Multiporphyrin Arrays

edited by Dongho Kim

Multiporphyrin Arrays

Fundamentals and Applications

Routledge
Taylor & Francis Group
LONDON AND NEW YORK

First published 2012 by Pan Stanford Publishing Pte. Ltd.

Published 2022 by Routledge
2 Park Square, Milton Park, Abingdon, Oxon OX14 4RN
605 Third Avenue, New York, NY 10017

Routledge is an imprint of the Taylor & Francis Group, an informa business

Publisher's Note

The publisher has gone to great lengths to ensure the quality of this reprint but points out that some imperfections in the original copies may be apparent.

British Library Cataloguing-in-Publication Data
A catalogue record for this book is available from the British Library.

Multiporphyrin Arrays: Fundamentals and Applications

ISBN 978-9-814-31660-6 (hbk)
ISBN 978-981-4364-28-7 (ebk)

Contents

Preface

Recent progresses in molecular photonics and electronics based on a variety of molecular species are ushering in a new era in the field of materials science. A photonic approach in the construction of molecular devices offers the advantages of ultrafast speed in the light-signal transmission as well as an easy access to control output performances. Furthermore, the molecular photonic devices circumvent connection problems, that is, a bottleneck that occurs when trying to combine electronic wires with macroscopic devices. With these versatilities, molecular photonics has drawn considerable attention in terms of the development of synthetic architectures for the absorption of light and the manipulation of excitation energy.

Natural photosynthetic light-harvesting complexes provide a source of inspiration in this regard, as excitation energy flows rapidly and efficiently among hundreds of bacteriochlorophyll pigments in a solid-state array for utilization as photoinduced electron transfer processes in reaction centers, which provides the fundamental reaction for energy conversion from light to chemical energy.

One of the most attractive building block elements for the preparation of biomimicking molecular photonic and electronic devices has been recognized as porphyrin molecules. A porphyrin provides a variety of features such as a rigid and planar geometry, high stability, intense electronic absorption, small HOMO-LUMO gap, and flexible tunability of the optical and redox properties by appropriate metallation. Owing to their fascinating and versatile physical properties and amenability to synthetic control, various synthetic strategies have been envisaged and developed for the construction of various types of covalently linked porphyrin arrays with linear, cyclic, and cross-linked geometries by using several types of shorter linkers via a meso position attachment. Various synthetic strategies involve the exploitation of a wide range of linkage motifs such as butadiyne, biphenyl, ethane, ethyne, enyne, furan, hexatriene, naphthalene, phenanthrene, phenylethene, and p-phenylene.

Although there exist opulent molecular architectures due to various linkers to adjoin the porphyrin moieties together, the requirements for ample electronic interactions between neighboring porphyrin pigments for efficient energy transfer and well-defined and rigid molecular structures to be void of any energy or charge transfer sink should be considered. In this sense, tremendous efforts have been made for the exploration of multiporphyrin arrays with a view of understanding the fundamental photophysical processes for a rational design of a wide range of molecular devices for photonic and electronic applications.

This book provides a comprehensive review of the fundamentals and applications of multiporphyrin arrays ranging from basic spectroscopic features to a wide range of promising applications such as molecular wires, switches, sensors, artificial photosynthetic devices, and dye-sensitized solar cells based on a variety of multiporphyrin architectures using covalent and/or noncovalent molecular assemblies. Of particular focus here are energy and electron transfer processes occurring in multiporphyrin arrays in various environments such as single-molecule level, composite materials, LB films, and solid surface to provide a better understanding of photofunctional molecular architectures. The key features addressed in this book include the fundamental electronic and structural description of a variety of multiporphyrin assemblies and a wide range of promising applications for molecular photonic/electronic devices.

Since this book is composed of contributions from internationally well-recognized and leading scientists in related fields, the scientific level of this book is very high to provide concise yet rich information on the fundamentals and applications of multiporphyrin arrays. This book is appropriate for advanced undergraduate and graduate-level students in the field of molecular electronic and photonic devices as well as researchers in nanotechnology, solar cell, molecular electronics and photonics, molecular recognition, sensors, artificial solar cell energy conversion, and single-molecule detection. Finally, I'd like to thank all the authors who have contributed to this book despite their busy schedules.

Dongho Kim

Chapter 1

Single-Molecule Photophysical Properties of Various Directly Linked Porphyrin Arrays

Dongho Kim[a] and Atsuhiro Osuka[b]
[a]*Department of Chemistry, Yonsei University,*
134 Sinchon-dong Seodaemun-gu, Seoul, 120-749, South Korea
[b]*Department of Chemistry, Graduate School of Science,*
Kyoto University, Sakyo-ku, Kyoto, 606-8502, Japan
dongho@yonsei.ac.kr; osuka@kuchem.kyoto-u.ac.jp

1.1 Introduction

Over the past decades, multiporphyrin arrays have been envisaged and prepared using several types of shorter linkers that are suitable for preparing linear or extended architectures via a *meso* position attachment. Various synthetic strategies involve the exploitation of a wide range of linkage motifs, such as butadiyne, biphenyl, ethene, ethyne, enyne, furan, hexatriene, naphthalene, phenanthrene, phenylethene, and p-phenylene [1–15]. Although there exist opulent molecular architectures due to various linkers to adjoin the porphyrin moieties together, the requirements for ample

Multiporphyrin Arrays: Fundamentals and Applications
Edited by Dongho Kim

www.panstanford.com

electronic interactions between neighboring porphyrin pigments for efficient excitation energy transfer (EET) and well-defined and rigid molecular structures to be void of any energy or charge sink should be considered. In this context, porphyrin arrays where constituent porphyrins are connected directly without any links can be promising molecular architectures for the application in molecular photonics and electronics.

This chapter deals with spectroscopic characterizations both at an ensemble and single-molecule level of various types of directly linked multiporphyrin arrays including directly *meso-meso* linked linear porphyrin arrays (**Z**$_N$), butadiyne–linked linear porphyrin arrays (**Z**$_N$**B**), *meso-meso*, β-β, β-β triply linked linear porphyrin arrays (**TB**$_N$), and two-dimensional cyclic porphyrin wheels (**C**$_N$**ZA** and **C**$_N$**ZB**) (Chart 1.1). In particular, single-molecule fluorescence

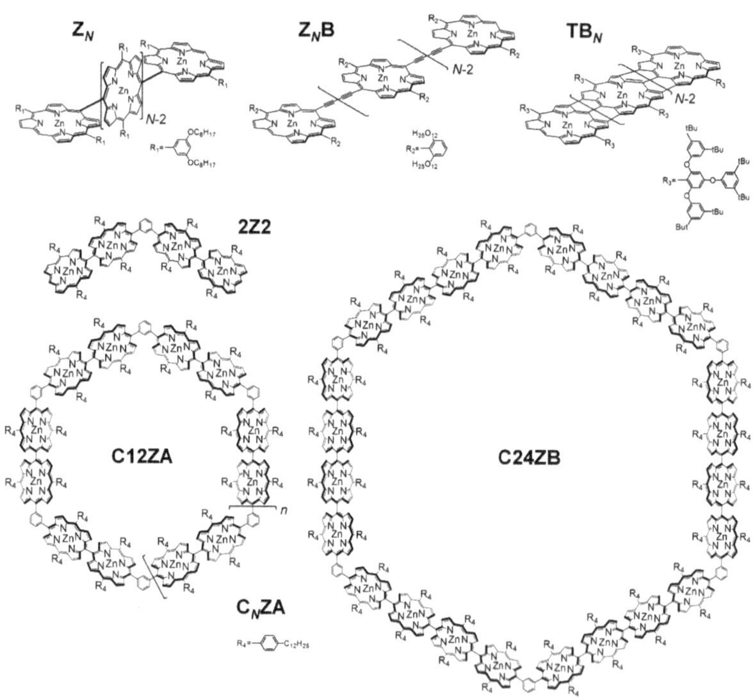

Chart 1.1. Molecular structures of various directly linked porphyrin arrays.

spectroscopy (SMFS) used in our researches has been instrumental, since it completely removes ensemble averaging by constructing a molecular distribution of physical quantities, from which we could obtain detailed information hidden in ensemble measurements: chemical and spectroscopic heterogeneities of single molecules and dynamics of these heterogeneities to a name a few [16–24]. Combination of SMFS and ultrafast time-resolved ensemble spectroscopic techniques provides comprehensive understanding on the excited-state dynamics such as excitation energy transfer and relaxation processes, and relationship of these processes with conformational heterogeneities that depend on molecular structure and size.

The order in this chapter comes from the dimensionality and type of linkages connecting neighboring porphyrins. Following this sequence, we have tried to describe the role of electronic couplings in determining photophysical properties of the porphyrin arrays since the electronic couplings depend largely on linkage motifs. Each four section in the following first illustrates the electronic couplings relevant for a particular series of porphyrin arrays and then accounts for the results of femtosecond transient absorption (TA) and transient absorption anisotropy (TAA) measurements and coincidence measurement to evaluate the EET process, the results of fluorescence intensity trajectories (FITs) and wide-field defocused imaging, which are related to single-molecule fluorescence dynamics, and finally the results of time-correlated single photon counting measurements performed both at the ensemble and single-molecule level to examine and compare conformational heterogeneities in different surrounding media.

1.2 Directly *Meso-Meso* Linked Linear Porphyrin Arrays

The molecular design of directly *meso-meso* linked linear porphyrin arrays as a model of light-harvesting antenna as well as a molecular photonic wire was envisaged to bring the porphyrin units closer for rapid energy transfer. For this purpose, zinc(II) 5,15-bis(3,5-bis(octyloxy)phenyl)porphyrin (**Z1**) and its directly *meso-meso* linked porphyrin arrays up to **Z128** ($\mathbf{Z}_N$, N represents the number of porphyrins) were synthesized (Chart 1.1) [25–28]. The fact that overall orthogonal conformation between the adjacent porphyrin units in the arrays disrupts π-electron conjugation over the array despite of the very short interporphyrin distance of 8.35 Å well resembles the facile energy migration processes in the biological light-harvesting complexes, where electronic delocalization is negligible due to a lack of direct bond linkage between individual pigment molecules [29–36].

1.2.1 *Exciton Coupling in Z_N*

1.2.1.1 Electronic nature of exciton split Soret band

The absorption spectra of $\mathbf{Z}_N$ normalized at ca. 413 nm which correspond to the high-energy Soret bands are shown in Fig. 1.1. With an increase in the number of porphyrin units, the Soret band at low energy is shifted to longer wavelength, while the Soret band at shorter wavelength remains nearly at the same position (ca. 413 nm), resulting in a progressive increase in the splitting energy. The relative intensities of split Soret bands also depend on the number of porphyrin units; the intensity of longer wavelength bands becomes increasingly stronger relative to that of shorter wavelength. On the other hand, the spectral shifts in the Q-band region are modest with gradual increase in intensity.

The systematic spectral changes of the Soret bands can be explained by the simple point-dipole exciton coupling theory developed by Kasha [37]. The Soret band of Zn(II) porphyrin has two

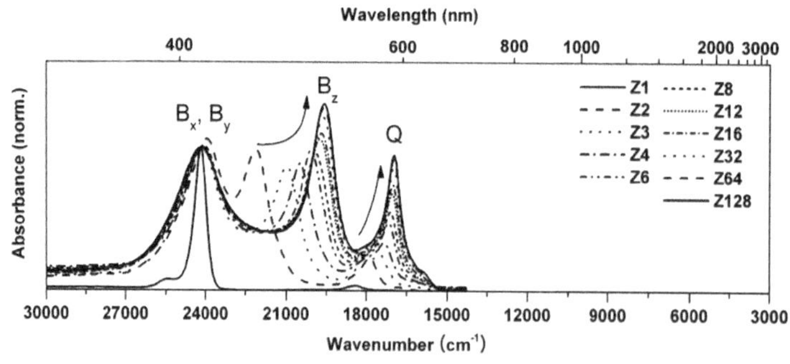

Figure 1.1. Steady-state absorption spectra of $\mathbf{Z}_N$ in THF. Reprinted with permission from [33].

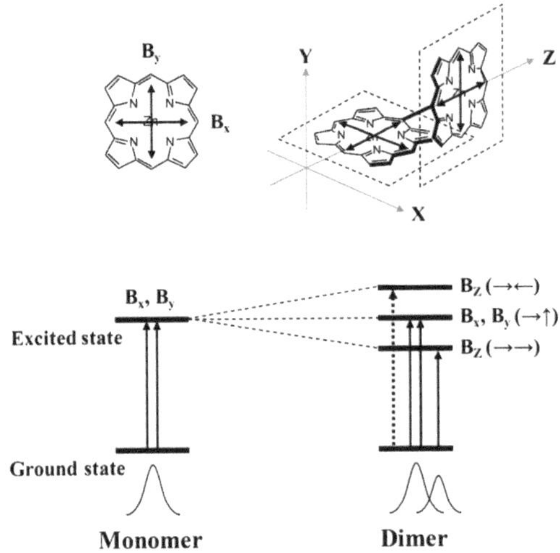

Scheme 1.1. Exciton coupling in $\mathbf{Z}_N$.

perpendicular components of B_x and B_y as depicted in Scheme 1.1. In a simple monomer they are degenerate, but in a porphyrin dimer they couple differently. In the case of **Z2**, only B_z transitions are parallel, and other dipole interactions should be zero for an averaged perpendicular conformation of **Z2**. Unperturbed Soret transitions observed at ca. 413 nm for all the arrays (**Z2-Z128**) suggest an orthogonal conformation and vice versa. And these bands become broader as the porphyrin arrays become longer, indicating the increasing conformational heterogeneities caused by the dihedral angle distribution, aggregate formation, and partial insolubility, if any, of longer arrays. Transitions are allowed to the lower energy of the two B_z states and the two unperturbed transitions B_x and B_y. Thus, the Soret band of **Z2** is split into a red-shifted B_z component and unperturbed B_x, B_y components (Scheme 1.1).

1.2.1.2 Photobleaching dynamics of **Z2** proves exciton coupling

Single-molecule photobleaching dynamics enables to confirm exciton coupling. Figure 1.2 shows typical FITs, that is, the fluorescence intensity of a single-molecule as a function of time, of **Z2** depending on the excitation wavelength.

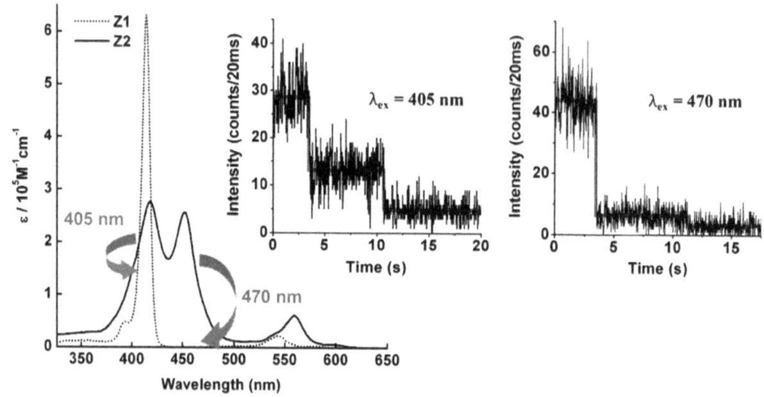

Figure 1.2. Typical fluorescence intensity trajectories (FITs) of **Z2** depending on the excitation wavelength. 405 and 470 nm correspond to the high-energy and low-energy Soret bands in the absorption spectrum of **Z2**, respectively. Arrows shown in the extinction coefficient spectra indicate the change in extinction coefficients at each excitation wavelength after the photobleaching of one porphyrin unit.

When **Z2** is photoirradiated at the wavelength of 405 nm corresponding to the high-energy Soret band, an extent of intensity drop from the first to second intensity level and that from the second to background level are similar with a ratio of 1.5 : 1. On the other hand, the FIT of **Z2** obtained by photoexcitation at 470 nm that corresponds to low-energy Soret band shows an order of magnitude intensity drop from the first to second level, with a concurrent increase in the fluorescence lifetime from 1.65 to 2.18 ns. The large difference found in the stepwise FITs arises from that, for the latter case, the breakdown of exciton coupling upon photobleaching of one porphyrin unit causes a significant drop in the oscillator strength to the almost negligible value, while it accompanies only a slight decrease in the oscillator strength for the former case, as schematically shown in the extinction coefficient spectra with the arrows (Fig. 1.2).

1.2.2 *Photobleaching Dynamics of Z_N*

For longer porphyrin arrays than **Z2**, in the cases when the FITs show the same number of steps as the constituent porphyrin units, the simultaneous analysis of the FITs and fluorescence lifetimes allows for observation of the photobleaching dynamics based on the photobleaching sequence of individual porphyrins [38]. Remnant FITs showing fewer steps than the number of porphyrin units, which may arise from the generation of a fluorescence trapping site that quenches excitation along the entire array [39], were excluded because they were not suitable for the purpose of this analysis.

As shown in Fig. 1.3, the photobleaching dynamics of **Z3** can be classified into two representative cases. In both FITs, the fluorescence lifetimes of the first emissive levels are shorter than that in **Z2**, which is clearly indicative of enhanced exciton delocalization. However, a striking difference is observed at the second emissive level. In Fig. 1.3a, the second emissive level in the FIT shows 11 times higher fluorescence intensity than that of the third level, and the fluorescence lifetime of 1.64 ns is intermediate between those of the first (1.59 ns) and third (2.07 ns) levels. Because this feature is similar to that found for **Z2**, we attribute it to the photobleaching of one of the terminal porphyrin units (Fig. 1.3a, schematic). On the other hand, when the central porphyrin unit is photobleached first (Fig. 1.3b, schematic), the remaining two porphyrins are separated from each other by approximately 17 Å, and the coupling strength diminishes by an order of magnitude. In this case, as shown in Fig. 1.3b, the second emissive level in the FIT carries only twice the intensity and a similar lifetime to the third level.

The photobleaching dynamics of **Z4** can be classified into four representative cases according to the characteristics of the second and third emissive levels in the FITs. Again, the first emissive levels show even shorter fluorescence lifetimes, from which we can infer the further augmentation of the exciton delocalization. For simplicity, we refer to the porphyrin units in **Z4** as p1–p4, starting from the left.

In Fig. 1.3c, the fluorescence intensities of the second and third emissive levels are 12 and 7 times higher, respectively, than that of the fourth level, and the respective fluorescence lifetimes of 1.62 and 1.70 ns are well matched with those observed for **Z2** and **Z3**. It is thought that the molecule exhibits trimer- and dimer-like behavior at the second and third levels, respectively, which can occur when the molecule follows the photobleaching sequences p4→p3→p1 (or p2) or p4→p1→p2 (or p3; Fig. 1.3c, schematic). On the other hand,

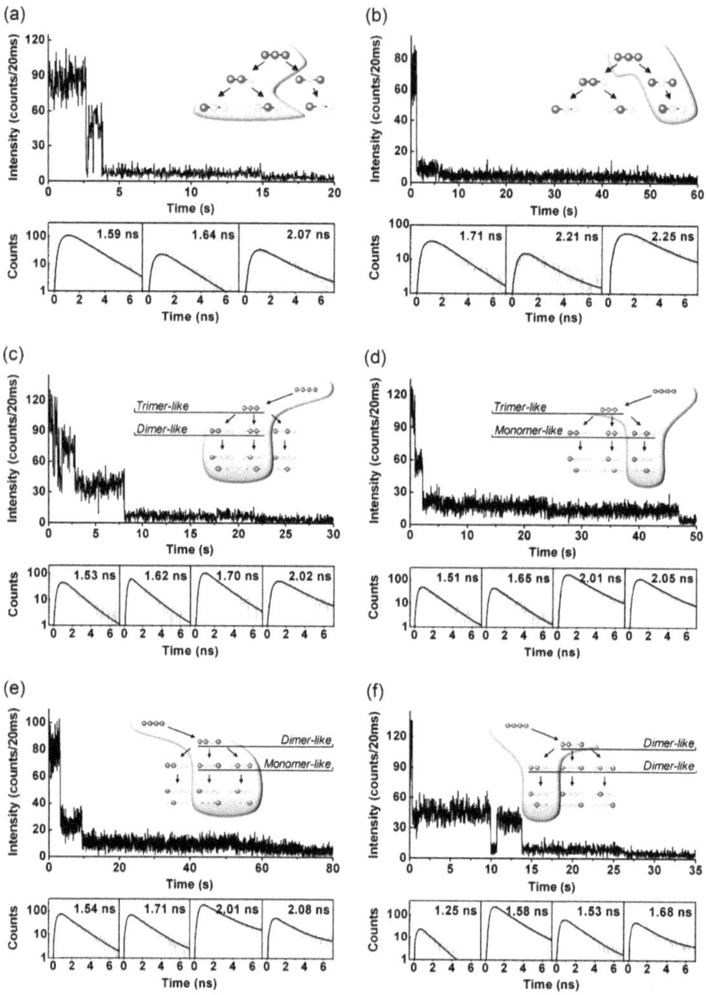

Figure 1.3. Representative FITs and fluorescence decay profiles corresponding to each emissive level of **Z3** molecules exhibiting (a) dimer-like and (b) monomer-like behavior at the second emissive level in the FITs, and of **Z4** molecules exhibiting (c) trimer- and dimer-like, (d) trimer- and monomer-like, (e) dimer- and monomer-like, and (f) dimer- and dimer-like behavior at the second and third emissive levels in the FITs. The schematic representations in each part display photobleaching sequences compatible with the experimental data. The L_S values for each molecule were calculated to be 2.7, 2.7, 3.6, 3.8, 3.7, and 3.7 (a–f, respectively). Reprinted with permission from [38].

the third emissive level of the FIT in Fig. 1.3d shows monomer-like behavior with only twice the intensity and a similar lifetime to the fourth level. We attribute this feature to the photobleaching sequence p4→p2→p1 (or p3; Fig. 1.3d, schematic). If one of the inner porphyrin units photobleaches first, the molecule shows dimer-like behavior at the second emissive level. In Fig. 1.3e, the second emissive level of the FIT has only half the intensity of that in Fig. 1.3d, with a slightly increased lifetime of 1.71 ns. Because the molecule exhibits dimer- and monomer-like behavior at the second and third levels, we assign this molecule the photobleaching sequences p3→p2→p1 (or p4) or p3→p1→p2 (or p4; Fig. 1.3e, schematic). As the last case, the FIT in Fig. 1.3f shows similar fluorescence intensities and lifetimes of 1.58 and 1.53 ns, respectively, at both the second and third emissive levels, which can be thought to arise from dimer-like behavior. In this case, the molecule might follow the photobleaching sequence p3→p4→p1 (or p2; Fig. 1.3f, schematic).

1.2.3 *Exciton Coherence Length of Z_N*

1.2.3.1 Exciton coherence length at the ensemble level

The coherence length of a strongly coupled molecular array is important in understanding the photoexcited-state dynamics and collective behavior of the transition dipole moments. When the pigments interact with each other and radiate in phase, the radiative decay rate becomes enhanced linearly with the number of pigments. According to the empirical formula for a one-dimensional linear array developed by Kakitani et al. [40–43], the exciton coherence length (N_C) is estimated by Eq. 1.1

$$N_C = 1.95 + 1.88V/\gamma \tag{1.1}$$

where V is the electronic coupling and γ is the homogeneous broadening. Using $V = 570\ \text{cm}^{-1}$ and $\gamma = 250\ \text{cm}^{-1}$ that were respectively obtained from $\mathbf{Z}_N$ and the spectral bandwidth in the Q-band of **Z1** at 103 K, we calculated the exciton coherence length (N_C) of $\mathbf{Z}_N$ to be ~6.

Spectroscopic observable directly related to the exciton coherence length is superradiance [44, 45]. The superradiance can be

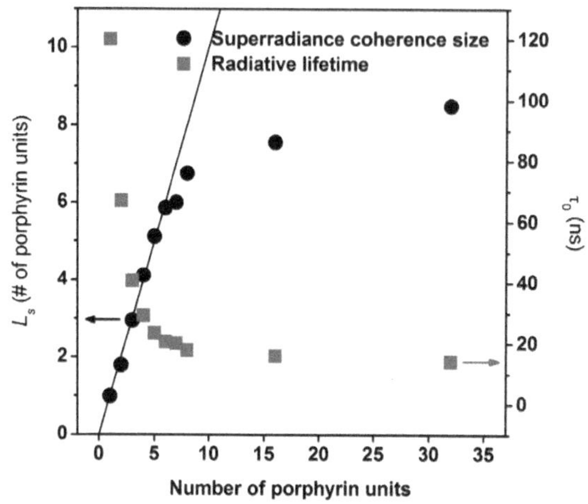

Figure 1.4. A plot of superradiance coherence sizes (L_S) (circles) and natural radiative lifetimes (τ_0) (squares) as a function of the number of porphyrin units. A straight line of function $y = x$ is inserted to show the linear increase of the L_S values up to **Z6**.

quantified using a superradiance coherence size, which is defined as the ratio of the radiative decay rate of the array to that of monomer, i.e., $L_S = k_{\text{rad, array}}/k_{\text{rad, monomer}}$, because the radiative decay rate ($k_{\text{rad}} = \Phi_F/\tau_F$; Φ_F = fluorescence quantum yield, τ_F = fluorescence lifetime) of the array increases when the constituent chromophores interact with each other and radiate in phase. The superradiance coherence size depends on the exciton dynamics and the geometry of molecular assemblies. However, when all dipoles in the assemblies are parallel, L_S provides a direct estimate of the exciton coherence length. Figure 1.4 shows the natural radiative decay rates and estimated superradiance coherence sizes of $\mathbf{Z}_N$ as a function of the number of porphyrin units. The deviation point from the linearity, that is, 6–7 porphyrin units seems to be reasonable as the exciton coherence length of $\mathbf{Z}_N$, which is also in a good agreement with the calculated exciton coherence length. Consequently, the EET in $\mathbf{Z}_N$ occurs in such a manner that ~6 porphyrin units within the porphyrin array behave as a coherently coupled exciton, whereas the light-signal transmission on a large scale resembles the hopping-like process of this exciton.

1.2.3.2 Exciton coherence length at the single-molecule level

The superradiance effect can also be examined for single molecules based on the photobleaching dynamics discussed in the section 1.2.2 [38]. In this approach, the radiative decay rates for the array and the monomer are calculated using fluorescence lifetimes at the first and last emissive levels, respectively, in the FITs and fluorescence quantum yields from bulk measurements, as it is impossible to measure them at the single-molecule level.

For the **Z2** molecule in Fig. 1.2, we obtained an L_S value of 1.8; for **Z3**, the L_S value was 2.7 for both of the molecules in Fig. 1.3a, b. The **Z4** molecules in Fig. 1.3c–f exhibit the L_S values of 3.6, 3.8, 3.7, and 3.7. The obtained L_S values for each array are slightly smaller than the ideal values of 2, 3, and 4, which would be expected for full exciton delocalization. The site-energy variation of individual chromophores (static disorder) and the exciton–phonon coupling (dynamic disorder), as we observe the fluorescence of thermalized excitons, might account for this discrepancy. Even so, this result indicates that the superradiance coherence size increases linearly as the array becomes longer. For longer arrays, however, the superradiance coherence size seems to show saturation behavior. In the FITs of single **Z5** and **Z6** molecules, the fluorescence lifetimes of the first emissive levels reach a saturated value of about 1.5 ns, which infers that the superradiance coherence size no longer increases linearly with the number of porphyrin units.

We carried out a statistical analysis by collecting 64 single-molecule datasets for each array. The probability of acquiring FITs for the analysis gradually decreases as the array becomes longer owing to the increased likelihood of generating fluorescence trapping sites that quench excitations along the entire array (Fig. 1.5a, red squares). In Fig. 1.5b, distributions of the superradiance coherence size show a progressive shift upward to **Z4**, after which it becomes saturated at a value around 4.5, which is clarified in a plot of the averaged L_S values (Fig. 1.5a, black circles). Thus, at the single-molecule level, 4–5 porphyrin units would be the best estimation of the exciton coherence length of $\mathbf{Z}_N$.

It should be noted that the individual porphyrin arrays retain coherent excitonic interactions over several porphyrin units, even

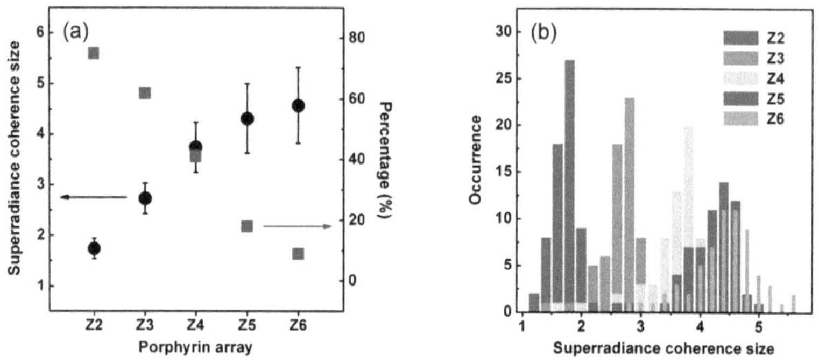

Figure 1.5. (a) A plot of the averaged superradiance coherence sizes (L_S) with standard deviations (black circles) and the percentage of FITs showing the same number of steps as porphyrin units (red squares). (b) Histograms for L_S of $\mathbf{Z}_N$, which were constructed by collecting 64 single-molecule datasets for each array. Reprinted with permission from [38]. See also Color Insert.

in the solid state at room temperature. Under the given experimental conditions, it is expected that the static and dynamic disorders greatly reduce the extent of exciton delocalization. Specifically, the static disorder plays an important role, because energy fluctuations between the chromophores become severe owing to the increase in the inhomogeneity of the surrounding medium. We believe that, for $\mathbf{Z}_N$, the covalent direct linkage with a short center-to-center distance of 8.35 Å facilitates dipole interactions among the chromophores, and the orthogonal geometry imposed by a large steric hindrance allows for the linearity to be void of any energy sink. These properties minimize the disorders and thus contribute to the preservation of coherence and to the facile EET of the arrays in the solid state.

1.2.4 *Conformational Heterogeneities in Z_N*

1.2.4.1 Conformational heterogeneities at the ensemble level

To transfer excitation energy over a long distance, it is of utmost importance to envisage well-arranged rigid molecular architectures. In this application, it is indispensable to minimize energy sinks, which may be formed by conformational heterogeneities in overall molecular structures, to realize high efficiency in the EET process. Although the large steric hindrance between the directly linked porphyrin moieties in $\mathbf{Z}_N$ helps to suppress conformational changes caused by the dihedral and tilt angle changes between the neighboring porphyrin planes, a summation of the dihedral angle distribution of $\pm15^\circ$ at ambient temperature is expected to give a large variation in linear structures especially in long porphyrin arrays [31].

The plot of the molar extinction coefficients of $\mathbf{Z}_N$ as a function of the number of porphyrin units exhibits a linear summation behavior up to **Z64**, indicating independent light-absorbing property for each porphyrin unit in the array [25, 34]. In contrast, the fluorescence quantum yields of $\mathbf{Z}_N$ show a maximum value at **Z16** and henceforth decrease as the arrays become longer (Table 1.1). The fluorescence decay measurements also show a similar trend. As shown

Table 1.1. Summary of various photophysical properties of $\mathbf{Z}_N$; Soret band splitting energies (ΔE), fluorescence quantum yields (Φ_F), fluorescence lifetimes (τ), and anisotropy values (r). Reprinted with permission from [34].

	ΔE (cm^{-1})	Φ_F^a	τ_{avg} (ns)^b	τ_1 (ns)^c	τ_2 (ns)^d	r^e
Z1	—	0.022	2.64			0.008
Z2	2060	0.034	1.83			0.03
Z3	3176	0.044	1.72			0.05
Z4	3692	0.055	1.65			0.16
Z6	4104	0.066	1.59			0.21
Z8	4283	0.074	1.55			0.29
Z12	4434	0.080	1.52			0.31
Z16	4499	0.088	1.50			0.31
Z24	4557	0.083	1.43	1.49 (95.4%)	0.14 (4.6%)	0.27
Z32	4569	0.062	1.40	1.48 (92.8%)	0.33 (7.2%)	0.24
Z48	4595	0.058	0.93	1.47 (61.1%)	0.08 (38.9%)	0.20
Z64	4595	0.033	0.51	1.47 (31.2%)	0.08 (68.8%)	0.17
Z96	4615	0.031	0.50	1.45 (27.7%)	0.14 (72.3%)	0.13
Z128	4615	0.012	0.49	1.42 (21.8%)	0.22 (78.2%)	0.12

[a] Relative fluorescence quantum yields determined with respect to $\Phi_F = 0.03$ of Zn^{II}TPP.
[b] Probed at the fluorescence maximum position with photoexcitation at 570 nm.
[c] Longer fluorescence lifetimes with their ratios.
[d] Shorter fluorescence lifetimes with their ratios.
[e] Maximum values at the fluorescence excitation anisotropy spectra.

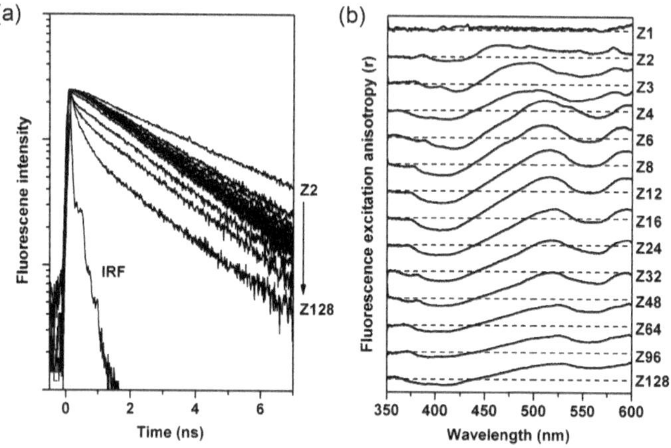

Figure 1.6. (a) Time-resolved fluorescence decay profiles and (b) steady-state fluorescence excitation anisotropy spectra (b) of $\mathbf{Z}_N$ taken in THF. Reprinted with permission from [33].

in Fig. 1.6a, starting from the fluorescence lifetime of 2.64 ns for **Z1**, the fluorescence lifetimes of $\mathbf{Z}_N$ decrease monotonically up to **Z16**. As the array length further increases, the fluorescence temporal profiles start to show double exponential decay in which the fast-decaying component increases gradually (Table 1.1).

To reveal the origin of this feature, we have recorded fluorescence excitation anisotropy spectra of $\mathbf{Z}_N$ (Fig. 1.6b) [33, 46]. The limiting anisotropy value of approximately −0.1 even in longer arrays such as **Z16** indicates a relatively large angle displacement between absorption and emission dipoles upon photoexcitation at ~413 nm, though this value is still smaller than the orthogonal orientation anisotropy value of −0.2. In longer arrays than **Z16**, this value starts to decrease gradually, but not so significantly. On the other hand, the anisotropy values above ~450 nm are positive. The maximum anisotropy value of ~0.3 above ~450 nm in **Z16**, which is still slightly smaller than 0.4 for the perfect in-plane orientation between absorption and emission dipoles, indicates a nearly parallel alignment of the absorption and emission dipoles upon photoexcitation at the low-energy Soret and Q-bands. The difference between the anisotropy values at ~413 nm and the entire visible region becomes larger upward to **Z16** and decreases in the arrays longer than **Z16** (Table 1.1). On the basis of our spectroscopic measurements, we conclude that the overall geometries of $\mathbf{Z}_N$ start to bend severely for >**Z16**.

1.2.4.2 Conformational heterogeneities at the single-molecule level

The spectral distributions of the single-molecule emission spectra of $\mathbf{Z}_N$ provide more information on the conformational heterogeneities contributing to nonradiative decay channels [47, 48]. Figure 1.7a shows the emission spectra of single **Z8, Z16**, and **Z32** molecules, and their spectra are compared to those taken in solution. Interestingly, as the number of porphyrin units increases the emission maximum position shifts to lower energy, resulting in the overall spectral distribution of the emission maximum of longer array, **Z32**, to be located in the lower energy region (Fig. 1.7b). This feature can be explained by that the energy variations in each constituent

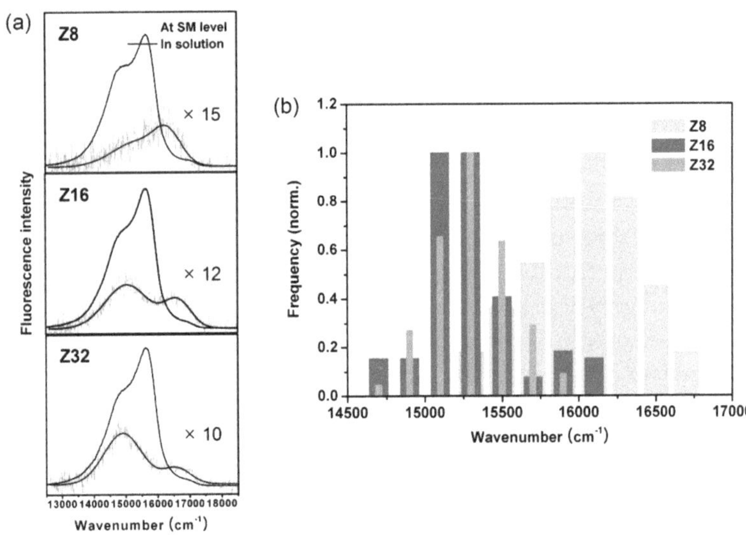

Figure 1.7. (a) Representative single-molecule fluorescence spectra of **Z8**, **Z16**, and **Z32**, with their ensemble-level fluorescence spectra taken in THF. (b) Histograms for fluorescence maximum positions of single-molecule fluorescence spectra of **Z8**, **Z16**, and **Z32**. Reprinted with permission from [47]. See also Color Insert.

porphyrin moiety in the array become larger due to the increase in conformational heterogeneities with increasing array length. The fluorescence may occur from the porphyrin unit that is lowest in energy in the array mediated by sequential energy transfer process. This observation at the singe-molecule level proves well the remorseless increase in conformational heterogeneities along with the length of the array.

1.3 Butadiyne-Linked Linear Porphyrin Arrays

Discrete molecular arrays with long π-conjugation pathways have long been of interest as organic conducting materials, near-infrared (near-IR) dyes, nonlinear optical materials, and molecular wires [49-51]. Porphyrins are intriguing building units to construct large π-conjugated molecular arrays because of their large polarizabilities and high optical oscillator strengths. Two types of synthetic strategies have been devised to develop conjugated porphyrin arrays, meso-alkyne-linked porphyrin arrays [52,53] and fused porphyrin arrays [54-56], both of which show unusual properties that result from strong π-conjugation. In this section, we describe single-molecule spectroscopic characteristics of butadiyne-linked linear porphyrin arrays (**Z**$_N$**B**) (Chart 1.1) [57], with main focuses on (i) a control of conformational dynamics of **Z2B** in its excited state and (ii) unique photobleaching dynamics distinguished from those of other multichromophoric arrays. Also, in the following section, the photophysical properties of fused porphyrin arrays, in which the electronic π-conjugation over the array is far stronger than butadiyne-linked porphyrin arrays, will be discussed.

1.3.1 *Control of excited-state conformational dynamics of Z2B*

For butadiyne-linked porphyrin arrays, conformations, particularly, those related to dihedral angles between adjacent porphyrins are one of the key issues, because the angular relationship between neighboring pigments mainly governs the electronic coupling in molecular arrays. As the butadiyne linkage potentially allows for

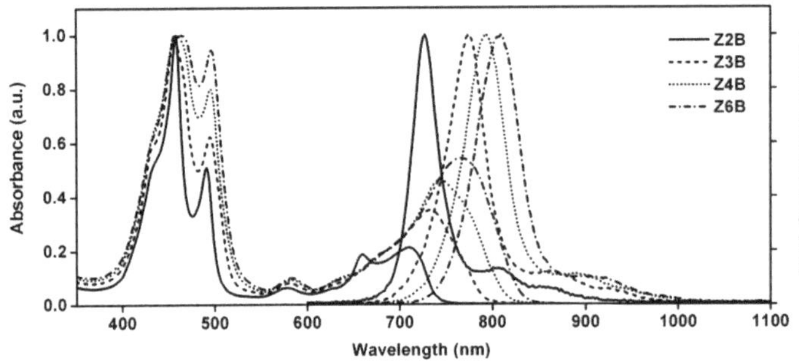

Figure 1.8. Steady-state absorption and emission spectra of $\mathbf{Z}_N\mathbf{B}$ in THF.

free rotation, in principle, a continuous distribution of the dihedral angles between adjacent porphyrins is possible. The π-conjugation should be strongest when the porphyrin moieties are coplanar and gradually decreases to a minimum when they are perpendicular. According to the previous report [58-61], butadiyne linked porphyrin dimer is in practice composed of two separable spectroscopic species; one is twisted conformer that is responsible for high energy B- and Q-bands in the absorption spectrum and the other is planar conformer for low energy B- and Q-bands (Fig. 1.8). While these two conformations coexist in the ground state because of the low barrier for rotation (0.67 kcal/mol), however, in the excited state, as the planar conformer is lower in energy (3.9 kcal/mol) than the twisted conformer, there is a strong bias toward the planar conformer via conformational dynamics [60].

In the solid-state, not only the conformations but also the conformational dynamics of **Z2B** depend on a concentration of host polymer matrix [62]. As displayed in Fig. 1.9a, the absorption spectrum of **Z2B** in the concentrated film shows a slight increase in the high energy B-band as compared with that measured in toluene, indicating that a portion of twisted conformer increases. This result suggests that an equilibrium between planar and twisted conformers shifts to a position where the portion of the twisted conformers is increased. It is thought that, in essence, equilibrium positions in concentrated films are determined ahead of the films being prepared by spin-coating procedure. This means that, already in solution that contains **Z2B**, PMMA, and toluene, PMMA polymer molecules interact with **Z2B** in competition with a tendency that **Z2B** is solvated by toluene molecules. Average molecular size of PMMA is much larger than those of **Z2B** and toluene molecules. Moreover, PMMA has irregular coiled conformation in apolar toluene solvent. We therefore speculate that an aspect of **Z2B** entangled by PMMA is rather complex, and that it is more likely to have twisted conformations rather than planar ones.

In Fig. 1.9b, as the concentration of PMMA increases, the high energy B-band gradually increases, indicating that the proportion of the twisted conformer becomes higher in dense polymer matrix. Similarly, in a series of fluorescence spectra of **Z2B**, the emission

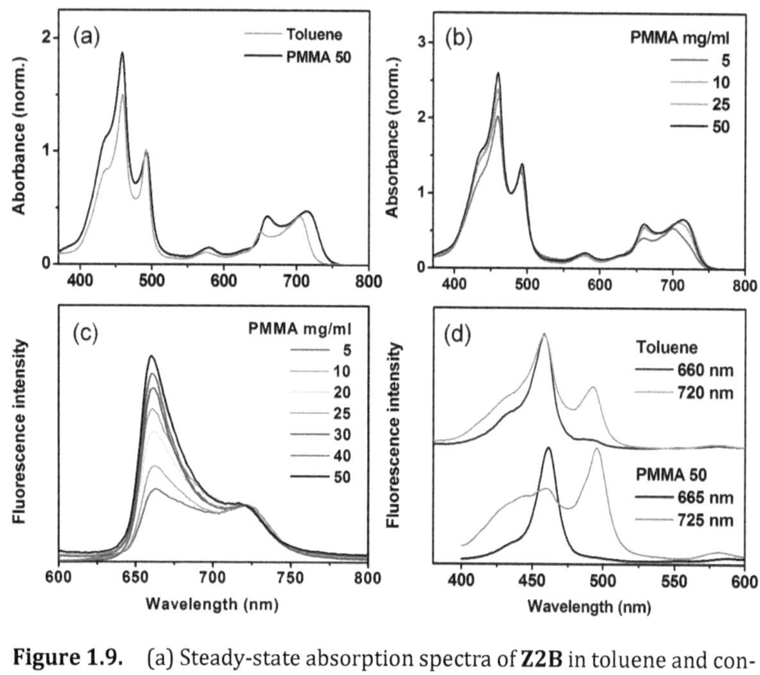

Figure 1.9. (a) Steady-state absorption spectra of **Z2B** in toluene and concentrated film prepared with 50 mg/ml PMMA. Steady-state absorption (b) and emission (c) spectra of concentrated films of **Z2B** prepared with a couple of different density of PMMA. (d) Fluorescence excitation spectra of **Z2B** in toluene and concentrated film prepared with 50 mg/ml PMMA. In the excitation spectra, the detection wavelengths of 660 (or 665) and 720 (or 725) nm correspond to the emission from the twisted and planar conformers, respectively. See also Color Insert.

band at 650 nm originating from the twisted conformer is considerably increased with increasing the concentration of PMMA (Fig. 1.9c). The magnitude of increase shows a linear relationship with the PMMA concentration. These results indicate that the excited-state planarization of twisted conformers is restricted in the solid-state because PMMA polymer molecules enclosing individual **Z2B** molecules may not provide a free space needed for such torsional motion. Figure 1.9d shows the fluorescence excitation spectra of **Z2B** in toluene and concentrated films. The excitation spectra monitored around 660 nm follows a spectral shape compatible with the high energy B-band of the absorption spectrum, irrespective of the environments that **Z2B** lies in. However, the excitation spectra monitored around 725 nm display different spectral features depending on the experimental condition. The emission peak coming from twisted conformers diminishes by a factor of 3 in the excitation spectrum of concentrated films, which is indicative of restriction of the excited-state planarization of twisted conformers.

Single-molecule photobleaching dynamics of **Z2B** also appears to be sensitive to the conformations [62]. Figure 1.10 shows the FITs of single **Z2B** molecules measured in the thin polymer films prepared with 5 and 50 mg/ml PMMA polymer matrix. Under the condition of low density of polymer matrix, one-step phtobleaching behavior is mainly observed (Fig. 1.10a). Surprisingly, 89% among 133 single molecules investigated exhibits this behavior. This result is quite distinctive, considering that the directly *meso-meso* linked porphyrin

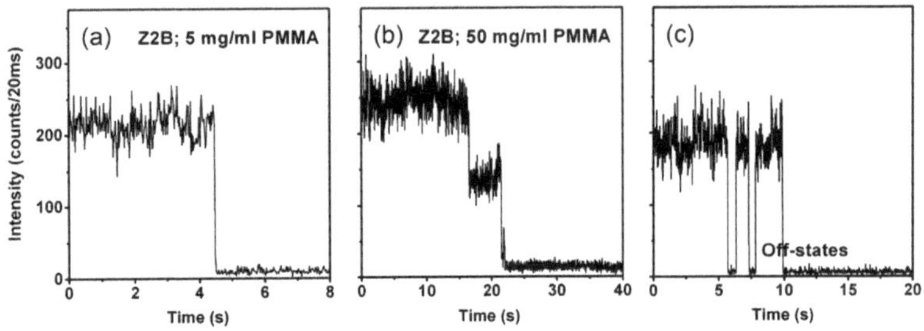

Figure 1.10. Typical FITs of **Z2B** embedded (a) in 5 mg/ml PMMA polymer film, (b) in 50 mg/ml PMMA polymer film, and (c) 10 mg/ml PMMA polymer film. The FITs were recorded with an excitation wavelength of 450 nm and irradiation power of 0.5 μW.

dimer, **Z2**, discussed in the previous section mostly shows two-step photobleaching behavior and that other multichromophoric systems with weak interchromophoric interaction usually show stepwise photobleaching behaviors in accordance with the number of chromophores [63-67]. On the other hand, as seen in Fig. 1.10b, single **Z2B** molecules in the high density of polymer matrix show two-step photobleaching behavior. In the statistical analysis, it appears that the ratio of two-step photobelaching behavior of **Z2B** gradually increases as the density of polymer matrix increases: 11% (15/133), 16% (23/146), 25% (33/134), and 35% (47/136) in 5, 10, 25 and 50 mg/ml PMMA, respectively. The two-step photobleaching behavior of **Z2B** comes from the twisted conformers, where the π-conjugation between the two porphyrins is so small that each porphyrin can be thought to be independent in electronical coupling. As confirmed in the steady-state spectroscopic measurements of the concentrated films of **Z2B**, the planarization of twisted conformer in the excited state becomes more restricted with increasing the density of polymer matrix. Thus, the relative portion of **Z2B** emitting in its twisted conformation increases in the higher polymer densities.

1.3.2 *Photobleaching dynamics of $Z_N B$: one quantum system*

Here we describe photobleaching dynamics of longer butadiyne-linked porphyrin arrays up to **Z6B**, which was examined for single molecules embedded in the polymer films prepared with 10 mg/ml PMMA [68]. For better understanding of the role of π-conjugation on single-molecule behaviors of $\mathbf{Z}_N\mathbf{B}$, the results of directly *meso-meso* linked porphyrin arrays, $\mathbf{Z}_N$, that exhibit strong dipole coupling over the array, will be taken into account together.

Figure 1.11 displays the representative FITs of two families of porphyrin arrays. As seen in Fig. 1.11a, $\mathbf{Z}_N$ photobleaches in a stepwise fashion, where the number of steps for each array is compatible with that of porphyrin units. The stepwise photobleachings may occur by successive photobleaching of emitting porphyrins that are lowest in energy at a given time, to which rapid energy transfer takes place owing to strong dipole coupling, indicating that the

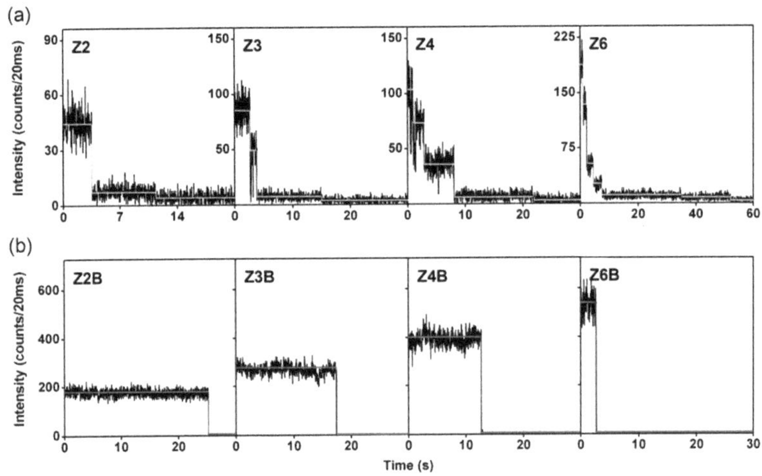

Figure 1.11. Comparison of photobleaching dynamics. Representative FITs (a) of $\mathbf{Z}_N$ and (b) of $\mathbf{Z}_N\mathbf{B}$. The lines involved in each set of FITs guide progress of the fluorescence intensities over time with clarifying discrete intensity levels.

excitonic dipole coupling facilitates strong communication over the whole array but enables each porphyrin to maintain its individual character (Scheme 1.2, upper). In contrast, the FITs of $\mathbf{Z}_N\mathbf{B}$ predominantly show one-step photobleaching behavior irrespective of the array length (Fig. 1.11b). This result is quite remarkable in that, even up to the longest array reaching ~80 Å in length, π-conjugation in the excited state is well elongated so that all $\mathbf{Z}_N\mathbf{B}$ are capable of acting as one quantum system (Scheme 1.2, lower).

Surprisingly, out of 120 single molecules investigated, more than half of $\mathbf{Z}_N\mathbf{B}$, that is, 65%, 56%, 51%, and 49% for **Z2B-Z6B**, appear to exhibit the collective one-step photobleaching behavior. For $\mathbf{Z}_N$, this behavior is observed only from 10-20% of the molecules. Furthermore, this bias in photobleaching dynamics of $\mathbf{Z}_N\mathbf{B}$ is clearly distinguished from the results found in single conjugated polymer chains [39], where only several hundredths of the total number of investigated chains have shown one-step photobleaching. The behavior of some chains as one quantum system has been explained by efficient Förster-type energy transfer to permanent photochemical traps that quench excitations along the entire chain. Traps are

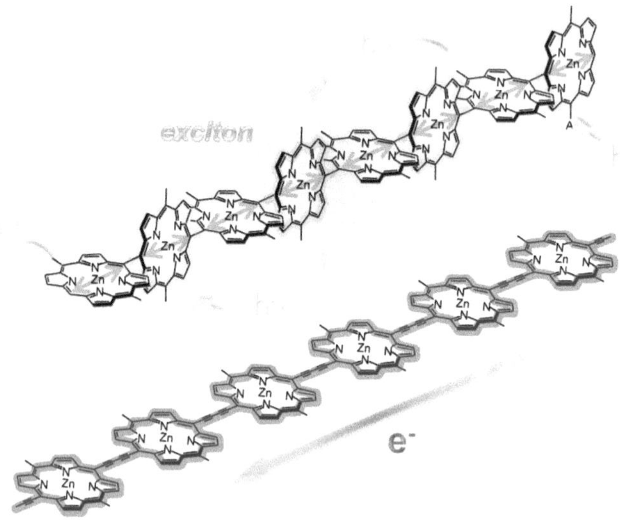

Scheme 1.2. Electronic couplings and energy/electron transfer mechanisms in $\mathbf{Z}_N$ and $\mathbf{Z}_N\mathbf{B}$.

assumed to be mostly generated by photo-oxidation. This explanation also accounts for the observed one-step photobleaching of $\mathbf{Z}_N$, because these arrays would show a similar type of energy transfer due to the strong dipole coupling. In contrast, in the case of $\mathbf{Z}_N\mathbf{B}$, we believe that each single array behaves as if it were one large delocalized single-molecule as a result of extensive π-conjugation achieved throughout the array, resulting in such a large population of arrays functioning as one quantum system.

Coincidence measurements were performed to verify whether $\mathbf{Z}_N\mathbf{B}$ acts as single photon emitter [69-71]. If a one-quantum system that consists of multiple chromophores is excited with an intense laser pulse, it is anticipated that only a single exciton is formed. This exciton may relax to the ground state in the form of fluorescent emission of a photon. After photon emission, the system must be re-excited and wait, on average, one fluorescence lifetime before another photon is emitted. Thus, time intervals between two consecutive detected photons conform to a multiple of the laser repetition rate. On the other hand, the interphoton time would be zero when two photons are detected simultaneously. In multichromophoric

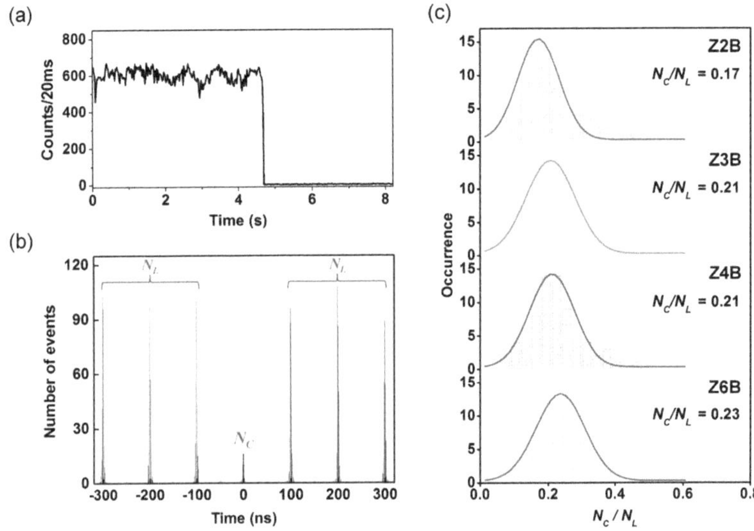

Figure 1.12. (a) The one-step photobleaching FIT and (b) the corresponding interphoton arrival time distribution of **Z4B**. In (b), only the central seven peaks are shown in the distributionfor clarity. (c) Histograms for the N_C/N_L values compiled from 70 single-molecule datasets for each array. All histograms were fitted with a Gaussian function.

systems, coincident photon pairs can be formed when constituent chromophores in the systems behave as independent emitters. In such a case, two or more excitons are generated per laser pulse, followed by simultaneous emission of multiple photons.

Figure 1.12 shows the one-step photobleaching FIT and corresponding interphoton arrival time distribution of **Z4B**. In Fig. 1.12b, the central peak, N_C, is attributed to the coincident photon-pairs. The lateral peaks, N_L, appear every ~100 ns in accordance with the multiple of the laser repetition rate of 10 MHz. It has been shown that a N_C/N_L value can be used to estimate the number of independently emitting chromophores [70]. The N_C/N_L values of 0 and 0.5 correspond to the single and two photon emitters, respectively. In practice, accidental simultaneous detection of two photons caused by contribution of background signals increases these values. As seen in Fig. 1.12b, the frequencies of lateral peaks are approximately five times higher than that of central peak, giving rise to the N_C/N_L

value of 0.18 [68]. Considering the S/B ratio of ~30.0 of the FIT displayed in Fig. 1.12a, the determined N_C/N_L value indicates that **Z4B** acts as the single photon emitter.

For all $\mathbf{Z}_N\mathbf{B}$, we have evaluated N_C/N_L values for the molecules showing the one-step photobleaching in the FITs; 70 single-molecule datasets were used for each array. As shown in Fig. 1.12c, the distributions of N_C/N_L values gradually shift to higher values in going from **Z2B** to **Z6B**. However, the magnitudes of the shifts are modest. The mean N_C/N_L values were determined to be 0.17±0.01, 0.21±0.02, 0.21±0.02, and 0.23±0.02 for **Z2B–Z6B**, respectively. It is observed that all the distributions are spread over the ranges; for example, the distribution of **Z2B** appears to be in the range between 0.04 and 0.44. In our coincidence measurements, the maximum S/B ratio of the FITs of **Z2B** was found to be 80.5 (minimum 7.5), resulting in the range of N_C/N_L values of 0.05–0.37. We assumed that **Z2B** is the single photon emitter. The similar calculations of the other arrays demonstrate that the calculated N_C/N_L ranges are roughly consistent with those observed from the experiments. We therefore conclude that $\mathbf{Z}_N\mathbf{B}$ acts as the single photon emitter as a natural result of the behavior of a one-quantum system.

1.3.3 *Wannier-Mott excitons in $Z_N B$ promote fluorescence intermittency*

We next examined fluorescence intensity fluctuations, which are mainly featured as collective off-states with a time scale ranging from tens of milliseconds to tens of seconds as exemplified by the FIT of **Z2B** (Fig. 1.10c) [68]. The long off-states are thought to arise from photo-induced electron transfer between the arrays and the surrounding polymer,[72-74] because (i) porphyrin radical cations are efficient quenchers of fluorescence due to their low-lying excited states and (ii) other conceivable causes of the off-states are unlikely as follows: triplet lifetimes of porphyrins are in μs-time scale and rotation of emitting dipoles is likely inhibited because of the high glass transition temperature of the polymer used ($T_g = 122°C$). Interestingly, the frequencies of long off-states differ dramatically in the two families of porphyrin arrays; while approximately 4% of $\mathbf{Z}_N$

exhibits this feature, the proportion grows by an order of magnitude in $\mathbf{Z}_N\mathbf{B}$: 37%, 39%, 45%, and 47% for **Z2B-Z6B**, respectively.

The observed large difference in the frequencies of long off-states can be well illustrated in terms of excitons, which are neutral, bound electron-hole pairs generated by photoexcitation. In $\mathbf{Z}_N$, as a lack of π-orbital overlaps between porphyrins prevents electron exchange along the array, each electron-hole pair is localized almost exclusively on a single porphyrin, forming a tightly bound Frenkel type exciton.[75] In contrast, the electron-hole pairs in $\mathbf{Z}_N\mathbf{B}$ develop into weakly bound Wannier-Mott type excitons[76,77] because the effective π-conjugation in the array accelerates the occurrence of electron exchange. Thus, the electron and hole are not necessarily localized in the same porphyrin at any given moment. As Coulombic attraction force is inversely proportional to the square of the distance between electron and hole, the force would decrease by a factor of $\sim$5 with consecutive addition of porphyrins in these arrays. Therefore, we conclude that the high frequencies of long off-states in $\mathbf{Z}_N\mathbf{B}$ originate from the nature of Wannier-Mott excitons that can easily dissociate into separate charges relative to Frenkel excitons. Thermal energies imparted by repetitive excitation of a single-molecule concomitant with single-molecule measurements may be operative to promote this dissociation.

1.4 Fused Linear Porphyrin Arrays

A straightforward strategy for maximizing π-overlap may be to hold the π-systems coplanar within a tape-like framework by fusing the units edge-to-edge, to make a covalently linked long, flat array. Scandium(III)-catalyzed oxidation of directly *meso-meso* linked porphyrin arrays with 2,3-dichloro-5,6-dicyano-1,4-benzoquinone (DDQ) led to efficient formation of triply *meso-meso*, β-β, β-β linked linear porphyrin arrays ($\mathbf{TB}_N$) [78]. These fully conjugated porphyrin arrays have planar tape-shaped structures and display drastically red-shifted absorption spectra that reach into the far-IR region, reflecting extensive π-conjugation. Here we illustrate energy relaxation dynamics of $\mathbf{TB}_N$ investigated by using femtosecond IR TA measurements [79]. Unfortunately, single-molecule studies on these arrays were unavailable due to the fact that the spectral region of absorption and emission of the arrays are far beyond the instrumental limitation. Note that $\mathbf{TB}_N$ should be distinguished from its previously reported analog, $\mathbf{T}_N$ [80, 81], which in some aspects has a problem concerning intermolecular aggregations caused by the flat structures although long alkyl chain substituents are attached in the peripheral positions. $\mathbf{TB}_N$ has more bulky substituents, (1,3,5-tris(3,5-di-tert-butylphenoxy)benzene), to circumvent this problem, which makes it possible to examine excited-state properties in aggregation-free condition.

1.4.1 *Energy Relaxation Dynamics in TB$_N$*

Figure 1.13 shows the steady-state absorption spectra of $\mathbf{TB}_N$ normalized at the high-energy Soret bands. The absorption bands are categorized into three distinct subgroups and are marked as bands I, II, and III, respectively. As the molecular length becomes longer, the bands II and III are continuously red-shifted upward to ~3500 cm^{-1}, while the band I remains at nearly the same position. It is remarkable that the bands III of all $\mathbf{TB}_N$ exhibit well-resolved vibrational structures. This feature is in sharp contrast with the bands III of $\mathbf{T}_N$, which appear to be broad and featureless in longer arrays with $N \geq 4$ [80]. Thus, it is inferred that $\mathbf{TB}_N$ is nearly free from aggregation-induced intermolecular π–π interaction owing to the bulky substituents.

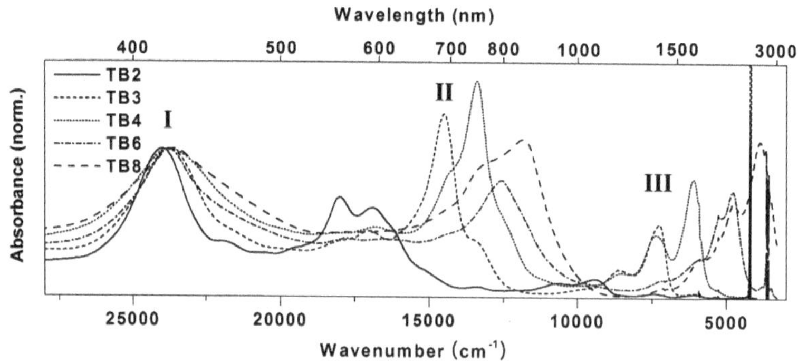

Figure 1.13. Steady-state absorption spectra of $\mathbf{TB}_N$ in $CHCl_3$.

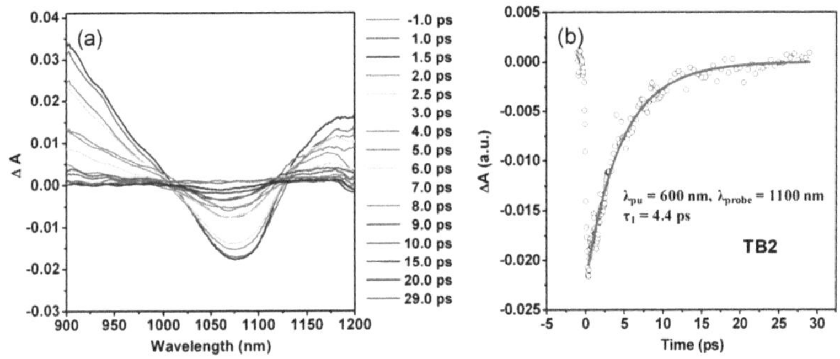

Figure 1.14. (a) Transient absorption spectra and (b) ground-state bleaching recovery dynamics of **TB2**.

To explore the excited-state dynamics of **TB**$_N$, we have carried out femtosecond TA measurements. We have adopted pump wavelengths that are in resonance with the bands III to solely observe energy relaxation dynamics occurring at the lowest excited states, except for **TB2**, for which we could not generate optical pulses near 1100 nm corresponding to the band III of **TB2**. Figure 1.14 shows a series of transient absorption spectra and a temporal profile of ground-state bleaching of **TB2** obtained with photoexcitation at 600 nm. The ground-state recovery process of **TB2** occurs with a single decay time constant of 4.4 ps. On the other hand, the temporal profiles of ground-state bleaching of **TB3**-**TB8** start to show double exponential decay, where fast time constants that take higher amplitudes are progressively shortened as the array becomes longer; 3.1, 1.3, 1.1, and 1.0 ps for **TB3**, **TB4**, **TB6**, and **TB8**, respectively (Fig. 1.15). These results are in accordance with the gradual decrease in the HOMO-LUMO energy gap with increasing array length, thus we demonstrate the fast time constants as the lowest excited-state lifetimes. In Fig. 1.15, the slow time constants appear to be almost the same as a value of ~9 ps, but their amplitudes increase in going from **TB2** to **TB8**. In the previous reports on porphyrin monomers [82–84], it has been revealed that vibrational relaxation process in porphyrin molecules occurs in the time scale of ~10 ps, which is well matched with the observed slow time constants for **TB**$_N$. Furthermore, as seen in the absorption spectra, **TB**$_N$ exhibits well-resolved vibrational structures even in the lowest absorption

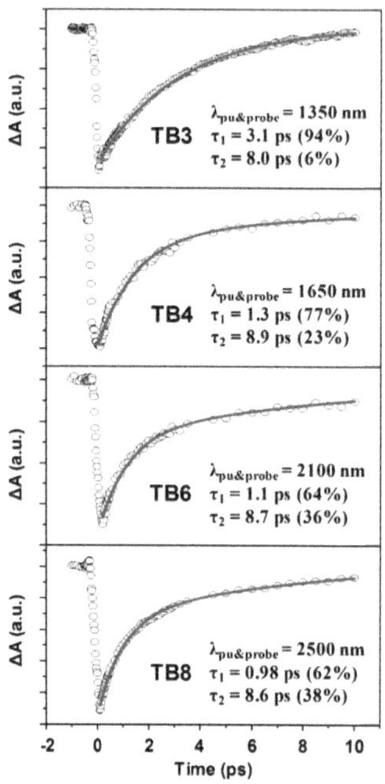

Figure 1.15. Ground-state bleaching recovery dynamics of **TB3–TB8** probed at the bands III after photoexcitation at the wavelength used for the probe.

bands. Accordingly, the slow recovery components originate from the vibrational relaxation process from S_1 to S_0 states and its contribution to excited-state dynamics becomes significant in the longer arrays.

Overall, upon photoexcitation of $\mathbf{TB}_N$, there exist two kinds of relaxation processes in the lowest excited-state manifolds. First, excited-state population relaxes to the ground state within ~5 ps through the electronic transition. Another process is the vibrational relaxation process including both intramolecular and intermolecular vibrational relaxations. As the number of porphyrin units increases in $\mathbf{TB}_N$, the HOMO-LUMO transition energy becomes lowered, accompanying the increase in the density of vibrational states. Thus, in the excited state, a competition between electronic and vibrational transitions becomes more enhanced in longer arrays because of their increased coupling with the vibrational manifolds.

In other words, the vibrational relaxation process is more activated when the electronic transition energy becomes extremely small, leading to the distinction between vibrational and electronic transitions. Based on these spectroscopic investigations, fused porphyrin arrays, $\mathbf{TB}_N$, have proven to be a good model to reveal vibronic coupling in large organic molecules and unique photophysical properties influenced by extensive π-conjugation.

1.5 Cyclic Porphyrin Wheels

In 1995, the crystal structure of light-harvesting antenna complex LH2 of purple bacteria *Rps. acidophila* was elucidated to be circularly arranged chromophoric assemblies [85–87]. LH2 consists of two wheel-like pigment arrays; B800 with 9 bacteriochlorophyll a (Bchl a) and B850 with 9 dimeric subunits (totally 18 pigments) of Bchl a. Since the advent of this wheel-like structure, the designs of light-harvesting antenna systems have focused on cyclic porphyrin wheels that allow for efficient EET. On the basis of the silver(I)-promoted oxidative coupling strategy, we have prepared two types of a series of extremely large yet discrete porphyrin rings, where directly *meso-meso* linked porphyrin dimer (**Z2**) and tetramer (**Z4**) are bridged by 1,3-phenylenes, respectively, to form cyclic structures $\mathbf{C}_N\mathbf{ZA}$ and $\mathbf{C}_N\mathbf{ZB}$ (Chart 1.1) [88–92]. The electronic interactions in these arrays are predominated by exciton coupling within the *meso-meso* linked porphyrin subunits and these subunits are weakly electronically coupled via through-space and through-bond interactions, which give rise to Förster-type EET in overall. In this section, singe-molecule fluorescence dynamics, rates and efficiencies of EET, and conformational heterogeneities with respect to molecular size are discussed based on the results of TCSPC, femtosecond TA and TAA, wide-field defocused imaging, and coincidence measurements.

1.5.1 *Exciton Coupling in C_NZA*

1.5.1.1 Exciton coupling in Z2, 2Z2, and C12ZA

Figure 1.16 shows steady-state absorption spectra of **Z2, 2Z2**, and **C12ZA**. With an increase in the number of porphyrin units, the

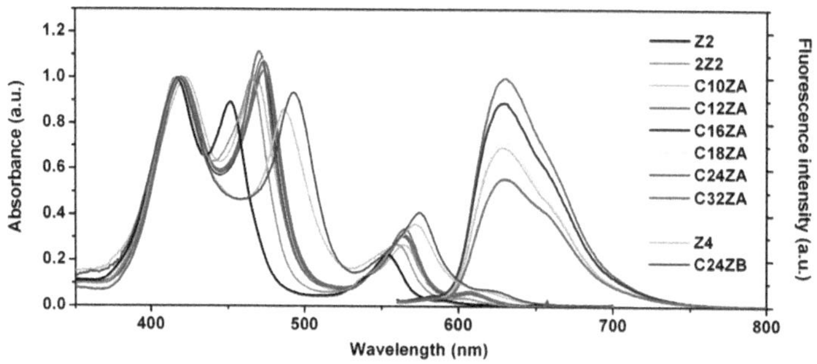

Figure 1.16. Steady-state absorption and emission spectra of **Z2**, **2Z2**, C_N**ZA**, **Z4**, and **C24ZB** in toluene. The absorption spectra were normalized to the high-energy Soret bands around 420 nm and the emission spectra were recorded with optical density of 0.1 at the excitation wavelength of 550 nm. Reprinted with permission from [94, 101]. See also Color Insert.

low-energy Soret band shifts to red, while the high-energy Soret band remains nearly at the same position as that of **Z1**. The split Soret bands of **Z2** are characteristic of excitonic dipole-dipole coupling between zinc(II) porphyrin monomers [30, 34]. The further split Soret bands observed for **2Z2** and **C12ZA** indicate additional dipole coupling between **Z2** subunits via the 1,3-phenylene spacer. In addition, relative intensities of the low-energy Soret bands increase in going from **Z2** to **2Z2** and **C12ZA**, which also reflects increased dipole coupling as the array becomes longer. These features in the absorption spectra are explained in terms of exciton coupling theory. As shown in Scheme 1.3, in **Z2**, two transition dipole moments along the long molecular axis (B_z) are coupled as a J-aggregate, giving rise to the red-shifted band (low-energy Soret band). On the other hand, B_x and B_y do not couple because of mutual orthogonal geometry, resulting in the same band position (high-energy Soret band) as that of **Z1**. The dipole moments of **2Z2** and **C12ZA** consist of the dipole moments (B_x, B_y) perpendicular to B_z along the long axis (*z*-axis). These dipole moments are coupled in a manner similar to that of **Z2**. For the B_z components, the in-phase arrangement of transition dipoles is attractive and leads to a lowering of energy, while the dipole couplings between B_x and B_y components cannot occur because of orthogonal geometries (Scheme 1.3).

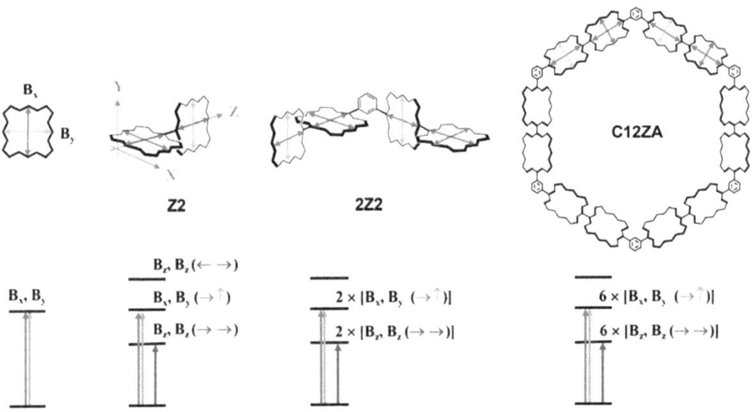

Scheme 1.3. Exciton coupling in **C12ZA**. See also Color Insert.

1.5.1.2 Exciton coupling in larger cyclic porphyrin wheels

As the number of **Z2** subunits increases from **C10ZA** to **C32ZA**, the exciton splitting energies corresponding to energy differences between high- and low-energy Soret bands become larger. The overall spectral features in the absorption spectra seem to be determined by the constituent subunit **Z2** in that additional exciton couplings observed for **C**$_N$**ZA** should be accumulated as the number of **Z2** subunits increases. In the absorption spectra, peak positions of the low-energy Soret bands of **C**$_N$**ZA** converge into a single position (Fig. 1.16), which is similar to the spectral feature observed in directly *meso-meso* linked linear arrays **Z2-Z128** [34]. However, in this case, such convergence of the low-energy Soret bands does not come from conformational flexibility as seen in the linear arrays but from exciton couplings among nonunidirectional transition dipole moments in circular geometries [92]. On the other hand, the Q-bands exhibit negligible red shift with slight intensification. Moreover, the fluorescence spectra of **C**$_N$**ZA** with photoexcitation at Q-band region are similar to each other (Fig. 1.16). These results infer that structural environment between two adjacent **Z2** remains relatively the same in all **C**$_N$**ZA**. In other words, three-dimensional orientation of **2Z2** moieties should be maintained in all the cyclic arrays.

When the subunit structure in cyclic porphyrin wheels is extended from **Z2** to **Z4**, a magnitude of red shift in the low-energy Soret band of **C24ZB** becomes smaller compared to that of **C12ZA** (Fig. 1.16). This result indicates a rather weak dipole coupling between **Z4** subunits probably due to the longer interchromophoric distance. The calculated dipole coupling energies between the neighboring subunits, **Z2**s in **C12ZA** and **Z4**s in **C24ZB**, respectively, are 943 and 38 cm^{-1} in the B-band and 140 and 11 cm^{-1} in the Q-band [93], which reinforces the strong dipole coupling in the cyclic arrays consisting of **Z2** subunits.

1.5.2 *Single-Molecule Fluorescence Dynamics of C12ZA*

Single-molecule FITs of **C12ZA** and its reference linear array **L12Z** exhibit structure-dependent behavior [94]. As displayed in Fig. 1.17a, the FIT of **L12Z** exhibits collective off-states. This feature can be explained in terms of existence of nonradiative decay channels, which are induced by the more flexible conformation that leads to more frequent kink structures. Such conformational flexibility is significantly reduced in the cyclic structure. The FIT of **C12ZA** shows a stepwise photobleaching behavior with lower occurrence of the off-states, indicating more efficient energy migration in this array (Fig. 1.17b). Since the stepwise photobleaching observed for

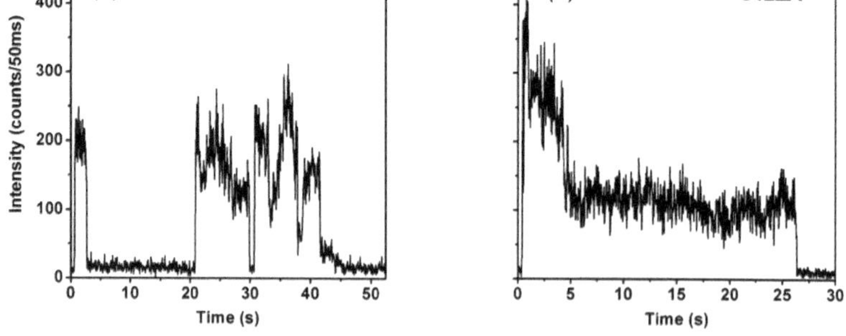

Figure 1.17. Typical FITs of 1,3-phenylene-bridged Zn(II) linear porphyrin array, **L12Z** (a) and cyclic array, **C12ZA** (b). Reprinted with permission from [94].

C12ZA is similar to the behavior seen in the FITs of other multi-chromophoric systems [63–67], the model proposed to describe the behavior of these systems, involving one of the chromophores acting as a fluorescent trapping site and energy migration toward this trapping site in combination with temporal evolution of the trapping site due to photobleaching, can be claimed here as well.

To confirm the above model, we performed wide-field defocused imaging [95, 96] for single **C12ZA** molecules embedded in a polymer film [48, 94]. The characteristic intensity distribution of the defocused images allows for the determination of three-dimensional orientation of emission dipole in a single-molecule, which may correspond to the fluorescence trapping site. There are six emitting dipoles in **C12ZA**. The two emission dipoles facing each other have the same emission pattern due to the symmetry of molecular system (identical angle ϕ in Fig. 1.18a), so a maximum of three differently oriented emitters are expected to appear. Note that if some of emission dipoles are not in-plane (the *xy* plane), in other words, if the angle θ between the *z* axis and the emission dipole (see Fig. 1.18a) is close to zero, the change in emission patterns cannot be observed clearly [95, 96].

The emission patterns of **C12ZA** as a function of time are displayed in Fig. 1.18b. It is noticeable that molecular rotational motion of **C12ZA** is prohibited in the PMMA polymer matrix at room

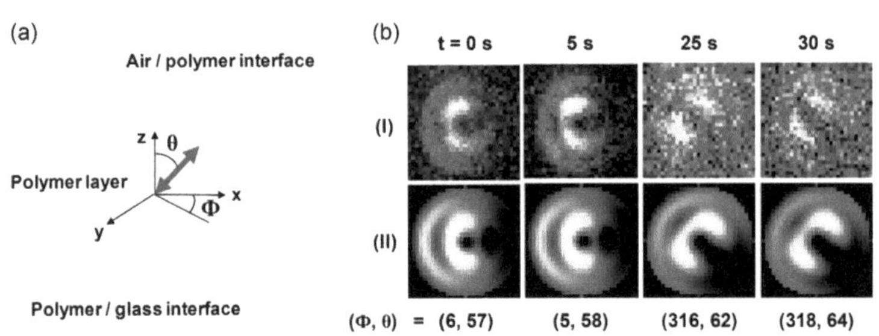

Figure 1.18. (a) Definition of angles: Φ indicates the in-plane (in the *xy* plane) angle, and θ indicates the out-of-plane angle (inclination angle between the optical dipole (bold arrow in red) and excitation optical axis (*z* axis)). (b) Snapshots of experimentally observed emission patterns of **C12ZA** (sequence I) and corresponding calculated patterns (sequence II) as a function of time. Reprinted with permission from [94]. See also Color Insert.

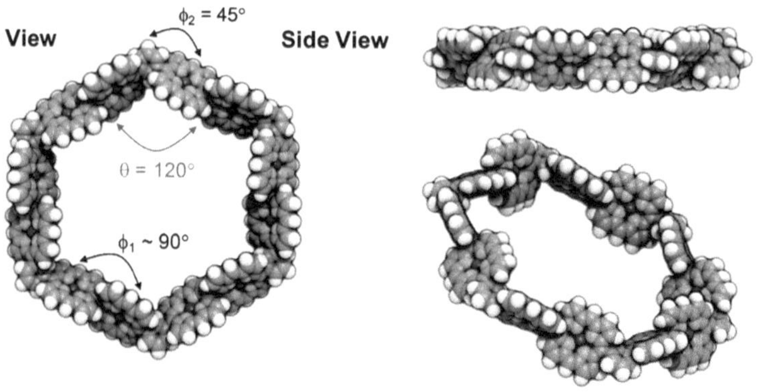

Figure 1.19. Three-dimensional structure of **C12ZA** optimized at the semiempirical PM3 level. Reprinted with permission from [102].

temperature due to the large size of the molecule with long side chains and the high glass transition temperature of PMMA (T_g = 122°C). Accordingly, the different patterns represent the different porphyrin units in **C12ZA** that serve as fluorescence trapping sites in the time course of the experiment. In the series of defocused images (Fig. 1.18b), we observed only two different patterns, probably due to the following reasons: (i) bleaching of the porphyrin units with different orientations during alignment or (ii) low count rate when the majority of the absorbing/emitting units are bleached. On the basis of fitted patterns, the relative angle between two emission dipoles was calculated to be ~60°, in excellent agreement with the value expected from the calculated molecular structure (Fig. 1.19). This experiment clearly proves the validity of the model that evokes emissive trapping sites and efficient energy migration toward these trapping sites.

1.5.3 *Excitation Energy Transfer in C_NZA*

1.5.3.1 Rate of excitation energy hopping in C12ZA

To explore the fast excitation energy hopping (EEH) process at the ensemble level, femtosecond TA and TAA measurements were conducted on **Z2**, **2Z2**, and **C12ZA** [97]. In the TA decays, **Z2** and **2Z2** reveal no power dependence and show only single decay

Table 1.2. Transient absorption decay parameters for **Z2**, **2Z2**, and **C12ZA** depending on pump power[a]. Reprinted with permission from [97].

Pump power (mW)	Fitted decay times[b]		
	τ_1 (ps)	τ_2 (ps)	τ_3 (ps)
	Z2		
1.0	1820 (100%)		
0.5	1820 (100%)		
0.2	1820 (100%)		
	2Z2		
1.0	1730 (100%)		
0.5	1730 (100%)		
0.2	1730 (100%)		
	C12ZA		
1.0	1.1 (22%)	6.2 (38%)	1660 (40%)
0.5	1.1 (20%)	6.2 (22%)	1660 (58%)
0.2	1.1 (14%)	6.2 (18%)	1660 (68%)

[a]The pump and probe wavelengths are 575 and 510 nm, respectively.
[b]Using the relation $\Delta OD(t) = A1\ \exp(-t/\tau_1) + A2\ \exp(-t/\tau_2) + A3\ \exp(-t/\tau_3)$, where $\Delta OD(t)$ is the transient absorption intensity, A the amplitude (noted in parentheses as the normalized percentage, i.e., $[Ai/(A1 + A2 + A3)] \times 100$), and τ the fitted decay time.

components that are in agreement with the S_1-state lifetimes (Table 1.2). On the other hand, the TA decays of **C12ZA** appear to be very sensitive to the pump power; when the pump power is increased, the contributions of relatively fast components, τ_1 and τ_2, are enhanced relative to the slowest component, τ_3 (Table 1.2). The pump power dependence of the TA decays is a strong indication of exciton–exciton annihilation [98, 99], because the excitation with high density of photons may generate two or more excitons in one cyclic array, followed by recombination between the excitons, which give rise to fast deactivation channels.

Figure 1.20 shows TAA decay profiles of **Z2**, **2Z2**, and **C12ZA**, reflecting fast depolarization channels resulting from the fast EEH. While **Z2** exhibits the single decay component with a time constant of 0.17 ps, **2Z2** and **C12ZA** exhibit the two decay components; 0.18 and 4.70 ps for **2Z2** and 0.16 and 1.22 ps for **C12ZA**. Since the time constants of fast components are well matched with that of **Z2**, these components are thought to arise from the EEH between porphyrin

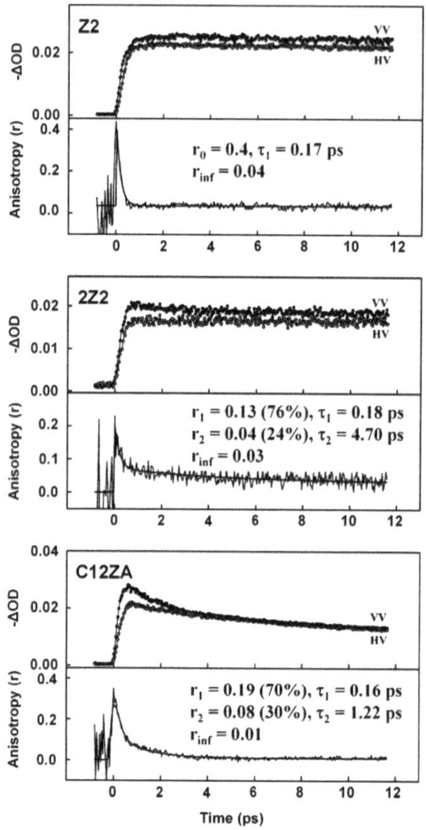

Figure 1.20. Transient absorption anisotropy decay profiles of **Z2**, **2Z2**, and **C12ZA** with their polarized transient absorption decays for parallel (VV) and perpendicular (HV) orientations between pump and probe beams. Insets show the deconvolution fitted anisotropy decay parameters with a train of 150-fs pump pulse. The pump and probe wavelengths were 575 and 510 nm, which correspond to Q-band pump and induced-absorption probe. Reprinted with permission from [97].

monomers via direct linkage. On the other hand, we attribute that the slow components result from the depolarization due to the EEH between **Z2** subunits via 1,3-phenylene spacer.

Because the EEHs of **Z2** and **2Z2** are reversible between porphyrin units, a simple equilibrium scheme should be introduced as follows

$$\mathrm{Por}_A^*\mathrm{Por}_B \Leftrightarrow \mathrm{Por}_A\mathrm{Por}_B^*$$

where Por is the porphyrin unit, and k_1 and k_{-1} are forward and backward reaction rate constants, respectively, which correspond to the EEH rate constants. Then, the EEH rate constant for the above reaction scheme can be defined by Eq. 1.2

$$\tau = 1/(k_1 + k_{-1}) \tag{1.2}$$

where τ is the measured anisotropy decay times of **Z2** and **2Z2**. According to Eq. 1.2, the EEH rate constant via the 1,3-phenylene spacer of **2Z2** was determined to be $1.06 \times 10^{11}\ \mathrm{s}^{-1}$ ($9.4\ \mathrm{ps}^{-1}$) with $\tau = 4.7$ ps, while the EEH rate constant via the direct linkage of **Z2** was determined to be $2.8 \times 10^{12}\ \mathrm{s}^{-1}$ ($0.36\ \mathrm{ps}^{-1}$) with $\tau = 0.18$ ps.

When the Förster-type incoherent energy hopping model is employed by assuming a random walk formalism of anisotropy decay and a migration-limited character of exciton–exciton annihilation, the analytical depolarization and exciton–exciton annihilation times are connected with the EEH time by Eqs. 1.3 and 1.4

$$\tau_{\mathrm{depolarization}} = \tau_{\mathrm{hopping}}/4(1 - \cos^2\alpha) \tag{1.3}$$

$$\tau_{\mathrm{annihilation}} = \tau_{\mathrm{hopping}}(N^2 - 1)/24 \tag{1.4}$$

where N is the number of effective hopping sites, α the angle between the adjacent transition dipoles, $\tau_{\mathrm{annihilation}}$ the slowest exciton–exciton annihilation time, and τ_{hopping} the inverse of the nearest neighbor energy hopping rate [100]. Equation 1.3 is understood by considering that the depolarization is complete when the transition dipole migrates through 90° and that how many hops are required for this rotation to be accomplished. On the other hand, Eq. 1.4 assumes that the exciton–exciton annihilation reflects the migration-limited exciton–exciton recombination along the whole cyclic array and how many hops are required for this recombination to be accomplished. The EEH time of **C12ZA** is calculated with the exciton–exciton annihilation and anisotropy decay times, given in Table 1.2 and Fig. 1.20. As **C12ZA** consists of six **Z2** subunits, the number of hopping sites would be $N = 6$. Introducing $N = 6$ and $\alpha = 60°$ to Eqs. 1.3 and 1.4, the EEH time between neighboring **Z2** subunits bridged by 1,3-phenylene spacer is calculated to be 3.66 and 4.25 ps, respectively. It is noteworthy that the two different experimental observables, exciton–exciton annihilation and anisotropy depolarization times, result in a consistent EEH time (4.0 ± 0.4 ps) within a small error range. The excitation energy migration, thus, is well described by the Förster-type incoherent energy hopping model, assuming the well-arranged cyclic structure of **C12ZA**.

1.5.3.2 Efficiencies of excitation energy transfer at the singe-molecule level

We have carried out coincidence measurements on the cyclic porphyrin wheels, **C10ZA**, **C12ZA**, **C16ZA**, and **C24ZB** and the reference linear molecule, **L12Z**, to evaluate EET efficiencies at the single-molecule level [48, 94, 101]. Refer to the section 1.3.2 for more details of the coincidence measurements. Figure 1.21b shows an interphoton time distribution of **C12ZA** constructed from photons that were detected at the first emissive level in the FIT. The distribution exhibits a very small central peak (N_C) with respect to lateral peaks (N_L), giving rise to the value of N_C/N_L ratio of 0.15. This result is commonly observed for a single-photon emitter, which infers that singlet-singlet annihilation readily occurs in **C12ZA** owing to the efficient EET within the array (Fig. 1.21a).

For statistical analysis, histograms of the N_C/N_L values were constructed and are as displayed in Fig. 1. 21c. First, it is apparent that the distribution for **L12Z** locates in the highest N_C/N_L value region with a mean value of 0.43 and is the broadest among all the distributions. The N_C/N_L value of 0.43 is close to the value expected for a two-photon emitter (0.5), indicating inefficient singlet-singlet annihilation in the linear array. The broad distribution is the result of large conformational heterogeneity because the **Z2** subunits in **L12Z** have rotational freedom with respect to 1,3-phenylene spacers, which may largely be reduced when the cyclic structure is formed. If we compare cyclic arrays **C12ZA** and **C24ZB**, the distribution of N_C/N_L values of **C24ZB** is slightly shifted to higher values relative to that of **C12ZA**, which reveals less efficient EET in the larger array. The relatively long interchromophoric distance between **Z4**

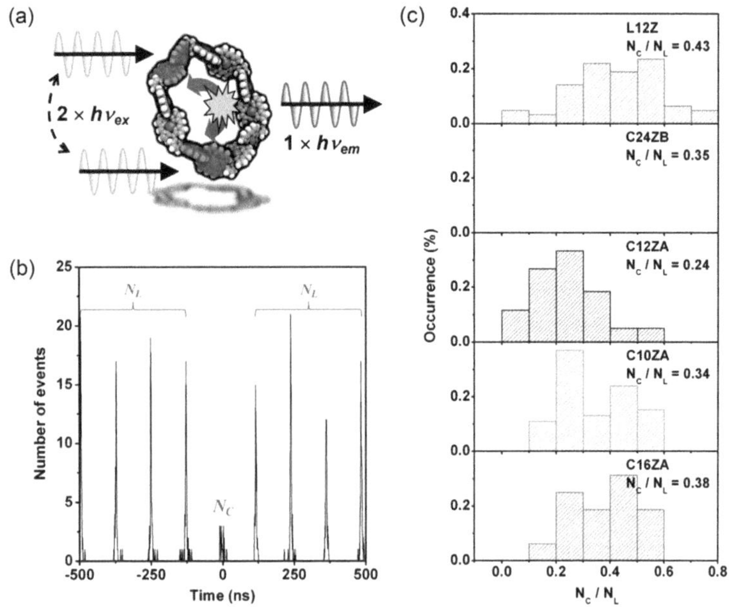

Figure 1.21. (a) Schematic representation of the singlet-singlet annihilation in **C12ZA**. (b) Interphoton arrival time distribution of **C12ZA**. (c) Histograms for the N_C/N_L values of **L12Z, C24ZB, C12ZA, C10ZA**, and **C16ZA** from top to bottom. Reprinted with permission from [94, 101].

subunits in **C24ZB** that reduces dipole coupling is thought to be responsible for this result. On the basis of these results, we have further focused on the examination of the cyclic porphyrin wheels, **C**$_N$**ZA**, consisting of **Z2** subunits.

In the histograms of the N_C/N_L values of **C10ZA**, **C12ZA**, and **C16ZA**, the distributions are shifted to higher values in the order of **C12ZA** (0.24) → **C10ZA** (0.34) → **C16ZA** (0.38) (Fig. 1.21c). At the same time, we could not find the N_C/N_L values smaller than 0.1 for **C10ZA** or **C16ZA**, but only for **C12ZA** (Fig. 1.21c). We attribute the less efficient singlet-singlet annihilation in **C10ZA** to considerable structural strain caused by the smaller ring size compared to **C12ZA**, which perturbs dipole coupling between **Z2** subunits by enforcing the dihedral angles and orientation factors to be changed unfavorably. On the other hand, in the case of **C16ZA**, the presence of less-favorable structural conformations that possess a large number of nonradiative decay channels resulting from conformational flexibility, as well as the longer interchromophoric distance between **Z2** subunits relative to that in **C12ZA** may disturb singlet-singlet annihilation to occur efficiently.

The small N_C / N_L value for **C12ZA** reveals that the overall structure of **C12ZA** is rigid enough without any significant distortion. The 1,3-phenylene spacer with an angle of 120° between the two linking C–C bonds at 1 and 3 positions of the phenyl group is well suited to accommodate six **Z2** subunits to form two-dimensional hexagonal structure (see Fig. 1.19). Thus, we conclude that **C12ZA** is the most promising material as an efficient light-harvesting antenna system among the investigated cyclic porphyrin wheels.

1.5.4 *Conformational Heterogeneities in C_NZA*

1.5.4.1 Conformational heterogeneities at the ensemble level

In the B800 and B850 of light-harvesting complex LH2 from *Rps. acidophila,* 9 and 18 Bchl a pigments are arranged symmetrically in a circular ring structure owing to the support of protein matrices [85–87]. However, as the cyclic porphyrin wheels are absent of such media to anchor constituent porphyrin units rigidly in a well-arranged manner, the relationship between molecular size and shape persistency is a significant issue that intimately relates to the EET efficiency. Increasing the number of chromophores in cyclic arrays is advantageous in absorbing capability, but will definitely accompany structural flexibility and heterogeneity, leading to energy sink.

The fluorescence temporal profiles of **C$_N$ZA** measured using TCSPC technique are depicted in Fig. 1.22, and their fitting parameters are listed in Table 1.3 [102]. Although all **C$_N$ZA** are photoexcited to higher electronic S_2 states, the observed decay times can be regarded as the S_1-state lifetimes since the internal conversion process from S_2 to S_1 states in **Z2** and **C$_N$ZA** is beyond the temporal resolution of TCSPC. While the fluorescence decays of **C10ZA** and **C12ZA** show single exponential decay, the other larger cyclic arrays exhibit double exponential decay. Fast components (<200 ps) in the fluorescence decays of **C16ZA** and larger arrays presumably result

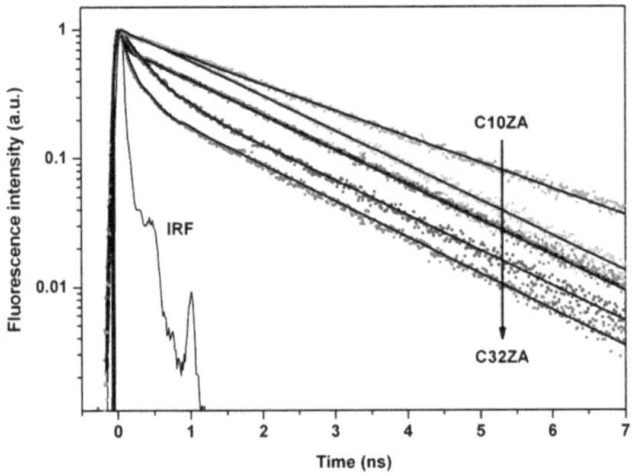

Figure 1.22. Fluorescence decay profiles of $\mathbf{C}_N\mathbf{ZA}$ in toluene. The excitation and detection wavelengths were 400 and 625 nm, respectively. Reprinted with permission from [102].

from the conformational heterogeneity contributed by the flexible dihedral angles between **Z2** subunits with respect to 1,3-phenylene spacers. Overall, as the molecular size increases, the average fluorescence lifetime gradually decreases in going from 1.73 ns (**C10ZA**)

Table 1.3. Fluorescence lifetimes[a] and fluorescence quantum yields[b] of **Z2**, **2Z2**, and $\mathbf{C}_N\mathbf{ZA}$ in Toluene. Reprinted with permission from [102].

Compounds	Fluorescence lifetime[a] (ns)			Φ_F[b]
	τ_{avg}	τ_1 (%)	τ_2 (%)	
Z2	1.82		1.82 (100)[c]	0.034
Z2Z	1.73		1.73 (100)[c]	0.040
C10ZA	1.73		1.73 (100)	0.025
			1.66 (100)[c]	
C12ZA	1.54		1.54 (100)	0.036
C16ZA	0.76	0.07 (55)	1.6 (45)	0.032
C18ZA	0.54	0.06 (69)	1.6 (31)	0.026
C24ZA	0.47	0.13 (75)	1.5 (25)	0.020
C32ZA	0.46	0.19 (79)	1.5 (21)	0.020

[a] Excitation and monitoring wavelengths were 400 and 625 nm, respectively. Relative amplitudes of each component are in parentheses.
[b] Relative fluorescence quantum yields obtained by comparison with respect to $\Phi_F = 0.033$ for ZnTPP in toluene. Excitation wavelength of 550 nm was used.
[c] Fluorescence lifetimes in THF.

to 0.46 ns (**C32ZA**). In addition, except for **C10ZA**, the overall fluorescence quantum yield (Φ_F) decreases with increasing molecular size because of the increased nonradiative decay channels in larger cyclic porphyrin wheels (Table 1.3). According to the fluorescence lifetimes and quantum yields, it is inferred that the number of conformational isomers increases as the size of $\mathbf{C}_N$**ZA** becomes larger.

1.5.4.2 Conformational heterogeneities at the single-molecule level

Figure 1.23a shows histograms of the fluorescence lifetimes at the initial emissive level in FITs of the cyclic porphyrin wheels, and Fig. 1.23b shows a typical FIT and fluorescence decay profile of single **C18ZA** molecule, illustrating the fluorescence lifetime at the initial emissive level in FIT [101]. The distribution of **C16ZA** is much wider ranging from 0.2 to 2.6 ns compared to those of smaller cyclic arrays, with the mean fluorescence lifetime of 0.84 ns. This feature can be explained by larger conformational freedom and, hence, larger conformational heterogeneity in **C16ZA** induced from the larger molecular size.

Remarkably, in the cases of **C18ZA**, **C24ZA**, and **C32ZA**, the distributions become narrow with a significant reduction in their fluorescence lifetimes. This is consistent with the trend of fluorescence lifetimes at the ensemble level, where the amplitudes of the fast decay components gradually increase as the cyclic arrays become larger. It is thought that the structural flexibilities in these larger cyclic arrays lead to a considerable variation in conformations, and most of them have a large number of nonradiative decay channels to shorten their S_1-state lifetimes.

Relative molecular brightness of the cyclic porphyrin wheels show a good correlation with the double exponential fluorescence decay abruptly appearing From **C16ZA**.

The relative molecular brightness at the single-molecule level is a compatible concept of fluorescence quantum yield and can be estimated by using initial fluorescence intensity in FIT and extinction coefficient of molecule. As the initial fluorescence intensity in FIT is related to the probability to emit fluorescence photons upon photoexcitation, a value obtained by dividing this intensity by molar

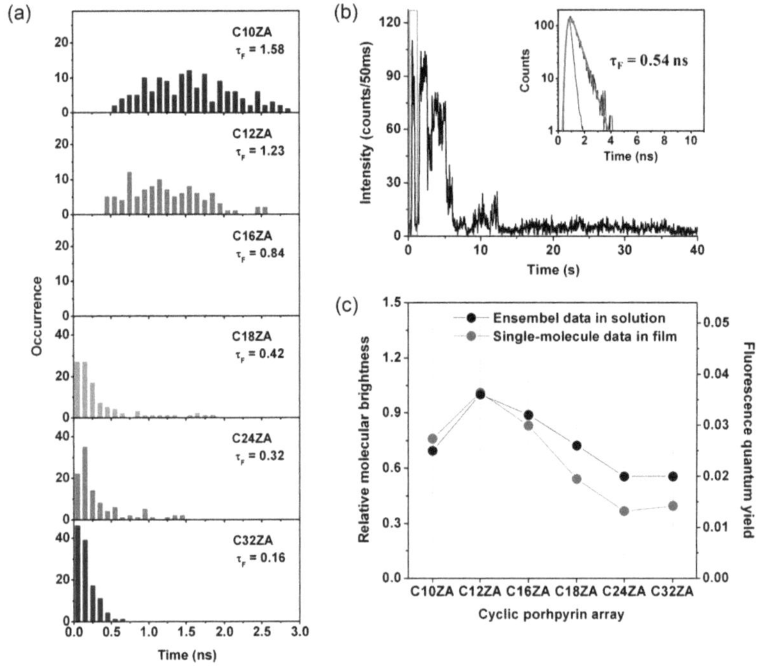

Figure 1.23. (a) Histograms of the fluorescence lifetimes of $\mathbf{C}_N\mathbf{ZA}$ at the initial emissive level in the FITs. (b) A typical FIT and fluorescence decay profile at the initial emissive level in the FIT of single **C18ZA** molecule. (c) A plot of average values of the relative molecular brightness at the single-molecule level and fluorescence quantum yields in toluene in the cyclic porphyrin wheels. The error range indicates the standard deviation of the relative molecular brightness for each cyclic porphyrin wheel. Reprinted with permission from [101].

extinction coefficient presents a fluorescence capability of single-molecule.

Figure 1.23c shows a plot of relative molecular brightness and fluorescence quantum yields of $\mathbf{C}_N\mathbf{ZA}$. In the cyclic porphyrin wheels, the molar extinction coefficients as a function of the number of **Z2** subunits exhibit a linear behavior, indicating higher absorptivity with increasing molecular size. In parallel, the initial fluorescence intensity also appears to increase as the cyclic array becomes larger. For **C10ZA** and **C12ZA**, for example, the average count rate in the initial emissive level increases from 2200 to 3900 s^{-1} and the extinction coefficient also increases from 1.59×10^5 to 2.06×10^5 M^{-1} cm^{-1}, which result in 16% increase in the relative molecular brightness in **C12ZA**. However, in larger cyclic arrays than **C12ZA**, the relative molecular brightness gradually decreases, because the count rate at the initial emissive level no longer increases linearly. These results are well matched with the ensemble-level fluorescence quantum yields as well as the very short excited-state lifetimes for the larger cyclic arrays. Therefore, in overall, conformational heterogeneity progressively increases as the cyclic porphyrin wheel becomes larger than **C12ZA**.

1.6 Conclusions and Future Outlook

A variety of porphyrin arrays have so far been investigated both at the ensemble and single-molecule level for applications in molecular photonics and electronics. In one-dimensional linear arrays, linkage motifs are one of key factors in determining angular relationship between neighboring porphyrins, which in turn principally determines electronic coupling and concomitant spectroscopic properties of the arrays. As seen in $\mathbf{Z}_N$, strong dipolar coupling renders the constituent porphyrin units to maintain individual characters in photobleaching dynamics. However, EET in these arrays occurs as the hopping-like process of coherently coupled exciton. On the other hand, $\mathbf{Z}_N\mathbf{B}$ has proven that extensive π-conjugation makes the arrays to perform as one quantum system underlying efficient EET. Surprisingly, this illustration is true for the longest array reaching ~80 Å in length. Also, through the femtosecond TA measurements on $\mathbf{TB}_N$, we have verified that, in the excited-state energy relaxation dynamics, π-conjugation promotes the contribution of vibrational relaxation process to be increased compared to that of electronic transition. In two-dimensional cyclic porphyrin wheels, structural rigidity with respect to molecular size plays an important role in EET efficiency. It has been demonstrated that as the molecular size becomes larger in $\mathbf{C}_N\mathbf{ZA}$, EET efficiency decreases in parallel with the increasing conformational flexibility that is thought to produce a large number of nonradiative decay channels. **C12ZA** has proven to be the most rigid and well-defined structure among the investigated cyclic arrays, where 1,3-phenylene spacers well accommodate six **Z2** subunits in two-dimensional hexagonal shape. In overall, in-depth understanding on the influences of electronic coupling and conformational heterogeneity on EET process in multiporphyrin molecular assemblies will be a platform in applying the porphyrin arrays as real molecular devices and in designing novel architectures with tuned electronic couplings to elicit desired photophysical properties.

Acknowledgement

This research was financially supported by the World Class University (R32-2010-000-10217) and Mid-career Researcher (2010-0029668) programs of the Ministry of Education, Science, and Technology (MEST) of Korea, and the AFSOR/AOARD grant (FA2386-09-1-4092). The work at Kyoto University was supported by Grants-in-Aid (A) (No. 19205006 (A) and 20108001 "pi-Space") for Scientific Research from MEXT of Japan.

References

1. Anderson, H. L. (1994) Conjugated porphyrin ladders, *Inorg. Chem.*, **33**, 972–981.
2. Arnold, D. P. and Nitschinsk, L. J. (1992) Porphyrin dimers linked by conjugated butadiynes, *Tetrahedron*, **48**, 8781–8792.
3. Arnold, D. P. and Nitschinsk, L. J. (1993) The preparation of novel porphyrins and bis(porphyrins) using palladium catalysed coupling reactions, *Tetrahedron Lett.*, **34**, 693–696.
4. Burrell, A. K., Officer, D. L. and Reid, D. C. W. (1995) Aldehyde-appended tetraphenylporphyrin: a new building block for porphyrin arrays, *Angew. Chem., Int. Ed. Engl.*, **34**, 900–902.
5. Lin, V. S.-Y. DiMagno, S. G. and Therien, M. J. (1994) Highly conjugated, acetylenyl bridged porphyrins: new models for light-harvesting antenna systems, *Science*, **264**, 1105–1111.
6. Lin, V. S.-Y. and Therien, M. J. (1995) The role of porphyrin-to-porphyrin linkage topology in the extensive modulation of the absorp-

tive and emissive properties of a series of ethynyl-and butadiynyl-bridged bis- and tris(porphinato)zinc chromophores, *Chem. Eur. J.*, **1**, 645–651.

7. Osuka, A. Maruyama, K. Yamazaki, I. and Tamai, N. (1990) Singlet excitation energy transfer in conformationally restricted zinc-free-base hybrid diporphyrins, *Chem. Phys. Lett.*, **165**, 392–396.
8. Osuka, A. Maruyama, K. Mataga, N. Asahi, T., Yamazaki, I. and Tamai, N. (1990) Geometry dependence of intramolecular photoinduced electron transfer in synthetic zinc-ferric hybrid diporphyrins, *J. Am. Chem. Soc.*, **112**, 4958–4959.
9. Osuka, A. Liu, B.-L. and Maruyama, K. (1993) An efficient one-pot synthetic procedure of multiple porphyrin-cyclization, *Chem. Lett.*, **22**, 949–952.
10. Ponomarev, G. V., Borovkov, V. V., Sugiura, K.-I., Sakata, Y. and Shul'ga, A. M. (1993) Synthesis and properties of cis-1,2-bis(octaethylporphyrinyl)ethylene, *Tetrahedron Lett.*, **34**, 2153–2156.
11. Rubtsov, I. V., Susumi, K., Rubtsov, G. I. and Therien, M. J. (2003) Ultrafast singlet excited-state polarization in electronically asymmetric ethynebridged bis[(porphinato)zinc(II)] complexes, *J. Am. Chem. Soc.*, **125**, 2687–2696.
12. Senge, M. O., Gerzevske, K. R., Vicente, M. G. H., Forsyth, T. P. and Smith, K. M. (1993) Models for the photosynthetic reaction center-synthesis and structure of porphyrin dimers with cis- and trans-ethene and skewed hydroxymethylene bridges, *Angew. Chem., Int. Ed. Engl.*, **32**, 750–753.
13. Senge, M. O., Vicente, M. G. H., Gerzevske, K. R., Forsyth, T. P. and Smith, K. M. (1994) Models for the photosynthetic reaction center: preparation, spectroscopy, and crystal and molecular structures of cofacial bisporphyrins linked by cis-1,2 and trans-1,2-ethene bridges and of 1,1-carbinol-bridged bisporphyrins, *Inorg. Chem.*, **33**, 5625–5638.
14. Vicente, M. G. H. and Smith, K. M. (1991) Vilsmeier reactions of porphyrins and chlorins with 3-(dimethylamino)acrolein to give *meso*-(2-formylvinyl)porphyrins: new syntheses of benzochlorins, benzoisobacteriochlorins, and benzobacteriochlorins and reductive coupling of porphyrins and chlorins using low-valent titanium complexes, *J. Org. Chem.*, **56**, 4407–4418.
15. Yang, S. I., Lammi, R. K., Seth, J., Riggs, J. A., Arai, T., Kim, D., Bocian, D. F., Holten, D. and Lindsey, J. S. (1998) Excited-state energy transfer and ground-state hole/electron hopping in *p*-phenylene-linked porphyrin dimers, *J. Phys. Chem. B*, **102**, 9426–9436.

16. De Schryver, F. C., Vosch, T., Cotlet, M., Van der Auweraer, M., Müllen, K., and Hofkens, J. (2005) Energy dissipation in multichromophoric single dendrimers, *Acc. Chem. Res.*, **38**, 514–522.

17. García-Parajó, M. F., Hernando, J., Mosteiro, G. S., Hoogenboom, J. P., van Dijk, E. M. H. P, and van Hulst, N. F. (2005) Energy transfer in single-molecule photonic wires, *ChemPhysChem*, **6**, 819–827.

18. Ha, T., Enderle, Th., Chemla, D. S., Selvin, P. R. and Weiss, S. (1996) Single molecule dynamics studied by polarization modulation, *Phys. Rev Lett.*, **77**, 3979–3982.

19. Lu, H. P. and Xie, X. S. (1997) Single-molecule kinetics of interfacial electron transfer, *J. Phys. Chem. B*, **101**, 2753–2757.

20. Moerner, W. E. and Orrit, M. (1999) Illuminating single molecules in condensed matter, *Science*, **283**, 1670–1676.

21. Moerner, W. E. (2002) A dozen years of single-molecule spectroscopy in physics, chemistry, and biophysics, *J. Phys. Chem. B*, **106**, 910–927.

22. Rigler, R., Orrit, M. and Basché, Th. (2002) *Single Molecule Spectroscopy*, 1st edn. Chemical Physics, Springer, Heidelberg.

23. Tinnefeld, P. and Sauer, M. (2005) Branching out of single-molecule fluorescence spectroscopy: challenges for chemistry and influence on biology, *Angew. Chem. Int. Ed.*, **44**, 2642–2671.

24. Xie, X. S. and Trautman, J. K. (1998) Optical studies of single molecules at room temperature, *Annu. Rev. Phys. Chem.*, **49**, 441–480.

25. Aratani, N., Osuka, A., Kim, Y. H., Jeong, D. H. and Kim, D. (2000) Extremely long, discrete *meso-meso*-coupled porphyrin arrays, *Angew. Chem. Int. Ed.*, **39**, 1458–1462.

26. Nakano, A., Osuka, A., Yamazaki, I., Yamazaki, T. and Nishimura, Y. (1998) Windmill-like porphyrin arrays as potent light-harvesting antenna complexes, *Angew. Chem., Int. Ed. Engl.*, **37**, 3023–3027.

27. Osuka, A. and Shimidzu, H. (1997) *Meso, meso*-linked porphyrin arrays, *Angew. Chem., Int. Ed. Engl.*, **36**, 135–137.

28. Yoshida, N. Shimidzu, H. and Osuka, A. (1998) Meso-meso linked diporphyrins from 5,10,15-trisubstituted porphyrins, *Chem. Lett.*, **27**, 55–56.

29. Aratani, N., Osuka, A., Cho, H. S. and Kim, D. (2002) Photochemistry of covalently-linked multi-porphyrinic systems, *J. Photochem. Photobiol. C: Photochem. Rev*, **3**, 25–52.

30. Cho, H. S., Song, N. W., Kim, Y. H., Jeoung, S. C., Hahn, S., Kim, D., Kim, S. K., Yoshida, N. and Osuka, A. (2000) Ultrafast energy relaxation dynamics of directly linked porphyrin arrays, *J. Phys. Chem. A*, **104**, 3287–3298.

31. Cho, H. S., Jeong, D. H., Yoon, M.-C., Kim, Y. H., Kim, Y.-R., Kim, D., Jeoung, S. C., Kim, S. K., Aratani, N., Shinmori, H. and Osuka, A. (2001) Excited-state energy transfer processes in phenylene- and biphenylene-linked and directly-linked zinc(II) and free-base hybrid diporphyrins, *J. Phys. Chem. A*, **105**, 4200–4210.

32. Kim, D. and Osuka, A. (2003) Photophysical properties of directly linked linear porphyrin arrays, *J. Phys. Chem. A*, **107**, 8791–8816.

33. Kim, D. and Osuka, A. (2004) Directly linked porphyrin arrays with tunable excitonic interactions, *Acc. Chem. Res.*, **37**, 735–745.

34. Kim, Y. H., Jeong, D. H., Kim, D., Jeoung, S. C., Cho, H. S., Kim, S. K., Aratani, N. and Osuka, A. (2001) Photophysical properties of long rodlike *meso-meso*-linked zinc(II) porphyrins investigated by time-resolved laser spectroscopic methods, *J. Am. Chem. Soc.*, **123**, 76–86.

35. Kim, Y. H., Cho, H. S., Kim, D., Kim, S. K., Yoshida, N. and Osuka, A. (2001) Photophysical properties of meso, meso-linked porphyrin arrays: steady-state and time-resolved fluorescence polarization, *Synth. Metal*, **117**, 183–187.

36. Min, C.-K., Joo, T., Yoon, M.-C., Kim, C. M., Hwang, Y. N., Kim, D., Aratani, N., Yoshida, N. and Osuka, A. (2001) Transient absorption anisotropy study of ultrafast energy transfer in porphyrin monomer, its direct *meso-meso* coupled dimer and trimer, *J. Chem. Phys.*, **114**, 6750–6758.

37. Kasha, M., Rawls, H. R. and El-Bayoumi, M. A. (1965) The exciton model in molecular spectroscopy, *Pure Appl. Chem.*, **11**, 371–392.

38. Yang, J., Yoo, H., Aratani, N., Osuka, A. and Kim, D. (2009) Determination of the superradiance coherence length of directly linked linear porphyrin arrays at the single-molecule level, *Angew. Chem. Int. Ed.*, **48**, 4323–4327.

39. Vanden Bout, D. A., Yip, W.-T., Hu, D., Fu, D.-K., Swager, T. M. and Barbara, P. F. (1997) Discrete intensity jumps and intramolecular electronic energy transfer in the spectroscopy of single conjugated polymer molecules, *Science*, **277**, 1074–1077.

40. Ha, J.-H., Cho, H. S., Song, J. K., Kim, D., Aratani, N. and Osuka, A. (2004) Excitonic coupling strength and coherence length in the singlet and triplet excited states of *meso-meso* directly linked Zn(II)porphyrin arrays, *ChemPhysChem*, **5**, 57–67.

41. Kakitani, T. and Kimura, A. (2002) Empirical formula of exciton coherent domain in oligomers and application to LH2, *J. Phys. Chem. A*, **106**, 2173–2179.

42. Kimura, A., Kakitani, T. and Yamato, T. (2000) Theory of excitation energy transfer in the intermediate coupling case. II. Criterion

for intermediate coupling excitation energy transfer mechanism and application to the photosynthetic antenna system, *J. Phys. Chem. B*, **104**, 9276–9286.

43. Kimura, A. and Kakitani, T. (2007) Advanced theory of excitation energy transfer in dimers, *J. Phys. Chem. A*, **111**, 12042–12048.
44. Pullerits, T., Chachisvilis, M. and Sundström, V. (1996) Exciton delocalization length in the B850 antenna of *Rhodobacter sphaeroides*, *J. Phys. Chem.*, **100**, 10787–10792.
45. Zhao, Y., Meier, T., Zhang, W. M., Chernyak, V. and Mukamel, S. (1999) Superradiance coherence sizes in single-molecule spectroscopy of LH2 antenna complexes, *J. Phys. Chem. B*, **103**, 3954–3962.
46. Ahn, T. K., Yoon, Z. S., Hwang, I.-W., Lim, J. K., Rhee, H. Joo, T., Sim, E., Kim, S. K., Aratani, N., Osuka, A. and Kim, D. (2005) Effect of conformational heterogeneity on excitation energy transfer efficiency in directly *meso-meso* linked Zn(II) porphyrin arrays, *J. Phys. Chem. B*, **109**, 11223–11230.
47. Park, M., Cho, S., Yoon, Z. S., Aratani, N., Osuka, A. and Kim, D. (2005) Single molecule spectroscopic investigation on conformational heterogeneity of directly linked zinc(II) porphyrin arrays, *J. Am. Chem. Soc.*, **127**, 15201–15206.
48. Yang, J. and Kim, D. (2009) Single molecule spectroscopic investigation on various multiporphyrin systems as molecular photonic devices, *J. Mater. Chem.*, **19**, 1057–1062.
49. Martin, R. E. and Diederich, F. (1999) Linear monodisperse π-conjugated oligomers: model compounds for polymers and more, *Angew. Chem. Int. Ed.*, **38**, 1350–1377.
50. Chen, J., Reed, M. A., Rawlett, A. M. and Tour, J. M. (1999) Large on-off ratios and negative differential resistance in a molecular electronic device, *Science*, **286**, 1550–1552.
51. Stegeman, G. and Likamwa, P. (1995) *Nonlinear Optical Materials and Devices for Applications in Information Technology* **289**, (ed. Miller, A., Welford, K. R., Daino, B.) Kluwer, Dordrecht, Netherlands pp. 285–320.
52. Arnold, D. P. and Heath, G. A. (1993) Voltammetric and UV to near-IR spectroelectrochemical characterization of the *meso-meso*-'buta-1,3-diyne-bridged octaethylporphyrin dimers {[M(OEP)](μ-C_4)[M(OEP)]} (M_2 = H_4, Co_2, Ni_2, Cu_2, Zn_2, Pd_2, Pt_2, Co/Ni, and Ni/Zn), in their neutral, mononegative, and dinegative oxidation states, *J. Am. Chem. Soc.*, **115**, 12197–12198.
53. Anderson, H. L. (1999) Building molecular wires from the colours of life: conjugated porphyrin oligomers, *Chem. Commun.*, 2323–2330.

54. Vicente, M. G. H., Jaquinod, L. and Smith, K. M. (1999) Oligomeric porphyrin arrays, *Chem. Commun.*, 1771–1782.

55. Tsuda, A., Nakano, A., Furuta, H., Yamochi, H. and Osuka, A. (2000) Doubly *meso*-β-linked diporphyrins from oxidation of 5,10,15-triaryl-substituted Ni^{II} and Pd^{II}- porphyrins, *Angew. Chem. Int. Ed.*, **39**, 558–561.

56. Tsuda, A., Furuta, H. and Osuka, A. (2000) Completely fused diporphyrins and triporphyrin, *Angew. Chem. Int. Ed.*, **39**, 2549–2552.

57. Lee, S. J., Mulfort, K. L., O'Donnell, J. L., Zuo, X., Goshe, A. J., Wesson, P. J., Nguyen, S. T., Hupp, J. T. and Tiede, D. M. (2006) Supramolecular porphyrinic prisms: coordinative assembly and solution phase X-ray structural characterization, *Chem. Commun.*, 4581–4583.

58. Kuimova, M. K., Botchway, S. W., Parker, A. W., Balaz, M., Collins, H. A., Anderson, H. L., Suhling, K. and Ogilby, P. R. (2009) Imaging intracellular viscosity of a single cell during photoinduced cell death, *Nature Chem.*, **1**, 69–73.

59. Kuimova, M. K., Balaz, M., Anderson, H. L. and Ogilby, P. R. (2009) Intramolecular rotation in a porphyrin dimer controls singlet oxygen production, *J. Am. Chem. Soc.*, **131**, 7948–7949.

60. Winters, M. U., Kärnbratt, J., Eng, M., Wilson, C. J., Anderson, H. L. and Albinsson, B. (2007) Photophysics of a butadiyne-linked porphyrin dimer: influence of conformational flexibility in the ground and first singlet excited state, *J. Phys. Chem. C*, **111**, 7192–7199.

61. Winters, M. U., Kärnbratt, J., Blades, H. E., Wilson, C. J., Frampton, M. J., Anderson, H. L. and Albinsson, B. (2007) Control of electron transfer in a conjugated porphyrin dimer by selective excitation of planar and perpendicular conformers, *Chem. Eur. J.*, **13**, 7385–7394.

62. Lee, J.-E., Yang, J. and Kim, D. (2011) Single-molecule fluorescence dynamics of butadiyne-linked porphyrin dimer: effect of conformational flexibility in host polymers, *Faraday Discuss.*, DOI: 10.1039/C1FD00082A.

63. Hernando, J., Hoogenboom, J. P., van Dijk, E. M. H. P., García-López, J. J., Crego-Calama, M., Reinhoudt, D. N., van Hulst, N. F. and García-Parajó, M. F. (2004) Single molecule photobleaching probes the exciton wavefunction in a multichromophoric system, *Phys. Rev. Lett.*, **93**, 236404.

64. Hofkens, J., Maus, M., Gensch, Th., Vosch, T., Cotlet, M., Köhn, F., Herrmann, A., Müllen, K. and De Schryver, F. C. (2000) Probing photophysical processes in individual multichromophoric dendrimers by single-molecule spectroscopy, *J. Am. Chem. Soc.*, **122**, 9278–9288.

65. Hofkens, J., Cotlet, M., Vosch, T., Tinnefeld, P., Weston, K. D., Ego, C., Grimsdale, A., Müllen, K., Beljonne, D., Brédas, J. L., Jordens, S., Schweitzer, G., Sauer, M. and De Schryver, F. C. (2003) Revealing competitive Förster-type resonance energy-transfer pathways in single bichromophoric molecules, *Proc. Natl. Acad. Sci. USA*, **100**, 13146–13151.

66. Lippitz, M., Hübner, C. G., Christ, Th., Eichner, H., Bordat, P., Herrmann, A., Müllen, K. and Basché, Th. (2004) Coherent electronic coupling versus localisation in individual molecular dimers, *Phys. Rev. Lett.*, **92**, 103001.

67. Vosch, T., Fron, E., Hotta, J.-I., Deres, A., Uji-i, H., Idrissi, A., Yang, J., Kim, D., Puhl, L., Haeuseler, A., Müllen, K., De Schryver, F. C., Sliwa, M. and Hofkens, J. (2009) Synthesis, ensemble, and single molecule characterization of a diphenyl-acetylene linked perylenediimide trimer, *J. Phys. Chem. C*, **113**, 11773–11782.

68. Yang, J., Lee, J.-E., Lee, C. Y., Aratani, N., Osuka, A., Hupp, J. T. and Kim, D. (2011) The role of electronic coupling in linear porphyrin arrays probed by single-molecule fluorescence spectroscopy, *Chem. Eur. J.*, **17**, 9219–9225.

69. Masuo, S., Vosch, T., Cotlet, M., Tinnefeld, P., Habuchi, S., Bell, T. D. M., Oesterling, I., Beljonne, D., Champagne, B., Müllen, K., Sauer, M., Hofkens, J. and De Schryver, F. C. (2004). Multichromophoric dendrimers as singlephoton sources: a singlemolecule study, *J. Phys. Chem. B*, **108**, 16686–16696.

70. Weston, K. D., Dyck, M., Tinnefeld, P., Müller, C., Herten, D. P. and Sauer, M. (2002). Measuring the number of independent emitters in single-molecule fluorescence images and trajectories using coincident photons, *Anal. Chem.* **74**, 5342–5349.

71. Hanbury-Brown, R. and Twiss, R. Q. (1956). Correlation between photons in two coherent beams of light, *Nature*, **177**, 27–29

72. Cichos, F., von Borczyskowski, C. and Orrit, M. (2007). Powerlaw intermittency of single emitters, *Curr. Opin. Colloid Interface Sci.* **12**, 272–284.

73. Hoogenboom, J. P., den Otter, W. K. and Offerhaus, H. L. (2006). Accurate and unbiased estimation of power-law exponents from singleemitter blinking data, *J. Chem. Phys.*, **125**, 204713.

74. Mirzov, O., Cichos, F., von Borczyskowski, C. and Scheblykin, I. G. (2004). Direct exciton quenching in single molecules of MEHPPV at 77 K, *Chem. Phys. Lett.*, **386**, 286–290.

75. Frenkel, J. (1931). On the transformation of light into heat in solids. II, *Phys. Rev.*, **37**, 1276–1294.
76. Wannier, G. H. (1937). The structure of electronic excitation levels in insulating crystals, *Phys. Rev.* **52**, 191–197.
77. Mott, N. F. (1938). Conduction in polar crystals. II. The conduction band and ultra-violet absorption of alkali-halide crystals, *Trans. Faraday Soc.*, **34**, 500–506.
78. Ikeda, T., Aratani, N. and Osuka, A. (2009) Synthesis of extremely π-extended porphyrin tapes from hybrid *meso-meso* linked porphyrin arrays: an approach towards the conjugation length, *Chem. Asian J.*, **4**, 1248–1256.
79. Kim, P., Ikeda, T., Lim, J. M., Park, J., Lim, M., Aratani, N., Osuka, A. and Kim, D. (2011) Excited-state energy relaxation dynamics of triply linked Zn(II) porphyrin arrays, *Chem. Commun.*, **47**, 4433–4435.
80. Cho, H. S., Jeong, D. H., Cho, S., Kim, D., Matsuzaki, Y., Tanaka, K., Tsuda, A. and Osuka, A. (2002) Photophysical properties of porphyrin tapes, *J. Am. Chem. Soc.*, **124**, 14642–14654.
81. Tsuda, A. and Osuka, A. (2001) Fully conjugated porphyrin tapes with electronic absorption bands that reach into Infrared, *Science*, **293**, 79–82.
82. Eom, H. S., Jeoung, S. C., Kim, D., Ha, J.-H. and Kim, Y.-R. (1997) Ultrafast vibrational relaxation and ligand photodissociation/photoassociation processes of nickel(II) porphyrins in the condensed phase, *J. Phys. Chem. A*, **101**, 3661–3669.
83. Rodriguez, J., Kirmaier, C. and Holten, D. (1989) Optical properties of metalloporphyrin excited states, *J. Am. Chem. Soc.*, **111**, 6500–6506.
84. Rodriguez, J., Kirmaier, C. and Holten, D. (1991) Time-resolved and static optical properties of vibrationally excited porphyrins, *J. Chem. Phys.*, **94**, 6020–6029.
85. Koepke, J., Hu, X., Muenke, C., Schulten, K. and Michel, H. (1996) The crystal structure of the light-harvesting complex II (B800–850) from *Rhodospirillum molischianum*, *Structure*, **4**, 581–597.
86. McDermott, G., Prince, S. M., Freer, A. A., Hawthornthwaite-Lowless, A. M., Papiz, M. Z., Cogdell R. J. and Isaacs, N. W. (1995) Crystal structure of an integral membrane light-harvesting complex from photosynthetic bacteria, *Nature*, **374**, 517–521.
87. Roszak, A. W., Howard, T. D., Southall, J., Gardiner, A. T., Law, C. J., Isaacs, N. W. and Cogdell, R. J. (2003) Crystal structure of the RC-LH1 core complex from *Rhodopseudomonas palustris*, *Science*, **302**, 1969–1972.

88. Aratani, N., Kim, D. and Osuka, A. (2009) Discrete cyclic porphyrin arrays as artificial light-harvesting antenna, *Acc. Chem. Res.*, **42**, 1922–1934.

89. Hori, T., Aratani, N., Takagi, A., Matsumoto, T., Kawai, T., Yoon, M.-C., Yoon, Z. S., Cho, S., Kim, D. and Osuka, A. (2006) Giant porphyrin wheels with large electronic coupling as models of light-harvesting photosynthetic antenna, *Chem. Eur. J.*, **12**, 1319–1327.

90. Hori, T., Peng, X., Aratani, N., Takagi, A., Matsumoto, T., Kawai, T., Yoon, Z. S., Yoon, M.-C., Yang, J., Kim, D. and Osuka, A. (2008) Synthesis of nanometer-scale porphyrin wheels of variable size, *Chem. Eur. J.*, **14**, 582–595.

91. Nakamura, Y., Aratani, N. and Osuka, A. (2007) Cyclic porphyrin arrays as artificial photosynthetic antenna: synthesis and excitation energy transfer, *Chem. Soc. Rev.*, **36**, 831–845.

92. Peng, X., Aratani, N., Takagi, A., Matsumoto, T., Kawai, T., Hwang, I.-W., Ahn, T., K., Kim, D. and Osuka, A. (2004) A dodecameric porphyrin wheel, *J. Am. Chem. Soc.*, **126**, 4468–4469.

93. Yoon, Z. S., Yoon, M.-C. and Kim, D. (2005) Excitonic coupling in covalently linked multiporphyrin systems by matrix diagonalization, *J. Photochem. Photobiol. C: Photochem. Rev.*, **6**, 249–263.

94. Park, M., Yoon, M.-C., Yoon, Z. S., Hori, T., Peng, X., Aratani, N., Hotta, J.-I., Uji-i, H., Sliwa, M., Hofkens, J., Osuka, A. and Kim, D. (2007) Single-molecule spectroscopic investigation of energy migration processes in cyclic porphyrin arrays, *J. Am. Chem. Soc.*, **129**, 3539–3544.

95. Bartko, A. P. and Dickson, R. M. (1999) Imaging three-dimensional single molecule orientations, *J. Phys. Chem. B*, **103**, 11237–11241.

96. Schroeyers, W., Vallée, R., Patra, D., Hofkens, J., Habuchi, S., Vosch, T., Cotlet, M., Müllen, K., Enderlein, J. and De Schryver, F. C. (2004) Fluorescence lifetimes and emission patterns probe the 3D orientation of the emitting chromophore in a multichromophoric system, *J. Am. Chem. Soc.*, **126**, 14310–14311.

97. Hwang, I.-W., Ko, D. M., Ahn, T. K., Yoon, Z. S., Kim, D., Peng, X., Aratani, N. and Osuka, A. (2005) Excitation energy migration in a dodecameric porphyrin wheel, *J. Phys. Chem. B*, **109**, 8643–8651.

98. Hwang, I.-W. Kamada, T. Ahn, T. K. Ko, D. M. Nakamura, T. Tsuda, A., Osuka, A. and Kim, D. (2004) Porphyrin boxes constructed by homochiral self-sorting assembly: optical separation, exciton coupling, and efficient excitation energy migration, *J. Am. Chem. Soc.*, **126**, 16187–16198.

99. Nakamura, Y., Hwang, I.-W., Aratani, N., Ahn, T. K., Ko, D., M. Takagi, A., Kawai, T., Matsumoto, T., Kim, D. and Osuka, A. (2005) Directly *meso-meso* linked porphyrin rings: synthesis, characterization, and efficient excitation energy hopping, *J. Am. Chem. Soc.*, **127**, 236–246.
100. Bradforth, S. E., Jimenez, R., van Mourik, F., van Grondelle, R. and Fleming, G. R. (1995) Excitation transfer in the core light-harvesting complex (LH-1) of *Rhodobacter sphaeroides*: an ultrafast fluorescence depolarization and annihilation study, *J. Phys. Chem.*, **99**, 16179–16191.
101. Yang, J., Park, M., Yoon, Z. S., Hori, T., Peng, X., Aratani, N., Dedecker, P., Hotta, J.-I., Uji-i, H., Sliwa, M., Hofkens, J., Osuka, A. and Kim, D. (2008) Excitation energy migration processes in cyclic porphyrin arrays probed by single molecule spectroscopy, *J. Am. Chem. Soc.*, **130**, 1879–1884.
102. Yoon, M.-C., Cho, S., Kim, P., Hori, T., Aratani, N., Osuka, A. and Kim, D. (2009) Structural dependence on excitation energy migration processes in artificial light harvesting cyclic zinc(II) porphyrin arrays, *J. Phys. Chem. B*, **113**, 15074–15082.

Chapter 2

Electron Transfer Through Butadiyne-Linked Porphyrin-Based Molecular Wires

Bo Albinsson[a] and Harry L. Anderson[b]

[a] *Department of Chemical and Biological Engineering, Chalmers University of Technology, Physical Chemistry, Kemivägen 3, SE - 412 96 Göteborg, Sweden*

[b] *Department of Chemistry, University of Oxford, Chemistry Research Laboratory, Mansfield Road, Oxford, United Kingdom*

balb@chalmers.se; harry.anderson@chem.ox.ac.uk

2.1 Introduction

Electron transfer is of fundamental importance in many areas of chemistry and biochemistry [1]. The ability to transfer charge efficiently over distances in the 10 nm range is crucial in various practical devices, such as organic transistors and solar cells—both dye-sensitized metal oxide solar cells and bulk-heterojunction blend devices [2]. Furthermore, in the context of the ever-decreasing dimensions of integrated circuits, it is interesting to study charge transport through individual molecules [3]. Organic π-conjugated oligomers and polymers are attractive candidates as molecular wires, because modern synthetic chemistry can be used to create

Multiporphyrin Arrays: Fundamentals and Applications
Edited by Dongho Kim

www.panstanford.com

almost any type of π-conjugated backbone, while non-covalent interactions and supramolecular chemistry can be used to modify the backbone conformation. Interactions of conjugated oligomers, both with each other and with the external environment, can also be controlled by self-assembly and non-covalent encapsulation [4].

Porphyrins and their derivatives occur widely in biological systems. For example chlorophylls, hemoglobin, and cytochromes are essential in biocatalytic processes. The synthesis and investigation of porphyrin-based molecular wires has been pioneered mainly by Crossley [5], Anderson [6, 7], Therien [8], and Osuka [9, 10]. The following five factors make porphyrins particularly suitable building blocks for the construction of molecular wires:

(i) *They are large aromatic macrocycles with small HOMO-LUMO gaps.* The delocalized electronic structures of porphyrins make them highly redox active and result in intense optical absorption bands across the whole of the visible spectrum.
(ii) *Porphyrins are remarkably stable*, as illustrated by the fact that they survive for millions of years in fossils. Bocian and coworkers have shown that porphyrins linked to silicon substrates are stable to extremes of temperature (400°C) for extended periods of time (1 hour), and survive large numbers (10^{12}) of redox cycles [11].
(iii) *Synthetic methodology is well established* for preparing conjugated porphyrin oligomers with lengths of 1–10 nm. This chemistry could be extended to target longer discrete oligomers and polydisperse polymers. Porphyrin chemistry is highly developed because of the tremendous effort that was invested in the synthesis of porphyrinoid natural products; for example the synthesis of vitamin B_{12}, completed by Woodward and Eschenmoser in 1972, is one of the greatest achievements in total synthesis [12].
(iv) *Porphyrins coordinate to a wide range of metal cations* providing a way to tune their redox potentials and photophysical behavior.
(v) *Coordination of axial ligands leads to scope for supramolecular assembly.* Axial ligands are generally labile enough to allow

reversible binding under ordinary conditions, leading to thermodynamic control. Supramolecular self-assembly can be used to dictate the three-dimensional shape of a porphyrin-based molecular wire [7].

This chapter focuses on molecular wires consisting of zinc porphyrin macrocycles linked at their *meso*-positions [5, 15] by butadiyne units. Oligomers of this type have the advantage that they are easily prepared by the oxidative Glaser-Hay coupling of the corresponding terminal acetylenes; this coupling reaction often proceeds in >95% yield at room temperature, after reaction times of about 30 minutes, even at concentrations of around 10^{-3} mol dm^3. Butadiyne-linked porphyrin oligomers also display strong π-conjugation, as indicated by a steep decrease in the optical π–π^* gap with increasing oligomer length [13]. In contrast, most other types of ostensibly π-conjugated connections to the *meso*-positions of porphyrins, such as vinylene or phenylene links, twist out of conjugation due to steric repulsion with the β-pyrrole hydrogens [14].

The presence of a coordinated metal simplifies the electronic absorption and NMR spectra of porphyrins, by avoiding the symmetry reduction caused by the two central hydrogen atoms in the free-base, and prevents accidental complexation of metal cations during synthetic transformations. Zinc is a convenient metal because of its simple closed-shell d_{10} electronic configuration. Zinc is easily and quantitatively inserted into free-base porphyrins using zinc(II) acetate, and in most cases it can be easily removed using strongly acidic conditions (e.g., trifluoroacetic acid), but zinc porphyrins are stable enough that accidental demetallation is rare, in contrast to magnesium(II), cadmium(II), and lead(II) porphyrins which are easily demetallated by traces of acid. The presence of a zinc cation enables one to increase the solubility of a porphyrin, and prevent aggregation, by using a coordinating ligand such as pyridine (e.g., 1% by volume of solvent) or tetrahydrofuran (as the solvent). Open-shell transition metal cations and heavier second or third row cations strongly perturb the photophysical behavior of porphyrins (reducing the singlet state lifetime), whereas the photophysics of zinc porphyrins is fairly similar to that of free-bases.

Electron transfer reactions can be very fast, and thus it is often beneficial to study photoinduced electron transfer in which either the donor or acceptor is excited to prepare an initial state with a driving force for electron transfer. The challenge of understanding natural photosynthetic processes has stimulated the investigation of photoinduced electron transfer in a great variety of synthetic donor–acceptor systems. The primary motivation of these studies has often been to efficiently create long-lived charge-separated states, i.e., to maximize the rate of charge separation and minimize the rate of charge recombination, so that the energy of the charge-separated state can be utilized to drive secondary redox reactions. The motivation for the work discussed here is completely different: We set out to investigate photoinduced electron transfer in conjugated porphyrin oligomers linked to donor and acceptor groups so as to learn about the ability of these molecular wires to mediate long-range charge transport.

2.2 Molecular Design

Most studies of photoinduced electron transfer in donor-bridge-acceptor systems rely on formation of the excited state of the donor (or the acceptor) to increase its reduction (or oxidation) potential. However, it would be difficult to adopt this strategy when testing electron transfer through a bridge consisting of a conjugated porphyrin oligomer because there are few available donors or acceptors that have lower energy excited states than a conjugated porphyrin oligomer. If the energy of the lowest singlet excited state of the bridge is smaller than that of the donor, then excitation of the donor will lead to energy transfer to the bridge, in competition with electron transfer. Thus we decided to adopt a different strategy, which relies on initial excitation of the bridge.

The general structure of the donor–acceptor systems, **Fc-P$_n$-C$_{60}$**, discussed in this chapter is shown in Fig. 2.1, together with a simplified energy level diagram for the processes that are expected to occur after excitation of the porphyrin oligomer [15]. The donor and acceptor, ferrocene and buckminsterfullerene, respectively, were chosen because their redox potentials make the long-range

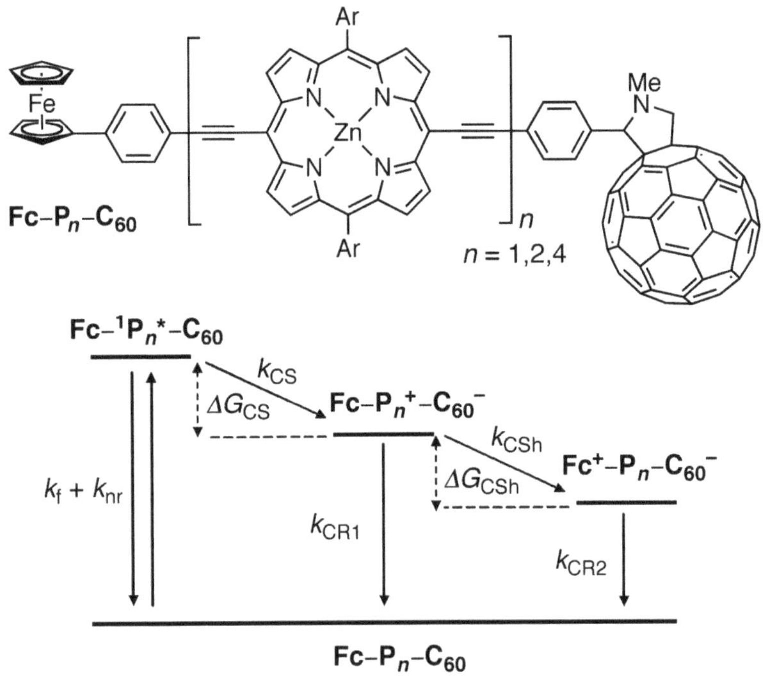

Figure 2.1. General structure of the triads and schematic energy level diagram showing the charge separation (CS), charge shift (CSh), and charge recombination (CR1 and CR2) reactions.

charge-separated state **Fc$^+$-P$_n$-C$_{60}$$^-$** lower in energy than the two possible intermediate states (**Fc-P$_{n^+}$-C$_{60}$$^-$** and **Fc$^+$-P$_n$$^-$-C$_{60}$**) for all oligomer lengths (i.e., for any value of *n*, Fig. 2.1, and quantitatively in Fig. 2.14). *Para*-phenylene spacers were used to link the porphyrin oligomer to the donor and the acceptor for synthetic convenience. Aryl side chains [Ar = 3,5-di(*t*-butyl)phenyl or 3,5-di(octyloxy)phenyl] were connected to the meso positions of the porphyrin units to provide high solubility in organic solvents and to hinder aggregation; these aryl groups are twisted out of conjugation with the porphyrin π-system so they have little effect on the photophysical behavior.

Many ferrocene-porphyrin-fullerene triads had been investigated previously by Fukuzumi and coworkers [16, 17], with the aim of preparing long-lived charge-separated states. This previous work demonstrated that porphyrin-fullerene systems have a small reorganization energy, placing the charge-separation process near the maximum-rate Marcus regime ($\lambda \approx -\Delta G^{\circ}$) [1]. The fullerene radical anion has a characteristic absorption at around 1000 nm, which provides a useful probe for charge-transfer processes.

The ferrocene donor has a convenient oxidation potential, and it is easy to handle during synthesis, but it has two significant disadvantages: (a) ferrocene has low lying excited states, which partially quench the excited states of the porphyrins by energy transfer [18], and (b) the UV-visible extinction coefficient of the ferrocinium radical cation is too low for the oxidized state of the ferrocene unit to be probed directly by transient absorption.

2.3 Electronic States and Spectra of Butadiyne-Linked Porphyrin Arrays

Butadiyne-linked porphyrin oligomers exist in a wide distribution of rotational conformers due to the low energy barrier for rotating adjacent porphyrins. Although the ground-state potential energy is quite insensitive to rotation of the porphyrins, the first excited state is found to depend more strongly on the dihedral angles and as a consequence the observed electronic absorption spectrum of a porphyrin oligomer is a mixture of contributions from different conformers. This effect was thoroughly studied in the butadiyne-linked porphyrin dimer (**P**$_2$) both experimentally and via quantum chemical methods [19]. **P**$_2$ can be used as the simplest model to understand the spectroscopy and photophysics of longer oligomers, as well as how the conformational flexibility in the ground and first excited state influence various important properties, such as electronic coupling and nonlinear absorption. A thorough understanding of the conformational dependence is a prerequisite for interpreting any measurement involving excited states of porphyrin oligomers.

The strong dependence of the absorption spectra on the dihedral angle is clearly demonstrated by titrating a bidentate pyridine ligand (**L**) to a solution of **P**$_2$. Figure 2.2 shows the absorption changes upon forming the strong 1:1 complex ($K \approx 10^7$–10^8 M^{-1}) of the ligand and **P**$_2$ in dichloromethane. The ligand effectively forces the dimer to stay planar or very nearly planar. The intensity of the B-band peak at 457 nm is reduced, whereas the intensity of the peak at 491 nm is increased. In the Q-band, the intensity of the peak farthest to the red is increased and that of the peak on the blue side decreased. The absorption of the ligand itself is in the UV, with a maximum at 340 nm. Thus, the final spectrum above 400 nm is almost exclusively that of the planar dimer, which exhibits two peaks in the B-band and one strong peak in the Q-band. It is similar to the spectrum of an aggregated dimer [20], as expected. The bending of the porphyrin dimer upon complexation has minor effects on the absorption spectra as shown in Fig. 2.2, and this is also supported by quantum mechanical calculations (*vide infra*). The ground-state absorption spectrum of disaggregated $\mathbf{P}_2$, free to rotate, is however a mixture of spectroscopic species. UV-vis titrations of $\mathbf{P}_2$ with this dipyridyl pyrrole ligand **L** were also performed in dichloromethane in the presence of pyridine in order to avoid the spectral shifts associated with complexation of the Zn-atoms and the spectral changes, apart from the red-shift of the Q-band, are very similar to those shown in Fig. 2.2.

$\mathbf{P}_2$ complexed with the dipyridyl pyrrole ligand **L** essentially gives the spectrum of the planar species. The other extreme, i.e., the spectrum of the twisted porphyrin dimer, could be extracted from the additional information available from fluorescence excitation spectra. In Fig. 2.3, the emission and excitation spectra of $\mathbf{P}_2$ at room temperature are shown. The excitation spectra are measured for two different emission wavelengths (727 and 666 nm) characteristic of emission from the thermally equilibrated (planar) and perpendicular (twisted) excited states [19]. Whereas recording the excitation spectra of the emission at 727 nm reproduced a spectrum resembling the full ground-state absorption spectrum, the emission at 666 nm did not reproduce the full spectrum but rather produced a spectrum that exhibits only two of its peaks. This spectrum is much more like that of a porphyrin monomer, albeit shifted to the red. This result indicates that the origin of the shoulder emission peak in the steady-state spectrum is due to a distinctive spectroscopic species. The excitation spectrum obtained from the emission at 666 nm mirrors the absorption spectrum of a twisted conformation.

Further evidence for conformational relaxation in the excited state comes from the temperature and time dependencies of the

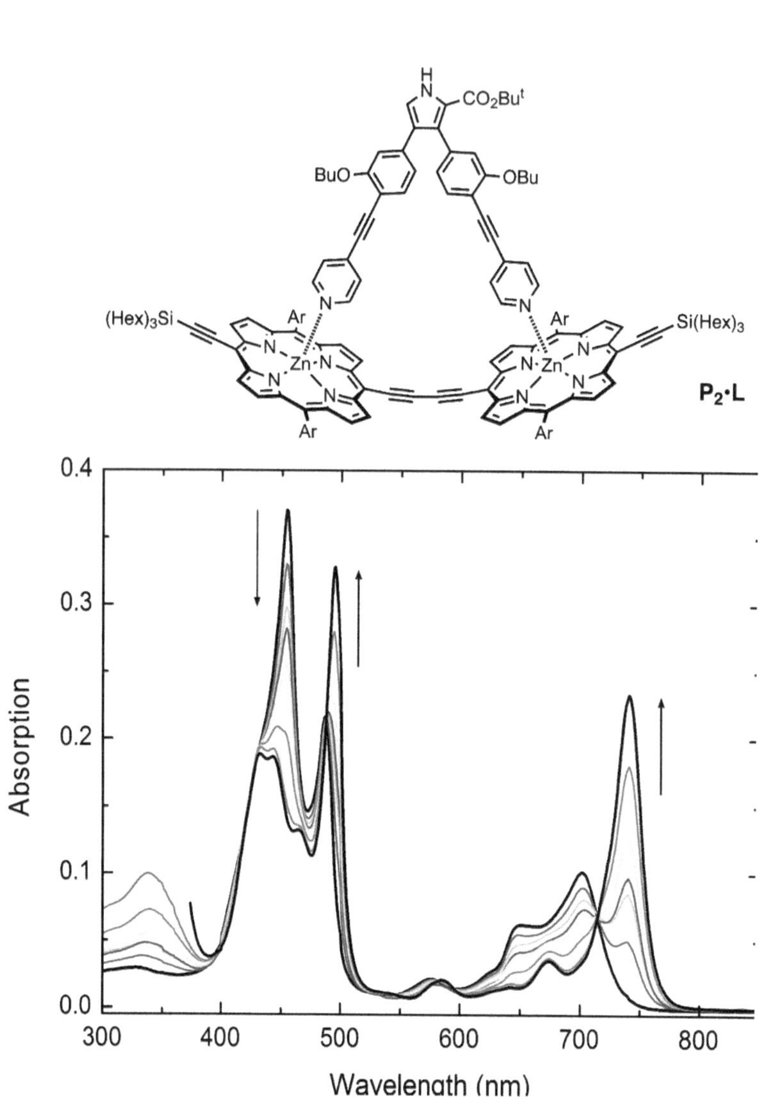

Figure 2.2. A solution of $\mathbf{P_2}$ in dichloromethane titrated by the dipyridyl pyrrole ligand **L** to form the complex $\mathbf{P_2 \cdot L}$. Isosbestic points are located at 487 and 715 nm. The concentration of $\mathbf{P_2}$ is approximately 0.7 μM.

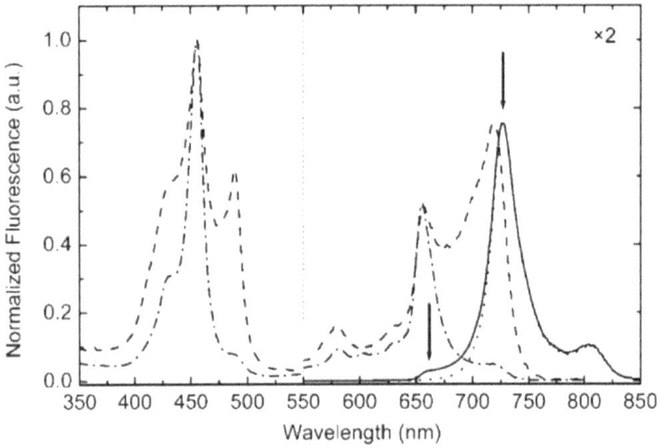

Figure 2.3. Emission and excitation spectra of $\mathbf{P}_2$ recorded in 2-methyl tetrahydrofuran (2-MTHF). Emission spectra were recorded using the excitation wavelengths 457 nm (solid black line) and 491 nm (dotted line). The emission wavelengths (indicated by arrows) used for the excitation spectra were 666 nm (dash-dot) and 727 nm (dash). The Q-band is slightly distorted due to scattering. Note that the excitation spectra in the right panel are scaled two times.

$\mathbf{P}_2$ emission spectrum [19]. If excited at either of the high energy peaks in the Soret or Q-band region, i.e., at 457 or 660 nm, respectively, a weak shoulder at 670 nm is noted in the emission (Fig. 2.3). This shoulder is intensified many orders of magnitude while the main emission at 730 nm is weakened upon cooling a sample of $\mathbf{P}_2$ in 2MTHF from room temperature to 90 K (*vide infra*, Fig. 2.10). According to the spectral assignments this high energy excitation populates the twisted conformer which at room temperature relaxes almost quantitatively to the planar conformer before emitting and returning to the ground state. However, as the temperature is lowered and the solvent viscosity is raised, the planarization is slowed down and at sufficiently low temperature all the emission comes from the initially populated twisted conformer. Another experiment that demonstrates the same process is time-resolved emission spectroscopy. Again, if the twisted conformer is populated by excitation into either of the high energy peaks and the emission spectrum is measured as a function of time a clear evolution from a blue shifted spectrum at short times corresponding to the twisted conformer to the emission spectrum corresponding to the relaxed planar species is observed. At room temperature in 2MTHF this process takes on average 100 ps, and since this is a much shorter time than the excited state lifetime (about 1.2 ns) the majority of the observed steady-state emission comes from the planar species.

2.4 Computational Modeling

Experimental results show clearly that the spectroscopic and photophysical properties of a butadiyne-linked porphyrin dimer (**P**$_2$) are dictated by its conformational flexibility. To support the experimental results and, more importantly, to generalize these findings to longer oligomers, a set of density functional theory (DFT: B3LYP/6-31G*) calculations of the ground and excited state properties of **P**$_2$ was undertaken. As expected, the ground-state potential energy varied very slightly with the dihedral angle and the barrier for rotating the two porphyrin planes relative to each other was only 0.7 kcal/mol and, thus, allowing all rotational conformers to be accessible at room temperature. The planar structure was the most stable conformer but the calculated barrier is within the accuracy of the theoretical method and should therefore be regarded as "small enough" to allow all conformers to be accessible. Electronic transitions for the different conformers were calculated with time-dependent DFT (TDDFT: B3LYP/6-31G*) and, as expected, varied quite extensively with the dihedral angle.

Calculated electronic transitions of 10 different conformations of **P**$_2$ are shown in Fig. 2.4. It is evident from these results that increasing the dihedral angle from 0° to 90° results in a blue shift of the Q band and an intensified Soret band transition at 420 nm. At a dihedral angle of 0°, the Soret band consists of two strong transitions, and the Q band consists of only one. This is in qualitative agreement with the spectrum of the planarized conformation displayed in Fig. 2.2. At a dihedral angle of 90°, a blue-shifted Q band is present, and only one major transition in the B band remains. This, in turn, reproduces the excitation spectrum in Fig. 2.3 that is believed to resemble the absorption spectrum of the twisted conformation.

The molecular structure of **P**$_2$ approximately belongs to the D_2 point group with three mutually perpendicular C_2 axes. The planar

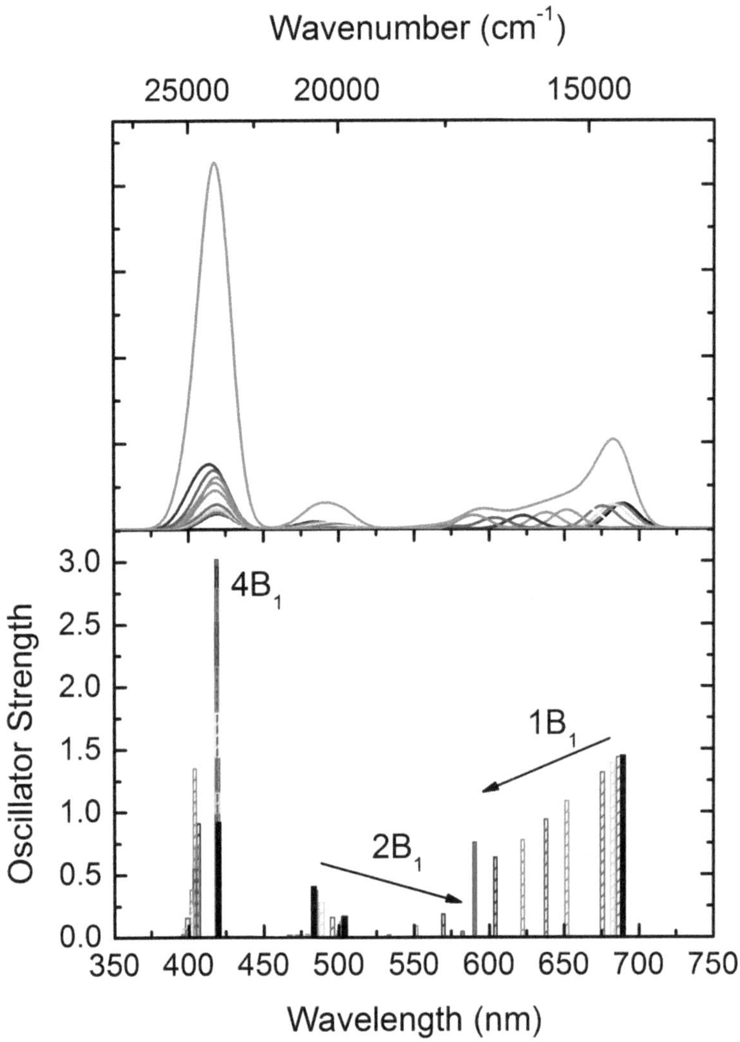

Figure 2.4. (Bottom) Calculated ground-state stick spectra at different dihedral angles; 0° (black, solid) and 90° (red, solid) represent the border cases and the intermediate dihedral angles are represented by hatched bars (10° (red), 20° (green), 30° (blue), 40° (cyan), 50° (magenta), 60° (yellow), 70° (dark green), 80° (navy blue)). (Top) Calculated spectra fitted to Gaussian components with half-widths at half-max arbitrarily set to 1000 cm^{-1}; 0°–90° in various colors and a simulated room temperature spectrum (dark cyan) based on the sum of the 10 equally weighted component spectra. See also Color Insert.

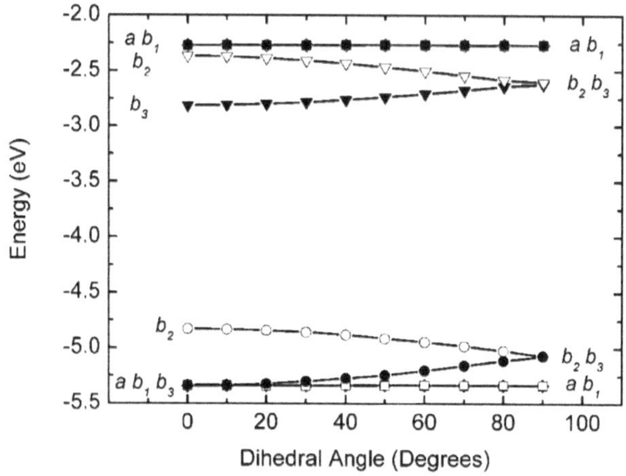

Figure 2.5. Calculated orbital energies for the four highest occupied and the four lowest unoccupied orbitals as a function of the dihedral angle (HOMO-3 is represented by filled diamonds, HOMO-2 by empty squares, HOMO-1 by filled circles, HOMO by empty circles, LUMO by filled triangles, LUMO+1 by empty triangles, LUMO+2 by filled squares, and LUMO+3 by empty diamonds). The orbital symmetries from HOMO-3 to LUMO+3 are a, b_1, b_3, b_2, b_3, b_2, a, and b_1, respectively. Graphical representations of the molecular orbitals are shown in Fig. 2.6.

and perpendicular conformers have additional symmetry elements raising the symmetries to the D_{2h} and D_{2d} point groups, respectively. In the D_2 point group, the π-orbitals transforms as the irreducible representations a, b_1, b_2, and b_3 (i.e., all available species in this point group) and these correlates with the a_u, b_{1g}, b_{2g} and b_{3u} species in the D_{2h} point group. In Fig. 2.5 the calculated orbital energies for the four highest occupied and four lowest unoccupied orbitals are shown as a function of the dihedral angle [21–23] and graphical representations of these orbitals are displayed in Fig. 2.6. It is quite obvious that the a and b_1 orbital energies do not depend on the angle between the porphyrin planes whereas the b_2 and b_3 orbitals depend strongly on this angle. This is easy to understand, since the a and b_1 π-orbitals are symmetric with respect to rotation around the z-axis (Fig. 2.6, long axis (z) along the butadiyne linker), and it is therefore not possible for these orbitals to spread over the butadiyne linker, which makes them localized on the porphyrin fragments.

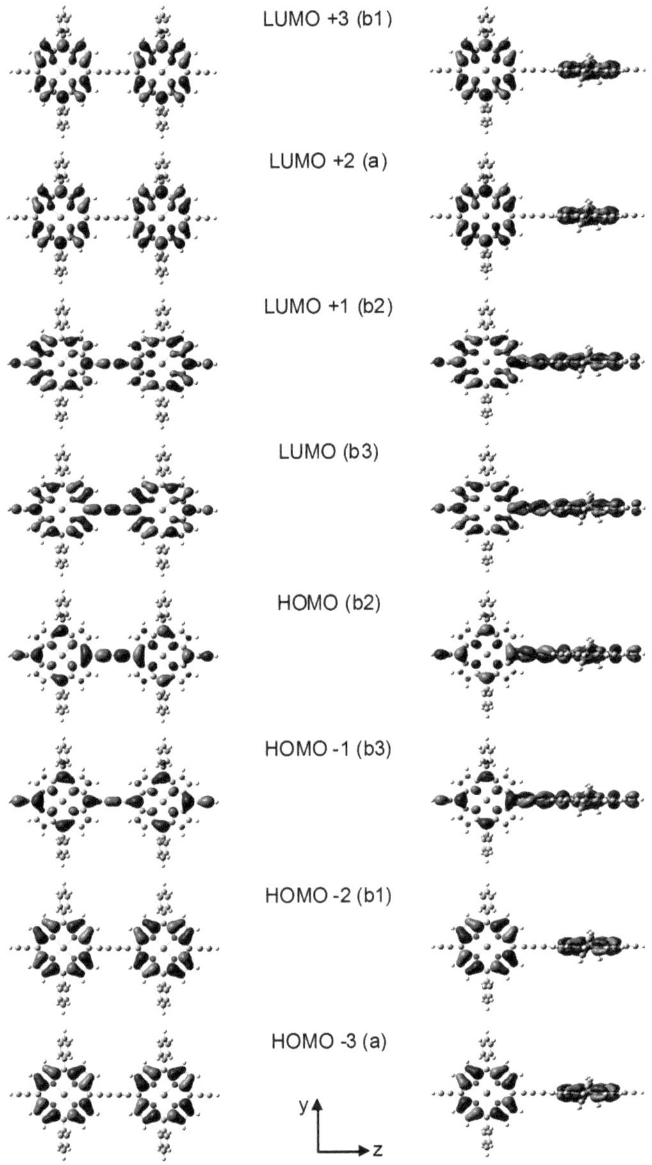

Figure 2.6. The four highest occupied and four lowest unoccupied orbitals represented at torsion angles of 0° (left column) and 90° (right column). Noted in parenthesis are the irreducible representations in the D_2 point group.

The b_2 and b_3 orbitals, however, are anti-symmetric with respect to z-rotation and will therefore delocalize over the butadiyne linker and couple the porphyrin fragments strongly and, as an inevitable consequence, their energies show strong dihedral angle dependence. Both the highest occupied and lowest unoccupied a and b_1 orbitals in Fig. 2.5 are accidentally (almost) degenerate for all dihedral angles. We can carry this quite general symmetry analysis further by noting that the b_2 and b_3 orbitals are symmetric and anti-symmetric, respectively, with respect to rotation around the y-axis (short in-plane axis for the planar configuration) and vice versa for the x-axis (short out-of-plane axis). This means that the π-orbitals belonging to the b_2 species are "anti-bonding" over the butadiyne bridge and the b_3 orbitals "bonding." As expected, the energy of the b_2 orbitals decreases as the dihedral angle increases, whereas the b_3 energies increase as they lose their π-overlap upon twisting. To correlate these changes in molecular orbital energies to spectral observations, we need to look at the possible electronic excited states that can be formed and which of those have strong, electric dipole allowed, transitions with the ground state (of symmetry A_1).

In the present calculation, transitions to excited states belonging to the irreducible representation B_1 are strong and long-axis polarized, while symmetry-allowed transitions to the other states of B_2 and B_3 symmetries are generally weak (Fig. 2.7). This is supported by other experiments, where the majority of the strong transitions observed in the absorption spectrum of $\mathbf{P}_2$ were shown to be long-axis polarized [24]. For this reason, only the states of B_1 symmetry will be considered. A correlation diagram of the B_1 states is shown in Fig. 2.7. The excited states of B_1 symmetry are composed of singly excited configurations where either an electron in a or b_1 orbitals is promoted to b_1 or a orbitals, respectively, and/or an electron in b_2 or b_3 orbitals is promoted to b_3 or b_2 orbitals, respectively. These electronic configurations can all contribute to the final state wavefunction of B_1 symmetry. Now, considering only the four highest occupied and four lowest unoccupied orbitals (see Fig. 2.5), the following single electron substitutions of B_1 symmetry are possible (approximately in increasing energy order): $b_2 \rightarrow b_3$, $b_3 \rightarrow b_2$, $a \rightarrow b_1$, and $b_1 \rightarrow a$. The lowest electronic transition, corresponding to the Q-band, has B_1 symmetry ($1B_1$) and is dominated

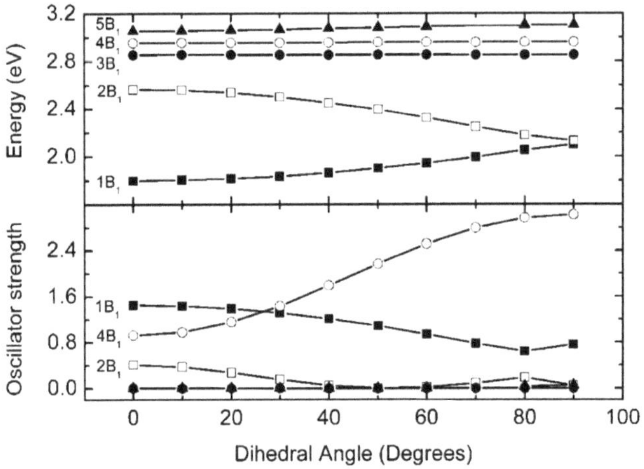

Figure 2.7. State correlation diagram of the calculated B_1 symmetry states as a function of the dihedral angle. $1B_1$ is represented by filled squares, $2B_1$ by empty squares, $3B_1$ by filled circles, $4B_1$ by empty circles, and $5B_1$ by filled triangles.

by the $b_2 \rightarrow b_3$(HOMO $\rightarrow$ LUMO) excitation for all dihedral angles, except at 90°. This transition shifts to higher energy and loses some intensity, as seen in the state correlation diagram (Fig. 2.7), which could easily be understood in terms of the orbital shifts displayed in Fig. 2.5. The next B_1 transition, calculated at 483 nm for the planar conformation, is part of the Soret band, but shifts to lower energy and loses intensity as the dihedral angle increases. At 90° the two lowest B_1 transitions almost coincide to form a pair of transitions that is expected to absorb at 590 nm (the Q-band of the twisted conformer). This $2B_1$ transition is dominated by the $b_3 \rightarrow b_2$ (HOMO-1 $\rightarrow$ LUMO+1) configuration and its red shift is understood from the decreased orbital splitting in Fig. 2.5.

Finally, the strong transition to the $4B_1$ state, corresponding to the high energy component of the Soret band, has an excitation energy that is almost independent of the dihedral angle. The distribution of electron configurations contributing to this transition is quite broad at all dihedral angles, but is dominated by the $a \rightarrow b_1$ and $b_1 \rightarrow a$ configurations, at least at large dihedral angles. Apparently, the configurations contributing to this transition either has no

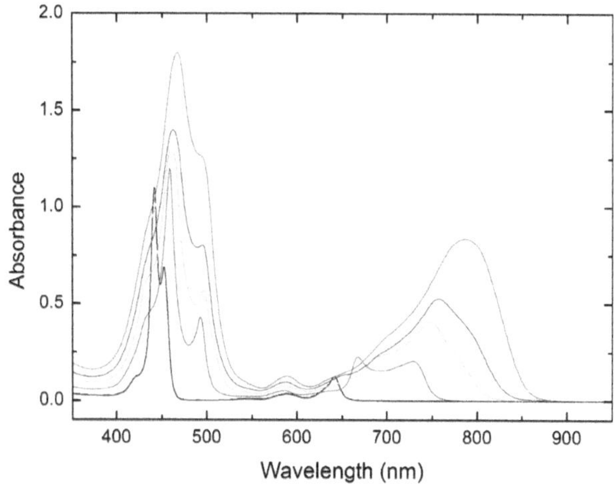

Figure 2.8. Ground-state absorption spectra of $\mathbf{P}_1$ (black), and $\mathbf{P}_2$ (red), $\mathbf{P}_3$ (green), $\mathbf{P}_4$ (blue), $\mathbf{P}_6$(cyan), and $\mathbf{P}_8$ (magenta) in THF with 1% pyridine. See also Color Insert.

dependence of the dihedral angle ($a \rightarrow b_1$ and $b_1 \rightarrow a$) or occur as a balanced mixture of configurations that increase ($b_2 \rightarrow b_3$) and decrease ($b_3 \rightarrow b_2$) in energy. Transitions to the $3B_1$ and $5B_1$ states included in Fig. 2.7 have very small transition moments at all dihedral angles and are therefore not considered here.

The detailed understanding of the electronic states and spectra of the porphyrin dimer could be used as a basis for understanding the linear oligomer spectra irrespective of their length. Figure 2.8 shows a series of absorption spectra for linear porphyrin oligomers ranging from one to eight subunits. It is evident that the lowest energy absorption band (Q-band) red shifts from 640 nm to 800 nm and gains in intensity as the oligomer gets longer. At the same time it appears that this band broadens and for the octamer it covers the region from the monomer Q-band around 600 nm to almost 900 nm. In contrast, the Soret band, that shows a double peak for the dimer at 450 and 495 nm, does not show any shift but only intensifies as the oligomer grows. All these observations could be understood by looking at the orbital composition of the lowest electronic excitations. The first electronic transition is expected to be related to the transition in $\mathbf{P}_2$ to the $1B_1$ state and this transition is dominated by the $b_2 \rightarrow b_3$(HOMO $\rightarrow$ LUMO) excitation. Even if symmetries for the oligomers longer than three subunits are different (lower) than $\mathbf{P}_2$, the approximate orbital composition of electronic transitions and orbital topology is expected to prevail and we will, for simplicity, adopt the same labels. The b_2 and b_3 orbitals are delocalized over the butadiyne linker and will therefore also have the possibility to be delocalized over all porphyrin subunits in the longer oligomers. The lowest electronic transition to the $1B_1$ excited state will, thus, be sensitive to the length of the oligomer and, furthermore, also be sensitive to the conformation. This is exactly what we see in the series of spectra in Fig. 2.8; the position, width, and shape of the Q-band of the longer oligomers are moving to lower energy and gets broader. For the longer oligomers the weight of the conformers with intermediate shifts increases because the number of combinations of dihedral angles giving a certain excitation energy increases making the band shape close to Gaussian for $\mathbf{P}_8$. The opposite is true for the Soret band that is dominated by the $a \rightarrow b_1$ and $b_1 \rightarrow a$ configurations. This band shows no shift in energy or shape as the oligomers grow longer. The only interaction possible for the a and b_1 orbitals is through space (i.e., exciton coupling), since delocalization over the butadiyne linkers is restricted by the symmetry properties of these orbitals (cf. Fig. 2.6).

2.5 Conformational Control of Charge Separation and Recombination in Butadiyne-Linked Porphyrins

Having a strong conformational dependence of both the spectroscopy and electronic communication in the porphyrin oligomers open up possibilities to control electron transport either by excitation wavelength selection or by chemically modulating the dihedral angle between individual porphyrin units. This was demonstrated by varying either the excitation energy or adding the bidentate ligand, **L**, to a solution of a donor–acceptor molecule comprised of $\mathbf{P}_2$ as electron donor and $\mathbf{C}_{60}$ as the acceptor (Fig. 2.9) [25]. As discussed above, the emission from the twisted conformer detected at 670 nm is significantly enhanced at the expense of the emission at 740 nm from the planar (relaxed) species if a solution of $\mathbf{P}_2$ is cooled

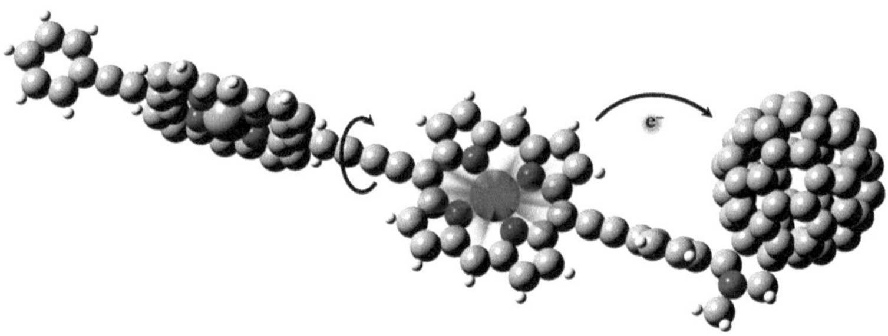

Figure 2.9. Electron transfer rates from $\mathbf{P_2}$ to $\mathbf{C_{60}}$ are controlled by the conformation of the porphyrin dimer.

down (Fig. 2.10a). In the $\mathbf{P_2}$-$\mathbf{C_{60}}$ compound where electron transfer from $\mathbf{P_2}$ to $\mathbf{C_{60}}$ is competing with structural relaxation and fluorescence the emission from the twisted conformer is strongly quenched at all temperatures whereas the quenched emission from the planar species increases for temperatures below 200 K. This experiment, shown in Fig. 2.10b, qualitatively demonstrates that the rates for electron transfer and their temperature dependencies are different for different conformations of $\mathbf{P_2}$. In order to get quantitative information about the difference in electron transfer rates for the different conformers of $\mathbf{P_2}$-$\mathbf{C_{60}}$, a set of time-resolved fluorescence experiments were conducted.

The fact that the two conformers stand in a mother-daughter relationship, makes steady-state fluorescence a less useful tool for quantitative analysis of the electron transfer kinetics that take place in $\mathbf{P_2}$-$\mathbf{C_{60}}$. However, the electron transfer rate can be studied by examining the fluorescence quenching efficiency of $\mathbf{P_2}$-$\mathbf{C_{60}}$ relative to $\mathbf{P_2}$, by way of fluorescence lifetime measurements, since electron transfer was shown to be the only additional deactivation pathway introduced by attaching the $\mathbf{C_{60}}$ electron acceptor [15]. A schematic energy level diagram for **Fc**-$\mathbf{P_2}$-$\mathbf{C_{60}}$ (Fig. 2.11) shows the two initial charge separation steps along with the subsequent charge recombination steps discussed below. With reference to this mechanism, the charge separation rate from the planar ($k_{CS//}$) and perpendicular ($k_{CS\perp}$) conformer can be determined by Eqs. 2a and 2b, respectively. In these equations, τ_{f} is the measured emission lifetime, τ_{f}^{0} is the intrinsic lifetime of unquenched $\mathbf{P_2}$, and k_{rot} is the rate constant

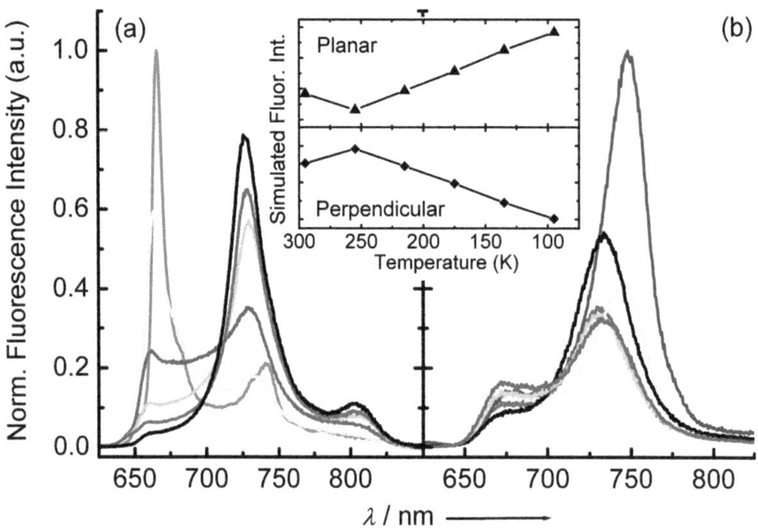

Figure 2.10. Emission spectra recorded at several temperatures obtained by exciting (a) $\mathbf{P_2}$ (2-MTHF) and (b) $\mathbf{P_2\text{-}C_{60}}$ (2-MTHF) at 460 nm. The spectra were measured at 295 K (black), 255 K (red), 215 K (green), 175 K (blue), 135 K (cyan), 115 K (purple, $\mathbf{P_2\text{-}C_{60}}$ only), and 95 K (magenta, $\mathbf{P_2}$ only). The inset shows the temperature variation of the integrated concentration of the planar and perpendicular species of $\mathbf{P_2\text{-}C_{60}}$ (top and bottom, respectively), estimated from the extracted rate constants. See also Color Insert.

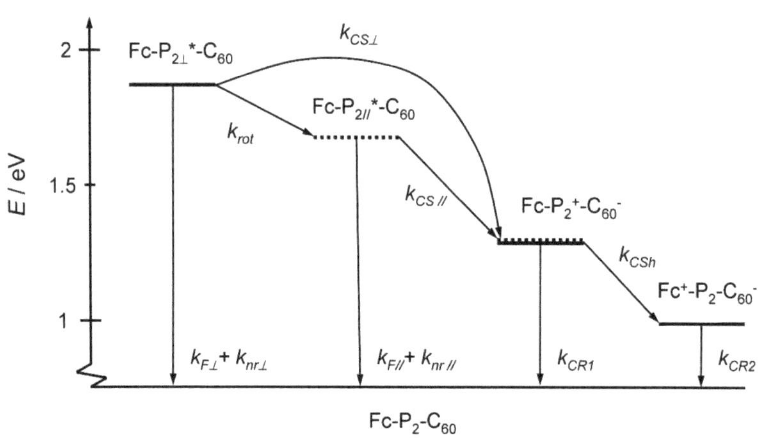

Figure 2.11. Relative energies of the states, electron transfer pathways and their respective rate constants.

for rotational relaxation of the excited perpendicular conformation.

$$k_{CS\perp} = (\tau_{f\perp})^{-1} - (\tau^{0}_{f\perp})^{-1} - k_{rot} \tag{2.1a}$$

$$k_{CS//} = (\tau_{f//})^{-1} - (\tau^{0}_{f//})^{-1} \tag{2.1b}$$

At room temperature, the rate for electron transfer from the twisted conformer was faster than the time-resolution of the experiment (<10 ps), while the charge separation rate for the planar species was estimated to be about (30 ps)$^{-1}$. At lower temperature both the electron transfer rates and the planarization is slowed down and so, for simplicity, the quantitative evaluation was made in the interval 200–100 K. The rate for electron transfer is related to several parameters, such as the electronic coupling (V), the inner and outer reorganization energy accompanying the redistribution of charge (λ_i and λ_o), and the thermodynamic driving force for the electron transfer process (ΔG°). Importantly, the rate for electron transfer is almost always temperature dependent. The Marcus model is often used to describe diabatic electron transfer in the high temperature limit (Eq. 2.2) [26–28], and by measuring the rate for electron transfer as a function of temperature, the unknown parameters of this expression can be experimentally estimated. To this end, the fluorescence lifetimes of **P**$_2$-**C**$_{60}$ were measured in 2-methyl tetrahydrofuran (2-MTHF) at low temperatures. The temperature range for this measurement was selected so that the short fluorescence lifetime of the perpendicular conformer could be properly resolved and therefore 2-MTHF was a suitable solvent. Moreover, the temperature was kept well above the glass temperature of the solvent ($T_g \sim 90$ K). In Fig. 2.12, the logarithms of the electron transfer rates $k_{CS//}$ and $k_{CS\perp}$, multiplied by $\sqrt{T}$ (see Eq. 2.2), are plotted against the inverse temperature. By linear regression of these data and from estimates of the driving forces for charge separation (ΔG°), the electronic couplings (V), and reorganization energies (λ) could be extracted (Table 2.1), assuming they are approximately constant in the temperature range used.

$$k_{CS} = \sqrt{\frac{\pi}{\hbar^2 \lambda\, k_B T}}\, |\, V\, |^2 \exp\left[-\frac{(\Delta G^\circ + \lambda)^2}{4\lambda\, k_B T}\right] \tag{2.2}$$

As Table 2.1 demonstrates, there is a significant difference in electronic coupling between the planar and perpendicular species,

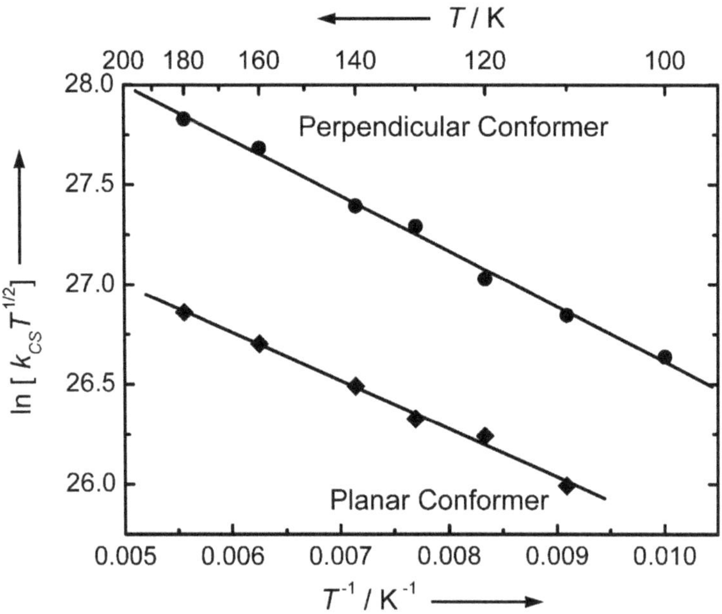

Figure 2.12. The rate constants for electron transfer from the perpendicular ($k_{CS\perp}$, circles) and planar ($k_{CS//}$, diamonds) conformer plotted against the inverse temperature. The sample was excited at 460 nm and the fluorescence lifetimes were measured at 670 and 740 nm, respectively. The lines are fits to a linear form of the Marcus equation (Eq. 2.2).

and surprisingly, the difference in driving force is of less importance to determine the electron transfer rates in this case. The slopes of the fitted lines in Fig. 2.12 are approximately equal signaling that the activation energies for the two charge separation reactions are similar but the offsets of these lines are significantly different. Since the offset of the temperature dependence is largely determined by the magnitude of the electronic coupling, it is this property that dictates the difference in charge separation rates in the **P**$_2$-**C**$_{60}$ system.

The charge-separated state in **P**$_2$-**C**$_{60}$ relaxes quite rapidly to the ground state by means of another electron transfer reaction–charge recombination. The rate of this process and other charge recombination and shift reactions were studied by femtosecond transient absorption since none of the states involved are luminescent. These processes are described in the next section.

Table 2.1. Driving forces for electron transfer (ΔG°), total reorganization energies (λ_{tot}), and the electronic couplings (V) for the process $\mathbf{P_2}$-$\mathbf{C_{60}} \rightarrow \mathbf{P_{2+}}$-$\mathbf{C_{60}}^-$.

Conformer	ΔG° [a] (eV)	λ_{tot} [b] (eV)	V [b] (cm^{-1})
Perpendicular	−0.54	0.78	32
Planar	−0.36	0.54	16

[a] Calculated from redox potentials (assuming that the oxidation potential of the perpendicular species is similar to that of the monomer $\mathbf{P_1}$) and excitation energies. [b] Estimated by linear regression of experimental data (see Fig. 2.12).

2.6 Long-Range Electron Transfer Mediated by Porphyrin Oligomers

By covalently attaching a secondary electron donor, ferrocene (**Fc**), to the $\mathbf{P}_n$-$\mathbf{C_{60}}$ systems a spatially extended charge-separated state, $\mathbf{Fc}^+$-$\mathbf{P}_n$-$\mathbf{C_{60}}^-$, could be formed. For this process to be efficient it is required that the charge shift reaction, $\mathbf{Fc}$-$\mathbf{P}_{n+}$-$\mathbf{C_{60}}^- \rightarrow \mathbf{Fc}^+$-$\mathbf{P}_n$-$\mathbf{C_{60}}^-$, is fast compared to the direct charge recombination, $\mathbf{Fc}$-$\mathbf{P_{2+}}$-$\mathbf{C_{60}}^- \rightarrow \mathbf{Fc}$-$\mathbf{P_2}$-$\mathbf{C_{60}}$. This was the case for the systems investigated but the overall quantum efficiency to form the fully charge-separated state dropped to only 25% for the $\mathbf{P_4}$ bridged system. Nevertheless, formation of this extended charge-separated state was efficient enough to make a systematic study of the distance dependence of the electron transfer rates (charge recombination of the extended charge-separated state) through the porphyrin oligomers as graphically depicted in Fig. 2.13.

Conjugated porphyrin oligomers feature long strongly coupled π-systems and small π-π^* gaps, so they are expected to be good mediators for long-range electron transfer [7, 10, 20]. For example Osuka and coworkers have shown that edge-fused porphyrin oligomers with lengths of up to 70 Å conduct charge efficiently across break-junctions [29]. Recently, Therien et al. have reported EPR measurements indicating that ethyne-linked porphyrin oligomers can mediate essentially barrierless hole transport over distances of up to 75 Å [30]. In this study we investigated the efficiency of electron transfer through porphyrin-based molecular wires by measuring the rate of charge recombination in a series of donor-bridge-acceptor molecules, $\mathbf{Fc}^+$-$\mathbf{P}_n$-$\mathbf{C_{60}}^-$, in which the

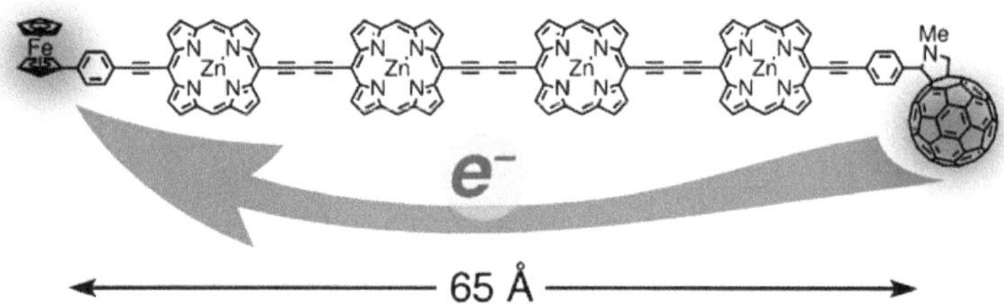

Figure 2.13. Charge recombination of the extended charge-separated state used to study the distance dependence of electron transfer through conjugated porphyrin oligomers.

electron donor ($\mathbf{C}_{60}{}^{-}$) is connected to the electron acceptor ($\mathbf{Fc}^{+}$) via a set of butadiyne-linked porphyrin oligomers ($\mathbf{P}_n$, $n = 1, 2$ or 4, Fig. 2.1). Surprisingly, the rate of electron transfer through the porphyrin tetramer bridge ($\mathbf{P}_4$, length: $R_{\text{DA}} = 65$ Å) is almost the same as that through the dimer ($\mathbf{P}_2$, length: $R_{\text{DA}} = 38$ Å); just comparing these two bridges gives an apparent attenuation factor of $\beta = 0.003$ Å^{-1}. Many ferrocene–porphyrin–fullerene triads have been investigated previously by Fukuzumi and coworkers, [16, 17] with the aim of preparing long-lived charge-separated states. The aim of this study was completely different: we were not primarily interested in the lifetimes of the charge-separated states but rather to investigate whether the intervening bridge acts as an efficient molecular wire. The energies of the states involved in the electron transfer process, calculated from electrochemical and spectroscopic data, are plotted in Fig. 2.14.

In the fully charge-separated state, the distances between $\mathbf{C}_{60^-}$ and $\mathbf{Fc}^{+}$ (R_{DA}) are approximately 24, 38, and 65 Å in **Fc-P$_1$-C$_{60}$**, **Fc-P$_2$-C$_{60}$**, and **Fc-P$_4$-C$_{60}$**, respectively, and the bridging porphyrin oligomers are in their electronic ground states. Figure 2.15 shows the transient absorption decays at wavelengths where, after an initial induction time, a clean single exponential $\mathbf{C}_{60^-}$ decay dominates the signal. The decay of the **Fc-P$_1$-C$_{60}$** system is evidently faster (0.64 ns) than both **Fc-P$_2$-C$_{60}$** and **Fc-P$_4$-C$_{60}$** (7.1 and 7.8 ns, respectively). In the lower panel of Fig. 2.15, the porphyrin ground-state recoveries of the corresponding systems are displayed. As expected from the model

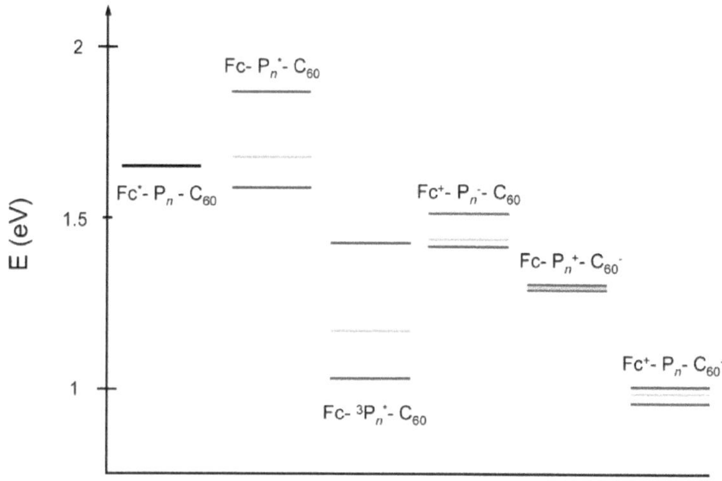

Figure 2.14. State energies estimated from electrochemical and spectroscopic data. Red lines are for $n = 1$, green lines for $n = 2$, and blue lines for $n = 4$. See also Color Insert.

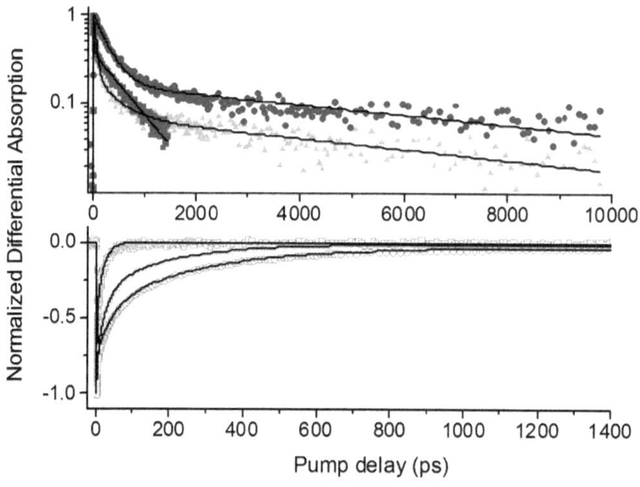

Figure 2.15. Transient absorption decays measured at 1000 nm (top), shown in logarithmic scale to emphasize the fact that they are single exponential decays at long delay times. The bottom panel shows the ground-state recovery of the oligomers (660 nm, 732 nm, and 756 nm). **Fc-P$_1$-C$_{60}$** is shown by red squares, **Fc-P$_2$-C$_{60}$** by green triangles, and **Fc-P$_4$-C$_{60}$** by blue circles. Note the different scales of the abscissa in the two panels. All measurements are in THF. See also Color Insert.

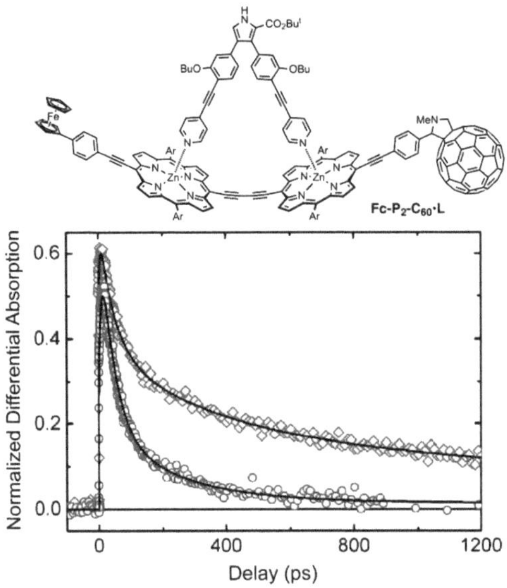

Figure 2.16. (**Bottom**) Transient absorption decays measured at 1020 nm for **Fc-P$_2$-C$_{60}$** in CH_2Cl_2 with 1% pyridine (squares) and in CH_2Cl_2 with stoichiometric amounts of the bidentate ligand, **L** (circles). Excitation is at 495 nm and the lines are in both cases 3 exponential fits to the measured data. (**Top**) Energy minimized structure of the **Fc-P$_2$-C$_{60}$** • L complex.

presented in Fig. 2.1, the porphyrin ground-state absorption recovers long before the charge-separated states have decayed. The ground-state recovery times agree well with the estimated decays of the porphyrin oligomers radical cations, further supporting the proposed model. The rate constants, k_{CR2}, for decay of the **Fc$^+$-P$_n$-C$_{60}$$^-$** ion pairs are all too fast for a significant contribution from a dynamic quenching process due to intermolecular electron transfer under the very dilute conditions of these experiments (concentration $\approx 10\ \mu$M).

Similar to the charge-separation process, the rate of charge recombination could be influenced by planarizing the porphyrin oligomer. Figure 2.16 shows how the lifetime of the fully charge-separated state of **Fc-P$_2$-C$_{60}$** decays about 10 times faster in the presence of the bidentate ligand, **L,** and the energy minimized structure of this complex is shown in the top panel [19]. This experiment clearly demonstrates the strong influence from conformational disorder on the electron transfer rate.

2.7 Discussion of Electron-Transfer Rates

The recombination rates, k_{CR2}, in $\mathbf{Fc^{+}}$-$\mathbf{P_2}$-$\mathbf{C_{60}}^{-}$ and $\mathbf{Fc^{+}}$-$\mathbf{P_4}$-$\mathbf{C_{60}}^{-}$ are remarkably fast when compared with those of similar systems [16, 31, 32], and this must be a consequence of the strong bridge-mediated electronic coupling. The distance dependence of k_{CR2} is also unusual, with almost no change in rate from dimer to tetramer.

Three mechanisms could contribute to charge-recombination from the fully charge-separated states ($\mathbf{Fc^{+}}$-$\mathbf{P}_n$-$\mathbf{C_{60}}^{-}$): (a) through-bond electron tunneling, (b) electron or hole hopping, or (c) recombination *via* a porphyrin oligomer triplet state [33–35]. Here we consider the interpretation of the distance dependence of k_{CR2} in terms of these three possible mechanisms.

2.7.1 *Tunneling*

Bridge-mediated through-bond tunneling is a mechanism in which bridge states are not populated during charge transfer. In the superexchange description [33], the rate of electron transfer decrease exponentially with donor–acceptor separation, according to Eq. 2.3,

$$k_{CR2} = k_0 e^{-\beta R_{DA}} \tag{2.3}$$

where R_{DA} is the length of the barrier and β is the attenuation factor. For a simple rectangular tunneling barrier, the value of β is related to the height of the barrier, ΔE, by Eq. 2.4 [36–38],

$$\beta = \frac{4\pi}{h}\sqrt{2\ m^{*}m_e\ \Delta E} \tag{2.4}$$

where m_e is the free electron mass, m^* is the effective (dimensionless) electron mass, and h is the Planck constant. This equation illustrates how β increases with the tunneling barrier, ΔE. Electron (or hole) transport is facilitated by the presence of a LUMO (or HOMO) on the bridge, even though the barriers to population of these orbitals (ΔE) are too high for thermally activated charge

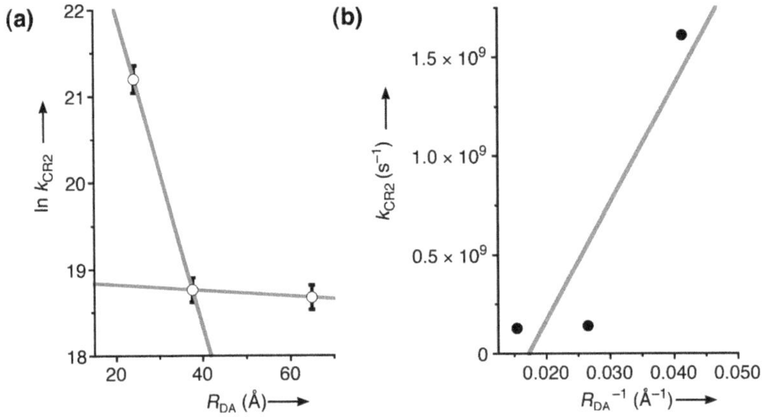

Figure 2.17. Distance dependence of the charge recombination rate, k_{CR2} in **Fc-P$_1$-C$_{60}$**, **Fc-P$_2$-C$_{60}$**, and **Fc-P$_4$-C$_{60}$**. (a) Semi-logarithmic plot of ln(k_{CR2}) vs. R_{DA} and (b) plot of k_{CR2} vs. $1/R_{DA}$.

injection into these orbitals. However, the practical utility of Eq. 2.4 is reduced by the fact that the effective mass, m^*, is difficult to estimate [37].

The logarithm of k_{CR2} is plotted against the fullerene–ferrocene distance in Fig. 2.17a. It is evident that k_{CR2} does not fall off exponentially with bridge length, and that the three points do not define a straight line. This non-exponential behavior indicates that the results do not fit a simple super-exchange tunneling model, or that length is not the only relevant parameter that changes on going from **Fc-P$_1$-C$_{60}$** to **Fc-P$_4$-C$_{60}$**. The slope of a line connecting the first two data points corresponds to $\beta = 0.18$ Å^{-1}, which is a value in the lower end of those typically found for conjugated bridge structures [31, 38–40], indicating a weak distance dependence, whereas connecting the second and third data points gives a line with practically no distance dependence ($\beta = 0.003$ Å^{-1}).

The electronic coupling elements, V can be estimated from the recombination rates, k_{CR2} and the free energy of reaction ΔG° using the Marcus equation for nonadiabatic electron transfer, Eq. 2.2 [26–28, 41]. Assuming a typical fullerene–ferrocene reorganization energy ($\lambda = 1.2$ eV) [16] gives $V = 3.3$, 0.92 and 0.85 cm^{-1} for recombination of **Fc$^+$-P$_1$-C$_{60}$$^-$**, **Fc$^+$-P$_2$-C$_{60}$$^-$** and **Fc$^+$-**

$\mathbf{P_4}$-$\mathbf{C_{60}}^-$, respectively. The coupling terms for the dimer and tetramer are similar to those reported in a 43 Å long phenylenethynylene bridge by Creager and coworkers (0.7 cm^{-1}) [42a], and a 40 Å phenylenethynylene bridge reported by Wiberg et al. (1 cm^{-1}) [42b], but much stronger than that reported for an analogous non-conjugated Fc-porphyrin-porphyrin-C_{60} system (5.6×10^{-5} cm^{-1}) [16]. The attenuation of tunneling rate with distance is often analyzed in terms of the distance dependence of the coupling, according to Eq. 2.5.

$$V = V_0\, e^{-\beta\, R_{\mathrm{DA}}/2} \tag{2.5}$$

A plot of ln V against R looks similar to Fig. 2.17a; connecting the first two points gives $\beta = 0.19$ Å^{-1} and $V_0 = 32$ cm^{-1}, while connecting the second and third point gives $\beta = 0.006$ Å^{-1} and $V_0 = 1.0$ cm^{-1}. It is not possible to calculate the barrier heights (ΔE) corresponding to these β-values by Eq. 2.4, due to the uncertainty in the effective mass (m^*); however, it seems likely that such small β values would correspond to barrier heights of less than the thermal energy at room temperature ($k_B T = 26$ meV). Such small barrier heights would be expected to result in thermal population of bridge states, under the conditions of the experiments, thus it is unrealistic to analyze these kinetic data in terms of a simple single-step tunneling mechanism. However, see Section 2.9.

2.7.2 *Hopping*

Incoherent hopping mechanisms involve the population of intermediate states, by thermally activated injection of an electron (or hole) into the LUMO (or HOMO) of the bridge [35, 43–47]. This requires a favorable energy difference between the states involved, and if they are resonant, or nearly so, a very weak distance dependence is generally found. The rate of hopping is expected to vary with bridge length, R_{DA} (or number of bridge hopping sites) according to Eq. 2.6,

$$k \propto R_{\mathrm{DA}}^{-x} \tag{2.6}$$

where $x = 2$ for a random diffusive hopping process (one-dimensional random walk) and $x = 1$ for a biased hopping process in which hops only occur toward the acceptor [48]. The values of

k_{CR2} are plotted against the reciprocal of the fullerene–ferrocene distance in Fig. 2.17b. The fit to this simple hopping model (k vs. R_{DA}^{-1}; Eq. 2.6; Fig. 2.17b) appears to be about as poor as to tunneling ($\ln k$ vs. R_{DA}; Eq. 2.3; Fig. 2.17a). A hopping mechanism might account for the experimental data, if one assumes that charge is fully delocalized over the conjugated porphyrin oligomer bridge, so that there is just one hopping site, regardless of the length of the bridge. From the electrochemical measurements, the barriers ΔE for hole transfer from Fc^+ to the bridge in **Fc^+-P_1-C_{60}**$^-$, **Fc^+-P_2-C_{60}**$^-$ and **Fc^+-P_4-C_{60}**$^-$, are 0.33, 0.31, and 0.30 eV, respectively (and the barriers to electron injection are even higher). A barrier of $\Delta E \approx 12k_BT$ would lead to a very unfavorable Boltzmann term in the Arrhenius equation, Eq. 2.7, so this model could only account for the observed behavior if the pre-exponential factor is extremely high ($A \approx 2 \times 10^{13}$ s^{-1}). Arrhenius pre-exponential factors of this magnitude are common in small molecule unimolecular reactions, but further analysis will be required to establish whether such values are plausible for thermally activated electron transfer in molecular wires.

$$k_{CR2} = A\, e^{-\Delta E/k_B T} \qquad (2.7)$$

2.7.3 *Recombination via Triplet States*

A third possible mechanism for charge recombination is that charges recombine by forming a bridge triplet state. For the tetramer system the oligomer triplet state and the charge shifted state are almost isoenergetic, suggesting that recombination *via* a triplet state might be viable. This process would result in long-lived ground-state bleaching, and this is observed to a small extent for the tetramer system, but the amount of bleaching does not increase on the time-scale of the decay of the **Fc^+-P_n-C_{60}**$^-$ state, and thus there is no evidence to indicate that the charge shifted state recombines to a bridge triplet state.

2.8 Conclusions and Outlook

The results discussed in this chapter show that butadiyne-linked porphyrin oligomers mediate charge transport efficiently over distances of at least 65 Å. Further experiments are currently in progress to test the mechanism of charge transport in these systems. A special feature of these molecular wires is that their conformations, and conformational homogeneity, can be controlled by supramolecular assembly, via coordination of axial ligands to the zinc porphyrin centers. In related work, time-resolved microwave conductivity experiments have been used to measure the mobility of holes along butadiyne-linked porphyrin polymers, both as single strands and as a double-strand 4,4'-bipyridyl ladder complex. In this case double-strand formation increased the charge mobility by an order of magnitude [49]. Theoretical analysis indicated that charge transport occurs by small polaron hopping mechanism and that double-strand formation increases the charge mobility both by increasing the orbital overlap along the backbone (by planarizing the wire) and by reducing the amount of energetic disorder [50]. Single molecule conductance measurements on butadiyne-linked porphyrin oligomers have confirmed the very shallow distance dependence observed during photoinduced electron transfer experiments [51]. The field of single molecular electronics will benefit from further studies in which direct electrical and photophysical measurements are compared for the same homologous group of molecules. In future we plan to extend these studies to investigate longer molecular wires and structures with highly redox active components.

2.9 Note Added in Proof

Since writing this chapter, two pieces of work have been carried out, and published [52, 53], which significantly alter our understanding of the systems discussed here.

A variable-temperature single-molecule conductance study of a series of butadiyne-linked porphyrin oligomers consisting of one to three porphyrin units, together with a detailed theoretical analysis, using density functional theory and an analytical model, strongly suggest that the mechanism of charge transport in these molecular wires is phase-coherent tunneling [52]. The previous conclusion that the very low attenuation factor in these systems ($\beta = 0.040$

$\pm$ 0.006 Å^{-1}) indicated an incoherent hopping transport mechanism, on the grounds that these β values must correspond to barrier heights of less than the thermal energy at room temperature, appears to be invalid because relationships such as Eq. 2.4 (which assume a free electron mass, a simple rectangular barrier and a barrier that is independent of length) are too crude.

We have also investigated the temperature-dependence of charge separation and recombination in the C_{60}-terminated oligomers $\mathbf{P_1}$-$\mathbf{C_{60}}$, $\mathbf{P_2}$-$\mathbf{C_{60}}$, $\mathbf{P_3}$-$\mathbf{C_{60}}$, $\mathbf{P_4}$-$\mathbf{C_{60}}$ and $\mathbf{P_6}$-$\mathbf{C_{60}}$ [53]. The temperature dependence of the charge-separation reaction revealed a complex process for the longer oligomers, in which a combination of (i) direct charge separation and (ii) migration of excitation energy along the oligomer followed by charge separation explained the observed fluorescence decay kinetics. The energy migration is controlled by the temperature-dependent conformational dynamics of the longer oligomers and thereby limits the quantum yield for charge separation. Charge recombination was also studied as a function of temperature through measurements of femtosecond transient absorption. The temperature dependence of the electron-transfer reactions could be successfully modeled using the Marcus equation through optimization of the electronic coupling (V) and the reorganization energy (λ). For the charge-separation rate, all of the donor-acceptor systems could be successfully described by a common electronic coupling, supporting a model in which energy migration is followed by charge separation. In this respect, the C_{60}-appended porphyrin oligomers are suitable model systems for practical charge-separation devices such as bulk-heterojunction solar cells, where conformational disorder strongly influences the electron-transfer reactions and performance of the device.

References

1. Barbara, P. F., Meyer, T. J., and Ratner, M. A. (1996) Contemporary issues in electron transfer research, *J. Phys. Chem.*, **100**, 13148–13168.
2. Cheng, Y.-J., Yang, S.-H., and Hsu, C.-S. (2009) Synthesis of conjugated polymers for organic solar cell applications, *Chem. Rev.*, **109**, 5868–5923.

3. McCreery, R. L. and Bergren, A. J. (2009) Progress with molecular electronic junctions: meeting experimental challenges in design and fabrication, *Adv. Mater.*, **21**, 4303–4322.
4. Frampton, M. J. and Anderson, H. L. (2007) Insulated molecular wires, *Angew. Chem. Int. Ed.*, **46**, 1028–1064.
5. Crossley, M. J. and Burn, P. L. (1987) Rigid, laterally-bridged bis-porphyrin systems, *J. Chem. Soc., Chem. Commun.*, 39–40.
6. Anderson, H. L., Martin, S. J., and Bradley, D. D. C. (1994) Synthesis and third order nonlinear optical properties of a conjugated porphyrin polymer, *Angew. Chem. Int. Ed. Engl.*, **33**, 655–657.
7. Anderson, H. L. (1999) Building molecular wires from the colours of life: conjugated porphyrin oligomers, *Chem. Commun.*, 2323–2330.
8. Lin, V. S.-Y, DiMagno, S. G., and Therien, M. J. (1994) Highly conjugated, acetylenyl bridged porphyrins: new models for light-harvesting antenna systems, *Science*, **264**, 1105–1111.
9. Tsuda, A. and Osuka, A. (2001) Fully conjugated porphyrin tapes with electronic absorption bands that reach into infrared, *Science*, **293**, 79–82.
10. Kim, D. and Osuka, A. (2004) Directly linked porphyrin arrays with tunable excitonic interaction, *Acc. Chem. Res.*, **37**, 735–745.
11. Liu, Z., Yasseri, A. A., Lindsey, J. S., and Bocian, D. F. (2003) Molecular memories that survive silicon device processing and real-world operation, *Science*, **302**, 1543–1545.
12. Eschenmoser, A. and Winter, C. E. (1977) Natural product synthesis and vitamin B_{12}, *Science*, **196**, 1410–1420.
13. Taylor, P. N., Huuskonen, J., Rumbles, G., Aplin, R. T., Williams, E., and Anderson, H. L. (1998) Conjugated porphyrin oligomers from monomer to hexamer, *Chem. Commun.* 909–910.
14. Frampton, M. J., Akdas, H., Cowley, A. R., Rogers, J. E., Slagle, J. E., Fleitz, P. A., Drobizhev, M., Rebane, A., and Anderson, H. L. (2005) Synthesis, crystal structure and nonlinear optical behavior of β-unsubstituted *meso-meso E*-vinylene-linked porphyrin dimers, *Org. Lett.*, **7**, 5365–5368.
15. Winters, M. U., Dahlstedt, E., Blades, H. E., Wilson, C. J., Frampton, M. J., Anderson H. L., and Albinsson, B. (2007) Probing the efficiency of electron transfer through porphyrin-based molecular wires, *J. Am. Chem. Soc.*, **129**, 4291–4297.
16. Imahori, H., Tamaki, K., Guldi, D. M., Luo, C., Fujitsuka, M., Ito, O., Sakata, Y., and Fukuzumi, S. (2001) Modulating charge separation and charge

recombination dynamics in porphyrin-fullerene linked dyads and triads: Marcus-Normal versus inverted region, *J. Am. Chem. Soc.*, **123**, 2607–2617.

17. Guldi, D. M., Imahori, H., Tamaki, K., Kashiwagi, Y., Yamada, H., Sakata, Y., and Fukuzumi, S. (2004) A molecular tetrad allowing efficient energy storage for 1.6 s at 163 K, *J. Phys. Chem. A*, **108**, 541–548.
18. Ishimura, K., Hada, M., and Nakatsuji, H. (2002) Ionized and excited states of ferrocene: Symmetry adapted cluster–configuration–interaction study, *J. Chem. Phys.*, **117**, 6533–6537. Rohmer, M.-M., Veillard, A., and Wood, M. H. (1974) Excited states and electronic spectrum of ferrocene, *Chem. Phys. Lett.*, **29**, 466–468. Armstrong, A. T., Smith, F., Elder, E., and McGlynn, S. P. (1967) Electronic absorption spectrum of ferrocene, *J. Chem. Phys.*, **46**, 4321–4328.
19. Winters, M. U., Kärnbratt, J., Eng, M., Wilson, C. J., Anderson, H. L., and Albinsson, B. (2007) Photophysics of a butadiyne-linked porphyrin dimer: influence of conformational flexibility in the ground and first singlet excited state, *J. Phys. Chem. C*, **111**, 7192–7199.
20. Anderson, H. L. (1994) Conjugated porphyrin ladders. *Inorg. Chem.*, **33**, 972–981.
21. Wilson, G. J. and D.P. Arnold, D. P. (2005) Time-dependent density functional molecular orbital and excited state calculations on bis(porphyrinyl)butadiynes in the monocationic, neutral, monoanionic, and dianionic oxidation states. *J. Phys. Chem. A*, **109**, 6104–6113.
22. Gouterman, M. (1959) Study of the effects of substitution on the absorption spectra of porphin, *J. Chem. Phys.*, **30**, 1139–1161.
23. Gouterman, M. (1961) Spectra of porphyrins, *J. Mol. Spectrosc.*, **6**, 138–163.
24. Drobizhev, M., Stepanenko, Y., Dzenis, Y., Karotki, A., Rebane, A., Taylor, P. N., and Anderson, H. L. (2005) Extremely strong near-IR two-photon absorption in conjugated porphyrin dimers: Quantitative description with three-essential-states model, *J. Phys. Chem. B*, **109**, 7223–7236.
25. Winters, M. U., Kärnbratt, J., Blades, H. E., Wilson, C. J., Frampton, M. J., Anderson, H. L., and Albinsson, B. (2007) Control of electron transfer in a conjugated porphyrin dimer by selective excitation of planar and perpendicular conformers. *Chem. Eur. J.*, **13**, 7385–7394.
26. Marcus, R. A. (1956) On the theory of oxidation-reduction reactions involving electron transfer. *J. Chem. Phys.*, **24**, 966–978.

27. Marcus, R. A. (1959) On the theory of electrochemical and chemical electron transfer processes, *Can. J. Chem.*, **37**, 155–163.
28. Marcus, R. A. and Sutin, N. (1985) Electron transfers in chemistry and biology, *Biochim. Biophys. Acta*, **811**, 265–322.
29. Yoon, D. H., Lee, S. B., Yoo, K.-H., Kim, J., Lim, J. K., Aratani, N., Tsuda, A., Osuka, A., and Kim, D. (2003) Electrical conduction through linear porphyrin arrays, *J. Am. Chem. Soc.*, **125**, 11062–11064. Kang, B. K., Aratani, N., Lim, J. K., Kim, D., Osuka, A., and Yoo, K.-H. (2005) Length and temperature dependence of electrical conduction through dithiolated porphyrin arrays, *Chem. Phys. Lett.*, **412**, 303–306.
30. Susumu, K., Frail, P. R., Angiolillo, P. J., and Therien, M. J. (2006) Conjugated chromophore arrays with unusually large hole polaron delocalization lengths, *J. Am. Chem. Soc.*, **128**, 8380–8381.
31. Pettersson, K., Wiberg, J., Ljungdahl, T., Mårtensson, J., and Albinsson, B. (2006) Interplay between barrier width and height in electron tunneling: photo-induced electron transfer in porphyrin-based donor-bridge-acceptor systems, *J. Phys. Chem. A*, **110**, 319–326. Eng, M. P., Ljungdahl, T., Mårtensson, J., and Albinsson, B. (2006) Triplet excitation energy transfer in porphyrin-based donor-bridge-acceptor systems with conjugated bridges of varying length: An experimental and DFT study, *J. Phys. Chem. B*, **110**, 6483–6491.
32. Giacalone, F., Segura, J. L., Martín, N., Ramey, J., and Guldi, D. M. (2005) Probing molecular wires: Synthesis, structural, and electronic study of donor–acceptor assemblies exhibiting long-range electron transfer, *Chem. Eur. J.*, **11**, 4819–4834. de la Torre, G., Giacalone, F., Segura, J. L., Martín, N., and Guldi, D. M. (2005) Electronic communication through π-conjugated wires in covalently linked porphyrin/C_{60} ensembles, *Chem. Eur. J.*, **11**, 1267–1280.
33. McConnell, H. (1961) Intramolecular charge transfer in aromatic free radicals, *J. Chem. Phys.*, **35**, 508–515.
34. Harriman, A., Khatyr, A., Ziessel, R., and Benniston, A. C. (2000) An unusually shallow distance-dependence for triplet-energy transfer, *Angew. Chem. Int. Ed.*, **39**, 4287–4290. Harriman, A., Rostron, S. A., Khatyr, A., and Ziessel, R. (2006) The mechanism of long-range electron exchange in molecular-scale photonic wires, *Faraday Discuss.*, **131**, 377–391.
35. Davis, W. B., Svec, W. A., Ratner, M. A., Wasielewski, M. R. (1998) Molecular-wire behaviour in *p*-phenylenevinylene oligomers, *Nature*, **396**, 60–63.

36. Vilan, A. (2007) Analyzing molecular current-voltage characteristics with the Simmons Tunneling Model: Scaling and linearization, *J. Phys. Chem. C.*, **111**, 4431–4444.

37. Joachim, C. and Magoga, M. (2002) The effective mass of an electron when tunneling through a molecular wire, *Chem. Phys.*, **281**, 347–352.

38. Albinsson, B., Eng, M. P., Pettersson, K., and Winters, M. U. (2007) Electron and energy transfer in donor-acceptor systems with conjugated molecular bridges, *Phys. Chem. Chem. Phys.*, **9**, 5847–5864.

39. Helms, A., Heiler, D., and McLendon, G. (1992) Electron transfer in bis-porphyrin donor-acceptor compounds with polyphenylene spacers shows a weak distance dependence, *J. Am. Chem. Soc.*, **114**, 6227–6238.

40. Sachs, S. B., Dudek, S. P., Hsung, R. P., Sita, L. R., Smalley, J. F., Newton, M. D., Feldberg, S. W., and Chidsey, C. E. D. (1997) Rates of interfacial electron transfer through π-conjugated spacers, *J. Am. Chem. Soc.*, **119**, 10563–10564.

41. Marcus, R. A. (1993) Electron transfer reactions in chemistry: Theory and experiment (Nobel Lecture), *Angew. Chem. Int. Ed. Engl.*, **32**, 1111–1121.

42. Creager, S., Yu, C. J., Bamdad, C., O'Connor, S., MacLean, T., Lam, E., Chong, Y., Olsen, G. T., Luo, J., Gozin, M., and Kayyem, J. F. (1999) Electron transfer at electrodes through conjugated "Molecular Wire" bridges, *J. Am. Chem. Soc.*, **121**, 1059–1064. Wiberg, J., Guo, L., Pettersson, K., Nilsson, D., Ljungdahl, T., Mårtensson, J., and Albinsson, B. (2007) Charge recombination versus charge separation in donor-bridge-acceptor systems, *J. Am. Chem. Soc.*, **129**, 155–163.

43. Berlin, Y. A., Grozema, F. C., Siebbeles, L. D. A., and Ratner, M. A. (2008) Charge transfer in donor-bridge-acceptor systems: Static disorder, dynamic fluctuations, and complex kinetics, *J. Phys. Chem. C*, **112**, 10988–11000.

44. Winters, M. U., Pettersson, K., Martensson, J., and Albinsson, B. (2005) Competition between superexchange-mediated and sequential electron transfer in a bridged donor–acceptor system, *Chem. Eur. J.*, **11**, 562–573.

45. Lambert, C., Nöll, G., Schelter, J. (2002) Bridge-mediated hopping or superexchange electron-transfer processes in bis(triarylamine) systems, *Nat. Mater.*, **1**, 69–73.

46. Weiss, E. A., Ahrens, M. J., Sinks, L. E., Gusev, A. V., Ratner, M. A., and Wasielewski, M. R. (2004) Making a molecular wire: Charge and spin transport through *para*-phenylene oligomers, *J. Am. Chem. Soc.*, **126**, 5577–5584.

47. Weiss, E. A., Tauber, M. J., Kelley, R. F., Ahrens, M. J., Ratner M. A., and Wasielewski, M. R. (2005) Conformationally gated switching between superexchange and hopping within oligo-*p*-phenylene-based molecular wires, *J. Am. Chem. Soc*, **127**, 11842–11850.
48. Jortner, J., Bixon, M., Langenbacher, T., and Michel-Beyerle, M. E. (1998) Charge transfer and transport in DNA. *Proc. Natl. Acad. Sci. USA*, **95**, 12759–12765.
49. Grozema, F. C., Houarner-Rassin, C., Prins, P., Siebbeles, L. D. A., and Anderson, H. L. (2007) Supramolecular control of charge transport in molecular wires, *J. Am. Chem. Soc.*, **129**, 13370–13371.
50. Kocherzhenko, A. A., Patwardhan, S., Grozema, F. C., Anderson, H. L., and Siebbeles, L. D. A. (2009) Mechanism of charge transport along zinc porphyrin-based molecular wires, *J. Am. Chem. Soc.*, **131**, 5522–5529.
51. Sedghi, G., Sawada, K., Esdaile, L. J., Hoffmann, M., Anderson, H. L., Bethell, D., Haiss, W., Higgins, S. J., and Nichols, R. J. (2008) Single molecule conductance of porphyrin wires with ultralow attenuation, *J. Am. Chem. Soc.*, **130**, 8582–8583.
52. Sedghi, G., García-Suárez, V. M., Esdaile, L. J., Anderson, H. L., Lambert, C. J., Martin, S. Bethell, D., Higgins, S. J., Elliott, M., Bennett, N., Macdonald, J. E., and Nichols, R. J. (2011) Long-range electron tunnelling in oligo-porphyrin molecular wires, *Nat. Nanotech.*, **6**, 517–523.
53. Kahnt, A., Kärnbratt, J., Esdaile, L. J., Hutin, M., Sawada, K., Anderson, H. L., and B. Albinsson, B. (2011) Temperature dependence of charge separation and recombination in porphyrin oligomer-fullerene donor-acceptor systems, *J. Am. Chem. Soc.* **133**, 9863–9871.

Chapter 3

Recent Applications of MCD Spectroscopy to Porphyrinoids

John Mack and Nagao Kobayashi
Department of Chemistry, Faculty of Graduate Science, Tohoku University, Sendai, 980–8578, Japan
nagaok@m.tohoku.ac.jp

Over the last 40 years, magnetic circular dichroism (MCD) spectroscopy has provided many key insights into the optical spectroscopy and electronic structure of porphyrinoid compounds [1, 2]. In this chapter, the use of the technique, since the start of the new millennium, to assign the optical spectra of newly synthesized porphyrinoid ligands and to elucidate their electronic structures is described in depth. The recent emergence of DFT theoretical treatments in commercially available software packages has made the calculation of detailed descriptions of the electronic structures of newly synthesized compounds routine. MCD spectroscopy provides key additional information about band polarization and state degeneracies, which can be used to validate time-dependent density functional theory (TD-DFT) calculations and to test the accuracy of density functional theory (DFT) descriptions of the electronic structure. In this chapter, the theoretical background to MCD spectroscopy and recent applications of the technique to

Multiporphyrin Arrays: Fundamentals and Applications
Edited by Dongho Kim

www.panstanford.com

radially symmetric and low-symmetry synthetic porphyrinoids and to bioinorganic complexes of transition metals, such as heme proteins, are described in detail. The radially symmetric synthetic porphyrinoids dealt with in this chapter include deeply saddled compounds such as tetraphenyltetraacenaphthoporphyrins and α-octaphenylphthalocyanines, ring-contracted subporphyrin and [14]triphyrin(2.1.1) compounds, and expanded porphyrinoids with both $4N$ and $4N+2$ π-electron systems such as gold hexaphyrins. The low-symmetry compounds covered include core-modified tetrabenzoporphyrins, corrolazines, fused-ring-expanded phthalocyanines, tetraazachlorins, azulene-fused porphyrins, azulenocyanines, and benzoporphycenes.

3.1 Introduction

The modern theoretical background to MCD spectroscopy is based primarily on the research of Buckingham, Stephens, and Schatz [3–5]. Unlike NMR and EPR spectroscopy, where there is resonance between microstates of the ground state, MCD spectroscopy is based on excitation to an excited electronic state due to the absorption of an incident photon of left or right circularly polarized light.

UV-visible absorption and MCD spectra are comprised of the same set of spectral bands but the band morphologies differ markedly. When a magnetic field is applied in the Faraday arrangement, parallel to the axis of light propagation, by mounting a magnet into the sample compartment of a CD spectrometer, there is a lifting of state degeneracies resulting in the formation of M_{J+1} microstates where M_J is the quantum number for total angular momentum, Fig. 3.1. Transitions in which $\Delta M_J = -1$ absorb right circularly polarized (rcp) photons, while those in which $\Delta M_J = +1$ absorb left circularly polarized (lcp) photons. The analysis of MCD spectroscopy is based on a normalized band shape function (usually a Gaussian-shaped curve), $f(\varepsilon)$, rather than the absorption of light at a discrete wavelength. According to the modified conventions recommended by Stephens, Piepho and Schatz [4, 5], the MCD

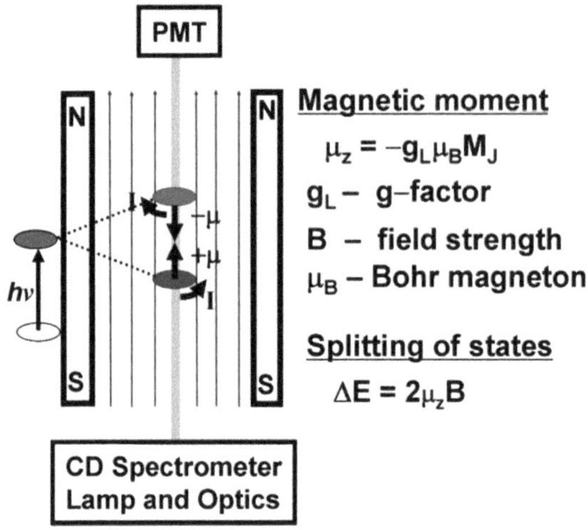

Figure 3.1. An MCD magnet's horizontal sample bore is surrounded by a solenoid so that a high magnetic field, typically of 1–7 tesla (T), is generated with field lines running from the south to north poles of the magnet parallel to the axis of light propagation so that the Faraday magnet alignment is obtained. A CD signal sensitivity of $\Delta A_{l-r} = 1 \times 10^{-6}$, is sufficient to detect the Zeeman splitting of states generated by the alignment of magnetic moments induced in the excited state by the circulation of charge either with or against the applied magnetic.

intensity equation for a fully allowed electronic transition is:

$$\Delta A_{l-r} = 152.5Bcl[\mathcal{A}_1(-df(\varepsilon)/d\varepsilon) + (\mathcal{B}_0 + \mathcal{C}_0/kT)f(\varepsilon)] \quad (3.1)$$

where B is the field strength in tesla, cl is the product of the concentration (mol.l^{-1}) and pathlength (cm). Information can be derived from the Zeeman splitting of states based on the three Faraday terms, $\mathcal{A}_1$, $\mathcal{B}_0$, and $\mathcal{C}_0$. $\mathcal{A}_1$ and $\mathcal{C}_0$ terms are observed in spectra of high symmetry complexes with orbitally degenerate ground and/or excited states. The derivative-shaped Faraday $\mathcal{A}_1$ term is temperature independent and identifies orbitally degenerate excited states, while the normally Gaussian-shaped $\mathcal{C}_0$ term is highly temperature dependent and identifies an orbitally degenerate ground state. The temperature-independent Faraday $\mathcal{B}_0$ term can exhibit a Gaussian shape of either sign and completely dominates the MCD spectra of low-symmetry compounds, which lack a threefold or higher axis of symmetry.

3.1.1 *Faraday Terms*

The Faraday $\mathcal{A}_1$ term, Fig. 3.2, arises from the $2g_J\mu_B B$ Zeeman splitting of the band centers of the lcp and rcp absorption bands, which arise for a transition into an orbitally-degenerate excited state, when a magnetic field is applied parallel to the axis of light propagation.

The equation for $\mathcal{A}_1$ term intensity, based on the conventions of Piepho and Schatz [5], is

$$\mathcal{A}_1 = \frac{1}{d_A}\sum < J|\mu_z|J > - < A|\mu_z|A > \left(\left[m_{-1}^{AJ}\right]^2 - \left[m_{+1}^{AJ}\right]^2\right) \quad (3.2)$$

$$\mu_z = -\mu_B(L_z + 2S_z) \quad (3.3)$$

where A and J refer to the ground and excited state and L_z, S_z, and m are the operators for orbital and spin angular momenta and the electric dipole moment, respectively. The sign of the derivative-shaped Faraday $\mathcal{A}_1$ term signal is determined by the difference in the total angular momentum associated with the ground and excited states. A positive Faraday $\mathcal{A}_1$ term is defined as one in which the lobe of positive intensity lies to higher energy, Fig. 3.2 [1].

Gaussian-shaped, temperature-independent $\mathcal{B}_0$ terms usually arise from field-induced mixing between excited states (labeled here

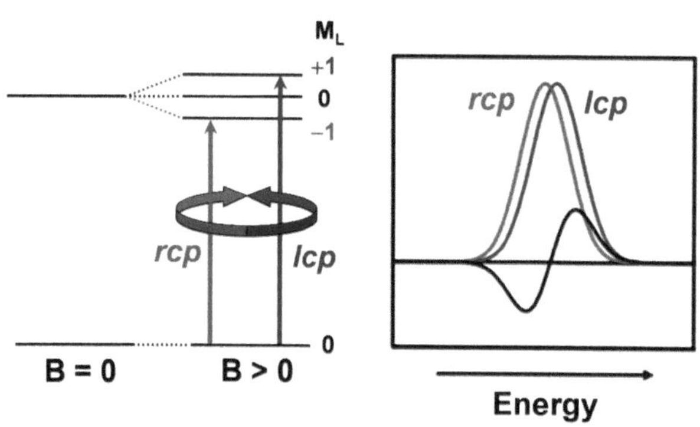

Figure 3.2. Selection rules for the absorption of circularly polarized light between an orbitally non-degenerate ground state and a degenerate excited state within a magnetic field applied parallel to the axis of light propagation. A temperature-independent derivative-shaped Faraday $\mathcal{A}_1$ term signal (black line) is observed due to the Zeeman splitting of the band centers of the lcp and rcp absorption bands.

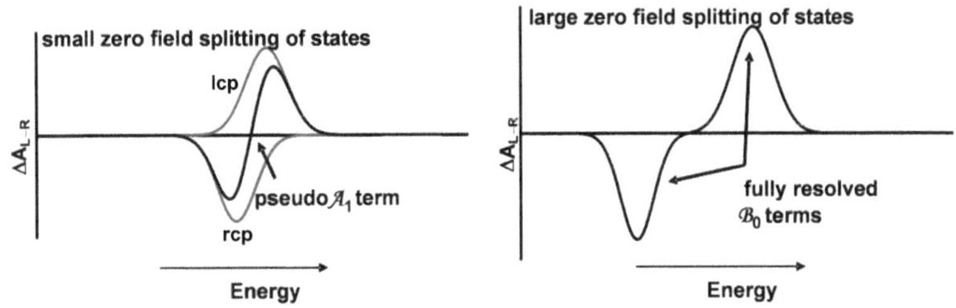

Figure 3.3. The appearance of coupled Faraday $\mathcal{B}_0$ terms depends upon the extent of the splitting of states at zero field relative to the bandwidths. When the ΔE_{KJ} energy value is small the coupled pair of oppositely signed $\mathcal{B}_0$ terms (black line) that arise from the differential absorbance of lcp and rcp can closely resemble a Faraday $\mathcal{A}_1$ term. This is sometimes referred to as a pseudo-$\mathcal{A}_1$ term.

as J and K) that are linked by a magnetic dipole transition moment:

$$\mathcal{B}_0 = \frac{2}{d_A}\Re e \sum \frac{< J|\mu_z|K >}{\Delta E_{KJ}} \left(\left[m_{-1}^{AJ}\right] \left[m_{+1}^{KA}\right] - \left[m_{+1}^{AJ}\right] \left[m_{-1}^{KA}\right] \right) \quad (3.4)$$

The key factor determining the intensity of $\mathcal{B}_0$ terms is the energy separation of the K and J excited states (ΔE_{KJ}), Fig. 3.3. When the separation is small relative to the spectral band widths, a so called pseudo-$\mathcal{A}_1$ term is observed. $\mathcal{B}_0$ term intensity sums to zero across the entire spectrum, since the intensity arises from the field-induced mixing of states In the spectra of radially symmetric porphyrinoids, the intensity of the $\mathcal{B}_0$ terms is usually significantly lower than that of the $\mathcal{A}_1$ or $\mathcal{C}_0$ terms.

Faraday $\mathcal{C}_0$ terms are observed for complexes with orbitally and/or spin degenerate ground states and will be described in more detail in Section 3.4. The Faraday $\mathcal{C}_0$ term exhibits strong 1/kT temperature dependence, Fig. 3.4, due to the Boltzmann population distribution across the degenerate ground state, Eq. 3.5, and can, therefore, be readily distinguished from $\mathcal{B}_0$ terms by recording spectra at both room and cryogenic temperatures. When the splitting of the microstates is small relative to the spectral band widths of the lcp and rcp absorption bands, the $\mathcal{C}_0$ term is essentially Gaussian-shaped, Fig. 3.4.

$$\mathcal{C}_0 = -\frac{1}{d_A} \sum < A|\mu_z|A > \left(\left[m_{-1}^{AJ}\right]^2 - \left[m_{+1}^{AJ}\right]^2 \right) \quad (3.5)$$

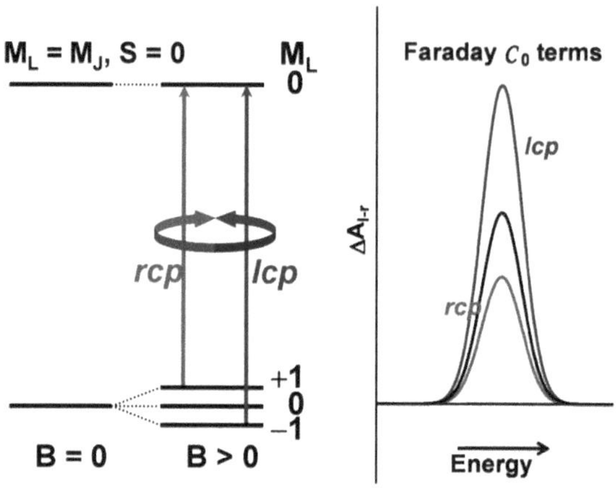

Figure 3.4. Selection rules (left) for the absorption of circularly polarized light based on a transition between an orbitally degenerate ground state and an orbitally non-degenerate excited state within a magnetic field applied parallel to the axis of light propagation. The $\mathcal{C}_0$ term signal (black line), generated by the differential absorbance of lcp and rcp (Right), exhibits $1/kT$ temperature dependence since the ground state microstates are populated according to a Boltzmann distribution. In this example, $S = 0$; so $M_L = M_J$. $\mathcal{C}_0$ terms, which arise when $S \neq 0$, are described in detail in Section 3.4.

3.1.2 *Gouterman's 4-Orbital Model*

Although TD-DFT calculations are now used routinely to calculate electronic absorption spectra, the three Faraday terms have not been implemented into commercially available DFT software packages. Older semi-empirical based approaches continue to play a key role in the analysis of MCD spectra and provides a readily accessible conceptual framework for the experimentalist. In the 1940s, Platt [6, 7], used molecular orbital theory to assign the bands observed in the electronic absorption spectrum of benzene based on the $M_L = 0$, ± 1, ± 2, 3 properties of the π-system MOs. Platt assigned allowed and forbidden B and L bands based on $\Delta M_L = \pm 1$ and ± 3 transitions, respectively, linking the orbitally degenerate HOMO ($M_L = \pm 1$) and LUMO ($M_L = \pm 2$). Since the magnetic and electric vectors of electromagnetic radiation can rotate a maximum of once per wavelength, only $\Delta M_L = 0, \pm 1$ transitions are electric dipole

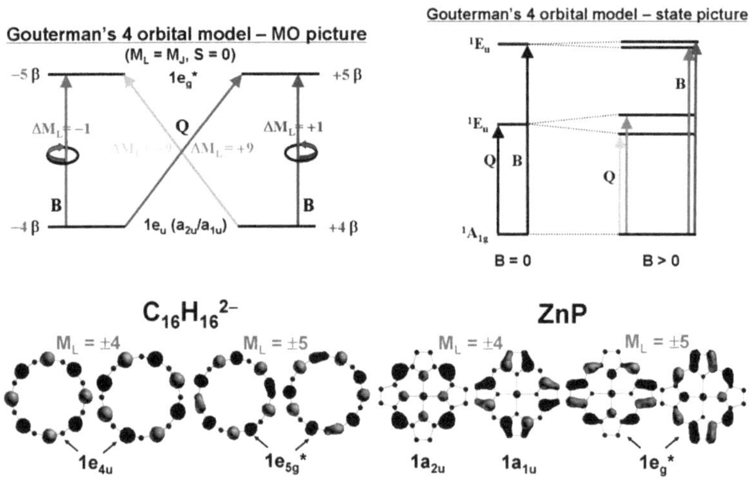

Figure 3.5. The origin of the Q and B bands in Gouterman's 4-orbital model [11] based on a consideration of only the four frontier π-MOs (top). The nodal patterns of the four key frontier π-MOs of $C_{16}H_{16}^{2-}$ and ZnP (bottom).

allowed. Gouterman and coworkers [8–12] subsequently demonstrated that the major spectral bands of metal porphyrin (MP) complexes could be assigned on a similar basis based on an $M_L = 0, \pm1$, ±2, ±3, ±4, ±5, ±6, ±7, 8 sequence of MOs in ascending energy associated with the 16 atom 18-π-electron system formed on the inner perimeter of the porphyrin ligand. The incorporation of the four pyrrole rings lowers the symmetry from D_{16h} to D_{4h}. Under D_{4h} symmetry, the LUMO ($1e_g$*) remains degenerate and has $M_L =$ ±5 nodal patterns, while the MOs with $M_L = \pm 4$ nodal patterns ($1a_{1u}$ and $1a_{2u}$) are non-degenerate. Within the band nomenclature of Gouterman's 4-orbital model [8–12], Fig. 3.5, there is an allowed B transition and a forbidden Q transition with $\Delta M_L = \pm1$ and ±9 properties, respectively.

The spectra of synthetic main group metal porphyrins contain a fully allowed B band at *ca.* 400 nm (also sometimes referred to as the Soret band) and a set of weaker Q bands between 500 and 650 nm (arising from the forbidden $\Delta M_L = \pm9$ transitions), Fig. 3.5 The x/y-polarized Q and B bands arise from transitions between the non-degenerate ground state ($^1A_{1g}$) and orbitally degenerate excited states (^1E_u). First derivative-shaped $\mathcal{A}_1$ terms, therefore, dominate the MCD spectra of metal porphyrin complexes, Fig. 3.3. Excited state magnetic moments can be derived from $\mathcal{A}_1/\mathcal{D}_0$ ratio, where $\mathcal{D}_0$ is the dipole strength of the corresponding absorption spectrum band. The allowed and forbidden nature of the B and Q bands was demonstrated on this basis [10]. In the context of free base porphyrinoid compounds, the symmetry is modified from D_{4h} to D_{2h} symmetry, so the x/y-polarized $\mathcal{A}_1$ terms are replaced by coupled and oppositely-signed x- and y-polarized Faraday $\mathcal{B}_0$ terms, Fig. 3.4. It should be noted that the Q bands of porphyrin complexes are only forbidden due to the accidental near degeneracy of the $1a_{1u}$ and $1a_{2u}$ MOs, since the allowed and forbidden intensity properties arise from the cancellation or intensification of near-equal contributions from the $1a_{1u} \rightarrow 1e_g{}^*$ and $1a_{2u} \rightarrow 1e_g{}^*$ one-electron transitions [11]. In the case of other porphyrinoids such as phthalocyanines (Pcs) there is a marked lifting of the degeneracy of the $1a_{1u}$ and $1a_{2u}$ MOs, since the incorporation of aza-nitrogens leads to a stabilization of the $1a_{2u}$ MO, while fused-ring-expansion causes a destabilization of the $1a_{1u}$ MO. This results in a mixing of the allowed and forbidden properties of the B and Q bands due to there being non-equal contributions from these one-electron transitions. The Q band, therefore, gains significant intensity and becomes the dominant spectral feature.

3.1.3 *Michl's (4N + 2)-Electron and 4N-Perimeter Models*

In the early 1950s, Moffitt [13, 14] developed a cyclic perimeter approach for analyzing the electronic structures of aromatic and heteroaromatic cyclic polyenes. Moffitt demonstrated that the nodal patterns of the frontier π-MOs of parent high symmetry hydrocarbons are retained even after substantial structural modifications [13, 14] and that the electronic structures of low symmetry cyclic polyenes can, therefore, be described in terms of perturbations to the structure of the parent hydrocarbon. In the 1970s, Michl [15–17] developed a general theory to account for the MCD spectroscopy of aromatic and heteroaromatic cyclic π-electron systems on this basis, which can be used to predict the sign sequences of $\mathcal{A}_1$ (or of coupled oppositely signed $\mathcal{B}_0$) terms based on a consideration of the excited state magnetic moments induced by Platt's forbidden and allowed

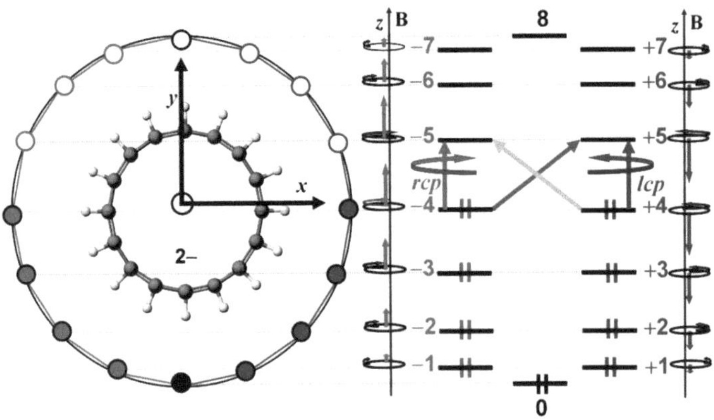

Figure 3.6. Michl's perimeter model (left) [15]. The circle is a diagrammatic representation of the clockwise (red) and counterclockwise (blue) motion of π-system electrons on the $C_{16}H_{16}^{2-}$ parent perimeter of porphyrinoid compounds based on the 16 $2p_z$ atomic orbitals of the inner ligand perimeter, generating the M_L = 0, ±1, ±2, ±3, ±4, ±5, ±6, ±7, 8 value for each complex wavefunction MO. The perspective is towards the light source. The magnitudes of the magnetic moments induced in the applied field within each MO (right) are plotted diagrammatically, based on the results of a set of LCAO calculations carried out by Michl [15]. The origin of the forbidden L and allowed B transitions linking the HOMO ($M_L = \pm 4$) and the LUMO ($M_L = \pm 5$) are indicated using arrows with the same color scheme adopted in Fig. 3.5 for the Q and B bands in Gouterman's 4-orbital model. See also Color Insert.

L and B transitions, Fig. 3.6. It was hoped that this would result in the MCD technique becoming a more widely used spectroscopic tool within organic chemistry.

Since the π-system forms a cyclic perimeter comprised of $2p_z$ atomic orbitals of carbon and nitrogen atoms, complex wavefunctions permit a description of the net electron circulation upon electronic excitation, which is not possible on the basis of the more commonly used real wavefunctions. A stack of MOs can be constructed for the $C_{16}H_{16}^{2-}$ parent perimeter of conventional radially symmetric tetrapyrrole porphyrinoids [1], Fig. 3.6. For each set of degenerate complex wavefunctions Ψ_N and Ψ_{-N}, the Ψ_{-N} MO involves electron circulation (based on a redistribution of electronic charge from one $2p_z$ atomic orbital to another) in a clockwise direction from the lowest energy $M_L = 0$ MO, while the Ψ_N MO involves electron circulation in a counterclockwise direction. When a magnetic field is applied parallel to the axis of light propagation, absorption of a photon of lcp and rcp results in either a $\Delta M_L = +1$ or a $\Delta M_L = -1$ transition, since the electric vector of the incident electromagnetic wave rotates clockwise or counterclockwise once per wavelength and the total angular momentum within the system must be conserved, there is either clockwise or anticlockwise charge circulation on the perimeter of the $\pi\pi^*$ excited state, based on the direction of rotation of the electric vector of the incident photon, Fig. 3.6.

In high symmetry cyclic polyenes, such as benzene, in which the π-MOs associated with the perimeter retain orbital degeneracy the orbital angular momentum (OAM) is no longer fully compensated by the motion of a second electron in the MO of the opposing handedness when an electron is promoted to the LUMO in an electric dipole allowed manner ($\Psi_N \rightarrow \Psi_{N+1}$ in the case of lcp and $\Psi_{-N} \rightarrow \Psi_{-N-1}$ in the case of rcp). When OAM is greater in the LUMO than in the HOMO, a negatively charged particle is effectively formed on the perimeter. The alignment of the magnetic dipole induced within the excited state can be visualized based on Ampere's rule for current flow in a solenoid. When the fingers of the left hand are curled around the loop (i.e., the convention would normally be to use the right hand band on conventional current flow with a positive charge carrier), the thumb points in the direction of the induced dipole moment. A positive magnetic dipole moment, which points along the z-axis of the applied field is the energetically favored alignment within the Zeeman splitting of states. Since the perspective is towards the detector rather than the source in this context, this corresponds to circulation induced by the absorption of a photon of incident rcp. This results in a derivative-shaped Faraday $\mathcal{A}_1$ terms signal with a $-$ve/$+$ve sign sequence. This is referred to as a positive $\mathcal{A}_1$ term since the higher energy component has a positive $\Delta A_{\text{l-r}}$ signal. Positive Faraday $\mathcal{A}_1$ terms, Fig. 3.7, are usually the dominant spectral feature of high symmetry cyclic polyenes, such as radially symmetric porphyrinoids, since the OAM associated with the circulation of negative charge in the LUMO level is usually greater than that associated with the circulation of positive charge in the HOMO level.

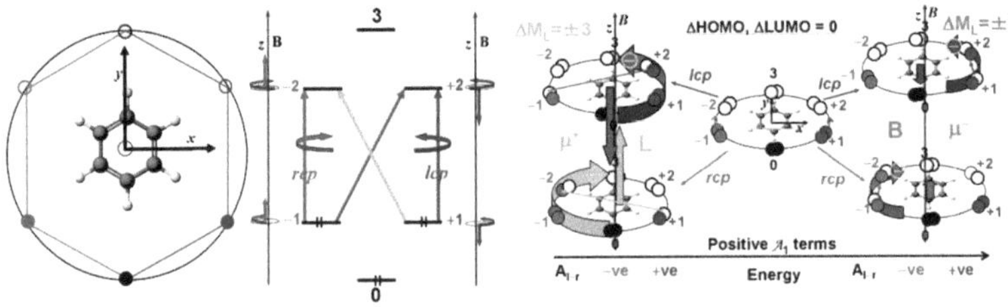

Figure 3.7. The sign orderings in ascending energy terms of the positive $\mathcal{A}_1$ terms associated with Platt's B and L bands [6] (left) is related to Ampere's rule (right-hand rule) for conventional current flow within a solenoid (right) in which the thumb is aligned with the magnetic dipole induced in the core of the current loop, when the fingers of the right hand are curled around the loop in the direction of conventional current flow. The left hand is used in this instance, since the charge carrier is negatively charged. The occupancy of the six complex MOs within Michl's perimeter model [15], Fig. 3.6, and the alignment of the μ^+ and μ^- magnetic dipoles, associated with the forbidden L and allowed B transitions, induced through the absorption of a photon of either lcp or rcp are indicated with colored circles and arrows, respectively, using the color scheme adopted in Fig. 3.5 for the Q and B bands in Gouterman's 4-orbital model. The resulting MCD sign sequence in ascending energy terms is shown based on the Zeeman splitting of states, which arises from the fact that a magnetic moment aligned with the applied field is energetically favored.

The induced $\pi\pi^*$ magnetic moments for Platt's electronically allowed B and forbidden L transitions for benzene (which would have $\Delta M_L = \pm 1$ and $\Delta M_L = \pm 3$ properties in the context of a free electron model approach) were referred to by Michl as the μ^- and μ^+ moments, respectively. Michl [15–17] predicted, based on an LCAO approach to the perimeter model, that the magnetic moments associated with the B and L bands (i.e., μ^- and μ^+) of a $C_{16}H_{16}^{2-}$ parent perimeter are -0.01 μ and -6.24 μ, respectively. Under the D_{16h} symmetry of the parent perimeter, the L band is fully electric dipole forbidden and can gain intensity only through vibrational borrowing from the allowed B band. When the structure is modified by incorporating pyrrole moieties to form a D_{4h} symmetry porphyrinoid there is a lifting of the degeneracy of the HOMO of the parent perimeter and a mixing of the allowed and forbidden properties of Q and B bands. The degree of mixing, α, and the magnitude of the induced magnetic dipole moments of the Q and B excited states ($\mu_{(Q)}$, $\mu_{(B)}$) can be estimated using Eqs. 3.6–3.8, based on the observed dipole strengths, $\mathcal{D}_0$, for the Q and B bands, which can be derived either empirically or from TD–DFT calculations by converting the predicted oscillator strength values.

$$\mathcal{D}_{0(Q)}/\mathcal{D}_{0(B)} = \tan^2\alpha \quad (3.6)$$

$$\mu_{(B)} = \pm(\cos^2\alpha)\mu^- \pm (\sin^2\alpha)\mu^+ \quad (3.7)$$

$$\mu_{(Q)} = \pm(\sin^2\alpha)\mu^- \pm (\cos^2\alpha)\mu^+ \quad (3.8)$$

When the symmetry of the parent perimeter is perturbed still further so that the fourfold axis of symmetry of radially-symmetric porphyrinoids is removed, the MCD spectrum is completely dominated by Gaussian-shaped $\mathcal{B}_0$ terms. Michl successfully demonstrated that the alignment of the magnetic moments and hence the sign sequence in the MCD spectrum is determined by whether the OAM of the incident lcp or rcp photon is conserved by the circulation of the positively charged hole left in the HOMO level or the negatively charged particle circulating in the LUMO level, based on the larger μ^+ moment associated with the forbidden L transition. In the context of porphyrinoids this is determined by the relative energy separations of the frontier π-MOs derived from the $1e_{4u}$ HOMO and $1e_{5g}$ LUMO of the $C_{16}H_{16}^{2-}$ parent hydrocarbon perimeter (Michl's terminology for this is the ΔHOMO and ΔLUMO values). When ΔHOMO $\gg$ ΔLUMO the OAM associated with the excited electron predominates and there is a $-$ve/$+$ve/$-$ve/$+$ve ΔA_{l-r} intensity pattern with ascending energy for the Faraday $\mathcal{B}_0$ terms associated with the Q and B bands. When ΔLUMO $\gg$ ΔHOMO circulation of the positive hole left within the HOMO level becomes the predominant factor. The magnetic moments induced by the absorption of a photon of lcp and rcp are reversed and the ΔA_{l-r} sign sequence for the Faraday $\mathcal{B}_0$ terms becomes $+$ve/$-$ve/$+$ve/$-$ve.

In the late 1990s, Michl and coworkers [18] developed a $4N$-perimeter model to analyze the electronic spectra of antiaromatic species and subsequently reported the analyses of several small cyclic π-electron systems on this basis [19–23]. Six frontier π-MOs derived from the HOMO, SOMO and LUMO of the parent hydrocarbon perimeter (referred to by Michl as h$-$, h$+$, s$-$, s$+$, l$-$, and l$+$)

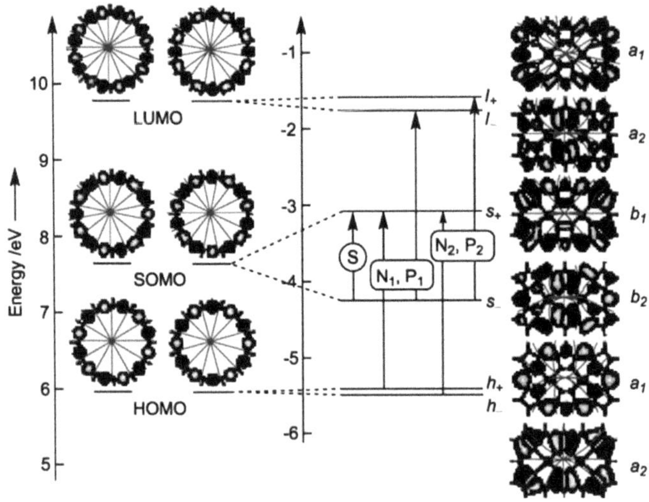

Figure 3.8. Isosurface plots of the frontier π-MOs, energy levels, and perimeter labels of an antiaromatic $C_{24}H_{24}^{4-}$ cyclic perimeter and an "unaromatic" Au^{III} hexaphyrin species [64]. Arbitrary nodal lines are drawn on the iso-surface plots. HOMO=s−, LUMO = s +.

play the key role in the low-energy singlet electronic states, Fig. 3.8. This is in marked contrast with the ($4N + 2$)-electron perimeter model where the L and B bands arise from transitions between the four frontier π-MOs derived from the HOMO and LUMO of the parent hydrocarbon perimeter [15–17]. Michl et al. used new terminology to describe even-electron cyclic $4N$-π-systems [18]. The term "antiaromatic" is reserved for π-systems that have very strong biradical-like character, while $4N$-electron perimeters in which the structural perturbations are strong enough to remove the biradical-like character are described as "unaromatic". The lowest energy band arises from the $s- \rightarrow s+$ transition, Fig. 3.8. The transition is of an intrashell nature and is electric-dipole forbidden. Although the transition is magnetic-dipole allowed, the absorption and hence also the MCD intensity of the S band is predicted to be zero [20]. The other four electronic transitions are intershell being derived from HOMO→SOMO and SOMO→LUMO one-electron transitions, Fig. 3.8. Two weakly (N_1, N_2) and two strongly (P_1, P_2) electric-dipole allowed transitions are predicted. The N and P transitions are analogous to the L and B transitions of aromatic systems. Configuration interaction between states arising from the h+ → s+ and s− → l− transitions causes cancellation and intensification of the transition dipole moments, resulting in weak and intense N_1 and P_1 bands, respectively. Weak N_2 and intense P_2 bands arise from the h−→s+ and s−→l+ transitions in a similar manner, Fig. 3.8.

It is important to note that Michl's $(4N + 2)$- and $4N$-perimeter models can not be applied to all porphyrinoid MCD spectra. Since the approach was originally developed for cyclic polyenes, the utility of these approaches is largely limited to free base porphyrinoids and the complexes of main group and closed shell d^{10} metals. The most accurate predictions are obtained when the visible region is almost completely dominated by Q and B bands with only limited configuration interaction between the B and higher energy $\pi\pi^*$ states. When there is scope for metal to ligand and ligand to metal charge transfer (MLCT and LMCT) bands, accurate predictions of the signs of Faraday $\mathcal{A}_1$ and $\mathcal{B}_0$ terms of the Q and B bands can no longer be made and in many instances temperature-dependent $\mathcal{C}_0$ terms will dominate the MCD spectra requiring a different theoretical approach which will be described in Section 3.4.

3.1.4 *TD-DFT Calculations of UV-Visible Absorption Spectra*

The recent emergence of DFT theoretical treatments [24] in commercially available software packages has made significantly more accurate descriptions of the ground state molecular structures of porphyrinoids available than those provided by the semi-empirical techniques used by Gouterman and Michl in the 1960s and 70s [11, 15–17]. This has made it easier for the experimentalist to derive calculated electronic absorption spectra for newly synthesized compounds, which can be validated by comparison with the observed UV-visible absorption and MCD spectra. It should be noted, that DFT calculations are usually not fully *ab initio*, since they tend to be based on exchange-correlation (x–c) functionals, which are partially empirically derived. A semi-empirical component based on the traditional Hartree–Fock (HF) approach is often introduced to derive predicted molecular property values, which are close to the observed experimental values. In the context of porphyrinoids the selection of the x–c functional can, therefore, have a significant impact on the predicted energies of the frontier π-MOs and hence on the energies of the main $\pi \rightarrow \pi^*$ spectral bands in TD-DFT calculations. Nemykin et al. [25] recently compared the MO energies predicted for the B3LYP geometry of ZnPc in a wide range of different calculations, ranging from fully semi-empirical methods to functionals such as BHandHLYP with a 50% HF component, functionals such as B3LYP in the Gaussian software package with a *ca.* 20% HF component, and finally to functionals such as BP86 in the Amsterdam Density Functional suite, which are based on the generalized gradient approximation (GGA) and have no HF component. The HOMO–LUMO energy gap was found to correlate linearly with the size of the HF component. The relative position of the $1a_{2u}$ π-MO from Gouterman's 4-orbital was also found to depend strongly on the selection of x–c functional. In the case of the semi-empirical PM3 and ZINDO/s methods and with calculations using "half-and-half" functionals, the $1a_{2u}$ MO is predicted to be HOMO$-$1 as would normally be anticipated in the context of Gouterman's 4-orbital model, while in BP86 GGA calculations it lies as low as HOMO$-$6. This is significant since even relatively minor differences in the relative energies of the frontier π-MOs can have a major impact on the predicted spectral properties of porphyrinoids, since they are largely determined by the OAM properties of the ground and excited states, Fig. 3.5.

Nemykin et al. also compared the effect of carrying out TD-DFT calculations for ZnPc with different x–c functionals and reported that those carried out with the GGA based BP86 functional provide support for an earlier assignment of a second electronic origin in the Q band region associated with an $n \rightarrow \pi^*$ transition out of a pyrrole and/or aza-nitrogen lone pair MO and into the LUMO of the π-system, which was originally made based on cryogenic temperature fluorescence [26, 27] and MCD measurements [28]. There has been some controversy over this in recent years, since $n\pi^*$ states are usually predicted to lie at considerably higher energies [29]. Nemykin and coworkers used a spectral band deconvolution analysis to argue that it is now possible to fully account for the optical properties of ZnPc based on the TD-DFT calculation carried out with the BP86 x–c functional. It should be noted, however, that Magyar and Tretiak [30] have demonstrated that TD-DFT calculations tend to underestimate the energy of charge transfer bands and concluded that BHandHLYP x–c functional with a 50% HF component provide the most accurate predictions for the energies of charge transfer states. Transitions linking the non-bonding MOs associated with the pyrrole and aza-nitrogen lone pair MOs and the π-system could reasonably be expected to fit this pattern. Although the assignment of a second electronic origin in the Q band region provides a good explanation for unusual features observed in the optical spectra of D_{4h} symmetry MPc complexes, it will also have to be consistent with the spectroscopy of the many low-symmetry Pcs that have subsequently been reported to stand the test of time [31].

Prior to Nemykin's research, Mack et al. [32] used a similar approach to analyze the optical spectra of a series of 17 different radially symmetric Zn(II) porphyrinoid compounds using the B3LYP functional to identify key trends in the calculated and observed optical spectra, Fig. 3.9, as part of a study on the optical spectroscopy of tetraphenyltetraacenaphthoporphyrins (TPTANPs), which had the most red-shifted B band to ever be reported for a radially symmetric porphyrinoid when first reported by Lash and coworkers

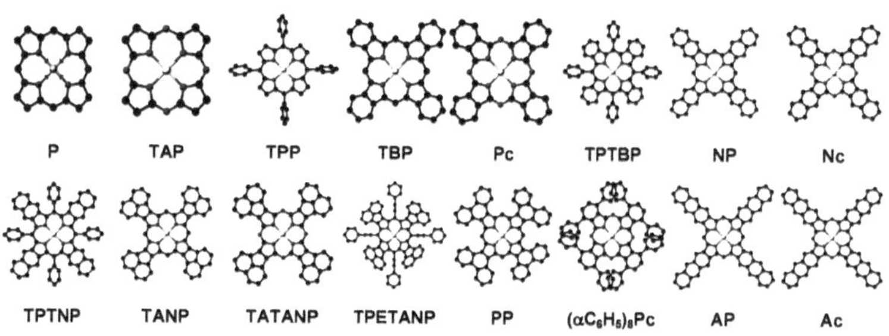

Figure 3.9. Molecular structures of the Zn(II) complexes of porphyrin (P), porphyrazine (TAP), tetraphenylporphyrin (TPP), tetrabenzoporphyrin (TBP), Pc, tetraphenyltetrabenzoporphyrin (TPTBP), naphthoporphyrin (NP), naphthalocyanine (Nc), tetraphenyltetranaphthoporphyrin (TPTNP), tetraacenaphthoporphyrin (TANP), tetraacenaphthotetraazaporphyrin (TATANP), tetraphenyltetraacenaphthoporphyrin (TPTANP), tetraphenylethynylacenaphthoporphyrin (TPETANP), tetraphenanthroporphyrin (PP), octa-α-phenylPc ($(\alpha C_6H_5)_8Pc$), anthracoporphyrin (AP) and anthracocyanine (Ac) based on DFT geometry optimizations using the B3LYP functional of the Gaussian 03 program with 6–31G basis sets (top).

[33, 34]. The band assignments were verified using the additional band polarization and state degeneracy information provided by MCD spectroscopy. The analysis of trends in the MCD data and TD-DFT calculations helped to resolve a controversy in the literature over the effect of ligand folding on the optical spectroscopy of sterically-rowded porphyrins, Fig. 3.10. Shelnutt and coworkers [35] have argued that red shifts observed in the Q and B bands are caused by the folding of the ligand and can be analyzed by determining the degree to which the π-system is distorted by doming, saddling and ruffling of the π-system. In contrast, Di Magno argued that a substituent effect was more important [36]. Although substituent effects can certainly be an important factor, Shelnutt's approach [35]

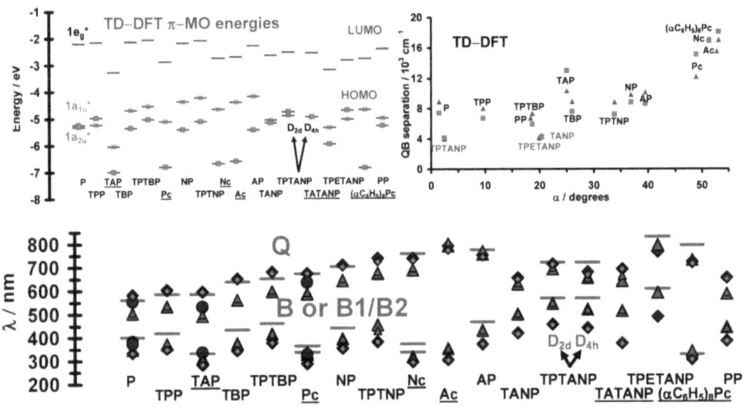

Figure 3.10. Trends in the energies of the four key frontier π-MOs associated with Gouterman's model [11] (top left) in TD-DFT calculations of 17 Zn(II) porphyrinoids, Fig. 3.9, based on optimizations with 6–31G basis sets [32]. Aza-substituted porphyrinoids are underlined. The experimentally observed (magenta squares) and calculated (blue diamonds and green triangles), energy separation of the Q and B bands (top right) plotted against the degree of mixing of the Q and B excited state magnetic moments (α) derived from oscillator strength values predicted by TD-DFT calculations carried out with the B3LYP functional using 6–31G basis sets. Data obtained from INDO/s and TD-DFT based on B3LYP DFT geometry optimizations with 6–31G and 6–31G(d) basis sets (bottom) appear as blue and magenta diamonds and green and red triangles, respectively. Data reported by Baerends et al. [29] based on calculations using the Amsterdam Density Functional software package are denoted by brown circles.

appears to be valid in certain contexts when ΔHOMO, ΔLUMO $\approx$ 0 and there is only limited configuration interaction between the B and higher energy $\pi\pi^*$ states, since folding of the ligand can modify the energies of the four key frontier π-MOs in Gouterman's 4-orbital model and hence the energies of the Q and B bands in a systematic manner. In the context of other porphyrinoids such as Pcs, however, where ΔHOMO $\gg$ 0, ΔLUMO $\approx$ 0 there is far greater configuration interaction between the B and higher energy $\pi\pi^*$ states and less scope for a red shift of the B band.

The study of a wide range of different porphyrinoids over the last decade with the B3LYP x–c functional has demonstrated that a significantly greater degree of complexity is usually predicted in the UV-region by TD-DFT calculations than is observed experimentally or is predicted based on older semi-empirical techniques such as ZINDO/s [32, 37]. This point will often be only of academic interest, however, since the main focus of current porphyrinoid research tends to be to fine tune the properties of the Q band to obtain compounds suitable for use in applications such as photodynamic therapy, solar cells and near-infrared (NIR) region dyes. This often entails shifting the band centre of the Q00 band to the red into the therapeutic window in the context of PDT research, to better fit the solar irradiance spectrum in the context of solar cells and to lie within the first telecom window in the context of NIR dyes. An ability to accurately predict the effects of different structural modifications on the properties of the Q band region often represents the key goal in molecular modeling, therefore, with the properties of bands in the UV region being only of secondary importance.

3.1.5 *TD-DFT Calculations of MCD Spectra*

The first examples of MCD spectra calculated using TD-DFT have recently been reported by Ziegler and coworkers based on the use of the BP86 functional of the Amsterdam Density Functional suite [38–47]. Analysis of induced magnetic moments values derived from the three Faraday terms would represent an additional spectroscopic basis for assessing the validity of DFT and TD-DFT based descriptions of the electronic structures and optical spectra of porphyrinoid complexes. In the case of MgP and ZnP, however, a second intense electronic band has been predicted in the B band region due to the presence of a $1b_{2u} \rightarrow 1e_g{}^*$ one-electron transition [47]. The sign sequence predicted for the two $\mathcal{A}_1$ terms in the B band region is opposite from that observed in the case of MgP. Coupled oppositely-igned $\mathcal{B}_0$ terms, due to the second close lying excited state, have to be invoked to account for this. This is problematic, since Nemykin and coworkers [25] have demonstrated that the analogous $1b_{2u}$ MO of ZnPc changes energy relative to the $1a_{2u}$ MO based on the size of the HF component used in the x–c functional. Given the impact of structural perturbations such as peripheral substitution, dication formation on the relative energies of the $1a_{2u}$ and $1b_{2u}$ MOs, it seems improbable that coupled oppositely-signed $\mathcal{B}_0$ terms could account for the wide range of positive Faraday $\mathcal{A}_1$ terms which have been reported in the B band regions of porphyrin MCD spectra. The problem may be related primarily to the selection of x–c functional rather than with the calculation of the Faraday $\mathcal{A}_1$ and $\mathcal{B}_0$ terms, however. Reimer and coworkers [48] recently demonstrated that calculations carried out with the BP86 and B3LYP functionals underestimate the energy of states associated with the $1b_{2u} \rightarrow 1e_g{}^*$ one-electron transition of free-base porphyrins and their fused-ring-expanded analogues and that more accurate predictions, which correspond more closely to Gouterman's 4-orbital model, are provided by a Coulomb-attenuated x–c functional (CAM-B3LYP). The CAM-B3LYP functional includes a long-range correction of the exchange potential, which incorporates an increasing fraction of HF exchange as the interelectronic separation increases.

3.2 Radially Symmetric Porphyrinoids

Until relatively recently the application of MCD spectroscopy to synthetic porphyrinoids focused primarily on radially symmetric ligands such as porphyrins and phthalocyanines [49]. Method of Moments analyses have been used to determine $\mathcal{A}_1/\mathcal{D}_0$ ratios and excited state magnetic moments and this confirmed the validity of Gouterman's 4-orbital model as the conceptual framework for describing the electronic structure and optical spectroscopy of a wide range of different porphyrinoids [11]. One of the key problems faced in the analysis of the optical spectra of fused-ring-expanded porphyrinoids such as Pcs is that broad envelopes of extensively overlapping vibrational and electronic spectral bands are observed in the UV region. Identifying the electronic origins can, therefore, be very challenging. In the 1980s and 90s, MCD spectroscopy provided the additional band polarization and state degeneracy information required to make detailed band assignments. Spectral band deconvolution studies by the Stillman group revealed that there are two overlapping intense Faraday $\mathcal{A}_1$ terms of almost equal magnitude and bandwidth in the B band region at *ca.* 350 nm. This led to a modification of the Q, B, N, L and C band nomenclature [12] that had earlier been proposed by the Gouterman group based on the results of semi-empirical calculations to include B1 and B2 bands [50–52]. In recent years the Kobayashi and Stillman groups have extended the use of spectral band deconvolution to a wide range of structurally modified porphyrinoids such as reduced symmetry π-systems, dimers, and corroles [2, 52–54]. After decades of research on porphyrinoids there is only limited scope for the synthesis of new planar radially symmetric tetrapyrrole porphyrinoid ligands. Recent research on radially symmetric porphyrinoids has focused on the effect of ligand folding due to steric crowding at the ligand periphery and on ring-contracted and ring-expanded compounds such as subporphyrins [55] and hexaphyrins [56]. MCD spectroscopy has provided important insights in each case.

3.2.1 *Non-Planar Porphyrinoids*

When bulky substituents are added to the porphyrin ligand at both the *meso*-carbons and the β-carbons of the pyrrole moieties, the ligand tends to fold due to steric crowding [35]. In most cases the main C_4 axis of symmetry is replaced by an S_4 symmetry axis due to saddling of the π-system. The degeneracy of the LUMO is retained on moving from D_{4h} to D_{2d} symmetry so $\mathcal{A}_1$ terms are still anticipated for the Q and B bands in the MCD spectra. Spectral deconvolution analyses of MCD and UV-visible absorption spectral data has been used to definitively identify the main

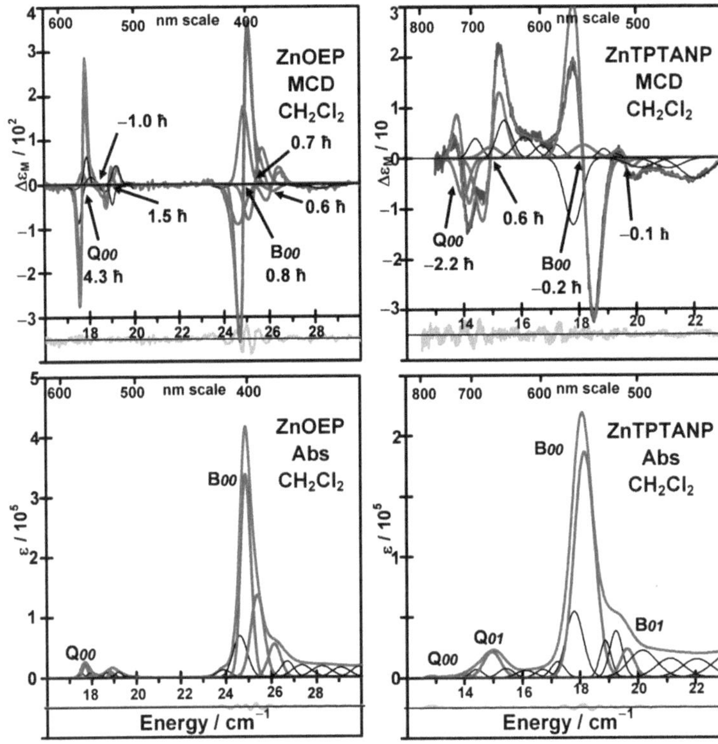

Figure 3.11. Spectral deconvolution analyses of the absorption and MCD spectra of zinc octaethylporphyrin (ZnOEP) and ZnTPTANP recorded in CH_2Cl_2 and the $\mathcal{A}_1/\mathcal{D}_0$ ratios associated with the $\mathcal{A}_1$ terms [32]. Purple, blue, green, black and red lines represent the calculated spectrum, the residual of the calculated fit, the experimentally observed spectral data, the component deconvoluted bands and bands associated with MCD $\mathcal{A}_1$ terms, respectively.

electronic bands on this basis. For example, Fig. 3.11, an almost fully forbidden Q00 band was found to lie at 717 nm well to the red of the value originally reported by Lash and coworkers [33, 34] for ZnTPTANP in what would appear on the basis of the UV-visible absorption spectrum alone to be a set of hot bands associated with the more intense Q01 vibrational band centered at 666 nm. The MCD spectra of M^{II}TPTANP complexes of main group and closed shell d^{10} metals were found to contain anomalous negative Faraday $\mathcal{A}_1$ terms (i.e., a +ve/−ve sign sequence in ascending energy terms), while the normal –ve/+ve sign sequence was observed for free base compounds and dication species [32]. Michl referred to cyclic polyenes such as TPTANPs in which ΔHOMO, ΔLUMO ≈ 0 as soft MCD chromophores [15–17], since small structural modifications can potentially reverse the alignment of the induced magnetic moments of the $\pi\pi^*$ excited states and hence the sign sequence observed for the Faraday $\mathcal{A}_1$ terms or coupled oppositely signed Faraday $\mathcal{B}_0$ terms associated with the main Q00 and B00 bands. The saddling of the ligand is believed to quench the OAM properties of the LUMO level to a greater extent than those of the HOMO level due to there being larger MO coefficients on the β-carbons of the pyrrole moieties and on the fused acenaphthalene rings. The OAM of the incident photon is, therefore, conserved by the circulation of the positively charged hole left in the HOMO, which reverses the alignment of the excited state magnetic moments and results in negative Faraday $\mathcal{A}_1$ terms.

Although it had long been known that steric crowding can result in the ruffling and saddling of porphyrin ligands, it has only been in the last decade that severely saddled Pc ligands have been reported. In 2000, Cook and coworkers [57] reported an X-ray structure for α-octaisopentylphthalocyanine with a dihedral angle of 32.0° between the planes formed by the saddled isoindole moieties due to steric hindrance between the bulky substituents at the α-positions. The following year Kobayashi and coworkers [58] reported that α-octaphenylphthalocyanine has dihedral angles of 46.0° and 41.9° due to a slight deviation from D_{2d} symmetry. Significant spectral changes are observed. A marked red shift of the Q band into the NIR region, Fig. 3.12, is observed due to a destabilization of the $1a_{1u}$ HOMO [59], but, unlike in the case of severely saddled porphyrins [32], TD-DFT results suggest that there is no red shift of the B1 and B2 bands. There is considerably greater configuration interaction between the B and higher energy $\pi\pi^*$ states, because ΔHOMO $\gg 0$ due to the stabilization of the $1a_{2u}$ MO of the Pc ligand caused by the large MO coefficients on the aza-nitrogens, Figure 3.5 These compounds, therefore, represent what Michl referred to as hard MCD chromophores [15–17], since there is no possibility of minor structural changes reversing the sign sequences of the $\mathcal{A}_1$ or $\mathcal{B}_0$ terms observed for the Q and B1 bands.

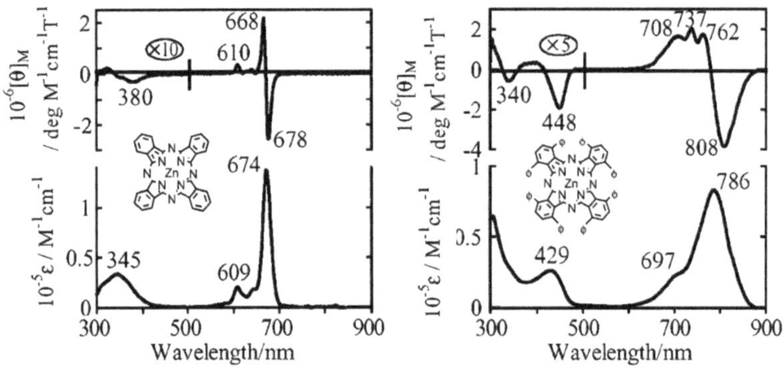

Figure 3.12. Electronic absorption and MCD spectra of ZnPc (right) and Znα(C_6H_5)$_8$Pc in pyridine (left) [59].

3.2.2 *Ring-Contracted Porphyrinoids*

Although boron subphthalocyanines (BsubPcs) were first reported as far back as 1972 [60], the first examples of boron subporphyrins (BsubPs) and boron subtribenzoporphyrins were reported by the Kobayashi [55] and Osuka [61] groups as recently as 2006. The optical spectroscopy of subporphyrinoids can be understood in terms of a $C_{12}H_{12}^{2-}$ parent perimeter with the key π-MOs arranged in an $M_L = 0, \pm1, \pm2, \pm3, \pm4, \pm5, 6$ MO sequence. The C_4 axis of conventional tetrapyrrole porphyrins is replaced by a C_3 axis. The LUMO has $M_L = \pm4$ nodal properties and remains degenerate, while the frontier π-MOs, derived from the $M_L = \pm3$ HOMO, which are very similar of the parent perimeter, have nodal patterns, which are very similar to those of the $1a_{1u}$ and $1a_{2u}$ MOs of tetrapyrrole porphyrins, Fig. 3.13. The major bands can, therefore, be assigned to B and Q transitions with $\Delta M_L = \pm1$ and ±7 properties, respectively, by analogy with Gouterman's 4-orbital model [11]. The electronic absorption and MCD spectra of six monomeric subPs and a μ-oxo dimer compound are displayed in Fig. 3.14. The intense absorption band in the 370–380 nm region is clearly similar to the B band of metal porphyrin complexes, Fig. 3.11, while the weaker absorption band in the 400–540 nm region is directly analogous to the Q band, since there is a relative intensification in the MCD spectrum based on the

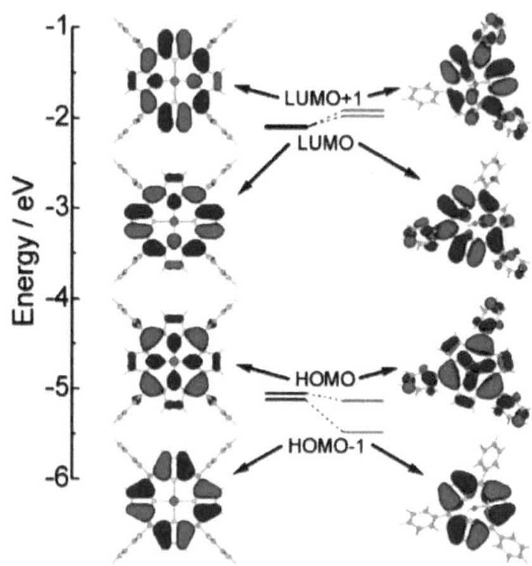

Figure 3.13. Nodal patterns of the four key frontier π-MOs of Zn^{II}TPP and B^{III}subP at an isosurface of 0.02 a.u. [55].

larger excited state magnetic moment associated with a forbidden transition with "$\Delta M_L = \pm 7$" properties.

The MCD spectra differ markedly from compound to compound when electron donating and withdrawing substituents are added to the phenyl groups on the three *meso*-carbons. The Q band regions of the MCD spectra can be divided into two distinct groups. SubPs with electron donating substituents exhibit an intense −ve/+ve pattern in ascending energy (i.e., positive Faraday $\mathcal{A}_1$ terms), while those with electron withdrawing groups are more complex with a concomitant decrease in intensity. In the B band region, the MCD spectra can be divided on a similar basis. The spectra of subPs with electron donating groups such as 4-methoxy and tolyl groups contain a −ve/+ve pattern as in the Q band region, while the spectra of subPs with electron withdrawing groups such as 3PysubP, 4PysubP, and TFsubP, Fig. 3.14, contain a +ve/−ve pattern (i.e., a negative Faraday $\mathcal{A}_1$ term) [1, 32]. In addition, careful inspection reveals that the −ve/+ve B band pattern observed for MOsubP changes gradually on going to TLsubP and then to PhsubP, Fig. 3.14. The relative intensity of the positive envelope at higher energy becomes weaker, so that, in PhsubP, only the negative trough is observed. On moving to 3PysubP, a positive MCD envelope develops at lower energy,

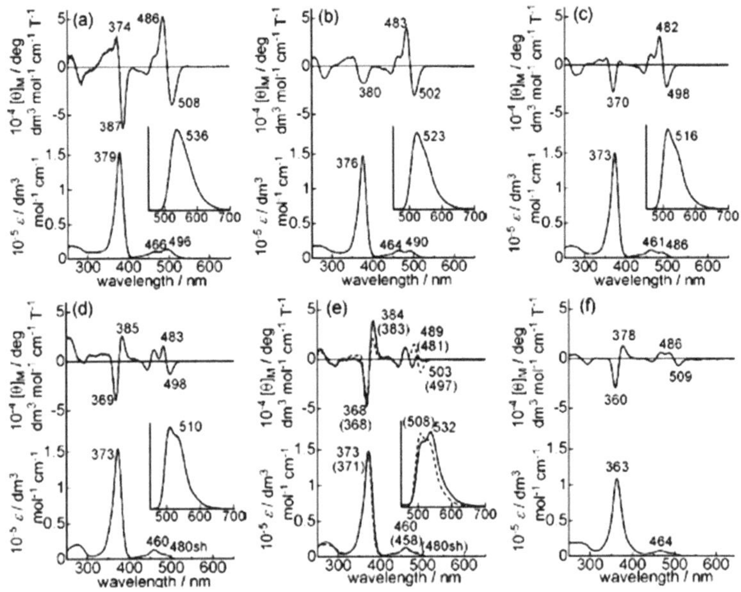

Figure 3.14. Electronic absorption, MCD, and fluorescence spectra of (a) tri(4-methoxyphenyl)subporphyrin (MOsubP), (b) tritolylsubporphyrin (TLsubP), (c) triphenylsubporphyrin (PhsubP), (d) tris{(4-trifluoromethyl)phenyl}subporphyrin (TFsubP), (e) tri(3-pyridyl)subporphyrin (3PysubP, broken lines) and tri(4-pyridyl)subporphyrin (4PysubP, solid lines), and (f) a μ-oxo dimer of 3PysubP ($(3PysubP)_2O$) recorded in $CHCl_3$ (absorption and MCD) and ethanol (fluorescence) [55].

and this intensifies in the case of TFsubP, to the point in the case of 4PysubP that the positive and negative MCD envelopes have almost equal intensities, Fig. 3.14. These drastic changes in the MCD spectra can only arise because the ligands are soft MCD chromophores with ΔHOMO, ΔLUMO ≈ 0. The spectra defy a straightforward explanation on the basis of Michl's perimeter model [15–17], however, and will probably provide a suitable test case for future implementations of the three Faraday terms into commercially available TD-DFT software. In contrast, the spectra of subPcs [62] and subPzs [63], contain only positive $\mathcal{A}_1$ terms, since these ligands are hard MCD chromophores due to the marked stabilization of the MO which is analogous to the $1a_{2u}$ MO of ZnP based on the large MO coefficients on the aza-nitrogens Fig. 3.13. This introduces large ΔHOMO values, while in contrast ΔLUMO = 0.

3.2.3 *Ring-Expanded Porphyrinoids*

Over the last decade there has been a growing research focus on the synthesis of ring-expanded porphyrinoids consisting of more than four pyrrole moieties. For example, the synthesis and characterization of *meso*-hexakis(pentafluorophenyl)-substituted hexaphyrin(1.1.1.1.1.1)s with multiple gold–carbon bonds has recently been reported [64], Fig. 3.15. Two-electron reduction of the aromatic Au^{III} hexaphyrins (Au_2-N) leads to reduced species with remarkable paratropic ring current effects (Au_2-R). The neutral and reduced forms can therefore be regarded as 26- and 28-π-electron systems. The electronic absorption spectrum of the heteroaromatic hexaphyrin ligand is dramatically changed upon

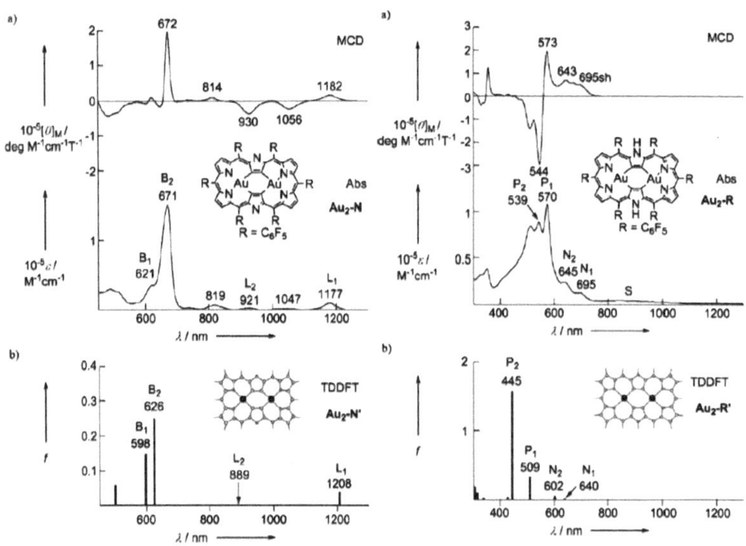

Figure 3.15. Electronic absorption and MCD spectra of aromatic (left) and "unaromatic" (right) Au^{III} hexaphyrin species (Au_2-N and Au_2-R) [64]. The main electronic bands are assigned based on the sign sequences in the MCD spectra, TDDFT calculations of model complexes and on Michl's $(4N + 2)$- [15–17] and $4N$- [18, 19] perimeter models.

reduction. Several NIR region absorption bands observed within the Au_2-N spectrum are clearly not present in the spectrum of Au_2-R, Fig. 3.15. Similar spectra were also observed when various peripheral substituents were added at the pyrrole β-positions [65]. An analysis [56] of the MCD spectra was used to validate the results of TD-DFT calculations. The main features of the optical spectra of Au_2-N and Au_2-R can be assigned using Michl's $(4N + 2)$- [15–17] and $4N$-perimeter models [18,19] based on the presence of L and B and S, N and P bands, respectively. Michl's perimeter model provides a readily accessible conceptual framework for rationalizing the key trends in what would otherwise appear to be highly complex spectra. Over the past decade, Waluk and Sessler have also used MCD spectral measurements and Michl's conventional $(4N + 2)$-perimeter model to analyze the electronic structures of pentaphyrin(1.1.1.0.0)s [66], cyclo[6]-, cyclo[7]-, and cyclo[8]pyrroles [67] and doubly N-confused sapphyrins [68].

3.3 Low-Symmetry Porphyrinoids

In recent years, research on porphyrinoid synthesis has increasingly focused on compounds, which lack a threefold or higher axis of symmetry. Over the past decade the Faraday $\mathcal{B}_0$ terms observed in the MCD spectra of these compounds have been used to identify the key electronic bands. There are a number of different types of structural modifications, which can be made to the porphyrin or phthalocyanine ligand. For example, the pyrrole nitrogens can be replaced by other heteroatoms such as sulfur and oxygen. An in depth study of the MCD spectroscopy of both planar and saddled core-modified tetrabenzoporphyrins (TBP) has recently been reported [37]. Modifications can also be made at the bridging atoms between the pyrrole moieties. Two key examples of this type of compound, which have been studied in depth using MCD spectroscopy, are firstly the [14]triphyrins(2.1.1) [69], novel subporphyrinoids with an extra *meso*-carbon, and secondly the corrolazines, analogues of the porphyrazine (Pz) ligand in which one of the aza-nitrogens is replaced by a direct bond between neighboring α-carbons [70, 71]. A third type of structural modification that has been studied is the reduction of a peripheral pyrrole bond to form chlorins, bacteriochlorins, and isobacteriochlorins. The MCD spectroscopy of these compounds was studied in depth in the 1970s and 1980s [72–74]. In the last decade there has been renewed interest in their tetraaza-analogues [75–84]. Finally, the π-system can be expanded by adding additional fused rings at the ligand periphery. MCD spectroscopy has provided valuable insights on the optical spectroscopy of partially benzo-fused Pz and Pc compounds [85–89], azulene-fused porphyrins [90], azulenocyanines [91], and benzoporphycenes [92].

3.3.1 *Core-Modified Tetrabenzoprophyrins*

When the Ono group [93, 94] prepared a series of core-modified tetrabenzoporphyrins in 2004, Fig. 3.16, it soon became clear that the OAM properties of the π-system are more significant than the changes to the geometry of the ligand in determining the optical properties, Fig. 3.17. The UV-visible absorption and MCD spectra of near planar ON3, SN3, S2N2, and OSN2 compounds (the

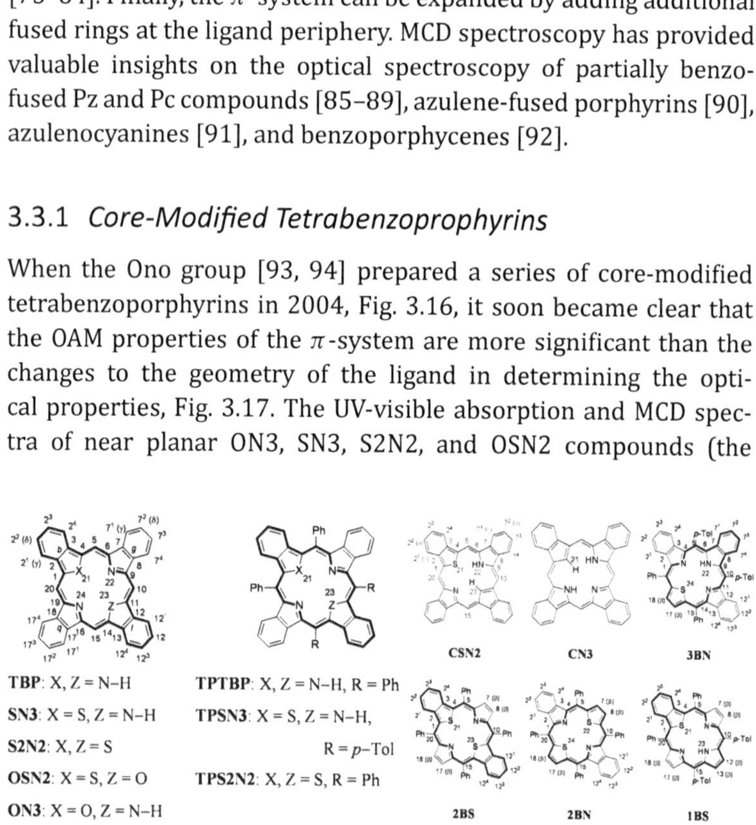

Figure 3.16. The molecular structures of tetraphenylporphyrin (TPP), tetrabenzo[*b*,*g*,*l*,*q*]porphyrin (TBP), 21-carba-, (CN3) 21-thia-, (SN3) 21,23-dithia- (S2N2) and 21-oxa-23-thia- (OSN2) tetrabenzo[*b*,*g*,*l*,*q*]porphyrins, 21-oxa- (ON3) and 21-carba-23–thia- (CSN2) tetrabenzo[*b*,*g*,*l*,*q*]porphyrins, tetraphenyltetrabenzo[*b*,*g*,*l*,*q*]-21-thiaporphyrin (TPSN3), tetraphenyltetrabenzo[*b*,*g*,*l*,*q*] -21,23-dithiaporphyrin (TPS2N2), tetraphenyldibenzo [*g*,*q*]-21,23-dithiaporphyrin (2BN), tetraphenyldibenzo [*b*,*l*]-21,23-dithiaporphyrin (2BS), tetraphenyltribenzo[*g*,*q*,*l*] -21-thiaporphyrin (3BN) and tetraphenylbenzo[*b*]-21-thiaporphyrin (1BS) (top) [95].

nomenclature used is based on the four core atoms of the inner ligand perimeter) are all very similar to those of TBP with only minor shifts observed in the band centers of the main Q and B bands [95]. The lone pairs in the $2p_z$ atomic orbitals of the protonated pyrrole nitrogens are directly replaced by another lone pair when oxygen and sulfur atoms are introduced. The nodal patterns of the four frontier π-MOs, Fig. 3.17, in Gouterman's 4-orbital model [11] are not significantly altered, therefore, so the optical spectra are broadly similar. Core modification with a carbon and a sulfur to form CSN2,

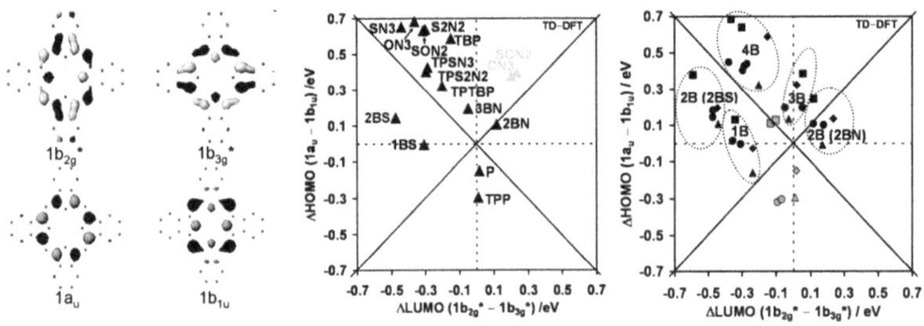

Figure 3.17. The nodal patterns of the four frontier π-MOs are shown at an isosurface value of 0.05 a. u. (left). The ΔHOMO and ΔLUMO values predicted for P, TPP, TBP, SN3, ON3, CN3, S2N2, OSN2, TPSN3, TPS2N2, 1BS, 3BN, 2BN 2BS and CSN2 (center) [95]. In the light gray shaded areas where ΔHOMO > ΔLUMO the normal −ve, +ve, −ve, +ve MCD sign sequence is anticipated on the basis of Michl's perimeter model [15–17] for the Q and B bands in ascending energy terms, while a +ve, −ve, +ve, −ve MCD sign sequence is anticipated when ΔLUMO > ΔHOMO. The effect of structural perturbations on the ΔHOMO and ΔLUMO values predicted (right). P, monobenzo[*b*]-21H,23H-porphyrin, opp-dibenzo[*b*,*g*]-21H, 23H-porphyrin, opp-dibenzo[*l*,*q*]-21H,23H-porphyrin, tribenzo[*b*,*g*,*l*]-21H, 23H-porphyrin, TBPs are denoted with diamonds. The corresponding 5,10,15,20-tetraphenyl analogues are denoted with triangles, while squares are used for 21-thia-23H-porphyrin, 21,23-dithiaporphyrin, monobenzo[*b*]-21-thia-23H-porphyrin, opp-dibenzo[*b*,*g*]-21H,23H-porphyrin, opp-dibenzo[*b*,*l*]-21,23-dithiaporphyrin, tribenzo[*g*,*l*,*q*]-21-thia-23H-porphyrin, SN3 and S2N2. The corresponding 5,10,15,20-tetraphenyl analogues, including 1BS, 2BN, 2BS, 3BN, TPS2N2 and TPSN3, are denoted with circles [95]. The compounds are identified based on the number of peripheral fused benzene rings and in the case of the opp-dibenzoporphyrins whether the two $2p_z$ atomic orbitals on the inner ligand perimeter containing lone pairs are located along the axis with fused benzene rings (2BS) or the axis with no benzene rings (2BN), Fig. 3.16.

however, has a significantly greater effect, Fig. 3.17, since a pyrrole nitrogen atom is protonated on the x-axis to compensate for the fact that the carbon $2p_z$ orbital does not contain a lone pair.

The anomalous MCD sign sequences observed for metal TPTANP complexes [32] led to the question of whether similar anomalous sign sequences would be observed for other non-planar porphyrinoids. The Ono, Kobayashi and Stillman groups [95] collaborated on a study of both planar and non-planar core-modified benzoporphyrins and tetraphenylbenzoporphyrins. The ΔHOMO and ΔLUMO values derived for the B3LYP optimized geometries of twelve core-modified compounds are shown in Fig. 3.17. The MCD spectra of most of the compounds exhibit the same sign sequence observed for the parent TBP and TPTBP compounds, Fig. 3.18. Anomalous +ve, −ve, −ve, +ve sequences are observed for the two non-planar tetraphenyldibenzoporphyrins 1BS and 2BS, Fig. 3.16, where ΔLUMO > ΔHOMO, Fig. 3.17, however. This is consistent with the MCD spectroscopy of the chlorins another type of porphyrinoid ligand with ΔLUMO > ΔHOMO, which was studied in detail in the 1980s by Djerassi and coworkers [72–74]. In instances where the mixing factor, α, is relatively small, Eq. 3.6, and |ΔLUMO −ΔHOMO| ≈ 0, the MCD sign sequence initially reverses only in the case of the Q band, since the μ^- moment remains the dominant factor in the case of the B band. The observed sign sequences, therefore, consistently match the pattern that would be predicted by Michl's perimeter model based on the ΔHOMO and ΔLUMO values. It seems safe to conclude that non-planarity usually does not reverse the sign sequences that would be anticipated for Faraday $\mathcal{B}_0$ terms when ΔHOMO ≠ ΔLUMO and a significant μ^+ contribution is introduced to the Q and B excited states, Eqs. 3.6–3.8. The anomalous +ve, −ve, +ve, −ve sign sequence observed for the Faraday $\mathcal{A}_1$ terms of ZnTPTANP [32] is believed to be caused by the formation of a soft MCD chromophore in which ΔHOMO, ΔLUMO ≈ 0 and $\alpha = 0$. The Q band is, therefore, fully forbidden in electronic terms, which eliminates the μ^+ contribution to the spectrum.

The UV-visible absorption spectra of 2BN and 2BS, Fig. 3.18, are markedly different despite the similar tetraphenyldibenzoporphyrin structures, Fig. 3.16. The weak absorption bands of 2BN at 639 and 704 nm are sufficiently separated from the bands at

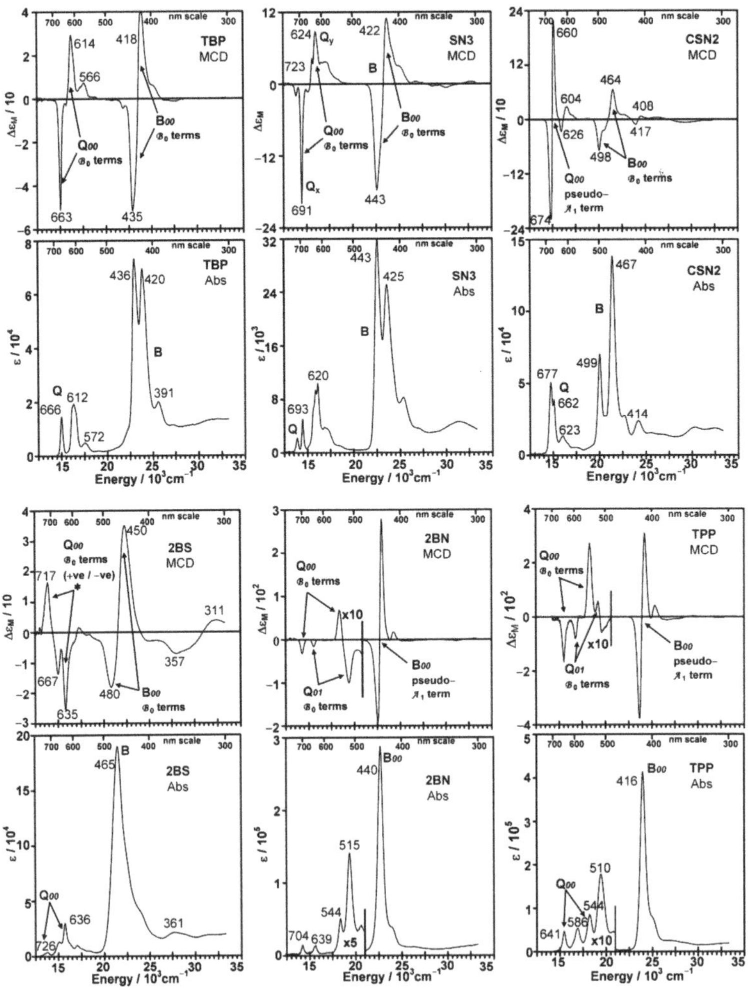

Figure 3.18. Absorption and MCD spectra of TBP, SN3, CSN2, 2BS, 2BN and TPP [95].

515 and 544 nm that their assignment would be problematic in the absence of the MCD spectral data. Both these bands appear as negatively signed Faraday $\mathcal{B}_0$ terms, as opposed to the coupled pair of oppositely signed bands observed for 2BS. When the ΔHOMO and ΔLUMO values predicted for TBP, S2N2, SN3, 1BS,

2BS, 2BN and 3BN, Fig. 3.16, the parent benzoporphyrins and their tetraphenyl analogs, and for non-phenylated analogs of 1BS, 2BS, 2BN and 3BN, are plotted it becomes clear that the pattern of partial benzo-substitution has the most significant effect on the ΔHOMO and ΔLUMO values, Fig. 3.17. This suggests that the differences in the spectra of 2BN and 2BS are best rationalized using a perimeter model approach. Despite the core modification, non-planarity and partial benzo-substitution of 2BN, the MCD spectrum is very similar to that of TPP, Fig. 3.18. This similarity arises because ΔHOMO $\approx$ ΔLUMO in both cases. The spectra of both 2BN and TPP can be expected to mimic that of a shared higher symmetry parent perimeter on this basis. The parent perimeter of free base core-modified tetrabenzoporphyrins is arguably best described as being $C_{18}H_{18}$, since there are two lone pairs localized on the protonated pyrrole nitrogens, along the y-axis as opposed to the situation in metal complexes where negative charge is spread evenly across the four pyrrole nitrogens to balance the charge on the divalent central metal. A key feature of $D_{(4N+2)}$h $(N > 1)$ symmetry is that the multiplication products of the LUMO and HOMO symmetries are both singly and doubly degenerate (for e_{1g} and e_{2u} in the case of the D_{6h} symmetry of benzene the products are e_{1u}, b_{1u} and b_{2u}), while in contrast in the case of $D_{(4N)h}$ $(N > 1)$ symmetry the products are doubly degenerate (for e_{4u} and e_{5g} under D_{16h} symmetry of $C_{16}H_{16}^{2-}$ the products are e_{1u} and e_{7u} [96]). Although the B bands of porphyrinoids derived from $C_{18}H_{18}$ perimeters can appear to be pseudo-$\mathcal{A}_1$ terms in the spectra of free base porphyrinoids such as TPP, Fig. 3.18, where ΔHOMO, ΔLUMO ≈ 0, the Q bands split markedly due to the inherent non-degeneracy of the L excited states of the parent perimeter. The spectra of 2BN and 2BS provide a good example of how Michl's perimeter model can provide a readily accessible conceptual framework for rationalizing key trends observed in the optical spectra of porphyrinoids which would otherwise defy straightfoward explanation.

3.3.2 *Triphyrins*

In 2008, the first serendipitous synthesis of a series of *meso*-aryl-substituted [14]triphyrin(2.1.1) and [14]benzotriphyrin(2.1.1)

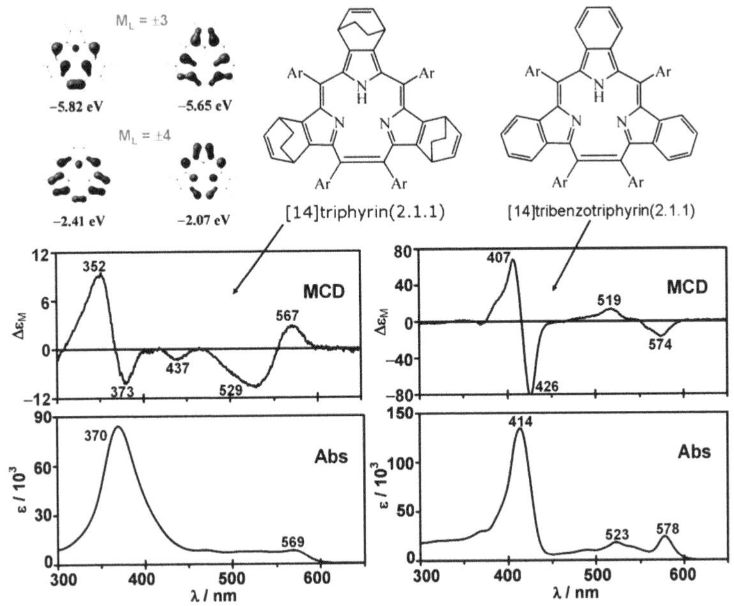

Figure 3.19. MCD and UV-visible absorption spectra in $CHCl_3$ at 298 K (bottom) and the frontier π-MOs (top) of [14]triphyrin(2.1.1) and [14]benzotriphyrin(2.1.1) [69].

compounds was reported [69]. These compounds were unprecedented in porphyrinoid research since they represented the first examples of heteroaromatic metal free ring-contracted porphyrinoids containing only the standard pyrrole and isoindoline moieties of porphyrins and tetrabenzoporphyrins. The properties of the π-system can be understood in terms of a $C_{14}H_{14}$ parent hydrocarbon perimeter with an $M_L = 0, \pm1, \pm2, \pm3, \pm4, \pm5, \pm6, 7$ sequence in ascending energy terms. The nodal patterns of the frontier π-MOs are broadly similar to those calculated for the BsubPs, Figs. 3.13 and 3.19. The insertion of an additional sp^2 hybridized *meso*-carbon eliminates the need for a small central boron atom to coordinate a ring-contracted dianion ligand, since a 14-π-electron system can be formed on the inner perimeter with a single inner proton. The central cavity of conventional subporphyrins is too small to accommodate two inner protons. The UV-visible absorption spectra of *meso*-aryl-substituted [14]triphyrin(2.1.1) and [14] benzotriphyrin(2.1.1), Fig. 3.19, contain intense bands at 370 and 414 nm, respectively, and weaker bands in the 500–600 nm region in a pattern similar to the B and Q bands of the porphyrins [49]. Since, in contrast with subporphyrins, there is no threefold axis of symmetry and hence no scope for degeneracy of the frontier π-MOs, the MCD spectra are dominated by coupled pairs of oppositely signed Gaussian-shaped Faraday $\mathcal{B}_0$ terms, similar to those observed for the Q and B bands of low-symmetry porphyrins and Pcs [49].

3.3.3 *Corrolazines*

Over the past decade there has been a growing research focus on the synthesis and properties of corrolazines (Czs) and tetrabenzocorrolazines (TBCzs) [70, 71, 97], Fig. 3.20. The removal of one of the aza-nitrogens of the Pz or Pc ligand, results in marked changes in the spectroscopic properties due to the loss of the main C_4 axis of symmetry and changes in the OAM properties [47, 71]. The four frontier π-MOs of the $C_{15}H_{15}^{3-}$ parent perimeter are almost identical to those of the porphyrins so Gouterman's 4-orbital model [11] and Michl's perimeter model [15–17] can still be used to rationalize the optical and redox properties, Fig. 3.20. Relatively small ΔLUMO values of 0.31 and 0.25 eV are predicted for the Cz and TBCz complexes, so only minor splittings of the Q00 and B00 bands are observed in the optical spectra [47], Fig. 3.21. The MO derived from the $1a_{2u}$ MO of MPc complexes is destabilized due to the large MO coefficients

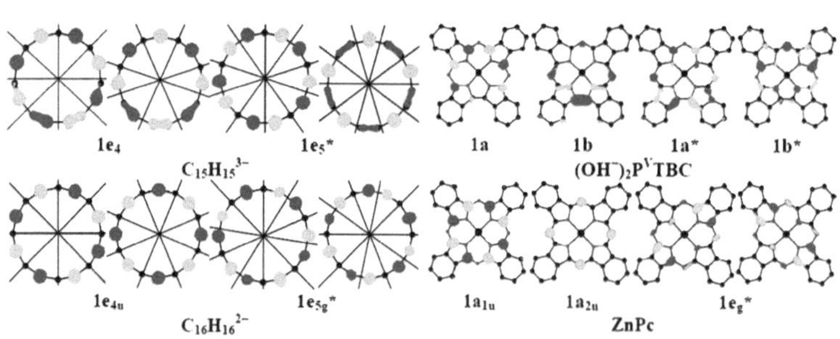

Figure 3.20. The nodal patterns of the four frontier π-MOs of $C_{15}H_{15}^{3-}$, $C_{16}H_{16}^{2-}$, $(OH^-)_2P^VTBC$ and ZnPc [47].

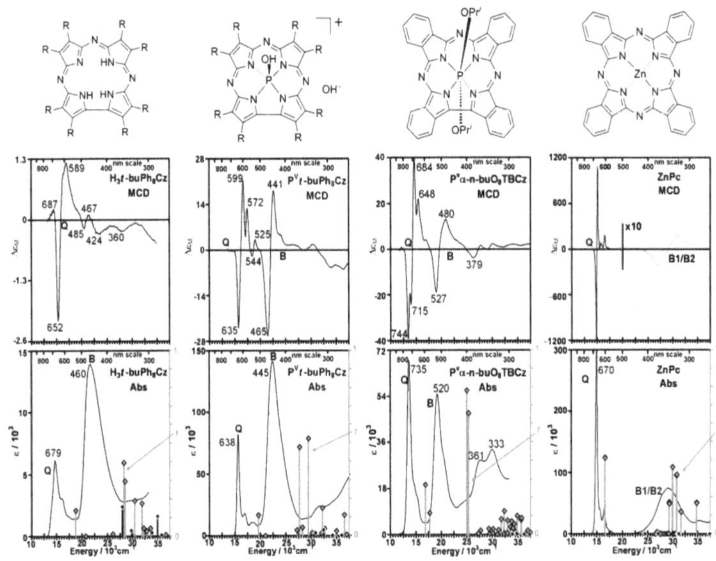

Figure 3.21. The molecular structures (top) and MCD and UV-visible absorption spectra (bottom) of $H_3(p\text{-}t\text{-buPh})_8Cz$, $(OH^-)_2P^V(p\text{-}t\text{-buPh})_8Cz$, $(OCH_3^-)_2P^V\alpha(n\text{-buO})_8TBCz$ and ZnPc [47]. Calculated TD-DFT absorption spectra for H_3Cz, $(OH^-)_2P^VCz$ and $(OH^-)_2P^VTBCz$ model compounds and for ZnPc are plotted against the right hand axes. Bands arising from transitions to nπ^* states are denoted by black diamonds.

on the aza-nitrogens so there is a marked decrease in the predicted ΔHOMO values from 1.15 and 1.82 eV for Pz and Pc compounds to 0.34 and 0.64 eV for the analogous Cz and TBCz compounds. As a consequence the optical spectra of TBCz complexes exhibit similarities to the spectra of TBP complexes, especially in the B band region, Figs. 3.18 and 3.21. The B band of $(OMe^-)P^V\alpha(n\text{-BuO})_8TBCz$ shifts to the red with an intensity maximum at 520 nm. A single pair of intense Faraday $\mathcal{B}_0$ terms is observed in the B band region of the MCD spectrum [47]. Since the ΔHOMO and ΔLUMO values are relatively similar and are both comparatively low, the OAM properties of the ligand are closer to those of the high symmetry parent hydrocarbon perimeter than is the case with Pc complexes and there is less interaction between the B and higher energy $\pi\pi^*$ states.

3.3.4 *Tetraazachlorins*

The Kobayashi and Luk'yanets groups have recently carried out an in depth study of the impact of the partial reduction of the peripheral pyrrole bonds of H_2Pz to form tetraazachlorin (H_2TAC), tetraazabacteriochlorin (H_2TAB) and tetraazaisobacteriochlorin ($H_2TA\textit{i}B$) compounds, Fig. 3.22 [75, 76]. The $1a_{1u}$ HOMO is destabilized, since there are large MO coefficients on these atoms in Pz π-systems, Figs. 3.5 and 3.23. In the case of H_2TAC and H_2TAB, there is a substantial splitting of the LUMO, since the perturbation of the π-system differs markedly along the x- and y-axes [77]. This results in increased splittings of the Qx and Qy bands of *ca.* 4500 cm^{-1} for H_2TAC and *ca.* 9000 cm^{-1} for H_2TAB, relative to the *ca.* 2300 cm^{-1} splitting observed in the spectrum of H_2Pz, Fig. 3.22. A marked red shift into the NIR region is observed for the Qx band due to the lifting of the degeneracy of the LUMO. The identification of the Qy bands based on the coupled pairs of Faraday $\mathcal{B}_0$ terms in the MCD spectrum made it clear that the average Q band energy remains almost constant, however. The adjacent pattern of the reduced bonds of $H_2TA\textit{i}B$ results in a smaller splitting of the Qx and Qy bands than is observed even in the case of H_2Pz, since the LUMO remains near degenerate because the perturbation to the structure is identical along the x- and y-axes. Similar Q band splittings have been reported when additional

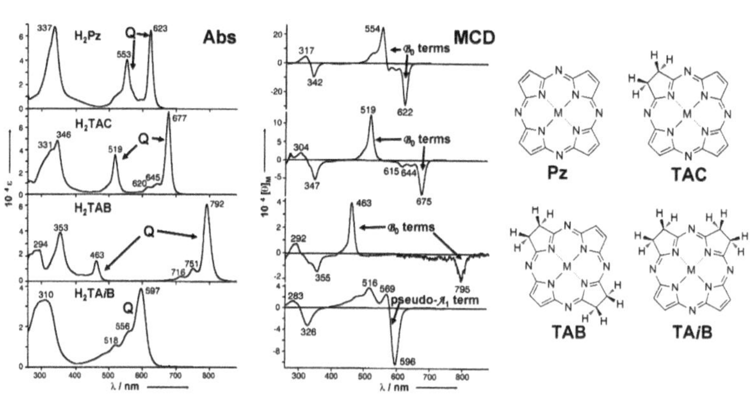

Figure 3.22. Electronic absorption (left) and MCD (center) spectra of H_2Pz, H_2TAC, H_2TAB and $H_2TA\textit{i}B$ in $CHCl_3$. The structures of Pz, TAC, TAB, and TA*i*B complexes (right) [77].

peripheral fused benzene rings are added to TAC and TAB compounds through the mixed condensation of tetramethylsuccinonitrile with phthalonitrile or 2,3-dicyanonaphthalene [77], and in the spectra of 2,3-naphthalene, pyrazine and 2,3- and 3,4-pyridine fused TAC, TAB and TAiB complexes [77–80] and C_{60} fused tribenzoTAC, dibenzoTAB, and dibenzoTA*i*B compounds [81–84].

3.3.5 *Partially Benzo-Fused Porphyrazines and Phthalocyanines*

Over the past decade the Kobayashi group has studied the effect of partial benzo-substitution of the Pz and Pc ligands making use of MCD spectroscopy to identify the main electronic bands [85–89]. These compounds can be viewed as having the same parent perimeters as Pz and Pc compounds, since the frontier π-MOs are located primarily on the inner ligand perimeter. Extension of the π-system with peripheral fused rings tends to destabilize the $1a_{1u}$ MO, since there are large MO coefficients on the peripheral β-carbon atoms of the pyrrole moieties of Pz complexes, Fig. 3.23. This results in a marked red shift of the lowest energy Q band, Fig. 3.24. The magnitude of the ΔLUMO value and hence of the splitting of the Q00 bands caused by the loss of the fourfold axis of symmetry, depends on the substitution pattern [85–89], Fig. 3.24. The splitting of the x- and y-polarized bands is most marked in the case of D_{2h} symmetry opp-di-substituted compounds, while the spectra of C_{2v} symmetry adj-di-substituted structures are typically very similar to those of D_{4h} symmetry MPz and MPc complexes [85–88], Fig. 3.24. This can be readily explained by the fact that the structural

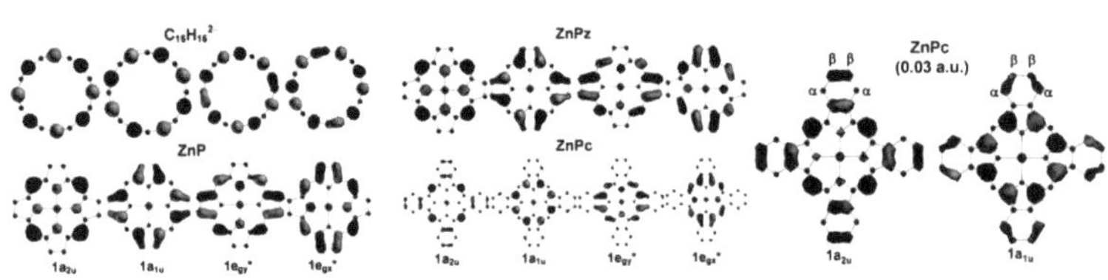

Figure 3.23. The frontier π-system MOs for $C_{16}H_{16}^{2-}$, ZnP, ZnPz, and ZnPc predicted by B3LYP geometry optimizations at an isosurface value of 0.04 a.u. for electrostatic potential (left) and the $1a_{1u}$ and $1a_{2u}$ MOs of ZnPc at an isosurface value of 0.03 a.u. (right).

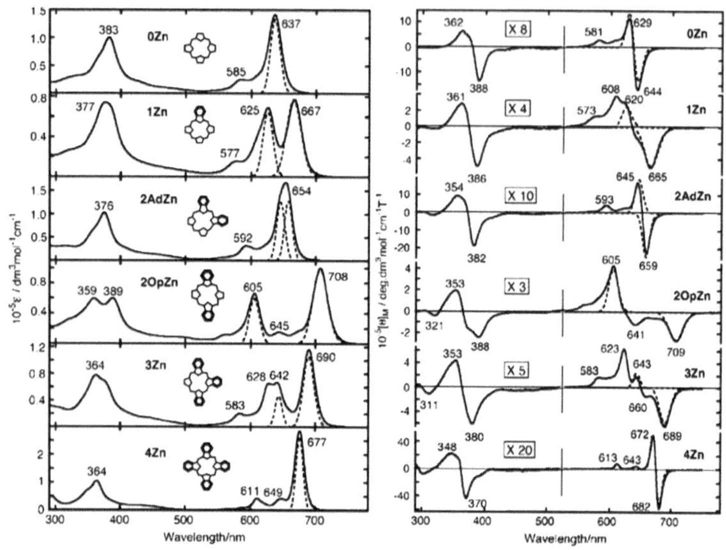

Figure 3.24. Electronic absorption (left) and MCD (right) spectra of a series of benzo-fused ZnPz compounds in toluene containing 1×10^{-2} M pyridine. The Q_x and Q_y bands evaluated by band deconvolution analyses are highlighted with dashed lines [88].

perturbation is the same along both the *x*- and *y*-axes in the case of the adj-disubstituted isomer but differs in the case of the opp-disubstituted isomer. The C_{2v} symmetry mono- [89] and tri- [87, 88] benzo-substituted structures of MPz complexes typically exhibit a pronounced splitting of the Q band, Fig. 3.24, while only a slight splitting is observed in the spectra of partially substituted MPc complexes [87]. Perturbations to the symmetry of Pz compounds tend to result in more marked spectral changes than those observed in the case of Pc and naphthalocyanine compounds, since the largest MO coefficients of the four key frontier π-MOs are located on the atoms on the inner ligand perimeter [86–88], Fig. 3.23.

3.3.6 *Azulene-Fused Porphyrins and Azulenocyanines*

In 2006, a series of azulene-fused porphyrins, Fig. 3.25, were synthesized by the Osuka group [98]. These compounds have large

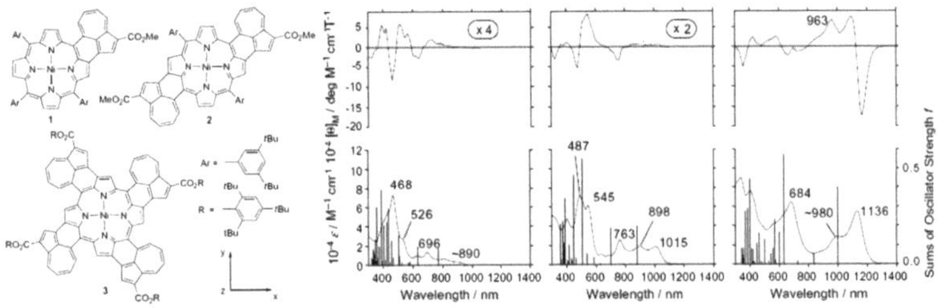

Figure 3.25. The molecular structures of mono- (1), opp-di- (2) and tetra-azulene-fused (3) π-expanded porphyrins (left). The electronic absorption, MCD and calculated TD-DFT spectra of, from left to right, 1–3 (right) [90].

two-photon absorption (TPA) cross-sections and intense absorption bands in the NIR region between 900 and 1200 nm. MCD spectroscopy was used, Fig. 3.25, to validate the results of TD-DFT calculations [90]. The DFT calculations for mono-azulene-fused porphyrins predict that ΔHOMO $\approx$ ΔLUMO, while the prediction for opp-di-azulene-fused porphyrins is that ΔHOMO $<$ ΔLUMO. The lowest energy bands of the mono-azulene-fused compound are, therefore, significantly less intense and a +ve, –ve sign sequence is observed in ascending energy terms in the coupled Faraday $\mathcal{B}_0$ terms in the MCD spectrum of the opp-di-fused compound. Pseuso-$\mathcal{A}_1$ terms are observed for both the B and Q bands in the MCD spectrum of tetra-azulene-fused porphyrins. The excited states clearly remain accidentally degenerate, therefore, despite C_2 molecular symmetry related to folding of the ligand. Muranaka et al. [91] recently reported the first synthesis of azulenocyanines, a structural analogue of phthalocyanines with peripheral azulene rather than benzene fused rings. Four isomers can be formed based on the relative orientation of the azulene moieties. The main product was isolated by HPLC and was found to have C_s symmetry. The MCD spectrum was used to validate the results of TD-DFT calculations

3.3.7 *Benzoporphycenes*

Recently, the first facile syntheses of free base di- and tetrabenzoporphycenes and their metal complexes were reported by Yamada and coworkers [92], based on retro-Diels-Alder reactions of the

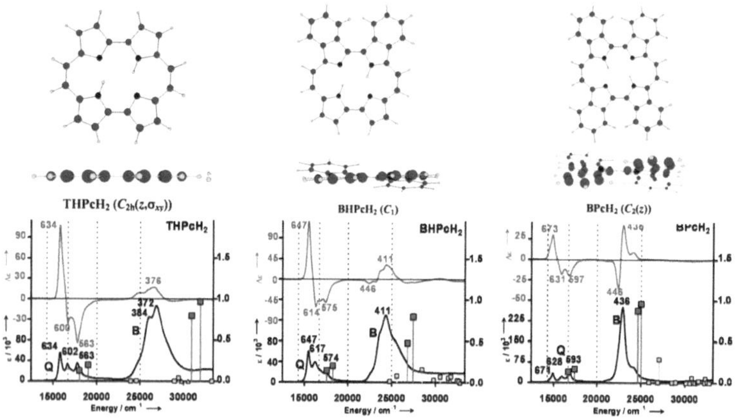

Figure 3.26. The molecular structures of porphycene ($THPcH_2$), dibenzoporphycene ($BHPcH_2$) and tetrabenzoporphycene ($BPcH_2$) (top). The electronic absorption, MCD and calculated TD-DFT spectra of $THPcH_2$, $BHPcH_2$ and $BPcH_2$ (bottom) [92].

corresponding bicyclo[2.2.2] octadiene-fused porphycenes (THPc). The main fourfold axis of symmetry, which is typically present in radially symmetric porphyrinoids, is removed. The addition of two adjacent benzo groups to form first dibenzoporphycene (BHPc) and then tetrabenzoporphycene (BPc), leads to a red shift of the Q and B bands as is also observed in the spectra of benzo-substituted Pzs, Fig. 3.24. A +ve/−ve sign sequence is observed in the Q band region, because ΔLUMO $>$ ΔHOMO. In the case of the B bands, a marked +ve/−ve sign sequence is observed in the spectra of $BHPcH_2$ and $BPcH_2$ but not for $THPcH_2$. The Q00 bands of BHPc and BPc are significantly weaker than those of THPc, Fig. 3.26, because the $|\Delta\text{LUMO} - \Delta\text{HOMO}|$ values are lowered upon substitution with fused benzene rings. As Djerassi and coworkers [72–74] demonstrated in the context of the chlorins in instances where α is relatively small, Eq. 3.1, and $|\Delta\text{LUMO} - \Delta\text{HOMO}| \approx 0$, the MCD sign sequence reverses only in the case of the Q band, since the μ^- moment remains the dominant factor in the context of the B band. Waluk and coworkers have also used Michl's perimeter model to analyze the MCD spectra of non-fused-ring-expanded porphycenes [99, 100].

3.4 Transition Metal Porphyrinoids

Key information about the degeneracy of the ground states of porphyrinoid complexes can be derived by determining whether the MCD spectrum contains temperature-dependent Faraday $\mathcal{C}_0$ terms. For example, in the MCD spectra of Fe(II) and Ru(II) porphyrinoids, the absence of temperature dependence is consistent with the presence of a low spin d^6 central metal configuration with $S = 0$, Fig. 3.27. The MCD spectra of low spin d^6 porphyrinoid complexes, such as $Fe^{II}Pc$ and $Ru^{II}Pc$ [49, 101, 102], are dominated by Faraday $\mathcal{A}_1$ terms and are similar to those of main group metal and closed shell d^{10} Pc complexes, complicated somewhat by the presence of metal to ligand and ligand to metal charge transfer (MLCT and LMCT) bands in addition to the $\pi \rightarrow \pi^*$ bands observed for main group metal complexes. For example, while the Q00 band is the lowest energy band observed in the spectra of low spin Fe(II)Pc complexes [101], the corresponding α-octaphenylphthalocyanine spectrum contains a broad absorption band envelope to the red of the Q00 band, which extends out to 1000 nm [54], Fig. 3.28. Spectral deconvolution analysis provides strong evidence for the presence of two MLCT transitions in the NIR region. A detailed analysis of a wide range of different D_{2d} symmetry transition metal complexes, based on the measurement of MCD spectra at both room and cryogenic

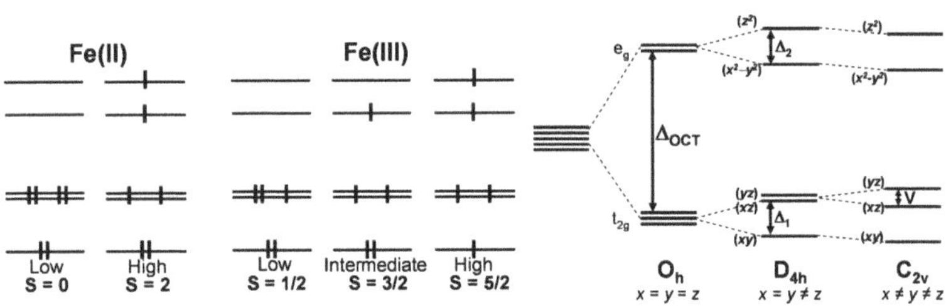

Figure 3.27. The low, intermediate, and high spin states for Fe(II) and Fe(III) based on the tetragonal environment formed by a planar porphyrinoid ligand (left). The crystal field splitting of Fe(II) and Fe(III) d orbitals in cubic O_h, tetragonal D_{4h} and rhombic C_{2v} environments (right). Only the latter two are typically observed for metal porphyrinoid complexes. The axial and rhombic ligand field parameters are denoted by Δ_1, Δ_2 and V.

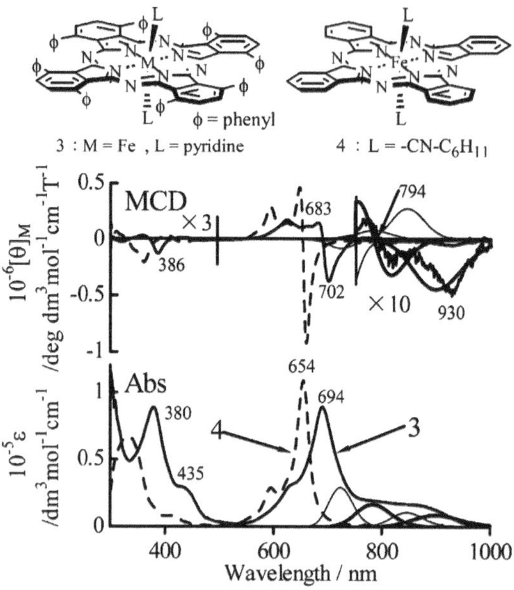

Figure 3.28. MCD (top) and absorption (bottom) spectra of $Fe^{II}\alpha(C_6H_5)_8Pc$, **3**, (solid lines) and $Fe^{II}Pc$, **4**, (dashed lines) in pyridine [54], respectively. Deconvoluted Gaussian-shaped bands are overlaid in the spectra.

temperatures, is still required to definitively assign these bands, however.

Since the mid-1960s, MCD spectroscopy has often been used to study heme-proteins, such as myoglobin and hemoglobin [103–106], and other biologically significant transition metal complexes. MCD spectroscopy has consistently provided key ground and excited state degeneracy information, which cannot easily be derived from an analysis of the UV-visible absorption spectra and theoretical calculations, alone. Major changes are observed in the MCD spectra when the oxidation or spin state changes due to the differing band morphologies of the three Faraday terms, while in contrast, only relatively minor changes are typically observed in the band centers and intensities of the UV-visible absorption spectra [10, 101, 103–107]. Since there is a partially filled set of *d* orbitals, Fig. 3.27, LMCT and MLCT bands are observed in addition to the $\pi \rightarrow \pi^*$ bands associated with

Gouterman's 4-orbital model. The redox chemistry of transition metal porphyrinoid complexes is significantly more complex than that of main group metal and closed shell d^{10} metal complexes. The spin state of Fe(II) and Fe(III) porphyrin complexes can be low, intermediate, or high, Fig. 3.27, depending on the nature of the axial ligation along the z-axis. The configuration interaction between the main $\pi\pi^*$ and CT states results in so called "fingerprint patterns", which can be used to derive information [103, 104] about several key points: (i) The nature and number of ligands attached to the iron. (ii) The spin and oxidation states of the iron. (iii) The ease with which the iron can be oxidized or reduced. And, (iv) whether axial ligands can approach and bind the iron? Recently the Stillman group has used this approach to explore the heme binding properties of iron-regulated surface determinant proteins in the human pathogen *Staphylococcus Aureus* [108–111], while Dawson and coworkers, who pioneered this approach, have reported similar MCD analyses for Fe-containing proteins [112].

There have been relatively few attempts in recent years to study the temperature dependence of Faraday $\mathbf{C}_0$ terms in porphyrinoid spectra, despite the fact that the theoretical approaches required to analyze these spectra were studied in depth in the 1980s and 90s. In the early 1990s, Thomson [113, 114] demonstrated that LMCT bands observed in the NIR region (1000 to 2000 nm) of the MCD spectra low spin d^5 Fe(III) hemes are an ideal probe for determining the extent of the rhombic splitting of the d_{yz} and d_{xz} orbitals, Fig. 3.27. Since $S = 1/2$, the ground and excited states of low spin d^5 complexes split into Kramers doublets within an applied magnetic field with $M_S = -1/2$ and $M_S = +1/2$. At high temperature and low field, the lcp and rcp bands tend to cancel resulting in a weak MCD signal, Fig. 3.29. As the temperature is lowered or the applied field is increased, the Boltzmann population of the higher sublevels decrease and the $\mathbf{C}_0$ term signal intensifies. When measurements are made at cryogenic temperatures (1–4 K) and high field strengths (4–7 T) in vitreous solutions or Shpol'skii matrices, only the lowest components of the ground state are populated and the MCD signal saturates in a distinctive derivative-shaped signal. Similar MCD spectra had earlier been reported by Stephens and coworkers for MCD spectra of cytochrome c recorded at room temperature [115]

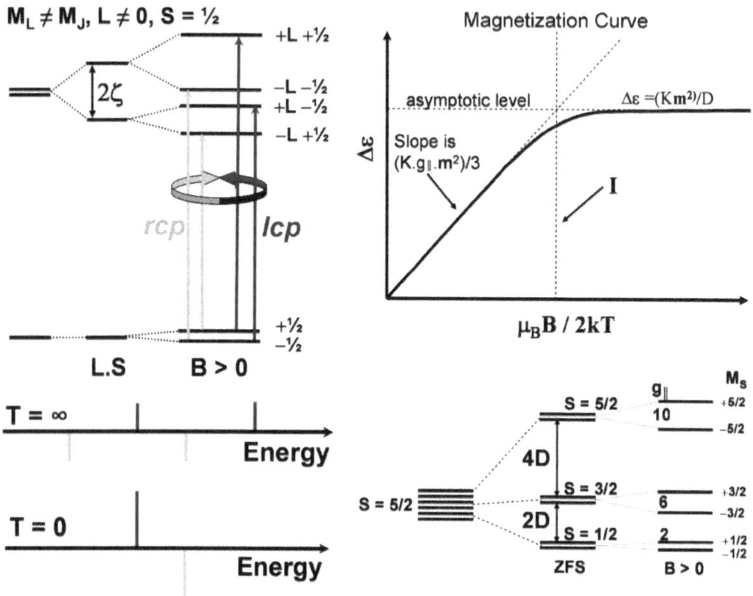

Figure 3.29. The origin of the Faraday C_0 term intensity arising from Kramers doublets formed by the ground and excited states of a paramagnetic complex with $S = 1/2$, an non orbitally degenerate ground state and an orbitally degenerate excited state (top left). Analysis of an MCD magnetization curve based on the approach developed by Schatz [118] and Thomson [119] (top right). The zero-field splitting of the ground state of a D_{4h} symmetry metal porphyrinoid complex with $S = 5/2$ based on the axial parameter, D (bottom right). When D is positive the $S = 1/2$ microstates are stabilized, while if D is negative the $S = 5/2$ microstates are stabilized.

and by Hatano and coworkers based on measurements between 138 and 283 K [116, 117]. Schatz [118] and Thomson [119] demonstrated that magnetization curves observed in plots of $\Delta\varepsilon$ against $\mu_B B/2kT$ can be used to derive the $g_{||}$ values for the ground states. Band polarization information can also be obtained since x/y- and z-polarized bands can be readily identified based on differences in the slopes of the linear portion of the magnetization plots, Fig. 3.29.

In the case of the $S = 5/2$ ground states of high spin Fe(III) complexes, there is a zero field splitting (ZFS) of the $M_S = \pm 1/2$,

±3/2, ±5/2 microstates due to spin–orbit coupling based on the axial (D) and rhombic (E) ZFS parameters within the spin Hamiltonian, Fig. 3.29. In the early 1980s, Stephens and coworkers [120] demonstrated that the magnitude of D could be derived by fitting a plot of $\Delta\varepsilon$ against T^{-1} based on cryogenic temperature MCD spectral data of a high spin Fe(III) tetraphenylporphyrin with a set of quadratic equations based on a least squares fitting approach, since the MCD signal is the sum of three Kramers doublets populated according to a Boltzmann distribution based on the magnitude and sign of D. This approach was subsequently adopted by the Stillman [121] and Sharonov [122] groups. Analysis of the $S = 2$ ground states associated with high spin d^6 configurations of Fe(II) are considerably more challenging than the Kramers doublet based systems, since multiple nested magnetization curves are observed based on the temperature and field strength values. During the 1990s, Solomon developed techniques for analyzing the MCD spectra of the non-Kramers doublet ground states of non-heme iron enzymes [123]. Since the turn of the century, there have been relatively few studies of the temperature-dependent $\mathcal{C}_0$ term properties of transition metal complexes, since very few research groups are currently actively carrying out MCD measurements at cryogenic temperatures [124–127]. It is possible that this field may undergo a renaissance in future, however, if advances in high temperature superconducting magnet technology make it possible to carry out measurements using liquid nitrogen thus eliminating the cost and difficulty associated with using liquid helium as a cryogenic coolant.

3.5 Conclusions

MCD spectroscopy continues to have an important role to play in porphyrinoid research. The additional band polarization and state degeneracy information, which it provides, can be used to definitively identify the main electronic bands within broad envelopes of extensively overlapping absorption bands. This information can be critical in validating the results of of TD–DFT calculations and in identifying key trends in the optical properties of porphyrinoid compounds, which can enable the rational design of novel compounds for use in specific practical applications. Once calculations for the three Faraday terms have been implemented into commercially available software packages the quantitative information that the MCD technique can provide about the OAM properties of ground and excited states will provide an additional test of the validity of DFT based descriptions of the electronic structures of porphyrinoids.

References

1. Mack, J., Stillman, M. J., and Kobayashi, N. (2007) Application of MCD spectroscopy to porphyrinoids, *Coord. Chem. Rev.* **251**, 429–453.
2. Kobayashi, N. and Nakai, K. (2007) Application of magnetic circular dichroism spectroscopy to porphyrins and phthalocyanines, *Chem. Commun.* 4077–4092.
3. Buckingham, A. D. and Stephens, P. J. (1966) Magnetic optical activity, *Ann. Rev. Phys. Chem.*, **17** 399–432.
4. Stephens, P. J. (1976) Magnetic circular dichroism, *Adv. Chem. Phys.*, **35** 197–264.
5. Piepho, S. B. and Schatz, P. N. (1983) *Group Theory in Spectroscopy with Applications to Magnetic Circular Dichroism*, Wiley, New York.
6. Platt, J. R. (1949) Classification of spectra of cata-condensed hydrocarbons, *J. Chem. Phys.*, **17** 484–495.
7. Platt, J. R. (1956) *Radiation Biology*, vol. 3, (ed. Hollander, A.), Chapter 2, McGraw-Hill, New York.
8. Gouterman, M. (1959) Spectra of porphyrins, *J. Mol. Spectrosc.*, **6**, 138–163.
9. Gouterman, M. (1959) Study of the effects of substitution on the absorption spectra of porphin, *J. Chem. Phys.*, **30**, 1139–1161.
10. McHugh, A. J., Gouterman, M., and Weiss Jr., C. (1972) Porphyrins XXIV. Energy, oscillator strength, and Zeeman splitting calculations (SCMO–CI) for phthalocyanine, porphyrins, and related ring systems, *Theoret. Chim. Acta*, **24**, 346–370.
11. Gouterman, M. (1978) Optical spectra and electronic structure, in *The Porphyrins,* vol. 3, part A, (ed. Dolphin, D.), Chapter 1 Academic Press, New York, 1–165.

12. Schaffer, A. M., Gouterman, M., and Davidson, E. R. (1973) Porphyrins XXVIII. Extended Hückel calculations on metal phthalocyanines and tetrazaporphins. *Theoret. Chim. Acta*, **30**, 9–30.
13. Moffitt, W. (1954) The electronic spectra of cata-condensed hydrocarbons, *J. Chem. Phys.*, **22** 320–333.
14. Moffitt, W. (1954) Configurational interaction in simple molecular orbital theory, *J. Chem. Phys.*, **22**, 1820–1829
15. Michl, J. (1978) Magnetic circular dichroism of cyclic π-electron systems. 1. Algebraic solution of the perimeter model for the A and B terms of high-symmetry systems with a $(4N + 2)$-electron [n]annulene perimeter, *J. Am. Chem. Soc.*, **100**, 6801–6811.
16. Michl, J. (1984) Magnetic circular dichroism of aromatic molecules, *Tetrahedron*, **4**, 3845–3934
17. Michl, J. (1980) Electronic structure of aromatic π-electron systems as reflected in their MCD spectra, *Pure Appl. Chem.*, **52**, 1549–1563
18. Höweler, U., Downing, J. W., Fleischhauer, J. and Michl, J. (1998) MCD of non-aromatic cyclic π-electron systems. Part 1. The perimeter model for antiaromatic $4N$-electron [n]annulene biradicals, *J. Chem. Soc., Perkin Trans.*, **2**, 1101–1117.
19. Fleischhauer, J., Höweler, U., and Michl, J. (1999) Magnetic circular dichroism of non-aromatic cyclic π-electron systems. 2. The perimeter model for high-symmetry 'unaromatic' and 'ambiaromatic' molecules derived from $4N$-electron [n]annulenes, *Spectrochim. Acta A*, **55**, 585–606.
20. Fleischhauer, J., Höweler, U., and Michl, J. (2000) MCD of nonaromatic cyclic π-electron systems. 3. The perimeter model for low-symmetry "Unaromatic" and "Ambiaromatic" molecules derived from $4N$-electron [n]Annulenes, *J. Phys. Chem. A*, **104**, 7762–7775.
21. Fleischhauer, J. and Michl, J. (2000) MCD of nonaromatic cyclic π-electron systems. 4. Explicit relations between molecular structure and spectra, *J. Phys. Chem. A*, **104**, 7776–7784.
22. Fleischhauer, J., Höweler, U., Spanget-Larsen, J., Raabe, G., and Michl, J. (2004) Magnetic circular dichroism of nonaromatic cyclic π-electron systems. 5. Biphenylene and its aza analogues, *J. Phys. Chem. A*, **108**, 3225–3234.
23. Fleischhauer, J., Raabe, G., Klingensmitj, K. A., Höweler, U., Chatterjee, R. K., Hafner, K., Vogel, E., and Michl, J. (2005) MCD of non-aromatic cyclic π-electron systems. Part 6: Pentalenes and heptalenes, *Int. J. Quantum Chem.*, **102**, 925–939.

24. Parr, R. G. and Wang, W. (1989) *Density-Functional Theory of Atoms and Molecules* Oxford University Press, Oxford.
25. Nemykin, V. N., Hadt, R. G., Belusludov, R. V., Mizuseki, H., and Kawazoe Y. (2007) Influence of molecular geometry, exchange-correlation functional, and solvent effects in the modeling of vertical excitation energies in phthalocyanines using time-dependent density functional theory (TDDFT) and polarized continuum model TDDFT methods: Can modern computational chemistry methods explain experimental controversies? *J. Phys. Chem. A*, **111**, 12901–12913.
26. Huang, T. H., Rieckhoff, K. E., and Voight, E. M. (1982) Shpol'skii effect and vibronic spectra of the phthalocyanines, *J. Chem. Phys.*, **77**, 3424–3441.
27. Huang, T. H., Rieckhoff, K. E., and Voight, E. M. (1981) New singlets in the phthalocyanines, *J. Phys. Chem.*, **85**, 3322–3326.
28. Mack, J. and Stillman, M. J. (1995) Band deconvolution analysis of the absorption and magnetic circular dichroism spectral data of ZnPc(−2) recorded at cryogenic temperatures, *J. Phys. Chem.*, **95**, 7935–7945.
29. Ricciardi, G., Rosa, A., and Baerends, E. J. (2001) Ground and excited states of zinc phthalocyanine studied by density functional methods, *J. Phys. Chem. A*, **105**, 5242–5254.
30. Magyar, R. J. and Tretiak, S. J. (2007) Dependence of spurious charge-transfer excited states on orbital exchange in TDDFT: Large molecules and clusters, *J. Chem. Theory Comput.*, **3**, 976–987.
31. Chidawanyika, W., Mack, J., Shimizu, S., Kobayashi, N., and Nyokong, T. (2009) Effect of peripheral fused ring substitution on the optical spectroscopy and electronic structure of metal phthalocyanine complexes, *J. Porphyrins Phthalocyanines*, **13**, 1053–1062.
32. Mack, J., Asano, Y., Kobayashi, N., and Stillman, M. J. (2005) Application of MCD spectroscopy and TD-DFT to a highly non-planar porphyrinoid ring system. New insights on red-shifted porphyrinoid spectral bands, *J. Am. Chem. Soc.*, **127**, 17697–17711.
33. Lash, T. D. and Chandrasekar, P. (1996) Synthesis of tetraphenyltetraacenaphthoporphyrin: A new highly conjugated porphyrin system with remarkably red-shifted electronic absorption spectra, *J. Am. Chem. Soc.*, **118**, 8767–8768.
34. Spence, J. D. and Lash, T. D. (2000) Porphyrins with exocyclic rings. 14.1 Synthesis of tetraacenaphthoporphyrins, a new family of highly conjugated porphyrins with record-breaking long-wavelength electronic absorptions, *J. Org. Chem.*, **65**, 1530–1539.

35. Haddad, R. E., Gazeau, S., Pécaut, J., Marchon, J.-C., Medforth, C. J., and Shelnutt, J. A. (2003) Origin of the red shifts in the optical absorption bands of nonplanar tetraalkylporphyrins, *J. Am. Chem. Soc.*, **125**, 1253–1268.
36. Wertsching, A. K., Koch, A. S., and Di Magno, S. G. (2001) On the negligible impact of ruffling on the electronic spectra of porphine, tetramethylporphyrin, and perfluoroalkylporphyrins, *J. Am. Chem. Soc.*, **123**, 3932–3939.
37. Seth, M., Ziegler, T., Banerjee, A., Autschbach, J., van Gisbergen, S. J. A., and Baerends, E. J. (2004) Calculation of the *A* term of magnetic circular dichroism based on time dependent-density functional theory I. Formulation and implementation, *J. Chem. Phys.*, **120**, 10942–10954.
38. Seth, M., Autschbach, J., and Ziegler, T. (2005). *Ab initio* calculation of the C/D ratio of magnetic circular dichroism, *J. Chem. Phys.*, **122**, 094112.
39. Seth, M. and Ziegler, T. (2006) Calculation of excitation energies of open-shell molecules with spatially degenerate ground states. II. Transformed reference via intermediate configuration Kohn–Sham time dependent density functional theory oscillator strengths and magnetic circular dichroism *C* terms, *J. Chem. Phys.*, **124**, 144105.
40. Krykunov, M., Banerjee, A., Ziegler, T., and Autschbach, J. (2005) Calculation of Verdet constants with time-dependent density functional theory: Implementation and results for small molecules, *J. Chem. Phys.*, **122**, 074105.
41. Seth, M., Autschbach, J., and Ziegler, T. (2007) Calculation of the term of magnetic circular dichroism. a time-dependent density functional theory approach, *J. Chem. Theory Comput.*, **3**, 434–447.
42. Krykunov, M., Seth, M., Ziegler, T., and Autschbach, J. (2007) Calculation of the magnetic circular dichroism *B* term from the imaginary part of the Verdet constant using damped time-dependent density functional theory, *J. Chem. Phys.*, **127**, 244102.
43. Seth, M., Krykunov, M., Ziegler, T., Autschbach, J., and Banerjee, A. (2008) Application of magnetically perturbed time-dependent density functional theory to magnetic circular dichroism: Calculation of *B* terms, *J. Chem. Phys.*, **128**, 144105.
44. Seth, M., Krykunov, M., Ziegler, T., and Autschbach, J. (2008) Application of magnetically perturbed time-dependent density functional theory to magnetic circular dichroism. II. Calculation of A terms, *J. Chem. Phys.*, **128**, 2341025.

45. Seth, M., Ziegler, T., and Autschbach, J. (2008) Application of magnetically perturbed time-dependent density functional theory to magnetic circular dichroism. III. Temperature-dependent magnetic circular dichroism induced by spin-orbit coupling, *J. Chem. Phys.*, **129**, 104105.

46. Peralta, G. A., Seth, M. and Ziegler, T. (2007) Magnetic circular dichroism of porphyrins containing M = Ca, Ni, and Zn. A computational study based on time-dependent density functional theory, *Inorg. Chem.*, **46**, 9111–9125.

47. Mack, J., Bunya, M., Lansky, D., Goldberg, D. P., and Kobayashi, N. (2008) The MCD spectroscopy of corrolazines and triazatetrabenzocorroles, *Heterocycles*, **76**, 1369–1380.

48. Cai, Z.-L., Crossley, M. J., Reimers, J. R., Kobayashi, R., and Amos, R. D. (2006) Density functional theory for charge transfer: The nature of the the N-bands of porphyrins and chlorophylls revealed through CAM-B3LYP, CASPT2, and SAC-CI Calculations, *J. Phys. Chem. B*, **110**, 15624–15632.

49. Mack, J. and Stillman, M. J. (2003) Electronic structures of metal phthalocyanine and porphyrin complexes from analysis of the UV-visible and magnetic circular dichroism spectra and molecular orbital calculations, in *The Porphyrin Handbook*, vol. 16, (ed. Kadish, K. M., Smith, K. M., and Guilard, R.), Chapter 103 Academic Press, New York, 43–116.

50. Stillman, M. J. and Nyokong, T. (1989) Absorption and magnetic circular dichroism spectral properties of phthalocyanines. Part 1: Complexes of the Dianion Pc(−2), in *Phthalocyanine. Principles and Properties*, vol. 1, (ed. Leznoff, C. C., and Lever, A. B. P.), Chapter 3 VCH Publications, New York, 133–289.

51. Stillman, M. J. (1993) Absorption and magnetic circular dichroism spectral properties of phthalocyanines. Part 2: Ring-oxidized and ring-reduced, in *Phthalocyanine. Principles and Properties*, vol. 3, (ed. Leznoff, C. C. and Lever, A. B. P.), Chapter 5 VCH Publications, New York, 227–296.

52. Mack, J. and Stillman, M. J. (2001) Assignment of the optical spectra of metal phthalocyanines through spectral band deconvolution analysis and ZINDO calculations, *Coord. Chem. Rev.*, **219–221**, 993–1032.

53. Keizer, S. P., Mack, J., Bench, B. A., Gorun, S. M., and Stillman, M. J. (2003) Spectroscopy and electronic structure of electron deficient zinc phthalocyanines *J. Am. Chem. Soc.*, **125**, 7067–7085.

54. Fukuda, T., Homma, S. and Kobayashi, N. (2003) A highly deformed iron(II) low-spin phthalocyanine which shows two MLCT transitions beyond the Q-band, *Chem. Commun.*, 1574–1575.
55. Takeuchi, Y., Matsuda, A., and Kobayashi, N. (2007) Synthesis and characterization of *meso*-Triarylsubporphyrins, *J. Am. Chem. Soc.*, **129**, 8271–8281.
56. Muranaka, A., Matsushita, O., Yoshida, K., Mori, S., Suzuki, M., Furuyama, T., Uchiyama M., Osuka, A., and Kobayashi, N. (2009) Application of the perimeter model to the assignment of the electronic absorption spectra of gold(III) hexaphyrins with [4n+2] and [4n] π-electron systems, *Chem. Eur. J.*, **15**, 3744–3751.
57. Chambrier, I., Cook, M. J., and Wood, P. T. (2000) Conformationally stressed phthalocyanines: the non-planarity of the 1,4,8,11,15, 18,22,25-octaisopentyl derivative, *Chem. Commun.*, 2133–2134.
58. Kobayashi, N., Fukuda, T., Ueno, K., and Ogino, H. (2001) Extremely non-planar phthalocyanines with saddle or helical conformation: Synthesis and structural characterizations *J. Am. Chem. Soc.*, **123**, 10740–10741.
59. Fukuda, T., Homma, S., and Kobayashi, N. (2005) Deformed phthalocyanines: synthesis and characterization of zinc phthalocyanines bearing phenyl substituents at the 1-, 4-, 8-, 11-, 15-, 18-, 22-, and/or 25-positions, *Chem. Eur. J.*, **11**, 5205–5216.
60. Meller, A. and Ossko, A. (1972) Phthalocyaninartige bor-komplexe, *Monatsh. Chem.*, **103**, 150–155.
61. Inokuma, Y., Kwon, J. H., Ahn, T. K., Yoo, M.-C., Kim, D., and Osuka, A. (2006) Tribenzosubporphines: Synthesis and characterization, *Angew. Chem. Int. Ed.*, **45**, 961–964.
62. Kobayashi, N., Ishizaki, T., Ishii, K., and Konami, H. (1999) Synthesis, spectroscopy, and molecular orbital calculations of subazaporphyrins, subphthalocyanines, subnaphthalocyanines, and compounds derived therefrom by ring expansion *J. Am. Chem. Soc.*, **121**, 9096–9110.
63. Stork, J. R., Brewer, J. J., Fukuda, T., Fitzgerald, J. P., Yee, G. T., Nazarenko, A. Y., Kobayashi, N., and Durfee, W. S. (2006) Chloro and hydroxo forms of a boron(III) subtriazaporphyrin macrocycle, *Inorg. Chem.*, **45**, 6148–6151.
64. Mori, S. and Osuka, A. (2005) Aromatic and antiaromatic gold(III) hexaphyrins with multiple gold-carbon bonds, *J. Am. Chem. Soc.*, **127**, 8030–8031.

65. Mori, S., Kim, K. S., Yoon, Z. S., Noh, S. B. Kim, D., and Osuka, A. (2007) Peripheral fabrications of a *Bis*-Gold(III) complex of [26]hexaphyrin(1.1.1.1.1.1) and aromatic versus antiaromatic effect on two-photon absorption cross section, *J. Am. Chem. Soc.*, **129**, 11344–11345.
66. Gorski, A., Lament, B., Davis, J. M., Sessler, J., and Waluk, J. (2001) Electronic states of a novel smaragdyrin isomer: polarized spectroscopy and theoretical studies, *J. Phys. Chem. A*, **105**, 4992–4999.
67. Gorski, A., Köhler, T., Seidel, D., Lee, J. T., Orzanowska, G., Sessler, J. L., and Waluk, J. (2005) Electronic structure, spectra, and magnetic circular dichroism of cyclohexa-, cyclohepta-, and cyclooctapyrrole, *Chem. Eur. J.*, **11**, 4179–4184.
68. Sessler, J. L., Cho, D.-G., Stępień, M., Lynch, V., Waluk, J., Yoon, Z. S., and Kim, D. (2006) Inverted sapphyrin: A new family of doubly n-confused expanded porphyrins, *J. Am. Chem. Soc.*, **128**, 12640–12641.
69. Xue, Z.-L., Shen, Z., Mack, J., Kuzuhara, D., Yamada, H., Okujima, T., Ono, N., You, X.-Z., and Kobayashi, N. (2008) A facile one-pot synthesis of *meso*-aryl-substituted [14]triphyrin(2.1.1), *J. Am. Chem. Soc.*, **130**, 16478–16479.
70. Kobayashi, N., Furuya, F., Yug, G.-C., Wakita, H., Yokomizo, M., and Ishikawa, N. (2002) Synthesis and characterization of phthalocyanines with direct Si-Si linkages, *Chem. Eur. J.*, **8**, 1474–1484.
71. Kobayashi, N., Yokoyama, M., Muranaka, A., and Ceulemans, A. (2004) Formation of silicon triazacorrole and tetrabenzotriazacorrole by the ring contraction of the corresponding tetraazaporphyrin ligands, *Tetrahedron Lett.*, **45**, 1755–1758.
72. Keegan, J. D., Stolzenberg, A. M., Lu, Y. C., Linder, R. E., Barth, G., Moscowitz, A., Bunnenberg, E., and Djerassi, C. (1982) Magnetic circular dichroism studies. 60. Substituent-induced sign variation in the magnetic circular dichroism spectra of reduced porphyrins. 1. Spectra and band assignments, *J. Am. Chem. Soc.*, **104**, 4305–4317.
73. Keegan, J. D., Stolzenberg, A. M., Lu, Y. C., Linder, R. E., Barth, G., Moscowitz, A., Bunnenberg, E., and Djerassi, C. (1982) Magnetic circular dichroism studies. 61. Substituent-induced sign variation in the magnetic circular dichroism spectra of reduced porphyrins. 2. Perturbed molecular orbital analysis, *J. Am. Chem. Soc.*, **104**, 4317–4329.
74. Djerassi, C., Lu, Y. C., Waleh, A., Shu, A. Y. L., Goldbeck, R. A., Kehres, L. A., Crandell, C. W., Wee, A. G. H., Kniezinger, A., Gaete-Holmes, R., Loew, G. H., Clezy, P. S., and Bunnenberg, E. (1984) Magnetic circular dichroism.

65. Sign variation in the magnetic circular dichroism spectra of free-base porphyrins having a single π-acceptor pyrrole ring substituent. Structure implications, *J. Am. Chem. Soc.*, **106**, 4241–4258.

75. Fukuda, T. and Kobayashi, N. (2008) Hydrogenated tetraazaporphyrins-old but new coremodified phthalocyanine analogues, *Dalton Trans.*, 4685–4704.

76. Makarova, E. A. and Lukyanets, E. A. (2009) Design and synthesis of tetraazachlorins, tetraazabacteriochlorins and tetraazaisobacteriochlorins, *J. Porphyrins Phthalocyanines*, **13**, 188–202.

77. Miwa, H., Makarova, E. A., Ishii, K., Luk'yanets, E. A., and Kobayashi, N. (2002) Spectroscopy, electrochemistry, and molecular orbital calculations of metal-free tetraazaporphyrin, -chlorin, -bacteriochlorin, and -isobacteriochlorin *Chem. Eur. J.*, **8**, 1082–1090.

78. Fukuda, T., Makarova, E. A., Luk'yanets, E. A., and Kobayashi, N. (2004) Synthesis and spectroscopic and electrochemical studies of novel benzo- or 2,3-naphtho-fused tetraazachlorins, bacteriochlorins, and isobacteriochlorins, *Chem. Eur. J.*, **10**, 117–133.

79. Makarova, E. A., Fukuda, T., Luk'yanets, E. A., and Kobayashi, N. (2005) Synthesis, spectroscopic, and electrochemical studies of 1,2-naphthalene-ring-fused tetraazachlorins, -bacteriochlorins, and -isobacteriochlorins: The separation and characterization of structural isomers, *Chem. Eur. J.*, **11**, 1235–1250.

80. Makarova, E. A., Dzyuina, E. V., Fukuda, T., Kaneko, H., Hashimoto, N., Kikukawa, Y., Kobayashi, N., and Lukyanets, E. A. (2009) Synthesis and spectroscopic and electrochemical studies of pyrazine- or pyridine-ring-fused tetraazachlorins, bacteriochlorins, and isobacteriochlorins, *Inorg. Chem.*, **48**, 164–173.

81. Fukuda, T., Masuda, S., and Kobayashi, N. (2007) Tetraazachlorin–fullerene conjugates: On-off control of electronic communication enabled by push-pull substituents, *J. Am. Chem. Soc.*, **129**, 5472–5479.

82. Fukuda, T., Masuda, S., Hashimoto, N., and Kobayashi, N. (2008) Metal effects on electronic structures of directly linked tribenzotetraazachlorin-fullerene conjugates, *Inorg. Chem.*, **47**, 2576–2580.

83. Ozoemena, K. I., Mamuru, S. A., Fukuda, T., Kobayashi, N., and Nyokong, T. (2009) Metal (Co, Fe) tribenzotetraazachlorin–fullerene conjugates: Impact of direct π-bonding on the redox behaviour and oxygen reduction reaction, *Electrochem. Commun.*, **11**, 1221–1225.

84. Fukuda, T., Hashimoto, N., Araki, Y., El-Khouly, M. E., Ito, O., and Kobayashi, N. (2009) Phthalocyanine–C_{60} fused conjugates exhibiting

molecular orbital interactions depending on the solvent polarity, *Chem. Asian. J.*, **4**, 1678–1686.

85. Kobayashi, N., Miwa, H., Isago, H., and Tomura, T. (1999) An adjacent dibenzotetraazaporphyrin: a structural intermediate between tetraazaporphyrin and phthalocyanine, *Inorg. Chem.*, **38**, 479–485.
86. Kobayashi, N., Miwa, H., and Nemykin, V. N. (2002) Adjacent versus opposite type di-aromatic ring-fused phthalocyanine derivatives: synthesis, spectroscopy, electrochemistry, and molecular orbital calculations, *J. Am. Chem. Soc.*, **124**, 8007–8020.
87. Kobayashi, N., Mack, J., Ishii, K., and Stillman, M. J. (2002) Electronic structure of reduced symmetry peripheral fused-ring-substituted phthalocyanines, *Inorg. Chem.*, **41**, 5350–5363.
88. Miwa, H., Ishii, K., and Kobayashi, N. (2004) Electronic structures of zinc and palladium tetraazaporphyrin derivatives controlled by fused benzo rings, *Chem. Eur. J.*, **10**, 4422–4435.
89. Kobayashi, N. and Fukuda, T. (2002) Mono-aromatic ring-fused versus adjacently di-aromatic ring-fused tetraazaporphyrins: regioselective synthesis and their spectroscopic and electrochemical properties, *J. Am. Chem. Soc.*, **124**, 8021–8034.
90. Nakai, K., Kurotobi, K., Osuka, A., Uchiyama, M., and Kobayashi, N. (2008) Electronic structures of azulene-fused porphyrins as seen by magnetic circular dichroism and TD-DFT calculations, *J. Inorg. Biochem.*, **102**, 466–471.
91. Muranaka, A., Yonehara, M., and Uchiyama, M. (2010) Azulenocyanine: A new family of phthalocyanines with intense near-IR absorption, *J. Am. Chem. Soc.*, **132**, 7844–7845.
92. Kuzuhara, D., Mack, J., Yamada, H., Okujima, T., Ono, N., and Kobayashi, N. (2009) Synthesis, structures, and optical and electrochemical properties of benzoporphycenes, *Chem. Eur. J.*, **15**, 10060–10069.
93. Shimizu, Y., Shen, Z., Okujima, T., Uno, H., and Ono, N. (2004) First synthesis of a series of core-modified tetrabenzoporphyrins, *Chem. Commun.*, 374–375.
94. Okujima, T., Komobuchi, N., Shimizu, Y., Uno, H., and Ono, N. (2004) An efficient synthesis of conjugation-expanded carba- and azuliporphyrins using a bicyclo[2.2.2]octadiene-fused tripyrrane, *Tetrahedron Lett.*, **45**, 5461–5464.
95. Mack, J., Bunya, M., Shimizu, Y., Uoyama, H., Komobuchi, N., Okujima, T., Uno, H., Ito, S., Stillman, M. J., Ono, N., and Kobayashi, N. (2008) Application of MCD spectroscopy and TD-DFT to nonplanar core-modified

tetrabenzoporphyrins: effect of reduced symmetry on nonplanar porphyrinoids, *Chem. Eur. J.*, **14**, 5001–5020.

96. Ceulemans, A., Oldenhof, W., Görller-Walrand, C., and Vanquickenborne,
L. G. (1986) Gouterman's "four-orbital" model and the MCD spectra of high-symmetry metalloporphyrins, *J. Am. Chem. Soc.*, **108**, 1155–1163.

97. Goldberg, D. P. (2007) Corrolazines: New frontiers in high-valent metalloporphyrinoid stability and reactivity, *Acc. Chem. Res.*, **40**, 626–634.

98. Kurotobi, K., Kim, K. S., Noh, S. B., Kim, D., and Osuka, A. (2006) A quadruply azulene-fused porphyrin with intense near-ir absorption and a large two-photon absorption cross section, *Angew. Chem. Int. Ed.*, **45**, 3947–3994.

99. Gorski, A., Vogel, E., Sessler, J. L., and Waluk, J. (2002) Magnetic circular dichroism of neutral and ionic forms of octaethylhemiporphycene, *Chem Phys.*, **282**, 37–49.

100. Gorski, A., Vogel, E., Sessler, J. L., and Waluk, J. (2002) Magnetic circular dichroism of octaethylcorrphycene and its doubly protonated and deprotonated forms, *J. Phys. Chem. A*, **106**, 8139–8145.

101. Ough, E. A. and Stillman, M. J. (1994) Analysis of the absorption and magnetic circular dichroism spectra of iron(II) phthalocyanine, *Inorg. Chem.*, **33**, 573–583.

102. Nyokong, T., Gasyna, Z., and Stillman, M. J. (1986) Photochemical formation of ruthenium phthalocyanine π-cation radical species, *Inorg. Chim. Acta*, **112**, 11–15.

103. Dawson, J. H. and Dooley, D. M. (1989) Magnetic circular dichroism spectroscopy of iron porphyrins and heme proteins, in *Iron Porphyrins*, part 3, vol. 4, (ed. Lever A. B. P. and Gray, H. B.), Chapter 1 VCH, London pp. 1–135.

104. Cheek, J. and Dawson, J. (2000) Magnetic circular dichroism spectroscopy of heme proteins and model systems, in *The Porphyrin Handbook*, vol. 7, (ed. Kadish, K. M., Smith, K. M., and Guilard, R.) Chapter 53 Academic Press, New York, **2000**, 339–369.

105. Kirk, M. L. and Peariso, K. (2003) Recent applications of MCD spectroscopy to metalloenzymes, *Curr. Opin. Chem. Biol.*, **7**, 22–227.

106. Solomon, E. I., Pavel, E. G., Loeb, K. E., and Campochiaro, C. (1995) Magnetic circular dichroism spectroscopy as a probe of the geometric and electronic structure of non-heme ferrous enzymes, *Coord. Chem. Rev.*, **144**, 369–460.

107. Mack, J. and Stillman, M. J. (1997) Assignment of the optical spectra of metal phthalocyanine anions, *Inorg. Chem.*, **36**, 413–425.

108. Mack, J., Vermeiren, C., Heinrichs, D. E., and Stillman, M. J. (2004) *In vivo* heme scavenging by *Staphylococcus aureus* IsdC and IsdE proteins, *Biochem. Biophys. Res. Commun.*, **320**, 781–788.

109. Vermeiren, C. L., Pluym, M., Mack, J., Heinrichs D. E., and Stillman, M. J. (2006) Characterization of the heme binding properties of *Staphylococcus aureus* IsdA, *Biochem.*, **45**, 12867–12875.

110. Pluym, M., Vermeiren, C. L., Mack, J., Heinrichs, D. E., and Stillman, M. J. (2007) Heme binding properties of *Staphylococcus aureus* Isd E, *Biochem.*, **46**, 12777–12787.

111. Heinrichs, D. E., Mack, J., Pluym, M., Vermeiren, C. L., and Stillman, M. J. (2007) Protoporphyrin IX and heme binding properties of *Staphylococcus aureus* Isd C, *J. Porphyrins Phthalocyanines*, **11**, 165–171.

112. Eakanunkul, S., Lukat-Rogers, G. S., Sumithran, S., Ghosh, A., Rodgers, K. R., Dawson, J. H., and Wilks, A. (1995) Characterization of the periplasmic heme-binding protein shut from the heme uptake system of *Shigella dysenteriae*, *Biochem.*, **44**, 13179–13191.

113. Thomson, A. J. and Gadsby, P. M. A. (1990) A theoretical model of the intensity of the near-infrared porphyrin-to-iron charge-transfer transitions in low-spin iron(III) haemoproteins. A correlation between the intensity of the magnetic circular dichroism bands and the rhombic distortion parameter of iron, *Dalton Trans.*, 1921–1928.

114. Gadsby, P. M. A. and Thomson, A. J. (1990) Assignment of the axial ligands of ferric ion in low-spin hemoproteins by near-infrared magnetic circular dichroism and electron paramagnetic resonance spectroscopy, *J. Am. Chem. Soc.*, **112**, 5003–5011.

115. Rawlings, R., Stephens, P. J., Nafie, L. A., and Kamen, M. D. (1977) Near-infrared magnetic circular dichroism of cytochrome *c'*, *Biochem.*, **16**, 1725–1729.

116. Kobayashi, N., Nozawa, T., and Hatano, M. (1981) Near-infrared magnetic circular dichroism studies on iron(III) horse heart cytochrome *c*, *Bull Chem. Soc. Jpn.*, **54**, 919–920.

117. Yamamoto, T., Nozawa, T., Kobayashi, N., and Hatano, M. (1982) Origins and spin dependence of near infrared magnetic circular dichroism of iron(III) hemoproteins, *Bull Chem. Soc. Jpn.*, **55**, 3059–3063.

118. Schatz, P. N., Mowery, R. L., and Krausz, E. R. (1978) M.C.D./M.C.P.L. saturation theory with application to molecules in $D_{\infty h}$ and its subgroups, *Mol. Phys.*, **35**, 1537–1557.

119. Thomson, A. J. and Johnson, M. K. (1980) Magnetization curves of haemoproteins measured by low-temperature magnetic-circular-dichroism spectroscopy, *Biochem. J.*, **191**, 411–420.

120. Browett, W. R., Fucaloro, A. F., Morgan, T. V., and Stephens, P. J. (1983) Magnetic circular dichroism determination of zero-field splitting in chloro(*meso*-tetraphenylporphinato)iron(III), *J. Am. Chem. Soc.*, **105**, 1868–1872.

121. Browett, W. R., Gasyna, Z., and Stillman, M. J. (1987) Temperature dependence and electronic transition energies in the magnetic circular dichroism spectrum of horeseradish peroxidase compound I, *J. Am. Chem. Soc.*, **110**, 3633–3640.

122. Oganesyan, V. S. and Sharonov, Y. A. (1997) A 4-term energy level scheme for the high-spin ferrous hemoproteins: Evidence for the ^5E_n and $^5B_2$ terms as the ground multiplets in hemoproteins with a histidine and a cysteine protein-derived heme ligand, respectively, *Spectrochim. Acta A*, **53**, 433–449.

123. Solomon, E. I., Brunold, T. C., Davis, M. I., Kemsley, J. N., Sang-Kyu Lee, S.-K. Lehnert, N., Neese, F., Skulan, A. J., Yang, Y.-S., and Jing Zhou, J. (2000) Geometric and electronic structure/function correlations in non-heme iron enzymes, *Chem. Rev.*, **100**, 235–350.

124. Praneeth, V. K. K., Neese, F., and Lehnert, N. (2005) Spin density distribution in five- and six-coordinate iron(II)-porphyrin no complexes evidenced by magnetic circular dichroism spectroscopy, *Inorg. Chem.*, **44**, 2570–2572.

125. Praneeth, V. K. K., Näther, C., Peters, G., and Lehnert, N. (2006) Spectroscopic properties and electronic structure of five- and six-coordinate iron(II) porphyrin no complexes: effect of the axial N-donor ligand, *Inorg. Chem.*, **45**, 2795–2811.

126. Paulat, F. and Lehnert, N. (2008) Detailed assignment of the magnetic circular dichroism and UV-vis spectra of five-coordinate high-spin ferric [Fe(TPP)(Cl)], *Inorg. Chem.*, **47**, 4963–4976.

127. Hu, C., Sulok, C. D., Paulat, F., Lehnert, N., Twigg, A. I., Hendrich, M. P., Schulz, C. E., and Scheidt, W. R. (2010) Just a proton: Distinguishing the two electronic states of five-coordinate high-spin iron(II) porphyrinates with imidazole/ate coordination, *J. Am. Chem. Soc.*, **132**, 3737–3750.

Chapter 4

Phthalocyanine–Porphyrin Heteroarrays: A Perfect Marriage Between Two Unique Macrocycles

Giovanni Bottari,[a] Gema de la Torre,[a] Dirk M. Guldi[b] and Tomas Torres,[ac]

[a] *Departamento de Química Orgánica, Universidad Autónoma de Madrid, 28049 Madrid, Spain*

[b] *IMDEA-Nanociencia, Facultad de Ciencias, Ciudad Universitaria de Cantoblanco, 28049 Madrid, Spain*

[c] *Friedrich-Alexander-Universität Erlangen-Nürnberg, Department of Chemistry and Pharmacy & Interdisciplinary Center for Molecular Materials (ICMM), Egerlandstrasse 3, 91058 Erlangen, Germany*

4.1 Introduction

Tetrapyrrolic macrocycles represent a class of compounds which are widespread in nature. Within the large family of porphyrinoid systems, porphyrins (Pors) [1] and their synthetic analogs phthalocyanines (Pcs) [2] deserve a special mention due to their unique and intriguing physicochemical properties. Both macrocycles, in fact, share important similarities such as a planar, 18-π-electron aromatic system and exhibit remarkable optical and electronic properties, which have to prompted their utilization in various disciplines ranging from materials science [3], catalysis [4], photovoltaics [5], or medicine [6] to mention a few.

Multiporphyrin Arrays: Fundamentals and Applications
Edited by Dongho Kim

www.panstanford.com

Owing to the presence in both macrocycles of a highly conjugated π-system, Pcs, and Pors show intense absorptions in the UV-visible region. Pors exhibit a strong Soret band around 400–450 nm together with a number of weaker Q bands around 500–600 nm, whereas Pcs show a medium strength B band at around 320–350 nm and a strong Q band around 700–800 nm. In this context, the preparation of Pc–Por heteroarrays as light harvesting systems is highly desirable, considering that the two macrocycles present intense, and complementary, optical absorption which cover a large portion of the UV-visible-near infrared component of the solar spectrum [7].

Moreover, both Pcs and Pors present interesting photophysical and electrochemical properties that can be easily modulated by the careful choice of the metal center and/or the peripheral substituents, so that efficient energy and/or photoinduced electron transfer processes are likely to take place in such Pc–Por heteroarrays after photoexcitation of one or both of the heteromacrocyclic components, thus making these systems interesting candidates for light harvesting, photovoltaic, or molecular photonic applications.

This chapter will revise the different synthetic strategies that have been pursued for the preparation of Pc–Por, as well as Pc-subphthalocyanine (SubPc) and Por-SubPc, heteroarrays in which the two heteromacrocycles have been covalently connected. A photophysical analysis of some of these systems will also be presented with the aim of rationalizing the effect that the structural and electronic features of the two heteromacrocycles which constitute these heteroarrays have on the photoinduced electron/energy transfer dynamics. Pc·Por heteroarrays assembled *via* supramolecular interactions have also been prepared and their photophysical properties studied with the aim of obtaining information of some of the fundamental steps occurring in natural photosynthesis, which also relies on supramolecular methods to assemble its components. The preparation and study of these supramolecular Pc·Por heteroarrays in solution, condensed phases or on metal surfaces will also be reviewed. The last part of this chapter will be devoted to the synthesis and study of multinuclear Pc and Pc–Por arrays connected, either covalently or *via* supramolecular interactions, to electroactive units such as C_{60} fullerene, anthraquinone, or perylenes as a way to prepare complex, multifunctional donor–acceptor (D–A) ensembles with improved optoelectronic properties.

4.2 Covalently Linked, Phthalocyanine–Porphyrin Heteroarrays

4.2.1 *Peripherally linked, Phthalocyanine–Porphyrin Heteroarrays*

Historically speaking, the first covalently linked Pc–Por heteroarrays were reported by Maillard and coworkers in 1986 [8]. The synthesis of dyads **1a,b** started with the *ipso* substitution reaction of 4-nitrophthalonitrile and monohydroxy-functionalized tetrapheny-porphyrin (TPP) **2a** (Scheme 4.1). The resulting Por-modified

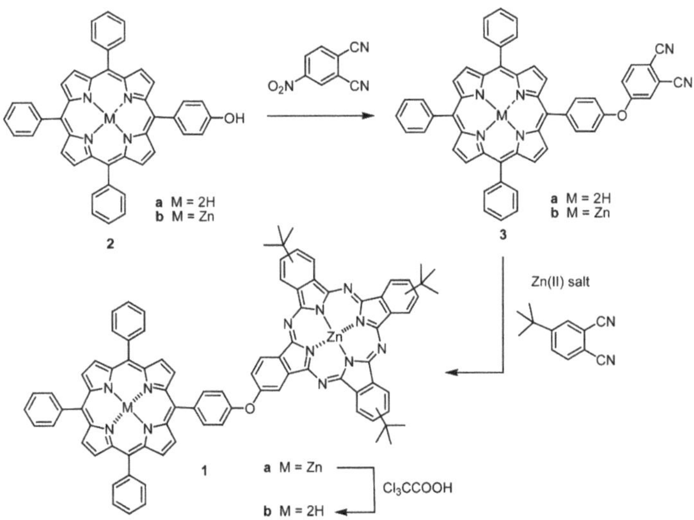

Scheme 4.1. Synthesis of oxo-bridged Pc–Por dyads **1a**,**b**.

phthalonitrile derivative **3a** was condensed with an excess of 4-*tert*-butylphthalonitrile in the presence of $ZnCl_2$ to afford dyad **1a**, in which each of the two macrocyclic cavities has a Zn(II) in its center. Selective demetallation of the TPP center in **1a** afforded the mixed dyad **1b**. The electronic absorption spectra of binuclear derivatives

1a,b in toluene are very similar to the sum of the spectra of the individual subunits, an equimolar mixture of either Zn(II) or metal-free TPP and tetra-*tert*-butyl Zn(II)Pc, thus indicating the lack of ground-state interactions.

The photophysics of the Zn(II) homodimetallic, covalently linked dyad **1a** in toluene is mainly governed by energy transfer reactions [9]. As the energy levels of the singlet and the triplet state of the Zn(II)Pc component lie below the corresponding Zn(II)Por levels, the excitation energy is only transferred from the Zn(II)Por to the Zn(II)Pc. Therefore, upon selective photoexcitation of the Zn(II)Por component at 424 nm in toluene solution, dyad **1a** exhibits a strong quenching of the Zn(II)Por-centered fluorescence, with concomitant emission of the Zn(II)Pc component at 690 nm. The energy transfer between the two chromophores is highly efficient, the quantum yield of the singlet–singlet energy transfer being close to in front of unity. Nanosecond absorption spectroscopy experiments revealed that the quantum yield of the triplet–triplet energy transfer process was also very high. Once the Zn(II)Pc excited states (i.e., singlet and triplet) are formed, their dissipative energy pathways are identical to those of the reference tetra-*tert*-butyl Zn(II)Pc.

A different scenario arises when the photophysical studies at dyad **1a** are performed in more polar solvents, such as DMSO. The polar environment favors the electron transfer mechanism on the expense of energy transfer; the fluorescence of both chromophores is quenched and the lifetime of both singlets is shortened. Moreover, a new short component (i.e., 0.33 ns) contributes to the deactivation of the Zn(II)Pc fluorescence, which can be assigned to the formation of a charge transfer (CT) state. The different behavior of dyad **1a** in DMSO was explained in terms of conformational changes and a large stabilization of the charge separated state in this polar solvent.

By using the self-cyclization reaction of Por-containing phthalonitrile **3b**, Pc-Por_4 pentads **4a,b** have been prepared by Nyokong and coworkers (Scheme 4.2) [10]. The Por-modified phthalonitrile derivative **3b** can be converted in the metal-free Pc derivative **4a** (10% yield) by treatment with lithium in octanol, and in **4b** by conversion of **3b** into the more activated diiminoisoindoline derivative and further treatment with $Zn(OAc)_2$ in order to obtain **4b** in 16% overall yield. A related Co(II)Pc-Co(II)Por_4

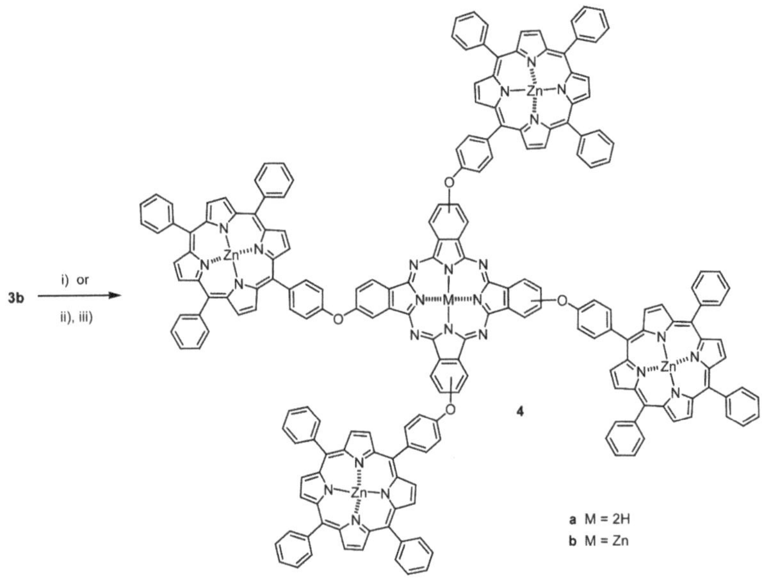

Scheme 4.2. Reaction conditions for the synthesis of pentads **4a**,**b**. (i) Li, octanol, 150C for **4a**; (ii) NH_3/NaOMe, methanol; and (iii) $Zn(OAc)_2{\cdot}2H_2O$, DBU, penthanol, 140C for **4b**.

heteropentamer containing four units of Co(II)Por linked to a central Co(II)Pc macrocycle *via* ether linkages has also been reported by the same authors [11]. A similar heteropentamer has also been recently prepared by *ipso* substitution reactions between Fe(II) and Zn(II) 1,8,15,22-tetranitroPcs and Por **2b** as the nucleophile [12].

The ether moiety is also the linkage between macrocycles in other Pc–Por heteroarrays reported by Nyokong and coworkers [13]. These authors have prepared Pc-Por_2 triads **5a**,**b** and Pc-Por_8 nonamers **6a**,**b** using a Por-containing dicyanobenzene **7** as a by key intermediate (Scheme 4.3). This compound was readily synthesized by reaction of Por **2a** with 4,5-dichlorophthalonitrile and further metallation of the Por macrocycles with $Zn(OAc)_2$. The preparation of H_2Pc-Zn(II)Por_2 triad **5a**, which is an unsymmetrically substituted Pc derivative, was undertaken by the "subphthalocyanine (SubPc) method", which involves the reaction of this preformed, lower Pc analogue with a phthalonitrile, or a moreactivated diiminoisoindoline, giving rise to a ring-expansion reaction to form

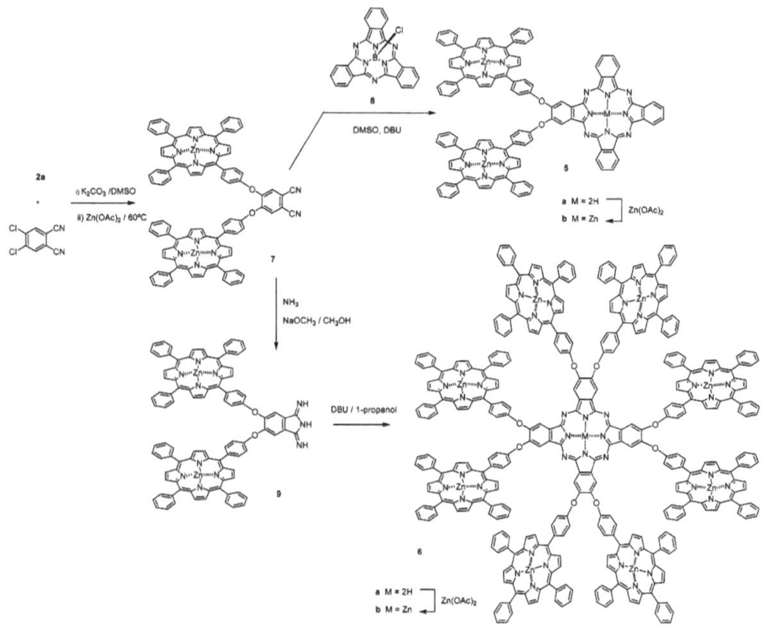

Scheme 4.3. Synthesis of Pc-Por_2 heterotrimers **5a**,**b** and "windmill" Pc-Por_8 heterononamers **6a**,**b**.

an unsymmetrical Pc molecule. In this case, the reaction of **7** with unsubstituted SubPc **8** afforded **5a** in 6% yield. The treatment of this derivative with $Zn(OAc)_2$ yielded the homotrimetallic Zn(II)Pc-Zn(II)Por_2 triad **5b**. On the other hand, the nonamer H_2Pc-$H_2$$Por_8$ **6a** was obtained in 14% yield by former conversion of the Por-containing phthalonitrile **7** into the more reactive diiminoisoindoline derivative **9** and subsequent self-condensation reaction. Zn(II)Pc-Zn(II)Por_8 **6b** was prepared by metallation of macrocycle **6a** with $Zn(OAc)_2$.

UV-vis measurements of these heteroarrays show that the B-band absorption of the Por moieties is not affected by the covalent linkage to the Pc moiety, whereas the Q-band of the latter is slightly blue-shifted with regard to the absorption of the unsubstituted, H_2Pc reference compound. The effect of the electronic interaction between the two chromophores is more pronounced in the excited state. Upon excitation of the Por moiety at 429 nm in triad

5a, the emission corresponding to the Zn(II)Por moiety at 657 nm disappears, and a 15 nm blue-shift in the emission peak of the H_2Pc component occurred, together with a remarkable enhancement of the Pc fluorescence intensity, suggesting that the excited Zn(II)Por component transfers energy to the H_2Pc unit. A similar intramolecular energy transfer process was observed in Zn(II)Pc-Zn(II)Por_2 triad **5b**, but with a higher efficiency (i.e., 30% in **5a** and 60% in **5b**). This effect was amplified in Pc-Zn(II)Por_8 derivatives **6a,b**, in which the efficiency of energy transfer can reach 90%.

A striking example of covalent, oxo-bridged Por–Pc heteroarrays was reported by the groups of Jiang, Weng and Li, who prepared and characterized Por-appended Eu(III) bis(phthalocyaninato) complexes **10a,b** and **11** (Scheme 4.4) [14]. These double-decker,

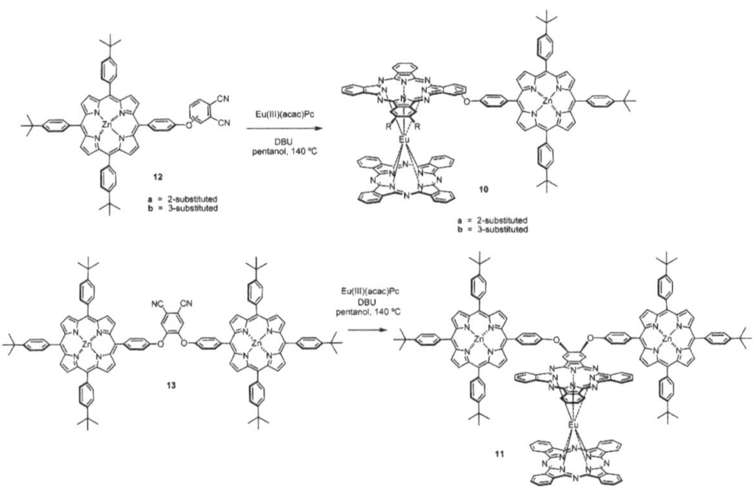

Scheme 4.4. Synthesis of Por-appended, Eu(III) bis(phthalocyaninato) complexes **10a,b** and **11**.

bisPc structures consist of a substituted Pc ligand, containing a different number of linked Por moieties at the peripheral (i.e., **10b** and **11**) or non-peripheral (i.e., **10a**) positions, and one unsubstituted Pc unit. Key elements for the preparation of heteroarrays **10a,b** and **11** are the Por-containing phthaloniriles **12a,b** and **13**, respectively. These precursors were subjected to one-pot, mixed cyclization with unsubstituted phthalonitrile, which takes place over the preformed mononuclear [Eu(III)(acac)(Pc)] (acac = acetylacetonate) which acts as template [15]. These reactions led to the formation of a mixture of products, from which the desired products **10a,b** and **11** could be isolated in 30–40% yield by silica gel and Biobeads column chromatography.

In general, the three heteroarrays **10a,b** and **11** show similar photophysical behavior, that is, the fluorescence of the Por moiety is effectively quenched by the double-decker Pc unit through an intramolecular photoinduced electron transfer process from the Por chromophore to the double-decker Pc. This process is thermodynamically favorable ($\Delta G = -0.11$ V) and takes place very quickly, in several hundreds of femtoseconds, while the recombination of the charge-separated state occurs in several picoseconds. The directionality of the electron transfer process was confirmed by time-resolved absorption experiments, which show a band at 620 nm in the femtosecond absorption spectra of **10a,b** and **11** which is ascribed to the monoanion of the bis(phthalocyaninato) unit (i.e., $[Eu(III)(Pc)_2]^-$). The effects of the number and, in particular, the position of the Por substituent(s) on the photophysical processes were also examined. In comparison with the β-substituted analogue **10b**, the α-substituted counterpart **10a** gives a slower electron transfer and recombination process, probably as a consequence of the increased separation between the Por chromophore and the $[Eu(III)Pc_2]$ unit. Nevertheless, the electron transfer and, especially, the recombination rates were even slower for tetrad **11** despite a similar Por-to-double-decker Pc separation to that of **10b**. The results can be reasonably rationalized by considering the intramolecular interaction between the two neighboring Por chromophores in **11**, which indeed stabilizes the charge-separated state due to delocalization of the positive charge on the two interacting Pors.

Very recently, a series of mixed (phthalocyaninato)(porphyrinato) yttrium double-decker complexes presenting one H_2Por macrocycle attached through an ester linkage at the *para*, *meta*, or *ortho* position of one *meso*-phenyl group of the Por ligand have been prepared [16]. Photophysical studies on these systems showed that the fluorescence of the H_2Por moiety attached at the *meta* and *ortho* position of one *meso*-phenyl group of the Por ligand is effectively quenched by the double-decker unit, whereas in the case of the *para*-isomer the fluorescence is only partially quenched. Density functional theory (DFT) calculations showed that the relative orientation between the H_2Por moiety and the double-decker unit is crossed for the *para*-isomer, while those of the *meta*- and *ortho*-isomers is open- and closed-shellfish-like, respectively, a finding that would support the different intramolecular fluorescent quenching efficiency observed for the three systems.

Following a related covalent, Pc–Por heteroatom-bridged structure, Cavaleiro, Torres, Guldi and coworkers have synthesized two types of NH-linked Pc–Por dyads **14a,b** (Scheme 4.5) and **15** (Scheme 4.6), in which the Por core is connected to the Pc either through the β-pyrrolic position or the meso phenyl

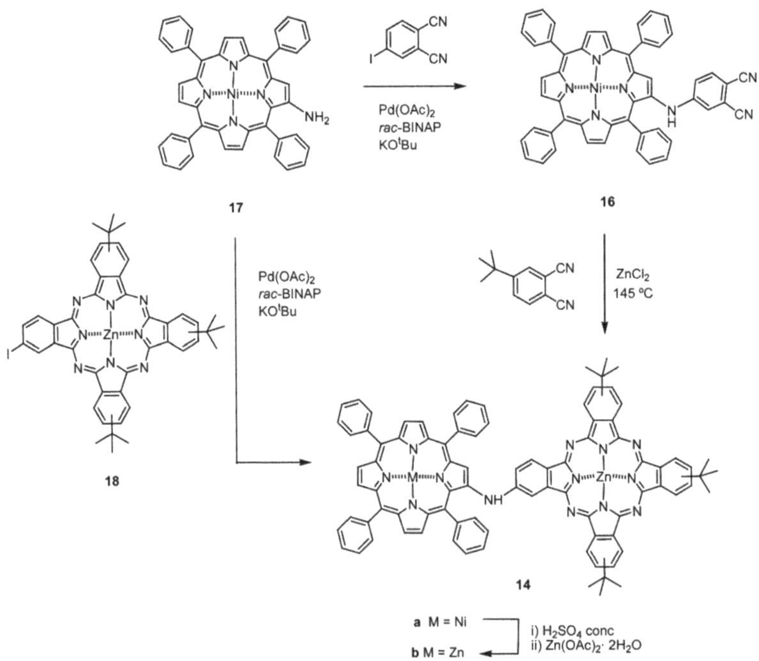

Scheme 4.5. Synthesis of *N*-linked Pc–Por dyads **14a,b**.

group, respectively [17]. These dyads have been prepared following two different procedures. In a former approach, 4-porphyrinylaminophthalonitrile **16** is a key intermediate for the synthesis of dyads **14a,b** (Scheme 4.5). This intermediate was

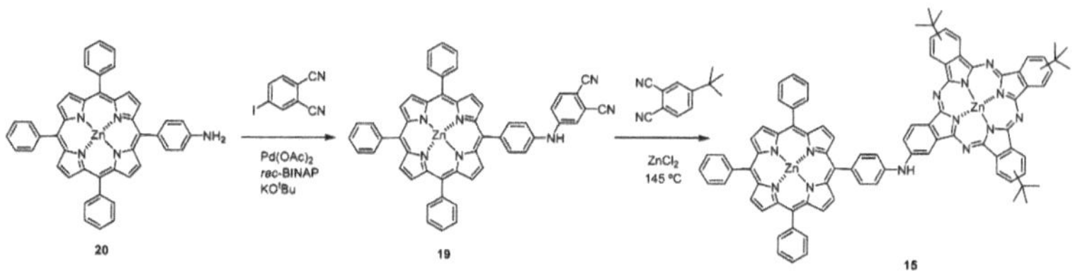

Scheme 4.6. Synthesis of *N*-linked Pc–Por dyad **15**.

obtained by a Buchwald-Hartwig amination reaction of Ni(II)Por **17** [18] and 4-iodophthalonitrile. Dyad **14a** was synthesized in 33% yield by statistical condensation of **16** with an excess of 4-*tert*-butylphthalonitrile. On the other hand, the same dyad was prepared in 70% yield using the above-mentioned Buchwald-Hartwig reaction conditions from the aminoPor **17** and iodoPc **18** [19]. Finally, in order to obtain Zn(II)Pc-Zn(II)Por dyad **14b**, the Por moiety of Zn(II)Pc-Ni(II)Por derivative **14a** was demetalated to the corresponding free base, and then subjected to metalation using $Zn(OAc)_2$.

Similar synthetic strategies were applied to the synthesis of dyad **15** (Scheme 4.6). In this case, porphyrinylaminophthalonitrile precursor **19** was prepared by a palladium-catalyzed coupling reaction of aminoporphyrin **20** with 4-iodophthalonitrile, and further subjected to cross condensation with an excess of 4-*tert*-butylphthalonitrile to give dyad **15** in 17% yield. The latter could also be prepared in 41% yield by coupling the Zn(II) aminoPor derivative **20** with iodophthalocyanine **18**. It is appropriate here to mention that aminoPor **20** has been used by Kobayashi to prepare Por_4-Pc pentads [20]. The condensation reaction of the tetra-anhydride of 2,3,9,10,16,17,23,24-octacarboxyPc with **20** gave a tetraPor-substituted Pc in 73% yield.

Ground state absorption spectra of **14b** and **15** show the characteristic absorptions of the individual constituents, namely strong absorptions assigned to the Por and Pc Soret- and Q-bands. A quite remarkable red-shift (i.e., *ca.* 23 nm in the case of **14b**) of the Zn(II)Pc based absorptions compared with those of a Zn(II)Pc model component suggests notable electronic interactions between the macrocycles. Likewise, a hypsochromic shift of the Zn(II)Por bands is also observed.

Results from steady-state and time-resolved fluorescence experiments with **14b** and **15** imply efficient Por-to-Pc energy transfer phenomena regardless of the linkage between the two macrocyclic units either through the β-pyrrolic (i.e., **14b**) or the *meso* (i.e., **15**) position of the Por core [21]. However, higher energy transfer efficiencies were noted for **14b**; while Por-centered S_2 deactivation efficiencies are around 25% for both **14b** and **15** (i.e., fluorescence upconversion measurements), the S_1 deactivation efficiencies (i.e., transient absorption measurements) of **14b** reach values of around 90%, which are nearly twice as high than in **15**. However, in terms of overall Zn(II)Pc fluorescence, **15** is more efficient to sensitize than **14b**, an observation that was rationalized in terms of a more pronounced red-shift of the Zn(II)Pc Q band in **14b**.

The direct linkage between the two chromophores in a Pc-Por heteroarray has also been explored by several authors; for instance Zn(II)Pc-Zn(II)Por dyads **21–23** in which the Pc moiety is directly linked to a β-pyrrolic position of the Por moiety have been synthesized (Scheme 4.7) [22]. The precursor in the

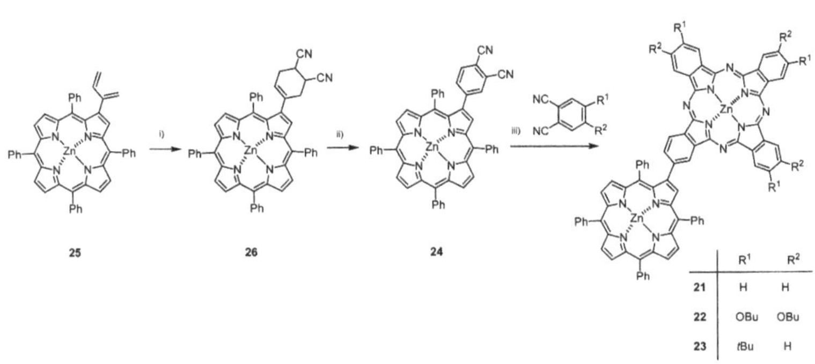

Scheme 4.7. Synthesis of Zn(II)Pc-Zn(II)Por dyads **21–23**. Reagents: (i) fumaronitrile, toluene, reflux, overnight; (ii) DDQ, toluene, reflux; and (iii) $ZnCl_2$, DMAE/*o*-dichlorobenzene, 145°C, Ar, 24 h.

synthesis of these dyads is the Por-phthalonitrile **24**, which was prepared in 75% yield by Diels-Alder reaction of diene **25** [23] with fumaronitrile, followed by the oxidation of the resulting adduct **26** with DDQ. The statistical condensation reaction of **24** with an excess of unsubstituted phthalonitrile, 4,5-dibutoxyphthalonitrile, or 4-*tert*-butylphthalonitrile, in the presence of $ZnCl_2$, yielded Zn(II)Pc-Zn(II)Por dyads **21–23**, respectively. As in other reported Pc-Por systems, the electronic absorption spectra of dyads **21–23** are very similar to those of an equimolar mixture of the corresponding subunits.

The most remarkable structural aspect of dyads **21–23** is that the free rotation through the direct linker is hindered and both chromophores are forced to locate at a very short distance. The short Pc–Por separation in dyads **21–23** leads to ultrafast energy transfer (*ca.* 10^{12} s^{-1}), as determined by femtosecond transient absorption measurements, from the highly energetic singlet state of the Zn(II)Por unit to the singlet state of the Zn(II)Pc, which evolves to populate the long-lived triplet (i.e., ^{3}Zn(II)Pc*) by intersystem crossing. Interestingly, the energy transfer seems to occur more efficiently in dyads with smaller energy gaps between the highest occupied molecular orbital (HOMO) and the lowest unoccupied molecular orbital (LUMO), namely **22** and **23**.

Van Lier and coworkers have used Pd-mediated (i.e., $Pd_2(dba)_3$ (dba = dibenzylidenacetone) and S-Phos ligand as catalytic complex) Suzuki cross-coupling methodologies to prepare Pc–Por dyads **27–29** (Scheme 4.8). In such dyads the two chromophores are directly linked, without and spacer, through the β-pyrrolic- (i.e., **27**), *meso*- (i.e., **28**), and *para*-position of the *meso*-phenyl group of TPP (i.e., **29**), through the coupling of a Zn(II)Pc-boronate synthon **30** with halogenated Por **31**, **32**, and **33**, respectively, with yields ranging from 30 to 65% [24]. Following the same methodology, related Zn(II)Pc-Zn(II)Por-Zn(II)Pc triads linked through the *meso*-position or the *para*-position of the *meso*-phenyl group of the central TPP have been prepared. The absorption features of dyads **27–29**, and related Pc_2–Por triads are very similar to those of the corresponding Zn(II)Pc and Zn(II)Por monomers.

Another example of directly linked Pc–Por heterodyads (H_2Pc-Zn(II)Por **34** and Zn(II)Pc-Zn(II)Por **35**) was reported by Kobuke and coworkers [25]. In these examples, the authors link an imidazolyl group at one of the Por *meso*-positions, while the opposite *meso* carbon is directly linked to the β-position of a Pc macrocycle. Since

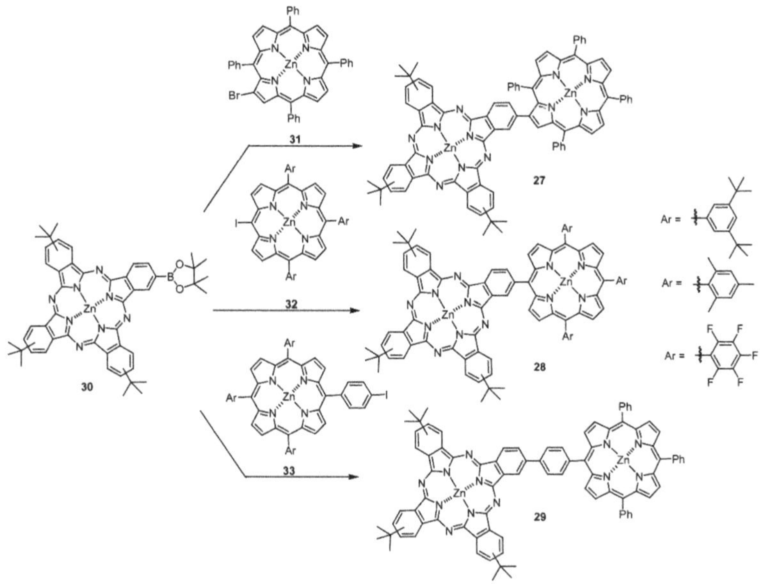

Scheme 4.8. Synthesis of Pc–Por dyads **27–29**. Reaction conditions: $Pd_2(dba)_3$, S-Phos, K_3PO_4, toluene, 90°C.

the two chromophores are in orthogonal geometry, intermolecular $\pi - \pi$ stacking of Por and Pc rings is avoided. However, the imidazolyl substituent affords the formation of stable, slipped cofacial coordination dimers of Pc–Por dyads in non-coordinating solvents, that is $(H_2Pc\text{-}Zn(II)Por)_2$ **36** and $(ZnPc\text{-}Zn(II)Por)_2$ **37**, through the coordination of the imidazol of **34** or **35** to the Zn(II) center of a complementary Por unit (Scheme 4.9), with extremely large

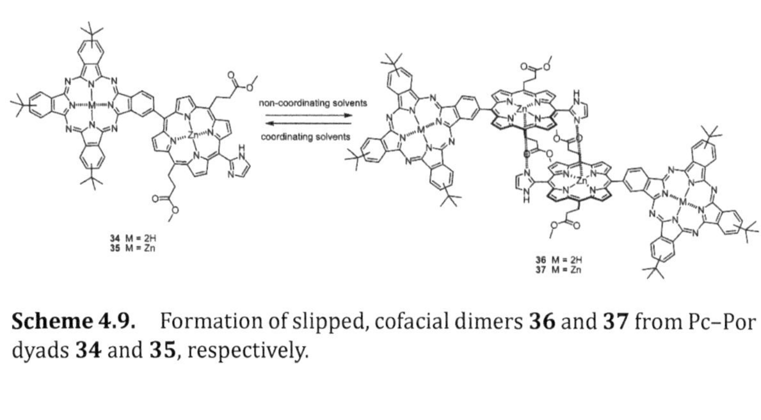

Scheme 4.9. Formation of slipped, cofacial dimers **36** and **37** from Pc–Por dyads **34** and **35**, respectively.

stability constants (10^{11} to 10^{12} M^{-1}). Self-assembled dimerization was observed in the absorption spectra as two split Soret bands of the Zn(II)Por moiety which cover the wavelengths from 400 to 460 nm. Q-bands of Pc moieties in **36** and **37** broaden significantly with respect to a mononuclear Zn(II)Pc model compound, and their λ_{max} are shifted to longer wavelengths by 7–10 nm. The formation of dimers **36** and **37** was also confirmed by ^{1}H NMR.

Heterodyad **35** and its slipped cofacial dimer **37** were investigated by means of femtosecond and picosecond transient absorption spectroscopies [26]. Worth of mention is the fact that, previously to this work, photoinduced electron transfer processes in Pc-Por heterodyads had been scarcely detected and, indeed, no direct proofs of the formation of charge separated species had been presented. As a general rule, this fact had been rationalized in terms of the very small energy gap between the S_1 state of the Zn(II)Pc component and the charge-separated state of the Zn(II)Pc-Zn(II)Por heterodyads. However, in the case of dimer **37**, a rapid energy transfer (i.e., in the subpicosecond timescale) from the singlet Zn(II)Por to the singlet Zn(II)Pc was followed by photoinduced charge separation and charge recombination with time constants of 47 and 510 ps, respectively. Typical absorption around 850 nm ascribed to the radical cation of Zn(II)Pc was detected in transient absorption experiments. On the other hand, in the monomer dyad **35** (existing in CH_2Cl_2 solution when 0.1 M 1-methylimidazole is added to avoid self-association), no clear charge separated state was observed. These results indicate that the dimeric slipped-cofacial arrangement allows for the effective production of the charge separated state as a consequence of a decrease in the reorganization energy of the dimer relative to that of the monomer system. The strong interaction between Zn(II)Pc and Zn(II)Por moieties (as demonstrated by electrochemical measurements of the former dyad **35**) also contributes to the fast charge separation process.

A related, straightforward approach to obtain cofacial stacks of Pc–Por dyads through complementary coordination consist in linking a Zn(II)Pc and a terminal imidazolyl moiety using metal-free, *meso*-substituted ethynylPor as a coplanar spacer [27]. Complementary coordination of the imidazolyl group to the central Zn(II) of a Pc subunit provides heterogeneous slipped-cofacial stacks of

H_2Por and Zn(II)Pc planes. In this case, the association constants for the coordination tetrads reach 10^{14} M^{-1}. The stacked tetrads show strong exciton coupling and ground-state charge-transfer properties between the H_2Por and Zn(II)Pc units, which can be modulated trough the electronic character of the peripheral substituents of the Zn(II)Pc subunit and by the solvent. Strong two- and three-photon absorptions of the supramolecular Zn(II)Pc-H_2Por tetrads have also been studied [28].

The previous example is not the pioneer report on the use of the ethynyl group as spacer in Pc–Por dyads. In fact, the first synthesis of Pc–Por heterodimers with a direct ethynyl linkage was reported in 2001 by Boyle [29]. Coplanar Zn(II)Pc-H_2Por and Zn(II)Pc-Zn(II)Por dyads were prepared by Shonogashira coupling of a Por functionalized with an ethynyl moiety in the *meso* position and Zn(II) tri-*tert*-butyliodoPc **18**. Lindsey and coworkers had also prepared a series of Pc–Por dyads **38a–c** linked through an ethynyl spacer which connects the Pc unit with the phenyl of a TPP unit (Scheme 4.10) [30]. To perform this synthesis, Mg(II) and Zn(II) ethynylPcs **39a,b** were prepared by

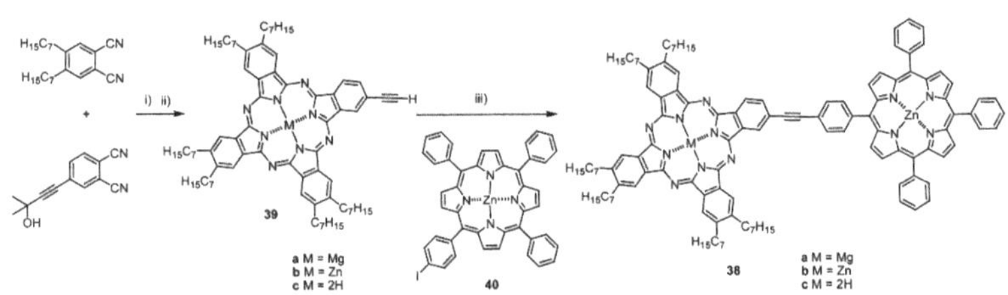

Scheme 4.10. Synthesis of Pc–Por dyads **38a–c**. (i) $MgCl_2$ or $Zn(OAc)_2{\cdot}2H_2O$, DBU, penthanol, 140°C, Ar; (ii) NaOH, toluene, 110°C; (iii) $Pd(OAc)_2$, tri-*o*-tolylphosphine, toluene/NEt_3 (5:1), 60°C, Ar.

statistical condensation between 4,5-diheptylphthalonitrile and 4-(3-hydroxy-3-methylbut-1-ynyl)phthalonitrile in the presence of 1,8-diazabicyclo[5.4.0]undec-7-ene (DBU) and the corresponding metal salt, and subsequent treatment of these compounds with NaOH in toluene to remove the carbinol protecting group. Additionally, metal-free ethynyPc **39c** can be obtained by treatment of the Mg(II) derivative **38a** with trifluoroacetic acid. Shonogashira coupling of **39a–c** with iodo-tetraphenylPor **40** using $Pd(OAc)_2$ and tri-*o*-tolylphosphine as catalyst afforded Pc–Por dyads **38a–c**. Additionally, the tetrapyrrol centers of the resulting dyads can be changed (to 2H, Mg(II), or Zn(II)) by selective metallation and demetallation reactions.

Transient absorption and steady-state emission experiments in dyads **38a–c** indicate that the energy transfer from the Por to the Pc moiety is very fast ($\leq$10 ps) and efficient ($\geq$90%), whereas photoinduced electron transfer involving the excited Pc and the Por unit is unfavorable. Hence, strong emission properties of the Pc ring (Φ_f = 0.37–0.75) are retained in these dyads.

Keeping the linear arrangement between chromophores of previous examples, Becker, Hammarstrom, Odobel and coworkers have reported on the synthesis and photophysical characterization of Zn(II)Pc-Sn(IV)Cl_2Por dyad **41**, in which the two macrocycles are connected trough phenylethynyl oligomers (Scheme 4.11) [31]. Iodo-substituted Zn(II)Pc **18** was connected to the monoprotected,

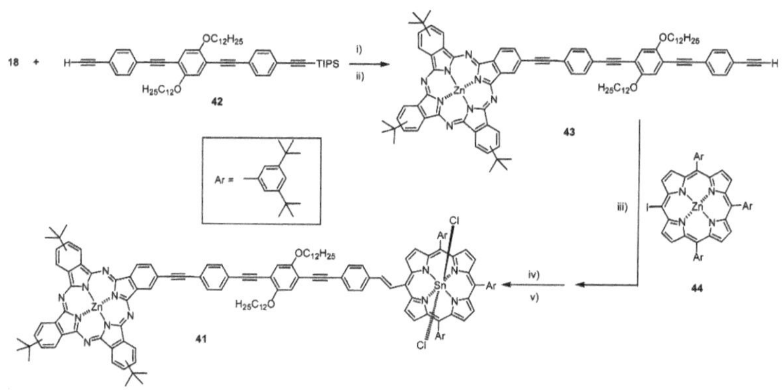

Scheme 4.11. Synthetic route to Pc-Por dyad **41**. Reagents and conditions: (i) piperidine, $Pd_2(dba)_3$-$CHCl_3$, $AsPh_3$, 40°C, 15 h; (ii) Bu_4NF, THF; (iii) THF, NEt_3, $Pd_2(dba)_3$-$CHCl_3$, $AsPh_3$, 60°C, 15 h; (iv) HCl, CH_2Cl_2; (v) pyridine, $SnCl_2$-$2H_2O$, 185°C, 3 h.

oligomeric bridge **42** using Heck ethynylation conditions. The triisopropyl protecting group was removed from the triple bond by treatment with tetrabutylammonium fluoride to give **43** and, then, iodo-substituted Zn(II)Por **44** was appended by a second Heck cross-coupling reaction. The resulting compound was selectively demetallated in the Por ring with hydrochloric acid and subsequently metallated by tin(II) chloride in refluxing pyridine. Surprisingly, during the demetallation/metallation process, the triple bond directly connected to the Por unit was reduced to a double bond to give dyad **41**.

The presence of the oligomeric phenylethynyl bridge reduces the symmetry of the Sn(IV)Cl_2Por and Zn(II)Pc units and leads to a conjugated π-system involving the bridge, thus resulting in red-shifted and broadened absorption bands as compared to reference compounds (i.e., tetra-*tert*-butyl Zn(II)Pc and tetramesityl Sn(IV)Cl_2Por), as well as in a split of the Zn(II)Pc-centered Q-bands. The steady-state fluorescence spectrum of dyad **41** in benzonitrile showed a strong quenching of both ^{1}Zn(II)Pc* and ^{1}Sn(IV)Por* emission relative to the corresponding reference compounds. Femtosecond transient absorption spectroscopy experiments indicate that, upon excitation at 680 nm, ^{1}Zn(II)Pc* is formed, which rapidly decays to the charge separated state. This can be clearly seen from the Sn(IV)Por ground state Soret bleach that grows in with a 13 ps time constant, while most of the Zn(II)Pc ground state bleach remains. At the same timescale there is a broad rise around 730 nm, which is ascribed to the absorption of Sn(IV)$Por^{\bullet -}$ and Zn(II)$Pc^{\bullet +}$ species, considering the simulated spectra constructed from the spectroelectrochemistry of oxidized Zn(II)Pc-oligomer **43** and reduced Sn(IV)Cl_2Por-oligomer reference compounds.

Ethynyl-type bridges are also present in Pc-Por_4 pentads synthesized by Lindsey (**45a–b**), in which four MPor moieties (M = Mg(II), Zn(II)) are linked to a central MPc (M = Mg(II), Zn(II)) in a star-shaped architecture [32]. The synthesis involves a straightforward cyclotetramerization of the Por-containing phthalonitrile **46** in the presence of the corresponding metallic salt (Scheme 4.12). Arrays with different metal centers (M = 2H, Zn(II), Mg(II)) are obtained by selective demetallation and metallation processes of the Por unit starting from **45a**. Following the same approach, these authors have also synthesized related star-shaped, light-harvesting arrays comprised of eight peripheral Pors and one Pc core by a streamlined synthesis with a high degree of convergence [33]. The synthesis

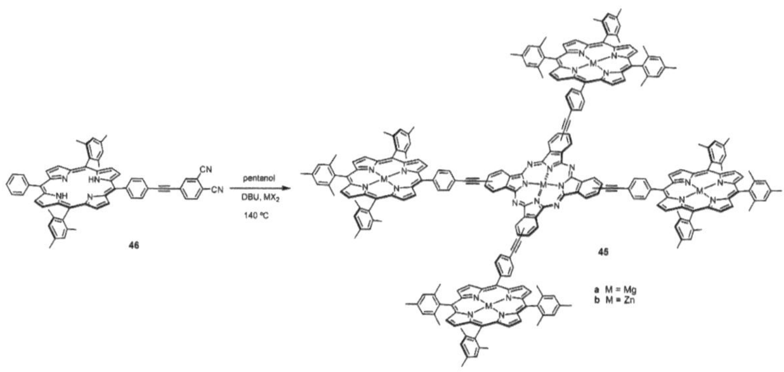

Scheme 4.12. Synthesis of Pc-Por$_4$ pentamers **45a,b**.

involves the formation of a Por dimer linked to a phthalonitrile moiety which was further cyclotetramerized to form the desired nonameric Pc-Por$_8$ array. By using a similar cyclotetramerization method, Nyokong and coworkers have prepared all-metal-free Pc-Por$_4$ heteropentamers consisting of four Por moieties connected through ethynyl bridges to the non-peripheral, α-positions of a central Pc core [34].

Compounds **45a,b** show intense absorptions in the blue and red region of the solar spectrum. Steady-state and time-resolved optical measurements on these compounds indicate that intramolecular singlet-excited-state energy transfer from the Por to the Pc moiety is extremely rapid (i.e., in the picosecond range) and efficient. For the mono-oxidized pentads, ground-state hole/electron hopping is rapid on the EPR time-scale, showing that electronic communication between the components is very efficient in these arrays.

Other bulky spacers have been incorporated in Pc–Por multicomponent systems, such as spirocyclic fused ring systems [35]. Highly rigid pentamers containing four Pc units linked to a central Por core have been synthesized, showing inhibited packing of the macrocycles in the solid state. Therefore, UV-vis absorption spectra of spin-coated films of these materials are almost identical to that obtained from the non-aggregated oligomers in dilute solution.

4.2.2 *Axially Bonded, Phthalocyanine–Porphyrin Heteroarrays*

The majority of the above-mentioned, covalently linked Pc–Por heteroarrays are obtained by multistep, and often cumbersome, organic reaction sequences carried out at the "outside" positions of the Por (either β-pyrrole or *meso*) and the Pc. In contrast, utilization of "inorganic" reactions, which can be readily conducted on the inner metalloid ion (i.e., metal–metal interaction, metal–ligand coordination, covalent-bond formation, etc.) appears to be an attractive and viable alternative for the facile construction of heteromacrocyclic Pc–Por arrays. In this section, we will overview the covalent, axial bonding strategy.

Dihydroxytin(IV) Pc is an oxophilic derivative which easily reacts with molecules holding hydroxyl moieties, giving rise to condensation reactions in the axial position of the Pc. By using dihydroxytin(IV) tetra-*tert*-butylPc **47** as a basal scaffolding unit and OH-containing, metal-free (i.e., **2a**) or Zn(II)TPPs (i.e., **2b**) as axial donor subunits, Pc-Por_2 trimers **48a–b** have been constructed (Scheme 6.13) [36].

A comparison of the UV-vis spectrum of Pc-Por_2 heterotrimers **48a,b** with the spectra of the corresponding Por and Pc precursors (**2a**,**b**) and **47**, respectively, shows that the extinction coefficients of the absorption bands of the heterotrimers are nearly equal to the sum of the extinction coefficients of the corresponding absorption bands of the constituent monomers. The features seen in the UV-vis spectra of these heteroarrays reflect the lack of any exciton coupling interactions between the individual macrocyclic units.

Unlike the ground-state properties, major differences were noticed between the singlet-state activities of heterotrimers **48a,b** and their corresponding individual constituents **2a**(**b**) and **47**. With excitation at 420 nm, that is, the λ_{max} of the Por absorption, quenched emission peaks were observed for both **48a** and **48b** relative to their individual constituents **2a,b** and **47**. This quenching was attributed to intramolecular energy transfer, competing with photoinduced electron transfer, from the singlet state of the axial Por to the central Sn(IV)Pc. On the other hand, a similar emission quenching observed with excitation at 700 nm is due to photoinduced

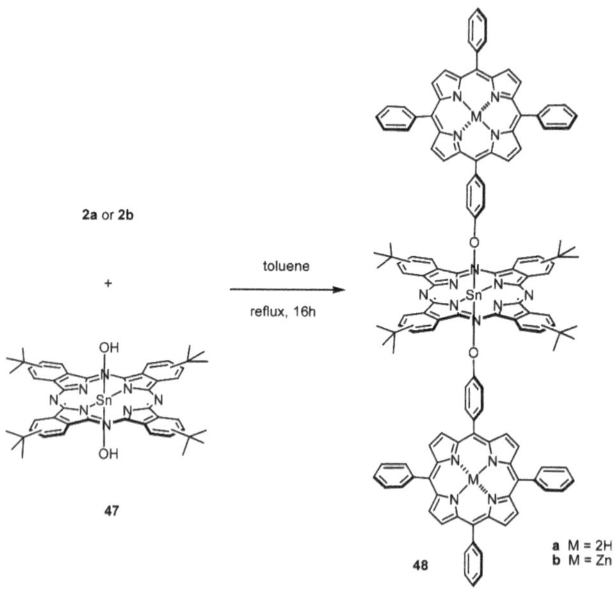

Scheme 4.13. Synthesis of axially linked Pc-Por_2 triads **48a,b**.

electron transfer from the ground state of the axial Por to the excited state of the central Sn(IV)Pc.

Si(IV)Pc-Por_2 triads **49a,b** have been synthesized following a related procedure, that is, the covalent axial bonding of two OH-containing TPP moieties (i.e., **2a,b**) to the central silicon(IV) atom of a Pc core (Figure 4.1) [37]. As in the case of Sn(IV)Pc-Por_2 triads, a competition between energy transfer and photoinduced electron transfer has been established. It has been shown that photoinduced electron transfer occurs between the photoexcited Por moieties (electron donor) and the Si(IV)Pc component (electron acceptor), but the rate of this process (1.4×10^{11} s^{-1}) is six times slower compared to the quenching rate of the Por-centered fluorescence (8.3×10^{11} s^{-1}). Fast depopulation of the first singlet excited state of the Por moieties and concomitant appearance of the Pc-centered fluorescence also takes place *via* energy transfer with a rate of $k = 6.9 \times 10^{11}$ s^{-1}. However, if the first singlet excited state of the Si(IV)Pc component is populated either directly or *via* energy transfer, the main quenching process is the hole transfer

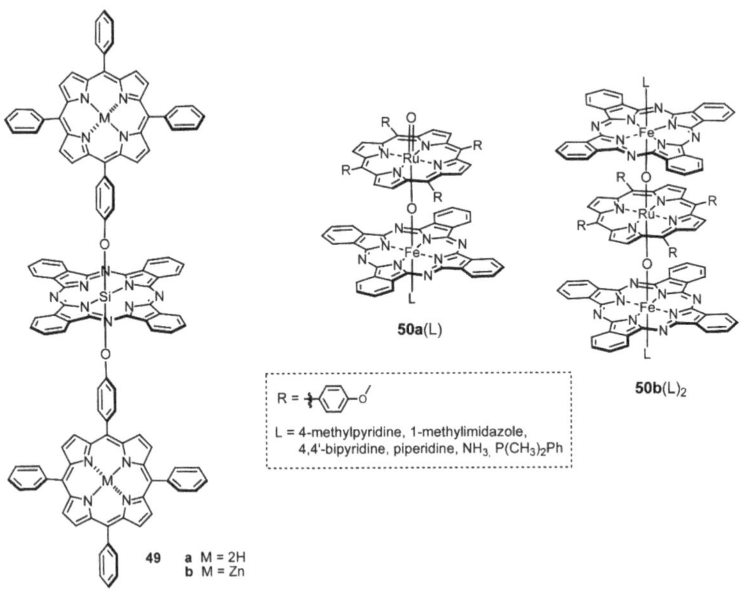

Figure 4.1. Molecular structures of multicomponent, Pc–Por systems **49a,b**, **50a**(L), and **50b**$(L)_2$.

to the charge-separated state. In polar solvents such as DMF, the charge-separated state is the lowest in energy, and charge recombination occurs directly to the ground state [38]. Using transient absorption spectroscopy, the lifetime of the charge-separated state was estimated to be 30 and 20 ps for **49a** and **49b**, respectively. In nonpolar solvents (i.e., toluene), the energy gap between the first singlet excited state of the Si(IV)Pc part and the charge-separated state is very small, and back hole-transfer occurs in both triads, resulting in the appearance of "delayed fluorescence" of the Si(IV)Pc component with a decay time similar to the lifetime of the charge-separated state (190 and 280 ps for **49a** and **49b**, respectively).

A striking approach towards the construction of heterochromophore systems through axial bonding consist in using a single atom link between the metal ions at the centers of the participating ligands, leading to close cofacial arrays in which exciton coupling effects can take place. Therefore, the absorption is expanded within these complex chromophoric systems. Examples of heterodyads linked in this way include those with μ-nitrido linkages e.g. of the type Fe(III)Pc-N-Fe(III)Por, Fe(III)Pc-N- Mn(III)Por, [39] or μ-oxo linkages as in Fe(III)Pc-O-Ru(III)Por [39]. A rare Cr(III)Pc(N_3)-O(H)-Mn(III)Por complex has been recently reported by Ercolani and coworkers [40]. More recently, a μ-oxo linked Si(IV)Pc-O-Ge(IV)Por dyad has been prepared by mixed condensation reactions between octakis(hexyl) dihydroxy Si(IV)Pc and octaethyl dihydroxy Ge(IV)Por [41]. This μ-oxo Pc-Por dyad exhibits absorption between 300 and 800 nm, with enhanced absorption intensity, relative to the monomer components, between 300 and 400 nm, together with a broad band centered at 746 nm. The latter presumably originates from exciton coupling effects that involve, at least, the transitions that originate from the Si(IV)Pc Q-band (centered at 705 nm in the octakis(hexyl) dihydroxy Si(IV)Pc monomer). This μ-oxo Pc-Por dyad exhibits a single fluorescence band ($\lambda_{max} = 784$ nm) upon excitation at various wavelengths (i.e., 345, 395, 460, 590, and 670 nm). Thus, both absorption and emission spectra indicate that the molecule acts as a unique, strongly coupled chromophore.

Diamagnetic hetero bi- (**50a**(L)) and trinuclear (**50b**$(L)_2$) (L = 4-methylpyridine (4-MePy), 1-methylimidazole (1-MeIm), 4,4'-bipyridine, piperidine, NH_3, $P(CH_3)_2Ph$)) oxo-bridged metal complexes based on Pc and Por have also been prepared (Figure 4.1) [42].

Formation of dinuclear complex **50a**(1-MeIm) proceeds through the oxidation of the metal center of a Fe(II)Pc$(1\text{-MeIm})_2$ complex by a Ru(II)PorO_2, in a process that requires the dissociation of the 1-MeIm ligand from the Fe(II)Pc coordination sphere. The binuclear complex **50a**(L) is formally a Fe(IV)-O-Ru(IV) and represents the first example of a Ru(II)Por complex containing both a terminal and a bridging oxo ligand.

Trinuclear complex **50b**$(4\text{-MePy})_2$ instead could be prepared by reacting Ru(II)PorO_2 with Fe(II)Pc$(4\text{-MePy})_2$. The presence of the labile 4-MePy ligand in fact gives rise to the formation of the highly reactive, binuclear species **50a**(4-MePy) which can coordinate to a Fe(II)Pc(4-MePy) which has a vacant coordination site, forming trinuclear oxo-bridged metal complex **50b**$(4\text{-MePy})_2$. Addition of other ligands (i.e., 4,4'-bipyridine, piperidine, NH_3, $P(CH_3)_2Ph$) to a chloroform solutions of trinuclear complex **50b**$(4\text{-MePy})_2$ allowed to generate various ligated derivatives of **50b**. The ligand displacement at the Fe(III) center of **50b**(4-MePy)$_2$ can be followed at room temperature by both ^{1}H NMR and near-IR spectroscopy following the order 4,4'-bipyridine < 4-MePy ~ piperidine < NH_3 < 1-MeIm < $P(CH_3)_2Ph$.

^{1}H NMR studies were also carried out on trinuclear **50b**(L)$_2$ suggesting an eclipsed conformation for the three macrocycles (i.e., one Por and two Pcs) in the complex, whereas UV-vis studies show a typical Por Soret band at 410 nm and a weak bathochromic shift for the Pc Q band at around 700 nm, whose low intensity can be due to the oxo-bridge complex containing face-to-face Pc and Por rings.

The trinuclear complex **50b**(L)$_2$ also undergoes slow reductive cleavage of the oxo-bridge in the presence of hydroquinones (H_2Q), phosphines, or phosphites as revealed by vis-IR and ^{1}H NMR analysis with a rate constant for the reduction (i.e., 10^{-6} s^{-1}) which was found to be independent of the concentration or nature of the reductant used. The proposed mechanism for this cleavage involves the dissociation of a Fe(II)Pc moiety from trinuclear complex **50b**(L)$_2$ as the rate determining step to generate a binuclear species, which is subsequently reduced to the monomeric Ru(II)Por and Fe(II)Pc complexes.

The synthesis and photophysical studies of related peripherally μ-oxo-bridged, non-metallated Por–Pc systems have also been tackled [43].

4.3 Supramolecular Phthalocyanine–Porphyrin Heteroarrays

Supramolecular interactions have widely been used as an efficient and "easy" method to prepare Pc–Por heteroarrays [7, 44].

In this sense, one of the most successful and versatile "supramolecular" approach that order has been used to prepare such heteroarrays is represented by the metal–ligand axial coordination. This method relies on the ability of nitrogenated ligands such as pyridine or imidazole to coordinate axially to the metal center of Pc or Por macrocycles.

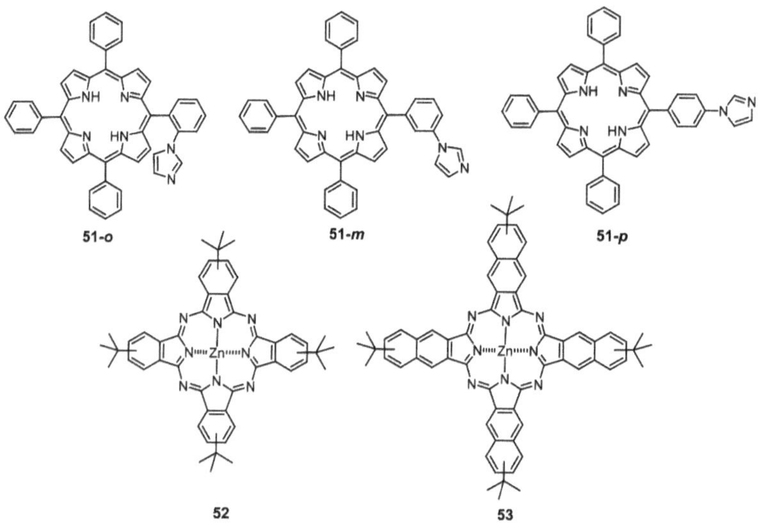

Figure 4.2. Molecular structures of imidazole-substituted, free-base Pors **51-*o***, **-*m***, **-*p***, Zn(II)Pc **52,** and Zn(II)Nc **53**.

Following this synthetic strategy, a series of Pc·Por and naphthalocyanine (Nc)·Por supramolecular ensembles have recently been prepared which are held together *via* the axial coordination of the imidazole moiety present on one of the *meso* positions of a free-base, Por macrocyle (i.e., **51**) to Zn(II)Pc **52** or Zn(II)Nc **53** (Figure 4.2) [45].

The substitution pattern of the imidazole unit at the H_2Por *meso* phenyl ring has been changed from *ortho* (i.e., **51-*o***), *meta* (i.e., **51-*m***) and *para* (i.e., **51-*p***) with the aim of obtaining Pc·Por and Nc·Por dyads in which the two macrocycles of the ensemble would present different relative orientation.

UV-vis studies on H_2Por **51-*o***, Zn(II)Pc **52** and Zn(II)Nc **53** in a non-coordinating solvent such as *o*-dichlorobenzene (*o*-DCB) show, in the case of H_2Por **51-*o***, the characteristic Por Soret band at around 432 nm (Pors **51-*m*** and **51-*p*** presented similar absorption spectra), whereas in the case of Pc **52** and Nc **53** their absorption maxima where located, respectively, in the visible (i.e., 681 nm) or infrared (IR) (i.e., 777 nm) region of the solar spectrum (Figure 4.3).

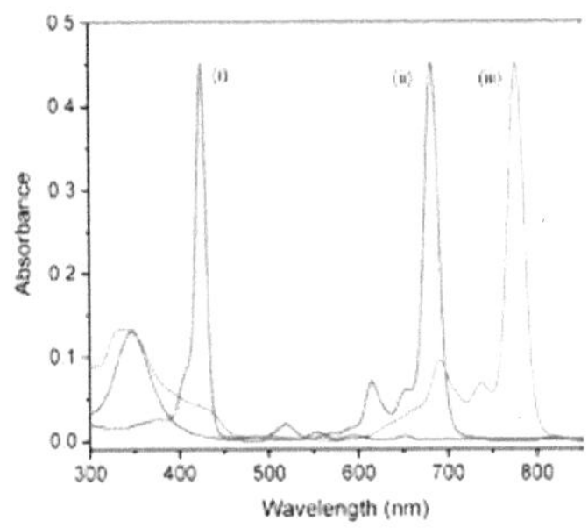

Figure 4.3. UV-vis absorption spectra in *o*-DBC of (i) imidazole-substituted, free-base Por **51-*o***, (ii) Zn(II)Pc **52,** and (iii) Zn(II)Nc **53**. Reproduced from Ref. [45] with permission of the American Chemical Society.

Addition of H_2Por **51-*p*** to a solution of Zn(II)Pc **52** led, beside the normal increase of the H_2Por absorption peaks during the titration, to some changes in the UV-vis absorption spectrum of the Pc, namely a diminishing of the 614 and 681 nm bands with isosbestic points at 609, 663, and 672 nm. A similar phenomenon (i.e., change in some of the Nc absorption peaks and presence of isosbestic points) was observed when titrating Zn(II)Nc **53** with H_2Por **51-*p***. Job plot experiments suggested the existence of a 1:1 complex formation between the H_2Por donor **51-*p*** and the acceptor macrocycles **52** or **53**, with association constant (K_{ass}) of 8.23×10^{-4} M^{-1} and 1.53×10^{-5} M^{-1}, respectively. The formation of a 1:1 complex was also observed when titrating Zn(II)Pc **52** or Zn(II)Nc **53** with H_2Por **51-*o*** or H_2Por **51-*m***, obtaining also in these cases K_{ass} around 10^{-4} to 10^{-5} M^{-1}. The K_{ass} values are higher in the case of the complexes formed between Pc **52** or Nc **53** and H_2Por **51-*p*** with respect to the same complexes formed with H_2Por **51-*o*** or H_2Por **51-*m***, probably due to the minor steric constraints present on the *para*-substituted H_2Por **51-*p***. Moreover, the binding constants for Zn(II)Nc **53** were found to be 2–3 times higher than those obtained for the corresponding Zn(II)Pc **52** binding, probably due to the fact that the Zn(II)Nc macrocycle is electron richer than the Zn(II)Pc one. Electrochemical studies on Pors **51**, Zn(II)Pc **52,** and Zn(II)Nc **53** revealed that these latter compounds were easier to oxidize than Pors **51**, making them, at least in principle, inappropriate candidates for the preparation of efficient D–A systems. Further DFT studies were used to evaluate the HOMO and the LUMO frontier orbitals of these Pc–Por and Nc–Por dyads. These studies showed that the HOMO orbital for all the dyads was fully localized on the Zn(II)Pc or Zn(II)Nc components while the LUMO was located on the Por macrocycle. In conjunction with the electrochemical experiments, these results show that electron transfer from the excited H_2Por to either the coordinated Zn(II)Pc or Zn(II)Nc is unlikely to take place. Finally, steady-state fluorescence studies were carried out on these axially bounded Pc–Por dyads revealing the occurrence, upon photoexcitation of the donor H_2Pors **51** at 518 nm which selectively excites the free base Por, of an energy transfer process from the Por to the acceptor Zn(II)Pc **52** or Zn(II)Nc **53** moieties, the extent of the energy transfer following the same order found for the K_{ass} of Pors **51** with either Zn(II)Pc **52** or Zn(II)Nc **53**, that is, H_2Por ***51-p*** > H_2Por ***51-m*** > H_2Por ***51-m***. Control experiments have demonstrated that the self-assembled nature of the Pc·Por dyads was crucial in order to give rise to this energy transfer process since when a H_2Por lacking the imidazole moiety was titrated with Zn(II)Pc **52** (or Zn(II)Nc **53**), only a slight increase of the Pc (or Nc) emission in the near-IR region was observed, implying the occurrence of little or no energy transfer.

Additional pump-probe and up-conversion spectral studies were carried out with these supramolecular dyad systems by performing transient absorption measurements in order to study the kinetics of the observed Forster's dipole–dipole energy transfer process (a Dexter's exchange mechanism should be excluded since it requires the presence of electronic communication *via* orbital overlap between the donor and acceptor species, a possibility excluded by spectroscopic and the DFT studies). Such time-resolved studies showed, upon photoexcitation, the disappearance of the Por excited-state features and the raising of the features typical of the singlet excited state of Zn(II)Pc in an ultrafast singlet-singlet energy transfer process with time constants on the order of 2–25 ps depending upon the type of dyad and following the order: *ortho* >*para* > *meta* substitution of the free-base Por.

A H_2Por macrocycle **54** substituted at its four *meso* positions with four pyridine groups has also been prepared and used for the construction of a Pc_4·Por supramolecular pentad (Figure 4.4) [46].

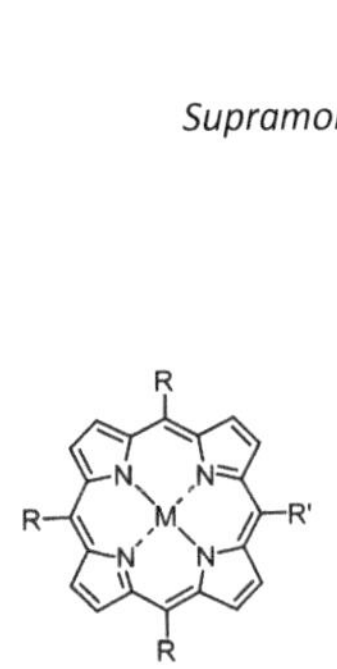

Compound	M	R	R'
54	H_2	C_5H_4N	C_5H_4N
56	H_2	Ph	Ph
57	H_2	Ph	C_5H_4N
58	Zn	C_6H_5Me	C_6H_5Me

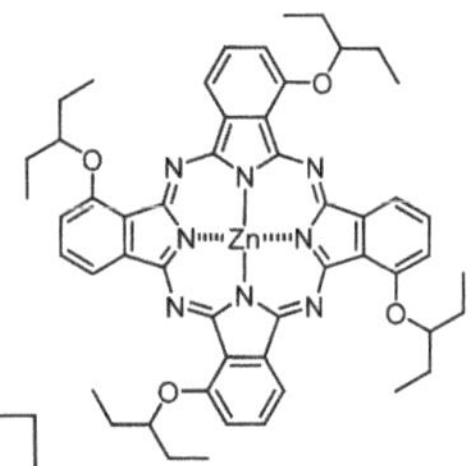

55

Figure 4.4. Molecular structures of Pors **54**, **56–58**, and Zn(II)Pc **55**.

^{1}H NMR experiments demonstrated that the addition of one equivalent of Zn(II)Pc **55** to a $CDCl_3$ solution of H_2Por **54** resulted in a change of some of the protons of the Por and the Pc macrocycles. A detailed analysis of these changes suggested the formation of an edge-to-face Pc·Por dyad in which one Pc unit is connected to one of the four pyridine moieties of Por macrocycle **54** in an arrangement in which Pc **55** is deshielded by the ring current generated by the Por macrocycle. Further addition of Zn(II)Pc **55** (i.e., 4 equivalents) to **54** in $CDCl_3$ resulted in a further downfield shift of some of the Zn(II)Pc protons, whereas the signals assigned to H_2Por **54** are further broadened and shifted upfield, suggesting, probably, the formation of higher aggregates such as a Pc_4·Por pentad **55**$_4$·**54**.

The UV-vis spectrum of Por **54** remained essentially identical upon addition of Pc **55** in $CHCl_3$ (and vice versa), thus, ruling out the presence of any electronic communication between the two hetero macrocycles in the ground state. On the contrary, a dramatic change in the fluorescence of Pc **55** (Por **54**) was observed upon addition of Por **54** (Pc **55**). In order to prove that the observed fluorescence quenching was actually due to the axial coordination of the pyridine unit of **54** to the Pc macrocycle, a control experiment was carried out which included mixing Pc **55** with a H_2Por lacking the pyridine units (i.e., H_2Por **56**). Steady-state fluorescence experiments revealed no observable changes in the fluorescence intensity of **56** upon addition of Zn(II)Pc **55**, thus providing strong evidence for the axial coordination of the pyridyl groups to the zinc(II) metal centre in ensemble $\mathbf{55}_4$·**56**.

^{1}H NMR and steady state fluorescence experiments were also carried out on mono pyridine-substituted H_2Por **57** and Zn(II)Pc **55**, which revealed the formation of a heterosupramolecular Pc–Por dyad. Analysis of the fluorescence data obtained by titrating Por **57** with Pc **55** allowed to calculate the K_{ass} for this edge-to-face Pc·Por dyad, which resulted to be 4.2×10^4 M^{-1} in $CHCl_3$.

In the attempt to prepare a face-to-face Pc·Por heterodimer, a mixture of Zn(II)Pc **55** and Zn(II)Por **58** and 4,4'-bipyridine (bipy) was studied. The ^{1}H NMR spectrum of this mixture recorded in $CDCl_3$ proved to be extremely difficult to interpret, probably due to the formation, beside the desired **58**·bipy·**55** systems, of the **58**·bipy·**58** and **55**·bipy·**55** ensembles. However, the fluorescence intensity of Zn(II)Por **58**, when excited at 515 nm in the presence of bipy in $CHCl_3$, gradually decreases during the titration with Zn(II)Pc **55**, whereas, in the absence of bipy, its intensity remained relatively unchanged, thus suggesting that the dinitrogenated ligand bipy plays a crucial role in joining **55** and **58** through coordination bonds.

Ru(II)Pcs have also been employed in the preparation of Pc–Por supramolecular heteroarrays such as **59** and **60** (Figure 4.5) [47].

Ru(II)(CO)Pc, one of the molecular component of Ru(II)Pc_2·H_2 Por heteroarray **59**, was prepared by metallation (i.e., $Ru_3(CO)_{12}$) of a free-base Pc bearing eight *n*-octyl groups at its non-peripheral

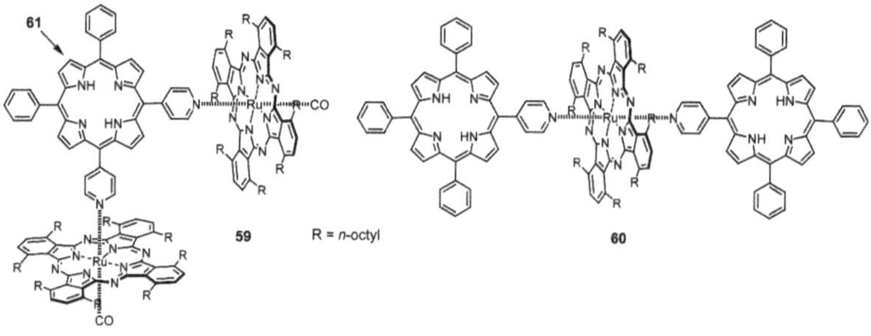

Figure 4.5. Molecular structures of Ru(II)Pc_2·H_2Por **59** and Ru(II)Pc·H_2 Por_2 **60**.

positions in refluxing benzonitrile. It is important to notice that such metallation reaction gives rise to the formation of two Ru(II)Pcs, one bearing a single, carbon monoxide ligand in the axial position coordinated to the Pc metal center, and the other having two labile benzonitrile ligands axially coordinated. Careful choice of the experimental conditions permitted the exclusive formation of one or the other Pc. In fact, when prolonged heating of the reaction in benzonitrile was carried out, the almost exclusive formation of Ru(II)Pc(benzonitrile)$_2$ was observed and the product was isolated in 96% yield, whereas treatment of this latter Ru(II)Pc with carbon monoxide led to the formation of Ru(II)(CO)Pc in 71% yield. Ru(II)Pc$_2$·H$_2$Por heteroarray **59** was finally obtained in 65% yield by reacting Ru(II)(CO)Pc with dipyridyl H$_2$Por **61**. In a similar fashion, Ru(II)Pc·H$_2$Por$_2$heteroarray **60** could be prepared in 70% yield by treatment of Ru(II)Pc(benzonitrile)$_2$ with two equivalents of monopyridyl H$_2$Por **57**.

Similar to the case of **55**$_4$·**54**, the formation of a Pc$_4$·Por pentad using Ru(II)Pc has also been achieved by mixing tetrapyridyl H$_2$Por **54** with Ru(II)(CO)Pc [48]. ^{1}H NMR experiments could provided some evidence of the axial coordination of the pyridine group(s) to the Ru(II)(CO)Pc in Ru(II)Pc$_4$·**54** such as a upfield shift of the characteristic singlet from the β-pyrrolic protons on the Por units, which in the case of this ensemble shifted more than 3 ppm upon complex formation.

More recently, a five-component, Pc·Por supramolecular heteroarray **62** has been reported, which is held together thanks to a combination of axial coordination and electrostatic interactions (Figure 4.6a) [49]. Such discrete, supramolecular assembly consists of a central diprotonated Por **63**$^{2+}$ and two Zn(II)Pc **64** unit connected by two 4-pyridinecarboxylate (4-PyCOO$^-$) "bridges".

Crucial for the preparation of supramolecular ensemble **62** is the non planar nature of the Pc and Por macrocycles. The saddle distortion of the dodecaphenyl Por facilitates the protonation of the pyrrole nitrogen atoms, thus allowing us to obtain the stable diprotonated Por **63**$^{2+}$ which can (i) act as an electron acceptor and (ii) interact *via* electrostatic interactions with groups bearing negative charges. On the other hand, the saddle-distorted octaphenyl Zn(II)Pc **64** (i) exhibits a lower oxidation potential relative to the Por **63** and

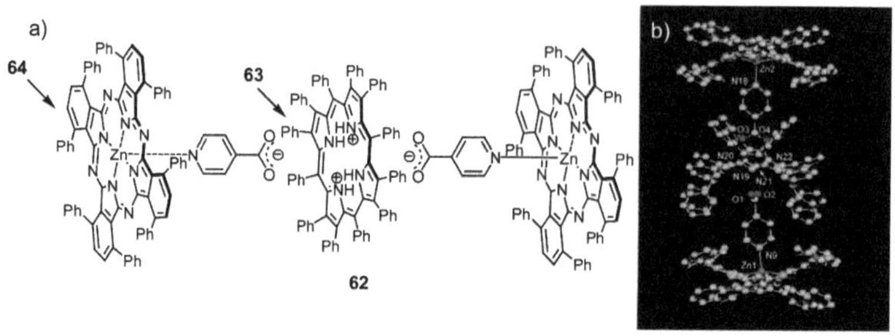

Figure 4.6. (a) Molecular structure and (b) crystal structure of the five-component, supramolecular triad **62** constituted of two Pc **64**, one Por **63**$^{2+}$, and two "bridging" 4-pyridinecarboxylate (4-PyCOO$^-$) units. Reproduced from Ref. [49] with permission of Wiley-VCH.

(ii) presents a metal center (i.e., a Zn(II) atom) able to coordinate axially to some nitrogenated ligand.

X-ray diffraction of single crystals of **62** grown by vapor diffusion of hexanes into a toluene mixture solution of [**63**$^{2+}$(4-PyCOO$^-$)$_2$] and Zn(II)Pc **64** allowed unambiguously to establish the structure of this supramolecular ensemble (Figure 4.6b). This structure contains one **63**$^{2+}$ macrocycle and two Zn(II)Pc **64** units connected by two 4-PyCOO$^-$ ligand "bridges". Each of the two charged 4-PyCOO$^-$ units coordinates to the Zn(II) metal center of **64** through the pyridine nitrogen atom, whereas its carboxylate group forms two-point, intermolecular hydrogen bonding with two pyrrole protons of the diprotonated, central macrocycle **63**$^{2+}$. Interestingly, the crystal packing of **62** exhibits a one-dimensional (1D) nanowire structure in which one of the crystallographic axis is due to the formation of intermolecular π–π interactions of the peripheral phenyl groups of **64**. Evidence of the formation of supramolecular complex **62** in solution was gathered from ^{1}H diffusion ordered spectroscopy (^{1}H DOSY) experiments in $CDCl_3$.

Femtosecond laser flash photolysis of **62** in benzonitrile was performed by photoexcitation at 410 nm, where both Por **63**$^{2+}$ and Pc **64** are excited. The formation of the singlet excited state of Pc **64** and Por **63**$^{2+}$ was observed immediately after the laser pulse (i.e., 4 ps) by the rise of the absorption maxima at 547 and 900 nm. At 150 ps, new absorption bands were observed at 547 and 1030 nm assigned

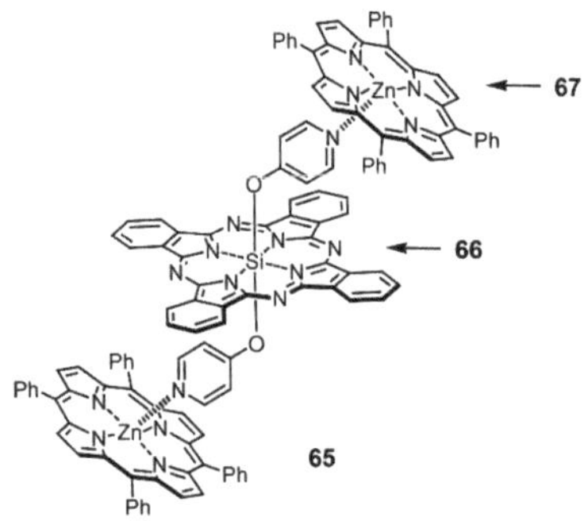

Figure 4.7. Molecular structure of supramolecular triad Pc·Por_2 **65** constituted by one Si(IV)Pc macrocycle **66** and two Pors **67**.

to the Por $\mathbf{63}^{+\bullet}$ and Pc $\mathbf{64}^{+\bullet}$ species, respectively. These bands suggest the occurrence in **62** of an intrasupramolecular electron transfer process from the Pc **64** moiety to Por $\mathbf{63}^{2+}$ to form a charge transfer state. Prior to this electron transfer, the fingerprint of an energy transfer event from photoexcited $^1(\mathbf{63}^{2+})^*$ to Pc **64** to give $^1\mathbf{64}^*$ was also observed. Finally, after 150 ps, the absorption at 1030 nm decays with a first-order rate constant to give an intrasupramolecular back electron transfer from Por $\mathbf{63}^{+\bullet}$ to Pc $\mathbf{64}^{+\bullet}$ to recover the ground state of **62**. The lifetime of the charge separated state of **62** was determined to be 667 ps.

Metal-ligand coordination between nitrogenated ligands and metallated Por units has also been used for the construction of Pc·Por heteroarrays such as **65** (Figure 4.7) [50].

Si(IV)Pc **66**, the central component of complex **65**, was prepared in 75% yield by ligand substitution of the commercially available Si(IV)Cl_2Pc with 4-hydroxypyridine in the presence of pyridine (i.e., the Si-O bond in **66** is "generally" considered as a covalent bond and not a coordinative one) followed by complexation with Zn(II)Por **67**.

^{1}H NMR studies on mixtures of **66** and **67** at different ratios showed an upfield shift for the signals for all the Pc and Por ring protons with respect to the unmixed solutions, in particular for the α-protons of **66** and the pyrrolic protons of **67**. Such changes were not observed when, in the same ^{1}H NMR experiment, Por **67** was replaced by its free-base analogous.

The broadening of the pyridyl protons' signals suggests that the complexation between **66** and **67** is rather weak and there is a rapid exchange, in the NMR timescale, between the coordinated and the free pyridyl groups. A Job's plot monitoring the shift of the ^{1}H NMR signal of the pyrrolic protons of **67** upon titration with **66** was done, showing a minimum when the molar fraction of **67** is about 0.65, thus suggesting the formation of the trinuclear supramolecular array **65**. The structure of this supramolecular assembly was unambiguously confirmed by X-ray diffraction analysis. A confirmation of the occurrence of a supramolecular interaction between **66** and **67** also in solution was gathered from steady-state fluorescence studies, which showed that the fluorescence of Por **67** decrease upon addition of Pc **66** in $CHCl_3$. This phenomenon is probably due to the occurrence of an electron transfer mechanism from Por **67** to Pc **66**, a process thermodynamically possible by looking at the redox potentials of the two hetero-macrocycles.

The interaction between pyridine-containing Pcs and Co(II)Pors have also been used for the preparation of Pc·Por heteroarrays, since, in principle, they should give rise to more stable heterodimers due to the stronger cobalt-pyridine binding interactions.

With this aim, the complexation of Si(IV)Pc **66** and tetraphenyl Co(II)Pors **68** was studied (Figure 4.8) [51].

Electronic absorption spectroscopy showed the occurrence of significant changes in the spectrum of Pc **66** upon addition of Co(II)Por **68a** attributable to the axially linked species. Job's plot analysis was used to infer the stoichiometry of the supramolecular

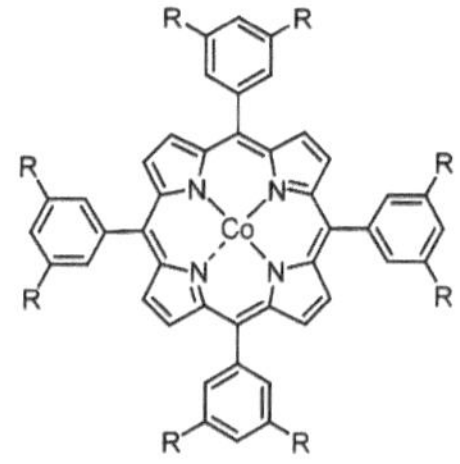

68 **a** R = H
b R = ^{t}Bu

Figure 4.8. Molecular structures of Co(II)Pors **68a,b**.

ensemble which resulted to be 1:2. Analysis of the UV-vis titration experiments allowed to calculate the K_{ass} for the **66**·**68a**$_2$ complex which was 1.2×10^{-4} M^{-1}.

The complexation of Si(IV)Pc **66** with Co(II)Por **68b**, as well as the 3-pyridyl analogue of **66** with Co(II)Pors **68a** and **68b**, was also investigated revealing the formation of Pc·Por supramolecular systems with identical stoichiometry (i.e., Pc·Por$_2$).

In such systems, the presence of the *tert*-butyl groups on the Por macrocycle seemed to not have any influence on the stability of the resulting Pc·Por$_2$complexes. Instead, the K_{ass} seems to be influenced by the position of the nitrogen atom on the pyridine ring (i.e., 4- *vs.* 3-substituted pyridine) of the Pc moiety. It was observed in fact that the complexes assembled using Si(IV)Pc **66** were more stable (i.e., three order of magnitude) with respect to the ones using the 3-pyridyl analogue of **66**. The stronger binding to the Co(II)Pors exerted by **66** with respect to its 3-pyridyl analogue can be attributed (i) to the fact that the 4-pyridinolato ligand is a better σ donor than the 3-pyridinlolato analogue and (ii) to the bigger steric congestion between the spatially close Pc and Por units when the Co(II)Por coordination occurs with the 3-pyridyl analogue of **66**.

A Si(IV)Pc **69** axially substituted with two cyclodextrin (CD) moieties was also prepared and complexed with a tetrasulfonated H$_2$Por **70** in aqueous media (Figure 4.9) [52].

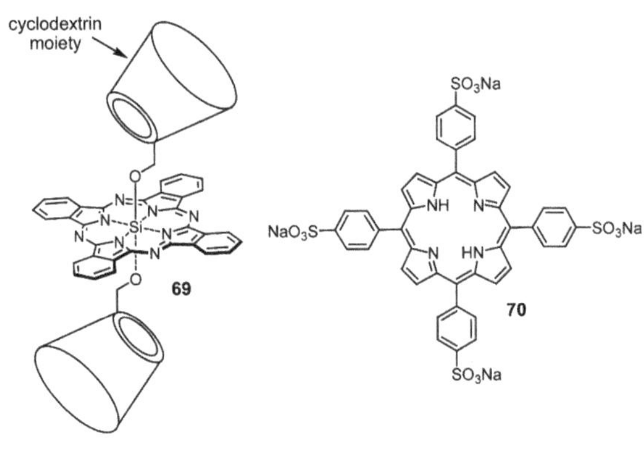

Figure 4.9. Molecular structures of cyclodextrin-substituted Si(IV)Pc **69** and tetrasulfonated H$_2$Por **70**.

Si(IV)Por **69** was obtained in 42% yield by treatment of Si(IV)Cl_2Pc with mono-6-hydroxy permethylated α-cyclodextrin. The complexation of **69** with tetrasulfonated H_2Por **70** was studied by ^{1}H NMR spectroscopy, which showed a splitting of the signals attributable to the Por macrocycle, which were also correlated, as demonstrated by ^{1}H NMR ROESY experiments, with the CD moiety, suggesting that H_2Por **70** is encapsulated in the CD cavity.

Further proofs of the interaction between the tetraanionic Por **70** and the CD moiety of **69** were obtained by UV-vis studies, which also allowed to determine the stoichiometry of the Pc·Por supramolecular complex that was found to be 1:1. This suggests the formation of a Pc·Por polymeric complex in which two CDs of two different Si(IV)Pc **69** are encapsulating two different sulfonate groups of the same H_2Por molecule in a head-to-tail fashion.

Photophysical experiments demonstrated that excitation at 610 nm of a water solution of **69** resulted, upon addition of tetraanionic H_2Por **70**, in a progressive quenching of the Pc fluorescence at 683 nm. This phenomenon is probably due to an electron transfer process taking place within the supramolecular complex, as confirmed by transient absorption experiments, which revealed the presence of a peak at 580 nm which could be attributed to the radical anion species of Pc **69**. The change in the fluorescence intensity of **69** upon addition of **70** was also used to determine the K_{ass} for this complexation, which resulted to be 2.1×10^{-8} M^{-1}. Similarly, excitation of the supramolecular complex **69·70** at 510 nm, where only H_2Por **70** has a weak absorption, resulted in a strong emission at 683 nm, which could be ascribed to the fluorescence of the Pc unit, thus suggesting an efficient singlet-singlet energy transfer process from the excited Por to the Pc component in the complex.

From these data it seems that the electron transfer mechanism is a way to depopulate the first excited-state of **69**, no matter if it is generated by direct excitation of **69** or through an energy transfer from the excited **70**. Finally, the *in vitro* photodynamic activity of this host-guest supramolecular system, as well as the one of its separate components Pc **69** and Por **70**, was also tested in HT29 human colon adenocarcinoma cells. These studies revealed that Pc **69** and complex **69·70** are highly photocytotoxic with a 50% growth-inhibitory ratio (IC_{50}) of 0.15 and 0.09 μM, respectively, although in the case of

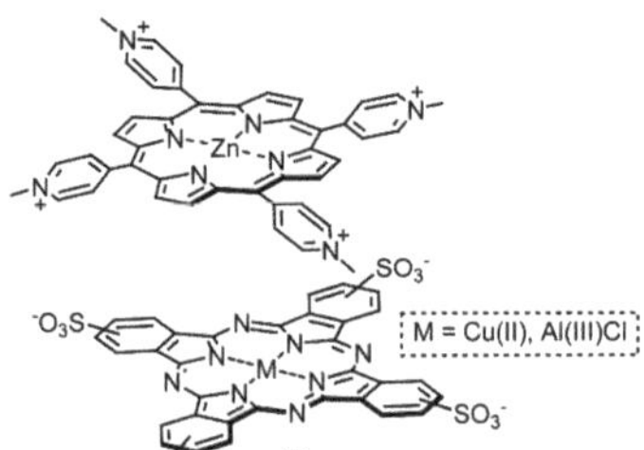

Figure 4.10. Molecular structure of Pc·Por **71**.

the complex the cellular uptake and localization properties resulted to be lower than that for Pc **69**.

Interactions between oppositely charged Pcs and Pors have also been exploited in order to drive the formation of Pc·Por supramolecular structures. In such systems, the charged substituents on the macrocycles provide a strong electrostatic force for the aggregate formation.

In this context, Pc·Por heterodimers **71** constituted by Zn(II)Por and Cu(II) or Al(III)Pcs bearing oppositely charged substituents (i.e., sulfonate and pyridinium groups, respectively) have been prepared (Figure 4.10) and the electrostatic formation of Pc·Por heterodimers in the liquid phase studied [53].

It was observed that the ground-state absorption spectra of such heterodimers (i.e., Cu(II)Pc·Zn(II)Por and Al(III)Pc·Zn(II)Por) were drastically changed with respect to those of the corresponding monomers, indicating the existence of a strong interaction between the two chromophores. Irrespective of the nature of the metal ions and of the peripheral substituents, the heterodimers present very similar ground-state spectra. The photophysical properties of the two heterodimers were investigated with the use of femto- and nano-second absorption spectroscopies, revealing the occurrence, upon photoexctitation of the Al(III)Pc·Zn(II)Por heterodimer at 565 or 620 nm, of an efficient electron transfer process from the Por to the Pc moiety. On the other hand, in the case of the Cu(II)Pc·Zn(II)Por heterodimer a very efficient intersystem conversion was observed, leading to the formation of a "triplet" excimer. The different fate of the photoexcited species in the two Pc·Por heterodimers has been rationalized in terms of the different thermodynamics of the transfer process in the two systems and the peculiar properties of the paramagnetic Cu(II)Pc.

Mixtures of a cationic tetrapyridinium Por and an anionic tetrasulfonated Pc have also been reported, which give rise to the formation, in polar solvents, of well-defined supramolecular assemblies, consisting either of heterodimers or heterotrimers [54]. This study showed that aggregation of these mixed complexes is governed by (i) the ability of the metal ions of the macrocycles to coordinate one or two axial ligands, (ii) the donor or acceptor character of the solvent, and (iii) the relative position of the substituents on the macrocycles.

It has also been demonstrated that molecular oxygen can bind to a Pc·Por heterodimer formed by a tetracationic Co(II)Por and an anionic tetrasulfonated Al(III)Pc [55]. In the ground state and in an oxygen-free solution, this Pc·Por heterodimer gives rise to the formation of a CT complex as inferred by UV-vis studies, which revealed the fingerprints of the Co(I)Por radical anion and Al(III)Pc radical cation species.

This partial CT favors the binding of O_2 to the Por cobalt metal center of the heterodimer, enhancing the CT from the Pc moiety to the oxygen through the Por macrocycle. As a consequence of the oxygen binding in fact a stable zwitterionic species, $O_2^{\bullet-}$Co(II)Por·Al(III)$Pc^{\bullet+}$ was formed as suggested by the shift of the Soret band of the Por moiety from the Co(I) to the Co(II) species while the Pc Q band shifts from 690 to 712 nm, the latter value being characteristic of the cation Al(III)Pc. The zwitterion can revert to the neutral form Al(III)Pc·Co(II)Por upon vigorous bubbling of the solution with an inert gas or under light stimuli. In such Pc·Por heterodimer it was observed that, upon photoexcitation, the oxygen-bound complex photoejects the O_2 molecule forming a long-lived triplet excited state. This latter species can then deactivate to the ground state by transferring its energy to nearby oxygen molecules, thus generating reactive singlet oxygen. The photodynamic properties of this zwitterionic Pc·Por complex were also studied against tumor cells *in vitro* obtaining modest results.

A detailed analysis on the formation and spectroscopic properties of several Pc·Por complexes consisting of negatively charged tetrasulfonatophenyl Por (**72a–c**) or tetracarboxyphenyl Por (**72d**)

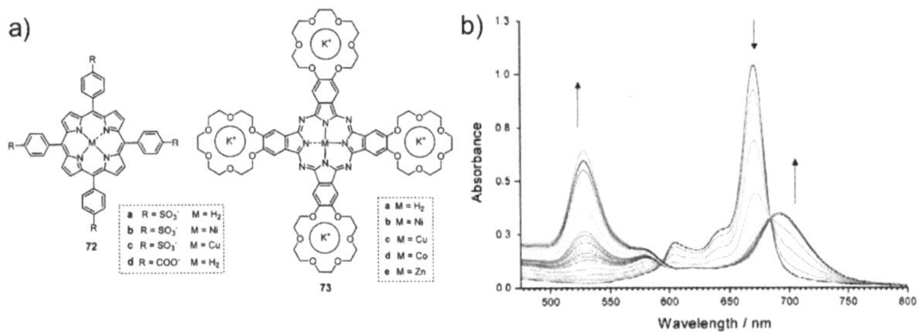

Figure 4.11. (a) Molecular structures of K^+-complexing, crown ether-substituted Pcs **73a–e** and tetraanionic Pors **72a–d**. (b) Evolution of the absorption spectrum of Ni(II)Pc **73b** upon addition of Ni(II)Por **72b** in DMSO. Figure 4.11b reproduced from Ref. [56] with permission of the American Chemical Society.

ion pairs with crown ether substituted Pcs **73** bearing potassium ions into its four crown ether cavities has also been carried out (Scheme 4.11a) [56]. The complexation of the potassium ions by the crown ether residues imparts a tetracationic nature to the Pc, thereby providing the potential for the K^+-doped Pcs to interact *via* electrostatic interactions with tetraanionic Pors.

The results of the titration of Ni(II)Pc **73b** with Cu(II)Por **72c** in DMSO caused in a small blue shift of the Por Soret band, complete disappearance of the crown ether-substituted Pc Q-band and the appearance of a new band in the red spectral region. Similar results were obtained when titrating Cu(II)Pc **73c** with Cu(II)Por **72c**, Ni(II)Pc **73b** with Ni(II)Por **72b** (Figure 4.11b), and Cu(II)Pc **73c** with Ni(II)Por **72b**. Interestingly, in the absence of K^+ ions, the spectral behavior seen in Fig. 4.11b was not observed, and the absorption spectrum of the Pc·Por mixtures was simply the sum of the spectra of the separate components. In contrast to the metal-containing Pc and Por systems described above, titration of the free-base Por **72a** and free-base (**73a**), Co(II) (**73d**) or Zn(II) (**73e**) Pcs, showed different behavior, whether in DMSO, ethanol, methanol, or water solutions. In the titration experiments, the isosbestic points were considerably smeared, and the absorption spectra continued to evolve even after several minutes following vigorous mixing of the components.

The method of continuous variation has also been used in order to determine the stoichiometry of the Pc·Por complexes in DMSO showing unexpected Pc-to-Por ratios of 2:1 for the systems Ni(II)Pc **73b**·Ni(II)Por **72b**, Ni(II)Pc **73b**·Cu(II)Por **72c**, Cu(II)Pc **73c**·Cu(II)Por **72c**, Cu(II)Pc **73c**·Ni(II)Por **72b**, and H_2Pc **73a**·H_2Por **72a**, whereas titration of H_2Por **72a** with Co(II)Pc **73d** in ethanol resulted in a 1:1 stoichiometry.

Spectroelectrochemical experiments were carried out with monomeric Pc molecules in order to examine the spectral properties of their electrochemically reduced and oxidized species. These studies showed a distinct maximum between 700 and 750 nm attributable to the radical cation of the Pc macrocycle. The fact that the absorption bands for both the Pc·Por complexes and the Pc radical cations are in the same spectral region (i.e., near 700 nm) suggests that the complexes have significant CT character, with a significant contribution from the Pc radical cation species. On the other hand, spectroelectrochemical observations on Ni(II)Por **72b** showed that its radical anion spectrum has a broad absorption band with a maximum near 770 nm, reduced in intensity and blue shifted (with respect to the neutral Por) Por Q-band at 525 nm, and an intense and red shifted (with respect to the neutral Por) Por Soret maximum at 412 nm. This information, coupled to the observation that the Pc radical species has also spectral intensity near 570 nm, is not inconsistent with the rationalization of the nature of the red band in the Pc·Por complexes. The observed red shift for the Soret band maximum of the reduced Por with respect to the Soret band of the neutral species is in agreement with the results obtained for H_2Pc **73a**·H_2Por **72a** and Co(II)Pc **73d**·H_2Por **72a** complexes where red shifts of the Soret maxima were observed. However, the blue shift of the Por B-band observed during the spectrophotometric titration for Ni(II)Pc **73b**·Ni(II)Por **72b**, Ni(II)Pc **73b**·Cu(II)Por **72c**, Cu(II)Pc **73c**·Cu(II)Por **72c** and Cu(II)Pc **73c**·Ni(II)Por **72b** systems seems to be at odds with the above argument. From these findings, it appears that another factor contributing to the spectral changes observed for the Pc·Por complex should be considered, that is, the formation of exciton coupling. A subtle balance between CT and exciton coupling interactions, which tend to generate the opposite spectral shifts, determine the position of the Soret band in the Pc·Por complex.

Luminescence studies were also carried out with these Pc·Por mixtures obtaining luminiscence only for the binary system H_2Pc **73a**·H_2Por **72a**, whereas in the case of the systems containing metallotetrapyrroles with transition metal centers no luminescence was observed upon photoexcitation. For heterosystem **73a**·**72a**, in fact, excitation at 408 nm (i.e., very close to the peak of the Por-like absorption of the complex and an isosbestic point observed in the spectrophotometric titration of the H_2Por **72a** with H_2Pc **73a**) showed that the emission spectrum of H_2Por **72a** diminishes in intensity as H_2Pc **73a** was added incrementally to the system. As the fluorescence of the Por decreased, a new emission band corresponding to the H_2Pc **73a**·H_2Por **72a** complex appeared near 765 nm, which increased in amplitude as the Por fluorescence diminished. Eventually, by adding sufficient Pc, the emission of the Por was completely suppressed and the sole emitting species was the excited state of the complex at 765 nm. An isoemissive point was observed at 740 nm, indicating that a single fluorescent product was formed throughout the titration. Direct excitation of the mixture of free-base tetrapyrroles into the absorption band at 715 nm resulted as well in an emission at 765 nm. From these data it seems that the excitation of the H_2Pc **73a**·H_2Por **72a** complex into its higher lying Por-like excited state results in the emission from the CT excited state.

Femtosecond transient absorption studies were also conducted with these Pc·Por systems, exciting at 400 nm, which populates the Por-localized S_2 state, or at 690 nm directly into the Pc·Por CT band [57]. These studies showed that, for all the examined Pc·Por systems, the transient features and kinetics were independent on the excitation wavelength. A common feature to all the Pc·Por complexes was the rapid conversion, upon photoexcitation at 400 nm, of the photogenerated S_2 Por excited state to the low-lying CT state. This latter state, visible in the ground state absorption spectrum of the Pc·Por complex, could also be accessed directly *via* excitation of the complexes near 700 nm. The low-lying CT state finally deactivates to the ground state through a metal-centered (d,d) state. However, differences could be found in the kinetics of the deactivation mechanism. In fact, whereas the lifetimes of the deactivation process for all the metal-based, Pc·Por systems was in the picosecond time regime (i.e., 5–10 ps), for the Cu(II)Pc **73c**·Ni(II)Por **72b** complex no significant decay of the transients was observed on the subnanosecond time scale (i.e., 200 ps). This could be due to the presence, in this latter system, of a d^9 Cu(II) center which (i) impedes the formation of an excited Cu-centered Pc state and (ii) perturbs the electronically excited Ni-center through the formation of spin–spin interactions. For the mixed, metal free-base Pc·Por complexes, the excited-state relaxation occurred through the metallotetrapyrrole component regardless of whether it was a Pc or a Por.

A solution mixture containing tetraanionic Por **72a** and a positively charged Pc has been studied in a polar solvent such as DMSO, yielding a well-defined supramolecular heterodimer Pc·Por and a supramolecular heterotrimer Pc·Por_2 [58]. These hybrid systems present absorption spectra considerably changed with respect to the corresponding monomers, which indicate the existence of a strong electrostatic interaction between these chromophores. The lifetimes of the first excited singlet states of these compounds were also measured and correlated with the singlet oxygen generation efficiency for the Pc·Por heterodimer and Pc·Por_2 heterotrimer.

The self-assembly of oppositely charged macrocycles have also been used to form multilayer films [59]. By alternating the deposition of charged Pcs and Pors onto an ionic substrate, a new type of functional, multilayer film was prepared which presented the Por Soret (at 440 nm) and the Pc Q (at 676 nm) bands red-shifted by *ca.* 500 and 200 cm^{-1}, respectively, compared to those in solution as a result of aggregation of the chromophores within the layered structure. The absorbance of these two bands also increases linearly with the number of layers, suggesting that the consecutive absorption is uniform and regular.

Self-assembled, amphiphilic Pc·Por systems have also been used to form Langmuir-Blodgett (LB) films. By alternating the deposition of fatty acids with that of a heteroarray of Zn(II) *meso*-tetrakis(4-*N*docosylpyridinium) Por and Cu(II) tetrasulfonated Pc, LB multilayers were formed [60]. Photophysical studies showed that, upon excitation of the Por moiety, long-lived radical ions are generated within the layers, differently from solution where only an energy transfer mechanism was observed. The difference in behavior can be attributed to the local high polarity experienced by the Pc·Por heteroarray in the condensed phase.

LB films have also been fabricated using an amphiphilic Zn(II)-tetrakis[(4-*N*,*N*-dimethyl-*N*-octadecylammonium)phenoxy]Pc and a *meso*-tetra(*p*-tolyl) Por [61]. A monolayer of these two compounds was prepared, spread on water from chloroform solution, and finally transferred onto a CaF_2 slide. Surface pressure-area isotherms and UV-vis spectroscopic studies suggested that, in the condensed phase, the Por macrocycle is located in the "pocket" created by the long alkyl chains of Pc unit, the latter macrocycle laying nearly parallel to the substrate surface.

A D-A heterocomplex based on a tetrapyridinium Zn(II)Por and a tetrasulfonated free-base Pc has also been prepared, organized in LB films and its photophysical properties studied [62]. This semi-amphiphilic Pc·Por heterocomplex displays the same optical properties in LB films as the corresponding system in DMSO, that is intense B- and Q-bands of the Pc and Por moieties, although in the case of the LB films, these bands were red-shifted and strongly depleted with respect to the same absorption bands of the species in DMSO, thus indicating the existence of some degree of electronic communication between the two chromophores in the condensed phase. A very small broadening of the bands in the solid film was also observed in the mixed Pc·Por complex, thus suggesting that the interaction between the neighboring molecules is negligible with respect to the intramolecular interaction. A study of this heterocomplex organized in LB films by using three complementary techniques such as four-wave mixing, transient absorption spectroscopy and transient photovoltage was also carried out. Transient absorption studies on this system revealed the formation, upon photoexcitation, of several transient species such as singlets, triplets and charge-separated species whose lifetimes could be calculated. The comparison of these lifetimes with those of the population of the excited states, contributing either to the non-linear optical response or to the photoelectrical response of the material, allowed to elucidate and connect the microscopic and the macroscopic properties of the material.

Week electrostatic interactions in solution between non-charged macrocycles such as an Al(III)ClPc and an Al(III)OHPor have also been studied [63]. These two systems have been chosen since Al(III)Pc exhibits low absorption in the Por Soret region, while at longer wavelengths it has a large overlap with the Al(III)Por emission, thus potentially providing an interesting energy transfer system. UV-vis and fluorescence studies on mixed solutions of the two macrocycles indicate the formation of a 1:1 Pc·Por heterodimer with a K_{ass} of 5.7×10^4 M^{-1}, a value higher than that of the respective Pc·Pc or Por·Por homodimers, but considerable lower that the one obtained for strongly interacting Pcs and Pors bearing oppositely charged substituents ($\sim 10^7$ M^{-1}) (*vide supra*). The time-resolved fluorescence data show that the fluorescence decay of Al(III)Por in the absence of the Pc macrocycle is biexponential. Upon addition of Al(III)Pc, the decay requires the use of three exponentials to fit the data, showing a clear increase in quantity of the shorter-lived (i.e., subnanosecond) component. The relative amount of longer-lived species (i.e., monomeric) is found to initially increase, probably linked to the dissociation of Por·Por homodimers, before exhibiting a decrease with increasing Al(III)Pc concentration. The rise of the shortest-lived component appears to be largely at the expense of the Al(III)Por dimer. As the overall fluorescence from the Por was also found to decrease upon titration with Al(III)Pc, this suggests a mechanism which involves the dissociation of the fluorescent Al(III)Por dimers and the formation of another species (i.e., Pc·Por heterodimer). When looking at the Pc emission in the Pc·Por system, it was possible to observe an enhancement in the Al(III)Pc fluorescence ($\lambda_{em} = 765$ nm) in the combined system with respect to that of the Al(III)Pc without Por. Considering that the excitation wavelength used (i.e., 400 nm) predominantly excites the Al(III)Por moiety, while the emission at 765 nm corresponds to the Pc macrocycle, sensitization of the Pc unit by the Por macrocycle in the Pc·Por mixed system possibly occurs, which could also explain the quenching of the Al(III)Por fluorescence.

Considering that the concentration of both Pc and Por in the mixture is low, it is unlikely that a Förster-type energy transfer process from monomers of Al(III)Por to Al(III)Pc is occurring. Instead the authors suggest either some sort of CT mechanism from Al(III)Por to Al(III)Pc in the mixed dimers or an electron exchange (Dexter) mechanism. The orbital energies of the Pc·Por system calculated by semiempirical PM3 model also support a favorable CT scenario from the excited Al(III)Por to the Al(III)Pc moiety, which then emits.

4.4 Phthalocyanines and Porphyrins on Metal Surfaces

More recently, the assembly of Pcs and Pors on metal surfaces has also been pursued as a method to obtain supramolecular nanoarchitectures based on these chromophores organized on solid supports. However, the design and preparation of these surface-supported, supramolecular systems represents not an easy task considering that it would require, in some cases, to rethink/modify some "old" concepts from the solution-based supramolecular chemistry and adapt/extend them to the realm of the supramolecular assembly on substrates.

In 2002, the first report on the formation of a bimolecular, self-assembled, Pc and Por-based 2D structure on a solid support appeared [64]. This study investigated the organization properties of hexadecafluoro Co(II)Pc and tetraphenyl Ni(II)Por on Au(111). Exposure of a gold substrate to the vapors of a 1:2 mixture of both Pc and Por resulted in the formation of films presenting highly ordered regions, as demonstrated by high resolution scanning tunnel microscopy (HRSTM). A closer inspection of these regions allowed to identify the presence of the two porphyrinoid species (i.e., Pc and Por) and to distinguish them at the molecular level as a result of the difference in the electronic configuration of the two transition metal ions (i.e., Co(II) and Ni(II)), finally responsible for the difference in tunneling current between the two macrocycles (Figure 4.12).

From the analysis of this STM image it is possible to notice that the Pc and the Por macrocycles are densely packed in a 1:1 ratio within the film and appear to have a well-defined structure, in which each Pc (Por) unit is surrounded by four Por (Pc) macrocycles. Moreover, it is important to notice that, in this system, the intermolecular interactions between Pc and Por give rise to a 2-D crystalline

Figure 4.12. High resolution STM image of hexadecafluoro Co(II)Pc and tetraphenyl Ni(II)Por adsorbed on Au(111) imaged under ultra high vacuum (UHV) conditions. The space-filling models of both the Pc and the Por compounds have been superimposed to the STM image. Reproduced from Ref. [64] with permission of the American Chemical Society. See also Color Insert.

architecture that presents a higher stability than that of the films resulting from evaporating each compound (i.e., Pc or Por) separately.

A more detailed study on these perfluorinated Co(II)Pc and Ni(II)Por, as well on unsubstituted Co(II)Pc, has also been carried out suggesting that (i) stabilizing H· · ·F interactions between adjacent Pc and Por macrocyles, (ii) attractions between the dipole moments normal to the surface at the periphery of each molecule and (iii) the destabilizing, but weak, quadrupole repulsions between the vertically oriented phenyl groups and the horizontally oriented fluorinated Pc ring are, possibly, among the driving forces responsible for the observed 2-D, Pc·Por organization [65].

Mixed adlayers of Co(II)Pc with Cu(II)TPP, Co(II)TPP, or octaethyl Cu(II)Por **74a** have also been prepared by immersing gold substrates (i.e., Au(111) or Au(100)) in a benzene solution containing both hetero-macrocycles, as revealed by cyclic voltammetry (CV) and *in situ* high-resolution STM studies [66]. It was observed that for the mixed adlayer consisting of Co(II)Pc and tetraphenyl Cu(II)Por molecules, the composition of the Pc·Por adlayer changed dramatically depending on the substrate (i.e., reconstructed Au(100)-(hex) substrate) immersion time, in a process in which the Co(II)Pc molecules displace the adsorbed Cu(II)Por molecules while increasing the substrate immersion time. A later study on the same Co(II)Pc·Cu(II)TPP system showed that ordered Pc·Por adlayer

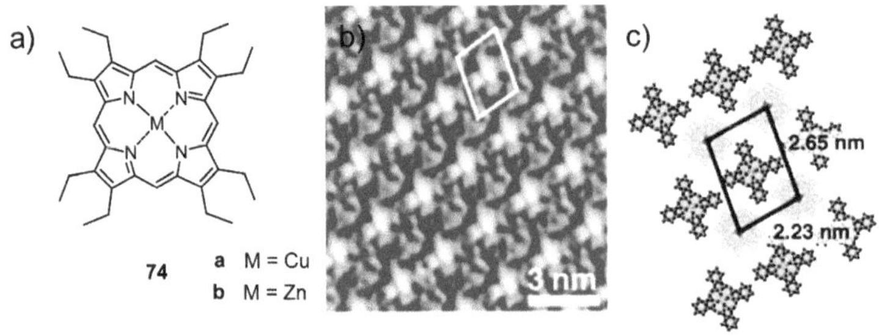

Figure 4.13. (a) Molecular structures of octaethyl Pors **74a,b**. (b) High-resolution STM image of alternate binary arrays of Co(II)Pc and Cu(II)Por **74a** on Au(111) in 0.1 M $HClO_4$ and (c) corresponding proposed model. Figure 4.13b reproduced from Ref. [68] with permission of the American Chemical Society. See also Color Insert.

structures could be obtained only on reconstructed Au(100)-(hex) surfaces, and not on Au(111), suggesting that the nature of the surface is a key factor in controlling the formation of Co(II)Pc and Cu(II)Por adlayers [67].

However, adlayers of Pc and Por alternately arranged on Au(111) substrates have been obtained by replacing Cu(II)TPP for octaethyl Cu(II)Por **74a** while maintaining the Co(II)Pc macrocycle (Figure 4.13) [68]. Similarly to the previous cases, it was observed that the composition of the Pc·Por mixed adlayer varied with the substrate immersion time, in a process in which the Cu(II)Por molecules were completely displaced by the Co(II)Pc molecules in the mixed solution after a prolonged modification.

Moreover, it was observed that, in such a system, the surface mobility and the molecular adlayer reorganization of Co(II)Pc and Cu(II)Por **74a** were accelerated by modulation of the scanning electrode potential, obtaining different surface structures at different electrode potentials.

More recently, two-component adlayers consisting of unsubstituted Zn(II)Pc and Zn(II)Por **74b** or Zn(II)TPP macrocycles have been prepared on Au(111) and used as bimolecular "chessboards" toward the supramolecular assembly of a third component (i.e., C_{60} fullerene) which is selectively trapped in the open spaces of the adlayers (Figure 4.14) [69].

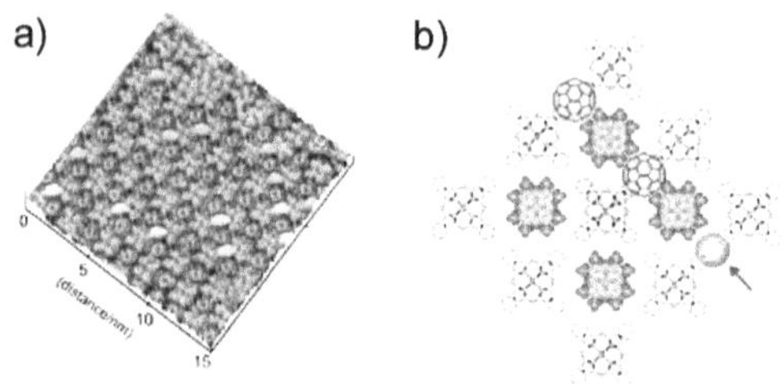

Figure 4.14. (a) Height shaded view of a STM image and (b) proposed model of C_{60} array in the bimolecular Pc·Por "chessboard" on Au(111). Reproduced from Ref. [69] with permission of the American Chemical Society. See also Color Insert.

Highly-ordered, but compositionally disordered, supramolecular organization of these macrocycles was also found on reconstructed Au(100)-(hex) surfaces although on this substrate no supramolecular assembly of C_{60} fullerene was found, suggesting that this organization is strongly dependent upon the bimolecular packing arrangement adopted by both porphyrinoids on the substrate.

4.5 Phthalocyanine-Subphthalocyanine and Porphyrin-Subphthalocyanine Heteroarrays

Among all the Pc derivatives, SubPcs [70] hold a privileged position. These lower Pc homologues are 14 π-electron aromatic compounds which comprise three N-fused diiminoisoindole units in a cone-shaped geometry, and a central boron atom axially substituted with a halogen atom or other type of ligands (i.e., alkoxy, etc...). SubPcs are very promising candidates in the fields of supramolecular chemistry [71], nonlinear optics [72], and also as photo- or electroactive systems [73]. In particular, SubPc-Pc systems are perfectly suited for the study of intramolecular energy/electron-transfer processes for several reasons. First, the absorption of each chromophore perfectly complements that of the other; Second, the energy level of their excited states and, therefore, of their optical transitions are very well suited for the efficient absorption and directional funneling, *via* energy-transfer processes, of photoexcitation energy; and finally, the redox gradient can be easily tuned by the introduction of different peripheral substituents on both macrocycles, and thus, control can be gained over the competition between photoinduced energy- and electron-transfer mechanisms.

Dyads **75a–c** [74], in which both units are covalently linked through their peripheral positions by a conjugated ethynyl spacer, have been synthesized by means of a Pd-mediated cross-coupling reaction between different monoiodo-SubPcs (**76a–c**) [75] and Pc **77** bearing a single ethynyl group (Scheme 4.14) [76]. A remarkable dependence of the reaction rate on the SubPc peripheral substitution was noticed; electron-poor aromatic rings react much faster (**76c** > **76a** > **76b**). After chromatographic separation, dyads **75a–c** were obtained in 53, 45, and 69% yield, respectively.

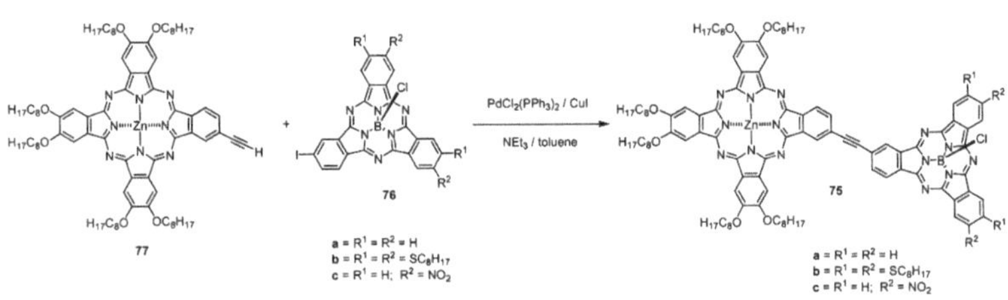

Scheme 4.14. Synthesis of Pc-SubPc dyads **75a–c**.

Steady-state absorption experiments show that, in all cases, the spectra of the Pc-SubPc dyads are not merely the sum of the spectra of their macrocyclic components; a small but clear bathochromic shift in both Pc and SubPc Q bands, and a more pronounced splitting of the Pc Q band are observed, which indicates that π conjugation is extended through the triple bond and confirms electronic communication.

As shown by cyclic and Osteryoung square-wave voltammetry experiments, the electronic characteristics of the SubPc moiety can be modulated by the introduction of different peripheral substituents. The reduction potential of this subunit is decreased by about 400 mV on going from thioether (i.e., **76b**) or no substituents (i.e., **76a**) to nitro groups (i.e., **76c**). Steady-state and time-resolved photophysical experiments demonstrated that, when the CT state lies high in energy, a quantitative singlet-singlet energy transfer mechanism from the excited SubPc to the Pc takes place.

On the contrary, stabilization of the radical pair by lowering the redox gap between electron donor and acceptor results in a highly efficient photoinduced electron-transfer process, even in solvents of low polarity such as toluene.

Covalent axial bonding of SubPc and Pc cores has also been undertaken by Cook and coworkers [41]. Cofacial Pc-SubPc dyad **77** and Pc-SubPc$_2$ triad **78** having a single atom link between the metal ions at the centres of the participating ligands have been prepared by condensation reaction of ClSubPc **8** and Si(IV)(OH)$_2$Pc **80** (Scheme 4.15). As far as **80** does not lead to self-condensation products, presumably because of steric interactions involving the hexyl chains, only dyad **77** and triad **78** are obtained in the reaction. However, only a 1% yield of triad **78** can be isolated from the mixture, even if an excess of ClSubPc **8** is employed in the reaction.

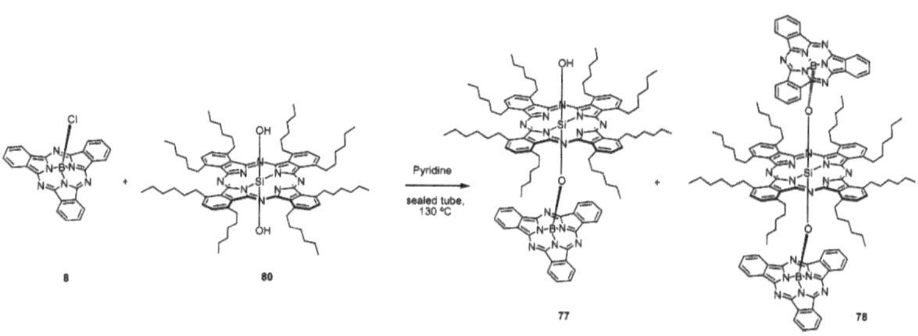

Scheme 4.15. Synthesis of μ-oxo Pc-SubPc dyad **77** and Pc-SubPc$_2$ triad **78**.

The principal absorption bands arising from the SubPc and Si(IV)Pc components are evident in the UV-vis absorption spectra of dyad **77** and triad **78**, but a small hypsochromic shift of the SubPc Q band and a small bathochromic shift of the Si(IV)Pc Q-band are observed with regard to the monomeric reference compounds **8** and **80**, respectively, which can be ascribed to differences in the axial environment more than to significant interactions of the π-electron orbitals of the component chromophores. The lack of interactions could be due to the cone-shaped structure of the SubPc unit, which avoids excitonic coupling even at such a short distance between the macrocycles. In accordance with this, the fluorescence spectra of dyad **77** and triad **78** show two bands at *ca.* 605 nm and *ca.* 780 nm, attributable to independent emissions from the SubPc and Si(IV)Pc, respectively.

Very recently, Shibata and Kobayashi have reported a related alkynyl-linked SubPc-Pc system **81** (Figure 4.15) but, in this case, the connection between both chromophores takes place between the peripheral position of the Pc core and an axial aromatic ligand at the boron center of the SubPc moiety [77]. The authors synthesized a trifluoroethoxy-coated SubPc holding an alkoxy moiety at its axial position. This latter group bears an iodo atom which was used to couple it to an appropriate ethynyl-containing Pc.

Therefore, functionalization of SubPcs can be performed either by derivatization at the peripheral positions or varying their axial ligands. Peripheral functionalization is sometimes difficult since the harsh reaction conditions for SubPc formation preclude the incorporation of a number of important functional groups. On the other hand, substitution of axial ligands of SubPcs is a feasible, modular approach towards the functionalization of these systems and, in particular, for the construction of multicomponent systems.

SubPc-Pc **82a,b** (Figure 4.15) and related Por-SubPc heterodyads have been constructed through axial coordination of SubPc macrocycles with a pyridiloxy substituent at the boron center and a series of Zn(II) and Ru(II)CO Pcs (i.e., **83a,b**) or Pors [78]. The complexation process was studied by ^{1}H NMR and fluorescence spectroscopic methods, which confirmed the 1:1 binding stoichiometry between the chromophoric units. The association constants, as determined by fluorescence titration, were generally higher for Ru(II)CO tetrapyrroles [$(2.5 - 4.7) \times 10^4$ M^{-1}] than for the Zn(II)Pc counterparts [$(0.3 - 1.8) \times 10^4$ M^{-1}]. UV-vis spectroscopy

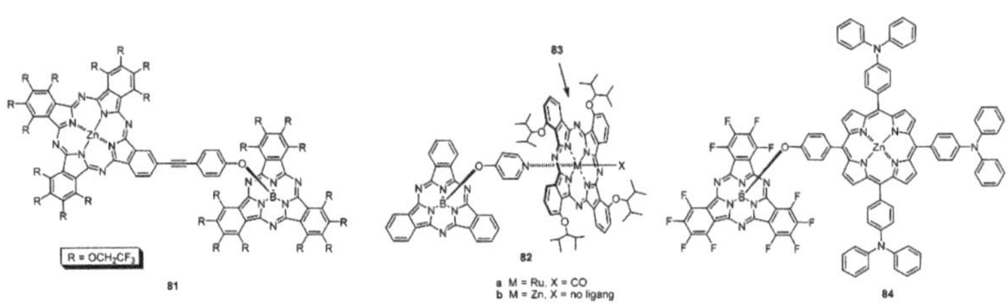

Figure 4.15. Molecular structures of Pc-SubPc (**81**, **82a,b**) and SubPc-Por (**84**) dyads.

experiments revealed negligible ground-state interaction between the two chromophores.

In the search of photoinduced interactions between strong donors, triphenylamine (TPA)-substituted Zn(II)Por (i.e., TPA_3 Zn(II)Por) and dodecafluoroSubPc (F_{12}SubPc) components, dyad **84** has been synthesized by means of the axial derivatization approach (Figure 4.15) [79]. Computational studies reveal that the HOMO is localized over the TPA entities and the Por macrocycle, while the LUMO is localized on the F_{12}SubPc moiety. The occurrence of fast (i.e., 10^{12} s^{-1}) and efficient charge-separation processes *via* the singlet excited state of TPA_3Zn(II)Por was confirmed by femtosecond transient absorption spectral measurements, both in polar and non-polar solvents. Nanosecond transient measurements with dyad **84** show that the radical–ion pair is extremely long-lived (i.e., 370 μs in benzonitrile), which is rationalized by the triplet spin character of the radical–ion pair.

4.6 Donor–Acceptor Multiporphyrinoid Homo- and Heteroarrays

4.6.1 *Covalently Linked Ensembles*

D–A multinuclear Pc systems have also been prepared with the aim of studying how the increased number of chromophoric units in these conjugates would affect some of the key factors that control the formation, and decide the fate, of the charge separated states (such as the electronic coupling between the donor and acceptor components or their structural arrangement).

A series of D–A triads (**85a–c**) have been prepared in which two Pc units have been covalently connected through vinyl spacers to the 1-, 5- (i.e., **85a**), 2-, 6- (i.e., **85b**) or 1-, 8- (i.e., **85c**) positions of an anthraquinone (AQ) unit, an excellent electron-acceptor moiety (Figure 4.16) [80, 81].

The photophysical and electrochemical properties of these triads have also been investigated suggesting that the substitution pattern at the AQ moiety exerts a profound influence over the aggregation status of the resulting Pc_2-AQ triads. Electrochemical

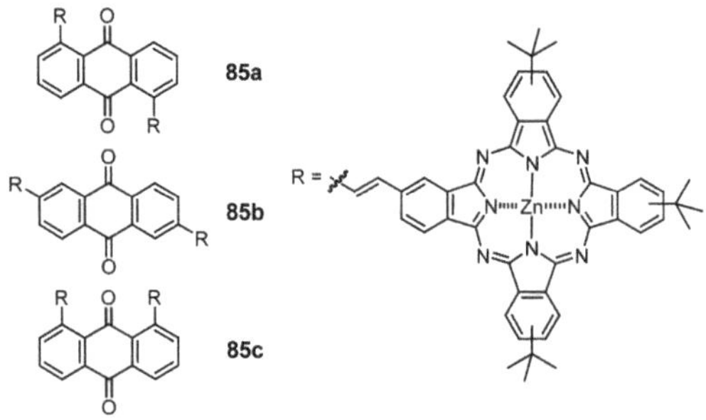

Figure 4.16. Molecular structures of Pc_2-AQ triads **85a–c**.

experiments carried out with these Pc-AQ ensembles evidenced the important effect of the solvent used (i.e., THF or *o*-DCB) as a way to modulate the inter- and/or intramolecular electronic interactions of the Pc and/or AQ subunits. Photophysical experiments demonstrated that, upon photoexcitation of the Pc units of **85a-c**, the formation of the $Pc^{\cdot+}$-$AQ^{\cdot-}$ state was observed. The kinetics of these electron transfer reactions could be modified as a function of the aggregation status of the Pc_2-AQ triads. These studies demonstrated that the triad topology (i.e., "packed" in the case of **85c** or "extended" in the case of **85a,b**) is able to drastically influence both the inter- and intramolecular interactions between the Pc donor and the AQ acceptor moieties.

En route toward the preparation of novel electron D–A conjugates, a multinuclear Pc_3-C_{60} fullerene tripodal architecture (**86**) containing a tetraphenylmethane core has been recently reported (Scheme 4.16) [82].

The starting compound used for the preparation of conjugate **86** was tetraiodo-substituted compound **87** which was reacted with 4-ethynylbenzaldehyde using Sonogashira cross-coupling reaction conditions to obtain aldehyde-containing **88**. Compound **88** was then subjected to a threefold cross-coupling reaction with ethynyl-containing Pc **89** to give a formyl-containing, tripod precursor which was then reacted with *N*-methyl glycine in the presence of C_{60} fullerene to obtain tetrad **86** in 22% yield. Photophysical studies carried out with tripod **86** demonstrated that charge separation occurs only in polar solvents such as benzonitrile or DMF, with radical ion

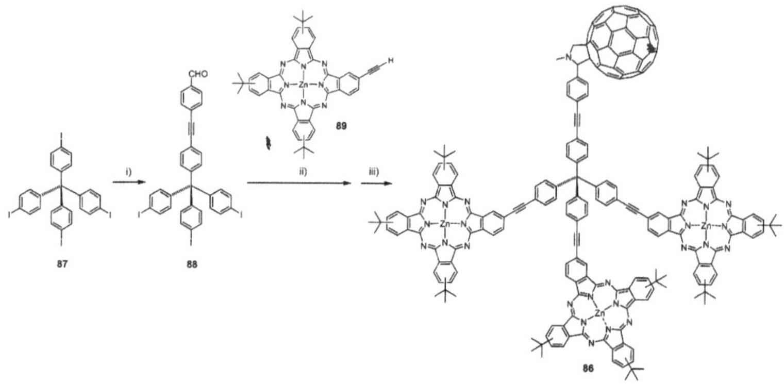

Scheme 4.16. Synthesis of Pc_3-C_{60} fullerene tripod **86**. (i) 4-ethynylbenzaldehyde, $Pd(PPh_3)_2Cl_2$, CuI, NEt_3, THF, RT. (ii) Pc **89**, $Pd_2(dba)_3$, $AsPh_3$, NEt_3, THF, RT. (iii) C_{60} fullerene, *N*-methylglycine, toluene, reflux.

pair state lifetimes of 1000 and 650 ps, respectively, whereas less polar solvents such as toluene, anisole, or THF are ineffective to trigger a CT event.

An ensemble containing two Pc macrocycles and two C_{60} fullerene units (i.e., **90**) has also recently been reported (Scheme 4.17) [83].

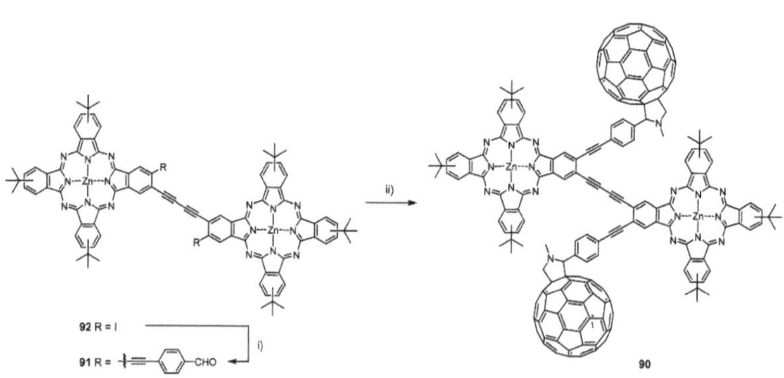

Scheme 4.17. Synthesis of Pc_2-$(C_{60})_2$ fullerene tetrad **90**. (i) 4-ethynylbenzaldehyde, $Pd(PPh_3)_2Cl_2$, CuI, NEt_3, THF, RT. (ii) $Pd_2(dba)_3$, $AsPh_3$, NEt_3, THF, RT. (iii) C_{60} fullerene, *N*-methylglycine, toluene, reflux.

The synthetic strategy used for the preparation of conjugate **90** consisted in the preparation of the diformyl-containing Pc precursor **91**, this latter compound being prepared from diiodo Pc **92** *via* palladium-catalyzed, cross-coupling reaction with 4-ethynylbenzaldehyde. The diformylPc **91** was then reacted with *N*-methyl glycine forming an azomethyne ylide species which gave rise to a 1,3 dipolar cycloaddition reaction over C_{60} fullerene to afford Pc_2-$(C_{60})_2$ tetrad **90**.

Photophysical analysis of ensemble **90** revealed relatively short charge separated state lifetimes (i.e., anisole = 2970 ps; THF = 1150 ps; benzonitrile = 180 ps) probably due to the close proximity of the Pc and the C_{60} subunits in the nanoconjugate, which could lead to a through-space deactivation dynamics.

D-A multi-Pc conjugates containing PDI as acceptor moiety have also been prepared. In this context, a covalently-linked Pc_2-PDI triad (**93**) was prepared following two different synthetic approaches, both of them involving a metal-catalyzed Sonogashira coupling reaction either between an ethynyl-containing Zn(II)Pc unit and a PDI unit substituted with two bromo atoms at its 1,7 positions (Strategy A, Scheme 4.18, 16% yield), or between a iodo-substituted Zn(II)Pc

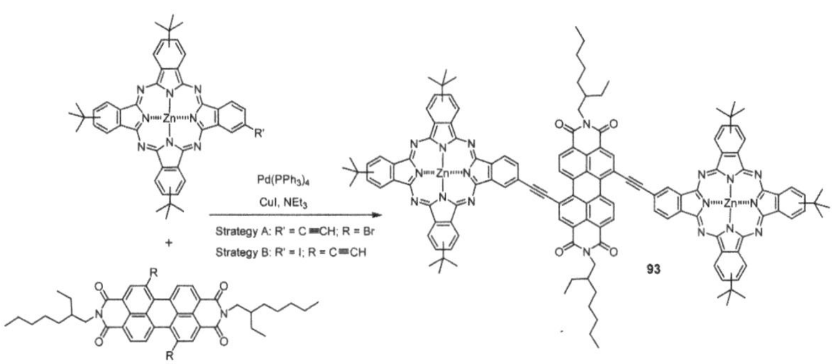

Scheme 4.18. Synthesis of Pc_2-PDI triad **93** following two different synthetic strategies.

and a bis(ethynyl) PDI moiety (Strategy B, Scheme 4.18, 33% yield), being the latter synthetic strategy the one that gave better results [84].

UV-vis analysis of the conjugated triad **93** and the Pc and PDI precursors accounted for a strong ground-state electronic communication between the two redox-active components in the triad system. Steady-state fluorescence was also used to shed light on the excited state interactions between the chromophores in the triad, revealing a strong quenching of the fluorescence (~ 250-fold), which occurred exciting either the Pc or the PDI moieties. Finally, transient absorption spectroscopy experiments demonstrated that, upon photoexcitation of the Zn(II)Pc component in triad **93**, the Pc singlet excited state features decayed rapidly (i.e., 7.2 ps) to yield a transient product that looked completely different from the excited states of the Zn(II)Pc triplet, the PDI singlet, or the PDI triplet, and that was finally attributed to the $Pc^{+\bullet}$-$PDI^{-\bullet}$-Pc charge-separated species.

In the same context, a Pc_2-PDI triad (**94**) has also been prepared in which the two Pc macrocycles have been covalently connected to the *N*-positions of a PDI moiety (Figure 4.17) [85].

The photoexcited properties of triad **94** were investigated by steady-state fluorescence measurements which showed, when exciting at 670 nm, a considerable quenching, with respect to a Zn(II)Pc reference compound, of the single-excited state of the Zn(II)Pc unit by the appended PDI chromophore.

By employing 490 nm as the excitation light, the emission peaks of $^1PDI^*$ in toluene were observed at 534 and 569 nm. Importantly, when exciting the PDI moiety at 490 nm, no Zn(II)Pc fluorescence was detected around 680 nm. This finding is helpful in ruling out the transfer of the singlet excited-state energies from $^1PDI^*$ to Zn(II)Pc

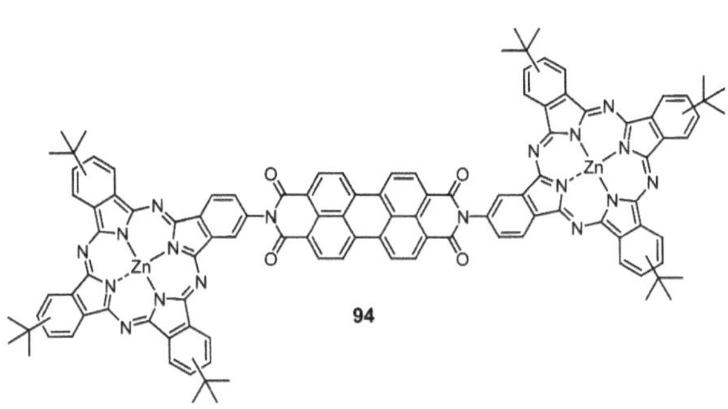

94

Figure 4.17. Molecular structure of Pc_2-PDI triad **94**.

in **94**, although this process would be feasible due to the higher energy level of ^{1}PDI* with respect to ^{1}Zn(II)Pc* by 0.31 eV.

These steady-state fluorescence observations were tracked by the fluorescence time profile measurements, which showed the shortening of the fluorescence lifetimes of Zn(II)Pc and PDI in **94** with respect to Zn(II)Pc and PDI reference compounds. These observations suggest that an efficient charge separation is taking place in triad **94** *via* both ^{1}Zn(II)Pc* and ^{1}PDI*. Complementary photophysical studies (i.e., nanosecond transient absorption) were carried out in toluene solutions of **94** applying a 532 nm laser light, which exclusively excited the PDI moiety. The recorded transient absorption spectra revealed the fingerprints of Zn(II)$Pc^{+\bullet}$ and $PDI^{-\bullet}$ transient species which decayed, through charge recombination, to the ground state with a lifetime of 25 ns in toluene and 20 ns in benzonitrile. Finally, molecular orbital calculations at B3LYP/3-21G levels were carried out with triad **94** showing that the majority of the HOMO is located in the Zn(II)Pc macrocycle whereas the majority of the LUMO is located on the PDI moiety. These findings support the photophysical results, that is, the occurrence of a charge-transfer state generated by transferring completely an electron from the Zn(II)Pc entity to the PDI moiety.

A linear array of chromophores consisting of a perylene monoimide (PMI) input unit, a free-base Por dimer transmission unit, and a free-base Pc output unit (**95**) has also been prepared for studies in artificial photosynthesis and molecular photonics (Scheme 4.19) [86].

The synthesis of **95** involved (i) the synthesis of two *trans*-ABCB Zn(II)Por building blocks bearing one phthalonitrile group (**96**) and one iodophenyl group (**97**), respectively, and having, each of them, one *meso*-position unsubstituted, (ii) oxidative, *meso*, *meso* coupling of Zn(II)Por monomers **96** and **97** to afford a Zn(II)Por dimer **98** bearing at its periphery one phthalonitrile group and one iodophenyl group, (iii) demetalation of **98** to obtain compound **99**, (iv) preparation of Por_2-Pc array **100** *via* a mixed cyclization involving **99** and 4-*tert*-butylphthalonitrile, and (v) Pd-mediated coupling of **100** with an ethynyl-substituted PMI to afford linear array **95** which strongly absorbs across the visible spectrum.

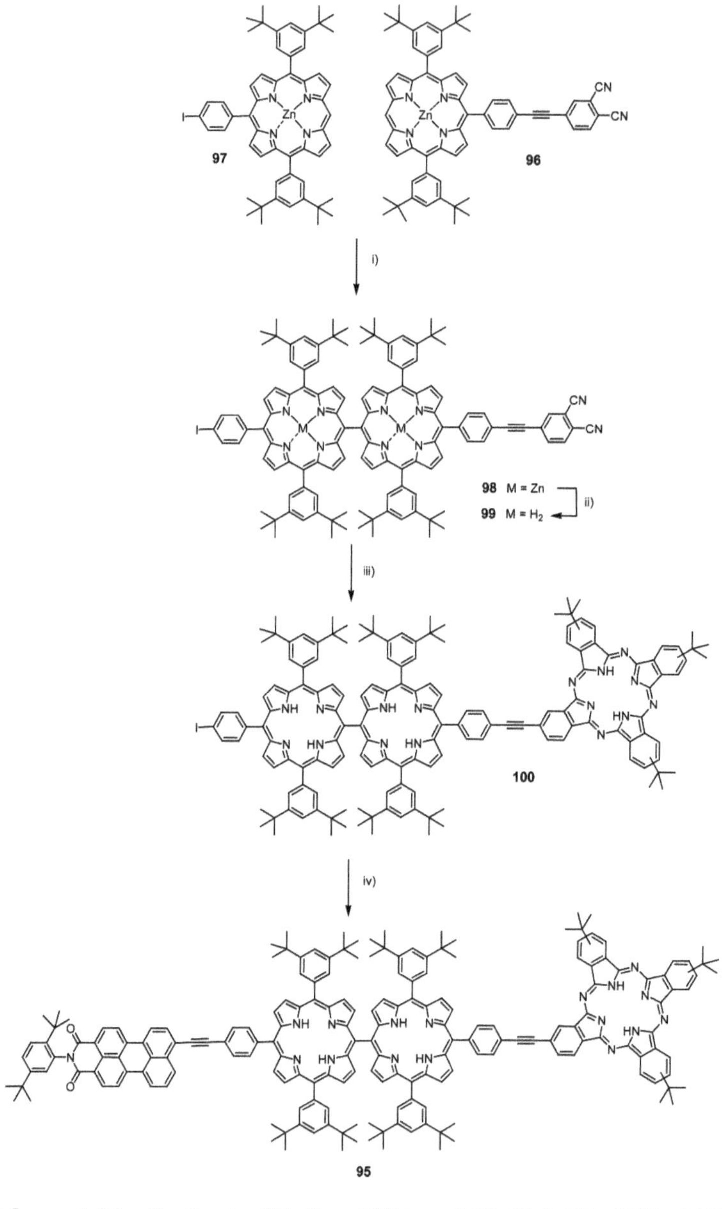

Scheme 4.19. Synthesis of Pc-Por_2-PMI tetrad **95**. (i) $AgPF_6$, $CHCl_3$, DMA. (ii) CH_2Cl_2/TFA. (iii) *tert*-butyl phthalonitrile, lithium turnings, penthanol, 145°C. (iv) ethynyl-substituted PMI, $Pd_2(dba)_3$, tris-*o*-toluenephosphine, toluene/NEt_3, 35°C.

Excitation of tetrad **95** at 490 nm, where the perylene unit preferentially absorbs (i.e., 60%) (the other 40% is represented by the Por absorption), results in fluorescence almost exclusively from the Pc macrocycle, with a Φ_f of 0.78, a value identical to the one of tetra-*tert*-butylPc reference compound (i.e., 0.77). These results suggest (i) quantitative energy transfer from the perylene and the bis-Por units to the Pc macrocycle, and (ii) intense emission from the Pc that is undiminished from that of the isolated chromophore. Transient absorption experiments were also carried out with **95** exciting at 529 nm where the perylene chromophore preferentially absorbs. These studies revealed that efficient (i.e., >99%) photoinduced energy transfer from the excited perylene to the ground state Pc component (*via* the intervening Por dimer) occurred.

Similarly, a triad system constituted of a Zn(II)Por input unit, a free-base Pc output unit, and ferrocene (Fc) as the redox-switching unit (**101**) has been repared (Figure 4.18) [87]. The synthesis of **101** involved a one-pot, Pd-coupling reaction between a Zn(II) tetraphenyl Por derivative bearing two iodo atoms, an ethynyl-substituted Fc compound, and an ethynyl-functionalized Pc. A mixture of trimeric, dimeric, and monomeric materials was obtained from the reaction, the purification of the desired triad **101** involving the use of chromatography procedures (i.e., silica gel and size exclusion chromatography).

Electrochemical studies showed that the potentials observed for array **101** were very similar to those observed for the individual components (i.e., Pc, Por and Fc), thus suggesting a

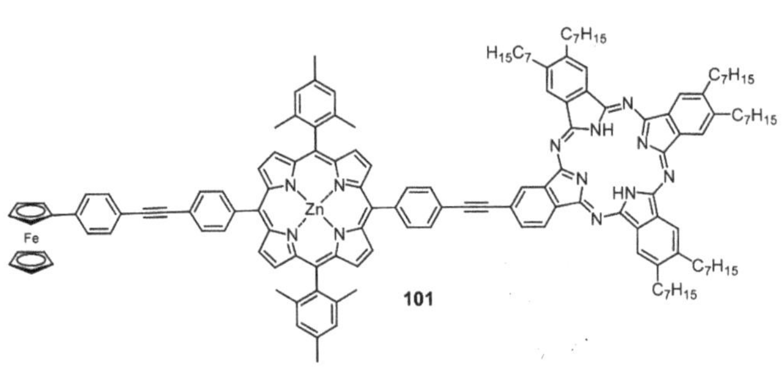

Figure 4.18. Molecular structure of Pc-Por-Fc triad **101**.

relatively weak coupling between the constituents of the array in the ground state. The fluorescence properties of triad **101** were also investigated. Excitation of triad **101** at the Pc absorption band (i.e., 662 nm) resulted in the Pc emission with a Φ_f of 0.79, thus suggesting that in such triad the location of the Fc moiety, quite distant from the Pc macrocycle, resulted in little, if any, quenching of its fluorescence. The fluorescence characteristics of triad **101** were next examined under conditions where the Fc unit was oxidized. In this case, oxidation of the Fc unit resulted in essentially complete quenching of the fluorescence emission from the Pc. This result indicates the Fc radical cation serves as an extremely efficient trap for the excited-state energy, acting as a viable redox-based switch. The quenching of the excited Pc output unit by the distant oxidized Fc redox-switching unit in **101** must occur *via* superexchange involving the central Zn(II)Por moiety. The alternative two-step mechanism involving first energy transfer from the excited Pc to the ground-state Zn(II)Por cannot occur because this step is uphill by about 0.17 eV.

Photoinduced intramolecular electron transfer in a AQ-substituted Pc–Por heterodimer (**102**, Figure 4.19) has been investigated by fluorescence quenching experiments and nanosecond flash photolysis techniques, giving rise to the formation of a charge separated species upon excitation of the Pc moiety [88].

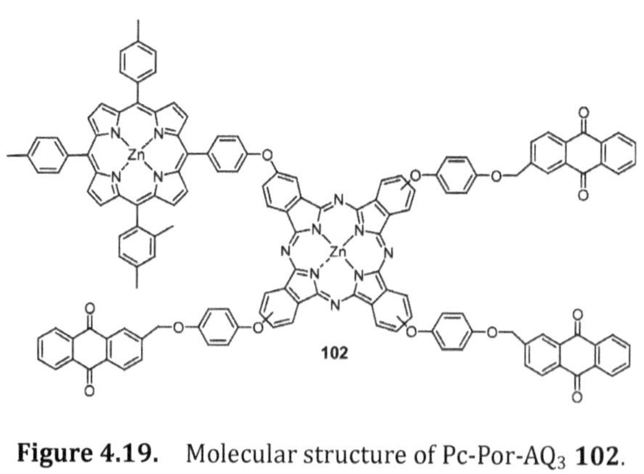

Figure 4.19. Molecular structure of Pc-Por-AQ_3 **102**.

4.6.2 *Systems Assembled* via *Supramolecular Interactions*

The preparation of D-A multiPc-based systems in which the active components assemble *via* non-covalent interactions is a promising strategy for the construction of functional systems. Such non-covalent systems in fact are expected to give rise to efficient and long-lived photoinduced charge separated processes, thus representing an attractive alternative to their covalently-linked counterparts.

In this context, a Pc_2-C_{60} supramolecular pseudorotaxane-like complex **103** has been prepared whose formation relies on the threading of a dibenzylammonium unit present on a fullerene derivative **104** through a dibenzo-24-crown-8 macrocycle substituted with two Pc units (**105**) (Figure 4.20) [89].

A stability constant of 1.9×10^4 M^{-1} was obtained for such pseudorotaxane complex as gathered analyzing the data of the progressive, non-linear quenching of the Zn(II)Pc fluorescence upon addition of fullerene derivative **104**.

Transient absorption measurements carried out with supramolecular complex **103** revealed the presence of three transient features, that is, the presence of an absorption band at 1040 nm, which is the characteristic absorption fingerprint of the one-electron reduced form of **104**, and two maxima at 500 and 860 nm, which correspond to the one-electron oxidized Zn(II)Pc radical cation. The lifetime of the charge separated state of **103** could also be determined obtaining a value of 1.3 μs. It is worth mentioning that

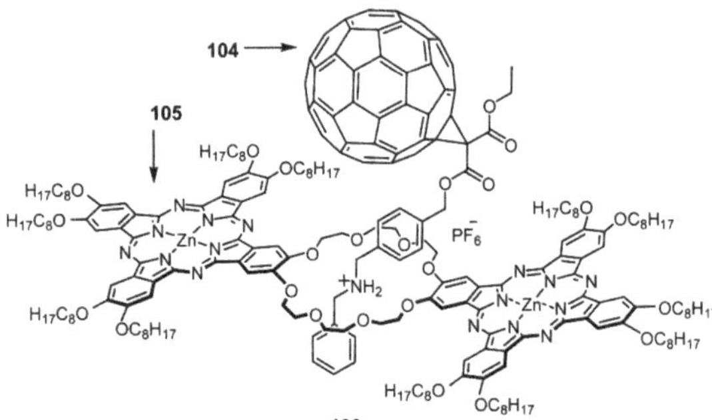

Figure 4.20. Molecular structure of [2]pseudorotaxane Pc_2-C_{60} triad **103**.

this supramolecular ensemble presents a stabilization of its radical ion pair state lifetime of more than two orders of magnitude with respect to covalently-linked Zn(II)Pc-C_{60} counterparts.

Metal-directed self-assembly has also been used to prepare D-A multiPc systems, since these metallo-supramolecular species present large energies associated with the metal-ligand bond formation giving rise to kinetically stable complexes.

The presence of Ru(II) in the central cavity of Pcs allows to link to the macrocycle either one or two ligands in its axial positions. In this context, metal coordination of pyridine derivatives to Ru(II)Pcs have been exploited to prepare D-A multiPc hybrids such as Ru(II)Pc_2-C_{60} **106** [90] and Ru(II)Pc_2-PDI **107** [91] in which the donor and the acceptor units bear an orthogonal geometry (Figure 4.21).

Ru(II)Pc_2-C_{60} **106** was prepared in 65% yield by reacting a tetra-*tert*-butylated phthalocyaninato ruthenium (II) derivative having a strongly ligating axial carbonyl moiety at one of the two axial Ru(II) coordination sites, with a highly elaborated *trans*-1-bis[*N*-(4-pyridyl)fulleropyrrolidine] ligand, which holds two pyridyl

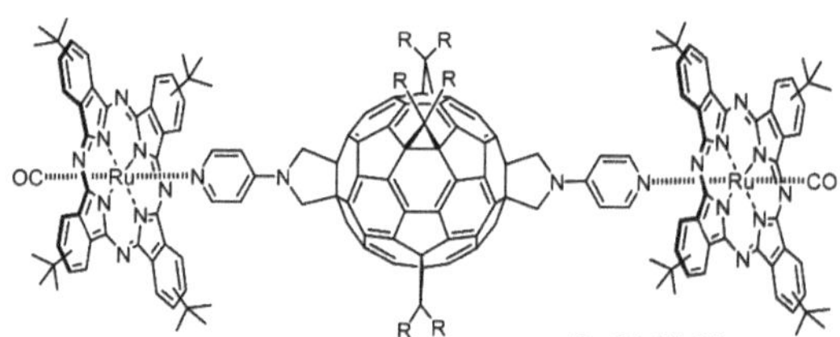

106

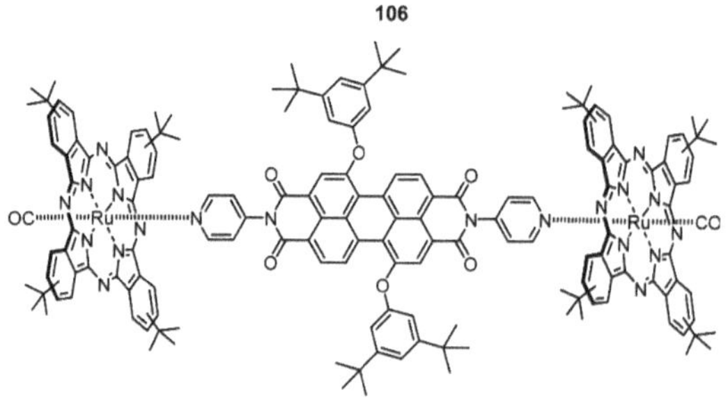

107

Figure 4.21. Molecular structures of supramolecular triads Ru(II)Pc_2-C_{60} **106** and Ru(II)Pc_2-PDI **107**.

ligands in an antiparalell arrangement. Electrochemical studies with **106** showed an anodic shift for the first three Pc-centered reduction peaks compared to a Ru(II)Pc reference derivative lacking the C_{60} fullerene moiety (i.e., easier to reduce), whereas the Pc-based oxidation peaks were shifted to higher potentials (i.e., more difficult to oxidize). Moreover, the fact that the first oxidation peak of **106** involves both Pc units implies that there is a lack of appreciable electronic coupling between the two Pc moieties.

In a photophysical context, photoexcitation of Ru(II)Pc_2-C_{60} **106** at 660 nm results in the formation of a charge separated state with lifetimes of 370 and 500 ps in toluene and anisole, respectively. Finally, the formed radical ion pair does not decay to the singlet ground state but to the Ru(II)Pc triplet excited state, which lies 0.64 eV below that of the charge separated state, thus offering a rapid deactivation pathway of the radical ion pair state. Thermodynamic considerations suggest that the dynamics of this process is nearly activationless.

A similar modular protocol has allowed the preparation of another D-A Ru(II)Pc-based hybrid (**107**) [91]. Compound **107** has been obtained through axial coordination of a perylenebisimide (PDI) moiety, bearing two 4-pyridyl substituents at its imido positions, to the Ru(II) metal centers of two Pcs.

A first insight into D–A interactions in **107** came from electrochemical measurements which showed that, for this triad, the peaks for the oxidative and reductive processes were shifted compared to the ones of its separated components (i.e., PDI and Ru(II)(CO)Pc) suggesting some electronic communication between the two chromophores in **107**.

Photophysical studies on triad **107** showed that photoexcitation of either chromophore (i.e., Pc or PDI) leads to a product, essentially identical for both pathways, characterized by transient maxima at 530 and 725 nm, plus transient minima at 580 nm and 650 nm.

This photoproduct was attributed to the photolytically generated, radical ion pair species [Ru(II)$Pc^{\bullet+}$-$PDI^{\bullet-}$-Ru(II)Pc] based on the radiolytically generated $PDI^{\bullet-}$ and Ru(II)$Pc^{\bullet+}$ species, which in the 300 and 900 nm range remarkably resemble those observed for photoexcited **107**. The lifetime of the photogenerated charge separated state in **107** (i.e., 115 ns) showed a significant stabilization compared to other D-A PDI-based hydribs [92] suggesting that the strongly exothermic charge recombination dynamics are buried deeply in the inverted region of the Marcus parabola.

References

1. (a) Burrell, A. K., Officer, D. L., Plieger, P. G., and Reid, D. C. W. (2001) *Chem. Rev.*, **101**, 2751–2796. (b) Beletskaya, I., Tyurin, V. S., Tsivadze, A. Y., Guilard, R., and Stern, C. (2009) *Chem. Rev.*, **109**, 1659–1713. (c) Drain, C. M., Varotto, A., and Radivojevic, I. (2009). *Chem. Rev.*, **109**, 1630–1658.
2. (a) McKeown, N. B. (1998) *Phthalocyanine Materials: Synthesis, Structure and Function.* Cambridge University Press, Cambridge. (b) Kadish, K. M., Smith, K. M., and Guilard, R. (eds) (2003). *The Porphyrin Handbook*, vol. 15, Academic Press, New York. (c) de la Torre, G., Vazquez, P., Agullo-Lopez, F., and Torres, T. (2004) *Chem. Rev.*, **104**, 3723–3750. (d) de la Torre, G., Claessens, C. G., and Torres, T. (2007) *Chem. Commun.* 2000–2015. (e) Claessens, C. G., Hahn, U., and Torres, T. (2008) *Chem. Rec.*, **8**, 75–97. (f) Rio, Y., Rodriguez-Morgade, M. S., and Torres, T. (2008) *Org. Biomol. Chem.*, **6**, 1877–1894. (g) Gonzalez-Rodriguez, D., and Bottari, G. (2009). *J. Porphyrins Phthalocyanines*, **13**, 624–636. (h) Martinez-Diaz, V. M., and Bottari, G. (2009) *J. Porphyrins Phthalocyanines*, **13**, 471–480. (i) de la Torre, G., Bottari, G., Hahn, U., and Torres, T. (2010) *Struct. Bond.*, **135**, 1–44.
3. For examples of Pcs, see: (a) Kadish, K. M., Smith, K. M., and Guilard, R. (eds) (2003). *The Porphyrin Handbook*, vol. 17, Academic Press, New York. (b) de la Torre, G. (2009) *J. Porphyrins Phthalocyanines*, **13**, 637–644. For examples of Pors, see: (c) Senge, M. O., Fazekas, M., Notaras, E. G. A., Blau, W. J., Zawadzka, M., Locos, O. B., and Mhuircheartaigh, E. M. N. (2007). *Adv. Mater.*, **19**, 2737–2774. (d) Elemans, J. A. A. W., van Hameren, R., Nolte, R. J. M., and Rowan, A. E. (2006) *Adv. Mater.*, **18**, 1251–1266. (e) Scandola, F., Chiorboli, C., Prodi, A., Iengo, E., and Alessio, E. (2006). *Coord. Chem. Rev.*, **250**, 1471–1496. (f) D'Souza, F., and Ito, O. (2009) *Chem. Commun.*, 4913–4928.
4. For examples of Pcs, see: (a) Kimura, M., Shirai, H. (2003) *The Porphyrin Handbook* (ed. Kadish, K. M., Smith, K. M., and Guilard, R.), vol. 19, Chapter 120, Academic Press, New York. (b) Zagal, J. H., Griveau, S., Silva, J. F., Nyokong, T., and Bedioui, F. (2010) *Coord. Chem. Rev.*, **254**, 2755–2791. For examples of Pors, see: (c) Simonneaux, G., and Le Maux, P. (2002)

Coord. Chem. Rev., **228**, 43–60. (d) Thordarson, P., Bijsterveld, E. J. A., Rowan, A. E., and Nolte, R. J. M. (2003) *Nature*, **424**, 915–918. (e) Rose, E., Andrioletti, B., Zrig, S., and Quelquejeu-Etheve, M. (2005) *Chem. Soc. Rev.*, **34**, 573–583.

5. For examples of Pcs, see: (a) Cid, J.-J., Yum, J.-H., Jang, S.-R., Nazeeruddin, M. K., Martinez-Ferrero, E., Palomares, E., Ko, J., Graetzel, M., and Torres, T. (2007). *Angew. Chem. Int. Ed.*, **46**, 8358–8362. (b) Martinez-Diaz, M. V., and Torres, T. (2010) *Handbook of Porphyrin Science* (ed. Kadish, K. M., Smith, K. M., and Guilard, R.), vol. 10, Academic Press, Singapore, pp. 141–181. (c) Martinez-Diaz, M. V., de la Torre, G., and Torres, T. (2010) *Chem. Commun.*, **46**, 7090–7108. (d) Bottari, G., de la Torre, G., Guldi, D. M., and Torres, T. (2010) *Chem. Rev.*, **110**, 6768–6816. For examples of Pors, see: (e) Imahori, H., and Fukuzumi, S. (2004) *Adv. Funct. Mater.*, **14**, 525–536. (f) Campbell, W. M., Burrell, A. K., Officer, D. L., and Jolley, K. W. (2004) *Coord. Chem. Rev.*, **248**, 1363–1379. (g) Nakamura, Y., Aratani, N., and Osuka, A. (2007) *Chem. Soc. Rev.*, **36**, 831–845. (h) Imahori, H., Umeyama, T., and Ito, S. (2009) *Acc. Chem. Res.*, **42**, 1809–1818. (i) Gust, D., Moore, T. A., and Moore, A. L. (2009) *Acc. Chem. Res.*, **42**, 1890–1898. (j) Bessho, T., Zakeeruddin, S., Yeh, C. Y., Diau, E. G., Grätzel, M. (2010) *Angew. Chem., Int. Ed.*, **49**, 6646–6649.
6. For examples of Pcs, see: (a) Ben-Hur, E., and Chan, W.-S. (2003) *The Porphyrin Handbook* (ed. Kadish, K. M., Smith, K. M., and Guilard, R.), vol. 19, Chapter 117, Academic Press, New York. For examples of Pors, see: (b) Kadish, K. M., Smith, K. M., and Guilard, R. (eds) (2003) *The Porphyrin Handbook*, vol. 14, Academic Press, New York. (c) Ethirajan, M., Chen, Y., Joshi, P., and Pandey, R. K. (2011) *Chem. Soc. Rev.*, **40**, 340–362.
7. Lo, P.-C., Leng, X., and Ng, D. K. P. (2007) *Coord. Chem. Rev.*, **251**, 2334–2353.
8. Gaspard, S., Giannotti, C., Maillard, P., Schaeffer, C., and Tran-Thi, T-H. (1986) *J. Chem. Soc., Chem. Commun.*, 1239–1241.
9. Tran Thi, T. H., Desforge, C., Thiec, C., and Gaspard, S. (1989) *J. Phys. Chem.*, **93**, 1226–1233.
10. Zhao, Z., Ogunsipe, A. O., Maree, M. D., and Nyokong, T. (2005) *J. Porphyrins Phthalocyanines*, **9**, 186–197.
11. Zhao, Z., Ozonema, K. I., Maree, M. D., and Nyokong, T. (2005) *Dalton Trans*, 1241–1248.
12. Osati, S., Safari, N., and Jamaat, P. R. (2010) *Inorg. Chim. Acta* **363**, 2180–2184.

13. (a) Zhao, Z., Maree, M. D., and Nyokong, T. (2005) *Dalton Trans*, 3732–3737. (b) Durmus, M., Chen, J. Y., and Zhao, Z. X., Nyokong, T. (2008) *Spectrochim. Acta Part A*, **70**, 42–49.
14. Bian, Y., Chen, X., Wang, D., Choi, C.-F., Zhou, Y., Zhu, P., Ng, D. K. P., Jiang, J., Weng, Y., and Li, X. (2007) *Chem. Eur. J.*, **13**, 4169–4177.
15. Ballesteros, B., de la Torre, G., Shearer, A., Hausmann, A., Herranz, M. A., Guldi, D. M., and Torres, T. (2010) *Chem. Eur. J.*, **16**, 114–125.
16. Zhang, X., Li, Y., Qi, D., Jiang, J., Yan, X., and Bian, Y. (2010) *J. Phys. Chem. B*, **114**, 13143–13151.
17. Soares, A. R. M., Martínez-Díaz, M. V., Bruckner, A., Pereira, A. M. V. M., Tomé, J. P. C., Alonso, C. M. A. (2007) *Org. Lett.*, **9**, 1557–1560.
18. Alonso, C. M. A., Neves, M. G. P. M. S., Tomé, A. C., Silva, A. M. S., and Cavaleiro, J. A. S. (2005) *Tetrahedron*, **61**, 11866–11872.
19. Maya, E. M., Vázquez, P., Torres, T. (1999) *Chem. Eur. J.*, **5**, 2004–2013.
20. (a) Kobayashi, N., Nishiyama, Y., Ohya, T., and Sato, M. (1987) *J. Chem. Soc., Chem. Commun.*, 390–392. (b) Kobayashi, N., Ohya, T., Sato, M., and Nakajima, S.-I. (1993) *Inorg. Chem.*, 1803–1808.
21. Hausmann, A., Soares, A. R. M., Martínez-Díaz, M. V., Neves, M. G. P. M. S., Tomé, A. C., Cavaleiro, J. A. S., Torres, T., and Guldi, D. M. (2010) *Photochem. Photobiol. Sci.*, **9**, 1027–1032.
22. Tomé, J. P. C., Pereira, A. M. V. M., Alonso, C. M. A., Neves, M. G. P. M. S., Tomé, A. C., Silva, A. M. S., Cavaleiro, J. A. S., Martínez-Díaz, M. V., Torres, T., Rahman, G. M. A., Ramey, J., and Guldi, D. M. (2006) *Eur. J. Org. Chem.*, 257–267.
23. Alonso, C. M. A., Neves, M. G. P. M. S., Tomé, A. C., Silva, A. M. S., and Cavaleiro, J. A. S. (2000) *Tetrahedron Lett.*, **41**, 5679–5682.
24. Ali, H., and van Lier J. E. (2009) *Tetrahedron Lett.*, **50**, 1113–1116.
25. Kameyama, K., Satake, A., and Kobuke, Y. (2004) *Tetrahedron Lett.*, **45**, 7617–7620.
26. Ito, F., Ishibashi, Y., Khan, S. R., Miyasaka, H., Kameyama, K., Morisue, M., Satake, A. and Kobuke, Y. (2006) *J. Phys. Chem. A.*, **110**, 12734–12742.
27. Morisue, M., and Kobuke, Y. (2008) *Chem. Eur. J.*, **14**, 4993–5000.
28. Morisue, M., Ogawa, K., Kamada, K., Ohta, K., and Kobuke, Y. (2010) *Chem. Commun.*, **46**, 2121–2123.
29. Sutton, J. M. and Boyle, R. W. (2001) *Chem. Commun.*, 2014–2015.
30. Yang, S. I., Li, J., Cho, H. S., Kim, D., Bocian, D. F., Holten, D., andLindsey, J. S. (2000) *J. Mater. Chem.*, **10**, 283–296.

31. Fortage, J., Göransson, E., Blart, E., Becker, H.-C., Hammarström, L., and Odobel, F. (2007) *Chem. Commun.*, 4629–4631.

32. Li, J., Diers, J. R., Seth, J., Yang, S. I., Bocian, D. F., Holten, D., and Lindsey, J. S. (1999) *J. Org. Chem.*, **64**, 9090–9100.

33. Li, J., and Lindsey, J. S. (1999) *J. Org. Chem.*, **64**, 9101–9108.

34. Zhao, Z., Poon, C.-T., Wong, W.-K., Wong, W.-Y., Tam, H.-T., Cheah, K.-W., Xie, T., and Wang, D. (2008) *Eur. J. Inorg. Chem.*, 119–128.

35. Makhseed, S., McKeown, N. B., Msayib, K., and Bumajda, A. (2005) *J. Mater. Chem.*, **15**, 1865–1870.

36. Giribabu, L., Kumar, C. V., and Reddy, P. Y. (2007) *Chem.-Asian J.*, **2**, 1574–1580.

37. Ermilov, E. A., Tannert, S., Werncke, T., Choi, M. T. M., Ng, D. K. P., and Röder, B. (2006) *Chem. Phys.*, **328**, 428–437.

38. Tannert, S., Ermilov, E. A., Vogel, J. O., Choi, M. T. M., Ng, D. K. P., and Röder, B. (2007) *J. Phys. Chem. B*, **111**, 8053–8062.

39. Floris, B., Donzello, M. P., and Ercolani, C. (2003) *The Porphyrin Handbook – Multiporphyrins, Multiphthalocyanines and Arrays*, ed. Kadish, K. M., Smith, K. M., and Guilard, R.), vol. 18, Academic Press, New York, Chapter 112.

40. Donzello, M. P., Bartolino, L., Ercolani, C., and Rizoli, C. (2006) *Inorg. Chem.*, **45**, 6988–6995.

41. Zhao, Z., Cammidge, A. N., and Cook, M. J. (2009) *Chem. Commun.*, 7530–7532.

42. Zobi, F., and Stynes, D. V. (2001) *Can. J. Chem.*, **79**, 795–801.

43. Tian, H.-J., Zhou, Q.-F., Shen, S.-Y., Xu, H.-J. (1993) *J. Photochem. Photobiol. A*, **72**, 163–168.

44. Tran-Thi, T. H. (1997) *Coord. Chem. Rev.*, **160**, 53–91.

45. Maligaspe, E., Kumpulainen, T., Lemmetyinen, H., Tkachenko, N. V., Subbaiyan, N. K., Zandler, M. E., D' Souza, F. (2010) *J. Phys. Chem. A*, **114**, 268–277.

46. Li, X.-Y., and Ng, D. K. P. (2000) *Eur. J. Inorg. Chem.*, 1845–1848.

47. Berber, G., Cammidge, A. N., Chambrier, I., Cook, M. J., and Hough, P. W. (2003) *Tetrahedron Lett.*, **44**, 5527–5529.

48. Cammidge, A. N., Berber, G., Chambrier, I., Hough, P. W., Cook, M. J. (2005) *Tetrahedron*, **61**, 4067–4074.

49. Kojima, T., Honda, T., Ohkubo, K., Shiro, M., Kusukawa, T., Fukuda, T., Kobayashi, N., and Fukuzumi, S. (2008) *Angew. Chem., Int. Ed.*, **47**, 6712–6716.

50. Choi, M. T. M., Choi, C.-F., and Ng, D. K. P. (2004) *Tetrahedron*, **60**, 6889–6894.
51. Leng, X., and Ng, D. K. P. (2007) *Eur. J. Inorg. Chem.*, 4615–4620.
52. Leng, X., Choi, C.-F., Lo, P.-C., Ng, D. K. P. (2007) *Org. Lett.*, **9**, 231–234.
53. Tran-Thi, T. H., Lipskier, J. F., Houde, D., Pepin, C., Keszei, E., and Jay-Gerin, J. P. (1992) *J. Chem. Soc., Faraday Trans.*, **88**, 2129–2137.
54. Lipskier, J. F., and Tran-Thi, T. H. (1993) *Inorg. Chem.*, **32**, 722–731.
55. Fournier, T., Liu, Z., Tran-Thi, T. H., Houde, D., Brasseur, N., La Madeleine, C., Langlois, R., Van Lier, J. E., and Lexa, D. (1999) *J. Phys. Chem. A*, **103**, 1179–1186.
56. Gusev, A. V., Danilov, E. O., and Rodgers, M. A. J. (2002) *J. Phys. Chem. A*, **106**, 1993–2001.
57. Gusev, A. V., and Rodgers, M. A. J. (2002) *J. Phys. Chem. A*, **106**, 1985–1992.
58. Agirtas, S., Ion, R.-M., and Bekaroglu, O. (2000) *Mater. Sci. Eng. C*, **7**, 105–110.
59. Sun, Y., Zhang, X., Sun, C., Wang, Z., Shen, J., Wang, D., and Li, T. (1996) *Chem. Commun.*, 2379–2380.
60. Tran-Thi, T. H., Lipskier, J. F., Simoes, M., and Palacin, S. (1992) *Thin Solid Films*, **210–211**, 150–152.
61. Liu, J., Xu, L., Shen, S., Zhou, Q., Li, T., and Xu, H. (1993) *J. Photochem. Photobiol., A: Chem.*, **71**, 275–278.
62. Tran-Thi, T. H., Fournier, T., Yu, A., Sharonov, N. Tkachenko, H. Lemmetyinen, P. Grenier, K. D. Truong, D. Houde, *Thin Solid Films* **1996**, *273*, 8–13.
63. Pereira, M. R., Ferreira, J. A., and Hungerford, G. (2005) *Chem. Phys. Lett.*, **406**, 360–365.
64. Hipps, K. W., Scudiero, L., Barlow, D. E., Jr. Cooke, M. P. (2002) *J. Am. Chem. Soc.*, **124**, 2126–2127.
65. Scudiero, L., Hipps, K. W., and Barlow, D. E. (2003) *J. Phys. Chem. B*, **107**, 2903–2909.
66. Suto, K., Yoshimoto, S., and Itaya, K. (2006) *Langmuir*, **22**, 10766–10776.
67. Suto, K., Yoshimoto, S., and Itaya, K. (2003) *J. Am. Chem. Soc.*, **125**, 14976–14977.
68. Yoshimoto, S., Higa, N., and Itaya, K. (2004) *J. Am. Chem. Soc.*, **126**, 8540–8545.
69. Yoshimoto, S., Honda, Y., Ito, O., and Itaya, K. (2007) *J. Am. Chem. Soc.*, **130**, 1085–1092.

70. Claessens, C. G., González-Rodríguez, D., and Torres, T. (2002) *Chem. Rev.*, **102**, 835–853.

71. (a) Claessens, C. G., and Torres, T. (2002) *J. Am. Chem. Soc.*, **124**, 14522–14523. (b) Claessens, C. G., and Torres, T. (2004) *Chem. Commun.*, 1298–1299.

72. (a) del Rey, B., Keller, U., Torres, T., Rojo, G., Agulló-López, F., Nonell, S., Marti, C., Brasselet, S., Ledoux, I., and Zyss, J. (1998) *J. Am. Chem. Soc.*, **120**, 12808–12817. (b) Martín, G., Rojo, G., Agulló-López, F., Ferro, V. R., García de la Vega, J. M., Martínez-Díaz, M. V., Torres, T., Ledoux, I., and Zyss, J. (2002) *J. Phys. Chem. B*, **106**, 13139–13145. (c) Claessens, C. G., González-Rodríguez, D., Torres, T., Martín, G., Agulló-López, F., Ledoux, I., Zyss, J., Ferro, V. R., and García de la Vega, J. M. (2005) *J. Phys. Chem. B*, **109**, 3800–3806.

73. (a) D. González-Rodríguez, T. Torres, M. A. Herranz, L. Echegoyen, E. Carbonell, D. M. Guldi, (2008) *Chem. Eur. J.*, **14**, 7670–7679. (b) D. González-Rodríguez, T. Torres, D. M. Guldi, J. Rivera, M. A. Herranz, L. Echegoyen, (2004) *J. Am. Chem. Soc.*, **126**, 6301–6313. (c) D. González-Rodríguez, E. Carbonell, D. M. Guldi, T. Torres, Angew. (2009) *Chem. Int. Ed.*, **48**, 8032–8036. (d) R. S. Iglesias, C. G. Claessens, T. Torres, G. M. A. Rahman, D. M. Guldi, (2005) *Chem. Commun.*, 2113–2115.)

74. Gonzalez-Rodriguez, D., Claessens, C. G., Torres, T., Liu, S. G., Echegoyen, L., Vila, N., and Nonell, S. (2005) *Chem. Eur. J.*, **11**, 3881–3893.

75. Claessens, C. G., González-Rodríguez, D., Torres, T., Mark, G., Schuchmann, H.-P., von Sonntag, C., Lye, J., MacDonald, J. G., Nohr, R. S. (2003) *Eur. J. Org. Chem.*, 2547–2551.

76. (a)Maya, E. M., García, C., García-Frutos, E. M., Vázquez, P., and Torres, T. (2000) *J. Org. Chem.* **65**, 2733–2739. (b) García-Frutos, E. M., Fernández-Lázaro, F., Maya, E. M., Vázquez, P., and Torres, T. (2000) *J. Org. Chem.*, **65**, 6841–6846.

77. Shibata, N., Das, B., Tokunaga, E., Shiro, M., and Kobayashi, N. (2010) *Chem. Eur. J.* **16**, 7554–7562.

78. Xu, H., and Ng, D. K. P. (2008) *Inorg. Chem.*, **47**, 7921–7927.

79. El-Khouly, M. E., Ryu, J. B., Kay, K.-Y., Ito, O., and Fukuzumi, S. (2009) *J. Phys. Chem. C* **113**, 15444–15453.

80. Gouloumis, A., Gonzalez-Rodriguez, D., Vazquez, P., Torres, T., Liu, S. G., Echegoyen, L., Ramey, J., Hug, G. L., and Guldi, D. M. (2006) *J. Am. Chem. Soc.*, **128**, 12674–12684.

81. Gouloumis, A., Liu, S. G., Vazquez, P., Echegoyen, L., and Torres, T. (2001) *Chem. Commun.*, 399–400.

82. Quintiliani, M., Kahnt, A., Vazquez, P., Guldi, D. M., and Torres, T. (2008) *J. Mater. Chem.*, **18**, 1542–1546.
83. Kahnt, A., Quintiliani, M., Vazquez, P., Guldi, D. M., and Torres, T. (2008) *ChemSusChem*, **1**, 97–102.
84. Jimenez, A. J., Spanig, F., Rodriguez-Morgade, M. S., Ohkubo, K., Fukuzumi, S., Guldi, D. M., and Torres, T. (2007) *Org. Lett.*, **9**, 2481–2484.
85. Chen, Y., Lin, Y., El-Khouly, M. E., Zhuang, X., Araki, Y., Ito, O., Zhang, W. (2007) *J. Phys. Chem. C*, **111**, 16096–16099.
86. Miller, M. A., Lammi, R. K., Prathapan, S., Holten, D., and Lindsey, J. S. (2000) *J. Org. Chem.*, **65**, 6634–6649.
87. Ambroise, A., Wagner, R. W., Rao, P. D., Riggs, J. A., Hascoat, P., Diers, J. R., Seth, J., Lammi, R. K., Bocian, D. F., Holten, D., Lindsey, J. S. (2001) *Chem. Mater.* **13**, 1023–1034.
88. Li, L., Shen, S., Yu, Q., Zhou, Q., and Xu, H. (1991) *J. Chem. Soc., Chem. Commun.*, 619–620.
89. Guldi, D. M., Ramey, J., Martinez-Diaz, M. V., de la Escosura, A., Torres, T., Da Ros, T., and Prato, M. (2002) *Chem. Commun.*, 2774–2775.
90. Rodriguez-Morgade, M. S., Plonska-Brzezinska, M. E., Athans, A. J., Carbonell, E., de Miguel, G., Guldi, D. M., Echegoyen, L., Torres, T. (2009) *J. Am. Chem. Soc.* **131**, 10484–10496.
91. Rodriguez-Morgade, M. S., Torres, T., Atienza-Castellanos, C., and Guldi, D. M. (2006) *J. Am. Chem. Soc.*, **128**, 15145–15154.
92. Prodi, A., Chiorboli, C., Scandola, F., Iengo, E., Alessio, E., Dobrawa, R., and Würthner, F. (2005) *J. Am. Chem. Soc.*, **127**, 1454–1462.

Chapter 5

Primary Photoevents in Self-Assembled Complexes: Multiporphyrin Arrays and "Semiconductor CdSe/ZnS Nanocrystal-Porphyrin" Composites

Eduard Zenkevich[a] and Christian von Borczyskowski[b]

[a] *Department of Information Technologies and Robototechnique, National Technical University of Belarus, Prospect Nezavisimosti 65 220013 Minsk, Belarus*

[b] *Institute of Physics, Center for Nanostructured Materials and Analytics, University of Technology Chemnitz, Reichenhainer Str. 70 D-09107 Chemnitz, Germany*

zenkev@tut.by; borczyskowski@physik.tu-chemnitz.de

It is clear that well-defined geometry is one of the important factors requisite for enhanced light-harvesting efficiency and/or ultrafast vectorial electron transfer in natural and artificial systems. With this idea in mind, we have elaborated the formation (in solutions and polymeric PMMA films at 293 K) various types of self-assembled nanoscale multiporphyrin arrays of controllable geometry and composition (up to 8 tetrapyrroles) using non-covalent twofold binding interactions of the meso-phenyl-bridged ZnOEP chemical dimers or trimers, $(ZnOEP)_2Ph$ or $(ZnOEP)_3Ph_2$, with di- and tetrapyridyl-substituted tetrapyrrole extra-ligands. The dynamics

Multiporphyrin Arrays: Fundamentals and Applications
Edited by Dongho Kim

www.panstanford.com

and mechanisms of relaxation processes (energy and charge transfer) in multiporphyrin arrays as a function of redox and photophysical properties of interacting subunits have been studied in solvents of various polarity and temperature (77–293 K), using static, time-resolved picosecond fluorescent and femtosecond pump-probe set-ups.

Inspired by work on self-assembled multiporphyrin arrays, we have succeeded in the direct labelling of semiconductor nanocrystals (NC) CdSe/ZnS with pyridyl-substituted porphyrins, $H_2P(m\text{-}Pyr)_n$ ($n = 1, 2, 3,$ or 4), based on the coordination of the pyridyl N lone pair with NC surface Zn or Cd atoms. Although part of the porphyrin-induced NC photoluminescence (PL) quenching could be explained by Foerster resonant energy transfer "NC→porphyrin," more than 85% of the PL quenching is caused by the other reason depending on the NC size (the charge tunnelling through the ZnS barrier in the conditions of quantum confinement).

In this review paper, we discuss our results on structural and photophysical properties for self-assembled multiporphyrin complexes of several types as well for "CdSe/ZnS NC-porphyrin" composites providing a characterization of the complex relaxation dynamics for the systems under study (including energy and charge transfer as well electron tunneling effects). The intention of this paper is not a thorough theoretical description of all the relaxation processes in these units, which would be too early given some open problems as well as some specific structural aspects. Rather the present work should be viewed as a more comparative characterization of the non-covalent self-assembly possibilities and the dynamics in these systems, thus providing the data for a further development of defined multicomponent structures for exploitation as artificial light-harvesting complexes, electro- and photochemical devices, nanosensors, etc.).

5.1 Introduction

5.1.1 *Photosynthetic Aspects and Applications in Nanotechnology*

At the moment, on the basis of x-ray data in combination with steady-state, time-resolved and single-molecule spectroscopy methods, it is well documented that photosynthesis is one of the finest piece of nanoscale molecular machinery where nature utilizes multicomponent arrays of tetrapyrrole molecules and other organic substances to channel energy and electron-transfer processes whose directionality and efficiencies are affected largely by the structure, properties, and orientation of interacting subunits [1–7]. In fact, the most complex and spectacular sets of tetrapyrrole containing organized arrays are found in photosynthetic objects that are capable for the directed very fast and efficient energy transfer (ET) among light-harvesting pigment–protein antenna complexes to the photochemical reaction center where the energy of excited states is converted into a stable transmembrane charge separation through a sequence of photoinduced electron transfer (PET) reactions. The elucidation of the mechanisms and dynamics of ET processes in light-harvesting antenna complexes in vivo as well as the intrinsic peculiarities of PET events, such as charge separation and charge recombination of the product ion pair state are the most fundamental and important problems. Noteworthy, despite intensive and wide research in this field, some aspects of the light collection and distant electron transfer reactions remain non-understood yet in great detail (for instance, the role of pigment–protein interactions and electronic couplings via bridge in the directed ET [8] and PET [6] processes, electronic couplings and optimized energy transfer in confined molecular assemblies [9], the relatively weak temperature dependence and high efficiency of charge separation). Additionally, there is an ongoing debate, as to whether PET is sequential for all steps, that is following the shortest pathway along the donor/acceptor array or whether so-called spectator states mediate the process via superexchange without participating in the transfer directly [10]. In spite of the fact that available experiments provide evidence for a sequential PET at room temperature [11] it cannot be excluded that the superexchange mechanism may play a role in various mutants. Theoretical investigations still allow for this possibility in bridged molecular systems [12, 13].

In this respect, multiporphyrin assemblies are of fundamental importance as models for mimicking and the detailed study of ET and PET processes taking place in the light-harvesting antenna complexes and the photosynthetic reaction centers of green plants and photosynthetic bacteria. In some sense, the real situation with native structurally and functionally correlated systems reflects their inner fundamental limitations which may be obviated by the disturbing the existing connections between individual parts of the objects *in vivo.* As neither the molecular photoprocesses mentioned above, nor more complex light-induced functions such as photoswitching of on/off receptor donor acceptor ability, is achieved in single molecules, but requires significant communication between interacting subunits, supramolecular assemblies are truly relevant biomimetic models for the fundamental energy and charge storage strategies observed in physiological photosynthetic systems.

From the other hand, the bottom-up construction of molecular/supramolecular devices including organic subunits offers a formidable challenge on the road toward modern nanotechnology [14–17]. This field is a new frontier of research that combines two seemingly incompatible objects: the building blocks of life and synthetic structures, both of them at a tiny, molecular-sized level. Its focus is on the development of powerful techniques and methods that merge the strengths of nanotechnology, working typically in the range of 1 to 100 nanometers, and biophysics, to generate a new type of 'bionanomaterial' which has some uniquely designed properties. In this respect, the interest in emerging nanostructures (huge part of them based on tetrapyrrolic macrocycles) is growing exponentially since they are not only good models for the mimicking the primary photoevents *in vivo* but seem to be considered as promising building blocks for advanced multifunctional nanocomposites with potential applications in improved drug delivery systems, photodynamic therapy, nanovoltaic cells, optoelectronic memory, multimolecular architectures for information storage and highly efficient catalysts, among several others [18, 19]. In fact, this direction is the supramolecular organic chemistry/photochemistry in solutions or at solid/liquid interfaces [20, 21], a highly interdisciplinary field of science covering the chemical, physical, and biological features of chemical species held together and organized by means of intermolecular binding interactions of various natures. The resulting crossover has provided novel principles and concepts in physico-chemistry such as molecular recognition, self-organization, regulation, cooperativity, replication. A significant interest of numerous scientific groups in this direction has been devoted to the design and investigation of tetrapyrrole compounds that fold or assemble predictably in order to form multicomponent well-defined arrays. A large body of interesting and important results obtained in this field of supramolecular chemistry has been presented and discussed in some comprehensive reviews [22–30] and numerous articles [e.g., 31–44].

The second direction of nanotechnology is based on the preparation (by lithographic methods, molecular beam epitaxy or conventional organometalic synthesis) and characterization of colloidal semiconductor nanocrystals (NC), e.g., CdSe or CdSe/ZnS and other II–VI systems with pronounced quantum confinement effects and thus tuneable colour and redox properties [45–48]. Such NCs become due to quantum size effects, quantum dots and are often strongly emissive [49]. NC optical properties arise from several phenomena (viz. quantum confinement of charge carriers, surface effects, and geometrical confinement of photons) and have turned these inorganic systems into promising materials for many applications [46, 48, 50], such as light-emitting diodes, lasers, charge-transfer composites, biology/medicine, etc.

In this respect, the combination of the two directions, that is the anchoring of organic molecules to tuneable wide gap semiconductor colloidal nanocrystals, is of considerable scientific and practical interest [51–55]. In heterocomposites of this type, the phenomena of photosensitized electron injection from organic subunits to semiconductor nanoparticles is used to activate charge separation with visible light quanta less than the semiconductor band gap energy [51]. In addition, the non-radiative energy transfer depending on absorption/emission properties and intercenter distances between coupled interacting nanoparticles–dye moieties strongly influences the dynamics and relaxation pathways in such complex systems [52–55]. The other driving force for the growing interest in the inorganic–organic nanocomposites is due to the use of NC as photoluminescence markers linked to biological relevant molecules [56], the development of new sensitizers of singlet oxygen for the photodynamic therapy [57, 58] (including ET process NC→ porphyrin [53, 59–61]), and the creation of new optical and laser materials [62, 63].

5.1.2 *Synthetic Approaches in the Formation of Multiporphyrin Arrays and "Semiconductor NC-Porphyrin" Nanocomposites*

At the moment, the number of photosynthetic porphyrin-containing models which mimic the primary photoinduced processes *in vivo* is vast, and descriptions of many of them may be found in reviews [22–30] and original papers. Beginning from pioneer investigations in this area [64] every following step has been a basis for the development of new strategies for the formation of larger scale complex systems. In fact, the formation of a variety of conformationally restricted, structurally and energetically well-defined multiporphyrin assemblies is based on two principally different synthetic strategies. One approach utilizes the covalent linkage through spacers of various nature and flexibility between supposedly essential components. The other, more common, approach is based on non-covalent interactions of various kinds (electrostatic interactions, hydrogen bonds, and the coordination chemistry of transition metals). Correspondingly, both synthetic methodologies, each with its advantages and disadvantages, continue to supply a large variety of new multicomponent arrays with pronounced ET and PET properties.

A new concept generally proposed first by J.-M. Lehn [20] for the formation of large functional supramolecular assemblies is based on the combination of these two principally different approaches. With all these ideas in mind, we realized a simple and yet potentially versatile strategy for the formation of highly organized and relatively rigid multimolecular tetrapyrrole assemblies in solutions and polymeric (PMMA) films [30, 65–76]. It includes a covalent stage where precursor molecular blocks such as Zn-porphyrin or Zn-chlorin chemical dimers with a covalent linkage of various nature (-CH_2-CH_2- or a phenyl ring in meso-position) as well as chemical trimers of Zn-octaethylporphyrins with phenyl spacers in meso-positions have been synthesized. Additionally, various types of electron acceptors (quinone, anthraquinone, pyromellitimide, NO_2 group) may be covalently linked to these dimers by flexible or rigid spacers. At the second stage dimeric molecular blocks (with/without electron acceptors) may be self-assembled with pyridyl containing porphyrin or chlorin extra-ligands through non-covalent binding interactions (two-fold extra-ligation effect following the "key-hole" principle). The key-hole organization principle is based on the complexation of central Zn ions of porphyrin chemical dimers or trimers with suitable extra-ligands (di- and tetrapyridyl-substituted tetrapyrrolic macrocycles) via two-fold non-covalent coordination [30, 65–68, 74]. Noteworthy, the matching geometry between nitrogen atoms in meso-pyridyl containing extra-ligands and Zn-Zn distance in the dimers plays the essential role in the formation of triads and pentads with relatively well-defined conformational rigidity. In the result, we have succeeded to form distinct arrays of variable structure with a controlled number of electronically interacting chromophores having given spectral, energetic and redox properties. It should be mentioned that the alternative principles using the combination of the two approaches have also been proposed and realized [77–80].

5.1.2.1 Self-assembled complexes with participation of tetrapyrrole molecules

The first part of this paper focuses mainly on the electronic excitation energy relaxation processes for some of self-assembled multiporphyrin complexes of known geometry and composition (up to eight tetrapyrrole macrocycles, the corresponding structures will be presented below) in solutions at 77–293 K. Details of the precursor molecular blocks preparation [Zn-octaethylporphyrin or Zn-octaethylchlorin chemical dimers with a covalent linkage of various nature ($-CH_2-CH_2-$ or a phenyl ring in meso-position), $(ZnOEP)_2$, $(ZnOEP)_2Ph$, $(ZnOEChl)_2$; chemical trimers of Zn-octaethylporphyrins with phenyl spacers in meso-positions, $(ZnOEP)_3Ph_2$; as well as di- and tetrapyridyl pyridyl-substituted porphyrin extra-ligands, $H_2P(m\wedge Pyr)_2$-(iso-PrPh)$_2$, H_2P(m-Pyr)$_4$, containing two adjacent or four meso-pyridyl rings with meta-position of nitrogens N, the formation of multiporphyrin complexes as well as experimental technique have been reported earlier [30, 65–67, 74–76]. Typically, at ambient temperature multiporphyrin complexes of various structure and composition are formed during a titration of the chemical dimer $(ZnOEP)_2Ph$ solution by the extra-ligand solution or by one-step mixing of the two solutions at appropriate concentrations. In the case of pair interactions, complexation constants K_C for the triad formation have been obtained by the conventional technique [30, 66, 67] on the basis of both absorption and fluorescence data measured in the titration experiments using the following expression:

$$K_C = (1 - \beta) \times \frac{1}{C_{D0} \cdot \beta (x + \beta - 1)} \qquad (5.1)$$

where C_{D0} is the initial concentration of the dimer $(ZnOEP)_2Ph$, $x = [\text{ligand}]/[(ZnOEP)_2Ph]$ is a molar ratio with [ligand] as the total experimental concentration of the extra-ligand in the solution; $\beta = I/I_0$ is the portion of the uncomplexed dimer molecules which can be estimated from the measurements of the uncomplexed dimer fluorescence; I and I_0 are the measured and initial integral fluorescence intensities, respectively, for the uncomplexed dimer at every step of the titration. K_C values for pentads may be derived upon the assumption that pentads are formed via the intermediate stage of the precursor triad formation [66, 67]. Due to two-fold extra-coordination, complexation constants for self-assembled triads with participation of the dimer $(ZnOEP)_2Ph$ and $H_2P(Pyr)_n$ have been estimated to be $K_C \sim 5\times 10^6\ M^{-1} \div 2\times 10^7\ M^{-1}$ (depending on the position of meso-pyridyl rings, opposite or adjacent) and are of two or three orders of magnitude larger than those obtained for the di-pyridinated dimer $(ZnOEP)_2$ Ph [30]. The corresponding K_C values for the formation self-assembled complexes with participation of the trimer $(ZnOEP)_3Ph_2$ have been estimated to be of $\sim(6.5 \pm 0.5) \times 10^7\ M^{-1}$ [74]. It means that the multiporphyrin systems under study are characterized by strong allosteric behavior showing that the first binding accelerates the second binding because of the chelate effect.

Two types of the triads with the same geometry will be discussed below [69–71]: the triad with porphyrin extra-ligand $H_2P(m{^\wedge}Pyr)_2$-

(iso-PrPh)$_2$ and the other one having 5,10-di(pentafluorophenyl)-15,20-dipyridyl free base porphyrin, $H_2P(m^\wedge Pyr)_2$-$(5FPh)_2$ known as a strong electron acceptor [81]. In addition, the chemical dimer $(ZnOEP)_2$ Ph may be covalently linked in meso-position to an electron acceptor *p*-benzoquinone (Q) or Pyromellitimide (Pim), having different redox properties and fixed at the same distance and geometry with respect to the dimer [30, 72]. For such tetrads, we like to concentrate on some aspects of interporphyrin and porphyrin-quinone PET competing with the non-radiative ET processes. The structures of the corresponding self-assembled triads and tetrads are shown in Fig. 5.1.

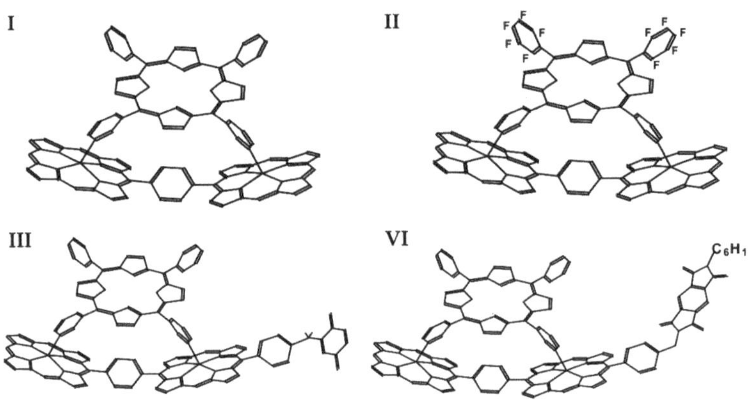

Figure 5.1. Mutual arrangement of the chemical dimer $(ZnOEP)_2Ph$, porphyrin extra-ligands in triads (**I, II**) and tetrads (**III, IV**) containing electron acceptors, quinone Q and pyromellitimide Pim. **I**: $(ZnOEP)_2$ Ph⊗$H_2P(m^\wedge Pyr)_2$-(iso-PrPh)$_2$; **II**: $(ZnOEP)_2$Ph⊗$H_2P(m^\wedge Pyr)_2$-$(5FPh)_2$; **III**: $(ZnOEP)_2$Ph-Q⊗$H_2P(m^\wedge Pyr)_2$-(iso-PrPh)$_2$; **IV**: $(ZnOEP)_2$Ph-Pim⊗$H_2P(m^\wedge Pyr)_2$-(iso-PrPh)$_2$. Optimized geometries have been calculated on the basis of HyperChem software package (release 4, semiempirical method PM3). For clarity, side alkyl substituents in pyrrole rings of the dimer and meso-phenyl rings of the extra-ligand are omitted. Central Zn ions in the dimer are coupled by co-ordination bonds with nitrogens of pyridyl rings (Pyr) of tetrapyrrole extra-ligands (H_2P is porphyrin). Two-fold interaction is formed via adjacent (^) pyridyl rings having nitrogens in meta- (m) positions. The symbol ⊗ is used in order to show what interacting subunits (dimer and extra-ligands) are bound together.

Using the above approach of the two-point coordination we have succeeded to bind two dimers $(ZnOEP)_2Ph$ via one molecule of tetrapyridyl-substituted $CuP(mPyr)_4$, i.e., to form pentads containing five tetrapyrrole macrocycles (Fig. 5.2, structure **V**, complexation constant $K_C = 1.2 \times 10^7\ M^{-1}$). At last, based on this

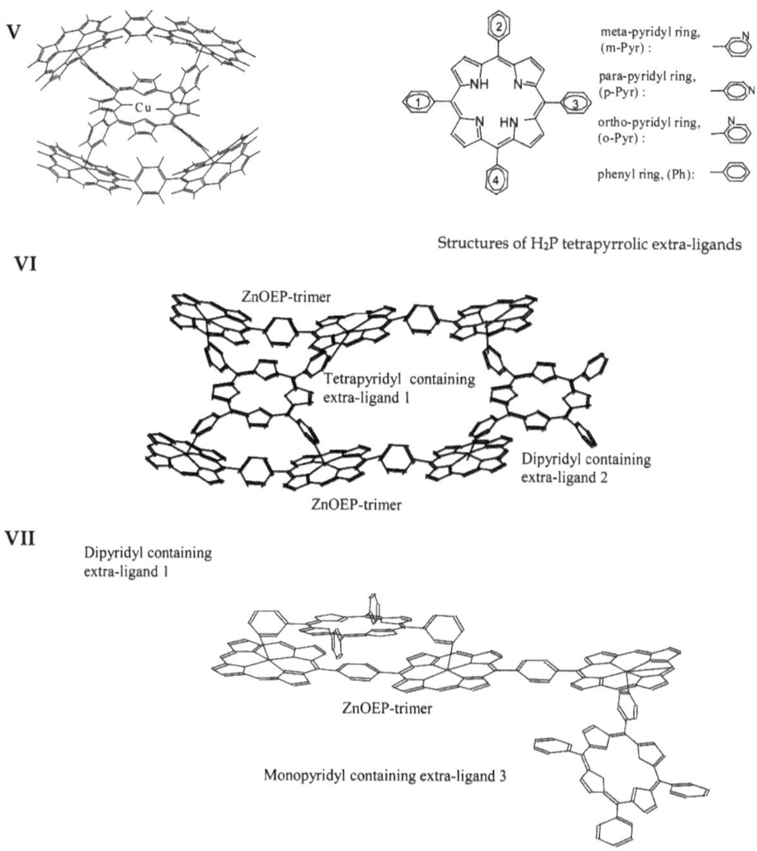

Figure 5.2. Mutual arrangement of the dimer $(ZnOEP)_2Ph$, trimer $(ZnOEP)_3Ph_2$ and various extra-ligands in multiporphyrin arrays with number of tetrapyrrole macrocycles $n \geq 5$. For clarity, β-ethyl substituents in $(ZnOEP)_2Ph$ and $(ZnOEP)_3Ph_2$ are omitted. **V**: – pentad $2(ZnOEP)_2Ph{\otimes}CuP(mPyr)_4$; **VI** and **VII**: self-assembled multiporphyrin complexes formed via two-point coordination and via the combination of two- and one-point extra-coordination.

strategy we could form multiporphyrin arrays with number of tetrapyrrole macrocycles $n \geq 5$ containing the trimer $(ZnOEP)_3Ph_2$ and various extra-ligands with given spectral, energetic, and redox properties (meso-pyridyl-substituted porphyrins, chlorins, tetrahydroporphyrins, and pentafluorophenyl porphyrins) [74]. Few structures based on the covalently linked trimer $ZnOEP)_3Ph_2$ and tetra-, di, and monopyridyl containing porphyrins are depicted in Fig. 5.2 (structures **VI** and **VII**). The detailed structural and photophysical description of various complexes based on the trimer has been presented in [74].

5.1.2.2 Self-assembled inorganic–organic nanocomposites containing semiconductor CdSe/ZnS nanocrystals and porphyrins

The tri-n-trioctylphosphine oxide (TOPO) caped highly monodisperse CdSe nanocrystals (NC) with an inorganic ZnS shell have been used in order to form semiconductor NC—porphyrin nanocomposites in toluene (or chloroform) at ambient temperature. Taking into account the existence of an inorganic ZnS shell covering CdSe NCs we have succeeded to realize the surface passivation of CdSe/ZnS NCs by meso-pyridyl-substituted porphyrins, $H_2P(m\text{-}Pyr)_n (n = 1, 2, 3, \text{ or } 4)$ and chlorins [53, 82, 83] as well as by other organic molecules with the corresponding anchoring groups [84–86] using the self-aggregation approach described above for multiporphyrin arrays (replacing Zn-porphyrin dimers with NCs) The colloidal TOPO-capped core/shell CdSe/ZnS NCs were obtained from Evident Technologies, Inc, Troy, NY, USA. The core diameters d_{CdSe} for NC vary between 2.1 and 5.2 nm, while (in most cases) two capping ZnS monolayers have been applied. Titration experiments were carried out by adding $H_2P(m\text{-}Pyr)_n$ molecules to the NC solution (with initial concentration C in the range $(1 \div 10) \times 10^{-7}$ M) at relative molar ratios x = [porphyrin]/[QD] in toluene, thus varying the number of porphyrin molecules on the NC surface. Details on the material properties, sample preparation and experimental setup are presented in our recent publications [53, 82–84].

Figure 5.3 shows a schematic presentation of such a heteronanoassembly consisting of a CdSe/ZnS nanocrystal with the

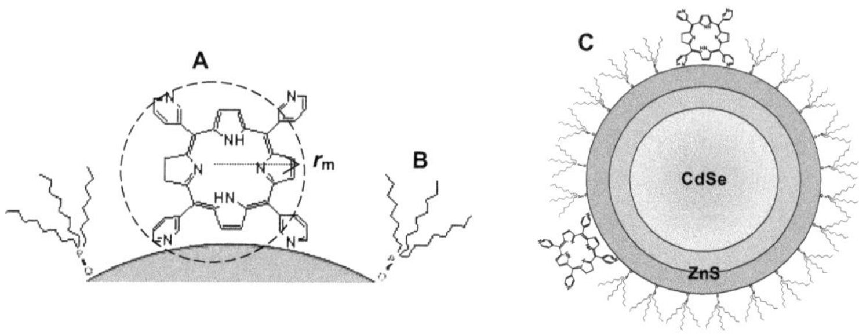

Figure 5.3. Structures of the porphyrin ligand (A), TOPO molecule (B) and schematic presentation of "NC-porphyrin" nanocomposite (C). The scales of CdSe core, ZnS shell, porphyrin and TOPO molecules as well as intercenter distances correspond to relative sizes of the main components of the arrays [53]: the thickness of one ZnS layer l = 5 Å; parameters for conical TOPO molecules $r_{\text{bottom}} = 5.5$ Å, $h_{\text{con}} = 9.9$ Å were used; $r_{\text{m}} = 7.5$ Å is the radius of porphyrin molecule with opposite pyridyl rings having nitrogens in meta-positions, $h = 10$ Å is the mean distance between meta-nitrogens of adjacent pyridyl rings (HyperChem software, release 4.0 geometry optimization with semiempirical PM3 method).

tri-*n*-octyl phosphine oxide (TOPO) surfactant layer and H_2P(m-Pyr)$_4$ molecules attached via their meso-pyridyl rings nearly perpendicular to the NC surface (based on both experimental results [53, 83] and theoretical calculations using HyperChem 7.0 and DFT method [60]). It follows from the detailed spectral observations [53, 82, 83] that at low molar ratios x all $H_2P(Pyr)_4$ molecules should be attached to the CdSe/ZnS surface. Thus from the experimental titration data on the basis of a bi-molecular reaction scheme (valid for a dynamic equilibrium between complexed and free constituents like in the case of multiporphyrin arrays, see Section 5.1.2) the values of the complexation constant K_{C} have been estimated for NC solutions ($C_{\text{NC}} = 5 \times 10^{-8}$ M, toluene, 293 K) as a function of various H_2P(m-Pyr)$_n$ extra-ligands [83]. The obtained results clearly show that for the given NC, K_{C} values are increased with the number of pyridyl rings. For $H_2P(Pyr)_4 K_{\text{C}} = 2.6\times 10^7$ M^{-1} being in a comparable range with respect to those characteristic for the two-fold extra-coordination in multiporphyrin arrays.

At last, it should be mentioned that among other reasons the tunability of the optical band gap of nanocrystals via the size dependent quantum confinement [46–50] (optical tunability) and the nearly unlimited possibilities of chemical engineering of controllable electronic properties of organic molecules [21–30] (chemical tunability) are stimulating and promising arguments to concentrate on the architecture of organic–inorganic nanoassemblies with respect to generation of new material properties.

5.2 Excited States Relaxation and Energy Transfer in Porphyrin Chemical Dimers

The comparative analysis of all self-assembled multiporphyrin complexes being studied [30, 65–68, 75, 76] has shown that the most pronounced spectral effects and energetic interactions upon their formation are observed for the phenyl-bridged dimers and trimers containing ZnOEP (Zn-octaethylporphyrin) and ZnTPP or ZnHTPP (Zn-tetraphenyl- or Zn-tetrahexyphenyl-porphyrin) monomeric components. Nevertheless, we apt to believe that for the comparison and understanding the photophysics of multiporphyrin arrays, our early results on pathways and mechanisms of excited states relaxation for chemical dimers (as "smallest" multiassemblies, Fig. 5.4) should be taken into account. Here, we present shortly some of them being obtained for geometry-restricted systems in a temperature range of 1.8 $\div$ 293 K (Fig. 5.4).

Main conclusions are the following [87–90]. The formation of intense UV absorption bands in the dimers is caused by two reasons: excitonic coupling and vibronic borrowing. Ground and first excited singlet/triplet states of subunits in the dimers with the inter-center distance $R = 0.55$–1.23 nm are involved into weak dipole–dipole/exchange interactions that result in pico-/nanosecond ET processes. In addition, interchromophoric and steric interactions manifest themselves in exchange d-π effects, ET induced NH-tautomerism, enhanced non-radiative deactivation of excited states, charge transfer etc.

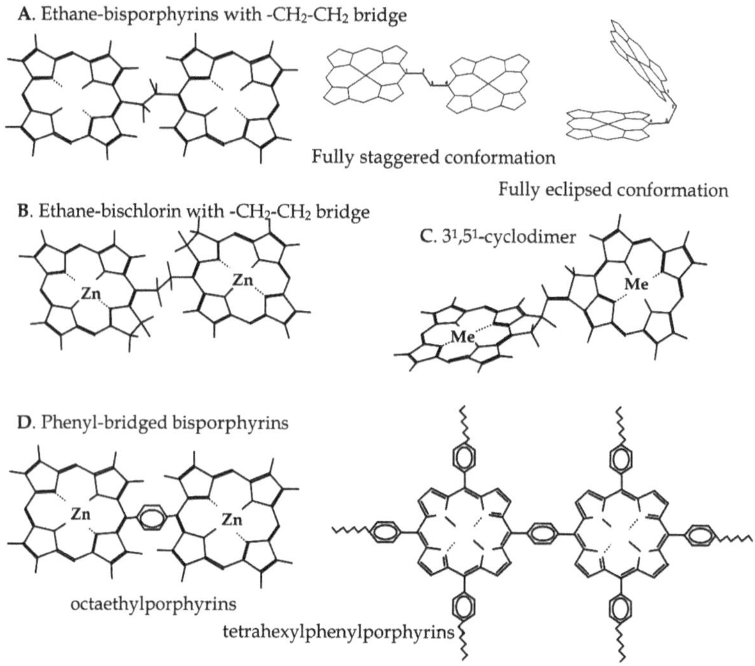

Figure 5.4. Structure of porphyrin and chlorine chemical dimers with various spacers. **A**. Ethane-bisporphyrins: homodimers $(OEP)_2$ and $(ZnOEP)_2$; heterodimers $(H_2+Cu)(OPP)_2$, $(Zn+Cu)(OEP)_2$ and $(H_2+Ni)(OEP)_2$. **B**. Ethane-bis-Zn-octaethylchlorin $(ZnOEChl)_2$with $-CH_2-CH_2$ bridge. **C**. 3^1,5^1-cyclodimer with covalent link via isocycle. **D**. Phenyl-bridged bisporphyrins: $(OEP)_2Ph$, $(ZnOEP)_2Ph$ and $(ZnHTPP)_2$.

5.2.1 *Interchromophoric Interactions and Energy Transfer in $-CH_2-CH_2-$Bridged Dimers*

Temperature NMR ^{1}H studies of ethane-bisporphyrins, $(ZnOEP)_2$, and $(OEP)_2$, revealed [91] that their conformers relative to the spacer $-CH_2-CH_2-$ range from the fully eclipsed (intercenter distance $d = 0.55$ nm) at ambient temperature to the fully staggered ($d = 1.06$ nm) at 77 K (Fig. 5.4A). Interchromophoric interactions in homodimers $(ZnOEP)_2$ and $(OEP)_2$ do not alter essentially their spectral and photophysical parameters with respect to the corresponding precursor monomeric compounds (Table 5.1) [91–94].

Table 5.1. Excited state parameters of porphyrin and chlorin chemical dimers and their monomeric subunits

Compound	φ_F 295 K	φ_F 77 K	τ_S(ns) 295 K	τ_S (ns) 77 K	τ_T(μs) 295 K	τ_T(μs) 77 K	τ_P (μs) 77 K	φ_P 77 K
Precursor monomers with corresponding side meso-substituents								
OEP-mesoCH_3	0.03	0.04	10.7	15.5	1000	7600	8600	0.0002
OEP-mesoPh	0.05	0.09	16.0	21.0	4.0	11100	13500	0.0002
ZnOEP-mesoCH_3	0.015	0.02[d]	1.7	2.3	1200	31400	35000	0.007
ZnOEP-mesoPh	0.02	0.03	1.6	2.3	5.8	32000	39000	0.03
CuOEP	—	—	—	—	0.120	65	72	0.14
CuOEP-mesoCH_3	—	—	—	—	0.025	160	150	0.04
CuOEP-mesoPh	—	—	—	—	0.025	40	48	0.05
TPP	0.09		12.2	16.0	1000	4900	5800	0.00006
ZnTPP	0,02	0.03	1.7	2.1	1200	30000	25000	0.007
Zn-chlorin-mesoCH_3	0.015	0.02	1.0	1.0	—	—	—	—
CuOEP-cycle	—	—	—	—	—	160	180	0.13
CuOEP-cycle=CH_2	—	—	—	—	—	220	200	0.02
OEP-cycle	0.05	0.10	16.0	22.4	—	19500	—	—
OEP-cycle=CH_2	0.04	0.09	15.0	21.1	—	23000	—	—
ZnOEP-cycle	0.015	0.02	1.9	2.9	—	72500	—	—
ZnOEP-cycle=CH_2	0.025	0.05	1.8	3.5	—	44500	—	—
Ethane-bisporphyrins and ethane-bischlorins with $-CH_2-CH_2-$ spacer								
$(OPP)_2$	0.025	0.035	8.2	11.6	1100	6300	7600	0.00002
$(ZnOPP)_2$	0.02	0.03	1.5	2.3	1150	33900	41000	0.015

(*Contd.*)

Table 5.1. *(Contd.)*

Compound	φ_F 295 K	φ_F 77 K	τ_S(ns) 295 K	τ_S (ns) 77 K	τ_T(μs) 295 K	τ_T(μs) 77 K	τ_P (μs) 77 K	φ_P 77 K
$(CuOPP)_2$	—	—	—	—	0.25	130	80	0.02
$(Cu+H_2)(OPP)_2$	0.0013	0.0018	—	—	0.9	350	300	0.003
$(Cu+Zn)(OEP)_2$	0.0002	0.0004	—	—	0.7	7900/1100	9200/1900	0.04
$(Ni+H_2)OEP)_2$	0.0001	-	0.4	—	—	—	—	—
$(ZnOEChl)_2$	0.015	0.02	1.1	1.2	—	—	—	—
Cyclodimer swith covalent link via isocycle								
$3^1,5^1$-cyclodimer	0.04	0.10	15.0	22.5	—	21600	—	—
Zn-$3^1,5^1$-cyclodimer	0.03	0.04	1.8	3.0	—	57400	—	—
Cu-$3^1,5^1$-cyclodimer	—	—	—	—	—	170	190	0.02
Chemical dimers with the phenyl spacer								
$(OEP)_2Ph$	0.06	0.08	11.2	16.0	2.45	3900	7600	0.00003
$(ZnOEP)_2Ph$	0.015	0.025	1.2	2.1	2,85	27000	38000	0.01
$(CuOEP)_2Ph$	—	—	—	—	0.015	0.28	0.24	0.01
$(Cu+H_2)(OEP)_2Ph$	0.002	0.0025	—	—	0.035	108	100	0.008
$(ZnHTPP)_2$	0.02	0,03	1.3	1.7	16.0	22300	14500	0.002

Notes: φ_F and τ_S are fluorescence quantum yield and lifetime values; τ_T and τ_P are excited triplet state decays measured by pump-probe and kinetic phosphorescence methods, correspondingly; φ_P is phosphorescence quantum yield; In the case of Cu-containing monomers and symmetrical dimers, the fluorescence was not observed within the limits of the experimental setup sensitivity ($\varphi_F < 10^{-5}$) in the temperature range of 77–295 K. At 295 K φ_F, τ_S, and τ_T values were measured for degassed solutions (TOL or MCH), at 77 K rigid glassy matrices (MCH + TOL, 3:1) were used.

The homodimer S_1 state is deactivated mainly through intersystem crossing to T_1 state where the excitation is lost via non-radiative processes (at 293 K). Dipole-dipole interactions between homodimer subunits ($1.5\ cm^{-1} \leq V_{12} \leq 11\ cm^{-1}$) lead to Foerster-type ET between them with a rate constant $k_{ET}^{SS} = (0.15 \div 13.5) \times 10^{10}\ s^{-1}$, and ET without quantum losses is much faster than the other deactivation processes in homodimers ($k_{\Sigma} = (0.9 \div 1.2) \times 10^{8}\ s^{-1}$). As a result, numerous "jumps" of excitation between the subunits within the S_1 state decay manifest themselves in the "collapse" of fluorescence excitation polarization spectra of homodimers compared to the corresponding individual monomers. In a temperature range of $77 \div 293$ K Foerster theory is still applicable [91–94].

At lower T (1.8–4.2 K) for the homodimer $(OEP)_2$, there is an energy gap between S_1 states of chemically identical halves caused by an essential spectral inhomogeneity in rigid glassy solution [95]. In these conditions the fluorescence excitation spectrum consists of two subbands: a structureless blue part and a quasi-line red one with an average splitting of 51 cm^{-1} (Fig. 5.5). These subbands are ascribed to the donor (D) "blue" and the acceptor (A) "red" halves of the homodimer. The anticorrelation between the D–A splitting and the 0–0 electronic transition frequency results in different values of fluorescence-absorption overlap integrals. The D fluorescence is effectively quenched by the directed singlet-singlet ET to A.

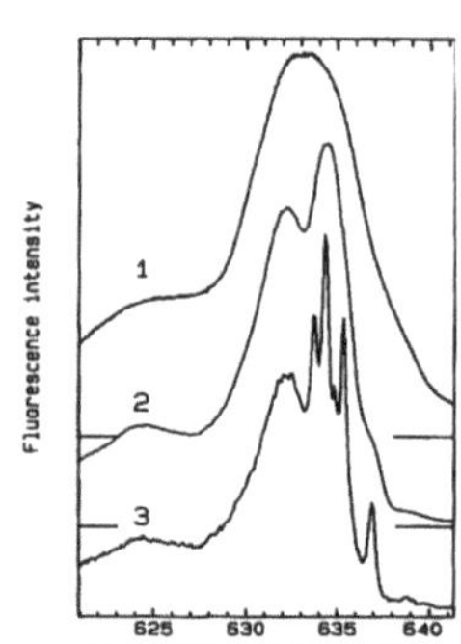

Figure 5.5. Fluorescence line narrowing in the fluorescence excitation spectrum of $(OEP)_2$: fluorescence recording through 650 nm cut-off filter (1) and through monochromator with slits of 1.1 nm (2) and 0.23 nm (3). Mixture of tetrahydrofurane-toluene (3:1), T = 1.8 K.

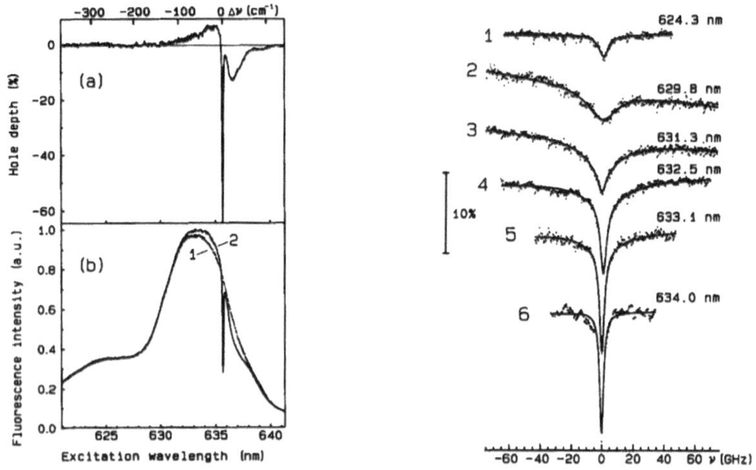

Figure 5.6. Fluorescence excitation spectra of $(OEP)_2$ before (1) and after (2) a saturation burning at $\lambda_{exc} = 635.7$ nm (b), the corresponding difference spectrum (a) and measured spectral holes at different excitation wavelengths (right part). Solid curves for spectral holes are the least-squares fits by one (1, 2, 6) and sum of two Lorentz components and a linear background. Mixture of tetrahydrofurane-toluene (3:1), $T = 1.8$ K, frequency jitter of 2 MHz.

Additionally, the efficiency and halfwidth of spectral hole burning (caused by the photoinduced NH-tautomerism [96]) have been studied as a function of λ_{exc} for the homodimer $(OEP)_2$ (Fig. 5.6). It follows from these results that for fully-staggered conformation at 1.8 K the effective ET rate constant $k_{ET} = 10^{11}$ s^{-1} is higher by an order of magnitude than that calculated using the Foerster approach. Thus, usual calculations of spectral overlap integrals according to Foerster theory are not applicable in this case [95].

Heterodimers of ethane-bisporphyrins, $(H_2{+}Cu)(OEP)_2$, (Zn+Cu) $(ZnOPP)_2$ and $(H_2{+}Ni)(OEP)_2$, containing a Cu or Ni ion in the center of one subunit, have some general features [92–94]. In fact, the emission of the heterodimers belongs to the H_2- or Zn-subunit. The singlet-singlet ET from Cu- or Ni-subunit to H_2- or Zn-subunit having a S_0-S_1 transitions is not detected while the fluorescence of the heterodimer is quenched strongly ($\varphi_D/\varphi_{D0} < 10^{-2}$, Table 5.1) compared to the corresponding homodimers. Additionally, the lifetime of the T_1 state (τ_T) of H_2- or Zn-subunits is decreased in the heterodimers. Different processes are responsible for the quenching in these dimers.

Picosecond pump-probe data reveal that excited states of the heterodimer $(H_2+Ni)(OEP)_2$ decay within 15 ± 5 ps at 293 K. In this dimer, the strong fluorescence quenching ($\varphi/\varphi_0 = 200$) and fast deactivation of the porphyrin free base subunit may be attributed to an exchange ET from the S_1-excited π-conjugated free base to low-lying d-levels of Ni ion of the other subunit. In the case of the heterodimers $(H_2+Cu)(OEP)_2$ and $(Zn+Cu)(OPP)_2$ the fluorescence quenching of free base or Zn-porphyrin halves ($\varphi/\varphi_0 \approx 40$, Table 5.1) is caused by exchange d-π effects, like in [64] and known before for monomeric porphyrins only [97]. In these dimers an interaction of the unpaired d-electron of central Cu ion in the Cu-porphyrin subunit with π-conjugated system of the other half leads to an increase of the intersystem crossing rate constant in the other half. As a result, the lifetimes of S_1 and T_1 states in the H_2- and Zn-subunits of the heterodimers are reduced compared to those for the corresponding homodimers. On the other hand, at 77 K phosphorescence of the Cu-subunit in $(H_2+Cu)(OEP)_2$ and $(Zn+Cu)(OPP)_2$ is not detected due to the directed exchange coupled T-T ET to the H_2- or Zn-subunit ($k_{ET}^{TT} \geq 2 \times 10^7$ s^{-1}). Later it has been also shown that T-T ET in hybrid (H_2P+CuP) dimers with rigid spacer gives rise to transient absorption spectra identical to that of T-T absorption of porphyrin free base half with respect to the intensity for individual H_2P monomer at the same experimental conditions [98]. Noteworthy, such exchange d-π effects are characteristic also for mixed polymeric ordered aggregates of chlorophyll and its analogs [99].

The conformational mobility of ethane-bischlorins $(ZnOEChl)_2$ about -CH_2-CH_2- bond is sterically hindered due to hydrated pyrrole rings of the subunits being in the vicinity of the spacer (Fig. 5.4B). Therefore, the fully staggered conformer is considered as the presumable one. In these dimers, the intensities of $S_0 \rightarrow S_1$ transitions of the subunit are comparable with those for the strong B-transitions forming the Soret band. Correspondingly, excitonic interactions ($V_{12} \approx 140$ cm^{-1}) of strong resonant Q-transitions in $(ZnOEChl)_2$ result in the splitting of Q(0,0) absorption band at 4.2–77 K (Fig. 5.7). Laser selective excitation into the shortwavelength splitted

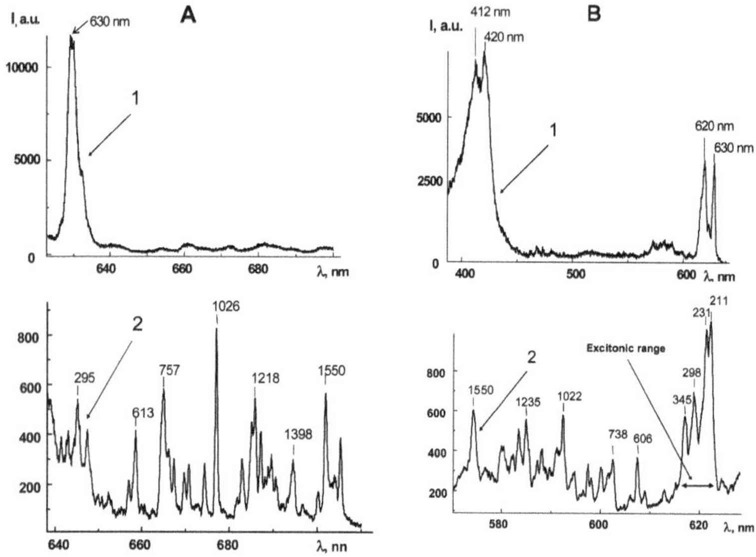

Figure 5.7. Fluorescence line narrowing in fluorescence (A) and fluorescence excitation spectra (B) for $(ZnOEChl)_2$ in mixture of tetrahydrofurane-toluene (3:1) at $T = 10$ K. A: 1 — non-selective excitation at $\lambda_{exc} = 620$ nm, 2 — selective excitation at $\lambda_{exc} = 630$ nm. B: 1 — non-selective registration at $\lambda_{reg} = 660$ nm, 2 — selective registration at $\lambda_{exc} = 631$ nm. Frequencies of the normal modes (in cm^{-1}) are indicated only for certain zero phonon lines.

component leads to the disappearance of zero phonon lines in fluorescence spectra. In addition, fluorescence excitation spectra at selective narrow ($\Delta\lambda \leq 0.01$ mn) registration within pure electronic and vibronic bands reflect the complex interplay between intramolecular relaxation modes and excitonic effects. In this dimer "hot" ET processes via a localized exciton lead to electronically excited states that are coherently delocalized over the two rings accompanied by the specific manifestation of intermolecular vibration modes [100]. Nevertheless energetic parameters of $(ZnOEChl)_2$ are almost unchanged relative to the corresponding monomeric compounds (see Table 5.1).

Noteworthy, very recently for chlorophyll trefoils the fluorescence dynamics dependence on the type of linkages has been studied by recording single-molecule fluorescence intensity trajectories and their corresponding lifetimes [44], that may be considered as the next step of selective spectroscopy methods with respect to fluorescence line narrowing and spectral hole burning experiments.

5.2.2 *Energy Transfer in Chemical Dimers of Cyclopentaneporphyrins*

$3^1,5^1$-cyclodimer (Zn-complex or free base) contain one molecule of ZnOEP-cycle (or OEP-cycle) as a potential energy donor *D* and one molecule of ZnOEP-cycle=CH_2(or OEP-cycle=CH_2) as an energy acceptor *A* which are covalently coupled via an isocycle with the intercenter distance of $d = 1.23$ nm and the angle of 78° between subunits planes (Fig. 5.4C). The rigid structure of the cyclodimers results in only a small overlap of π-electronic systems of their subunits. However, weak dipole–dipole ($V_{12} \leq 3$ cm^{-1}) interactions of the cyclodimer subunits as well as the energy gap $\Delta E = E(S_1^D) - E(S_1^A)$ give rise to an effective quenching of the *D* emission due to ET ZnOEP-cycle*→ ZnOEP-cycle [101–104]. In Zn-cyclodimers ET can involve differently polarized S_0-S_1 and S_0-S_2 transitions of the dimer subunits [104, 105], since they are close in energy scale (Fig. 5.8).

At 293 K in Zn-cyclodimers, experimental ET quantum efficiency is of $\Phi_{DA} = 0.99$ and $\Phi_{DA} = 0.95$ at 77 K as calculated by both from *D* fluorescence quenching and A fluorescence sensitization. It means the lack of quantum losses during ET in this case. On the basis of calculated spectral overlap integrals ($J = 7.6 \times 10^{-16} \div 3.6 \times 10^{-14}$ cm^3·mol^{-1}), and estimated orientation factors ($k^2 = 0.07 \div 2.0$) using both computer-aided simulation of cyclodimer structure and experimental data on linear dichroism and fluorescence polarization [105], the main paths of ET in Zn-cyclodimers may be presented in the following way. At 77 K only the S_1 level of *D* (ZnOEP-cycle) plays an essential role in ET with a rate constant of $k_{ET}(S_1^D \rightarrow S_1^A) = 6.6 \times 10^{10}$ s^{-1}, while at 293 K the thermally activated S_2 state of *D* is involved in ET as well: $k_{ET}(S_1^D \rightarrow S_1^A) = 4.1 \times 10^{10}$ s^{-1} and $k_{ET}(S_2^D \rightarrow S_1^A) = 1.1 \times 10^{10}$ s^{-1}. A good agreement of experimental and calculated data for the *D* fluorescence quenching via ET at 77 K indicates that Foerster model is still applicable to weakly interacting porphyrins at the intercenter distance of 1.23 nm.

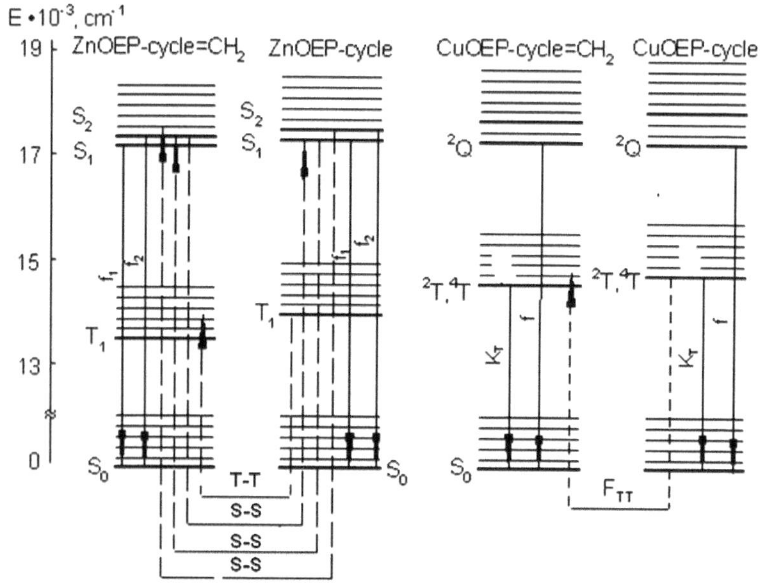

Figure 5.8. Diagram of the electronic energy deactivation main paths in Zn- and Cu-complexes of 3^1,5^1-cyclodimer.

The effective ET in Zn-cyclodimers was observed also at 4.2 K [106]. Under these conditions, the selective laser excitation within the band of the $S_0 \rightarrow S_1$ transition of *A* leads to the dimer fluorescence spectrum consisting of zero phonon lines, while the excitation into the $S_0 \rightarrow S_1$ transition of the donor subunit (ZnOEP-cycle) leads to disappearance of fluorescence line narrowing. The latter effect may be treated as the non-correlation of $S_{0D} \rightarrow S_{1D}$ and $S_{1A} \rightarrow S_{0A}$ transitions for different donor–acceptor pairs in ET processes (rate constant $k_{EM} \geq 10^{10}$ s^{-1}) caused by inhomogeneous broadening of electronic levels at 4.2 K.

For cyclodimers consisting of free bases, ET processes have interesting features due to existence of two NH-tautomers with different absorption and fluorescence spectra for *D* and *A* subunits [101, 107, 108]. In spite of the fact that the photoinduced NH-tautomerism does not compete with the deactivation of S_1 and T_1 states in the cyclodimers at 77 K due to the best resonance conditions for main tautomers of *D* and *A* ($J = 8 \times 10^{-15}$ cm^3·mol^{-1}), the S-S ET in the dimers results in the directed photoinduced transformations of NH-tautomers in *A* upon selective excitation of the *D* subunit.

In the case of Cu-cyclodimers, fast intersystem crossing to the trip-dublet/trip-quartet ^{2,4}T states with the rate constant of $k_{\rm ISC} = 3 \times 10^{13}$ s^{-1} in both subunits (Fig. 5.8) prevents S-S ET D→A (calculated rate constant is of $k_{\rm ET}^{\rm theor} < 4 \times 10^{10}$ s^{-1}). Nevertheless, a strong quenching of *D* phosphorescence is observed for these dimers at 77 K [109]. The detection of the *D* bands in phosphorescence excitation spectra of the *A* subunit leads to the conclusion that this quenching is due to an exchange ET between T levels of interacting macrocycles in the Cu-cyclodimer. In this case, the experimental value of the T-T ET rate constant ($k_{\rm ET}^{\rm TT} = 2\times 10^8$ s^{-1}) is much higher than the rate constant of the donor T state deactivation ($k_{\rm T}^{0} = 6.7\times 10^3$ s^{-1}).

5.2.3 *Steric Hindrance Effects and Excited State Relaxation in Phenyl-Bridged Dimers*

Numerous literature data show that in a lot of cases multiporphyrin arrays of various types contain an inserted phenyl spacer between interacting components [see e.g., 22, 27–30, 65, 68, 81, 110–113], like in homodimers $(ZnHTPP)_2$ or $(ZnOEP)_2Ph$ (Fig. 5.4) and the heterodimer [ZnOEP-Ph-ZnP(iso-PrPh)$_3$] in our studies. Correspondingly, it has been found that steric interactions of the phenyl spacer with the connected subunits as well as the π-electronic nature of the linkage may influence on mechanisms and the efficiency of the electron/energy transfer [110–113]. Thus, for the decision whether or not the inserted phenyl ring acts as a stepping stone in ET and PET processes, it would be desirable to test the role of a phenyl ring itself in the deactivation of porphyrin excited S_1 and T_1 states. Typically, the addition of bulky substituents of various nature in meso-positions of tetrapyrrole macrocycle leads to the formation of sterically encumbered porphyrins in which deformations from planarity are induced by several reasons followed by a pronounced decrease of fluorescence quantum yields ($\varphi_{\rm F}$) and decay ($\tau_{\rm S}$) shortening as well by essential changes of optical, photophysical and redox properties of tetrapyrrole compounds [114–118].

In this respect, we found for the first time that mono- and di-meso-phenyl substitution in octaethylporphyrins (free bases and corresponding Zn-complexes) as well as the formation of octaethylporphyrin chemical dimers with a meso-phenyl spacer, $(ZnOEP)_2Ph$ and $(OEP)_2Ph$, lead to a strong shortening of triplet lifetimes (by ~300–500 times) at room temperature in degassed liquid solutions without any strong influence on spectral-kinetic parameters of S_0 and S_1 states (Table 5.1) [119–123]. Based on experimental spectral-kinetic data and theoretical analysis (semiempirical PM3, DFT, TD-DFT calculations), it was concluded that being planar in S_0 and S_1 states, mono-meso-phenyl-substituted octaalkylporphyrins in the triplet excited state may occur in the highly nonplanar conformation characterized by the out-of-plane displacement of the single C_{m1}–C_1 bond and the increased overlap of the porphyrin macrocycle and the meso-phenyl ring. For the sterically strained porphyrins, the transition into this nonplanar conformation is accompanied by the essential decrease of the triplet-singlet energy gap $\Delta E(T_1\text{-}S_0)$ that leads to the experimentally detected drastic reduction of the triplet state decays for this type molecules (see Table 5.1). Later, the same experimental tendencies and ideas were additionally discussed in [124, 125].

It follows from above results and the analysis of data presented in Table 5.1, that interchromophoric interactions in phenyl-bridged homodimers $(ZnHTPP)_2$ or $(ZnOEP)_2Ph$ do not change practically their spectral and photophysical parameters with respect to the corresponding precursor monomeric compounds. The only principal feature is that in the case of $(ZnOEP)_2Ph$ the triplet state decay is of $\tau_T^0 = 2.85$ μs in degassed liquid solutions at 295 K (because of sterical reasons, not excitonic interactions) while for $(ZnHTPP)_2$ $\tau_T^0 = 1160$ μs at the same conditions (because of lack of sterical hindrances). The intermediate case is characteristic for the heterodimer [ZnOEP-Ph-ZnP(iso-PrPh)$_3$], where excitonic interactions of strong non-resonant B-transitions are negligible while steric interactions of the phenyl bridge should lead to the triplet state decay shortening for ZnOEP subunit. But this effect seems to be hardly detected because of the efficient singlet-singlet ET ZnOEP-Ph*→ZnP(iso-PrPh)$_3$ at $d = 1.22$ nm with the rate constant of $k_{ET} = 1.3 \times 10^{10}$ s^{-1} which is essentially higher than the intersystem rate constant $k_{ISC} \approx 6 \times 10^8$ s^{-1} for ZnOEP-Ph subunit (Fig. 5.9). In addition, for this heterodimer in methylcyclohexane at 295 K, the photoinduced electron transfer does not compete with ET as far as the energy of charge-transfer state E(CT) = 2.57 eV is essentially higher the energies of locally excited S_1 states of both subunits (2.14 eV for ZnOEP-Ph and 2.09 eV for ZnP(iso-PrPh)$_3$) [75, 76].

Finally, we found [120, 121] that the rate constant of the photoinduced electron transfer depends essentially on steric hindrance effects and electronic properties of the bridge in meso-phenyl-substituted octaalkylporphyrins having covalently linked electron-accepting NO_2-group in para-, meta-, and ortho-positions of the phenyl ring. The highest value of the PET rate constant (k^S_{PET} = 9.5×10^9 s^{-1}) was obtained for monomeric ortho-NO_2-substituted molecules. In this case sterical interactions of bulky β-C_2H_5 substituents and NO_2 group favour the overlap of molecular orbitals for interacting D and A subunits (the electronic coupling term is V = 130–190 cm^{-1} in dimethylformamide) resulting in a direct through-space PET from the porphyrin locally excited S_1 state to the low-lying charge-transfer state of the radical ion pair. In the dimer OEP-Ph-OEP-Ph(o-NO_2), the reduction of k^S_{PET} by three times is connected with the existence of the competing non-radiative S-S energy transfer between covalently linked porphyrin macrocycles with the rate constant of $k^{SS}_{ET} = (1.9\text{–}5.0) \times 10^9$ s^{-1}.

The presented results evidently show that steric hindrance effects and the dynamic non-planarity of meso-phenyl-substituted porphyrins and their chemical dimers with the phenyl spacer in the excited triplet state have be taken into account in the analysis of photophysical properties of locally excited states and photochemical events in systems of this type or more complex multiporphyrin arrays.

5.3 Competition Between Energy and Photoinduced Electron Transfer Processes in Self-Assembled Porphyrin Triads

Here, we will start with triads of type **I** (the corresponding structure is depicted in Fig. 5.1) as an example of our approach in the analysis of pathways and mechanisms of electronic excitation relaxation dynamics for multiporphyrin complexes [66, 68–70, 72].

5.3.1 *Steady-State and Time-Resolved Data*

Absorption spectra of the triads of this type are a linear combination of the corresponding di-pyridinated dimer $(ZnOEP)_2Ph$ (extra-ligation effects) and the corresponding porphyrin extra-ligand, with only small differences in wavelength maxima and band shapes (Fig. 5.8A). Thus, the interaction between the two subunits is weak in the ground state, and they retain their individual identities.

In the triad **I**, $(ZnOEP)_2Ph \otimes H_2P(m^\wedge Pyr)_2$-(iso-PrPh)$_2$, the dimer fluorescence does show strong quenching (fluorescence decay is shorten from $\tau_{SD}^0 = 1.15$ ns down to $\tau_{SD} \leq 1.7$ ps for the triad in toluene at 295 K), and fluorescence spectra of the triads mainly consist of the porphyrin extra-ligand fluorescence bands (Fig. 5.9B). As far as there is a good spectral overlap between the dimer $(ZnOEP)_2Ph$ fluorescence and the extra-ligand $H_2P(m^\wedge Pyr)_2$-(iso-PrPh)$_2$ absorption bands it seems to be reasonable to connect the

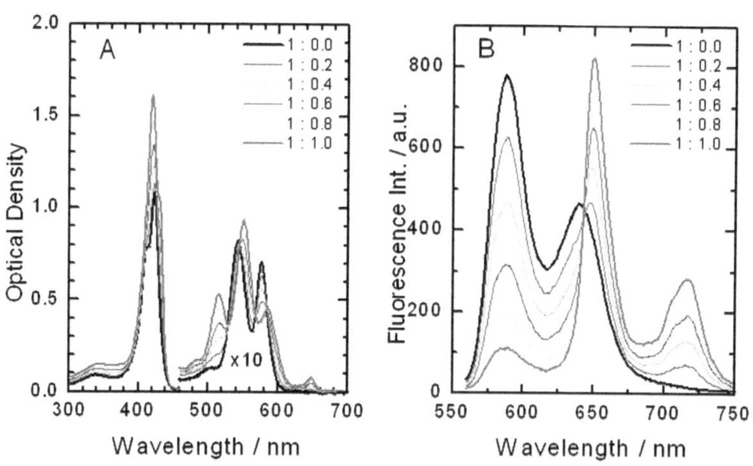

Figure 5.9. Absorption (A) and fluorescence (B, $\lambda_{ex} = 546$ nm, isosbestic point) spectra of the dimer $(ZnOEP)_2Ph$ with increasing amounts of extra-ligand $H_2P(m^\wedge Pyr)_2$-(iso-PrPh)$_2$ (toluene, 295 K). The molar ratio x = [ligand]/[dimer] varies from $x = 0.0$ to 1.0 (0.0, 0.2, 0.4, 0.6, 0.8, 1.0). See also Color Insert.

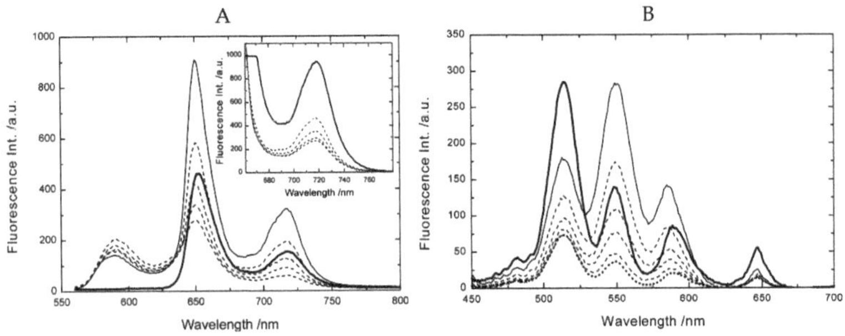

Figure 5.10. Corrected fluorescence (A, $\lambda_{ex} = 546$ nm) and fluorescence excitation (B, $\lambda_{em} = 720$ nm) spectra of the triad $(ZnOEP)_2$ $Ph\otimes H_2P(m^\wedge Pyr)_2$-(iso-PrPh)$_2$ in toluene (thin solid lines) and upon addition of 3, 6, 9, 17 vol% of acetone (thin dashed lines). Bold line spectra represent the pure extra-ligand emission and fluorescence excitation spectrum in toluene at the same concentration as in triad and at the same excitation conditions. Inset: the same experimental solutions at $\lambda_{ex} = 650$ nm (within longwavelength Q-band of the extra-ligand).

above quenching with ET dimer*→ligand. Quite obviously ET from the dimer to an extra-ligand has to lead to the sensitized emission of the last. For example, Fig. 5.10B shows that the fluorescence excitation spectrum of the triad $(ZnOEP)_2Ph\otimes H_2P(m^\wedge Pyr)_2$-(iso-PrPh)$_2$ in pure toluene (monitored at 720 nm where the majority of the emission comes from the extra-ligand) contains 549 nm and 587 nm bands belonging in the absorption to the dimer. It follows from these results that the singlet-singlet ET efficiency was estimated to be $\Phi =$ 70–80% in pure toluene [70, 71]. As was described in Section 5.2 for porphyrin chemical dimers with various structures and properties and discussed for self-assembled multiporphyrin complexes [71, 74], at temperature range of 77–295 K and intercenter distances between porphyrin $D - A$ moieties of $r_{DA} \leq 0.82$ nm one may apply Foerster inductive-resonant model for the quantitative description of the non-radiative electron-vibrational ET.

On the basis of optimized geometry of the triads of type **I** (HyperChem software package, release 4, semiempirical methods AM1 and PM3), experimental data for D and A and using Foerster inductive-resonant model [35] we calculated the corresponding ET parameters: (i) orientational factors k^2= [Cos(μ_D, μ_A) — 3 Cos(μ_D, $\mathbf{r}_{DA}$)·

$\mathrm{Cos}(\mu_A, \mathbf{r}_{DA})]^2 = 0.47 \div 0.75$, (ii) spectral overlap integrals $J(\nu) = 1.2 \cdot 10^{-14} \div 5.9 \cdot 10^{-14}$ cm^3 M^{-1}, (iii) critical ET distances $R_0^{\mathrm{theor}} = 1.68 \div 2.023$ nm. Correspondingly, for the systems under consideration theoretical values of ET rate constants $[k_{ET} = (1/\tau_S)(R_0^{\mathrm{theor}}/r_{DA})^6]$ have to be in the range of $k_{EM} = (6.7 \times 10^{10}) \div (1.4 \times 10^{11}\ \mathrm{s}^{-1})$. It means that the S_1state lifetime of the dimer has to be shortened to values of 10–8 ps.

However, experimental femtosecond pump-probe results for the triad $(ZnOEP)_2Ph \otimes H_2P(m^\wedge Pyr)_2$-(iso-$PrPh)_2$ (Fig. 5.11A) [70] do not coincide with the theoretical estimations. It follows from the detailed analysis of these data [70] that the non-radiative relaxation of the dimer S_1 state in the triads takes place within 1.7 ps that is essentially faster than 13–15 ps. In addition, the formation of charge transfer (CT) states is clearly detected (especially appearance of the Zn-porphyrin cation absorption at 670 nm) thus indicating the existence of the photoinduced electron transfer. Hence, in order to analyze the interplay between ET and PET processes in the triad under study one should take into account that at the constant geometry of the triad ET is hard dependent on the temperature (77–293 K)

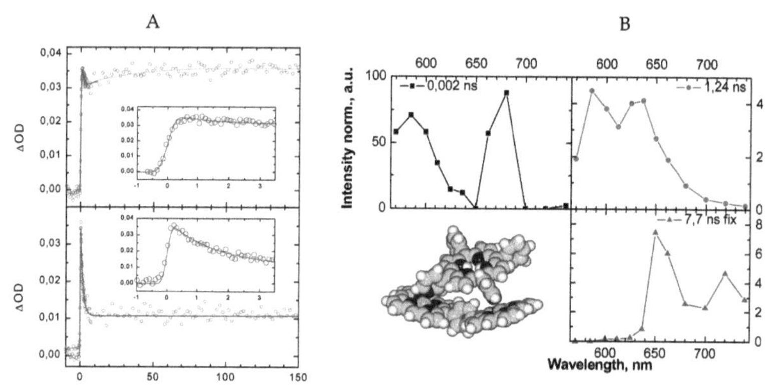

Figure 5.11. Time-resolved femtosecond pump-probe (A) and picosecond fluorescence (B) data for the triad $(ZnOEP)_2Ph \otimes H_2P(m^\wedge Pyr)_2$-(iso-$PrPh)_2$ in toluene at 293 K. A: Time evolution of transient absorbance at $\lambda_{pump} = 555$ nm and $\lambda_{probe} = 670$ nm (top), 510 nm (bottom). B: Decay-associated spectra ($\lambda_{ex} = 546$ nm) derived from a global analysis of 12 time correlated single photon counting (TCSPC) fluorescence measurements. See also Color Insert.

and polarity of the solvent [126] while rate constants of the non-adiabatic ET are sensitive to these parameters and are dependent on redox properties of interacting $D - A$ reactants [127].

Correspondingly, main fluorescent features in the triad of type **I** being obtained during "polarity-temperature" experiments are as follows (Figs. 5.10A, 5.10B and 5.11B) [69–72, 75]. (1) In nonpolar toluene at 295 K, the extra-ligand fluorescence quantum efficiency is reduced by 1.5–2 times as compared to the individual $H_2P(m^\wedge Pyr)_2$-(iso-PrPh)$_2$ (Fig. 5.10A, bands at 650 nm and 714 nm, and inset). (2) In non-degassed toluene at ambient temperature, the fluorescence decay of extra-ligand in the triad **I** is reduced noticeably ($\tau =$ 7.7 ns) with respect to that for individual uncomplexed porphyrin ($\tau_0 = 9.5$ ns). (3) The increase of the solvent polarity by a subsequent acetone addition to toluene leads to the decrease of the extra-ligand fluorescence intensity in the triad while the emission of the individual extra-ligand does not show fluorescence quenching upon acetone addition. (4) The decay time shortening increases upon the solvent polarity rise ($\tau_S = 5.5$ ns for $H_2P(m^\wedge Pyr)_2$-(iso-PrPh)$_2$ in the triad in toluene + 7 vol % of acetone), the last tendency correlates with a pronounced decrease of the extra-ligand fluorescence quantum yield. (5) At 295 K in toluene, fluorescence excitation spectrum of the triad **I** detected at extra-ligand fluorescence bands ($\lambda_{det} \geq$ 720 nm, toluene, 293 K) is „triad-like", clearly showing the existence of absorption bands of the dimer (549 nm and 587 nm). Upon addition of 17 vol% of acetone admixture in toluene the dimer emission in the triad remains strongly quenched but the form of the excitation spectrum of the triad becomes almost identical to that of the individual extra-ligand $H_2P(m^\wedge Pyr)_2$-(iso-PrPh)$_2$. It indicates that the sensitization effect due to the singlet-singlet ET process is absent, though usually through-space S-S ET in multiporphyrin arrays is hardly dependent on the solvent polarity [128]. (6) Fluorescence intensity of complexed H_2P is decreased upon the temperature lowering.

5.3.2 *Mechanisms and Pathways of Relaxation Dynamics*

In the result, on the basis of the detailed experimental data in the combination with theoretical estimations the whole scheme of the dynamics of locally excited S_1 states for the components in triad **I** may be presented by Scheme 5.1.

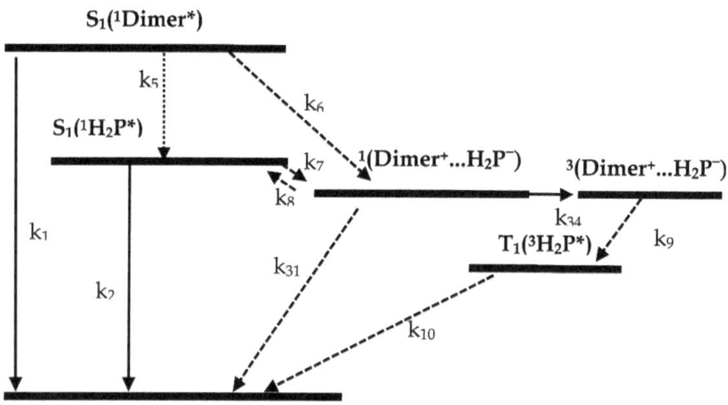

Scheme 5.1.

The non-radiative deactivation of the dimer locally excited S_1-state is caused by competing S-S ET and PET processes (rate constants k_5 and k_6). The competition between S-S ET and PET channels depends on photophysical and redox parameters of the subunits and may be driven by temperature and polarity of surrounding. ET transfer parameters have been evaluated in Section 5.3.1. Our estimations have shown [69–72] that in toluene at 295 K, the energy of charge transfer (CT) state for the photoinduced electron transfer $(ZnOEP)_2Ph \rightarrow$extra-ligand $E(\text{IP}) = 1.9$ eV being very close to locally excited S_1 state of the extra-ligand. This situation explains the existence of stimulated fluorescence of the extra-ligand in pure toluene at 295 K as a result of simultaneous realization of S-S ET, PET and repopulation processes. The increase of the solvent polarity (addition of acetone) leading to CT state lowering manifests itself in the extra-ligand fluorescence additional quenching and the absence of sensitizing effect, which may hint to ET process presumably involved in the later case. It should be mentioned also that the photoinduced hole transfer from the excited state of the extra-ligand to the dimer (rate constant k_7) may lead to the singlet radical ion pair state formation. In its turn, the fast repopulation of the extra-ligand S_1 state is caused by the effective thermally activated charge recombination (rate constant k_8) from the singlet radical ion pair state to the close-lying locally excited singlet state.

At last, we have shown for the triads of type **I** [70, 72], that the direct intersystem crossing $S_1 \rightsquigarrow T_1$ in the extra-ligand subunit is extremely low in comparison with the processes discussed above. The population of the extra-ligand locally excited T_1 state may take place from the upper-lying triplet radical ion pair state $^3(\text{Dimer}^+ \ldots H_2P^-)$ formed via the spin rephasing between the singlet and triplet radical ion pairs (rate constant k_{34}).

Our theoretical results [69, 71] have shown that the generalized Haken-Strobl-Reineker formalism provides both qualitative and quantitative description of time-resolved and steady-state properties such as fluorescence quenching in triads of type **I** due to competing ET and PET processes. From the theoretical point of view coherent or incoherent-type of electron or exciton transport in the systems under consideration depends on the coupling to the relevant vibrational modes. For the triads containing the dimer and the extra-ligand H_2P, the dynamics of the population of excited states $|1\rangle = |\text{Dimer}^* - H_2P\rangle$, $|2\rangle = |\text{Dimer}^+ - H_2P^-\rangle$, and $|3\rangle = |\text{Dimer} - H_2P^*\rangle$ is described by the equation of motion for the relevant reduced density matrix [129] with neglecting of the vibrational substructure of the electronic states [130]. It is seen from Fig. 5.12 that calculated dependencies of the extra-ligand H_2P population in

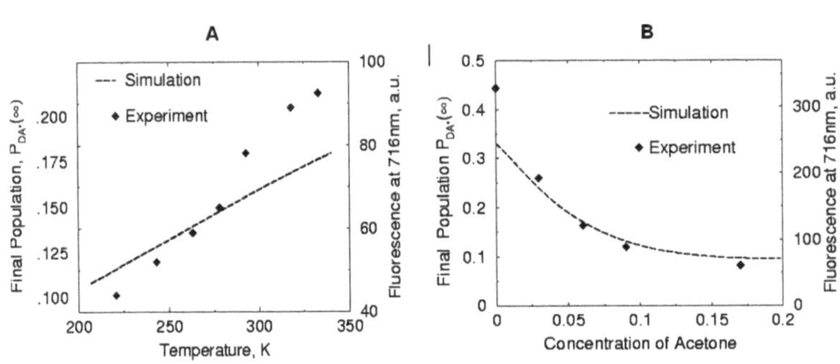

Figure 5.12. Experimental data and theoretical simulation on temperature (A) and solvent polarity (B) dependence of the triad $(ZnOEP)_2Ph\otimes H_2P(m^\wedge Pyr)_2$-(iso-PrPh)$_2$ fluorescence band at 716–720 nm in toluene at 295 K.

S_1 state $P_{DA^*}(\infty)$ on temperature and solvent polarity are in a reasonable agreement with the observed fluorescence changes.

Our simulations display that at low acetone concentrations (0–7 vol%) the quasi-thermal equilibrium population $P_{DA^*}(\infty)$ is reached in one-step (due to ET process presumably) and a reaction rate constant can be found with an one-exponential fit. In the case of high acetone concentration (10–20 vol%) the energy detuning between states $|2\rangle$ and $|3\rangle$ becomes larger. Correspondingly, in addition to ET process, the hole transfer $|3\rangle \rightarrow |2\rangle$ takes place, thus the equilibrium population $P_{DA^*}(\infty)$ is reached in two steps. The reaction occurs with the help of a sequential transfer, which is described by two rates (increase and decrease) and the one-exponential fit for a reaction rate constant can not be used in this case.

5.3.3 *Ultrafast PET in Triad Containing Fluorinated Extra-Ligand*

It was found [70] that the dynamics of electronic excitation energy deactivation in triads **I** and **II** of the same geometry (Fig. 5.1) differs significantly. While the absorption spectra of the titration experiments for the dimer $(ZnOEP)Ph_2$ with $H_2P(m^\wedge Pyr)_2$-(iso-PrPh)$_2$ and $H_2P(m^\wedge Pyr)_2$-$(5FPh)_2$ ligands are very similar but great differences are observed in fluorescence spectra of the triad **II** in comparison to the triad **I**. In contrast to triad **I**, extremely low fluorescence of the fluorinated extra-ligand $H_2P(m^\wedge Pyr)_2$-$(5FPh)_2$ is observed for the triad **II**. At 295 K in toluene, the fluorescence excitation spectrum (monitored at 717 with appropriate signal amplification) is identical in shape to that of the individual $H_2P(m^\wedge Pyr)_2$-$(5FPh)_2$, indicating that no fluorescence is sensitized via $(ZnOEP)Ph_2$absorption. At this conditions, femtosecond pump-probe measurements reveal the formation of Zn-porphyrin cation band at $\lambda_{rec} = 670$ nm with the time constant 0.7 ($\pm$0.1) ps. (Fig. 5.13).

These data indicate that for the triad **II** an extremely fast PET takes place on a time scale of 0.7 ps. (free Gibbs energy $\Delta G^0 \approx -0.25$ eV, energy of charge transfer state $E(\text{IP}) = 1.66$ eV, $r_{DA} = 8.8$ Å) [70]. No fluorescence is sensitized via the dimer absorption, and correspondingly S-S ET $(ZnOEP)_2Ph^* \rightarrow H_2P(m^\wedge Pyr)_2$-$(5FPh)_2$ is not realized at 295 K. Noteworthy, in rigid glassy matrices and PMMA

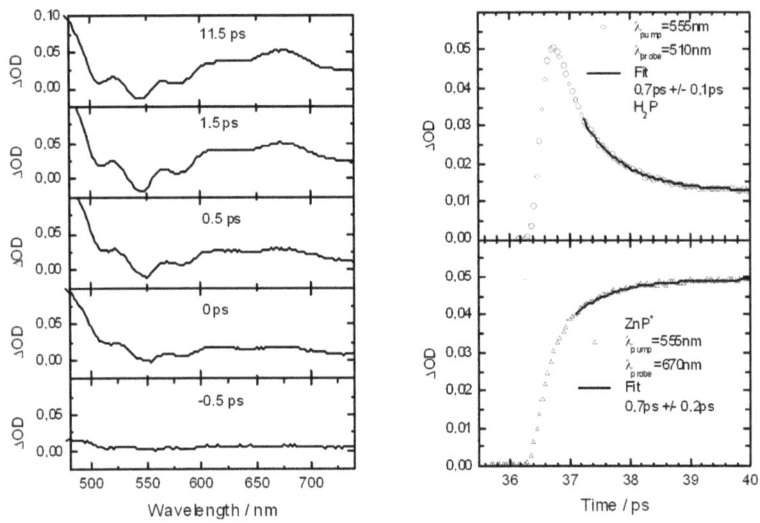

Figure 5.13. Femtosecond pump-probe data for the triad **II**$(ZnOEP)_2$ Ph⊗H_2P(m^Pyr)$_2$-(5FPh)$_2$with fluorinated extra-ligand in toluene at 295 K. **A:** Time-resolved transient absorption spectra at various delay times with respect to the exciting pump pulse at 400 nm. **B:** Time evolution of the transient absorbance for the fluorinated triad formed by the excitation at 555 nm and measured at 510 nm (top) and 670 nm (bottom).

thin films at 120–77 K this PET remains still effective (rate constant is $k_{PET} \sim 10^{11}$ s^{-1}) and competes with the singlet-singlet EM process from the dimer to the extra-ligand [70]. Some reasons for the effective low-temperature ET in this case may be considered: (i) The fluorinated porphyrin free base is strongly electron withdrawing and stabilizes a negative charge on its π-conjugated macrocycle; (ii) The coordination of the electron-donating pyridyl rings helps to stabilize a positive charge on the Zn-dimer and thus lowers the energy of the radical ion pair state ^{1}(Dimer$^+$...Lig$^-$). In the triad **II** PET is adiabatic at room temperature, while in rigid solution at 77 K electron quantum tunneling may take place. It should be noted that PET processes in many biological objects (including photosynthetic reaction centers [2]) appear to be operative at cryogenic temperatures, while only few model systems [131] with this property (including our discussed here) have been described.

Because of fast PET in the triad **II** the direct population of the locally excited triplet T_1 state of the fluorinated extra-ligand can take place via intersystem crossing. In this case, the effective formation of the extra-ligand low-lying T_1 state in the triad ($\tau_T = 6.4$ μs in degassed solution) takes place from the upper-lying triplet or singlet radical ion pair states. Noteworthy, the triplet decay time of the non-fluorinated extra-ligand in the triad **I** ($\tau_T = 1.3$ ms in degassed toluene at 293 K) is practically the same with respect to that for pure H_2P at the same conditions. In later case, the locally excited H_2P triplet state ($E_T = 1.44$ eV) lies essentially lower that CT state. Therefore in the triad **I** the upper-lying CT state does not influence on the deactivation of the locally excited T_1-state of the extra-ligand. In contrast to this, for the triad **II** the triplet decay time of $H_2P(m^\wedge Pyr)_2$-$(5FPh)_2$ is shorten essentially due to the perturbation influence of closer-lying CT state.

The comparative analysis of relaxation processes in triads of various composition (Zn-homo- and heterodimers, porphyrin, chlorine and tetrahydroporphyrin extra-ligand) and geometry one may find in our earlier publications [30, 72, 73, 75].

5.4 Triads with Covalently Linked Electron Acceptors of Non-Porphyrin Nature

Here, we present the dynamics of relaxation processes in self-assembled triads of the same geometry composed of the dimer $(ZnOEP)_2Ph$, with covalently linked electron acceptor *A* (para-benzoquinone, Q, or pyromellitimide, Pim), and di-meta-pyridyl-substituted tetrapyrrole extra-ligand with adjacent pyridyl rings (porphyrin, H_2P, chlorin, Chl, or tetrahydroporphyrin, THP) (Triads of **III** and **IV** types, Fig. 5.1). Like for the triads discussed in Section 5.3, steady-state absorption spectra of the triads with *A*'s are essentially a linear combination of the dipyridinated dimer $(ZnOEP)_2Ph$, extra-ligand and the corresponding acceptor [72]. These observations provide good evidence for the absence of ground–state interactions between subunits in the triads. In order to carry out the quantitative analysis of relaxation processes in these self-assembled complexes consider the main energetic parameters for individual covalently linked systems $(ZnOEP)_2$Ph-Pim and $ZnOEP)_2$Ph-Q.

5.4.1 *Photoinduced PET in Dimers Covalently Linked to Electron Acceptors*

In contrast to the pure dimer $(ZnOEP)_2$Ph (fluorescence decay $\tau_S^0 =$ 1.21 ns in toluene at 293 K), essential fluorescence quenching is observed for the Pim containing dimer ($\tau_S = 135$ ps according to TCSPC data [72]). This quenching becomes stronger ($\tau_S = 54$ ps) upon pyridine addition leading to the well-known formation of pyridynated complexes of the dimer [28, 72]. It should be mentioned that namely later data are of interest with respect to the self-assembled triads as far as ZnOEP components are ligated [30, 65–68]. Femtosecond transient absorption shows a noticeable spectral dynamics (Fig. 5.14). The rising absorption near 670 nm and 715 nm can be ascribed to ZnP^+ and Pim^- species, respectively, by referring to the corresponding spectra obtained by electrochemical oxidation of ZnOEP [132, 133] and reduction of Pim [134]. Transient absorption decays measured at 670 nm and 715 nm coincide within experimental error manifesting the existence of an one-step PET (^{1}Dimer*...Pim)→(Dimer$^+$...Pim$^-$). Addition of pyridine to toluene solutions leads to the accelerated formation of the radical ion pair CT state with respect to that found for toluene solutions. The same tendency upon pyridine action has been observed for other covalently linked D-A systems containing Zn-porphyrins [28, 135]. A reasonable agreement between experimental decays obtained by TCSPC method (presented above) and femtosecond pump-probe technique (depicted in Fig. 5.14) for the locally excited S_1 state of the dimer in $(ZnOEP)_2$Ph-Pim seems to be considered as a real background for PET in this case. The same conclusion has been done for $(ZnOEP)_2$Ph-Q ($\tau_S = 34 - 25$ ps in toluene, and $\tau_S = 27 - 17.6$ ps upon pyridine addition [72]). Thus, based on TCSPC data, the corresponding rate constants k_{PET} have been calculated using the formula

$$k_{PET} = (\tau_S)^{-1} - (\tau_{S0})^{-1} \tag{5.2}$$

In toluene at 293 K, $k_{PET} = 6.6\times 10^9$ s^{-1} for $(ZnOEP)_2$Ph-Pim and $k_{PET} = 2.9\times 10^{10}$ s^{-1} $(ZnOEP)_2$Ph-Q.

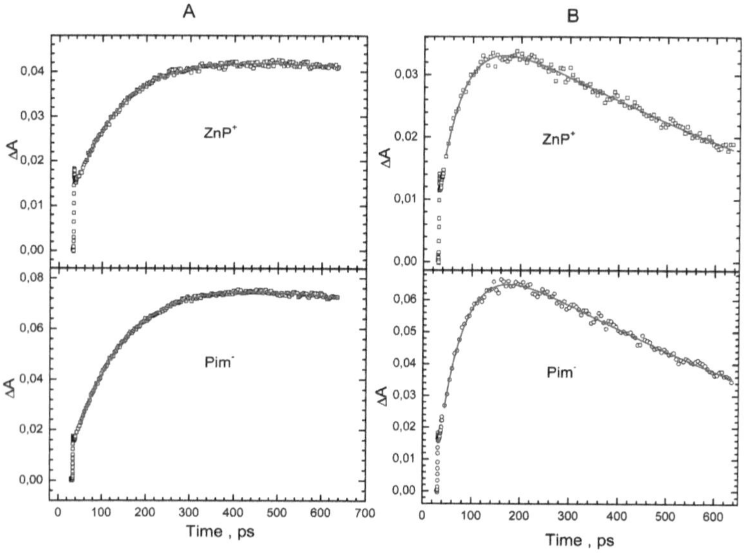

Figure 5.14. Time evolution of the transient absorbance for $(ZnOEP)_2$Ph-Pim in toluene (A) and toluene+pyridine (B) at 293 K formed by the excitation at $\lambda_{pump} = 540$ nm and measured at 670 nm (ZnP^+) and 715 nm (Pim^-). The kinetics were deconvoluted and fitted using the Marquardt algorithm. Two-exponential fit $I(t) = A_1 \cdot \exp(-t) + A_2 \cdot \exp(-t/\tau_2) + \delta(t)$ gives the following: $\tau_1 = 111$ ps ($A_1 = -0.033$), $\tau_2 = 6037$ ps ($A_2 = 0.046$) for ZnP^+ and $\tau_1 = 126$ ps ($A_1 = -0.073$), $\tau_2 = 3435$ ps ($A_2 = 0.087$) for Pim^- in toluene; $\tau_1 = 52.9$ ps ($A_1 = -0.033$), $\tau_2 = 1078$ ps ($A_2 = 0.056$) for ZnP^+ and $\tau_1 = 5$ ZnOEP)$_2$Ph-A_1 4.3 ps ($A_1 = -0.073$) and $\tau_2 = 914$ ps ($A_2 = 0.102$) for Pim^- in toluene + pyridine.

At high temperatures, the semiclassical Marcus theory of endergonic or moderately exergonic non-adiabatic PET occurring within the "normal" region predicts the following expressions for the rate constant k_{PET} [127]:

$$k_{PET}^{S} = \frac{2\pi}{\hbar} \cdot \frac{V^2}{(4\pi\lambda k_B T)^{1/2}} \cdot \exp\left(-\frac{\Delta G^*}{k_B T}\right) \tag{5.3}$$

$$\Delta G^* = \frac{\left(\Delta G^0 + \lambda\right)^2}{4\lambda}. \tag{5.4}$$

Here k_B is Boltzmann's constant, T is the temperature, h is Planck's constant, V is the electronic coupling term between the electronic wave functions of the reactant and product states, $\lambda = \lambda_{in} + \lambda_{ext}$

is the Gibbs reorganization energy determined by the nuclear λ_{in} and solvent λ_{ext} reorganization energies, ΔG^0 is the Gibbs free energy of the PET reaction, ΔG^* is the Gibbs activation energy. For porphyrin macrocycles, the term, λ_{in}, involving vibrational energy changes between the reactant and product states was estimated to be $\lambda_{in} \approx 0.3$ eV [22, 23, 28, 30]. The solvent-dependent term λ_{ext} for the surrounding medium treated as a dielectric continuum is expressed as:

$$\lambda_{solv} = \frac{e^2}{4\pi\varepsilon_0}\left[\frac{1}{2r_D} + \frac{1}{2r_A} - \frac{1}{r_{DA}}\right] \cdot \left[\frac{1}{\varepsilon_{op}} - \frac{1}{\varepsilon_{st}}\right] \tag{5.5}$$

where r_D and r_A are D and A radii, respectively, $\varepsilon_{op} = n^2$ is the optical dielectric constant, n is the refraction index and ε_{st} is the static dielectric constant of the solvent.

On the bases of direct TCSPC data for $(ZnOEP)_2$Ph-Pim in toluene and toluene + pyridine in a temperature range of 273–196 K nuclear and electronic factors have been evaluated: $\lambda = 0.343$ eV, $V = 1.18$ meV (toluene), and $\lambda = 0.417$ eV, $V = 2.12$ meV (toluene/pyridine). Taking into account that nuclear factors λ do not differ significantly for the dimers $ZnOEP)_2$Ph-Pim and $(ZnOEP)_2$Ph-Q of the same geometry, and using different values of Gibbs free energy of PET ΔG^0 (Table 5.7) we estimated the corresponding values for the quinone-substituted dimer: $V = 2.35$ meV (toluene) and $V = 4.22$ meV (toluene/pyridine). We have shown additionally that according to the Landau-Zener criteria $\pi^2 V^2 / h\omega(2\lambda k_B T)^{1/2} < 1$ [133], the dimer S_1 state quenching is due to a non-adiabatic PET in toluene. Upon pyridine addition PET processes for $(ZnOEP)_2$Ph-Pim may be assigned to the "normal" region of the Marcus dependence, $\log k_{ET} = f(-\Delta G^0)$, while for $ZnOEP)_2$Ph-Q PET reaction is probably near the turning point beyond which the system goes into the "inverted" region [72].

5.4.2 *Superexchange Electron Transfer in Triads with Covalently Linked Electron Acceptors*

Comparative studies of fluorescence properties for the triads **III**, $H_2P(m^\wedge Pyr)_2$-(iso-PrPh)$_2$⊗$(ZnOEP)_2$Ph-Q and **IV**, $H_2P(m^\wedge Pyr)_2$-(iso-PrPh)$_2$⊗$(ZnOEP)_2$Ph-Pim, reveal the complex behaviour both for the dimer and extra-ligand depending also on the polarity of the surrounding (addition of polar acetone or acetonitrile to toluene) [30, 72]: (1) In toluene at 293 K, upon the triad self-assembly the initial fluorescence of the *A*-containing dimers $(ZnOEP)_2$Ph-Q or $(ZnOEP)_2$Ph-Pim (being strongly quenched due to PET process dimer → *A* as discussed in Section 5.4.1), does show a remarkable additional quenching. This quenching increases at the solvent polarity rise. (2) The fluorescence quantum efficiency of complexed extra-ligands is smaller essentially with respect to that found for the same extra-ligands in the corresponding triads without *A*'s. The observed quenching becomes stronger by 1.5–2.0 times for both triads **III** and **IV** 9 vol % of polar solvent. (3) For Pim containing triad in toluene at 293 K fluorescence excitation spectra reveal a small sensitization effect with respect to that observed for the triad **I** of the same geometry but without *A* (Fig. 5.1). The polarity increase leads to the full disappearance of the sensitization effect (e.g., the energy transfer absence). (4) For Q containing triad even in toluene at 293 K is almost identical in shape to that of the individual extra-ligand. It means that the S-S ET process Zn-dimer*→extra-ligand is slow compared to other pathways (including PET) of electronic energy deactivation of the locally excited S_1 state of the dimer $^1(ZnOEP)_2Ph^*$.

TCSPC measurements get a more quantitative insight into the dynamics of relaxation processes in the systems under consideration (Fig. 5.15 and Table 5.2). Noteworthy, a small but noticeable

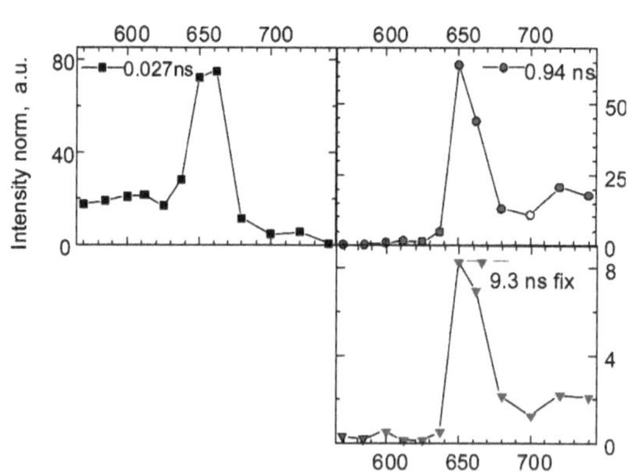

Figure 5.15. Decay-associated spectra of the triads $(ZnOEP)_2$Ph-Q ⊗ $H_2P(m^\wedge Pyr)_2$-(iso-PrPh)$_2$ (toluene, 293 K, λ_{ex} = 546 nm) derived from global analysis of 12 TCSPC time-resolved fluorescence measurements.

shortening of fluorescence decays was detected for the same extra-ligands in triads without *A*'s that was attributed to a photoinduced hole transfer from the extra-ligand to the dimer [70–72] (see Sections 5.3.1 and 5.3.2).

Table 5.2 shows also that for the *A*-containing triads of the same geometry but having extra-ligands of various nature the extra-ligand fluorescence decay shortening decreases in the following sequence: $H_2P(m{\wedge}Pyr)_2$-(iso→PrPh)$_2$ → $H_2Chl(m{\wedge}Pyr)_2$ → $H_2THP(m{\wedge}Pyr)_2$. Femtosecond pump-probe data for the triad $(ZnOEP)_2$Ph-Pim⊗$H_2P(m{\wedge}Pyr)_2$-(iso-PrPh)$_2$ reveal that the primary step, the non-radiative deactivation of the dimer locally excited S_1 state is faster and takes place within $\tau_1 = 0.9$ ps [30, 72]. This short component is accompanied by a relatively longer one ($\tau_1 = 5.4$ ps) that may be attributed to fast competing relaxation processes with participation of the other excited states of various nature. In fact, the non-radiative relaxation processes for both the dimer and extra-ligand S_1 states are faster in *A*-containing triads with respect to those found for *A*-free triads [70].

$$\Delta G_S = \frac{e^2}{4\pi\varepsilon_0}\left[\left(\frac{1}{2r_D}+\frac{1}{2r_A}-\frac{1}{r_{DA}}\right)\frac{1}{\varepsilon}-\left(\frac{1}{2r_D}\frac{1}{\varepsilon_D}+\frac{1}{2r_A}\frac{1}{\varepsilon_A}\right)\right] \tag{5.6}$$

Finally, using schematic energy level diagram for A-containing triads based on the obtained experimental data (Scheme 5.2), one may discuss possible pathways, which are responsible for the observed

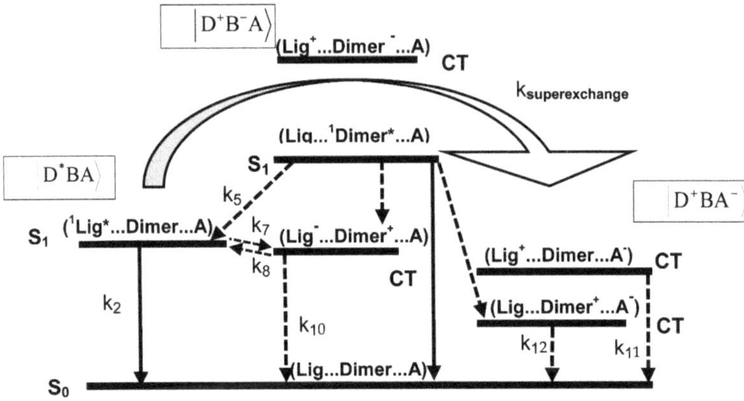

Scheme 5.2.

Table 5.2. Measured and estimated parameters for superexchange PET in self-assembled triads of the same geometry with electron acceptors Pim or Q and various extra-ligands (toluene, 293 K)

Triad No	$E(S_1^D)$,[(a)] eV	r_{DB},[(b)] nm	r_{DA},[(b)] nm	E_D^{OX},[(c)] eV	$E(D^+B^-A)$,[(d)] eV	τ_{S0}^D,[(e)] ns	τ_S^D,[(e)] ns	$k_{PET}/10^8$,[(f)] s^{-1}
1	1.91	0.91	2.42	1.10	3.08	7.7	2.67	2.5
2	1.91	0.91	2.08	1.10	3.08	7.7	0.95	9.2
3	1.89	0.82	1.80	1.07	3.05	6.6	1.24	6.5
4	1.66	0.82	1.80	1.09	3.07	4.3	1.04	7.3

Notes: Triad composition: No. 1—$H_2P(m\hat{}Pyr)_2$-(iso-PrPh)$_2$⊗$(ZnOEP)_2$Ph-Pim, No. 2—$H_2P(m\hat{}Pyr)_2$-(iso-PrPh)$_2$⊗$(ZnOEP)_2$Ph-Q, No 3.—$H_2Chl(m\hat{}Pyr)_2$⊗$(ZnOEP)_2$Ph-Q, No 4.—$H_2THP(m\hat{}Pyr)_2$⊗$(ZnOEP)_2$Ph-Q.

[(a)] Energy levels of excited S_1-states of extra-ligands (D) were determined on the basis of fluorescence and absorption Q(0,0) bands.

[(b)] Intercenter distances r_{DB}, r_{DA} and molecular radii $r_D = 5.5$ Å, $r_A = 3.5$ Å (Pim) or 3.3 Å (Q) were estimated from optimized structures of the triads (Fig. 5.1).

[(c)] Redox potentials for extra-ligands E_D^{OX} and coordinated dimer $(ZnOEP)_2$Ph, $E_A^{RED} = 0.63$ V (in DMF vs. SCE) were extracted from literature data (cited in our papers [68–73]).

[(d)] The energy of a bridge level $E(D^+B^-A)$,was estimated by $E(D^+B^-A) = e(E_D^{OX} - E_A^{RED}) + \Delta G_S$, [72] with dielectric constants of ε (toluene) $= 2.38$ and $\varepsilon'_A = \varepsilon'_D = 36.7$ (DMF).

[(e)] τ_{S0}^D and τ_S^D values correspond to fluorescence decays for extra-ligands in triads without and with additional A's, respectively.

[(f)] rate constants k_{PET} were calculated by $k_{PET} = (\tau_S^D)^{-1} - (\tau_{S0}^D)^{-1}$. Correction term for the position of CT state in toluene at 293 was estimated from [22, 28, 29].

shortening of the dimer and the extra-ligand locally excited S_1-states. Really, after the direct excitation S_1-state (Lig...^{1}Dimer...A) of the dimer $(ZnOEP)_2Ph$ may be deactivated due to the following non-radiative processes:

(i) One-step PET (Lig...^{1}Dimer*...A) $\xrightarrow{k_9}$ (Lig...Dimer$^+$...A$^-$), $k_9 = 6.6 \times 10^9$ s^{-1} for Pim and $k_9 = 2.9 \times 10^{10}$ s^{-1} for Q;

(ii) one-step ET (Lig...^{1}Dimer*...A) $\xrightarrow{k_6}$ (Lig$^-$...Dimer$^+$...A);

(iii) Non-radiative S-S ET, (Lig...^{1}Dimer*...A) $\xrightarrow{k_5}$ (^{1}Lig* ...Dimer...A). Rate constants $(k_5+k_6) = 1/1.7$ ps $= 5.9 \times 10^{11}$ s^{-1}[69, 70]. Thus, $(k_5 + k_6) \gg k_9$, and the direct one-step PET process (i) is low probable with respect to processes (ii) and (iii). In addition, $(k_5 + k_6) \gg k_1 = 1/\tau_S^0 = 1/1.15ns = 8.7 \times 10^8$ s^{-1}; thus no detectable fluorescence of the dimer $(ZnOEP)_2Ph$ is observed in the triads. Upon the solvent polarity increase the sequential PET (ii) becomes dominant in the non-radiative deactivation of the dimer S_1 state with respect to the process (iii).

It follows from scheme 1.2 that at 293 K the extra-ligand S_1 state (^{1}Lig*...Dimer...A) could arise from both direct photoexcitation and fast exothermic energy transfer (Lig...^{1}Dimer...A)→ (^{1}Lig*...Dimer...A) or via thermally activated charge recombination (Lig$^-$...Dimer$^+$...A) $\xrightarrow{k_8}$ (Dimer...^{1}Lig*...A). Once formed, the excited S_1 state of the extra-ligand may decay via two non-radiative processes:

(iv) Bridge-mediated long-distance superexchange PET, discussed in [12, 13, 68] (^{1}Lig*...Dimer...A) $\xrightarrow{k_{superexchange}}$ (Lig$^+$ ...Dimer...A$^-$), where $(ZnOEP)_2Ph$ dimer is a bridge;

(v) (v) photoinduced hole transfer (^{1}Lig*...Dimer...A) $\xrightarrow{k_7}$ (Lig$^-$...Dimer$^+$...A).

Superexchange PET occurs because of coherent mixing of the three or more states of the system [12, 13] (shown as $|D^*BA\rangle$, $|D^+B^-A\rangle$ and $|D^+BA^-\rangle$ in Scheme 5.2). Being not directly populated a high-lying "spectator" state $|D^+B^-A\rangle$ mediates the distant ET from a donor state $|D^*BA\rangle$ to CT state $|D^+BA^-\rangle$. Within this model the charge separation rate constant $k_{super} \sim (V_{12} \cdot V_{23}/\delta E)$. V_{12} and V_{23} are the electronic coupling terms for PET processes $|D^*BA\rangle \rightarrow |D^+B^-A\rangle$ and $|D^+B^-A\rangle \rightarrow |D^+BA^-\rangle$ respectively, both being essentially lower than the energy differences between the relevant system states. δE is the energy difference of $|D^+BA^-\rangle$ and the crossing point of the potential energy curves of $|D^*BA\rangle$ and $|D^+BA^-\rangle$ along the reaction coordinate. As is seen from Table 5.2 that superexchange PET rate constants are hard dependent on the extra-ligand nature and the triad organisation (because of a strong electronic coupling term V_{12}). At the same time the decrease in PET rate constant for Pim - containing triad (kPET $= 2.5 \times 10^8$ s^{-1}) with respect to that for the triad with Q ($k_{PET} = 9.2 \times 10^8$ s^{-1}) may be due to the increase of rDB and rDA distances as well as changes of the energy of $|D^+B^-A\rangle$ and $|D^+BA^-\rangle$ states.

At last, it should be mentioned that hole transfer pathway (v) leads to the formation of CT state (Lig$^-$ ... Dimer$^+$) which is not a final CT state in *A*-containing triads. Indeed, both PET processes to low-lying CT states, (Lig$^-$ ... Dimer$^+$ ... A) $\rightarrow$ (Lig ... Dimer$^+$ ... A$^-$) and (Lig$^-$ ... Dimer$^+$ ... A) $\rightarrow$ (Lig$^+$ ... Dimer ... A$^-$), may be considered as superexchange PET processes mediated by coherent mixing of the corresponding upper-lying CT states.

Thus, self-organised nanoscale porphyrin triads containing additional *A*'s of non-porphyrin nature show complex ET and PET dynamics depending on the geometry of the complex, redox and photophysical properties of interacting subunits as well as on the solvent polarity. The non-radiative deactivation of locally excited S states in the triads includes multistep PET processes of various type (sequential PET, hole transfer, and long-range superexchange PET to *A*) thus mimicking the primary charge transfer *in vivo*.

5.5 Excited State Properties of Multiporphyrin Complexes with Number of Macrocycles $N \geq 5$

The replacement of the chemical dimers by trimers in self-assembly with pyridyl containing di- and tetrapyrrole extra-ligands (porphyrin, pentafluorophenyl-substituted porphyrin, Cu-porphyrin, tetrahydroporphyrin) increases a variety of possible (and controllable) geometries and compositions of the complexes under study. In its turn, it may lead to novel pathways of electronic excitation energy dynamics depending on physico-chemical characteristics of interacting subunits and properties of surrounding. Some of the arrays under study will be discussed below, while the comparative analysis for these complexes of a given structure and composition has been presented recently in our papers [74, 76].

5.5.1 *Exchange d-π Effects in Self-Assembled Complexes Containing Cu-Porphyrins*

Titration experiments have shown (Fig. 5.16A, B) that during pentad **V**, $2(ZnOEP)_2Ph \otimes CuP(mPyr)_4$, formation the dimer $(ZnOEP)_2Ph$ is complexed practically completely at a molar ratio of x = $[CuP(mPyr)_4]/[ZnP\text{-dimer}] \approx 0.5$, in contrast to $x \approx 1$ obtained for the corresponding triads [66]. An allosteric effect found for the Zn_2-gable-porphyrins [137] and a lack of evidence on higher aggregate formation for the systems under study may be considered as an additional support of the pentads formation. Almost the same spectral position of the Soret band components and practically doubled extinction coefficients of the pentad compared to those of the triads [67] lead to the conclusion that the intense B-transitions of two dimers $(ZnOEP)_2Ph$ bound in the same pentad are not involved into strong coupling. Like in the case for other pentads of the same geometry [67] the formation of the complex $2(ZnOEP)_2Ph \otimes CuP(mPyr)_4$ are characterized by a slight red shift of the dimer and ligand visible electronic bands which is larger with respect to that for the corresponding triads [66]. This shift and differences may be explained by a larger distortion of tetrapyridyl-containing extra-ligands from planarity in the pentad relative to that for dipyridyl-containing porphyrins in the triad, such a correlation is known for sterically hindered non-planar porphyrins [114–118]. Figure 5.16B shows that the dimer fluorescence is strongly quenched in the pentad $2(ZnOEP)_2Ph \otimes CuP\text{-}(m\text{-}Pyr)_4$. Noteworthy, at $x = 0.71$ the rest of the non-red-shifted fluorescence belongs to non-complexed dimer $(ZnOEP)_2Ph$ exclusively.

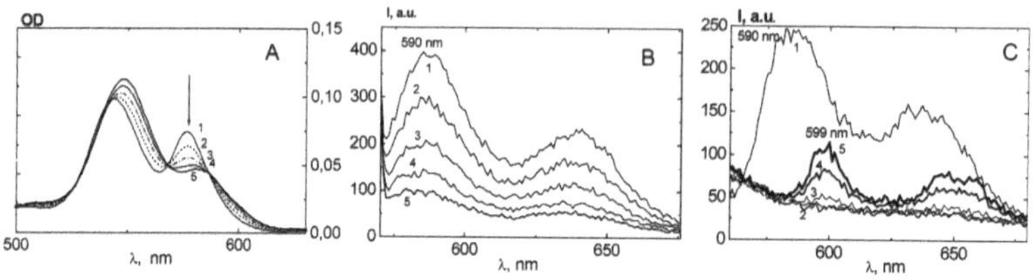

Figure 5.16. Q-band absorption (A), fluorescence (B) spectra of the dimer $(ZnOEP)_2Ph$ upon titration by tetra-pyridyl containing extra-ligand, $CuP(mPyr)_4$, in toluene at 293 K, and temperature dependence of the pentad $2(ZnOEP)_2Ph \otimes CuP(mPyr)_4$ emission spectra upon temperature lowering (C). **A**: Absorption spectra transformations at various molar ratio $x = [CuP(mPyr)_4]/[ZnOEP)_2Ph)]$: 0.14 (1), 0.29 (2), 0.43 (3), 0.57 (4), and 0.71 (5). **B**: Fluorescence intensity decrease (λ_{ex} = 567 nm, isosbestic point) of the uncomplexed dimer (λ_{max} = 590–591 nm) upon the sequential increase of molar ratio x: 0.14 (1), 0.29 (2), 0.43 (3), 0.57 (4), and 0.71 (5). At $x = 0.71$, the rest of the non-redshifted fluorescence belongs to non-complexed dimer. **C**: Fluorescence spectra (λ_{ex} = 542 nm) of the pentad $2(ZnOEP)_2Ph \otimes CuP(mPyr)_4$ ($x = 0.71$) in toluene (thin transparent layers) at various temperatures: 300 K (1, belongs to the rest of the non-complexed dimer), 220 K (2, additional quenching is due to the full complexation), 190 K (3), 165 K (4), and 150 K (5). The red-shifted fluorescence bands at $T \leq 210$ K (increasing upon temperature lowering) belong to the emission of the dimer extra-ligated by $CuP(mPyr)_4$ in pentad.

In Section 5.2.1, we described that for Cu-porphyrin containing hybrid dimers (H_2P+CuP), (ZnP+CuP) with a relatively close proximity of interacting halves the fluorescence lifetime shortening and quantum yield decrease detected for ZnP or H_2P counterparts are caused mostly by the enhancement of the intersystem crossing in these halves [64, 92, 98]. This enhancement is explained by the existence of the exchange coupling of unpaired d-electron of central Cu(II) ion of one half of the dimer with π-conjugated electronic system of the other half (exchange d-π effects [64, 92, 98]. In some cases this coupling may govern by through-bond interactions via spacer [138]. At ambient temperature an additional fluorescence quenching for ZnP half in hybrid (ZnP+CuP) dimers may be caused by the thermally activated singlet-singlet ET ZnP→CuP [92]. At last, the formation CT states has been discussed as a possible reason of ZnP fluorescence quenching in the dimers of such composition [139–141].

Correspondingly from above physico-chemical backgrounds the dimer fluorescence in pentad $2(ZnOEP)_2Ph\otimes CuP(mPyr)_4$ may be caused by several reasons (Scheme 5.3):

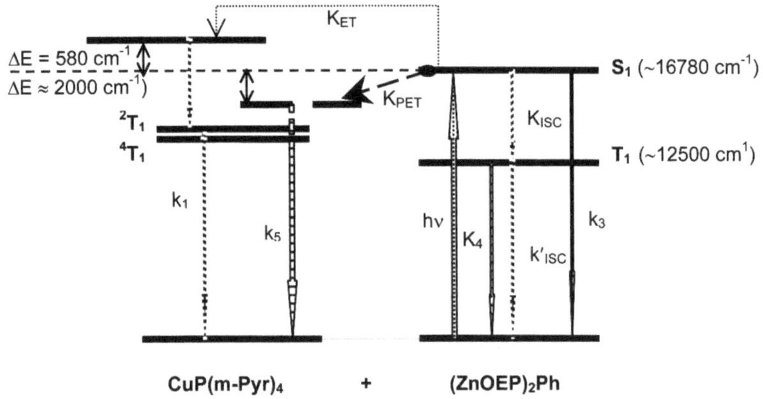

Scheme 5.3.

(i) the photoinduced electron transfer, PET [Zn-dimer*...CuP]→ [Zn-dimer$^+$...CuP$^-$]; (ii) thermally activated ET from the dimer S_1 state to ^2Q state of $CuP(mPyr)_4$; (iii) exchanged d-π effects leading to an increase of the intersystem crossing rate constant in the dimer subunits caused by an interaction of Cu unpaired d-electron with π-conjugated systems of the dimers.

From the thermodynamic background, PET becomes possible if the free energy of the process $\Delta G^0 < 0$ [127]. With respect to the pentad, Gibbs free energy of PET reaction was calculated according to [76]

$$\Delta G^0 = e(E_D^{OX} - E_A^{RED}) + \Delta G_S - E(S_1^D) \qquad (5.7)$$

The oxidation potential for coordinated dimer $(ZnOEP)_2Ph$ was taken to be $E_D^{OX} = 0.63$ V, like for pyridinated ZnOEP in dimethylformamide (DMF) *vs.* SCE [28, 110, 111]. Reduction potential for $CuP(mPyr)_4$ $E_A^{RED} = -1.07$ V (in DMF vs. SCE) has been extracted from literature data taking into account that pyridyl substituents increase the reduction potential of porphyrins and chlorins [74, 76]. According to Eq. (5.6) correction term for the position of CT state in toluene at 293 K was calculated to be $\Delta G_S = 0.287$ eV, with donor and acceptor radii of $r_D = r_A = 0.55$ nm and intercenter distance of $r_{DA} = 0.88$ nm estimated the pentad **V** optimized structure. Thus, in toluene at ambient temperature $E(IP) = e(E_D^{OX} - E_A^{RED}) + \Delta G_S = 1.99$ eV, $\Delta G^0 = -0.14 eV < 0$, and PET at $r_{DA} = 8.8$ Å might become rather effective and fast (within tens of picoseconds) in the present case.

At room temperature (kT = 209 cm^{-1}) thermally activated inductive-resonance ET from the dimer S_1 state to ^2Q state of CuP(mPyr)$_4$ (energy gap $\Delta E \approx 580$ cm^{-1}, see Scheme 5.3) is still possible with the rate constant $k_{ET} \leq 3.7 \times 10^9$ s^{-1}. In fact, like in hybrid (ZnP+CuP) dimers [87, 91], this process in the pentad may lead to the additional fluorescence quenching of (ZnOEP)$_2$Ph moiety.

Obviously, temperature lowering and transition to rigid solutions would lead to the switching off two processes in the system under discussion: (i) energy migration (due to Boltzmann factor $\exp(-\Delta E/kT)$ and decrease of spectral overlap integral leading to k_{ET} diminishing), and (ii) photoinduced electron transfer due to the destabilization of the ion pair [Zn-dimer$^+$... CuP$^-$] state [22]. From the other hand, exchanged d-π effects are hardly dependent on temperature [91, 97, 98, 138–141]. Experimental findings show (see Fig. 5.16C) that at $T \leq 210$ K the fluorescence of the complexed dimer (ZnOEP)$_2$Ph in the pentad 2(ZnOEP)$_2$Ph⊗CuP-(m-Pyr)$_4$ remains still strongly quenched like that found at 293 K, but a small pronounced increase of red-shifted fluorescence bands belonging to the dimeric subunit in the pentad (like di-pyridinated Zn-dimer [66, 74]) is observed. It means that in a temperature range of 150 K ÷ 293 K the strong quenching of (ZnOEP)$_2$Ph emission in CuP(mPyr)$_4$ containing pentads is caused by exchanged d-π effects presumably.

The manifestation of exchanged d-π effects is also observed in the system (ZnOEP)$_3$Ph$_2$⊗CuP(mPyr)$_4$⊗H$_2$P(m^Pyr)$_2$ characterized by the structure **VI** (see Fig. 5.2). Basically the existence of CuP(mPyr)$_4$ in this complex leads to the fluorescence quenching for both (ZnOEP)$_3$Ph$_2$ and H$_2$P(m^Pyr)$_2$ subunits [74]. Indeed, PET H$_2$P(m^Pyr)$_2$ →CuP(mPyr)$_4$ is impossible due to thermodynamic reasons ($\Delta G^0 > 0$). Energy transfer H$_2$P(m^Pyr)$_2$ →CuP(mPyr)$_4$ is low probable due to a high energy gap $\Delta E \approx 1950$ cm^{-1}. Thus, H$_2$P(m^Pyr)$_2$ and (ZnOEP)$_3$Ph$_2$ fluorescence quenching is caused mostly by the enhancement of the intersystem crossing due to exchange d-π effects. As for the complex $(ZnOEP)_3Ph_2 \otimes CuP(mPyr)_4 \otimes H_2P(m\wedge Pyr)_2$, an additional reason of the trimer emission quenching has to be taken into account. Recently, we have shown [71, 72] that in the triads with participation of the dimer $(ZnOEP)_2Ph$ and $H_2P(m\wedge Pyr)_2$ (structure **I**, Fig. 5.2) there are two charge transfer processes leading to the non-radiative deactivation of the locally excited S_1 states in nano-, picosecond time scale [$(1\text{Dimer} * \ldots \text{Lig}) \longrightarrow (\text{Dimer}^{+} \ldots \text{Lig}^{-})$, $(\text{Dimer} \ldots {}^{1}\text{Lig}^{*}) \longrightarrow (\text{Dimer}^{+} \ldots \text{Lig}^{-})$] and the ET process $(ZnOEP)_2Ph \rightarrow H_2P(m\wedge Pyr)_2$. Some details of this quenching have been analyzed for various systems in [74].

5.5.2 *Relaxation Pathways in Multicomponent Complexes*

Depending on the physico-chemical parameters of interacting subunits as well as on their mutual geometry in multicomponent self-assembled complexes (structures **VI** and **VII** in Fig. 5.2) the competition on of the main two channels (ET and PET) have been realized governed by the solvent polarity and temperature [74]. Table 5.3 collects experimental parameters and theoretical values characterizing ET processes in multiporphyrin arrays of VI and VII structures based on the trimer $(ZnOEP)_3Ph_2$ and various extra-ligands.

Finally, it should be mentioned that the formation of multiporphyrin complexes may be governed by temperature [74]. For instance, the complex $(ZnOEP)_3Ph_2 \otimes H_2THP(m\text{-}Pyr)_2\text{-}(Ph)_2 \otimes H_2P(5F\text{-}Ph)_3\text{-}(p\text{-}Pyr)_1$ (structure **VII,** Fig. 5.2) may be formed at low temperature only because of small complexation constant for monopyridyl containing fluorinated extra-ligand (L_2) at ambient temperature ($K_C < 10^3\ M^{-1}$). Typically, at 293 K the complexed trimer is strongly quenched due to ET/PET processes $(ZnOEP)_3Ph_2 \rightarrow H_2THP(m\text{-}Pyr)_2\text{-}(Ph)_2$ (ligand L_1) in the complex $(ZnOEP)_3Ph_2 \otimes H_2THP(m\text{-}Pyr)_2\text{-}(Ph)_2$, while fluorinated extra-ligand L_2 does fluoresce not being connected to the above complex. Transition to 77 K manifests itself in a pronounced decrease of the ligand L_1 fluorescence band and full disappearance of the ligand L_2 emission. These facts are indicative of the complexation of the fluorophenyl-substituted mono-pyridyl containing extra-ligand L_2 with the trimer at low temperature. Again, the majority of the trimer

Table 5.3. Spectral-kinetic and ET parameters for interacting subunits in multiporphyrin arrays (toluene, 293 K, refractive index $n = 1.506$)

No	System and D-A pair	τ_D^0, ns	φ_D^0	ε_A, $M^{-1}\cdot cm^{-1}$	J, 10^{-14} $cm^3\cdot M^{-1}$	k^2	R_0^{theor}, Å	r_{DA}, Å	k_{ET}, 10^{10} s^{-1}
1	Dimer* $(ZnOEP)_2$-h	1.10	0.012	24900 (588)**	3.95	1.002	21.1	12.5	2.1
2	Trimer $(ZnOEP)_3$-h_2	0.9	0.01	47900 (574)	7.37	1.002	22.7	12.5	3.95
2a	Trimer $(ZnOEP)_3$-h_2 (Pyridin)	1.25	0.016	24900 (588)	5.30	1.002	23.3	12.5	3.35
3	$(ZnOEP)_2Ph \otimes H_2P(m^{\wedge}Pyr)_2$-$(iso\text{-}PrPh)_2$	1.10	0.012	9100 (654)	4.18	0.760	20.3	8.8	13.7
4	$(ZnOEP)_3Ph_2 \otimes H_2P(m^{\wedge}Pyr)_2$-$(iso\text{-}PrPh)_2$	0.9	0.01	9100 (654)	4.35	0.772	19.9	8.8	14.8
5	$(ZnOEP)_3Ph_2 \otimes H_2THP(m\text{-}Pyr)_2$-$(Ph)_2$	0.9	0.01	130000 (739)	6.15	0.525	19.8	8.2	22.0
6	$(ZnOEP)_3Ph_2 \otimes H_2THP(mPyr)_4$	0.9	0.01	130000 (739)	6.15	0.772	21.1	8.8	21.1
7	$(ZnOEP)_3Ph_2 \otimes H_2P(mPyr)_4$	0.9	0.01	9100 (654)	4.35	0.772	19.9	8.8	14.8
8	$(ZnOEP)_2Ph \otimes CuP(mPyr)_4$	1.10	0.012	2500 (576)	0.13	0.760	11.1	8.8	0.37
9	$(ZnOEP)_3Ph_2 \otimes H_2P(m^{\wedge}Pyr)_2$-$(iso\text{-}PrPh)_2$ (L_2)	0.9	0.01	9100 (654)	4.35	0.772	19.9	8.8	14.8
10	$(ZnOEP)_3Ph_2 \otimes H_2THP(m\text{-}Pyr)_2$-$(Ph)_2$ (L_1)	0.9	0.01	130000(739)	6.15	0.525	19.8	8.2	22.0
11	$(ZnOEP)_3Ph_2 \otimes H_2P(Ph)_3$- $(p\text{-}Pyr)_1(L_2)$	0.9	0.01	10900(654)	4.18	0.472	18.2	9.1	7.1
12	$(L_2)H_2P(Ph)_3$-$(p\text{-}Pyr)_1 \otimes (ZnOEP)_3Ph_2 \otimes H_2THP$ $(m\text{-}Pyr)_2$-$(Ph)_2$ (L_1)	9.1	0.07	130000 (739)	7.42	0.671	29.1	18.4	0.17

Table 5.3.

13	$(ZnOEP)_3Ph_2 \otimes H_2THP(m\text{-}Pyr)_2\text{-}(Ph)_2$ (L_1)	0.9	0.01	130000 (739)	6.15	0.525	19.8	8.2	22.0
14	$(ZnOEP)_3Ph_2 \otimes H_2P(5F\text{-}Ph)_3\text{-}(p\text{-}Pyr)_1$ (L_3)	0.9	0.01	600 (641)	1.22	0.472	14.9	9.1	2.1
16	$(ZnOEP)_3Ph_2 \otimes H_2P(m\text{-}Pyr)_2\text{-}(5F\text{-}Ph)_2$ (L_1)	0.9	0.01	600 (641)	1.85	0.525	16.6	8.2	6.3
17	$(ZnOEP)_3Ph_2 \otimes H_2THP(Ph)_3\text{-}(p\text{-}Pyr)_1$ (L_3)	0.9	0.01	130000 (739)	6.15	0.472	19.9	9.1	12.1

Notes: Abbreviations (L_1), (L_2) and (L_3) in D-A pairs correspond to extra-ligands 1, 2, and 3 in multiporphyrin arrays (structures **VI** and **VII**, Fig. 5.2).
φ_D^0 and τ_D^0 are experimentally measured fluorescence quantum yield and lifetime values for individual donor in the presence of O_2 in solutions.
*Both ZnOEP macrocycles are extra-ligated. Decimal extinction coefficients ε_A were measured with an average total error of 15%.
**corresponding λ_{max} in nm is indicated in brackets. Intercenter distances r_{DA} were estimated from optimised structures of the complexes (HyperChem software package, release 4, methods AM1, PM3).
Energy migration parameters have been calculated in the frame of Foerster inductive-resonant model [126]:
Critical transfer distance $R_0^{theor} = \frac{9000 \cdot \ln 10 \cdot k^2 \cdot \varphi_D^0}{128\pi^5 n^4 N_A} \times \int_0^\infty f_D(\nu)\varepsilon_A(\nu)\frac{d\nu}{\nu^4}$.

N_A is Avogadro number. Spectral overlap integrals $J = \int_0^\infty f_D(\nu)\varepsilon_A(\nu)\frac{d\nu}{\nu^4}$ have been calculated on the basis of experimental acceptor absorption and donor fluorescence spectra. Orientational factors $k^2 = [\mathrm{Cos}(\mu_D, \mu_A) - 3\,\mathrm{Cos}(\mu_D, \mathbf{r}_{DA}) \cdot \mathrm{Cos}(\mu_A, \mathbf{r}_{DA})]^2$, where (μ_D, μ_A) is the angle between transition dipole moment vectors of the D and A subunits, $(\mu_D\ \mathbf{r}_{DA})$ and $(\mu_A, \mathbf{r}_{DA})$, denote the angles between the dipole vectors of D and A and the direction D→A, respectively. k^2 values have been calculated on the basis of optimised structures of the complexes and oscillator models for ZnP-dimers and extra-ligands. ET theoretical rate constants have been calculated according to $k_{EM} = (1/\tau_S^0)(R_0^{theor}/r_{DA})^6$.

fluorescence at 77 K is quenched due to PET $(ZnOEP)_3Ph_2 \rightarrow H_2PF$ and $(ZnOEP)_3Ph_2 \rightarrow H_2THP$ competing with possible ET processes $(ZnOEP)_3Ph_2{}^* \rightarrow H_2THP$, $(ZnOEP)_3Ph_2{}^* \rightarrow H_2PF$. In its turn, the emission quenching for H_2THP moieties may be also connected with the manifestation of charge transfer interactions in the conditions of increased overlap of donor and acceptor electronic wave functions.

5.6 Exciton Relaxation and Electron Wave Tunneling in Self-Assembled "Semiconductor Nanocrystal–Porphyrin" Composites

Typical transformations of absorption and photoluminescence (PL) spectra of the NC solution as a function of added monomeric tetra-meso-pyridyl-substituted porphyrins $H_2P(m\text{-}Pyr)_4$ at well-defined molar ratios $x = [H_2P]/[QD]$ are depicted in Fig. 5.17. It is seen that upon molar ratio x increase, besides NC absorption bands ($\lambda_{max} = 556$ nm) a linear increase of the porphyrin absorption bands takes place. Noteworthy, in all cases. NC absorption remains constant, while the PL ($\lambda_{max} = 585$ nm) is considerably quenched upon titration by $H_2P(m\text{-}Pyr)_4$ molecules.

Time-resolved PL measurements show that emission of pure NC without porphyrin ligand is characterized by a non-exponential decay (Fig. 5.18) that is typical for these objects reflecting the

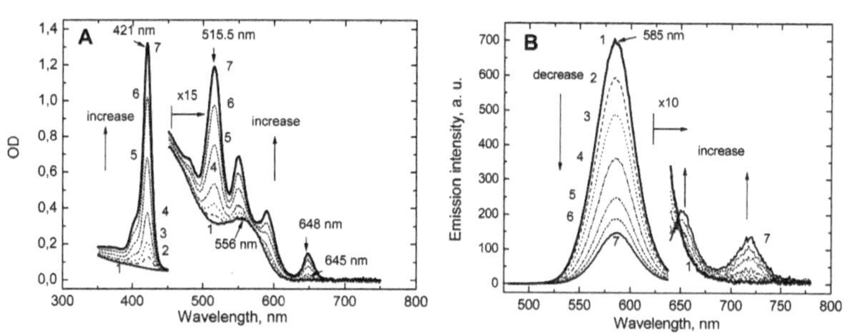

Figure 5.17. Absorption (A) and emission (B, $\lambda_{ex} = 465$ nm) spectra of CdSe/ZnS NC (diameter of CdSe core $d = 3.8$ nm, 3 ZnS monolayers) and $H_2P(m\text{-}Pyr)_4$ molecules in toluene at 295 upon molar ratio x = [porphyrin]/[NC] increase: 0.0(1); 1.15 (2); 2.3 (3); 4.5 (4); 9.1 (5); 13.5 (6); 17.8 (7).

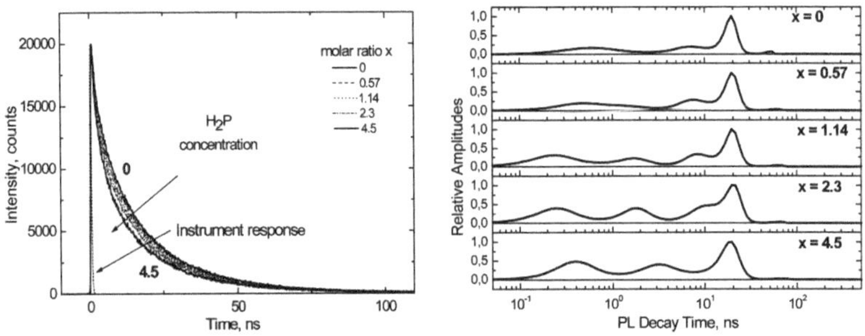

Figure 5.18. Time-resolved PL traces and decay amplitude distributions for CdSe/ZnS QD (diameter of CdSe core $d = 3.8$ nm, 3 ZnS monolayers) upon molar ratio x increase of H_2P-(m-Pyr)$_3$ in toluene (295 K, $\lambda_{exc} = 575$ nm, $\lambda_{em} = 585$ nm).

complex exciton dynamics in the conditions of quantum confinement [46–48, 50]. The interaction with anchored porphyrin molecules manifests itself in the appearance and rise of two additional short time components ($\sim$7 ns and $\sim$700 ps). In this conditions fluorescence parameters for H_2P(m-Pyr)$_4$ molecules (spectral position, quantum efficiency, decay) remain unchanged with respect to individual monomers.

All these facts indicate that in "NC-monomeric porphyrin" nanosize composites PL quenching of CdSe counterpart is a dynamic process caused by the increased non-radiative relaxation channels in the excited states of a nanocrystal. The increase of the solvent polarity (an addition of 32 vol % of polar acetone to toluene) enhances an additional decrease of NC PL by more than one order of magnitude [53].

It should be mentioned that in most cases the formation of "NC-organic dye" composites is followed by NC PL quenching which is commonly interpreted as being due to photoinduced charge transfer (CT) [142, 143] and/or Foerster-type resonant excitation ET) NC$\rightarrow$dye [52, 144]. Though in most cases ample qualitative evidence for the presence of such quenching processes is given, only a few reports unravel quantitatively, that the PL quenching can uniquely be assigned to CT and/or ET [52]. In fact, PL quenching is may be induced by other processes than ET, which might be related to the inherent photoinduced blinking of single NCs, which has been for the first time described systematically as being due to photoinduced self-trapping of charges in the dielectric medium of the environment of NCs [145].

Usually in the ET case, the direct verification of the energy migration process as a real reason of PL quenching is the comparison of the experimental values of ET efficiencies via the donor (NC) PL quenching and the sensibilization of the acceptor (dye) fluorescence. Such a comparison has been carried out recently by us for "NC-porphyrin" composites [53, 146]. Firstly, it has been definitely shown that the porphyrin fluorescence enhancement is of order of 10% being much smaller as compared to the corresponding quenching efficiency values found from NC photoluminescence. Secondly, in the case of one type of CdSe/ZnS NC ($d = 2.6$ nm and 2 ZnS monolayers), upon the replacement of H_2P(m-Pyr)$_4$ molecules by tetra-pyridyl-substituted CuP(m-Pyr)$_4$ and tetrahydroporphyrin, THP(m-Pyr)$_4$ derivatives (thus changing the overlap integral values $J(\nu) = \int_0^\infty f_D(\nu)\varepsilon_A(\nu)\frac{d\nu}{\nu^4}$ by factor of 2.5) the quenching efficiency of NC photoluminescence remains nearly the same [83, 84]. So, the contribution of ET NC→porphyrin to the total PL quenching seems to be minor and hence is negligible.

Taking into account the results discussed in previous sections, one should conclude that in the case of the photoinduced charge (hole or electron) transfer processes in "NC-porphyrin" composites the porphyrin ligand fluorescence should be also diminished. Nevertheless, as we mentioned before, fluorescence parameters (efficiency φ_F and decay τ) for H_2P(m-Pyr)$_4$ molecules upon complexation with NCs remain the same practically with respect to those measured for individual ligands and the same conditions [83, 84]. In addition, the comparative titrations of the same NC solutions by H_2P(m-Pyr)$_4$ and THP(m-Pyr)$_4$ (holes acceptor) ligands as well as by H_2P(m^Pyr)$_2$(Ph)$_2$ and electron acceptors, H_2P(m^Pyr)$_2$(5FPh)$_2$ or H_2P(m^Pyr)$_2$(Anthraquinone)$_2$, gives practically the same curves for NC photoluminescence quenching [83, 84]. Thus, non-dependence of NC PL quenching efficiency on redox properties of porphyrin ligands and the absence of the porphyrin fluorescence quenching in "NC-porphyrin" composites rules out the dominant role of usual photoinduced charge

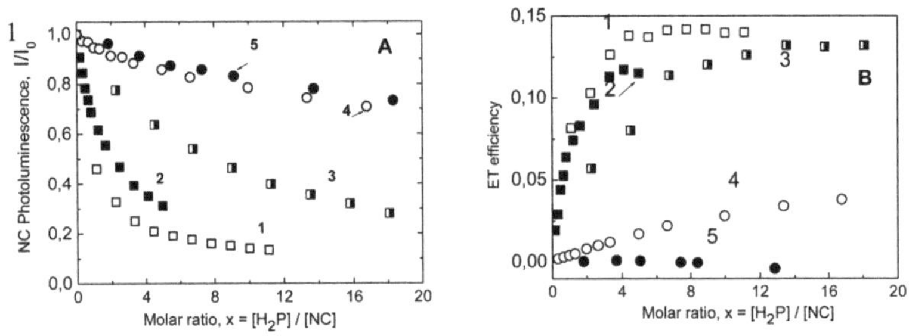

Figure 5.19. Relative integrated PL emission intensities $I(\mathrm{x})/I_0$ (A) and efficiency of resonant energy transfer (B) for CdSe/ZnS NCs of various sizes upon titration by H_2P(m-Pyr)$_4$ as function of the molar ratio $x = [H_2P$(m-Pyr)$_4]/[QD]$ in toluene at 295 K. Legend: (1): $d_{\mathrm{CdSe}} = 3.46$ nm/0 ML ZnS; (2): $d_{\mathrm{CdSe}} = 2.14$ nm/2 ML ZnS; (3): $d_{\mathrm{CdSe}} = 3.00$ nm/2 ML ZnS, (4): $d_{\mathrm{CdSe}} = 4.08$ nm/2 ML ZnS; (5): $d_{\mathrm{CdSe}} = 5.16$ nm/2 ML ZnS. Efficiencies of resonant energy transfer (ET) have been calculated on the basis of approach [53] from the sensitized integrated porphyrin fluorescence intensity.

transfer processes with participation of molecular orbitals of porphyrin macrocycle in NC PL quenching for the systems under study.

We have shown in recent experiments that a third mechanism clearly distinct from CT or ET may cause ligand-induced PL quenching (the so-called non-FRET [146]). This process has not been considered in most of the related publications and is connected with the extension of the wave function of the exciton to the outside of the NC. In this study, the experimental background was based on the comparative analysis of the NC PL quenching by one type of porphyrin H_2P(m-Pyr)$_4$ molecule for NCs of different sizes (as well as having different but known number of ZnS monolayers).

It is seen from Fig. 5.19A that at the same molar ratios x the NC PL quenching is more effective for smaller NCs. For the analysis of the PL quenching curves as a function of x, Stern–Volmer formalism has been modified as follows

$$\frac{I_0}{I} = 1 + \int_0^{\infty} K(x) \cdot dx, \tag{5.8}$$

where I and I_0 represent NC PL in presence (I) and absence (I_0) of H_2P(m-Pyr)$_4$, respectively. Stern–Volmer parameter $K(x) =$

Table 5.4. Photoluminescence quenching rate constants k_q for NCs of various CdSe core diameters d_{CdSe} upon titration by $H_2P(m\text{-}Pyr)_4$ (toluene, 295 K). The number of ZnS monolayers is $n = 2$ for all NCs

d_{CdSe} (nm)	4.3	5.2	6.3	7.3
$< K_{SV}(x) >$	0.65 ± 0.10	0.115 ± 0.015	0.055 ± 0.007	0.020 ± 0.003
k_q, ns^{-1}	0.041 ± 0.016	0.0057 ± 0.0023	0.0027 ± 0.0016	0.0015 ± 0.0009

$\langle K_{SV}(x)\rangle$ is expressed by the first derivative of the experimental data I_0/I as the result of few experimental titration curves for NC of every size. In parallel to this, experimental ET efficiencies have been evaluated from sensitized emission of $H_2P(m\text{-}Pyr)_4$ molecules being attached to NC surface (Fig. 5.19B). The corresponding quenching rate constants k_q have been determined by $k_q = K(x)/\tau_0$ (collected in Table 5.3), where τ_0 is mean intrinsic (non-exponential) PL lifetime measured for every NC.

Data presented in Table 5.3 show that for CdSe/ZnS nanocrystals with two ZnS monolayers k_q values follow a monotonous function drastically decaying with the NC core diameter. From the physico-chemical point of view, we conjecture that upon interaction of $H_2P(m\text{-}Pyr)_4$ molecule with NC surface, the electron wave function may be locally modified (via inductive and/or mesomeric effects [84]) forming a surface local state capable to trap the electron of the photogenerated exciton. The general scheme of our model is shown in Fig. 5.20. We consider the behaviour of the electron wave function at the interface to the functional pyridyl group of the attached porphyrin. In particular, for the quantitative calculation of the probability density function $\psi^2(r)$ we have applied the particle-in-a-box model proposed for a core/shell NC according to [147, 148] solving the Schrödinger equation for an idealized NC with respect to the confinement Eigen energies. The single-carrier envelope wave functions ψ_a in a spherical core/shell NC are determined by the Schrödinger equation

$$\left(\frac{\hbar^2}{2}\frac{\partial^2}{\partial r^2}\left(\frac{1}{m_a^*(r)}+V_a(r)\right)\right)\psi_a = E_a\psi_a \tag{5.9}$$

where the index $a = e, h$ represents electron and hole, respectively. Solutions were obtained by applying continuity relations for the single-carrier wave functions at the interfaces between the CdSe

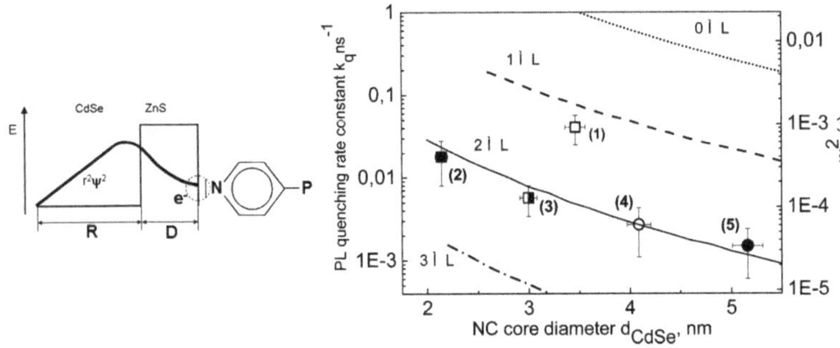

Figure 5.20. Scheme of the NC PL quenching model and the comparison of experimental PL quenching rate constants k_q with calculated size-dependent curves for of the probability density functions $\psi^2(R + D)$ of a 1s electron at the outer interface between the ZnS shell and the environment for CdSe/ZnS NCs of various sizes in toluene at 295 K. Legend: (1): $d_{CdSe} = 3.46$ nm/0 ML ZnS; (2): $d_{CdSe} = 2.14$ nm/2 ML ZnS; (3): $d_{CdSe} = 3.00$ nm/2 ML ZnS, (4): $d_{CdSe} = 4.08$ nm/2 ML ZnS; (5): $d_{CdSe} = 5.16$ nm/2 ML ZnS. Model: Upon excitation, an electron–hole pair (exciton) is created in the CdSe core of the QD. The electron is delocalized over the core and the ZnS shell. As a result of the finite ZnS energy barrier E, the electron can tunnel to the ZnS surface (and the environment). Upon interaction with the lone-pair of a pyridyl group of a porphyrin molecule, the electron becomes partly localized in the vicinity of the attachment site. Calculations: A barrier energy of 4 eV between the ZnS shell and the matrix. Effective masses have been taken as 0.11 for the CdSe core and 0.5 for the ZnS shell, respectively. The amplitude of $\psi^2(R + D)$ has been adjusted to the experimental value at $d_{CdSe} = 4.08$ nm/2 ML ZnS.

core ($i = 1$), ZnS shell ($i, j = 2$), and matrix ($j = 2$), respectively, by assuming that

$$\psi_i\left(r_{ji}\right) = \psi_j\left(r_{ji}\right) \text{ and } \left.\frac{1}{m_i^*}\frac{d}{dr}\psi_i\left(r\right)\right|_{r=r_{ji}} = \left.\frac{1}{m_j^*}\frac{d}{dr}\psi_j\left(r\right)\right|_{r=r_{ji}} \quad (5.10)$$

To account for the local character of this interaction at $r = R + D$, where $R = 1/2\ d_{CdSe}$ corresponds to the core radius and D to the ZnS shell thickness, respectively, $r^2\psi^2(r)$ has to be normalized with respect to r^2 that is with respect to the total QD surface area. This approximation holds since we are only considering point-like interactions at the position of one nitrogen lone pair orbital. Hence the

PL quenching rate becomes

$$k_q(r) = C\psi^2(r) \tag{5.11}$$

The comparison of rate constants k_q (points with bars) and the calculated (and scaled) probability densities $\psi^2(r)$ (lines) shows a good correlation (Fig. 5.20). Thus, one point-like charge density perturbation caused by organic linker group or chromophore at NC interface forces the electron of the delocalized exciton of nanocrystal to become localized. These results reveal also that single functionalized porphyrin molecules can be considered as one of the probes for the complex interface physics and dynamics of colloidal semiconductor nanocrystals.

5.7 Conclusions

Here, we demonstrated that by means of a non-covalent two-fold extra-ligation in conjunction with Zn-porphyrin chemical dimers and trimers it is possible to create by predictable manner a variety of multiporphyrin structures that are capable of undergoing photodriven fast effective intra-complex energy transfer and photoinduced charge separation. The cooperative existence and competition between PET and ET depends on the temperature and polarity of surrounding and results in a complex relaxation dynamics of the locally excited S_1 and T_1 states of interacting counterparts.

In addition, self-assembly principles (based on the non-covalent two-fold extra-ligation) elaborated for the formation of porphyrin triads, pentads and multicomponent arrays may be successfully extended to anchor in a systematic and directed way π-conjugated tetrapyrrole molecules on semiconductor CdSe/ZnS nanocrystal surfaces in solutions. With respect to two-fold extra-coordination, complexation constant K_C values for "NC-porphyrin" composites are in a comparable range with those obtained for various multiporphyrin complexes. Depending on CdSe size (optical tunability), ZnS layer thickness, spectral and anchoring properties of porphyrin moieties (chemical tunability), the photoluminescence quenching nanocrystals in "NC-porphyrin" composites with closely contacting organic and inorganic components can be understood in terms of a tunnelling of the electron (of the excited electron–hole pair) followed by a (self-) localization of the electron or formation of trap states.

Our preliminary results have shown also that the above supramolecular approach can be successfully employed to create nanosize heterocomposites consisting of CdSe/ZnS NCs with multichromophore organic component (poprhyrin triads) anchored on NC surface [149, 150]. The presented results show that design and self-assembly of organic/inorganic moieties into functional superstructures with vectorial relaxation dynamics are perspective for nanotechnology and supramolecular electronics.

Acknowledgements

This work was supported by Volkswagen Foundation (Priority Program "Physics, Chemistry and Biology with Single Molecules"), Belarus Complex Program for Scientific Research (Nanotech 6.18), German Academic Exchange Service (DAAD, grant No A/08/08573, EZ), and Belarusian Republican Foundation for Fundamental Research (grant No ϕ10CO-005). We like to thank also our collaborators Dr. A. Shulga, Dr. Habil. E. Sagun, Dr. A. Stupak, Dr. V. Knyukshto (B.I. Stepanov Institute of Physics NAS, Minsk, Belarus), Prof. F. Cichos (Molecular Nanophotonics, University of Leipzig, Germany), Dr. T. Blaudeck (Institute for Print and Media Technology, Chemnitz University of Technology, Germany) and Dr. E. Petrov (Centre of Biotechnology, Dresden University of Technology, Germany) for help and fruitful discussions.

References

1. Michel-Beyerle, M.-E. (ed.) (1996) The reaction center of photosynthetic bacteria. *Structure and Dynamics*, Springer-Verlag, Berlin-Heidelberg.
2. Cogdell, R. J., Gall, A., and Koehler, J. (2006) The architecture and function of the light-harvesting apparatus of purple bacteria: from single molecules to in vivo membranes, *Quarterly Reviews of Biophysics*, **39** (3), 227–324.

3. Cheng, Y. C. and Fleming, G. R. (2009) Dynamics of light harvesting in photosynthesis *Annual Rev. Phys. Chem.* **60** (1), 241–262.
4. Gall, A., Sogalia, E., Gulbinas, V., Ilioaia, O., Robert, B., and Valkunas, L. (2010) Spectral dependence of energy transfer in wild-type peripheral light-harvesting complexes of photosynthetic bacteria, *Biochim et Biophys. Acta*, **1797** (8), 1465–1469.
5. Valkunas, L., Trinkunas, G., Chmeliov, J., and Ruban, A.V. (2009) Modeling of exciton quenching in photosystem II, *Phys. Chem. Chem. Phys.*, **11**, 7576–7584.
6. Ivashin, N., and Larsson, S. (2008) Trapped water molecule in the charge separation of a bacterial reaction center, *J. Phys. Chem. B.*, **112** (38), 12124–12133.
7. Freiberg, A., and Trinkunas, G. (2009) Photosynthesis in silico: Understanding complexity from molecules to ecosystems, in *Unraveling the Hidden Nature of Antenna Excitations* (ed. Laisk, A., Nedbal, L., and Govindjee), Springer Science+Media B.V., Chapter 4, pp. 55–82.
8. Yang, S. I., Lammi, R. K., Seth, J., Riggs, J. A., Arai, T., Kim, D., Bocian, D. F., Holten, D., and Lindsey, J. S. (1998) Excited state energy transfer and ground-state hole/electron hopping in p-phenylene-linked porphyrin dimers, *J. Phys. Chem. B.*, **102** (47), 9426–9436.
9. Jordanides, X.J., Scholes, G.D., and Fleming, G.R. (2001) The mechanism of energy transfer in the bacterial photosynthetic reaction center, *J. Phys. Chem. B.*, **105** (8), 1652–1669.
10. Bixon, M., Jortner, J., and Michel-Beyerle, M. E. (1995) A kinetic analysis of the primary charge separation in bacterial photosynthesis. Energy gaps and static heterogeneity *Chem. Phys.*, **197** (1), 389–404 and references therein.
11. Schmidt, S., Arlet, T., Hamm, P., Huber, H., Nagele, T., Wachteveitl, J., Meyer, M., Scheer, H., and Zinth, W. (1994) Energetics of the primary electron transfer reaction revealed by ultrafast spectroscopy on modified bacterial reaction centers, *Chem. Phys. Lett.*, **223** (1–2), 116–120.
12. Davis, W. B., Ratner, M. A., and Wasielewski M. R. (2001) Conformational gating of long distance electron transfer through wire-like bridges in donorbridgeacceptor molecules, *J. Am. Chem. Soc.*, **123** (32), 7877–7886.
13. Kilin, D., Kleinekathofer, U., and Schreiber, M. (2000) Electron transfer in porphyrin complexes in different solvents, *J. Phys. Chem. A.*, **104** (23), 5413–5421.

14. Petty, M. C., Bryce, M. R, and Bloor, D., (eds) (1995) *An Introduction to Molecular Electronics London,* Edward Arnold, a division of Holder Headline PLC, Melbourne, Auckland, p. 320.
15. Mansoori, G. A. (2005) *Principles of Nanotechnology. Molecular-Based Study of Condensed Matter in Small Systems,* University of Illinois at Chicago, Chicago.
16. Nicolini, C. (2010) *Nanotechnology and Nanobiosciences,* "Nanoscale Materials", Pan Stanford Series on Nanobiotechnology, Chapter 1, vol. 1.
17. Mann, S. (2008) Life as a nanoscale phenomenon, *Angew. Chem. Int. Ed.,* **47** (29) 5306–5320.
18. Mozer, A. J., Griffith, M. J., Tsekouras, G., Wagner, P., Wallace, G. G., Mori, S., Sunahara, K., Miyashita, M., Earles, J. C., Gordon, K. C., Du, L., Katoh, R., Furube, A., and Officer, D. L. (2009) Zn-Zn porphyrin dimer-sensitized solar cells: toward 3-D light harvesting, *J. Am. Chem. Soc.,* **131** (43) 15621–15623.
19. Kadish, K., Smith, K. M., and Guilard, R. (eds) (2010) *Handbook of Porphyrin Science (With Applications to Chemistry, Physics, Material Science, Engineering, Biology and Medicine*), Volumes 1 "Supramolecular Chemistry", 4 "Phototherapy, Radioimunotherapy and Imaging", 10 "Catalysis and Bio-Inspired Systems", World Scientific Publishing UK Ltd., Abingdon, UK.
20. Lehn, J.-M. (1990) Perspectives in supramolecular chemistry - from molecular recognition towards molecular information processing and self-organization. *Angew. Chem., Int. Ed. Engl.,* **29** (11), 1304–1319.
21. Balzani, V., and Scandola F. (1991) *Supramolecular Photochemistry,* Ellis Horwoord, New York, London, Toronto, Sydney, Tokyo, Singapore.
22. Wasielewski, M. R. (1992) Photoinduced electron transfer in supramolecular systems for artificial photosynthesis, *Chem. Rev.,* **92** (3), 435–461.
23. Imamura, T., and Fukushima, K. (2000) Self-assembly of metallopyridylporphyrin oligomers. *Coord. Chem. Rev.,* **198** (1), 133–156.
24. Wojaczynski, J., and Latos-Grazynski, L. (2000) Poly- and oligometalloporphyrins associated through coordination *Coord. Chem. Rev.,* **204**, 113–171.
25. Burrell, A. K., Officer, D. P., Plieger, G., and Reid, D. C. W. (2001) Synthetic routes to multiporphyrin arrays, Chem. Rev., **101** (9), 2751–2796.
26. Chambron, J.-C., Heitz, V., and Sauvage, J.-P. (2000) Noncovalent multiporphyrin assemblies, in *The Porphyrin Handbook* (ed. Kadish, K.M.,

Smith, K.M., and Guilard, R.). Vol. 6. Academic Press, New York, Chapter 40, pp. 1–41.

27. Ogoshi, H., Mizutani, T., Hayashi, T., and Kuroda, Y. (2000) Porphyrins and metalloporphyins as receptor models in molecular recognition, in *The Porphyrin Handbook* (ed. Kadish, K. M., Smith, K. M., and Guilard, R.), vol. 6, Academic Press, New York, Chapter 46, pp. 279–340.
28. Gust, D., and Moore, T. A. (2000) Intermolecular photoinduced electrotransfer reactions of porphyrins, in *The Porphyrin Handbook* (ed. Kadish, K. M., Smith, K. M., and Guilard, R.), vol. 6, Academic Press, New York, Chapter 57, pp. 153–190.
29. Fukuzumi, S. (2000) Electron transfer chemistry of porphyrins and metalloporphyrins, in *The Porphyrin Handbook* (ed. Kadish, K. M., Smith, K. M., and Guilard, R.), vol. 6, Academic Press, New York, p. 115.
30. Zenkevich, E. I., von Borczyskowski, C. (2002) Multiporphyrin self-assembled arrays in solutions and films: Thermodynamics, spectroscopy and photochemistry, in *Handbook of Polyelectrolytes and Their Applications* (ed. Tripathy, S. K., Kumar, J., and Nalwa, H. S.), American Scientific Publishers, USA, vol. 2, Chapter 11, pp. 301–348.
31. Liu, Z., Yasseri, A. A., Lindsey, J. S., and Bocian, D. F. (2003) Molecular memories that survive silicon device processing and real-world operation, *Science*, **302** (5650), 1543–1544.
32. Peng, X., Aratani, N., Takagi, A., Matsumoto, T., Kawai, T., Hwang, I.-W., Ahn, T. K., and Osuka A. (2004) A dodecameric porphyrin wheel, *J. Am. Chem. Soc.*, **126** (14) 4468–4469.
33. Harvey, P. D., Stern, C., Gros, C. P., and Guilard, R. (2010) Through-space energy transfers in light-harvesting systems and cofacial bisporphyrin diads, *J. Porphyrins and Phthalocyanines*, **14** (1), 55–63.
34. Albinsson, B., and Mårtensson J. (2010) Excitation energy transfer in donor–bridge–acceptor systems, *Phys. Chem. Chem. Phys.*, **12** (27), 7338–7351.
35. Kim, Y. H., Jeong, D. H., Kim, D., Jeoung, S. C., Cho, H. S., Kim, S. K., Aratani, N., and Osuka, A. (2001) Photophysical properties of long rodlike *mesomeso*-linked zinc(II) porphyrins investigated by time-resolved laser spectroscopic methods, *J. Am. Chem. Soc.*, **123** (1), 76–86.
36. Martelli, C., Canning, J., Khoury, T., Skivesen, N., Kristensen, M., Huyang, G., Jensen, P., Neto, C., Jing Sum, T., Hovgaard, B. M., Gibson, B. C. and Crossley, M. J. (2010) Self-assembled porphyrin microrods and observation of structure-induced iridescence, *J. Mater. Chem.* **20** (12), 2310–2316.

37. Foubert, P., Vanoppen, P., Martin, M., Gemsch, T., Hofkens, J., Hesler, A., Seeger, A. R., Taylor, M., Rowan, A. E., Nolte, R. J. M., and De Schryver, F. C. (2000) Mechanical and optical manipulation of porphyrin rings at the submicrometer scale, *Nanotechnology*, **11** (1), 16–23.

38. Balaban, T. S., Bhise, A. D., Bringmann, G., Bürck, J., Chappaz-Gillot, C., Eichhöfer, A., Fenske, D., Götz, D. C. G., Knauer, M., Mizoguchi, T., Mössinger, D., Rösner, H., Roussel, C., Schraut, M., Tamiaki, H., and Vanthuyne, N. (2009) Mimics of the self-assembling chlorosomal bacteriochlorophylls: regio- and stereoselective synthesis and stereoanalysis of acyl(1-hydroxyalkyl)porphyrins, *J. Am. Chem Soc.*, **131** (40), 14480–14492.

39. Kaunisto, K., Chukharev, V., Tkachenko, N. V., Efimov, A., and Lemmetyinen, H. (2009) Energy and electron transfer in multilayer films containing porphyrin-fullerene dyad, *J. Phys. Chem. C*, **113** (9), 3819–3825.

40. Holten, D., Bocian, D. F., and Lindsey, J. S. (2002) Probing electronic communication in covalently linked multiporphyrin arrays. A Guide to the rational design of molecular photonic devices, *Acc. Chem. Res.*, **35** (1) 57–69.

41. Berera, R., van Stokkum, I. H. M., Kodis, G., Keirstead, A., Pillai, S., Herrero, C., Palacios, R. Vengris, E. M., van Grondelle, R., Gust, D., Moore, T. A., Moore, A. L., and Kennis, J. T. M. (2007) Energy transfer, excited-state deactivation and exciplex formation in artificial carotenophthalocyanine light harvesting antennas, *J. Phys. Chem. B.,* **111** (24), 6868–6877.

42. Xu, H., Ermilov, E. A., Röder, B., Ng, D. K. P. (2010) Formation and energy transfer property of a subphthalocyanine–porphyrin complex held by host–guest interactions. *Phys. Chem. Chem. Phys.* **12** (27), pp. 7366–7370.

43. Maligaspe E., Kumpulainen, T., Lemmetyinen, H., Tkachenko, N. V., Subbaiyan, N. K., Zandler, M. E., and D'Souza, F. (2010) Ultrafast singlet-singlet energy transfer in self-assembled via metal-ligand axial coordination of free-base porphyrin-zinc phthalocyanine and free-base porphyrin-zinc naphthalocyanine dyads, *J. Phys. Chem. A*, **114** (1), 268–277.

44. Lee, J.-E., Yang, J., Gunderson, V. L., Wasielewski, M. R., and Kim, D. (2010) Fluorescence dynamics of chlorophyll trefoils in the solid state studied by single-molecule fluorescence spectroscopy, *J. Phys. Chem. Lett.*, **1** (1), 284–289.

45. Grundman, M., Christen, J., Ledentsov, N. N., Boehrer, J., Bimberg, D., Ruvimov, S. S., Werner, P., Richter, U., Goesele, U., Heydenreich, J., Usti-

nov, V. M., Egorov, A. Yu., Zhukov, A. E., Kop'ev, P. S., and Alferov, Z. I. (1995) Ultranarrow luminescence lines from single quantum dots, *Phys. Rev. Lett.*, **74** (20), 4043–4046.

46. Gaponenko, S. V. (1998) *Optical Properties of Semiconductor Nanocrystals*, University Press, Cambridge.
47. Klimov, V. (2000) Optical Properties of Nanocrystals, in *Handbook of Nanostructured Materials and Nanotechnology* (ed. Nalwa, H. S.), Academic Press., USA, vol. **4**, 451–527.
48. Rogach, A. L. (ed.) (2008) *Semiconductor Nanocrystal Quantum Dots (Synthesis, Assembly, Spectroscopy and Applications)*, Springer-Verlag, Wien, p. 369.
49. Talapin, D. V., Haubold, S., Rogach, A. L., Kornowski, A., Haase, M., and Weller, H. (2001) A novel organometallic synthesis of highly luminescent CdTe nanocrystals, *J. Phys. Chem. B*, **105** (12), 2260–2263.
50. Schmid, G. (ed.) (2004) *Nanoparticles: From Theory to Applications*, Wiley-VCH, Weinheim, Germany.
51. Wang, M., Moon, S-J., Xu, M., Chittibabu, K., Wang, P., Cevey-Ha, N-L., Humphry-Baker, R., Zakeeruddin, SM. and Graetzel, M. (2010) Efficient and stable solid-state dye-sensitized solar cells based on a high-molar-extinction-coefficient sensitizer, *Small*, **6** (2), 319–324.
52. Clapp, R. Medintz, I. L. and Mattoussi H. (2006) Foerster resonance energy transfer investigations using quantum-dot fluorophores, *ChemPhysChem* **7** (1), 47–57
53. Zenkevich, E., Shulga, A., Cichos, F., Petrov, E., Blaudeck, T., and von Borczyskowski, C. (2005) Nanoassemblies designed from semiconductor quantum dots and molecular arrays, *J. Phys. Chem. B.*, **109** (18), 8679–8692.
54. Dayal, S., Krolicki, R., Lou, Y., Qiu, X., Berlin, J., Kenney, M., and Burda C. (2006) Femtosecond time-resolved energy transfer from CdSe nanoparticles to phthalocyanines. *Applied Physics B*, **84** (1–2), 309–315.
55. Van Embden, J. L., Sader, J.E., Davidson, M., and Mulvaney P. (2009) Evolution of colloidal nanocrystals: theory and modelling of their nucleation and growth, *J. Phys. Chem.*, **113** (37), 16342–16355
56. De, M., Ghosh, P.S., and Rotello, V. M. (2008) Applications of nanoparticles in biology, *Adv. Mater.*, 20 (22), 4225–4241.
57. Bakalova, R., Ohba, H., Zhelev, Z., Nagase, T., Jose, R., Ishikawa, M., and Baba, Y. (2004) Quantum dot anti-CD conjugates: are they potential

photosensitizers or potentiators of classical photosensitizing agents in photodynamic therapy of cancer?, *Nano Lett.*, 4 (9), 1567–1573.

58. Samia, A. C. S., Dayal, S., and Burda, C. (2006) Quantum dot-based energy transfer: perspectives and potential for applications in photodynamic therapy. *Photochemistry and Photobiology*, **82** (3), 617–625.
59. Shi, L., Hernandez, B., and Selke, M. (2006) Singlet oxygen generation from water-soluble quantum dot-organic dye nanocomposites, *J. Am. Chem. Soc.*, **128** (19), pp. 6278–6279.
60. Kilin D. S., Tsemekhman K., Prezhdo O. V., Zenkevich E. I., and von Borczyskowski C. (2007) Ab initio study of exciton transfer dynamics from a core-shell semiconductor quantum-dot to a porphyrin-sensitizer, *J. Photochem. Phobiol A: Chemistry*, **190** (2–3), 342–354.
61. Lihitkara, N. B., Singha, S. B., Singhb, J., Srivastavab, O. N., Naikc, R. H., and Kulkarnid, S. K. (2009) Optical investigations of interaction between zinc tetra phenyl porphyrin and CdSe nanoparticles, *Chem. Phys. Lett.*, **483** (4–6), 227–232.
62. Coe-Sullivan, S., Woo, W.-K., Steckel, J. S., Bawendi, M., and Bulovic, V. (2003) Tuning the performance of hybrid organic/inorganic quantum dot light-emitting devices, Organic *Electronics*, **4** (2–3), 123–130.
63. Katoh, R., Furube, A., Yoshihara, T., Hara, K., Fujikash,i G., Takano, A., Murata, S., Arakawa, H., Tachiya, M. (2004) Efficiencies of electron injection from excited N3 dye into nanocrystalline semiconductor (ZrO2, TiO2, ZnO, Nb2O5, SnO2, In2O3) films, *J. Phys. Chem. B*, , **108** (10), 4818–4822.
64. Schwarz, F. P., Gouterman, M., Muljiami, Z., and Dolphin, D. (1972) Energy transfer between covalently linked metal porphyrins, *Bioinorg. Chem.*, **2** (2), 1–32.
65. Rempel, U., von Maltzan, B., and von Borczyskowski, C. (1993) Temperature and solvent dependent charge transfer in self-organized porphyrin-quinone compounds, *Pure and Appl. Chem.*, **65** (8), 1681–1685.
66. Chernook, A. V., Shulga, A. M., Zenkevich, E. I., Rempel, U., and von Borczyskowski, C. (1996) Complexation and interchromophoric interactions in self-organized porphyrin and chlorin triads, *J. Phys. Chem.*, **100** (5), 1918–1926.
67. Chernook, A. V., Rempel, U., von Borczyskowski, C., Zenkevich, E. I., and Shulga, A. M. (1996) Formation and optical properties of self-organized pentameric porphyrin arrays, *Chem. Phys. Letters*, **254** (3–4), 229–241.

68. Rempel, U., Meyer, S., von Maltzan, B., and von Borczyskowski, C. (1998) Energy transfer and distance independent charge separation in self-organised porphyrin-quinone aggregates, *J. of Luminesc.*, **78** (2), 97–110.
69. Zenkevich, E. I., Bachilo, S. M., Shulga, A. M., Rempel, U., Willert, A., and von Borczyskowski, C. (1998) Electron transfer in porphyrin multimolecular self-organized nanostructures, *Molecular Crystals and Liquid Crystals*, **324**, 169–176.
70. Bachilo, S. M., Willert, A., Rempel, U., Shulga, A. M., Zenkevich, E. I., and von Borczyskowski, Ch. (1999) Efficient low temperature charge transfer in self-assembled porphyrin aggregate, *J. Photochem. Photobiol. A: Chemistry*, **126** (1–3), pp. 99–109.
71. Zenkevich, E. I., Willert, A., Bachilo, S. M., Rempel, U., Kilin, D. S., Shulga, A. M., and Ch. von Borczyskowski. (2001) Competition between electron transfer and energy migration in self-assembled porphyrin triads, *Materials Science and Engineering C.*, **18** (1–2), 99–111.
72. Zenkevich, E. I., von Borczyskowski, C., Shulga, A. M., Bachilo, S. M., Rempel, U., and Willert, A. (2002) Self-assembled nanoscale photomimetic models: structure and related dynamics, *Chemical Physics*, **275** (1–3), 185–209.
73. Sagun E. I., Zenkevich, E. I., Knyukshto, V. N., Shulga, A. M., Starukhin, D. A., and von Borczyskowski, C. (2002) Interaction of multiporphyrin systems with molecular oxygen in liquid solutions: extra-ligation and screening effects, *Chemical Physics*, **275** (1–3), 211–237.
74. Zenkevich, E. I., von Borczyskowski, C., Shulga, A. M. (2003) Structure and excited states properties of mutiporphyrin arrays formed by supramolecular design, *J. of Porphyrins and Phthalocyanines*, **7** (11), 731–754.
75. Sagun, E. I., Zenkevich, E. I., Knyukshto, V. N., Shulga, A. M., Ivashin, N. V. (2009) Relaxation processes in self-assembled triads based on porphyrin Zn-heterodimer, *Optics and Spectroscopy*, **107** (6), 1011–1025.
76. Sagun, E. I., Zenkevich, E. I., Knyukshto, V. N., Shulga, A. M., and Ivashin, N. V. (2010) Self-assembly effects and mechanisms of relaxation processes in porphyrin pentads, *Optics and Spectroscopy*, **108** (4), 590–607.
77. Hunter, C. A., and Hyde, R. K. (1996) Photoinduced energy and electron transfer in supramolecular porphyrin assemblies, *Angew. Chem., Int. Ed. Engl.*, **35**, pp.1936–1954.

78. Tamiaka, H., Miyatake, T., Tanikaga, R., Holtzwart, A. R., and Schaffner, K. (1996) Self-assembly of an artificial light-harvesting antenna: energy transfer from a zinc chlorin to a bacteriochlorin in a supramolecular aggregate *Angew. Chem. Int. Ed. Engl.*, **35**, 772–779.

79. Webb, S. I., and Sanders, J. K. M. (2000) Synthesis and recognition properties of a ruthenium(II)bis(zinc) cyclic porphyrin trimer, *Inorg. Chem.*, **39** (26), 5912–5919.

80. Winters, M. U., Dahlstedt, E., Blades, H. E., Wilson, C. J., Frampton, M. J., Anderson, H. L., and Albinsson, B. (2007) Probing the efficiency of electron transfer through porphyrin-based molecular wires, *J. Am. Chem. Soc.*, **129** (14), 4291–4297.

81. Gust, D., Moore, T. A., Moore, A. L., Gao, F., Luttrull, D., DeGraziano, J. M., Ma, X. C., Makings, L. R., Lee, S.-J., Trier, T. T., Bittersmann, E., Seely, G. R., Woodward, S., Bensasson, R. V., Rougue, M., De Schryver, F. C., and Van der Auweraer, M. (1991) Long-lived photoinitiated charge separation in carotene-diporphyrin triad molecules, *J. Am. Chem. Soc.*, **113** (10), 3638–3649.

82. Zenkevich, E., Blaudeck, T., Abdel-Mottaleb, M., Cichos, F., Shulga, A., and von Borczyskowski, C. (2006) Photophysical properties of porphyrin-semiconductor nanoassemblies, *Internat. Journal of Photoenergy* (ISSN: 1110-662X), **2006**, Article 1D90242, 1–6.

83. Zenkevich, E. I., Blaudeck, T., Shulga, A. M., Cichos, F., and von Borczyskowski, C. (2007) Identification and assignment of porphyrin-CdSe heteronanoassemblies, *J. of Luminescence*, **122–123**, 784–788.

84. Zenkevich, E. I., Sagun, E. I., Yarovoi, A. A., Shulga, A. M., Knyukshto, V. N., Stupak, A. P., and von Borczyskowski, C. (2007) Photoinduced relaxation properties in complexes based on semiconductor nanocrystals CdSe and organic molecules, *Optics and Spectroscopy*, **103** (6), 998–1009.

85. Kowerko, D., Schuster, J., Amecke, N., Abdel-Mottaleb, M., Dobrawa, R., Würthner, F., and von Borczyskowski, C. (2010) FRET and ligand related NON-FRET processes in single quantum dot-perylene bisimide assemblies, *Phys. Chem. Chem. Phys.*, **12**, 4112–4123.

86. Kowerko, D., Krause, S., Amecke, N., Abdel-Mottaleb, M., Schuster, J., and von Borczyskowski, C. (2009) Identification of different donor-acceptor structures via FRET in quantum-dot-perylene bisimide assemblies, *International Journal of Molecular Science*, **10** (12), 5239–5256.

87. Zenkevich, E. I., Shulga, A. M., Sagun E. I., von Borczyskowski, C., Rempel, U., and Chernook, A. V. (1997) Multimolecular systems on the base of tetrapyrrolic compounds: structure, photonics, applications, in *Successes in Porphyrin Chemistry* (ed. Golubchikov, O.A.) St-Petersburg Pubs, St-Petersburg, vol. 2, Chapter 12, 270–315.

88. Zenkevich, E. I., Shulga, A. M., Chernook, A. V., Gurinovich, G. P., and Sagun, E. I. (1991) Covalently linked porphyrin dimers as model systems of the photosynthetic special pair: spectroscopy, energetics and photochemistry, in *Light in Biology and Medicine*, Plenum Press, New York and London (ed. Douglas, R. H., Moan, J., and Ronto, G.), **2**, 337–344.

89. Zenkevich, E. I., Shulga, A. M., Chernook, A. V., Sagun, E. I., and Gurinovich, G. P. (1995) Dipole-dipole and exchange energy transfer in different types of porphyrin chemical dimers, *Proceedings of the Indian Acad. Sci., Chemical Sciences*, **107** (6), 795–802.

90. Zenkevich, E. I., Knyukshto, V. N., Shulga, A. M., Kuzmitzkii, V. A., Gael, V. I., Levinson, E. G., and Mironov, A. F. (1997) Spectroscopy and photophysics of covalent ether-bonded porphyrin-chlorin heterodimers, *J. of Luminescence*, **75** (3), 229–244.

91. Gurinovich, G. P., Zenkevich, E. I., Sagun, E. I., and Shulga, A. M. (1984) Spectral-luminescent properties and energetics of covalently-linked dimers of some porphyrins, *Opt. Spectrosc.*, **56** (6), 1037–1043.

92. Gurinovich, G. P., Zenkevich, E. I., Shulga, A. M., Sagun, E. I., Mauring, K., and Suisalu, A. (1984) Deactivation of electronic excitation in porphyrin chemical dimers, *J. Appl. Spectroscopy*, **41** (3), 446–450.

93. Zenkevich, E. I., Shulga, A. M., Gurinovich, G. P., Sagun, E. I. (1985) Polarized luminescence and optical activity of covalently-linked dimers (ethane-bisporphyrins), *J. Appl. Spectroscopy*, **42** (2), 207–213.

94. Zenkevich, E. I., Shulga, A. M., Chernook, A. V., Sagun, E. I., and Gurinovich, G. P. (1989) Intermolecular interactions in complexes of ethane-bisporphyrins with transition metal ions, *Khimicheskaya Fizika*, **8** (6), 842–853.

95. Mauring, K., Suisalu, A., Kikas, J., Zenkevich, E. I., Chernook, A. V., Shulga, A. M., and Gurinovich G. P. (1995) Energy transfer in ethane-bisporphyrins studied by fluorescence line narrowing and spectral hole burning, *J. of Luminesc.*, **64** (1–6), 141–148.

96. Zenkevich, E. I., Shulga, A. M., Chernook, A. V., and Gurinovich, G. P. (1984) Spectral peculiarities of NH-tautomerism in isocycle-containing porphyrins and their covalently linked dimers, *Chem. Phys. Lett.*, **109** (3), 306–311.

97. Kim, D., Holten, D., and Gouterman, M. (1984) Evidence from picosecond transient absorption and kinetic studies of charge-transfer states in copper(II) porphyrins, *J. Am. Chem. Soc*, **106** (6), 2793–2798.

98. Asano-Someda, M., and Kaizu, Y. (1999) Highly efficient triplettriplet intramolecular energy transfer and enhanced intersystem crossing in rigidly linked copper(II) porphyrinfree base porphyrin hybrid dimers, *Inorg. Chem.*, **38** (10), 2303–2311.

99. Zenkevich, E. I., Losev, A. P., Kochubeev, G. A., and Gurinovich, G. P. (1978) Regularities of intermolecular interactions and energy transfer in mixed associates of chlorophyll and its analogs, *J. Molec. Struct.*, **45**, 423–436.

100. Zenkevich, E. I., Starukhin, A. S., and Shulga, A. M. (1999) Energy transfer studies in porphyrin and chlorin chemical dimers by methods of site-selection spectroscopy at 1.8–10 K, *J. Appl. Spectroscopy*, **64** (4), 500–504.

101. Zenkevich, E. I., Shulga, A. M., Sagun, E. I., Gurinovich, G. P., and Chernook, A. V. (1985) Energetics and photophysical processes in covalently linked porphyrin dimers, in *Teubner-Texte zur Physik*, (ed. Feller, K.H.), BSB B. G.Teubner Verlagsgesselschaft, Leipzig, Band 4, pp. 297–300.

102. Zenkevich, E. I., Shulga, A. M., Sagun, E. I., Chernook, A. V., and Gurinovich, G. P. (1985) Photophysics of cyclopentaneporphyrin covalently-linked dimers of cyclopentaneporphyrins, *J. Appl. Spectroscopy*, **43** (3), 455–461.

103. Zenkevich, E. I., Shulga, A. M., Chernook, A. V., and Gurinovich, G. P. (1986) Electronic excitation energy migration in chemical dimers of cyclopentaneporphyrins, *J. Appl. Spectroscopy*, **45** (6), 984–991.

104. Zenkevich, E. I., Chernook, A. V., Shulga, A. M., Sagun, E. I., and Gurinovich, G. P. (1989) Electronic structure in covalently-linked dimers of Zn-cyclopentaneporphyrins and energy transfer, *Khimicheskaya Fizika*, **8** (7), 891–901.

105. Chernook, A. V., Zenkevich, E. I., and Shulga, A. M. (1991) Experimental grounding of oscillator model for cyclopentaneporphyrins and their chemical dimers, *J. Appl. Spectroscopy*, **55** (3), 445–451.

106. Starukhin, A., Zenkevich, E., Shulga, A., and Chernook, A. (1996) Fluorescence line narrowing study of cyclopentaneporphyrin chemical dimers at 4.2 K, *J. of Luminescence*, **68** (6), 313–323.

107. Zenkevich, E. I., Shulga, A. M., Filatov, I. V., Chernook, A. V., and Gurinovich, G. P. (1985) NH-tautomerism and visible absorption spectra

of porphyrins with nonsymmetrical substitution. Oscillator model and MO calculations, *Chem. Phys. Lett.*, **120** (1–2), 63–68.

108. Gurinovich, G. P., Zenkevich, E. I., and Shulga (1986) The role of NH-tautomerism in the formation of excited states. Porphyrins with non-symmetrical substitution, in *Proceedings of the International Conference on Excited States and Dynamics of Porphyrins* (ed. Gouterman, M. Rentzepis, P. M., and Straub, K. D.), *Am. Chem. Soc. Series*, Washington, Book 321, pp. 74–93.

109. Zenkevich, E. I., Chernook, A. V., Shulga, A. M., Pershukevich, P. P., Gurinovich, G. P., and Sagun, E. I. (1991) Role of dipole-dipole and exchange energy processes in electronic excitation deactivation for chemical dimers of cyclopentaneporphyrins, *Khimicheskaya Fizika*, **10** (9), 1183–1191.

110. Gust, D., Moore, T. A., Moore A. L., Devadoss, C., Liddell, P. A., Hermant, R., Nieman, R. A., Demanche, L. G., DeGraziano, J. M., and Gouni, I. (1992) Triplet and singlet energy transfer in carotene-porphyrin dyads: role of the linkage bonds, *J. Am. Chem. Soc.*, **114** (10), 3590–3603.

111. Kuciauskas, D., Liddell, P. A., Hung, S.-C., Lin, S., Stone, S., Seely, G. R., Moore, A. L., Moore, T.A., and Gust, D. (1997) Structural effects on photoinduced electron transfer in carotenoidporphyrinquinone triads, *J. Phys. Chem. B.*, **101** (3), 429–440.

112. Andreasson, J., Kajanus, J., Martensson, J., and Albinsson, B. (2000) Triplet energy transfer in porphyrin dimers: comparison between π- and σ-chromophore bridged systems, *J. Am. Chem. Soc.*, **122** (40), 9844–9845.

113. Wilson, T. M., Hori, T., Yoon, M-C., Aratani, N., Osuka, A., Kim, D., and Wasielewski, M R. (2010) Rapid intramolecular hole hopping in *mesomeso* and *meta*-phenylene linked linear and cyclic multiporphyrin arrays, *J. Am. Chem. Soc.*, **132** (4), 1383–1388

114. Gentemann, S., Medforth, C. J., Ema, T., Nelson, N. Y., Smith, K. M., Fajer, J., and Holten, D. (1995) Unusual picosecond $^1(\pi, \pi^*)$ deactivation of ruffled nonplanar porphyrins *Chem. Phys. Lett.*, **245** (4–5), 441–447.

115. Barkigia, K. M., Nurco, D. J., Renner, M. W., Melamed, D., Smith, K. M., and Fajer, J. (1998) Multiconformational surfaces in porphyrins: previews into excited-state landscapes, *J. Phys. Chem B.*, **102** (1), 322–326.

116. Chirvony, V. S., van Hoek, A., Galievsky, V. A., Sazanovich, I. V., Schaafsma, T. J., and Holten, D. (2000). Comparative study of the photophysical properties of nonplanar tetraphenylporphyrin and octaethylporphyrin diacids, *J. Phys. Chem. B*, **104** (42), 9909–9917.

117. Retsek, J. L., Gentemann, S., Medforth, C. J., Smith, K. M., Chirvony, V. S., Fajer, J., and Holten, D. (2000) Photoinduced evolution on the conformational landscape of nonplanar dodecaphenylporphyrin: picosecond relaxation dynamics in the $^1(\pi\pi^*)$ excited state, *J. Phys. Chem. B*, **104** (40), 6690–6693.

118. Chirvony, V. S., Avilov, I. V., Panarin, A. Yu., Malinovskii, V. L., and Galievsky, V. A. (2007) The triplet state decay kinetics and deactivation funnel geometry of a series of nonplanar saddle-shaped porphyrins, *Chem. Phys. Lett.*, **434** (1), 116–120.

119. Knyukshto, V. N., Shulga, A. M., Zenkevich, E. I., Sagun, E. I., and Bachilo, S. M. (1998) Unusual dynamic relaxation of triplet-excited meso-phenyl-substituted porphyrins and their chemical dimers at room temperature, *Chem. Phys. Lett.*, **297** (1–2), 97–108.

120. Knyukshto, V. N., Zenkevich, E. I., Sagun, E. I., Shulga, A. M., and Bachilo, S. M. (1999) Pathways for photoinduced electron transfer in *meso*-nitro-phenyl-octaethylporphyrins and their chemical dimers, *Chem. Phys. Lett.*, **304** (3–4), 155–166.

121. Knyukshto, V. N., Sagun, E. I., Shulga, A. M., Bachilo, S. M., Zenkevich, E. I. (1999) Dynamic relaxation of triplet porphyrin states with mesoaryl substitution and of their chemical dimers, *Chem. Phys. Reports*, **18** (5), 855–872.

122. Zenkevich, E. I., Knyukshto, V. N., Sagun, E. I., Shulga, A. M., and Bachilo, S.M. (2000) Unusual pathways of triplet state dynamic relaxation in meso-arylsubstituted porphyrins and their dimers at 295 K, *J. of Fluorescence*, **10** (1), 55–68.

123. Avilov, I. V., Zenkevich, E. I., Sagun, E. I., and Filatov, I. V. (2004) Quantum-chemical investigation of the conformational dynamics of the mono-meso-phenyl-substituted octaalkylporphyrins in the triplet excited state, *J. Phys. Chem. A.*, **108** (26), 5684–5691.

124. Andreasson, J., Zetterqvist, H., Kajanus, J., Martensson, J., and Albinsson, B. (2000) Efficient non-radiative deactivation and conformational flexibility of *meso*-diaryloctaalkylporphyrins in the excited triplet state, *J. Phys. Chem. A.*, **104** (41), 9307–9314.

125. Kyrychenko, A., Andreasson, J., Martensson, J., and Albinsson, B. (2002) Sterically induced conformational relaxation and structure of *meso*-diaryloctaalkyl porphyrins in the excited triplet state: experimental and DFT studies, *J. Phys. Chem. B.*, **106** (48), 12613–12622.

126. Foerster, Th. (1965) Delocalized excitation and excitation transfer, in *Modern Quantum Chemistry* (ed. Sinanoglu, O.), Academic Press, New York, London, vol. **3**, pp. 93–118.

127. Marcus, R. A. (1993) Electron transfer reactions in chemistry, *Rev. Modern Phys.*, **65** (3), 99–610.
128. Hsiao, J.-S., Krueger, B. P., Wagner, R. W., Johnson, T. E., Delaney, J. K., Mauzerall, D. C., Fleming, G. R., Lindsey, J. S., Bocian, D. F., and Donohoe, R. J. (1996) Soluble synthetic multiporphyrin arrays. 2. Photodynamics of energy-transfer processes, *J. Am. Chem. Soc.*, **118** (35), 11181–11189.
129. Reineker, P. (1982) Stochastic Liouville equation approach: Coupled coherent and incoherent motion, optical line shapes, magnetic resonance phenomena, in *Springer Tracts in Modern Physics. Exciton Dynamics in Molecular Crystals and Aggregates* (ed. Kenkre, V. M., and Reineker, P.), Springer, Berlin, vol. 94, pp. 111–226.
130. Schreiber, M., Kilin, D., and Kleinekathoefer, U. (1999) Comparison of two models for bridge-assisted charge transfer, *J. of Luminescence*, **83–84** (1–2), 235–240.
131. Wasielewski, M. R., Johnson, D. J., Swec, W. A., Kersey, K. M., and Minsek, D. W. (1988) Achieving high quantum yield charge separation in porphyrin-containing donor-acceptor molecules at 10 K, *J. Am. Chem. Soc.*, **110** (21), 7219–7221.
132. Furhop, J.-H., and Mauzerall, D. (1969) The one-electron oxidation of metalloporphyrins, *J. Am. Chem. Soc.*, **91** (15), 4174–4181.
133. Johnson, D. J., Niemczyk, M. P., Minsek, D. W., Wiererrechy, G. P., Svec, W. A., Gaines III, G. L., and Wasielewski, M. R. (1993) Photochemical electron transfer in chlorophyll-porphyrin-quinone triads: the role of the porphyrin-bridging molecule, *J. Am. Chem. Soc.*, **115** (13), 5692–5701.
134. Osuka, A., Marumo, S., Mataga, N., Taniguchi, S., Okada, T., Yamazaki, I., Nishimura, Y., Ohno, T., and Nozaki, K. (1996) A stepwise electron-transfer relay mimicking the primary charge separation in bacterial photosynthetic reaction center, *J. Am. Chem. Soc.*, **118** (1), 155–168.
135. Rempel, U., von Maltzan, B., and von Borczyskowski, C. (1995) Competition between charge transfer via superexchange and thermally activated energy transfer in porphyrinheterodimerquinone systems, *Chem. Phys. Lett.*, **245** (2–3), 253–261.
136. Liu, J., and Bolton, J. R. (1992) Intramolecular photochemical electron transfer. 7. Temperature dependence of electron-transfer rates in covalently linked porphyrin-amide-quinone molecules, *J. Phys. Chem.*, **96** (4), 1718–1725.

137. Tabushi, A., Kugimiya, S., Kinnaird, M. G., and Sasaki, T. (1985) Artificial allosteric system. 2. Cooperative 1-methylimidazole binding to an artificial allosteric system, zinc-gable porphyrin-dipyridylmethane complex, *J. Am. Chem. Soc.*, **107** (14), 4192–4199.

138. Toyama, N., Asano-Someda, M., Ichino, T., and Kaizu, Y. (2000) Enhanced intersystem crossing in gable-type copper(II) porphyrinfree base porphyrin dimers: evidence of through-bond exchange interaction, *J. Phys. Chem. A.*, **104** (21), 4857–4865.

139. Mialocq, J. C., Giannotti, C., and Maillard, P. (1984) Energy transfer in "covalently linked" and "face-to-face" bisporphyrins, *Chem. Phys. Lett.*, **112** (1), 87–93.

140. Hugerat, M., Levanon, H., Ojadi, E., Biczok, L., and Linschitz, H. (1991) Multiple decay pathways and electron transfer in excited ion-paired zinc-copper porphyrins: laser photolysis and time-resolved EPR spectroscopy, *Chem. Phys. Lett.*, **181** (5), 400–406.

141. Sagun E. I. (1994) Quenching of singlet oxygen luminescence by chemical dimers of porphyrins, *Journal of Applied Spectroscopy*, **60** (1–2), 82–85

142. Schmelz, O., Mews, A., Basché, T., Herrmann, A., and Müllen, K. (2001) Supramolecular complexes from CdSe nanocrystals and organic fluorophors, *Langmuir*, **17** (9), 2861–2865.

143. Boulesbaa, A., Issac, A., Stockwell, D., Huang, Z. Huang, J. Guo, J. and Lian, T. (2007) Ultrafast charge separation at CdS quantum dot/rhodamine B molecule interface, *J. Am. Chem. Soc.*, **129** (49), 15132–15133.

144. Orlova, A. O., Maslov, V. G., Baranov, A. V., Gounko, I. and Byrne, S. (2008) Spectral-luminescence study of the formation of QD–sulfophthalocyanine molecule complexes in an aqueous solution, *Optics and Spectroscopy*, **105** (5), 726–800.

145. Issac, A., von Borczyskowski, C., and Cichos, F. (2005) Correlation between photolumi-nescence intermittency of CdSe quantum dots and self-trapped states in dielectric media, *Phys. Rev. B*, **71** (16), 161302 (1–4).

146. Blaudeck, T., Zenkevich, E., Cichos, F., and von Borczyskowski, C. (2008) Probing wave functions at semiconductor quantum dot surfaces by non-FRET photoluminescence quenching, *J. Phys. Chem. C*, **112** (51), 20251–20257.

147. Haus, J. W., Zhou, H. S., Homma, I., and Komiyama, H. (1993) Quantum confinement in semiconductor heterostructure nanometer-size particles, *Phys. Rev. B*, **47** (3), 1359–1365.

148. Dabbousi, B. O., Rodriguez-Viejo, J., Mikulec, F. V., Heine, J. R., Mattoussi, H., Ober, R, Jensen, K. F., and Bawendi, M. G. (1997) CdSe/ZnS core-shell quantum dots: synthesis and characterization of a size series of highly luminescent nanocrystallites, *J. Phys. Chem. B*, **101** (46), 9463–9475.

149. Zenkevich, E. I., and von Borczyskowski, C. (2009) Photoinduced relaxation processes in self-assembled nanostructures: multiporphyrin complexes and composites "CdSe/ZnS quantum dot-porphyrin", *Macroheterocycles*, **2** (3), 207–221.

150. Zenkevich, E. I., and von Borczyskowski, C. (2009) Assembly principles and relaxation processes in nanosized heterogeneous complexes: multiporphyrin structures and CdSe/ZnS nanocrystals, *High Energy Chemistry*, **43** (7), 570–576.

Chapter 6

Structural and Functional Mimics of Light-Harvesting Systems in Bacterial Photosynthesis

Yoshiaki Kobuke
Institute of Advanced Energy, Kyoto University,
Gokasho, Uji, Kyoto 611-0011, Japan
kobuke@iae.kyoto-u.ac.jp

6.1 Introduction

When I first looked at the historic report on the X-ray crystallographic analysis of light-harvesting complex LH2 from *Rhodopseudomonas* purple bacteria, I was greatly astonished [1]. In B850, many bacteriochlorophylls are organized systematically into a large macroring. Soon, cryomicroscopy reported the structure of LH1 as a similarly organized ring [2]. The difference between these two antennae was the number of bacteriochlorophylls, 18 and 32 for LH2 and LH1, respectively. Later, the LH1 structure was discussed from the viewpoint of the forbidden nature of the optical transition of a true circular structure and also the lack of an exit site for the reduced quinone from the LH1–reaction center

Multiporphyrin Arrays: Fundamentals and Applications
Edited by Dongho Kim

www.panstanford.com

composite, and another structure of an ellipse form [3] or another having an open mouth was also reported [4, 5]. Irrespective of these differences, the most important feature of these antenna systems comes from the ring architecture constructed from many members of bacteriochlorophylls in a supramolecular fashion by the conducting action of polypeptides. Later on, atomic force microscopy (AFM) studies elucidated the closely packed assembly of the LH1–reaction center composite and LH2 in the membrane [6–9]. Furthermore, the control of the numbers and species of participating antennae were found to be dependent on light intensities [10, 11]. The natural working system was so nicely designed and seemed to be far beyond the power of synthetic organic chemists.

There are several reasons of my astonishment. First, the construction unit of the ring is composed of a bacteriochlorophyll dimer, which is arranged in a slipped cofacial fashion. This unit, as will be explained later in detail, has been reported as the core element of the photosynthetic reaction center, called a "special pair" [12, 13]. A synthetic model of the special pair had been prepared by complementary coordination of imidazolyl-to-Zn(II), mimicking the coordination of the histidyl unit to the Mg(II) center in the bacteriochlorophyll in the reaction center [14]. Nature therefore employs the *identical* methodology for constructing the reaction center and the light-harvesting system by using the *identical* construction parts, bacteriochlorophyll and imidazolyl side chain of the histidyl unit in the transmembrane helix, and the *identical* supramolecular methodology for structure formation, coordination of His-to-Mg(II). Once I confirmed these points, the construction of an artificial antenna system mimicking the way of nature became my major target in chemistry. Such analysis sufficiently raised my feeling as "I started first. Who else can trace our way of mimicking nature?" However, it was really tough work, partly because I had only a few years' experience on porphyrin chemistry, although I had started research in chemistry around 30 years before. The other reason for the difficulty is of an intrinsic nature associated with supramolecular chemistry itself, i.e., the difficulty of structural proof in spite of relatively easy structure creation. This problem was overcome by the gradual development of newer methodologies associated with general advancement in this field. In this chapter, fortunately, the author will have the privilege of concentrating on his own methodology, because many excellent chemists in this field are well placed to extend their own fields in this book.

6.2 Chromophore Arrangements in the Photosynthetic System

The natural photosynthetic system looks like a factory for mass production. It may not be a smart idea to develop numerous small factories covering the total conversion system from dilute light energy to oxidation–reduction products. It is much more efficient to reduce the number of central total production sites and to widely distribute simpler factories devoted to the limited function of transferring the incoming dilute energy to a central production site. Nature, I feel, is always a most ingenious chemist and has designed a very sophisticated molecular system: the reaction center produces electrons and holes by the action of light energy delivered from a large number of light-harvesting devices of simple structure in the surrounding space. In natural photosynthetic bacteria, light energy is captured efficiently by chromophores distributed widely in the membrane and transferred efficiently to the reaction center. Bacteriochlorophyll-*a* or -*b* with extinction coefficients as large as 10^5–10^6 M^{-1} cm^{-1} are major components and are organized with the aid of polypeptide chains. The absorption band is tuned for undertaking an energy-transfer cascade by interactions with polypeptides and chromophores themselves. The absorption band of monomeric bacteriochlorophyll-*a* appears at 770 nm and shifts to 800 and 850 nm in LH2 and further to 870–880 nm in LH1. To design our molecular system, the construction principle of the bacterial photosynthetic system is described below in more detail.

Figure 6.1 shows the first X-ray crystallographic structure reported as membrane-bound antenna complexes in 1995 [1]. The complex is composed of two types of bacteriochlorophyll assemblies, B800 and B850. The former is a ring assembly composed of nine bacteriochlorophylls in a plane parallel to the membrane surface. The latter ring illustrated in Fig. 6.1 has a barrel-type structure provided by the assembly of nine bacteriochlorophyll dimers, where

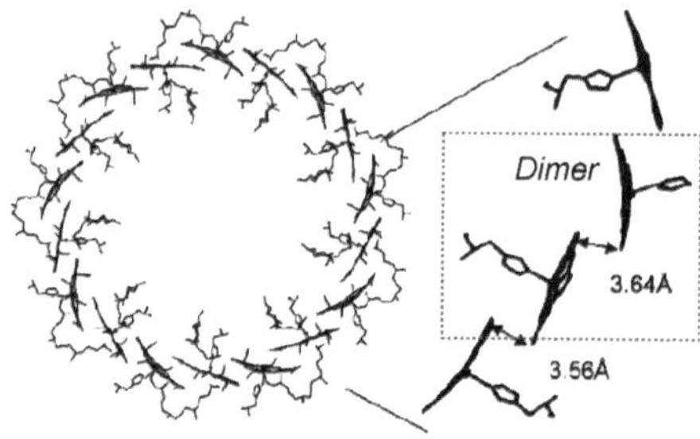

Figure 6.1. Arrangement of chlorophylls in B850 of the light-harvesting complex II (LH2) of photosynthetic purple bacteria [1].

each chromophore is coordinated by an imidazolyl side chain from transmembrane helices of α- and β-polypeptides, located inside and outside the chromophores, respectively, to the central Mg(II) ion in the bacteriochlorophyll. Each chlorophyll unit is partially overlapped in this dimer unit and each dimer unit is again partially overlapped with neighboring dimer units, placing all of the units in an almost identical environment. The chlorophyll plane is perpendicularly arranged in the lipid membrane. As a consequence of this partial overlap of the chlorophyll π-orbitals, the absorption maximum is shifted to a longer wavelength at 850 nm. This shift is appropriate for receiving excitation energy from B800, where the absorption maximum appears at 800 nm as a result of an arrangement of almost independent chromophores separated from each other in the plane.

Another important antenna assembly is LH1, which contains a barrel structure similar to B850 of LH2 [2, 5]. In this case, the barrel is composed of 15 bacteriochlorophyll dimers with an open mouth (Fig. 6.2). The barrel structure is maintained again by the same principle, i.e., coordination from the imidazolyl side-chain of the histidyl residue in the transmembrane helices located inside and outside the bacteriochlorophyll rings. Reflecting the closer interaction of each dimer unit, the absorption maximum is shifted further to 870–880 nm. This larger LH1 ring can accommodate the reaction-center system in its central cavity of around 70 Å in diameter.

The core unit in the reaction-center complex is composed of a bacteriochlorophyll dimer, a so-called special pair, where two bacteriochlorophylls are contacting at a van der Waals distance

Figure 6.2. Arrangement of chlorophylls in the light-harvesting complex I (LH1) of photosynthetic purple bacteria [5].

with a small slipping distance, i.e., in a slipped cofacial arrangement with mutual overlap of one pyrrole ring. The structural basis is almost the same as the dimer unit employed for B850 in LH1 and LH2. It is noteworthy that the geometrical arrangement of the special pair is generated again by coordination of imidazolyl ligands of the histidyl residues in the L and M subunits to the central Mg(II) ions in the bacteriochlorophylls. This structure was elucidated early in 1984 as the first successful X-ray crystal structure of a membrane protein (Fig. 6.3) [12, 13].

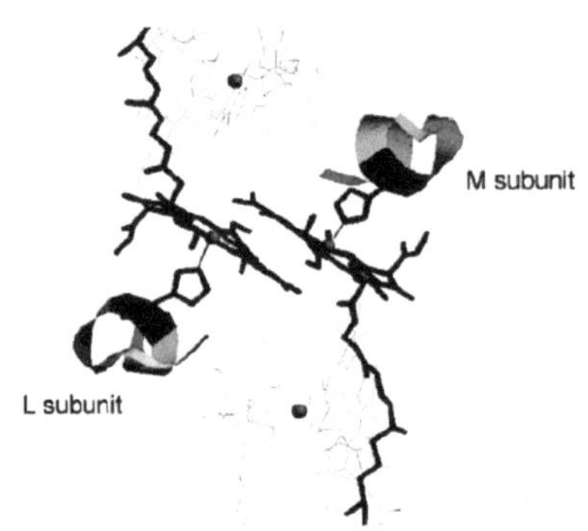

Figure 6.3. Coordination mode of the special pair in photosynthetic purple bacteria [12, 13].

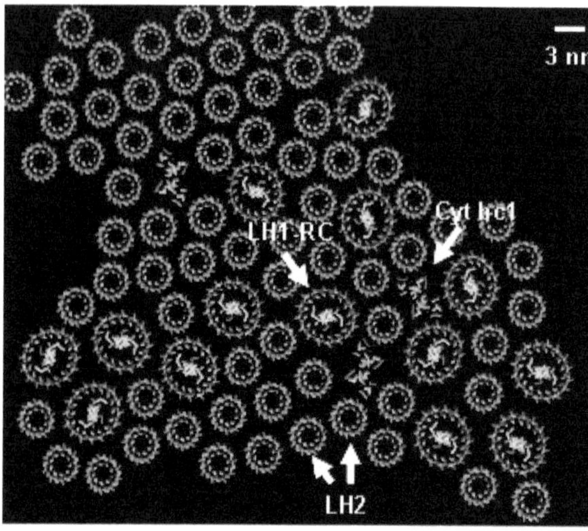

Figure 6.4. Molecular organization of the photosynthetic apparatus containing LH1, LH2, reaction center, and cytochrome-bc1 Scheuring, S. Sturgis, J N. Prima, V., Bernadac, A. Lévy D. and Rigaud, J.-L. (2004) *Proc. Natl. Acad. Sci. U. S. A.*, **101**, 11293–11297. Copyright (2004) National Academy of Sciences, U.S.A.

The whole arrangement of the bacterial photosynthetic system in the membrane elucidated by AFM is illustrated in Fig. 6.4, in which large numbers of LH2 are closely packed with limited numbers of LH1–reaction center composite along with the cytochrome-bc1 complex [15]. Light energy reaching this membrane is captured primarily by the most abundant LH2 antenna system. Excitation energy captured at the B800 site is transferred rapidly within the planar ring (ca. 0.5 ps) and then transferred to B850 (1.2 ps) according to the energy gradient and delocalized over the B850 ring [16, 17]. The exciton delocalization length in the B850 ring was much debated and contradictory views were given soon after the crystal structure was obtained, but now it is generally accepted that the exciton delocalization process is time dependent, leading to the rather different numerical values of the exciton length in the literature [18–22]. The energy-delocalization and energy-hopping rates are fast (0.1–0.2 ps) compared with its lifetime (ca. 1 ns) and the ring is regarded as the "energy storage ring". The excitation energy stored in one ring is then transferred (ca. 5 ps) to the neighboring LH2 rings, and finally to the LH1 ring. The interring LH2–LH1 energy transfer is reported to occur on a time scale of 3–5 ps. The last step in the excitation-energy-transfer chain occurs from LH1 to the reaction center in the central cavity. This is the slowest step in the whole series of energy-transfer processes and occurs on the time scale of 30–40 ps. Then, ejection of an electron from the special pair initiates the oxidation–reduction sequences of photosynthesis.

Another important antenna assembly is known as chlorosome in green photosynthetic bacteria [23]. Chlorosomes with a cylindrical rod structure are attached to the cytoplasmic side on the inner cell membrane as the peripheral light-harvesting system. The assembly is constructed primarily from interactions between bacteriochlorophylls themselves, with little participation of protein. It is interesting that pigment–pigment interactions solely create the higher and functional structure. Unfortunately, a clear and unambiguously accepted structural model does not exist and prevents synthetic approaches on a definite structural basis. Among light-harvesting antennae determined by X-ray crystallography, plant and cyanobacteria, in a close mutual structural similarity, employ totally different arrangements of chlorophylls [24–27]. However, their whole architecture seems to lack structural regularity, in contrast to those in purple bacteria, which makes it extremely difficult to extract the simple principle that provides efficient utilization of light energy [28].

6.3 Methodology for Photosynthetic Mimics

As a chromophore for constructing light-harvesting systems, metal complexes of porphyrin and phthalocyanine occupy a central position in view of their excellent UV-vis absorption, fluorescence properties, and varieties of possible synthetic modifications. Covalent linking of chromophores is outside our interest and supramolecular chromophore-assembly formation is the focus of this chapter. There are several methods for supramolecular assembly formation.

Even though molecular assembly is intended, I would like to avoid the structural ambiguity and discuss the situation from a clear molecular basis. For this purpose, very strong intermolecular interactions are desirable for the assembly formation. From this viewpoint, metal coordination provides the best method compared with other intermolecular forces such as hydrogen bonds, ionic, and π–π stacking interactions. Other intermolecular forces are also used in combination with metal coordination, but supermolecules of discrete structure in solution are hardly expected to form using their weak intermolecular forces alone [29, 30].

According to the biomimetic methodology, Mg(II) may be the choice for the central metal ion, but its use is limited by its unstable nature toward heat, acid, and oxidizing agents, as well as the synthetic difficulty upon insertion to the free base. Zn(II) as a d^{10} metal ion is an excellent candidate considering all of its properties. Differences between Mg(II) and Zn(II) are seen in the decrease in coordination number and fluorescence quantum yield. Mg(II) accepts a sixth coordination, although with a much decreased stability constant. This property may be the reason for the complexity of the debated organized structures, especially on assemblies in the chlorosome. In contrast, Zn(II) hardly accepts a sixth coordination and its structure can be clearly discussed on the basis of monodentation of the ligand because the other four sites are already occupied as the central metal ions in macrocyclic ligands such as porphyrin or phthalocyanine. As a metal ion of d^{10} electronic configuration without an odd electron, Zn(II) cannot provide a strong ligand field and shows only a limited strength for the stability constant. This property, on the other hand, has merits that are similar to what is observed in nature. The coordination complex has a dynamic nature and allows ligand-exchange reactions giving rise to the formation of the most stable equilibrium product. The absence of an odd electron in a d^{10} metal ion is desirable for reactions occurring in the singlet state without suffering fast intersystem crossings. Of course, various transition metal ion complexes of porphyrin and phthalocyanine show the reverse properties in their higher stability, higher coordination number, rigid structure, and many other properties. Their importance cannot be stressed too much in the field of photochemical and electrochemical studies, but the singlet nature in biomimetic light-harvesting events puts the chemistry outside the scope of this chapter.

As discussed above, Zn(II) as a d^{10} metal ion cannot provide large stability constants considering the pentacoordination, i.e., monodentation from the ligand outside the macrocycle [29]. However,

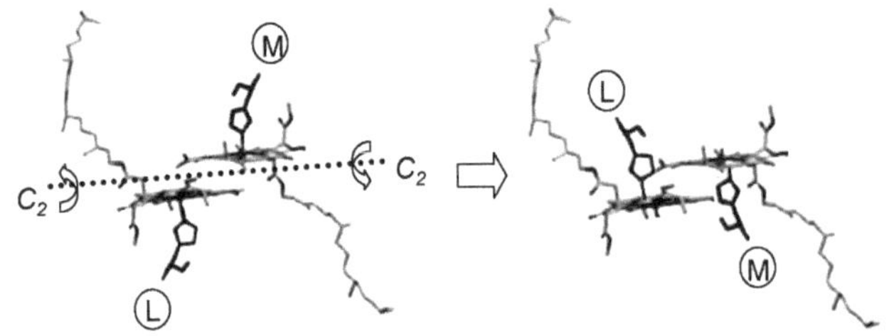

Figure 6.5. C_2 operation on the special pair to produce a complementary dimer of imidazolyl-coordinated porphyrin.

we need a stability constant larger than 10^8–10^{10} M^{-1} in solution, so that 1:1 molar addition of ligand to the macrocycle generates more than 90–99% complex, respectively, under the dilute solution of 10^{-6} M solution. This micromolar concentration is commonly used because of the large extinction coefficient (10^5–10^6 M^{-1} cm^{-1}) of the porphyrin Soret band. Such a demand is difficult to achieve unless somewhat unusual geometrical conditions are satisfied. Very fortunately, we can achieve such a case by complementary coordination. The basic idea is illustrated in Fig. 6.5. If L and M subunits were subjected to a C_2 rotation, each imidazolyl ligand may switch the position close to the chlorophyll in the counterpart subunit. The connection between imidazolyl and subunit may then be changed to a direct bond to the chromophore. According to this scheme, *N*-methylimidazolyl-substituted Znporphyrin may give rise to the formation of a dimer of the special-pair-type geometry. Here, three replacements were made from a synthetic point of view: imidazolyl to *N*-methylimidazolyl, chlorophyll (Mg(II) in it) to porphyrin with Zn(II).

Examination of the crystal structure proved the formation of the slipped cofacial complementary dimer [31, 32]. The extremely large stability constant, in the order of 10^{11} M^{-1} in $CHCl_3$ largely comes from its complementarity. When one imidazolyl ligand in the first porphyrin coordinates to Zn(II) in the second porphyrin, this imidazolyl ligand faces Zn(II) in the first porphyrin juxtaposed for coordination, thereby contributing to decreasing the unfavorable entropy on complexation. At the same time, the enthalpy term arising from π–π interaction between two large porphyrin π-networks in close contact contributes significantly to enhance the stability constant.

Moreover, as a nitrogen ligand, *N*-methylimidazole has an excellent advantageous position. Because no stabilization effect through back donation of electrons from metal to ligand is expected for ligands without d-orbitals, the stability constant is primarily determined by the electron-donating power of the ligand. In a series of aromatic amines, the stability constants increase with an increase in the pK_a values of the amines. The more basic *N*-methylimidazole, having a larger pK_a value, has order of magnitude larger stability constants compared with pyridine derivatives [29, 33–36]. Furthermore, compared with pyridyl ligands of six-membered rings, the α-CH unit in the five-membered imidazole ring with a smaller internal angle has a longer separation distance from the porphyrin π-plane, and the steric repulsion between the α-hydrogen atom and porphyrin π-plane is decreased on ligand coordination. When pyridine or thiazole of lower basicity was employed as the porphyrin substituent, much higher concentrations were required to obtain reasonable complexation percentages in solution. This proves the importance of the factors discussed above.

The photosynthetic reaction proceeds through a singlet excited state and its conversion to the triplet state is protected by several mechanisms. To transfer the excitation energy preventing various decay processes within their lifetime, chromophores must be placed at close interchromophore distances. The close contact of chromophores, however, sometimes induces energy-sink formation, where the excitation energy is lost through internal conversion to vibrational states. Because random assembly tends to contain this type of unfavorable defect, a regular arrangement of the chromophores is most desirable.

In such assembly formation, we must consider the types of strong chromophoric interactions. According to Kasha's theory, face-to-face arrangement of the chromophores induces a blue shift of the absorption band, whereas head-to-tail interaction shifts the band to longer wavelength [37]. The excited state in the face-to-face arrangement tends to lose its energy through internal conversion to vibrational energy and to decay without fluorescence emission. On the other hand, excitons in a head-to-tail orientation can maintain the excited state longer and transfer the energy within its lifetime. Therefore, regular assembly in a head-to-tail orientation is desirable for enabling singlet energy transfer.

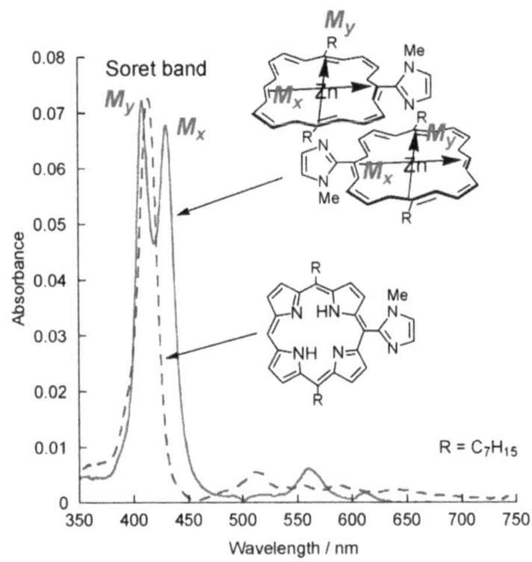

Figure 6.6. UV-vis absorption spectra of a slipped cofacial porphyrin dimer and interaction modes of the transition dipole moment. See also Color Insert.

Because porphyrin has two orthogonal in-plane transition dipoles, the head-to-tail and face-to-face transition dipoles, M_x and M_y, respectively, in the slipped cofacial arrangement generate red and blue shifts of the Soret band, respectively (Fig. 6.6). The large split Soret band (1035 cm^{-1} in the case of *N*-methylimidazolyl-appended Znoctaethylporphyrin) is a characteristic feature of strong exciton coupling of the complementary dimer.

6.4 Light-Harvesting Antenna Ring

Based on the above structural unit, two *N*-methylimidazolylZnporphyrin units were connected through a *meta*-phenylene spacer to adjust the internal angle between two porphyrin planes to 120° in the hope that the successive complementary coordination may lead to a hexagonal ring as the strain-free product [38, 39]. However, this view is too simple and optimistic because there is a variety of routes leading to products other than the desired cyclization. Actually, a broad GPC elution curve, suggesting formation of by-products, was obtained after the introduction of Zn(II) to *m*-phenylene-linked bis(*N*-methylimidazolyl-Znporphyrin **1**. A

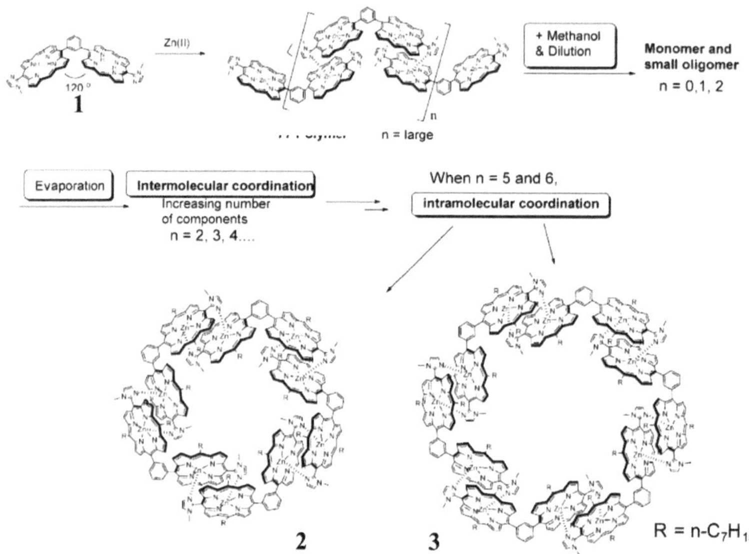

Figure 6.7. Zn(II) insertion to porphyrin leading to a linear oligomer followed by reorganization to macrocyclic 5- and 6-mer formation.

reorganization procedure utilizing the dynamic nature of imidazolyl-to-Zn(II) coordination was developed to obtain macrocyclic products selectively. The whole product mixture after metal insertion was subjected to equilibrium conditions by dissolving in $CHCl_3$/MeOH (1:1, v/v) followed by solvent evaporation. Coordination organization under high dilution conditions (3.5 μM) leads to the formation of macrocyclic products because the growing uncoordinated *N*-methylimidazolyl-Znporphyrin terminals try to find the counterpart in their own linear chain because of the lack of monomer supply in the dilute solution (Fig. 6.7). Therefore, the smallest ring formation is favored unless the angle strain significantly overrules this factor.

The chromatogram after this treatment became much sharper and showed two peaks with almost the same intensities (Fig. 6.8). These two peaks were separated easily with a preparative GPC column. The products were stable in the absence of coordinating solvents and the first and second peaks were assigned tentatively to macrocyclic hexamer **2** and pentamer **3**, respectively, from their

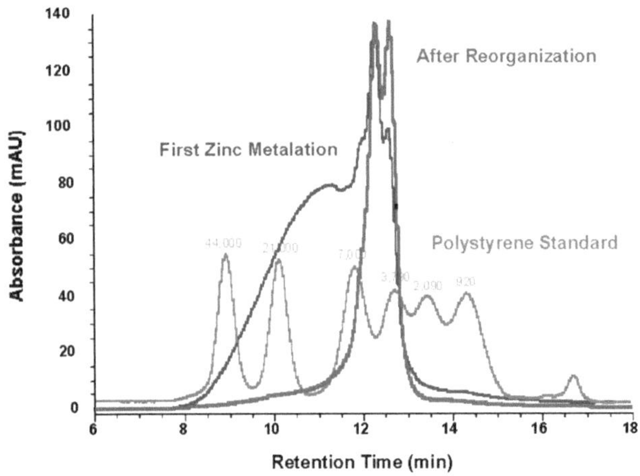

Figure 6.8. GPC elution curves immediately after Zn(II) insertion (blue), after reorganization (red), and of polystyrene standards (purple). Reprinted with permission from Takahashi, R., and Kobuke, Y. (2003) *J. Am. Chem. Soc.*, **125**, 2372–2373. Copyright 2003 American Chemical Society. See also Color Insert.

molecular weight–elution time calibration curve. It is concluded that the reorganization procedure eliminated almost all of the oligomeric species of higher molecular weights and afforded macrorings of the least angle strain (hexamer) accompanied by its smaller (pentamer) analogue, while excluding larger (heptamer) and smaller (tetramer) ring products [38, 39].

Although structure formation according to supramolecular methodology is excellent in view of providing even great molecular architectures with reduced synthetic effort, the structure proof does not generally proceed in a straightforward manner. The biggest problem in many cases is the lack of exact molecular weight information from mass spectrometry, even though various mild ionization techniques have been developed in the past decades. In the above macroring cases, trials of electrospray ionization mass spectrometry, which is known to be the mildest ionization method, gave monomeric peaks of bisporphyrin in overwhelmingly high intensities accompanied by tiny peaks corresponding to the dimer without any evidence for higher assembly formation. STM gave a somewhat hopeful sign suggesting vague images of the macroring at

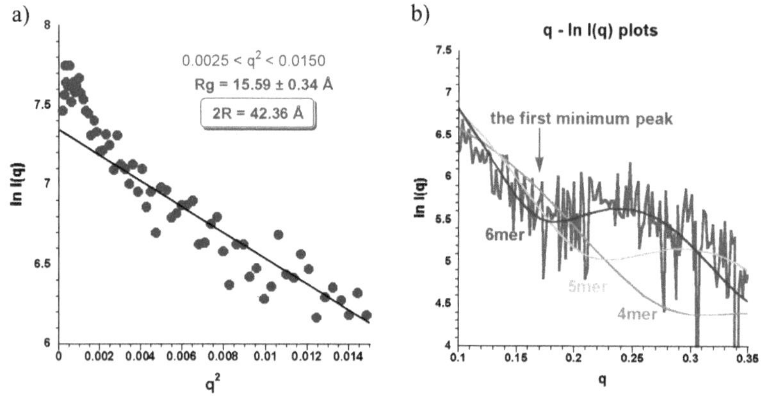

Figure 6.9. (a) Guinier and (b) SAXS plots for the isolated first peak. Reprinted with permission from Takahashi, R., and Kobuke, Y. (2005) *J. Org. Chem.*, **70**, 2745–2753. Copyright 2005 American Chemical Society.

this research stage. Synchrotron solution small angle X-ray scattering (SAXS) experiments have been employed as a powerful tool for elucidating molecular sizes in the field of biological studies. The plot of scattering intensities as a function of the square of the scattering vector (Guiner analysis, Fig. 6.9(a)) afforded a radius of gyration (R_g) of 15.59 $\pm$ 0.34 Å for the first peak [40]. The molecular diameters were calculated as 42.36 and 40.26 Å depending on the sphere and cylinder approximations, respectively. These values match well with the estimated value of 41 Å from a molecular mechanics calculation for the cyclic hexamer. From the similar plot for the second peak, an R_g value of 11.12 $\pm$ 0.66 Å was determined and then the diameter of 28.96 Å was obtained according to the cylinder approximation. This diameter agrees reasonably with the value of 30 Å calculated for a cyclic pentamer. In the plot of scattering intensities of the first peak in the wide-angle region (Fig. 6.9(b)), a wave profile characteristic of a hollow cylindrical structure was observed. Its minimum peak appeared at 0.17 Å^{-1}, corresponding closely to the calculated value of 0.18 Å^{-1}, compared with other cyclic oligomers, which were 0.23 and 0.30 Å^{-1} for the cyclic pentamer and tetramer, respectively [41]. A similar plot for the second peak was of relatively low resolution because of the smaller molecular size, but was roughly estimated to be 0.18–0.26 Å^{-1}. It might be concluded that the first peak should be a hexameric and the second peak a pentameric macroring. SAXS studies introduced a unique methodology for evaluating the size of molecular assemblies in solution, although the scattering intensities are relatively low in organic solvents compared with the case of biological material in aqueous solution.

In the above study, various methods of structural analysis were combined to elucidate the structure with confidence. Even so, precise and direct molecular weight information from mass spectrometry is desirable. The idea of connecting the complementary dimer pair by a C═C bond by an olefin metathesis reaction using a Grubbs catalyst has emerged [42]. On the basis of molecular model analysis, two pairs of substituents introduced at the 10,20 and 10′,20′-*meso* positions should be close to each other in the complementary dimer. The metathesis reaction of *N*-methylimidazolyl-Znporphyrin appended with 10,20-bis(allyloxycarbonylethyl) substituents **4** was found to give metathesized product **5** almost quantitatively (Fig. 6.10) [43, 44]. Because the reaction does not proceed in the free-base porphyrin at all under the same conditions, strapping over the porphyrin plane does not take place. Then, the supramolecular assembly formation and reorganization reactions for bisporphyrin with bisalkenyl substituents were undertaken in a similar way and the whole reaction mixture was subjected to the metathesis reaction. Two predominant peaks were observed at m/z 7910 and 6952, corresponding to hexamer **6** (C-P6) and pentamer **7** (C-P5), respectively [45]. This means that all of the allyloxy units, 24 in the hexamer and 20 in the pentamer, were connected in a pairwise fashion. It is now safely concluded that supramolecular organization of bisporphyrin with an internal angle of 120° leads to almost equal amounts of cyclic hexamer and pentamer by successive complementary coordination. Combined information on molecular weight and ring formation is confidently supported by GPC, STM, SAXS, and MS analyses.

Next, two *N*-methylimidazolyl-Znporphyrin units were connected through a *meta*-diethynylphenylene spacer to explore further the supramolecular cyclic pentamer and hexamer formation. Introduction of an ethynyl moiety is expected to increase the electronic interaction between complementary dimer units after assembly formation, in contrast to the above cases where the phenylene spacer inhibits π-conjugation between the porphyrin dimers [46].

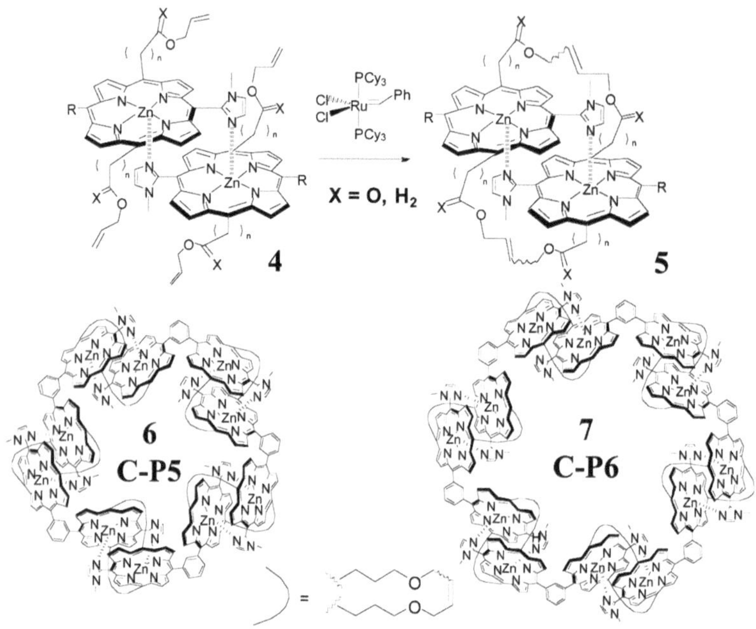

Figure 6.10. Ring-closing metathesis fixation of complementary dimer and macrocyclic 5- and 6-mers.

Furthermore, the presence of an ethynyl moiety allows free rotation around the ethyne axis and induces a red shift through π-conjugation with an increased transition dipole along this axis. The synthesis of target porphyrin and the introduction of Zn(II) followed by reorganization and metathesis linking according to the established method led to macrocyclic pentamer and hexamer in a similar way. The supramolecular structural proof was undertaken by using not only GPC, MS analyses as above, but also clear NMR in this case. An STM study is discussed later, but the π-conjugated spacer allowed the analysis with much higher resolution. The same characteristics enabled almost perfect NMR spectral assignment. The essence is illustrated in Fig. 6.11 for macrocyclic pentamer N-EP5 **10** and hexamer N-EP6 **11**, along with starting monomer **8** and dimer **9** for comparison. The dimer spectrum shows a very characteristic higher field shift of Im_4 and Im_5 by the ring-current effect from the porphyrin on coordination. At the same time, $\beta_1 \gg \beta_2$ protons are shifted to higher field, while β_3 and β_4 move to lower field, again

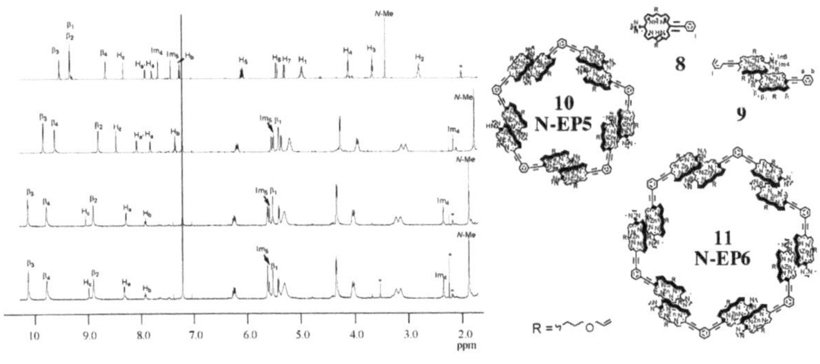

Figure 6.11. ^{1}H NMR spectra of *meta*-diethynylphenylene-linked Zn(II)porphyrin dimer: (a) free base **8**, (b) dimer **9**, (c) N-EP5 **10**, and (d) N-EP6 **11**.

because of the anisotropic shift caused by the porphyrin. The most characteristic points are that both pentamer and hexamer gave surprisingly simple spectra, exhibiting only a single set of porphyrin protons, similar to that of the reference dimer **9**. The large differences are seen only for the downfield shift of H_a, H_b, and H_c of the phenylene unit. This difference originates from the downfield shift by the porphyrin dimer connected at the other side of the phenylene bridge and supports the macroring formation. This simple NMR indicates that all 10 or 12 porphyrins are identical without discrimination of inner- and outer-ring porphyrins of macrocyclic arrangements. This is completely different from *m*-phenylene-linked bisporphyrin cases, where the NMR spectra were too complex to analyze because there are several geometrical in–out porphyrin isomers because of the high rotational barrier between the porphyrin and the phenylene bond. In the present case, no isomers exist because porphyrins can rotate freely around the ethyne bond.

Introduction of an ethyne moiety between the porphyrin and phenylene units leads to a red shift of the absorption spectra of these macrorings **18** and **19** compared with those of phenylene-linked macrorings **6** and **7**. Furthermore, the Q-band intensities of **18** and **19** are significantly higher than those of **6** and **7**. Because the emission spectra of **6** and **7** overlap well with the absorption spectra of **18** and **19**, the latter macrorings are suitable for accepting excitation energy from the formers, in a manner similar to the relationship of LH2 and LH1.

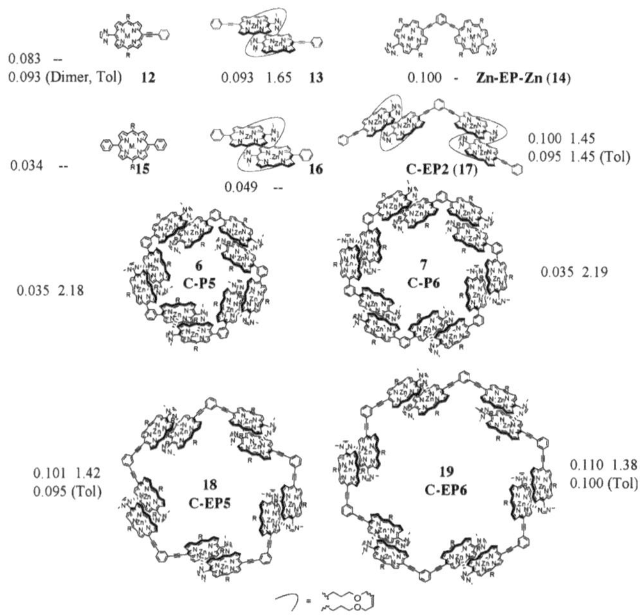

Figure 6.12. Fluorescence quantum yield (left column) and lifetime (ns, right column) of porphyrin macrorings, reference monomer, and dimers. Data in pyridine unless otherwise described.

As one of the important criteria for light-harvesting antenna molecules, assembly formation should not induce quenching of the excitation energy. The steady-state fluorescence quantum yields and lifetimes were summarized for this series of porphyrin macrorings along with the reference monomer and complementary dimer (Fig. 6.12).

All of the ethynyl-linked porphyrins show inherently high fluorescence quantum yields, compared with those of Znporphyrins, normally being 3–4%. Dimer formation never lowers the fluorescence quantum yields, but rather increases the value. When the number of dimers increases from one to five or six, the quantum yields are also increased. These are very important characteristics for light-harvesting properties.

The preparation of macrorings larger than a hexagon may be useful for elucidating the effect of the energy-transfer rate on the ring size. For this purpose, two imidazolylporphyrin units were

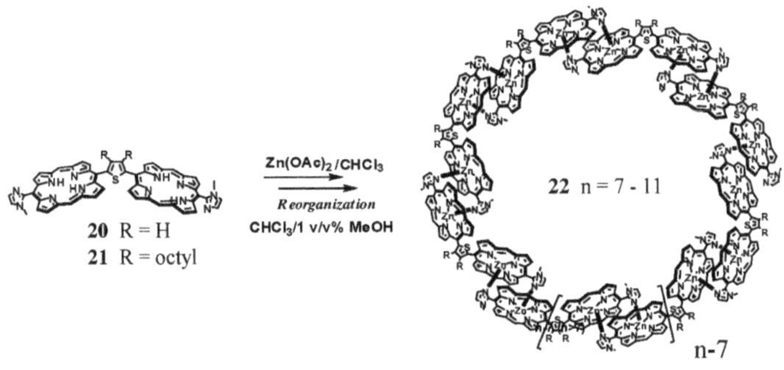

Figure 6.13. Thiophene-bridged porphyrin dimer **20**, **21** and macrorings **22**.

connected through thiophene and 2,3-dioctylthiophene to give **20** and **21**, respectively [47]. Molecular orbital calculations (AM1) predict the internal angle between two porphyrins as 154° and 148°, suggesting cyclic 14- and 11-mers, respectively, as macrorings with the least steric strain.

Zn insertion and reorganization showed the conventional conversion to macrorings under equilibrium conditions (Fig. 6.13). Then, the converged sample was analyzed by GPC with a combination of metathesis connection of allylic ether side chains of each complementary coordinated porphyrin pair. In the case of **20**, macrorings were distributed widely starting from the 7-mer to even larger than the 16-mer and significant peak overlaps made further analysis difficult. In contrast, **21** gave a much narrower GPC elution diagram. The mass spectra of the crude converged sample after the metathesis reaction showed the formation of macrorings **22** ranging from 7-mer to 11-mer, and each peak was separated by recycled GPC (Fig. 6.14). The peaks of the macrorings before metathesis corresponded to the calibration line obtained from the sample after metathesis.

Then, to determine the exact distribution ratio before the metathesis reaction, the GPC peak was deconvoluted by a Gaussian function approximation to give a maximum distribution at the 8-mer as follows: 7-mer = (27 ± 2%), 8-mer = (36 ± 2%), 9-mer = (18 ± 1%), 10-mer = (11 ± 2%), and larger than 10-mer = (7 ± 3%). This size population is apparently shifted

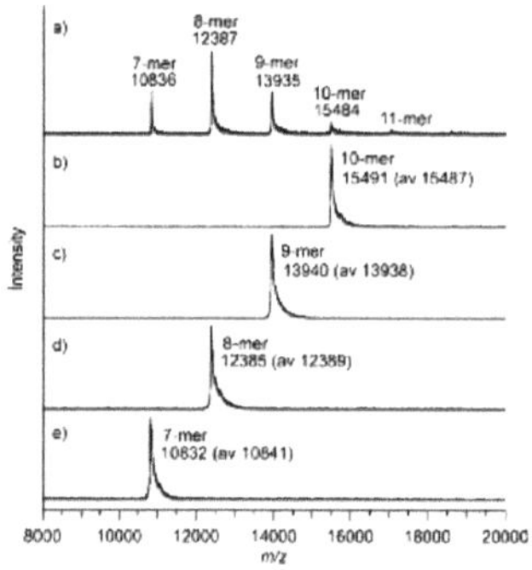

Figure 6.14. Mass spectra of metathesis-linked macroring **22** of (a) mixture and (b)–(e) isolated fractions.

toward macrorings of smaller unit numbers than the expected formation of 11-mer as the least steric strain product expected from the internal angle (148°) of the bisporphyrin unit. Under the high dilution conditions, terminal imidazolyls try to find counterparts in their own linear chains rather than those in other molecular chains, leading to the formation of smaller rings. To make quantitative use of this general tendency, free-energy differences were calculated between the 8-mer and the cyclic n-mer ($n = 7, 9$, and 11). Results calculated from semi-empirical MO, AM1 for ΔH and a rigid particle association model for the entropy term are illustrated in Fig. 6.15.

The entropy term decreases almost linearly for larger macrorings and favors the smaller macrorings. The enthalpy term strongly disfavors the smallest macrorings of the 7-mer. As a result, ΔG gets a minimum at the 8-mer, illustrating the relative free-energy order of 7-mer $\geq$ 8-mer $<$ 9-mer $<$ 11-mer, showing agreement with the experimentally obtained order of formation.

In the UV-vis spectra of a series of macrorings, the peak maxima of the longer wavelength Soret bands are systematically shifted to longer wavelengths in the order of increased ring size. The interaction energy for all of the porphyrin units in the macroring was calculated by Eq. 6.1 derived from excitonic coupling theory:

$$E = \Sigma 2\boldsymbol{m}_{mx}\boldsymbol{m}_{nx}\kappa_{mx,nx}/(\boldsymbol{R}_{mx,nx})^3, \tag{6.1}$$

where $\boldsymbol{m}_{mx}$ and $\boldsymbol{m}_{nx}$ are the x-components of the transition dipole moments of the m- and n-th dimer units, $\boldsymbol{R}_{mx,nx}$ is the center-to-center distance between $\boldsymbol{m}_{mx}$ and $\boldsymbol{m}_{nx}$, and $\kappa_{mx,nx}$ is the orienta-

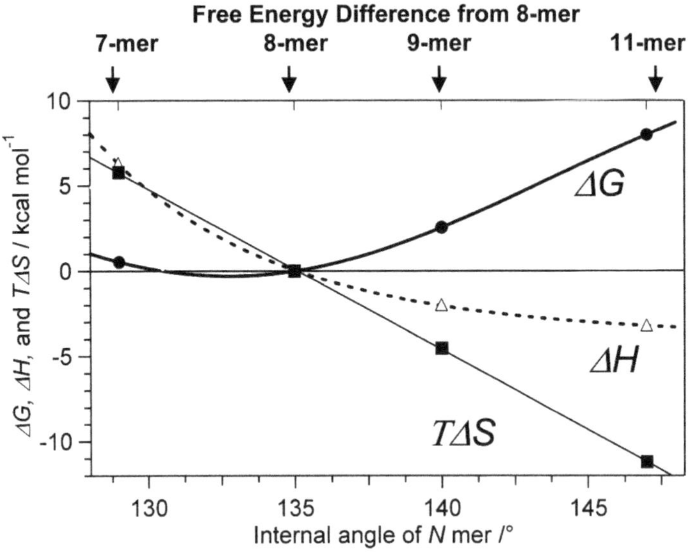

Figure 6.15. Calculated free energy changes of n-mers from 8-mer.

tion factor of $\boldsymbol{m}_{mx}$ and $\boldsymbol{m}_{nx}$, defined as $1 - 3\cos^2\theta$, with θ being the angle between the two dipoles. The total excitonic coupling energy was calculated as shown in Table 6.1 and increased with increasing ring size. Faster excitonic energy-transfer rates are expected upon increasing the value of the unit number. The preliminary results of the excitation-energy hopping rates show agreement with the expected general tendency.

Table 6.1. Total excitonic coupling energies of **22** based on equation 6.1. Transition dipoles are shown for the case of 8-mer

N-mer	E (10^{-6} m_0^2)
7	−1451
8	−1810
9	−1981
10	−2299
11	−2491
12	−2788

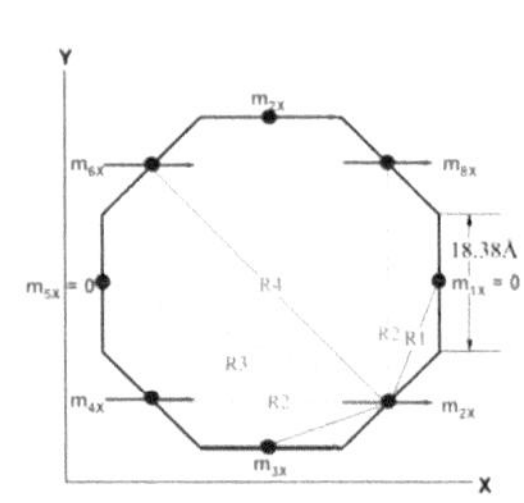

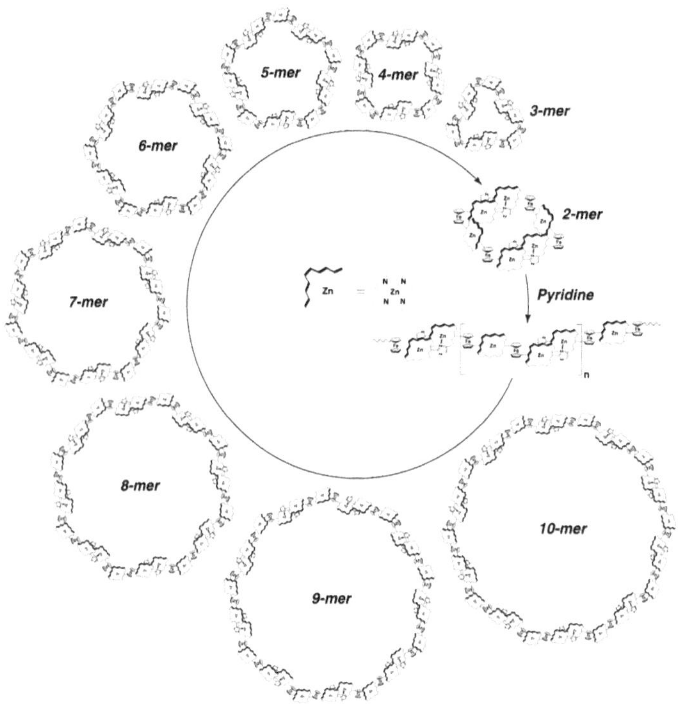

Figure 6.16. Formation of giant macrocycles by complementary coordination of ferrocene-bridged tris-Znporphyrin.

All of the previous systems employed connectors of fixed internal angles. When the connector was changed to a hinge-like unit of adjustable angle, the ring size could be controlled freely depending on the reorganization conditions from linear to large and small rings. For this purpose, two *N*-methylimidazolylporphyrin units were connected to the central porphyrin through ferrocenyl spacers [48]. Upon Zn(II) insertion, the trisporphyrin was spontaneously dimerized by complementary coordination. The hinge-like motion of the ferrocenyl unit allowed the smallest ring formation. The dimer was dissolved first in pyridine to dissociate into the monomeric trisporphyrin–pyridine complex. On evaporation of pyridine, high molecular weight polymer was obtained (through GPC observation), where the pyridyl ligands were gradually replaced by complementary imidazolyl coordination and the concentration exceeded

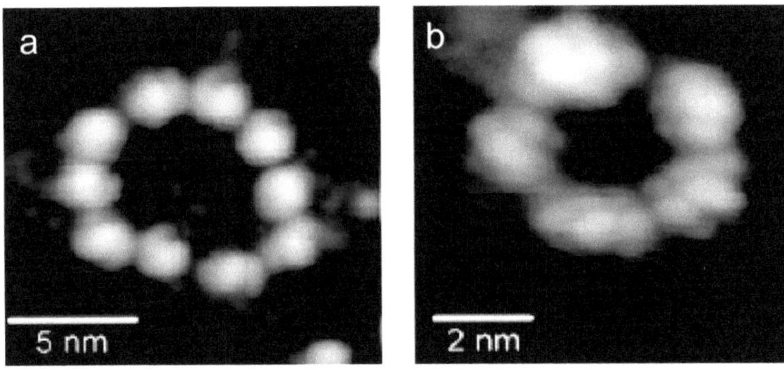

Figure 6.17. HRSTM images of ferrocene-bridged macrorings: (a) 10-mer and (b) 5-mer. Reprinted with permission from Shoji, O., Tanaka, H., Kawai, T., Kobuke, Y. (2005) *J. Am. Chem. Soc.*, **127**, 8598–8599. Copyright 2005 American Chemical Society.

the critical value above which only linear polymer increases [49]. On standing the dilute solution of the polymer in $CHCl_3$ containing 0.5% EtOH, peaks of longer retention times in GPC gradually increased and at least nine separate peaks were observed at the final equilibrium. All of the peak components other than the dimer were separated and identified as a series of macrorings composed of 3–10 trisporphyrin units after the ring-closing metathesis reaction (Fig. 6.16). It may be noteworthy that the largest ring contains 30 porphyrin units and 20 ferrocene units, and 40 alkenyl *meso*-substituents were metathesized in a pairwise fashion to give satisfactorily exact mass numbers.

High-resolution STM (HRSTM) images of these macrorings were obtained using pulse injection to the substrate under ultra high vacuum conditions [50, 51]. Clear round spots corresponding to the complementary coordination dimer were arranged in a circular form. Figure 6.17 shows 10 and 5 complementary dimer units arranged into macrorings [52].

In view of the HRSTM images, the macrorings prepared from *m*-diethynylphenylene-linked bisporphyrin gave a picture with excellent resolution [53]. In this case, free rotation around the ethynyl bond allowed the flat molecular structure to make maximum contact with the substrate. The pentamer-rich sample (98%

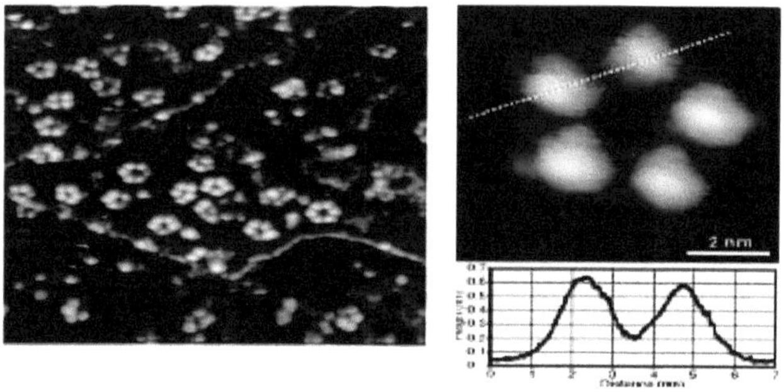

Figure 6.18. HRSTM images of diethynylphenylene-bridged macrorings: (a) 5-mer and 6-mer and (b) 5-mer and its profile along the line. Satake, A., Tanaka, H., Hajjaj, F., Kawai, T., and Kobuke, Y. (2006) *Chem. Commun.*, 2542–2544. Reproduced by permission of The Royal Society of Chemistry.

pentamer **18** and 2% hexamer **19**) was deposited on a Au(111) surface by a pulse injection method under high vacuum. Many clear spots arranged into five- or six-membered rings were detected (Fig. 6.18a). Analysis shows that each bright spot corresponds to imidazolyl-coordinated porphyrin dimer and each dark node corresponds to a diethynylphenylene moiety. One of the pentameric circles was analyzed in detail and is shown in Fig. 6.18(b).

6.5 Linear Antenna Array

The Mg(II) complex of imidazolylporphyrin was tested first to extend the number of complementary coordinations organized by using the sixth coordination (Fig. 6.19) [54], despite the limited stability constant mentioned above. Examination of the NMR spectrum of *N*-methylimidazolyl-appended Mg(II)octaethylporphyrin in $CHCl_3$ (2 mM) showed extension of the complementary coordination up to a mean assembly number of 2.4, which corresponds to the composition of 60% dimer and 40% trimer. Inspection of the electrospray ionization mass spectrum showed the formation of larger assemblies up to a heptamer **23** as dicationic peaks. It was confirmed that the Mg(II) metal center can accept the sixth coordination from an imidazolyl ligand to some extent but, at the same time, the formation of a much larger assembly cannot be expected according to this methodology.

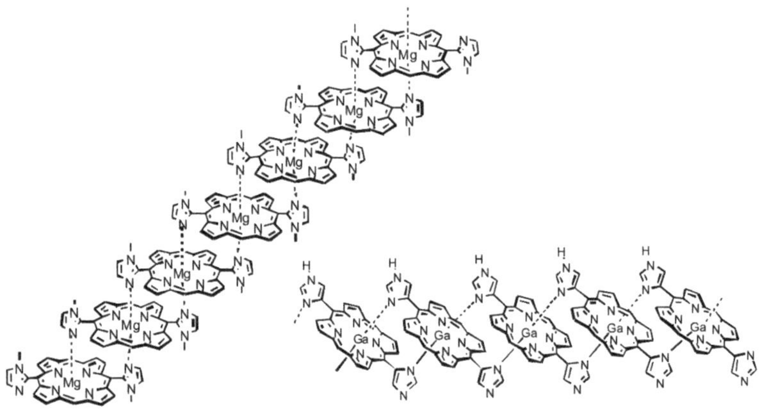

23 **24**

Figure 6.19. Staircase array of Mg(II) and Ga(III) imidazolylporphyrins.

ImidazolylGa(III)porphyrin is interesting for another type of coordination organization, where the extra positive charge on the metal center is expected to induce the dissociation of imidazolyl ligand into an imidazolate anion [55]. Then, Ga(III)porphyrin is expected to accept anionic and neutral nitrogens in the imidazolate anion to extend the linear structure via a staircase structure formation **24** (Fig. 6.19). In accord with this expectation, rod-like assemblies of 200–300 nm in length and 5–6 nm in height could be detected by AFM observation [56].

To explore the formation of a much larger linear assembly, 5-*N*-methylimidazolyl-Znporphyrin was subjected to a coupling reaction at the C-15 *meso*-position to give **25** [57]. The *meso–meso* linking forces an almost orthogonal orientation of two porphyrins because of the steric repulsion of the two C=H bonds *β* to the *meso*-carbon. Complementary coordination at 5- and 5′-*N*-methylimidazolyl substituents located at the terminal ends of this bisporphyrin is hardly affected and extends the molecular growth to both sides. Because the substituent at the other *meso*-positions (10- and 20-C) was chosen as the *n*-heptyl group primarily to prevent scrambling of the *meso*-substituents during porphyrin synthesis, the resulting linear molecular array was covered by long alkyl substituents extending up, down, left, and right of the molecule and showed excellent

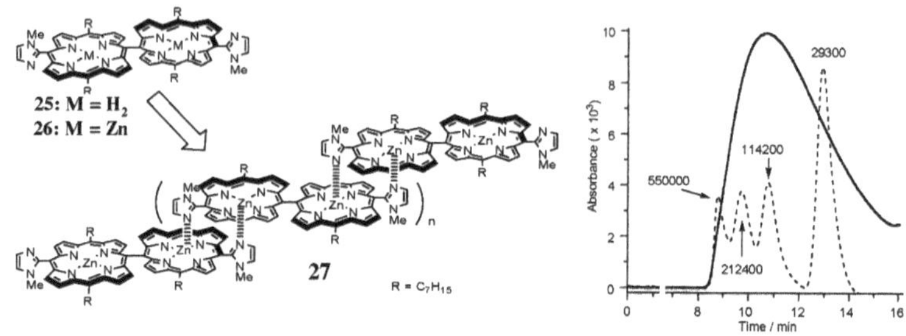

Figure 6.20. Formation of a one-dimensional linear array **27** from *meso–meso*-linked bis(*N*-methylimidazolyl-Znporphyrin) **26** and its GPC elution curve.

solubility in $CHCl_3$. This enabled the formation of soluble material even with a large molecular weight arising from the large stability constant of the complementary imidazolyl-to-Znporphyrin coordination. The GPC elution curve of the bisporphyrin assembly is shown in Fig. 6.20, along with those of polystyrene mixtures of standard molecular weights. The first elution edge appeared near the exclusion limit (5.0×10^5 Da) and the peak maximum was around 1.0×10^5 Da, which corresponds to 80 molecular units of bis(*N*-methylimidazolyl-Znporphyrin) with a molecular weight of 1298. Because a molecular mechanics calculation using Cerius2 afforded the molecular length of 1.43 nm for the bisporphyrin unit, the molecular length corresponds to 550 and 110 nm for its longest and distribution-top array, respectively.

Complementary coordination induces a characteristic splitting of the Soret band. The Soret band of dimeric mono-*N*-methylimidazolyl-Znporphyrin **28** appears at 409 and 430 nm by face-to-face and head-to-tail exciton interactions, respectively. Because the two molecular planes of bis(imidazolylporphyrin) are orthogonal to each other, only the transition dipoles along the axis passing through the *meso–meso* bond interact with each other to produce a red shift giving the split Soret bands at 412 and 453 nm. In the case of bis(*N*-methylimidazolyl-Znporphyrin) **27**, the Soret band at the longer wavelength is shifted further to 490 nm because of the cumulative head-to-tail exciton interactions, while the shorter band is slightly blue shifted to 409 nm because face-to-face interactions

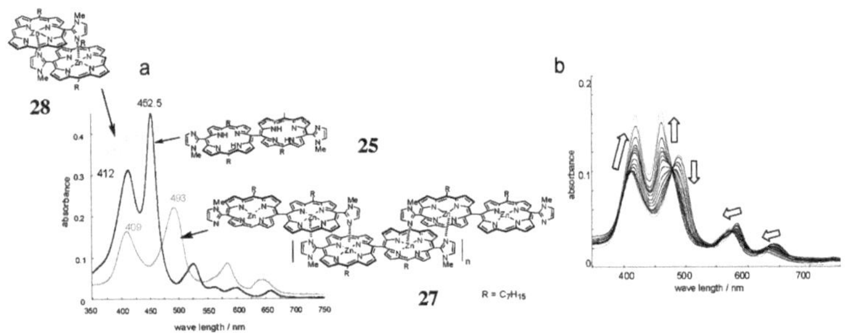

Figure 6.21. (a) Comparison of absorption spectra of polyporphyrin **27**, free-base bisporphyrin **25**, and complementary dimer **28**. (b) UV-vis spectral change on MeOH addition.

in the *meso–meso* dimer are orthogonal to each other and cannot be conveyed through the molecular array extension (Fig. 6.21(a)).

Although the complementary coordination of imidazolyl-to-Znporphyrin is very stable in nonpolar solvents, its structure can be broken by the addition of coordinating solvents. Formation and breaking of the coordination organization is most easily monitored by the absorption spectral change of the Soret band. When MeOH was added to the $CHCl_3$ solution of bis(*N*-methylimidazolyl-Znporphyrin), the longer Soret band shifted to shorter wavelength, passing through an isosbestic point, indicating that the equilibrium shifted to the array of low molecular weight (Fig. 6.21(b)). When MeOH was removed from the solution, the original UV absorption spectrum was recovered, suggesting that the cleavage and regeneration of the linear array structure can be controlled reversibly by the amount of external reagent such as MeOH. Then, MeOH was added to a mixture of bis(*N*-methylimidazolyl-Znporphyrin) **27** and *N*-methylimidazolyl-Znporphyrin **28** in $CHCl_3$ and then evaporated after a while. The GPC elution curve before MeOH addition showed the corresponding two peaks. The elution curve in $CHCl_3$ after the MeOH treatment showed several peaks, which were separated and identified as oligomers terminated by mono(*N*-methylimidazolyl-Znporphyrin) unit **29** ($n = 1$, 2, and 3) (Fig. 6.22). The ratio **28**/**27** determines the molecular length.

When the terminal imidazolyl-Znporphyrin is immobilized on a substrate surface such as a gold or indium–tin oxide electrode

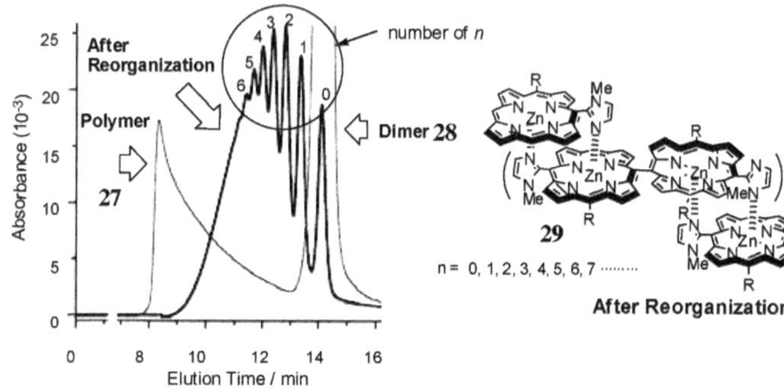

Figure 6.22. GPC chart of a mixture of poly(bisporphyrin) **27** and dimer **28** before (solid line) and after (bold line) reorganization.

by using thiolate or carboxylate, addition of bis(imidazolyl-Znporphyrin) starts the array formation *in situ*, which may serve as a molecular wire or light-harvesting unit for molecular electronics or energy-conversion applications. This is discussed in Section 6.9.

A butadiynylene-linked bis(*N*-2-ethylhexylimidazolyl-Znporphyrin) **30** was prepared with the aim of obtaining better photonic and electronic communication between porphyrins [58]. It was spontaneously organized into a large molecular weight polymer **31** by complementary coordination. The NMR spectrum was simple and similar to that of mono-*N*-methylimidazolyl-Znporphyrin dimer because no geometrical isomer is formed on successive complementary coordination because of the free rotation around the diyne bond. The long array structure allowed molecular wiring between gold nanoparticles using *N*-methylimidazolyl-Znporphyrin as a connector molecule, which was modified on gold by dithiolate bonds. The STM allowed the observation of single molecular wire bridging gold nanoparticles (Fig. 6.23).

The concept of one-dimensional array formation was extended further to develop a two-dimensional network system. Any defect once generated in the one-dimensional array system may cut off the energy/electron/hole transport along the linear line, but the construction of a higher network can find other connected routes in the system to convey the signal. For this purpose, four *N*-methylimidazolyl-Znporphyrin units were connected to the central porphyrin to give **32**, so that the complementary coordination may develop in four directions (Fig. 6.24) [59]. Because straightforward

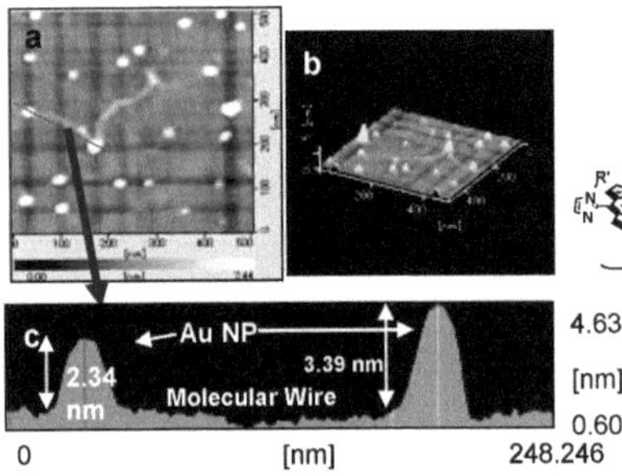

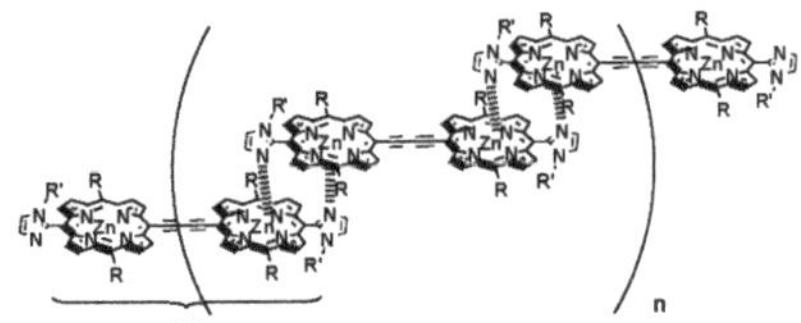

30

31

Figure 6.23. Linear array from butadiynylene-bridged bisporphyrin and AFM images of molecular wiring (thin white line) connecting gold nanoparticles (dots): (a) 2D image, (b) 3D image, and (c) profile along the line in (a). Satake, A., Fujita, M., Kurimoto, Y., and Kobuke, Y. (2009) *Chem. Commun.*, 1231–1233. Reproduced by permission of The Royal Society of Chemistry.

Zn(II) introduction resulted in a material that was insoluble in any solvent, Zn(II) insertion and coordination organization were undertaken in a stepwise fashion. As the first step, Zn(II) insertion was conducted in the presence of enough pyridine to suppress *in situ* imidazolyl coordination. After complete conversion to Znporphyrin, pyridine was removed gradually from nitrobenzene solution to grow a two-dimensional network system composed of complementary coordination units **33** in a soluble form. An AFM study of the solid after complete solvent evaporation elucidated a plate-like structure with a length of over 390 nm and a height of 2.0–2.4 nm. The two-dimensional network area corresponds to the assembly of ca. 130 porphyrin pentamer units. Because coordination occurs randomly on both the top and bottom of the porphyrin plane, the observed

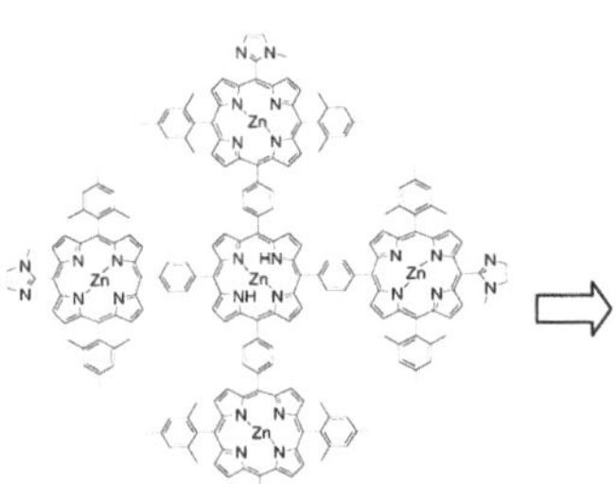

32

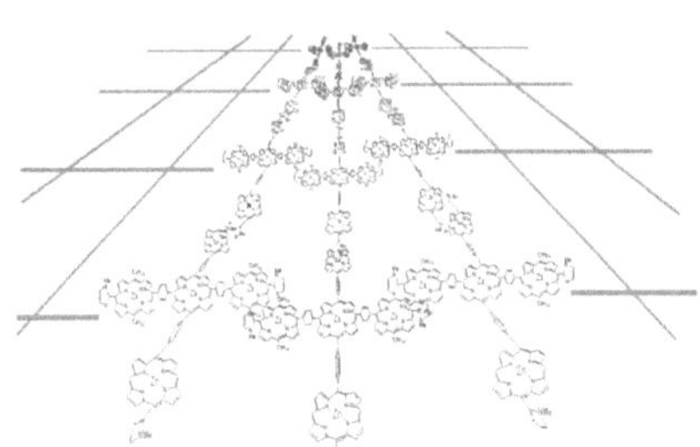

33

Figure 6.24. Formation of the two-dimensional porphyrin array **33**.

height is reasonable considering 6–7 porphyrin layers on average. It is worthwhile commenting that solid material self assembled on a glass plate still retains significant luminescence intensity after such a large two-dimensional assembly formation.

6.6 Intraring Excited Energy Transfer

The intraring energy-transfer rate was determined first for the macrocyclic pentamer C-P5 (**6**) and hexamer C-P6 (**7**), both of which were connected by a ring-closing metathesis reaction so as to be stable without reorganization under concentrated concentrations for detecting weak signals of transient species [60]. The complementary dimer **34**, bisporphyrin monomer coordinated by pyridine **35**, and bisporphyrin dimer **36** were used as the reference compounds. To detect the fast excitation-energy-transfer process, pump-power dependent transient absorption (TA) with a femtosecond time resolution and transient absorption anisotropy (TAA) decays were combined with nanosecond time-resolved fluorescence decay and fluorescence anisotropy decay [60–65]. The TAA decay reflects directly the depolarization of an initially localized exciton to weakly coupled chromophores according to the excitation-energy transfer. Under multiphoton absorption conditions, annihilation occurs by energy hopping and this process can be extracted through pump-power dependent excitation of the multichromophores.

The TAA decay of **34** showed a single time component of 180 fs. In contrast, the TAA decays of **35** and **36** were fitted by two time constants of 190 ps and 4.7 ps, and 230 ps and 4.6 ps, respectively (Table 6.2). The fast ~200 fs components are almost the same and assigned to the in-plane dipole equilibrium in the Znporphyrin plane. The slower components were assigned to the excitation-energy transfer (EEH) between Znporphyrin monomers in **35**, and the two cofacial Znporphyrin dimers in **36**. In the case of C-P5 and C-P6, the slower TAA decay time was obtained as 2.11 and 1.62 ps for C-P5 and C-P6, respectively. These values for multichromophores must be calculated to obtain the elemental EEH rate between neighboring chromophores and will be shown below. At the same time, the transient absorption showed pump-power dependency for C-P5 and

Table 6.2. Energy-hopping rates of porphytin macrorings along with the rates of reference porphyrins

			Hopping time const/ps		
Compound	No.	Distance (Å)	Anisotropy	Annihiration	Mean
C-P5	**6**	15.5	7.6	8.3	8.0
C-P6	**7**	16.1	4.9	5.7	5.3
C-EP5	**18**	19.9	19.5	22.0	21.0
C-EP6	**19**	20.7	12.6	13.0	12.8

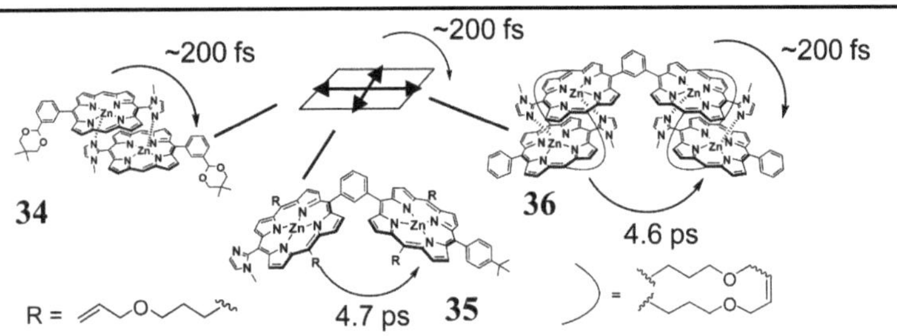

C-P6, but not for the reference compounds **34** or **35**. This observation is in accord with the fact that the exciton–exciton annihilation between the complementary dimer units occurs only in the multichromophoric systems of C-P5 and C-P6. The fitted decay times of the slowest process for C-P5 and C-P6 were 8.27 and 8.28 ps, respectively.

From these exciton–exciton annihilation and anisotropy depolarization time constants, the EEH processes were analyzed based on the Förster-type energy-transfer model assuming migration-limited exciton–exciton annihilation and a random-walk process for the anisotropy decay. Equations 6.2 and 6.3 were obtained using Fleming's model to deduce $\tau_{\text{annihilation}}$, as the slowest exciton–exciton annihilation time and τ_{hopping}, as the inverse of the nearest neighbor energy-hopping rate [61].

$$\tau_{\text{depolarization}} = \tau_{\text{hopping}}/4(1 - \cos^2(2\pi/N)) = \tau_{\text{hopping}}/4(1 - \cos^2\alpha) \tag{6.2}$$

$$\tau_{\text{annihilation}} = \{(N^2 - 1)/24\} \times \tau_{\text{hopping}} \tag{6.3}$$

where N is the number of EEH sites and α is the angle between the neighboring transition dipoles.

Using the values $N = 5$, $\alpha = 72°$, and $\tau_{depolarization} = 2.11$ ps for C-P5, and $N = 6$, $\alpha = 60°$, and $\tau_{depolarization} = 1.62$ ps for C-P6, the $\tau_{hopping}$ values from anisotropy depolarization experiments were obtained as 7.6 and 4.9 ps for C-P5 and C-P6, respectively. Similarly from Eq. 6.3, annihilation experiments derived the values $\tau_{hopping} = 8.3$ and 5.7 ps for C-P5 and C-P6, respectively. The time constants obtained from the two different experiments must be the same within experimental errors, and the final EEH times are determined to be 8.0 ± 0.5 ps for C-P5 and 5.3 ± 0.6 ps for C-P6. Considering the fluorescence lifetimes of ~2.2 ns for C-P5 and C-P6, rapid energy transfer takes place within the ring 300–400 times during their excited-state lifetimes. These dynamic and static properties along with barrel-type structures, which were constructed by characteristic coordination of the imidazolyl ligand to create partially overlapped porphyrin π-planes, support this excellent B850-type light-harvesting antenna model.

Larger macrorings C-EP5 (**18**) and C-EP6 (**19**) were synthesized to examine the relationship between the macroring structure and EEH time constants [66]. In these molecules, two porphyrin units are connected through a 1,3-diethynylphenylene moiety expecting strong electronic coupling between two complementary cofacial dimer units. By using the above time-resolved anisotropic spectroscopies, the EEH times were evaluated to be 21.0 ps for C-EP5 and 12.8 ps for C-EP6. Comparison of the EEH times for the C-P and C-EP series suggests that the distance between the chromophores is most important, even though the introduction of the ethyne bond induces coplanarity of the chromophores and their enhanced electronic communication. These data are summarized in Table 6.2.

It is interesting to note that the EEH rates of the hexameric macrorings of C-P6 and C-EP6 are faster than those of the corresponding pentameric rings. The Förster-type energy rate constant k_{ET} is expressed by Eq. 6.4:

$$k_{EEH} = (8.8 \times 10^{-25} \kappa^2 \Phi / n^4 R^6 \tau) J \tag{6.4}$$

where n is the refractive index of the solvent, R is the center-to-center distance between donor and acceptor, τ is the fluorescence lifetime of an energy donor, κ is the orientation factor, Φ is the fluorescence quantum yield, and J is the spectral overlap integral.

Because we assume planar transition dipole interactions for pentagonal and hexagonal rings, κ is calculated to be 1.65 for C-P5 and C-EP5 and 1.75 for C-P6 and C-EP6. In terms of the orientation factor, hexamers may lead to a faster exciton energy-hopping rate by a factor of $(1.75/1.65)^2 = 1.12$. The center-to-center distances between the two complementary dimer units are calculated from molecular modeling to be 15.5, 16.1, 19.9, and 20.7 Å for C-P5, C-P6, C-EP5, and C-EP6, respectively. Therefore, the pentamers are more favorable in terms of distances by a factor of $(1.04)^6 = 1.27$. Introducing the fluorescence quantum yields and the lifetimes for C-P5, C-P6, C-EP5, and C-EP6 listed in Fig. 6.12, the relative EEH rate constants, $k_{5\text{-mer}}/k_{6\text{-mer}}$, are calculated to be 1.12 and 1.00 for the C-P and C-EP series, respectively. The larger orientation factor of the hexamers is cancelled or outweighed primarily by the unfavorable distance factor. In contrast, the EEH rates obtained experimentally are faster for hexamers C-P6 and C-EP6 than for pentamers C-P5 and C-EP5. The treatment assuming only neighboring exciton interactions may be modified to include next neighbors and more distant exciton interactions. A series of macrorings from thiophene-bridged bisporphyrin **22** are showing a promising sign on this point, but we must await further studies for conclusive results, as described previously.

It is interesting that *meso–meso* coupled bis(*N*-methylimidazolyl-Znporphyrin) does not lose its fluorescence intensity even after formation of huge molecular weight polymer **27** by successive complementary coordination. The lack of extra quenching on assembly formation means that this linear porphyrin array can serve as a linear light-harvesting system. However, evaluation of the energy-transfer rate along the linear array is not as easy as in the case of a ring system, where all of the chromophores are degenerate without chain terminals. To achieve energy-transfer efficiency in the linear porphyrin array, a fluorescence-quencher molecule was introduced at the polymer chain terminals. *N*-methylimidazolylMn(III)porphyrin was introduced by complementary heterodimerization at the terminal Znporphyrin to afford **37** [67]. The intermediate strength of heterodimer formation compared with strong Zn–Zn and weak Mn–Mn homodimer formation allowed the introduction of a controlled amount of Mnporphyrin at the molecular terminals, thereby controlling the Zn–Zn homodimer array

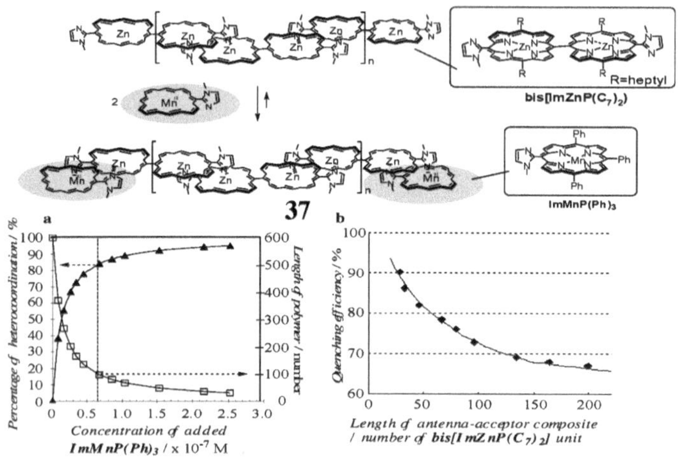

Figure 6.25. Formation of Mn(III)porphyrin-terminated linear Znporphyrin array. (a) Percentage of heterocoordination (left scale) and mean polymer length (right scale). (b) Fluorescence quenching efficiency as a function of polymer length.

length. Because the fluorescence is quenched at the terminal Mnporphyrins, energy-transfer efficiency along the Zn–Zn porphyrin array can be monitored by measuring the quenching efficiency of various array lengths. Figure 6.25 shows, for example, that the fluorescence from the array of 130 bisporphyrin units is quenched by 70%. In other words, the energy excited at any porphyrin in the region of $130 \times 0.7/2 = 45$ bisporphyrin units from the terminal is transported without loss along the array to the terminal Mnporphyrin and quenched.

6.7 Antenna–Reaction Center Composite

In the bacterial photosynthetic system, the reaction center is accommodated in the large cavity of the LH1 antenna to receive effectively the energy transferred from the light-harvesting system. A synthetic approach was designed according to the following idea (Fig. 6.26). Two *N*-methylimidazolyl-Znporphyrin units were connected through bis(*m*-phenylene)-substituted Znporphyrin

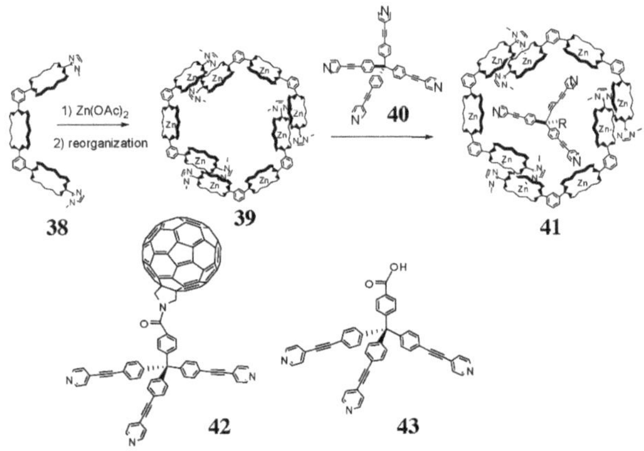

Figure 6.26. Formation of a macroring from trisporphyrin and incorporation of the tripodal ligand.

to afford **38** [68]. A similar complementary coordination took place using the terminal imidazolyl-Znporphyrin part. The self-organization under dilute conditions produced a cyclic trimer with a high selectivity. The competitive process leading to a smaller dimeric ring is strongly prohibited by the large angle strain from hexagon (120°) to square (90°). Because coordination organization takes place at the terminal porphyrin unit, the central Znporphyrin unit provides a multitopic and cooperative coordination site toward the tripodal ligand approaching from inside the ring. By using tetrapyridyl ligand **40**, UV and fluorescence titrations conducted in toluene showed a very large stability constant of 8×10^8 M^{-1}, suggesting cooperative 1:1 complexation with participation of three pyridyl units. In polar benzonitrile, the UV-vis spectrum showed coordination of benzonitrile prior to the pyridyl titration, but still gave a large stability constant of 1.2×10^7 M^{-1}, in keeping with the nature of cooperative binding.

An antenna ring capable of incorporating a multidentate ligand into its cavity with a large binding constant has thus been prepared. The fourth arm of the above tetrapyridyl ligand can be substituted by an appropriate energy/electron-accepting group without affecting the essential nature of the tridentate ligation for constructing a desired antenna–reaction center model. Without doubt, a fullerene unit may be the best choice for such an acceptor in view of its excellent electronic characteristics. With the help of a CPK molecular model, tripod ligand C_{60}–Tripod **42** was designed and synthesized. It is anticipated that the fullerene unit is extruded out of the barrel supported by cooperative coordination of the tripodal pyridyl ligand inside the ring.

UV-vis and fluorescence titrations were conducted in toluene and in benzonitrile as well, considering that photoinduced electron-transfer experiments are usually conducted in polar solvents, particularly in benzonitrile, to stabilize the charge-separated species. Upon titration with C_{60}–tripod, the Soret and Q band showed characteristic changes of pyridyl coordination, namely, a slight red shift with a decrease in the intensities. The fluorescence from the ring was quenched efficiently by 85% with addition of an equimolar amount of C_{60}–Tripod and by 95% with the addition of 4 equiv. Combined with Job's plot, the titration curve showed a clear 1:1 complexation, giving the binding constant of 3.1–3.4 $\times$ 10^8 M^{-1} in benzonitrile by UV-vis and fluorescence titrations. This binding constant is approximately 10 times larger than those compared with the tetrapodal ligand **40** (1.2 $\times$ 10^7 M^{-1}) or tripodal ligand **43** (2.1 $\times$ 10^7 M^{-1}) without the fullerene unit as its precursor. This result suggests that the fullerene contributes significantly to enhancing the binding of C_{60}–Tripod **42** in the macroring even compared with pyridyl coordination. At this stage, the initial idea that three pyridyl arms interact inside the porphyrin macroring by tripodal coordination and the fullerene moiety is extruded out of the ring (Fig. 6.27(a)) may be modified. To establish a plausible structure, photophysical properties of the macroring/C_{60}–Tripod composite were investigated.

Time-resolved fluorescence decay profiles of the macroring/C_{60}–Tripod composite showed a biexponential decay with time constants of 164 ps (62%) and 2100 ps (38%) [69]. Fluorescence from the macroring is thus efficiently quenched with a high rate constant of 5.6 $\times$ 10^9 s^{-1}. The fluorescence quenching efficiency was as high as 0.97 or 0.92 from steady-state and dynamic experiments, respectively. In spite of this observation, the nanosecond transient absorption spectra of the macroring/C_{60}–Tripod composite did not show any sign of the presence of the charge-separated species in the anticipated region of 700–1200 nm. The absorption peak at 700–900 nm, corresponding to the triplet excited state of

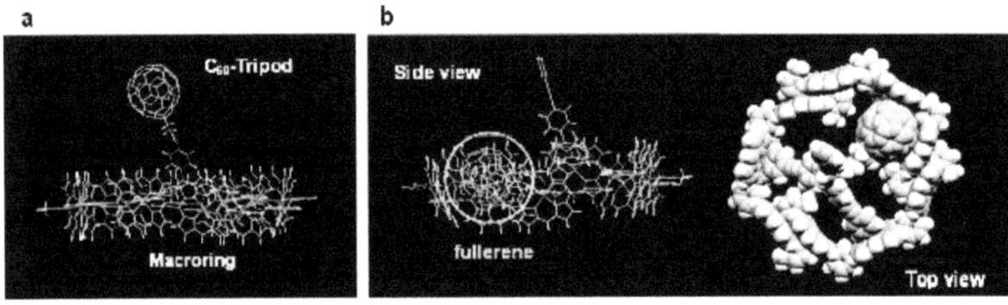

Figure 6.27. Specific formation of macrocyclic trimer from trisporphyrin and incorporation of a tetrapodal pyridyl ligand in the pore. Reprinted with permission from Kuramochi, Y., Satake, A., Itou, M., Ogawa, K., Araki, A., Ito, O., and Kobuke, Y. (2008) *Chem. Eur. J.*, **14**, 2827–2841.

porphyrin observed after 50 μs with the tripod guest without C_{60}, disappeared in agreement with this behavior. The transient absorption spectra in the subpicosecond time region showed a broad and featureless band in the 600–800 nm region extending over 1000 nm. The lifetime of this band was 77 ps ($k = 1.3 \times 10^{10}$ s^{-1}). These results suggest that the fullerene unit of C_{60}–Tripod has fallen down into the inner surface of the macroring, making direct contact with the porphyrin through π–π interaction, as shown by pictures created with Material Studio® (Accelrys) in Fig. 6.27(b). In such a geometry, a charge-transfer state seems to be generated, showing a broad and featureless transient absorption band in the region of 600–1000 nm as reported in the porphyrinZn–fullerene π-stack dyad [70]. The decay time of this dyad (38 ps) measured in benzonitrile is similar to the present lifetime of the macroring–C_{60}–Tripod composite (77ps).

Although the conclusion outlined in the above section must be accepted as the most reasonable explanation, the idea was still not fully convincing and the behavior of guest accommodation into this macroring was examined in more detail by preparing further fullerenyl guest derivatives [71]. The molecular structures and stability constants in benzonitrile of guest molecules containing multipyridyl and fullerenyl units are summarized in Fig. 6.28.

Introduction of the third pyridyl arm in **45** to bipyridyl ligand **44** enhanced the stability constant by ca. 10 times. At first glance, this effect may be reasonable as a result of incremental pyridyl coordination, but the 10-fold enhancement is rather small considering that cooperative pyridyl coordination should contribute to increase

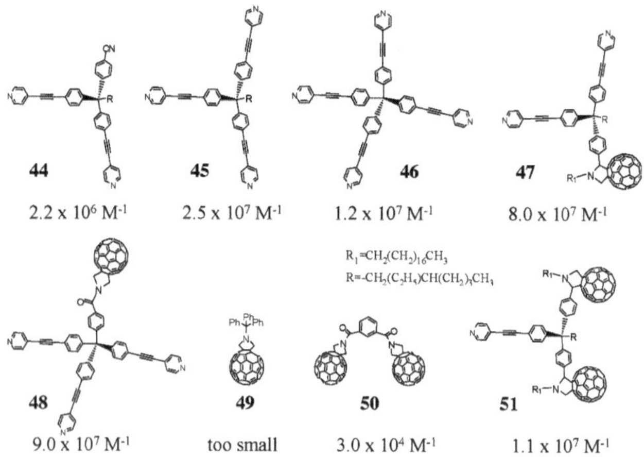

Figure 6.28. Stability constants of selected guest ligands with the macroring from trisporphyrin.

the value by a factor of 10^3. Substitution of one of the pyridyls into the fullerenyl arm in tripyridyl or tetrapyridyl ligands enhanced the stability constant by a factor of 3.2 and 7.5 when the stability differences were compared between **47/45** and **48/46**, respectively. This is rather unusual considering the small stability constants of fullerenyl units for **49** or **50** based on the weaker π–π interaction compared with pyridyl coordination.

To understand the apparently unusual behavior of the stability constants, thermodynamic parameters for the formation between macroring **39** and **44**, **45**, **47**, and **48** were examined by variable-temperature UV-vis titrations in benzonitrile. The results are shown in Table 6.3. Reflecting strong cooperative coordination of multiple pyridyl units inside the macroring **39**, bis- and trispyridyl ligands

Table 6.3. Thermodynamic parameters for complex formation of macrorind with ligands

Ligand No.	ΔH^o (kj mol^{-1})	ΔS^o (j mol^{-1} j^{-1})	Solvent
44	−24.4	40.9	Benzonitrile
45	−36.3	19.8	Benzonitrile
45	−75.7	−66.5	Chloroform
47	−32.8	41.1	Benzonitrile
48	−32.3	43.8	Benzonitrile

44 and **45** exhibited large negative ΔH° values. At the same time, however, positive and somewhat large ΔS° values seemed unusual for such strong ligand coordination. Introduction of an additional pyridyl unit in **45** induced a 1.5-fold increase in ΔH° with a loss of half of the ΔS° value for **44**. At this point, it was confirmed that the three-pyridyl coordination enhances the binding by the enthalpy term, and that significant loss of ligand flexibility is accompanied by the rigid three-point complex formation. Even so, the overall entropy change was favorable, suggesting that significant numbers of solvent molecules must be released on complexation. This behavior was observed characteristically in benzonitrile, because the same guest **45** gave large negative ΔH° and ΔS° values measured in chloroform. Examination of thermodynamic parameters for fullerenyl ligands **47** and **48** gave additional important features for the complex formation with macroring **39**. Interestingly, the enthalpy gains for both ligands were even larger than that for **44**, by -8.7 and -8.2 kJ·mol^{-1} for **47** and **48**, respectively. These ΔH° increments must suggest that the fullerenyl unit contributes to the π–π interaction between the fullerene moiety and porphyrin in addition to two-pyridyl coordination. In these cases, ΔS° values remain almost the same as that of **44**, in contrast to the decrease when compared with the case for ligand **45**. Therefore, the entropy loss to be associated with three-point interaction of two-pyridyl coordination and fullerenyl π–π interaction must be compensated by the release of benzonitrile molecules that interacted with the fullerenyl part prior to the complexation. Ligands containing two pyridyl and one fullerenyl units, **47** and **48**, interacted more strongly than trispyridine ligand **45** by virtue of three-point interactions of two pyridyl coordinations and one π–π interaction of fullerene and porphyrin rather than three pyridyl coordinations. The result explains satisfactorily that the fullerenyl unit instead of pyridyl interacts favorably with the porphyrin π-plane by the configuration in which the pyridine has fallen inside the macroring. It is now fully understood that the resulting close contact of fullerene with porphyrin induced a strong charge-transfer interaction leading to no observation of stable charge-separated species.

The above conclusion forced us to modify the structure of the guest molecule so that it could achieve the original idea of

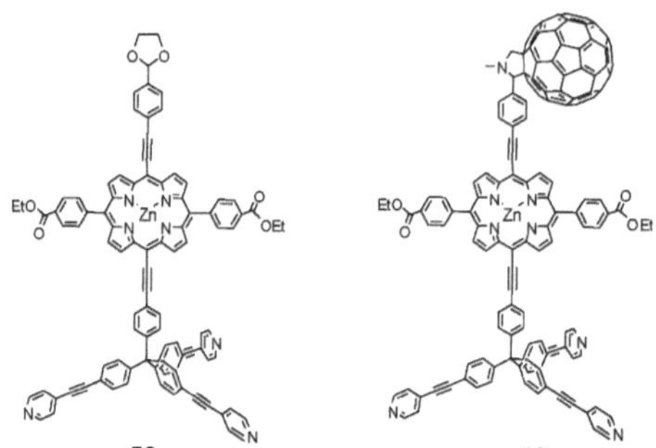

Figure 6.29. Tripodal ligands with an ethynyl-linked porphyrin–fullerene unit and their precursors.

demonstrating the photosynthetic process, light harvesting followed by charge separation. The improvement must be made at the point where the fullerenyl unit could not contact directly any porphyrin part in the macroring. For this purpose, we designed a new guest molecule C_{60}–ZnP–Tripod **53**. The target molecule **53** along with its synthetic precursor ZnP–Tripod **52** contains some tricks (Figure 6.29) [72].

The basic strategy to bind strongly into the macroring cavity is not changed from the idea of the tripodal pyridyl ligand. The fourth arm should be extended from the tetrahedral carbon making the molecule long enough to never fall down into the cavity. As the first component in this arm, diethynylporphyrin was selected as the key element that accepts energy from the macroring and initiates charge separation. The absorption spectrum of this unit is shifted to a longer wavelength by the presence of two conjugated ethynyl linkers to create an overlap with the fluorescence emission spectrum of the macroring **39** large enough to expect Förster-type energy transfer. At the terminal end, a fullerenyl unit was introduced expecting electron transfer from the ZnP unit. Consideration of the oxidation–reduction potentials of ZnP and C_{60} units and the S_1 energy level of ZnP suggests photoinduced electron transfer from ZnP to C_{60} moieties in benzonitrile. The molecular mechanics calculation (Material Studio® created the most probable structure of C_{60}–ZnP-Tripod **53**/Macroring **39** composite depicted in Fig. 6.30. The fluorescence of **53** was quenched by 99.3% compared with that of **52**, indicating

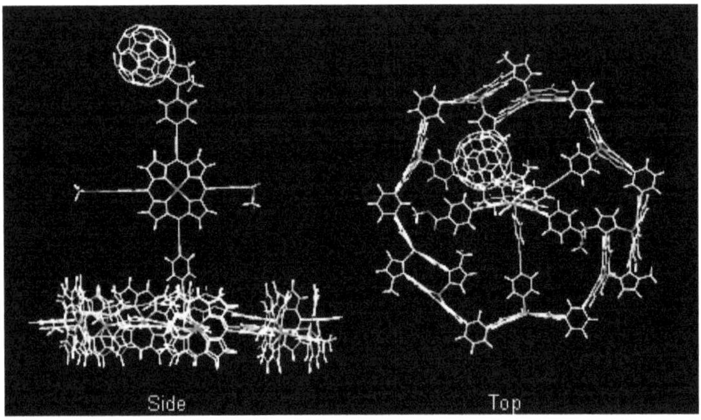

Figure 6.30. Molecular model of the C_{60}–porphyrin–tripod ligand **53**/macroring **39** composite optimized by molecular mechanics calculations. Reprinted with permission from Kuramochi, Y., Sandanayaka, A. S. D., Satake, A., Araki, Y., Ogawa, K., Ito, O., and Kobuke, Y. (2009) *Chem. Eur. J.*, **15**, 2317–2327.

efficient photoinduced electron transfer from the excited singlet state of bisethynylporphyrin to C_{60} units in this newly designed guest molecule. Photodynamic measurements gave consistent evidence. Transient absorption afforded a peak at 1020 nm, characteristic of the anion radical of the fulleropyrrolidine moiety, along with a broad peak at 400–900 nm, assignable to the cation radical of ZnP. From the comparison of fluorescence lifetimes of **52** and **53**, the electron-transfer rate constant in this donor–acceptor pair was estimated to be 8.5×10^9 s^{-1}.

The final step must be the successful incorporation of **53** into the cavity of **39** followed by excitation-energy transfer from macroring **39** to the ZnP moiety in **53** and succeeding processes. UV-vis and fluorescence titrations in benzonitrile suggested the incorporation of **53** into **39** with tripodal pyridyl coordination, because the 1:1 stability constant of 1.6×10^7 M^{-1} was similar to those with **52** (1.4×10^7 M^{-1}) and other tri- or tetrapodal pyridyl ligands. In the titration experiments with **52**, selective excitation of the macroring at 567 nm induced a decrease in the fluorescence from the macroring at 623 nm, with a concomitant increase in the fluorescence from the ZnP part in **52**, indicating the excitation-energy transfer from

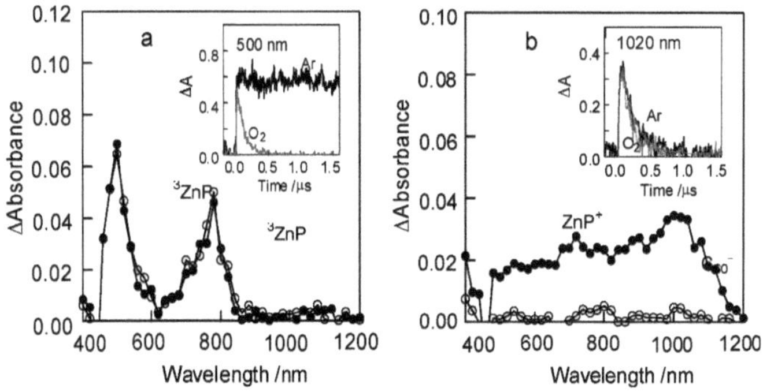

Figure 6.31. Nanosecond transient absorption spectra of the composite formed by mixing the macroring with (a) ZnP–Tripod or (b) C_{60}–ZnP–Tripod in benzonitrile. The spectra were recorded at 0.1 μs (•) and 1.0 μs (○) after predominant irradiation at the macroring. Inset: absorption time profile at (a) 500 nm and (b) 1020 nm in Ar- and O_2-saturated benzonitrile. Reprinted with permission from Kuramochi, Y., Sandanayaka, A. S. D., Satake, A., Araki, Y., Ogawa, K., Ito, O., and Kobuke, Y. (2009) *Chem. Eur. J.*, **15**, 2317–2327.

39 to the Zn–P unit. The efficiency was determined to be 80%. The energy-transfer rate constant was determined from the fluorescence decay process to be 4.3×10^9 s^{-1}, which is consistent with the calculated rate constant based on the Förster-type through-space mechanism.

When titrated with **53**, only a fluorescence decrease from the macroring was observed without emission from the ZnP part. The fluorescence quenching efficiency (83%) by **53** was close to the energy-transfer efficiency by **52** (80%), suggesting that the dominant quenching mechanism in the 1:1 composite of **53** and **39** is energy transfer from **39** to the ZnP part in **53** followed by electron transfer from the excited ZnP to the C_{60} part. The fluorescence decay rate constant for this composite (5.0×10^9 s^{-1}) from time-resolved fluorescence spectra was similar to the **39–52** composite, again supporting energy transfer to the ZnP moiety as the predominant quenching route. At the same time, only the decay of the macroring was detected without a rise in the fluorescence of ZnP. Therefore, electron transfer from ZnP to C_{60} is faster than energy transfer from the macroring to ZnP [73].

Comparison of the time-resolved absorption spectra of **52–39** and **53–39** composites demonstrated a clear difference (Fig. 6.31). In the former case, two sharp peaks appeared at 500 and 780 nm, which were similar to those observed for **52** alone and ascribed to the excited triplet state of the ZnP moiety. In this case, energy transfer from the macroring to the ZnP moiety is followed by intersystem crossing to the triplet state. In the latter case, the characteristic peak of the C_{60} radical anion at 1020 nm was accompanied by bands in the 400–900 nm region, as expected for the cation radical of ZnP. Lack of observation of the triplet excited state suggests that the excitation energy transferred from the macroring to ZnP induces photoinduced electron transfer from the excited singlet state of ZnP to C_{60} with high efficiency. Unfortunately, the electron-transfer rate in the C_{60}–ZnP-Tripod/Macroring **53–39** could not be determined by direct excitation of the ZnP moiety because of the instability of the laser power at wavelengths required for selective excitation at the ZnP moiety [74]. However, the rate constant determined for C_{60}–ZnP–Tripod **53** itself, 8.5×10^9 s^{-1}, may be a reasonable measure. On the other hand, the decay profile at 1020 nm determined the charge-recombination rate constant as 5.0×10^6 s^{-1}. All of the rate constants for energy and electron-transfer reactions involved in this antenna–reaction center model are listed in Table 6.4, along with their quantum efficiencies.

In summary, we could present a total photosynthetic model combining light harvesting and charge separation processes as shown in Fig. 6.32. The light energy collected by a macroring composed

Table 6.4. Energy-Transfer Rate Constants (k_{En}) and Efficiency (Φ_{En}), and Charge-Separation and Charge-Recombination Rate Constants (k_{CS} and k_{CR}), and Quantum Yields of Charge-Separation (Φ_{CS})

		39 → ZnP		ZnP → C_{60}		
Compound		k_{En}/s^{-1}	Φ_{En}	k_{CS} /s^{-1}	Φ_{CS}	k_{CR} /s^{-1}
C_{60}– ZnP-Tripod	**53**	–	–	8.5×10^9	0.94	3.4×10^6
ZnP-Tripod	**52/39**	4.3×10^9	0.90	–	–	–
C_{60}–ZnP-Tripod	**53/39**	5.0×10^9	0.91	a	a	5.0×10^6

[a]Not determined.

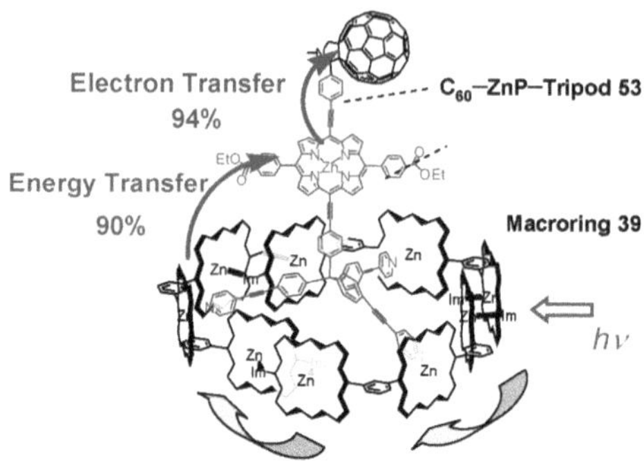

Figure 6.32. Energy transfer from the antenna ring to the acceptor porphyrin followed by electron transfer to the fullerene unit. See also Color Insert.

of nine Znporphyrins was transferred within the ring and then to the central acceptor porphyrin, ZnP, which is stably bound in 1:1 composition even in micromolar concentrations by tripodal cooperative coordination to the macroring Znporphyrin. It is remarkable that simple mixing of the components in micromolar concentrations leads to the formation of the desired three-dimensional architecture. Photoinduced charge separation occurs from Znporphyrin to the C_{60} moiety. The charge-recombination rate is slower than the forward charge separation rate by a factor of ca. 2000.

6.8 Interring Excited Energy Transfer

On considering the light-harvesting system, not only energy transfer in an individual ring but also interring energy-transfer processes, LH2–LH2 and LH2–LH1, are important. However, construction of such a system is much more difficult compared with the case of an isolated ring. In a biomimetic approach, it is desirable that some, preferably definite, structural principle be extracted from nature, but no such information is available for the interring cases. In spite of this, surprisingly short interring energy-transfer times, such as 3–5 ps for LH1–LH2 and 3–10 ps for LH2–LH2 have been reported [17–21].

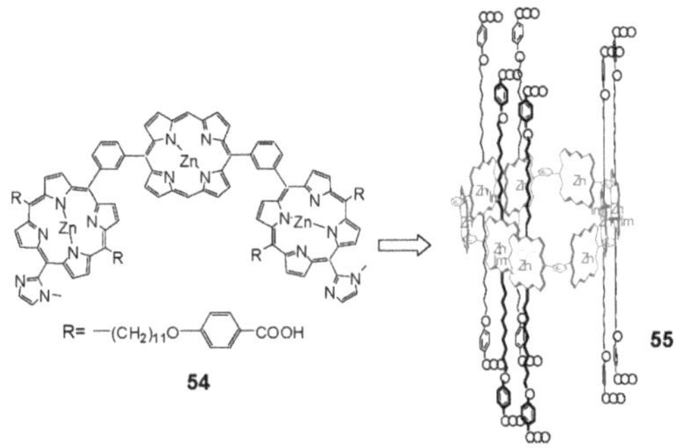

Figure 6.33. Membrane incorporation model of light-harvesting antenna. See also Color Insert.

As the first approach, we tried to assemble macrorings in a membrane through weak molecular interactions in a lipid. The second approach employed hydrogen bonding to assemble macrorings into a somewhat discrete nanotube structure. To incorporate the cyclic assembly from trisporphyrin into a bilayer membrane of a liposome, four amphiphilic *p*-benzoyloxyundecanyl groups were substituted at the two *meso*-positions of each terminal porphyrin **54** [75]. The molecular design intends that a total of 12 carboxylate anions in the macroring **55** may locate near the membrane surface after placing the central porphyrin moiety in the inner core of the membrane with an orientation of the plane perpendicular to the membrane surface (Fig. 6.33). Intermolecular forces operative in the membrane may tend to isolate macrorings composed of a large π-electronic plane from simple alkyl chains of the lipid component, thereby assembling macrorings in the membrane.

Investigation of the absorption and fluorescence spectra suggested that the macroring **55** was incorporated quantitatively in the small unilamellar vesicle membrane keeping the trimeric structure as suggested in the solution studies at relative **55**/lipid mole ratios of 1/1500 and 1/7500. Then, the energy-transfer process was monitored by fluorescence quenching using a tripodal pyridyl ligand bearing the C_{60} energy/electron-acceptor unit **42**. The relative fluorescence intensities of the macroring **55** at various quencher/macroring **42**/**55** ratios are shown in Fig. 6.34 in comparison with the plot undertaken in a homogeneous $CHCl_3$ solution.

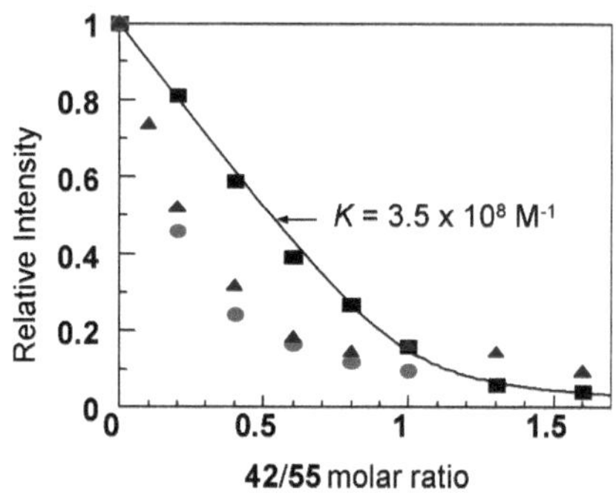

Figure 6.34. Fluorescence quenching of the macroring on titration with tripodal ligand **42** in SUV at a mole ratio of macroring **55**/lipid: 1/1500 (•), 1/7500 (▲), compared with that in $CHCl_3$ solution (■).

For a 0.2 mole ratio, the quenching efficiency was 54% in the lipid case 1/1500. The efficiency is threefold compared with 19% in a homogeneous solution. Because 1:1 complexation between macroring and quencher occurs similarly in the bilayer membrane with $K = 3.5 \times 10^8$ M^{-1}, this means that three macrorings can transfer energy to one another within their fluorescence lifetime and that the trapping of one quencher into one of the macrorings quenches the other two simultaneously. When the macroring concentration was reduced to 1/7500, the quenching efficiency was slightly decreased because the mean number of macrorings capable of rapid energy transfer was reduced.

The overall diagram involving interring energy-transfer quenching is presented in Fig. 6.35. Formally, the construction of LH1–LH2–reaction center composite is successfully demonstrated. However, the number of LH2 participating the energy-transfer event is small because of slow interring energy-transfer rates and/or the lack of appropriate mediator connecting antenna rings.

6.9 Organic Photovoltaic Cell

Dye-sensitized and organic thin film solar cells are interesting fields of application where artificial photosynthetic devices such as light harvesting, charge separation, hole and electron transport are

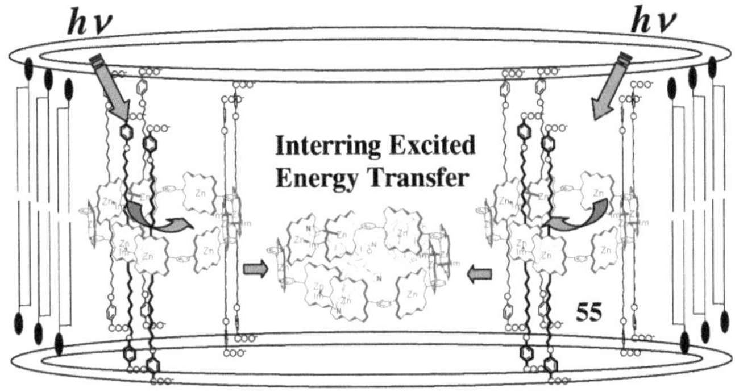

Figure 6.35. Schematic diagram of interring excitation-energy transfer among antenna rings in a bilayer membrane. See also Color Insert.

tested for their usefulness. A self-assembled monolayer (SAM) on a solid electrode or semiconductor has been used in principle to achieve efficient photocurrent generation. However, any available chromophores are not sufficient to absorb the incoming light efficiently by a single layer, and solid materials having large surface areas, usually 1000 times larger than a flat substrate, must be used to overcome the problem. This solution is associated with a hard and rigid substrate. At the same time, this choice is reasonable considering the limitation of organic materials, which are inferior to semiconductors in exciton and hole-migration properties, and therefore the charge separation takes place better at the substrate surface without transferring exciton and hole. This concept is different from photosynthesis, which employs efficient exciton migration by a light-harvesting system, as described in the previous sections.

The light-harvesting system developed here is capable of transferring excitons a few tens of nanometers without loss and it is interesting to explore the possibility in photovoltaic cells. The first approach was undertaken by sequential elongation of the porphyrin array starting from imidazolylporphyrin attached to a gold surface by a thiolate substituent [76–80]. Complementary coordination of *meso–meso*-linked bisimidazolylporphyrin was undertaken by repeated cycles, as shown in Fig. 6.36(a): (1) soaking of porphyrin solution in CH_2Cl_2 with MeOH to dissociate the otherwise extensively developed coordination, (2) rinsing with CH_2Cl_2 to

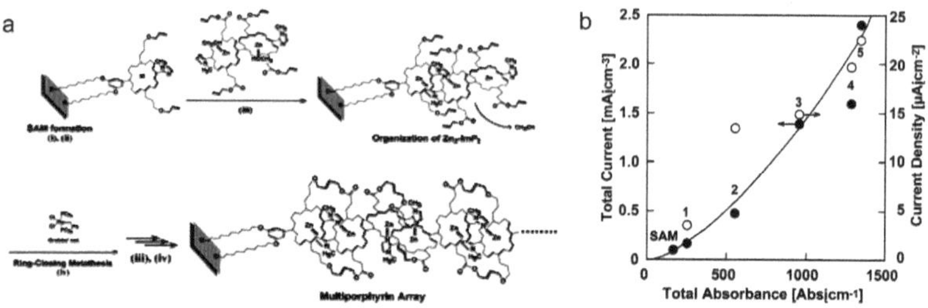

Figure 6.36. (a) Multiporphyrin array formation on a Au surface and metathesis linking. (b) Dependence of photocurrent on surface grafting.

organize the imidazolyl-to-Zn(II) coordination, (3) soaking in a dilute CH_2Cl_2 solution of Grubbs catalyst to immobilize the allylether side chains by a ring-closing metathesis reaction, and (4) successive washing with excess CH_2Cl_2, MeOH, and H_2O, followed by drying. This series of procedures was repeated until multiporphyrin layers were accumulated, as detected by the gradual increase in the characteristic porphyrin absorption. Photocurrent measurement (Fig. 6.36b) in aqueous solution containing methylviologen showed a small initial increase for the first two accumulation cycles, but significant increments for the last three, in accord with strong quenching of the excited singlet states of Znporphyrin in the vicinity of the gold surface. The action spectrum resembles those of porphyrin-assembled electrodes over the whole range of 350–750 nm. Photocurrent quantum efficiencies at the peak maximum (480 nm) were 0.4% after five deposition cycles. The significant increase in the photocurrent on accumulated deposition indicates that the multiporphyrin array works as a light-harvesting antenna on the gold electrode.

Organic thin film solar cells have received increasing attention as the newest class of photovoltaic cell that enables simple processing of charge separating, hole and electron transporting materials between electrodes without the use of a liquid electrolyte. Bulk-heterojunction-type solar cells employ a phase-separated blend of electron donor and acceptor [81–84]. To improve the photon/power conversion efficiency, an efficient electron-donor/acceptor combination is of central importance. As a candidate for such donor material, the polymeric porphyrin array developed here seems attractive

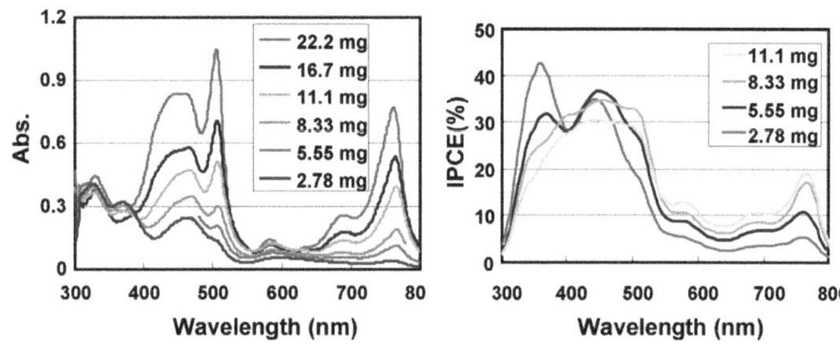

Figure 6.37. (a) Absorption spectra and (b) IPCE from the butadiynylene-linked bisporphyrin array **31** at various concentrations. See also Color Insert.

in view of (1) the coverage of a wide absorption range of visible light, (2) spontaneous formation of a complementary coordination polymer from soluble monomeric porphyrin, (3) no quenching of the singlet excited state on polymer formation, (4) very long (a few tens of nanometers) exciton diffusion length, probably associated with excellent hole transport properties, (5) excellent charge-separation properties, (6) facile synthetic modification to control HOMO/LUMO levels, solubility, and others.

Butadiynylene-linked bis(*N*-methylimidazolyl-Znporphyrin) **31** was used in combination with PCBM as an active layer [58]. The absorption spectra and incident-photon-to-electron conversion efficiency (IPCE) of various concentrations are illustrated in Fig. 6.37 [85]. The coverage of a wide absorption range and moderate IPCE values are promising for further investigation.

6.10 Phthalocyanine

Phthalocyanine shows a much higher absorption band in the Q-band region at longer wavelength than porphyrin because of the nondegenerate Q_x and Q_y bands. In a similar manner to porphyrin, the *N*-methylimidazolyl substituent was introduced at one of the β-carbons of phthalocyanine [86]. A very stable complementary coordination dimer of phthalocyanine was obtained on Zn(II) and Mg(II) metal introduction with stability constants of 1.4×10^{11} M^{-1} and 1.1×10^{12} M^{-1}, respectively, for the dimerization. These large stability

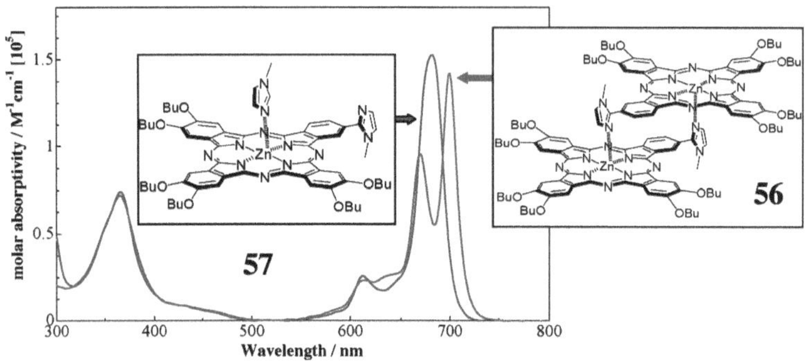

Figure 6.38. Absorption spectra of phthalocyanine dimer and monomer.

constants reflect the slipped cofacial interaction of the large π-electronic plane of phthalocyanine.

The absorption spectrum of the dimer **56** shows split Q bands at 669.5 and 699 nm compared with 679 nm for the monomer **57** (Fig. 6.38). The apparent splitting energy of 630 cm^{-1} is quite large considering the Q-band region absorption. The fluorescence quantum yields of the dimers from Zn(t-Bu)$_3$-, Zn(OBu)$_6$-, and Mg(t-Bu)$_3$-phthalocyanines maintained the high values of phthalocyanine, being 0.26, 0.45, and 0.76, respectively. These values for the dimer were almost the same as those for the monomer, showing once again the merit of using complementary dimer formation. It is a remarkable property considering that, in most cases, metal-assisted organization of phthalocyanine in the past resulted in quenching of the singlet excitation energy because of face-to-face stacking interactions [87–90].

In another approach to the use of phthalocyanine as an antenna chromophore, it was introduced as a *meso*-substituent of *N*-methylimidazolyl-Znporphyrin. On forming the complementary dimer **58** from the porphyrin part, two phthalocyanine units were connected through the porphyrin dimer [91–93]. The absorption spectrum covers a wide visible region by the sum of two chromophores, because the porphyrin dimer is characterized by the large split Soret bands and phthalocyanine by Q bands with large extinction coefficients, both exceeding the extinction coefficient of natural chlorophylls as shown in

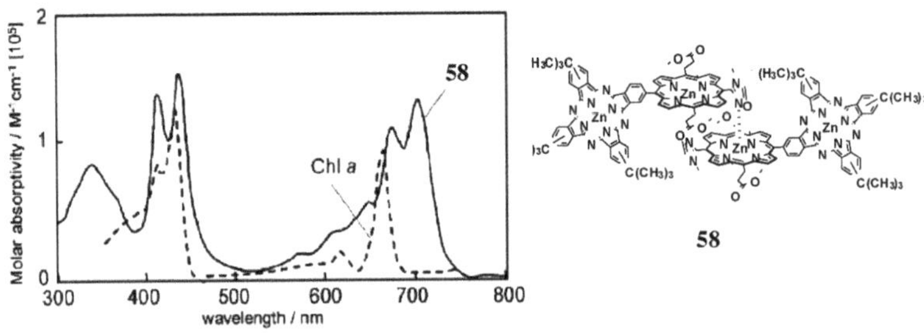

Figure 6.39. Comparison of absorption spectra of the complementary dimer from phthalocyanine-substituted porphyrin and chlorophyll *a*.

In Fig. 6.39, the energy excited at any absorption area falls into the S_1 state of phthalocyanine through efficient energy transfer and emits fluorescence with a quantum yield as high as 0.71. The overall feature of the united porphyrin–phthalocyanine system is characterized by an excellent antenna composite having broad and large absorption bands at both S_2 of porphyrin and S_1 of phthalocyanine and, furthermore, by a high fluorescence quantum yield from the lowest excited state of phthalocyanine.

6.11 Conclusions

The structure of light-harvesting antenna complexes from photosynthetic purple bacteria was elucidated as assemblies of bacteriochlorophylls in a barrel form. Their structure is produced by coordination of an imidazolyl side chain from transmembrane helices arranged in a doubly cylindrical fashion inside and outside the barrel. The principle of this structure formation is identical to that of the special pair in the photosynthetic reaction center. Keeping these points in mind, complementary coordination of imidazolyl-to-Zn in *N*-methylimidazolyl-Znporphyrin was designed to create artificial reaction center and antenna complexes. This methodology produced important structural characteristics: very large stability constants for dimerization, amounting to 10^{11} M^{-1} in non-polar solvents, and slipped cofacial structure, in which the porphyrin π-orbitals are coplanar and partially overlap each other leading to split Soret bands. Connection of two such units leads to macrorings of various sizes and a linear array of planar or orthogonal dimers depending on the connecting bond unit. The large stability constants for the dimerization afforded stable macroring products and polyporphyrin arrays of huge molecular weights. At the same time, addition of coordinating solvents destroyed the structure partially or totally depending on the nature and the amount, whereas elimination of the solvent regenerated the coordination structure. This property allowed reorganization of the formed structures leading to exclusive formation of the macroring or modification of terminal porphyrin units. The photophysical properties of macrorings and linear arrays constructed from the complementary coordination dimer units were characterized most typically by "no loss of excitation energy" on formation of the assembly. The fluorescence quantum yields and fluorescence lifetimes were almost identical to, or even larger than, the parent monomer. The excitation-energy hopping between neighboring porphyrin dimers occurred through the Förster mechanism with time constants of 5–20 ps, showing 100–200 times energy migration in the rings within the excited state lifetimes. In the case of a linear porphyrin array, these properties allowed excitation-energy transfer from the site separated by 45 bisporphyrin units, corresponding to 60 nm distance. Compared with the fast intraring energy-transfer rates, interring rates were relatively slow and only a few rings could be incorporated in the energy-transfer event during the excited state lifetime. A tripodal pyridyl ligand appended with another porphyrin of a lower energy level and fullerene units was introduced into the inner cavity of the macroring to construct an LH1-reaction center mimic. The selectively excited antenna porphyrin transferred the energy to the inner porphyrin, where charge separation took place, thereby presenting a formal mimic of the antenna LH1–reaction center composites. Furthermore, an LH2–LH1–reaction center mimic was demonstrated by incorporating the macroring into a bilayer membrane, where interring energy transfer followed by energy/electron transfer occurred to the tripodal acceptor ligand, thereby presenting the formal LH2–LH1–reaction center composite system. The antenna effect was demonstrated in a photovoltaic cell using complementary imidazolyl-to-Zn dimer formation. Phthalocyanine was also suggested to be an excellent antenna chromophore by its own coordination dimerization motif or by conjugation with porphyrin.

Acknowledgments

Naturally, this work came about as a result of collaboration with many people. I would like to thank all of the people who collaborated in this whole series of research. This research was initiated by Dr. Hidekazu Miyaji who, as a fresh student, opened the door to me by proposing the first research in this field. Complementary coordination of imidazolyl-to-Zn(II) became our tool throughout the photosynthetic work. The second big barrier appeared in organizing bisimidazolyl-Znporphyrin into a macrocyclic hexamer. The synthesis of bisporphyrin was an extremely tough problem. Dr. Ryo-ichi Takahashi overcame the difficult task by scrutinizing all the points and trying to find solutions with never-ending enthusiasm in chemistry. The structure proof was also a tough job. X-ray small angle diffraction analysis, kindly proposed by Prof. Mikio Kataoka (next door laboratory at NAIST), was successfully applied. Not only GPC analysis but also covalent linking of the coordinated dimer pair by a metathesis reaction proposed by Assistant Prof. Akiharu Satake eliminated all of the ambiguous points in the structure proof. Before accomplishing the ring assemblies, a linear supramolecular coordination polymer of bisimidazolyl-Znporphyrin was obtained by a research associate, the later Assistant Prof. Kazuya Ogawa. Newly developed characterization methods along with reorganization techniques in this new field were indispensable for all of the later activities. The first energy-hopping process in the porphyrin macroring was analyzed by Prof. Dongho Kim (Yonsei University, Editor of this book), who clearly demonstrated the most important photophysical properties.

At this point, I regret to consider the page limitation for acknowledgments. I remember all of the collaborators in my photosynthetic work with highest respect and gratitude as memorable contributors.

Students in my laboratory: Mr. Jiro Kondo, Dr. Naoto Nagata, Mr. Motoki Murakami, Dr. Hidekane Ozeki, Ms. Aya Ryuno, Mr. Satoshi Morikawa, Dr. Kazuya Kameyama, Mr. Kazuhiro Kobayashi, Mr. Masataka Aikawa, Mr. Keiji Maruo, Mr. Akira Tanaka, Mr. Satoshi Momozawa, Mr. Shigeru Yamatsu, Dr. Hiromichi Nakagawa, Dr. Fatin Hajjaj, Ms. Sachiko Ushikusa, Mr. Jun-ichi Tanihara, Mr. Daisuke Furutsu, Dr. Yusuke Kuramochi, Ms. Saori Okada, Mr. Hirokazu Matsuda, Mr. Ryo Yokoyama, Mr. Takafumi

Fujita, Mr. Toshimasa Sugimura, Mr. Masakuni Fujita, Dr. Zafer Uyer, and Dr. Kaori Fujisawa.

Technician: Ms. Noriko Haruta.

Research associates: Dr. Akihiro Nomoto, Dr. Chusaku Ikeda, Dr. Atsushi Ohashi, Dr. Osami Shoji, Dr. Mitsuhiko Morisue, and Dr. Dipak Kalita.

Professors in other universities: Prof. Kei-taro Yoshihara (Japan Institute of Science and Technology, Energy Transfer), Prof. Hiroshi Miyasaka (Osaka University, Energy Transfer), Prof. Osamu Ito (Tohoku University, Electron Transfer), Prof. Tomoji Kawai (Osaka University, STM), and members in these laboratories.

Finally, I would like to acknowledge financial support from Nara Institute of Science and Technology, CREST (Creative Research for Evolutional Science and Technology, Prof. Hideki Sakurai as a Project Leader), Grants-in-Aid from the Ministry of Education, Culture, Sports, Science and Technology of the Japanese Government (Prof. Fumio Toda (Ehime University), Prof. Toyoki Kunitake (Kyushu University), Prof. Isao Saito (Kyoto University), and Prof. Tasuku Ito (Tohoku University) as Project Leaders involved in the research areas). Without these links, neither compounds nor characteristics would have been achieved.

References

1. McDermott, G. Prince, S. M., Freer, A. A., Hawthornthwaite-Lawless A. M., Papiz, M. Z., Cogdell, R. J., and Isaacs, N. W. (1995) *Nature*, **374**, 517–521.
2. Karrasch, S. Bullough, P. A., and Ghosh, R. (1995) *EMBO J.*, **14**, 631–638.
3. Jamieson, S. J., Wang, P. Qian, P. Kirkland, J. Y., Conroy, M. J., Hunter, C. N., and Bullough, P. A. (2002) *EMBO J.*, **21**, 3927–3935.
4. Jungas, C., Rananck, J.-L., Rigaud, J. L., Joliot, P., and Vermeglio, A. (1999) *EMBO J.*, **18**, 534–542.
5. Roszak, A. W., Howard, T. D., Southall, J., Gardiner, A. T., Law, C. J., Isaacs, N. W., and Cogdell, R. J. (2003) *Science*, **302**, 1969–1972.
6. Scheuring, S., Francia, F., Busselez, J., Melandri, B. A., Rigaud, J.-J., and Levy, D. (2004) *J. Biol. Chem.* **279**, 3620–3626.

7. Scheuring, S. Sturgis, J N. Prima, V., Bernadac, A. Lévy D. and Rigaud, J.-L. (2004) *Proc. Natl. Acad. Sci. U. S. A.*, **101**, 11293–11297
8. Bahatyrova, S., Frese, R. N., Siebert, C. A., Olsen, J. D., van derWerf, K. O. Van Grondelle, R. Niederman, R. A. Bullough, P A. Otto, C., and Hunter, C. N. (2004) *Nature*, **430**, 1058–1062
9. Scheuring, S. Lévy, D. and Rigaud, J.-L. (2005) *Biochim. Biophys. Acta*, **1712**, 109–127.
10. Green, B. R. and Parson, W. W. (eds.) (2003) Light-harvesting antennas in photosynthesis, in *Advances in Photosynthesis and Respiration* (ed. Govindjee), vol. 13, Kluwer Academic Press, Dordrecht, Netherland.
11. Andrews, D. L. (ed.) (2005) *Energy Harvesting Materials*, World Scientific, Hackensack, NJ.
12. Deisenhofer, J., Epp, O., Miki, K., Huber, R., and Michel, H. (1984) *J. Mol. Biol.*, **180**, 385–398.
13. Deisenhofer, J., and Michel, H. (1989) *Angew. Chem. Int. Ed., Engl.*, **28**, 829–847.
14. Kobuke, Y., and Miyaji, H. (1994) *J. Am. Chem. Soc.* **116**, 4111–4112.
15. Scheuring, S. Sturgis, J N. Prima, V., Bernadac, A. Lévy D. and Rigaud, J.-L. (2004) *Proc. Natl. Acad. Sci. U. S. A.*, **101**, 11293–11297
16. Robert, B., Cogdell, R. J., and van Grondelle, R. (2003) The light havesting system of purple bacteria, in *Light-Harvesting Antennas in Photosynthesis* (ed. Green, B. R., and Parson, W. W.), Chapter 5, Kluwer Academic Press, pp. 169–194.
17. Hu, X., Ritz, T., Damjanovic, A., and Schulten, K. (1997) *J. Phys. Chem. B*, **101**, 3854–3871.
18. Hess S., Akesson E., Cogdell R. J., Pullerits T., and Sundstrom V (1995) *Biophys J.*, **69**, 2211–2225.
19. Shreve A. P., Trautman J. K., Frank H. A., Owens T. G., and Albrecht A. C (1991) *Biochim Biophys Acta* **1058**, 280–300
20. Jimenez R., Dikshit S. N., Bradforth S. E., and Fleming G. R (1996) *J Phys Chem.*, **100**, 6825–6834.
21. Hess S., Chachisvilis M., Timpmann K., Jones M. R., Hunter C. N., and Sundstrom V (1995) *Proc Natl Acad Sci USA.*, **92**, 12333–12337.
22. Beekm, P., van Moan, L., Murik, F., Jones, M. R., Visser, H. M., Hunter, C. N., and van Grondelle, R. (1994) *Biochemistry*, **33**, 3143–3147.
23. Pšenčík, J. Ikonen, T. P. Laurinmäki, P. Merckel, M. C. Butcher, S. J. Serimaa R. E. and Tuma, R. (2004) *Biophys. J.*, **87**, 1165–1172.

24. Zouni, A. Witt, H.-T. Kern, J., Fromme, P., Kraus, N., Saenger W. and Orth, P. (2001) *Nature*, **409**, 739–744.
25. Jordan, P., Fromme, P., Witt, H. T., Klukas, O., Saenger, W., and Krauss, N. (2001) *Nature*, **411**, 909–917.
26. Liu, Z., Yan, H., Wang, K., Kuang, T., Zhang, J., Gui, L., An, X., and Chang, W. (2004) *Nature*, **428**, 287–292.
27. Nelson, N., and Ben-Shem, A. (2005) *Bioessays*, **27**, 914–922.
28. Kobuke, Y. (2006) *Structure & Bonding*, **121**, 49–104.
29. Satake, A., and Kobuke, Y. (2005) *Tetrahedron*, **61**, 13–41.
30. Kobuke, Y. (2006) *Eur. J. Inorg. Chem.*, 2333–2351.
31. Inaba, Y., Nomoto, A., and Kobuke, Y., unpublished result.
32. Satake, A., and Kobuke, Y. (2007) *Org. Biomol. Chem.*, **5**, 1679–1691.
33. Summers, J. S., and Stolzenberg, A. M. (1993) *J. Am. Chem. Soc.*, **115**, 10559–10567.
34. Imai, H., Nakagawa, S., Kyuno, E. (1992) *J. Am. Chem. Soc.*, 1992, **114**, 6719–6723.
35. Brewer, C. T., and Brewer, G. J. (1990) *Chem. Soc. Dalton Trans.*, 843–847.
36. Bhyrappa, P., Krishnan, V., and Nethaji, M. J. (1993) *Chem. Soc. Dalton Trans.*, 1901–1906.
37. Kasha, M., Rawls, H. R., and El-Bayyoumi, M. A. (1965) *Pure Appl. Chem.*, **11**, 371–392.
38. Takahashi, R., and Kobuke, Y. (2003) *J. Am. Chem. Soc.*, **125**, 2372–2373.
39. Takahashi, R., and Kobuke, Y. (2005) *J. Org. Chem.*, **70**, 2745–2753.
40. Svergun, D. I., Barberato, C., and Koch, M. H. J. (1995) *J. Appl. Crystallogr.*, **28**, 768–773.
41. Pickover, C. A., and Engelman, D. M. (1982), *Biopolymers*, **21**, 817–831.
42. Review: Trnka, T. M., and Grubbs, R. H. (2001) *Acc. Chem. Res.*, **34**, 18–29.
43. Ikeda, C., Satake, A., and Kobuke, Y. (2003) *Org. Lett.*, **5**, 4935–4938.
44. Ohashi, A., Satake, A., and Kobuke, Y. (2004) *Bull. Chem. Soc. Japan*, **77**, 365–374.
45. It might be better to use another compound name in addition to the systematic numbering for macrorings in this section by defining, for example C-P5 as C=metathesized, P=phenylene-bridged, and 5=5-mer. Other notations include N, EP as non-metathesized, diethynylphenylene-bridged, respectively.

46. Hajjaj, F., Yoon, Z. S., Yoon, M.-C., Park, J., Satake, A., Kim, D.-H., and Kobuke, Y. (2006) *J. Am. Chem. Soc.*, **128**, 4612–4623.
47. Fujisawa, K., Satake, A., Hirota, S., and Kobuke, Y. (2008) *Chem. Eur. J.* **14**, 10735–10744.
48. Shoji, O., Okada, S. Satake, A. and Kobuke, Y. (2005) *J. Am. Chem. Soc.*, **127**, 2201–2210.
49. Chi, X., Guerin, A. J., Haycock, R. A., Hunter, C. A., and Sarson, L. D. (1995) *J. Chem. Soc. Chem. Commun.*, 2563–2565.
50. Tanaka H., and Kawai T (1997) *J Vac Sci Technol B* **15**, 602–604.
51. Tanaka H., and Kawai T (2003) *Surf Sci.*, **539**, L531–L536.
52. Shoji, O., Tanaka, H., Kawai, T., Kobuke, Y. (2005) *J. Am. Chem. Soc.*, **127**, 8598–8599.
53. Satake, A., Tanaka, H., Hajjaj, F., Kawai, T., and Kobuke, Y. (2006) *Chem. Commun.*, 2542–2544.
54. Kobuke, Y., and Miyaji, H. (1996) *Bull. Chem. Soc. Japan*, **69**, 3563–3569.
55. Nagata, N., Kugimiya, S.-I., Fujiwara, E.-I., and Kobuke, Y. (2004) *Trans. Materials Res. Soc. Japan*, **29**, 911–914.
56. Nagata, N., Kugimiya, S., Fujiwara, E., and Kobuke, Y. (2003) *New J. Chem.*, 743–747.
57. Ogawa, K., and Kobuke, Y. (2000) *Angew. Chem. Int. Ed.*, **39**, 4070–4073.
58. Satake, A., Fujita, M., Kurimoto, Y., and Kobuke, Y. (2009) *Chem. Commun.*, 1231–1233.
59. Morikawa, S., Ikeda, C., Ogawa, K., and Kobuke, Y. (2004) *Lett. Org. Chem.*, **1**, 6–11.
60. Hwang, I.-W., Ko, D. M., Ahn, T. K., Kim, D.-H., Ito, F., Ishibashi, Y., Khan, S. R., Nagasawa, Y., Miyasaka, H., Ikeda, C., Takahashi, R., Ogawa, K., Satake, A., and Kobuke, Y. (2005) *Chem. Eur. J.*, **11**, 3753–3761.
61. Bradforth, S. E., Jimenez, R., van Mourik, F., van Grondelle, R., Fleming, G. R. (1995) *J. Phys. Chem*, **99**, 16179–16191.
62. Trinkunas, G. Herek, J. L. Polívka, T. Sundström, V., Pullerits, T. (2001) *Phys. Rev. Lett.* **86**, 4167–4170
63. Brüggemann, B., and May, V. (2004) *J. Chem. Phys*, **120**, 2325–2336.
64. Müller, M. G., Hucke, M. Reus, M. and Holzwarth, A. R. (1996) *J. Phys. Chem*, **100**, 9537–9544.

65. Brüggemann, B., Herek, J. L., Sundstrcm, V., Pullerits, T., and May, V. (2001) *J. Phys. Chem. B*, **105**, 11391–11394.
66. Hajjaj, F., Yoon, Z. S., Yoon, M.-C., Park, J., Satake, A., Kim, D.-H., and Kobuke, Y. (2006) *J. Am. Chem. Soc.*, **128**, 4612–4623.
67. Furutsu, D., Satake, A., and Kobuke, Y. (2005) *Inorg. Chem.*, **44**, 4460–4462.
68. Kuramochi, Y., Satake, A., and Kobuke, Y. (2004) *J. Am. Chem. Soc.*, **126**, 8668–8669.
69. Kuramochi, Y., Satake, A., Itou, M., Ogawa, K., Araki, A., Ito, O., and Kobuke, Y. (2008) *Chem. Eur. J.*, **14,** 2827–2841.
70. Trachenko, N. V., Lemmetyinen, H., Sonoda, J., Ohkubo, K., Sato, T., Imahori, H., and Fukuzumi, S. (2003) *J. Phys. Chem. A*, **107**, 8834–8844.
71. Uyar, Z., Satake, Y., Kobuke, Y., and Hirota, S. (2008) *Tetrahedron Letters*, **49**, 5484–5487.
72. Kuramochi, Y., Sandanayaka, A. S. D., Satake, A., Araki, Y., Ogawa, K., Ito, O., and Kobuke, Y. (2009) *Chem. Eur. J.*, **15**, 2317–2327.
73. The slipped-cofacial dimeric and pyridyl-coordinated monomeric porphyrin units of **39** have similar lowest excited energy levels and are expected to transfer rapidly the excitated energy among the porphyrin units as reported for macrorings composed of solely dimeric units, C-P5 (**6**) and C-P6 (**7**), where hopping times of 8.0 and 5.5 ps are obtained, respectively.
74. The difficulty came from the instability of the laser power at ~460 and 600 nm, wavelengths that are required for selective excitation at the central ZnP in **53**/**39** composite.
75. Nagata, N., Kuramochi, Y., and Kobuke, Y. (2009) *J. Am. Chem. Soc.*, **131**, 10–11.
76. Nomoto, A., and Kobuke, Y. (2002) *Chem. Commun.*, 1104–1105.
77. Nomoto, A., Mitsuoka, H., Ozeki, H., and Kobuke, Y. (2005) *Chem. Commun.*, 1074–1075.
78. Morisue, M., Yamatsu, S., Haruta, N., and Kobuke, Y. (2005) *Chem. Eur. J.*, **11**, 5563–5574.
79. Morisue, M., Haruta, N., Kalita, D., and Kobuke, Y. (2006) *Chem. Eur. J.*, **12**, 8123–8135.
80. Morisue, M., Kalita, D., Haruta, N., and Kobuke, Y. (2007) *Chem. Comm.*, **23**, 2348–2350.

81. Peet, J., Heeger, A. J. and Bazan, G. C. (2009) *Acc. Chem. Res.*, **42**, 1700–1708.
82. Chen, J., and Cao, Y. (2009), *Acc. Chem. Res.*, **42**, 1709–1718.
83. Inganäs, O., Zhang, F., and Andersson, M. R. (2010) *Chem. Soc. Rev.*, **39**, 1731–1739.
84. Heeger, A. J. (2010) *Chem. Soc. Rev.*, **39**, 2354–2371.
85. Yoshikawa, O., Yoshikawa, S., and Kobuke, Y. unpublished results.
86. Kameyama, K., Morisue, M., Satake, A., and Kobuke, Y. (2005) *Angew. Chem. Int. Ed.*, **44**, 4763–4766.
87. Kaneko, Y., Arai, T., Tokumaru, K., Matsunaga, D., and Sakuragi, H. (1996) *Chem. Lett.* 345–346
88. Ishii, K., Abiko, S., Fujitsuka, M., Ito, O., and Kobayashi, N. (2002) *J. Chem. Soc. Dalton Trans.* 1735–1739
89. Li, X.-Y. and Ng, D. K. P. (2001) *Tetrahedron Lett.* **42**, 305–309
90. Kimura, M., Kuroda, T., Ohta, K., Hanabusa, K., Shirai, H., and Kobayashi, N. (2003) *Langmuir* **19**, 4825–4830.
91. Kameyama, K., Satake, A., and Kobuke, Y. (2004) *Tetrahedron Letters,* **45**, 7617–7620.
92. Morisue, M., and Kobuke, Y. (2008) *Chem. Eur. J.*, **14**, 4993–5000.
93. Ito, F., Ishibashi, Y., Khan, S. R., Miyasaka, H., Kameyama, K., Morisue, M., Satake, A., Ogawa, K., and Kobuke, Y. (2006) *J. Phys. Chem., A,* **110**, 12734–12742.

Chapter 7

Light Harvesting, Excitation Energy/Electron Transfer, and Photoregulation in Artificial Photosynthetic Systems

Yuichi Terazono, Thomas A. Moore, Ana L. Moore, and Devens Gust
Department of Chemistry and Biochemistry, Center for Bioenergy and Photosynthesis, and Center for Bio-Inspired Solar Fuel Production, Arizona State University, Tempe, AZ 85287 USA
gust@asu.edu

7.1 Introduction

This chapter summarizes recent research from our laboratory on multichromophoric molecules that perform as components of artificial photosynthetic systems. The basic scientific principles underlying the design of these molecules were drawn from natural photosynthetic energy conversion, but the materials used to prepare them are synthetic. The systems consist of chromophores and electron and energy donors and acceptors that have been chemically linked to form artificial reaction centers and antennas, and to mimic some aspects of photosynthetic regulation and photoprotection.

Multiporphyrin Arrays: Fundamentals and Applications
Edited by Dongho Kim

These chromophores include porphyrins, fullerenes, carotenoid polyenes, antenna moieties, and photochromic materials. We begin with an introduction to artificial photosynthesis and a simple example. More complex systems including multichromophore arrays and systems that incorporate photoregulatory elements are then discussed. We have chosen examples based on a single organic framework—hexaphenylbenzene—as an organizing unit for the various chromophores, donors, and acceptors.

7.2 Natural and Artificial Photosynthesis

Although sunlight illuminates the earth with an abundance of energy, it is diffuse ($\leq$1000 W/m^2). To be useful to humans and other life forms, sunlight must be converted to other forms of energy. When living organisms began to obtain their energy from sunlight, they faced both these problems, and solved them. Nature's solution is photosynthesis. Photosynthesis provides most of the energy used by organisms and, through fossil fuels, is the source of most of the technological energy used by humans. Burning fossil fuels is not a sustainable energy source, but sunlight can in principle fill all the energy needs of humanity many times over. Realizing the promise of solar energy requires new technologies to convert solar energy to electrical and chemical energy inexpensively, efficiently, and safely. Photosynthesis can provide blueprints for this new technology—blueprints for artificial photosynthesis.

The idea of using the basic science underlying photosynthesis to design systems for artificial photosynthesis is not a new one [1]. However, in recent years much progress has been made. This is due in part to the work of biologists and biochemists who are uncovering the intimate details of how the natural process works. Before giving examples of artificial photosynthetic molecules, we briefly discuss the natural system.

Photosynthesis begins with the absorption of sunlight, and this is carried out almost entirely by photosynthetic antennas. These antennas are arrays of chromophores that are designed to efficiently capture light in the visible spectral region, rapidly and efficiently transfer the resulting excitation energy to other antenna molecules, and ultimately move excitation to the photosynthetic reaction center. A variety of antenna structures are known, and the type and arrangement of chromophores is determined to a large extent by where the organism lives. Different environments provide different qualities of light (intensity, range of wavelengths), and therefore demand different antennas. Many antenna systems consist of chlorophylls and their relatives. The absorption properties of chlorophylls may be tuned over a fairly broad range by structural changes and by changes in the surrounding protein environment, but their absorption still does not effectively cover the entire useful solar spectrum. Therefore, organisms use chromophores such as carotenoid polyenes, phycobilins and phycocyanins as antennas for some wavelengths. A variety of antenna architectures are also found, ranging from cyclic, protein-associated structures lying in the photosynthetic membranes to rod-like structures that are almost pure pigment. [2] This variability in structure provides fertile ground for those searching for inspiration for artificial antenna design.

No matter what the chromophore composition of an antenna, absorbed energy must be transferred between chromophores and ultimately out of the antenna unit. Photosynthesis uses singlet excitation energy, which wastes less of the energy stored in the photon than triplet excitation. However, unlike triplet states, singlet states are short lived, and decay by the unimolecular processes of internal conversion, intersystem crossing, and fluorescence. Thus, the energy transfer processes must be fast enough to compete effectively with this relaxation. When the chromophores are separated by tens of Å singlet-singlet energy transfer is usually adequately described by the Förster theory [34] but at very short distances more sophisticated theories must be used [5–9].

Photosynthetic antennas not only absorb sunlight and move the excitation energy to reaction centers, but also perform important photoregulatory and photoprotective functions [10–14]. These are necessary because under some conditions, excitation energy is delivered to reaction centers more rapidly than the high-energy chemical species generated by reaction centers can be used by the organism. We discuss photoprotection and the possible need for similar artificial processes later in this review.

The photosynthetic reaction center receives excitation energy from the antennas and performs the vital function of converting excitation to electrochemical potential. This is done by the process of photoinduced electron transfer. An excited state chlorophyll electron donor transfers an electron to an adjacent chlorophyll acceptor, creating an initial charge-separated state that preserves some of the photon energy as chemical energy. The lifetime of this initial state is very short, as the electron can readily hop back to the now-positively charged chlorophyll. A short lifetime does not permit utilization of the energy in the charge-separated state to power the processes in the organism that produce useful biological energy. Therefore, reaction centers rapidly move the electron away from the initial acceptor and through a series of additional donor–acceptor species to generate a final charge-separated state in which the electron and positive charge are spatially well separated. This multistep electron transfer reduces the electronic coupling interaction between the two radical ions to a point where recombination occurs too slowly to interfere with the ensuing chemistry of energy conversion. Each of the electron transfer steps is, of course, exergonic, but this is the energetic price paid by the reaction center to slow recombination to yield a ground or excited state of the initial donor and waste the harvested energy.

The photosynthetic reaction center is thus a kind of nanoscale photovoltaic device, in that it converts light into electrochemical potential energy. This energy is used by the later "dark" reactions of photosynthesis. The nature of these dark reactions depends to a degree on the organism in question. For example, oxygenic photosynthetic organisms use the oxidizing potential of the reaction center chlorophyll donor to oxidize water to molecular oxygen and hydrogen ions. The reduction potential of the transferred electron is used in part to generate reducing equivalents stored in NADH and capable of reducing carbon dioxide to carbohydrate or other kinds of stored energy. Biology has developed excellent (but not perfect) enzymatic catalysts for these kinds of processes.

A useful theoretical framework for understanding electron transfer is provided by the work of Marcus, Hush, and Levich [15–17], and elaborated by many others. The major factors that control electron transfer rates are thermodynamic driving force, electronic coupling, and reorganization energy.

7.3 Artificial Photosynthetic Reaction Centers

Using photosynthesis as a guide, it is clear that the minimum requirements for an artificial photosynthetic reaction center are an electron donor that absorbs light in the visible region, an energetically suitable electron acceptor, and an electron coupling interaction between them that allows photoinduced electron transfer to compete with decay of the excited donor by other pathways. Typical early artificial reaction centers consisted of porphyrins (P) covalently linked to quinone (Q) electron acceptors [18–21]. Excitation of the porphyrin generates the first excited singlet state ^{1}P-Q, which undergoes photoinduced electron transfer to yield a $P^{\bullet+}$-$Q^{\bullet-}$ charge-separated state. These states can be formed in high quantum yield and preserve a large fraction of the photon energy as electrochemical potential. However, they recombine in a few hundred ps or less, making the harvesting of the stored energy by other chemical processes impractical.

Turning again to the photosynthetic blueprint, a useful way to prolong the lifetime of charge separation is to incorporate additional donors or acceptors into the artificial reaction center. The earliest example of a donor–porphyrin–acceptor artificial reaction center is a carotenoid (C) porphyrin (P) quinone triad (C-P-Q) [22, 23]. Excitation of the porphyrin of this molecule yields C-^{1}P-Q, which evolves into C-$P^{\bullet+}$-$Q^{\bullet-}$. Competing with recombination of this state is electron donation from the carotenoid to yield $C^{\bullet+}$-P-$Q^{\bullet-}$, which has a lifetime of hundreds of ns. This multistep electron transfer strategy for generating long-lived charge-separated states has been used in many other systems, including those with additional donor and acceptor species [21, 24–31]. As an extreme example, excitation of a molecule consisting of a linear assembly of carotene, two porphyrins and two quinones produces a final charge-separated state with a lifetime of 55 μs and a quantum yield of 83% [32].

7.3.1 *Triads Containing Fullerene, C-P-C_{60}*

Since the first report of a covalently linked porphyrin-fullerene [33], fullerenes have become very popular as electron acceptors in artificial reaction centers because of their convenient range of

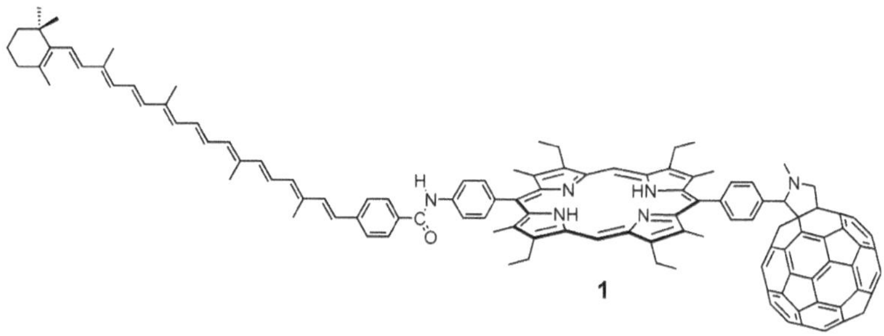

1

Figure 7.1. Carotene-porphyrin-fullerene triad 1.

redox potentials, their low solvent reorganization energies for electron transfer, and the low sensitivity of their ions to solvent effects [29, 33–43]. The first triad using this architecture is shown in Fig. 7.1 [35].

In 2-methyltetrahydrofuran solution **1** undergoes photoinduced electron transfer to yield C-P$^{\bullet+}$-C$_{60}^{\bullet-}$, which evolves into C$^{\bullet+}$-P-C$_{60}^{\bullet-}$ with an overall quantum yield of 0.14. This state decays by charge recombination to yield the carotenoid triplet state with a time constant of 170 ns. Even in a glass at 77K, C$^{\bullet+}$-P-C$_{60}^{\bullet-}$ is formed with a quantum yield of $\sim$0.10 and again decays mainly by charge recombination to give ^{3}C-P-C$_{60}$. Subsequent C-P-C$_{60}$ triads have quantum yields approaching unity [44].

Triad **1** differs from the quinone-based analogs in that it undergoes charge recombination to yield a triplet state, rather than the ground state, and charge separation still occurs in glassy solvents at low temperatures. These phenomena are due to the low reorganization energies of the three chromophores and the fact that the energies of the charge-separated states are relatively insensitive to the solvent dielectric constant. In the quinone systems mentioned above, the decrease in solvent dielectric constant that occurs upon freezing destabilizes the charge-separated states to the point that photoinduced electron transfer does not occur. The low-temperature electron transfer and charge recombination to a triplet observed for **1** are phenomena that are also observed in natural photosynthetic reaction centers, but rather rare in synthetic systems. The rate of charge recombination to yield the triplet state in

1 is affected by magnetic fields, even at low field strength [45]. This unusual phenomenon may be related to mechanisms by which birds and other animals navigate [46].

7.4 Antenna-Reaction Center Complexes

7.4.1 *Porphyrin-Based Antenna-Reaction Center Complex*

As discussed earlier, most of the light used in photosynthesis is not absorbed by the reaction centers, but rather by antenna systems. There has been considerable progress made in the preparation of synthetic antenna systems and antenna-reaction center complexes. Antennas consisting of porphyrins are in some ways analogs of the types of natural antennas based mainly on chlorophyll. Many porphyrin-based antenna arrays have been reported [29, 47–64]. Figure 7.2 shows a porphyrin antenna array linked to a porphyrin-fullerene artificial reaction center (**2**) [47, 48]. The four zinc porphyrins are designed to act as antenna chromophores for the freebase porphyrin, which absorbs light at longer wavelengths than the zinc analogs. The freebase porphyrin and fullerene are arranged to favor photoinduced electron transfer as in triad **1** to yield a charge-separated state.

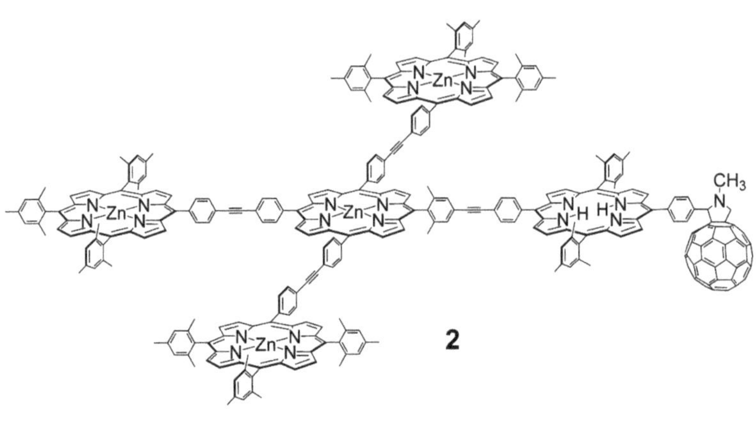

2

Figure 7.2. Hexad artificial photosynthetic antenna-reaction center

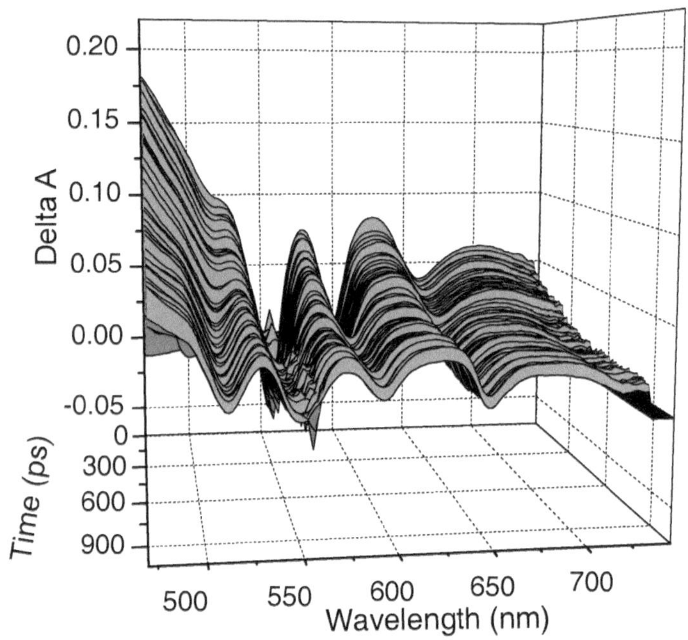

Figure 7.3. Typical transient absorption data for a model compound of **2** that lacks the fullerene moiety. Global analysis of the data yielded decay-associated spectra with exponential lifetime components of 9.5 ps, 160 ps, and 11.0 ns.

Discovering the pathways of energy and electron transfer in a molecule this complex requires spectroscopic investigations of the molecule and relevant model systems using a variety of techniques. For **2**, these techniques included time-resolved fluorescence spectroscopy and transient absorption spectroscopy on the ps and ns time scales. Figure 7.3, for example, shows a transient absorption data set obtained after excitation of a 2-methyltetrahydrofuran solution of a model compound for **2** that lacks the fullerene moiety. These data were analyzed globally to extract spectral components with various lifetimes. From these kinds of data, it was possible to determine time constants for the energy and electron transfer events that follow excitation of chromophores in **2**.

The analysis reveals that light absorbed by a peripheral zinc porphyrin yields the first excited singlet state, which transfers excitation to the central zinc porphyrin to yield $(P_{ZP})_3$-$^1P_{ZC}$-P-C_{60} with a time constant of 50 ps. In addition to exchanging excitation energy with the essentially isoenergetic peripheral zinc porphyrins $(P_{ZP})_3$-$^1P_{ZC}$-P-C_{60} donates energy to the attached freebase porphyrin with a time constant of 32 ps, giving $(P_{ZP})_3$-P_{ZC}-^{1}P-C_{60}. Photoinduced electron transfer from this state yields $(P_{ZP})_3$-P_{ZC}-$P^{\bullet+}$-$C_{60}^{\bullet-}$ ($\tau = 25$ ps). In $(P_{ZP})_3$-P_{ZC}-$P^{\bullet+}$-$C_{60}^{\bullet-}$, charge shift from the free base to the central zinc porphyrin, which is more easily oxidized than the free base, occurs with a time constant of 380 ps and is significantly faster than decay of $(P_{ZP})_3$-P_{ZC}-$P^{\bullet+}$-$C_{60}^{\bullet-}$ to the ground state. The resulting $((P_{ZP})_3$-$P_{ZC})^{\bullet+}$-P-$C_{60}^{\bullet-}$ decays to the ground state with a lifetime of 240 ns. In benzonitrile, the lifetime is 25 μs.

The fullerene of **2** also absorbs light. The absorption bands extend throughout the visible out to about 705 nm, but the extinction coefficients are low relative to those of the porphyrins. When $(P_{ZP})_3$-P_{ZC}-P-$^1C_{60}$ is formed, it decays by photoinduced electron transfer ($\tau = 75$ ps) to give the same $(P_{ZP})_3$-P_{ZC}-$P^{\bullet+}$-$C_{60}^{\bullet-}$ state that is produced by porphyrin excitation. All of the energy and electron transfer processes in **2** are rapid compared to competing pathways, and so the yield of $((P_{ZP})_3$-$P_{ZC})^{\bullet+}$-P-$C_{60}^{\bullet-}$ is ~90%.

7.4.2 *Non-Porphyrin Antenna Chromophores*

7.4.2.1 Wheel-shaped antenna-reaction center complex

As mentioned above, chlorophylls, and by extension the porphyrins, do not have high extinction coefficients throughout the entire visible spectral region, and therefore photosynthetic organisms use other sorts of auxiliary antenna chromophores. For example, carotenoid polyenes absorb strongly in the region around 450 nm. It is possible to construct artificial antenna systems that employ carotenoids [65–72] but the unusual photophysical properties of carotenoids make it difficult to achieve effective energy transfer from them to porphyrins and chlorophylls. The transition between the carotenoid ground state and the first excited singlet state is electric-dipole-forbidden, and this short-lived state is not a good energy donor by the Förster mechanism. Absorption into higher-lying states is strongly allowed, but these states are very short lived, which makes a high yield of energy transfer difficult to obtain.

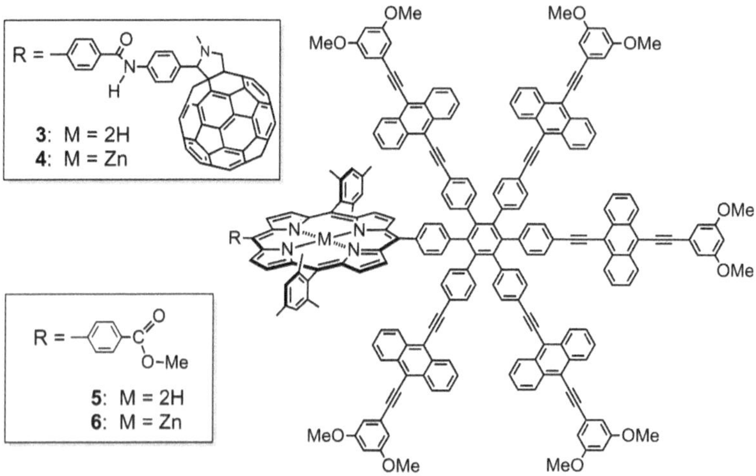

Figure 7.4. Antenna-reaction center heptads **3** and **4** and antenna hexads **5** and **6**.

Molecular heptads **3** and **4** (Fig. 7.4) make use of five bis(phenylethynyl)anthracene (BPEA) antenna chromophores to absorb light in spectral regions similar to carotenoids and transfer excitation energy to a porphyrin-fullerene artificial reaction center [40, 73]. The components of the heptads are organized by a hexaphenylbenzene core. The rigid core constrains the interchromophore separations and orientations, both holding the BPEA and porphyrin units in an arrangement conducive to rapid energy transfer and preventing the chromophore quenching that can occur if the moieties are too closely associated [74]. Hexaphenylbenzene is a useful choice because it has points of attachment for a large number of peripheral substituents on the six phenyl rings surrounding the central benzene ring, is constrained to a rigid conformation in which the six peripheral rings are nearly perpendicular to the central ring, and can be readily synthesized with a number of well-defined substitution patterns. In addition to the two heptads, model compounds **5** – **15** were prepared for use in uncovering the various energy and electron transfer pathways (Figs. 7.4 and 7.5).

As the photochemistry of the heptads is relatively complex, we first discuss that of some of the model compounds. Figure 7.6 shows

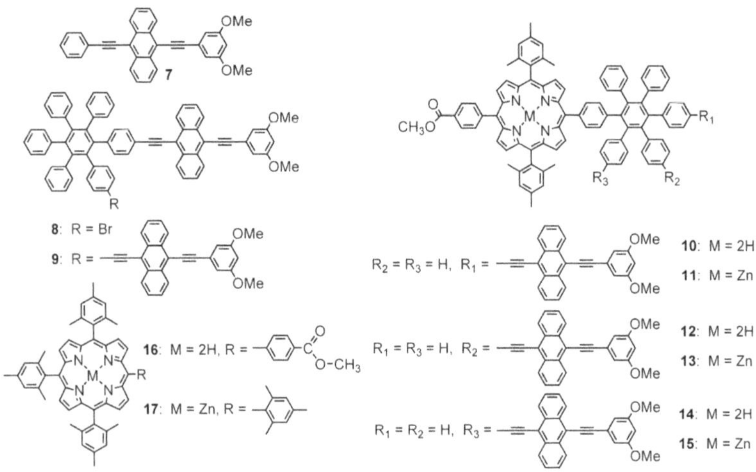

Figure 7.5. Structures of model compounds for heptads **3** and **4**.

the absorption and emission spectra of BPEA models **7**, **8**, and **9** in 2-methyltetrahydrofuran. Linking the BPEA unit to the hexaphenylbenzene core has little effect on the BPEA absorption spectrum, leading mainly to a small shift to longer wavelengths. Adding a second BPEA unit, as in **9**, does not lead to significant spectra shifts, indicating that the BPEA units do not interact in ways that strongly perturb the spectra. This is also seen in time-resolved fluorescence studies of **8** and **9**, which yield excited singlet state lifetimes of 2.80 and 2.83 ns, respectively. However, the two BPEA units can exchange singlet excitation energy. Stimulated emission anisotropy decay studies yielded a time constant of 0.4 ps for this transfer process.

Next, we examine the effect of adding the porphyrin chromophore to the hexaphenylbenzene bearing a BPEA unit. Figure 7.7a shows the absorption spectra of freebase porphyrin **16**, porphyrin-BPEA dyads **10** and **14**, and hexad **5**. The results show that the spectrum of the hexad is similar to the sum of the spectra of the porphyrin and the five BPEA units, with little perturbation due to the linking of the chromophores. Figure 7.7b shows similar results for the zinc-containing series of compounds: porphyrin **17**, porphyrin-BPEA dyads **11** and **15**, and hexad **6**.

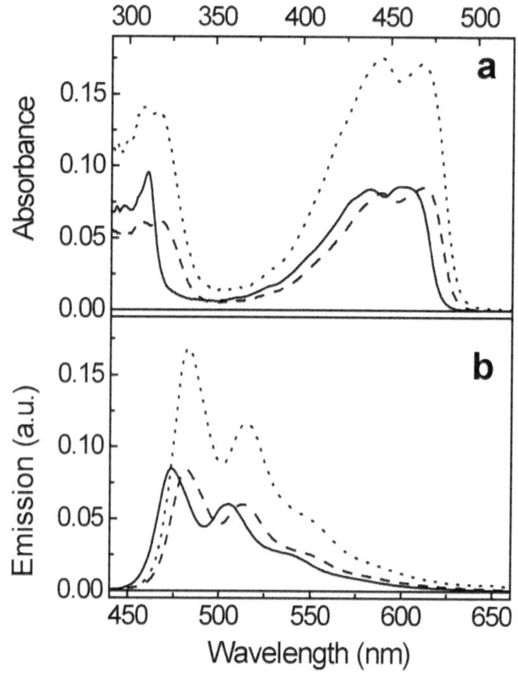

Figure 7.6. Absorption (a) and emission (λ_{ex} = 400 nm) spectra of model BPEA antenna **7** (solid), **8** (broken), and **9** (dotted) in 2-methyltetrahydrofuran solution.

Figure 7.7 also shows the strong absorption of the hexads in the ca. 430–490 nm region due to the BPEA units. As seen in the spectra of model porphyrins **16** and **17**, porphyrins have negligible absorption in this region, and augmentation by antennas is useful.

Of course, the extra absorption provided by the BPEA moieties is not beneficial to an artificial photosynthetic reaction center unless the excitation energy in the BPEA units is transferred to the porphyrin. Transfer was investigated using steady-state fluorescence excitation spectroscopy. Figure 7.8 shows the porphyrin fluorescence of hexad **6** as a function of the wavelength of the exciting light superimposed on the absorption spectrum of the hexad. The two spectra are nearly identical, indicating that the light absorbed by the BPEA units gives rise to emission from the porphyrin. Thus, energy transfer from all BPEA units to the porphyrin occurs with

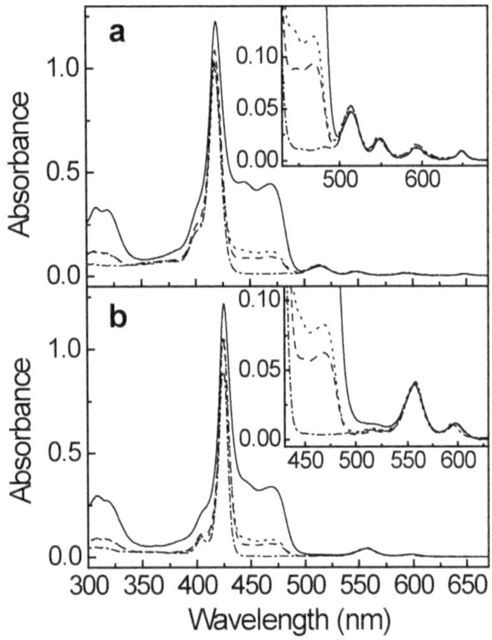

Figure 7.7. Absorption spectra in 2-methyltetrahydrofuran of (a) **5** (solid), **10** (dotted), **14** (dash), and porphyrin **16** (dash-dot) and (b) **6** (solid), **11** (dotted), **15** (dash), and porphyrin **17**(dash-dot).

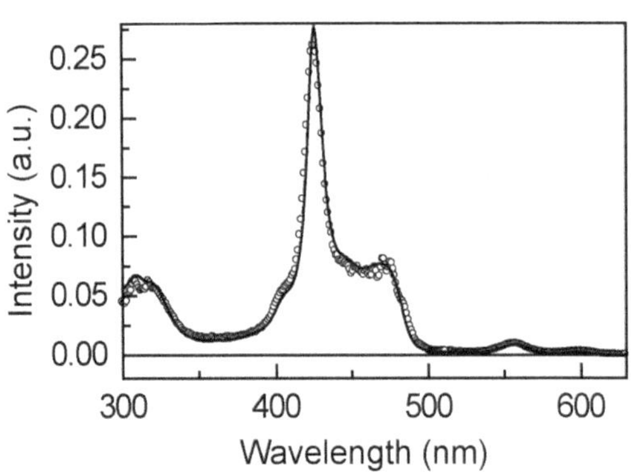

Figure 7.8. Steady-state absorption (solid) and corrected fluorescence excitation ($\lambda_{em} = 650$ nm) spectra of zinc hexad 6.

efficiencies close to 100%. Similar results were found for freebase hexad **5** These findings prove that energy transfer from all BPEA units to the porphyrins in the zinc and freebase hexads occurs much faster than decay of the BPEA excited states by other pathways. Rate constants for energy transfer from BPEA units at each position on the hexaphenylbenzene core to the porphyrin were determined by transient absorption studies of the hexads and model compounds discussed above [40].

Turning now to heptads **3** and **4**, the absorption spectra are essentially identical to those of hexads **5** and **6**, except for very weak absorption by the C_{60} moiety throughout the visible, with a long-wavelength maximum at 704 nm. The porphyrin fluorescence of both heptads is strongly quenched relative to that of the corresponding hexads. This quenching is due to photoinduced electron transfer, as occurs in the other artificial reaction centers discussed earlier. Electron transfer was confirmed by transient absorption spectroscopy, which showed absorbance ascribed to the porphyrin radical cation in each heptad, plus an absorbance at 1015 nm which is characteristic of the fullerene radical anion. The lifetimes of the $BPEA_5$-$P^{\bullet+}$-$C_{60}^{\bullet-}$ charge-separated states, determined by nanosecond transient absorption spectroscopy, are 8.9 ns for **3** and 15.3 ns for **4**. The quantum yields for the formation of charge-separated states are 80% for **3** and 96% for **4**.

The various transient spectroscopic studies of the heptads and the related model compounds allowed extraction of the rate constants for all of the energy and electron transfer processes. Figure 7.9 schematically illustrates the processes, and Table 7.1 gives the measured values.

The Förster theory was used to estimate singlet-singlet energy transfer rate constants for the heptads, based on molecular-mechanics-minimized structures. The results are shown in Table 7.1. It is clear that the theory gives transfer rate constants that are comparable to those observed, suggesting that the energy transfer occurs mainly by this mechanism. However, there are some anomalies. The Förster theory predicts that energy transfer from a *meta*-BPEA unit to the porphyrin will be faster than that from the *para*-BPEA, but the opposite is in fact observed. This suggests that the through-bond Dexter mechanism may also play a role.

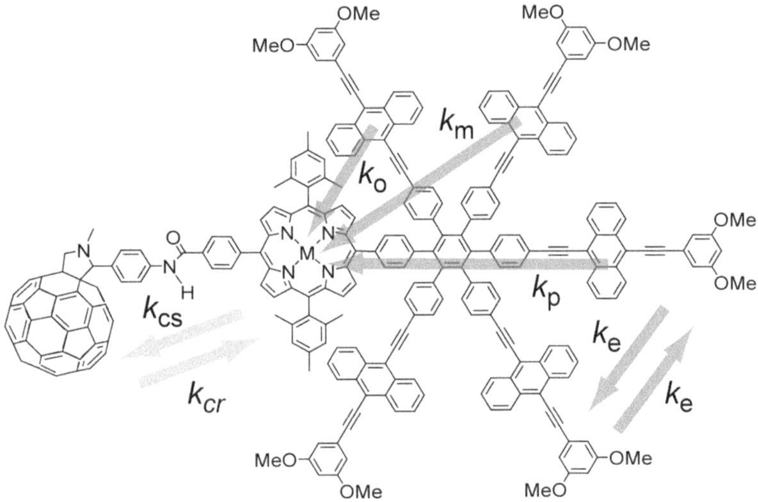

Figure 7.9. Singlet energy (red) and electron (blue) transfer pathways in heptads **3** and **4**. See also Color Insert.

The spectroscopic results show that heptads **3** and **4** function very well as artificial antenna-reaction center systems. Energy transfer from all of the antenna BPEA units to the porphyrin is essentially quantitative. The electron transfer quantum yield is not unity, but is still very high. The slightly lower overall yield for generation of the charge-separated state in these molecules as compared to **1** and **2** is due to slower photoinduced electron transfer from the porphyrin

Table 7.1. Energy and electron transfer rate constants for heptads **3** and **4**.

	Heptad **3**		Heptad **4**	
Rate constant (s^{-1})	exp.	calcd.	exp.	calcd.
k_e	2.5×10^{12}	1.9×10^{12}	2.5×10^{12}	1.9×10^{12}
k_o	1.1×10^{12}	[a]3.8×10^{12}	8.3×10^{11}	2.3×10^{12}
k_m	1.3×10^{11}	[a]1.3×10^{12}	1.4×10^{11}	1.3×10^{11}
k_p	2.5×10^{11}	[a]1.9×10^{11}	1.8×10^{11}	6.7×10^{10}
k_{cs}	4.0×10^{8}	[a]8.8×10^{10}	1.2×10^{10}	
k_{cr}	1.1×10^{8}	7.4×10^{10}	6.5×10^{7}	

[a]Values are reported for the two isomers that differ in the disposition of hydrogen atoms on the porphyrin nitrogen atoms.

to the fullerene, which results in turn from the longer chemical linkage between the moieties. The longer linkage reduces the electronic coupling, and thus the transfer rate.

7.4.2.2 Antenna-reaction center complexes using self-assembly

Antenna-reaction center complexes **3** and **4** illustrate the use of non-porphyrin accessory chromophores to increase the portion of the solar spectrum that an artificial reaction center can use. However, the combination of the porphyrin and BPEA units still does not fully exploit the visible spectral region. To solve this problem, antenna-reaction center complex **18** was designed and constructed (Fig. 7.10). The antenna portion of the complex, **19**, consists of a hexaphenylbenzene core that organizes two blue-light-absorbing BPEA antennas, two green-light-absorbing borondipyrromethene antennas (BDPY), and two porphyrins, whose Q-bands absorb in the red spectral region out to about 650 nm. As discussed in detail below, the BPEA and BDPY units are designed to transfer excitation energy to the porphyrins, yielding porphyrin first excited singlet states. Antenna **19** lacks a good electron acceptor unit, and thus

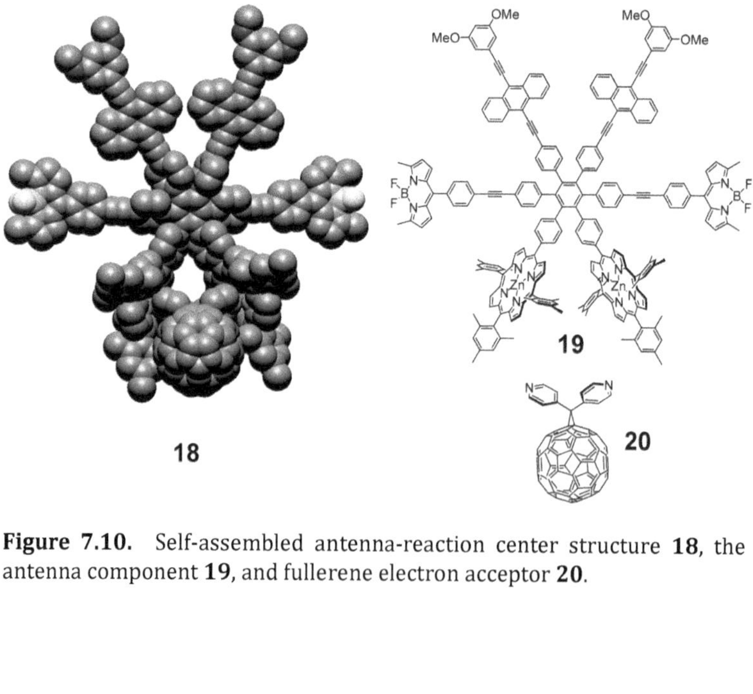

Figure 7.10. Self-assembled antenna-reaction center structure **18**, the antenna component **19**, and fullerene electron acceptor **20**.

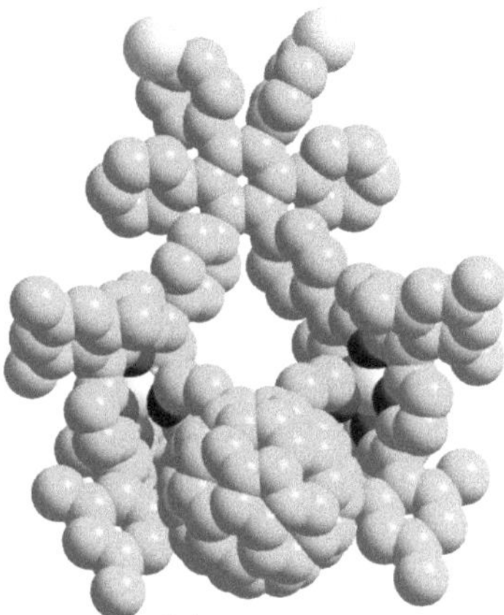

21

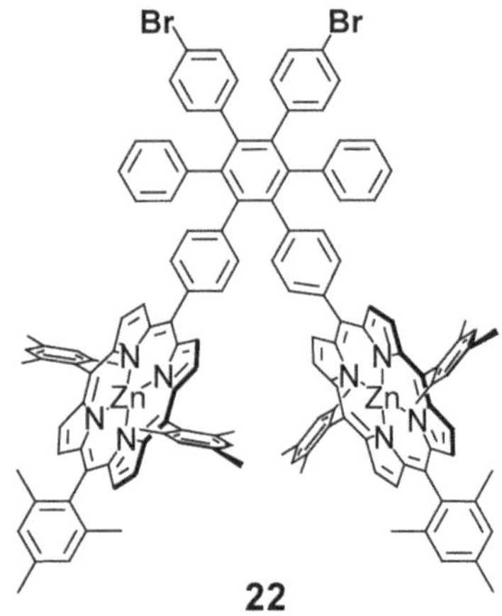

22

Figure 7.11. Model compounds for heptad **18**.

a reaction center function. This was added by self-assembly of a fullerene bearing two pyridyl groups (**20**). The nitrogen atoms of the pyridyl groups coordinate the two zinc atoms of **19** to assemble the final antenna-reaction center complex **18**. As with **3** and **4**, several model systems (Fig. 7.11) were constructed in order to help select the final design for **18** and to allow us to extract estimates of the various energy and electron transfer rate constants in the complete system.

The initial model was designed to evaluate self-assembly and photoinduced electron transfer in the reaction center portion of the heptad [75]. Triad **21** is assembled from porphyrin dyad **22** and dipyridyl-fullerene **20**. The binding constant for **20** to **22** in 1,2-difluorobenzene was determined by spectrophotometric titration, and found to be $7.3 \times 10^4 \pm 3 \times 10^3 M^{-1}$, which is considerably larger than binding constants for complexation of pyridine itself to zinc porphyrins ($\sim$5000 M^{-1}) [76].

Transient spectroscopic studies of **21** were carried out in 1,2-difluorobenzene containing a minimal amount of CS_2 for solubility purposes. Excitation of the porphyrin moieties at 600 nm with $\sim$100 fs laser pulses results in the formation of the porphyrin first excited singlet state, which decays by photoinduced electron transfer to form the $PP^{\bullet+}$-$C_{60}^{\bullet-}$ charge-separated state with a rate constant of $3.3 \times 10^{11}\ s^{-1}$. This state was identified by the absorption of the zinc porphyrin radical cation in the 650 nm region and the fullerene radical anion near 1040 nm. The $PP^{\bullet +}$-$C_{60}^{\bullet -}$ state decayed to the ground state with a time constant of 230 ps. This model compound demonstrated that the dipyridyl fullerene associates with the diporphyrin **22** with a reasonably high binding constant, and that the resulting complex **21** demonstrates very rapid photoinduced electron transfer from the excited porphyrin to the fullerene.

Having established that the reaction center portion of heptad **18** was viable, we investigated models for the antenna moieties of the compound. Some of the relevant molecules are shown in Fig. 7.12. Figure 7.13 shows the absorption and fluorescence emission spectra in 2-methyltetrahydrofuran of model BPEA **8**, of model BDPY **23**, and of dyad **25** which features both a BPEA and a BDPY moiety bound to a hexaphenylbenzene ring. The BPEA absorbs strongly in the 450 nm region, and the BDPY around 515 nm. The zinc porphyrins of **18** have Soret absorptions at 418 nm and Q-bands at 557 and 598 nm. Thus the BDBY and BPEA antennas absorb well in a region where the porphyrin is not effective, as desired for antennas. The spectrum of dyad **25** in the visible is essentially a linear combination of the spectra of **8** and **23**. Figure 7.13 shows that the fluorescence emission of the BPEA overlaps well with the absorption of BDPY, indicating that singlet energy transfer from BPEA to BDPY is thermodynamically favorable. The fluorescence quantum yield of the BPEA chromophore is 0.94, whereas that of the BDPY is 0.04. When dyad **25** is excited at 400 nm, where both chromophores have nearly equal extinction coefficients, the emission spectrum (Fig. 7.13b) shows mainly BDPY emission, with only a trace of emission from BPEA. This verifies efficient singlet-singlet energy transfer from BPEA to BDPY.

Figure 7.14 shows the absorption spectrum of the antenna hexad **19** and associated model compounds. It is very clear in these spectra that the absorbance of the BPEA and BDPY units completely fills the region between ~430 nm and ~540 nm, where porphyrin absorption is essentially zero.

Turning now to the final self-assembled heptad molecule **18**, transient spectroscopic techniques were used to tease out the various energy and electron transfer steps and their rate constants. This

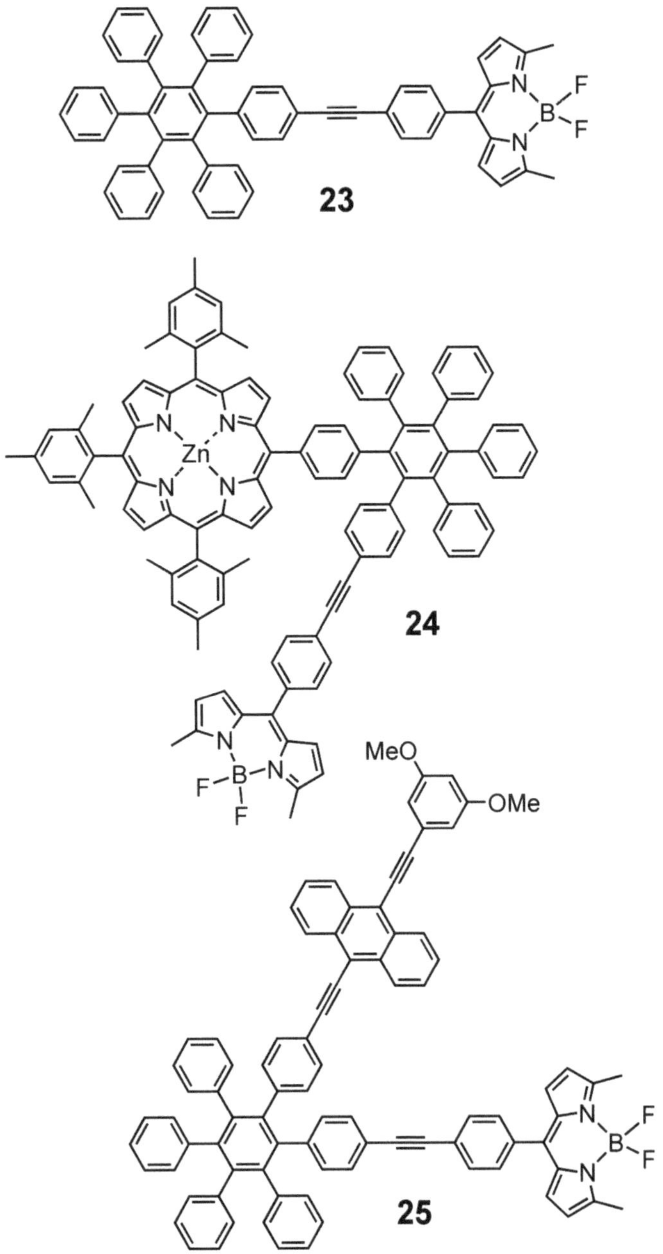

Figure 7.12. Model compounds for heptad **18**.

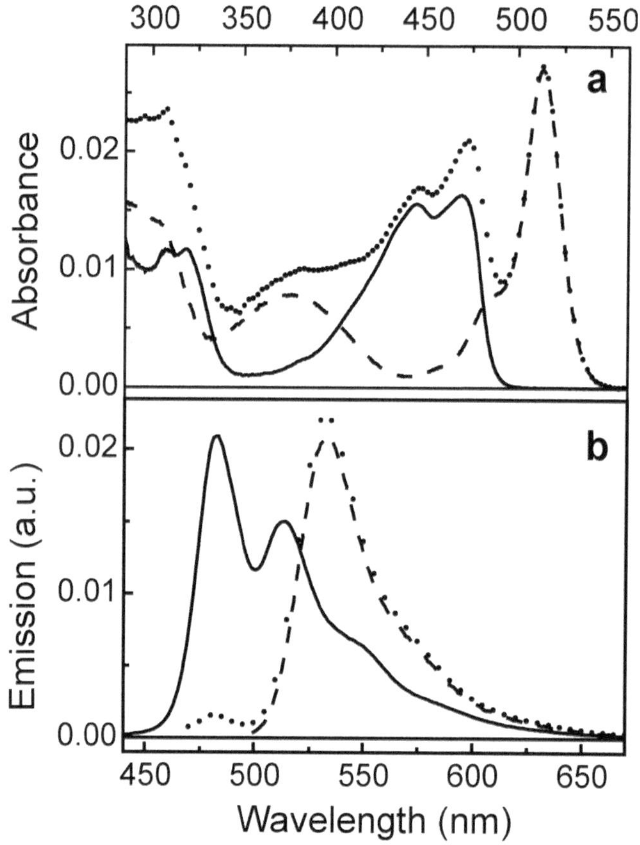

Figure 7.13. Absorption (a) and fluorescence (b) spectra of BPEA **8** (solid), BDPY **23** (dash), and dyad **25** (dotted).

was done by studying both **18** itself and the various model compounds. The various pathways are shown in Fig. 7.15, and the relevant time constants are given in Table 7.2.

Energy transfer from BPEA to BDPY and directly to the porphyrins is very fast relative to decay of ^{1}BPEA by other photophysical processes, and the quantum yield of energy transfer is essentially unity. In **18** and hexad **19** energy transfer from BPEA to the porphyrins occurs on the ca. 5 ps time scale. Two sets of BDPY conformations were observed, corresponding to different orientations of the boron-containing ring relative to the attached phenyl ring. These

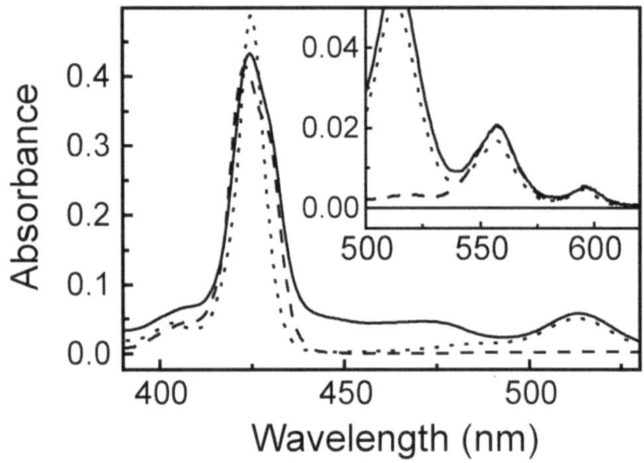

Figure 7.14. Absorption spectra in 2-methyltetrahydrofuran of zinc porphyrin hexad **19** (——), a zinc porphyrin dyad similar to **22** (- - -), and BDPY-zinc porphyrin dyad **24** (• • • •).

have different energy transfer efficiencies. Transfer from ^{1}BPEA to the most efficient set of BDPY conformations occurs on the ca. 5 ps time scale, whereas transfer to a slightly less efficient set of conformations of BDPY is slower (14 ps). Thus, in these molecules, decay of ^{1}BPEA by the 3 faster pathways dominates, and the yield of ^{1}BDPY from ^{1}BPEA by the slower process is $\sim$0.10. Energy transfer from ^{1}BDPY to the porphyrins is also multi-exponential, and the fastest process (4–5 ps) is essentially quantitative. Transfer by the slower process is slightly less effective, but the quantum yield is still $\sim$0.96. Calculations based on molecular models and using the Förster treatment yield rate constants consistent with those in the Table. Thus the two kinds of chromophores are highly effective antennas for the zinc porphyrins in **18**, transferring energy both directly and by a stepwise mechanism whereby excitation migrates around the hexaphenylbenzene from BPEA to BDPY and on to the porphyrin.

Turning to electron transfer in these compounds, it was discovered that in zinc hexad **19**, where there is no fullerene acceptor, photoinduced electron transfer from the porphyrin first excited singlet states to the BDPY, acting as an acceptor to form an undesired charge-separated state, is slow but relatively efficient (step 6 in Fig. 7.15).

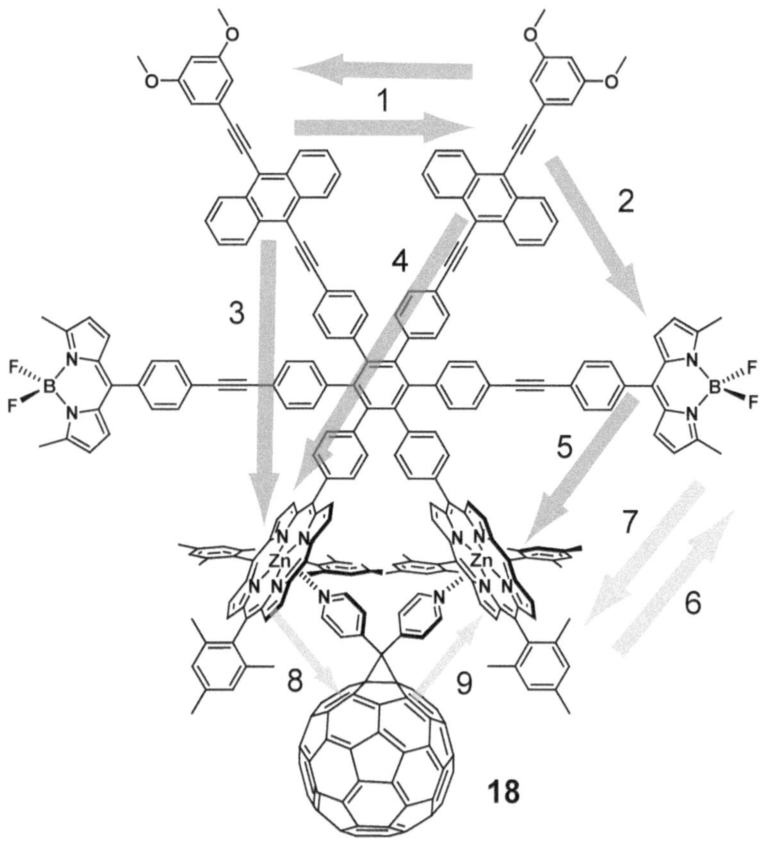

Figure 7.15. Singlet excitation energy (red arrows) and electron (blue arrows) transfer pathways observed in heptad **18** and model compounds. The relevant time constants are shown in Table 7.2. See also Color Insert.

In heptad **18**, on the other hand, photoinduced electron transfer to the fullerene (step 8) is extremely rapid, as expected from the studies of model **21**, and occurs with a quantum yield of essentially unity (Table 7.2). The yield of photoinduced electron transfer to the BDPY is insignificant. Electron transfer to the fullerene is substantially faster than has been observed for some other donor–acceptor systems formed by association of a fullerene-bearing nitrogenous group with zinc porphyrins [77–81]. This can be ascribed to good electronic coupling between the porphyrin and fullerene due to the

Table 7.2. Time constants for transfer processes in heptad **18** and model **19** (ps).

	Hexad **19**		Heptad **18**
Process	2-MeTHF[a]	1,2-DFB[b]	1,2-DFB[c]
Step 1	0.4	0.4[d]	0.4
Step 2	5–14	5–13	5 - 13
Step 3	7	7[d]	7
Step 4	6	6	6
Step 5	4–20	-15	2–15
Step 6	570	230	230
Step 7	4800	1500	1500
Step 8			3
Step 9			230

[a] 2-Methyltetrahydrofuran
[b] 1,2-Difluorobenzene
[c] 1,2-Difluorobenzene containing excess fullerene **20**.
[d] Estimated from results in 2-methyltetrahydrfuran.

short covalent linkage and the close spatial approach of the moieties. Of course, this same electronic coupling leads to a charge recombination time constant, 230 ps, that, although 70 times larger than the time constant for charge separation, is still smaller than those observed for many porphyrin-fullerene systems with weaker coupling.

7.5 Self-Regulation of Artificial Photosynthetic Systems

As mentioned in the Introduction, most of the light that drives photosynthesis is absorbed by antennas. Under conditions of bright sunlight, it is possible for antennas to harvest photons at a rate that drives charge separation in reaction centers faster than subsequent dark reactions can use the resulting electrochemical energy to produce fuels or other useful molecules. When this occurs, the charge-separated species can destroy parts of the photosynthetic apparatus or generate harmful species such as singlet oxygen, which in turn attack tissues. To minimize this kind of damage, photosynthetic organisms employ a variety of photoprotective, photoregulatory and self-repair mechanisms. One of these is nonphotochemical quenching (NPQ), a process whereby excess excitation energy in the antenna system is degraded to harmless heat instead of being transferred to reaction centers [10–12, 82,83]. In cyanobacteria, NPQ is carried out by orange carotenoid protein, which is associated with the phycobilisome antennas [84]. Upon absorption of blue-green light, orange carotenoid protein is photoconverted to a red-absorbing form. This form thermally reverts to the orange form. Thus, the ratio of the orange and red forms is controlled by the light level and temperature. It has been proposed that the carotenoid in the red form quenches chlorophyll excited states in the cyanobacterial antenna, thus preventing delivery of excess excitation to reaction centers and down-regulating the photosynthetic process.

Artificial photosynthetic systems may also be able to benefit from photoprotection, photoregulation, and self-repair. Indeed, many conceivable nanoscale systems or devices will likely require components that not only carry out a function, but also regulate that function in response to external stimuli. Chemists are only beginning to focus on this aspect of molecular design.

Drawing again on photosynthesis as a model, we have designed and studied molecular pentad **26**, which functions as an artificial antenna-reaction center complex with a built-in light-activated control unit (Fig. 7.16) [85]. The reaction center component consists of a free-base porphyrin and a fullerene linked via an amide as in heptads **3** and **4**. The porphyrin bears two BPEA moieties at *meso*-positions. These were shown to be excellent antennas for porphyrins in compounds **3**, **4**, and **18**. The hexaphenylbenzene unit bears a photochromic control unit in addition to the porphyrin. Photochromes are molecules that exist in two metastable forms that can be interconverted by light, or light and heat. The dihydroindolizine photochrome (DHI) in **26c** is thermally stable, but when it absorbs light in the blue spectral region it photoisomerizes to the open chain, zwitterionic betaine (BT) to give **26o**. The betaine thermally isomerizes to the DHI form, and also photochemically isomerizes to that form, although the quantum yield is not high.

The light absorption properties of the DHI and BT forms of the photochrome differ significantly, and this is key to the operation of the control unit. This is illustrated by model dihydroindolizine **27** (Fig. 7.17). The absorption spectra of this molecule in

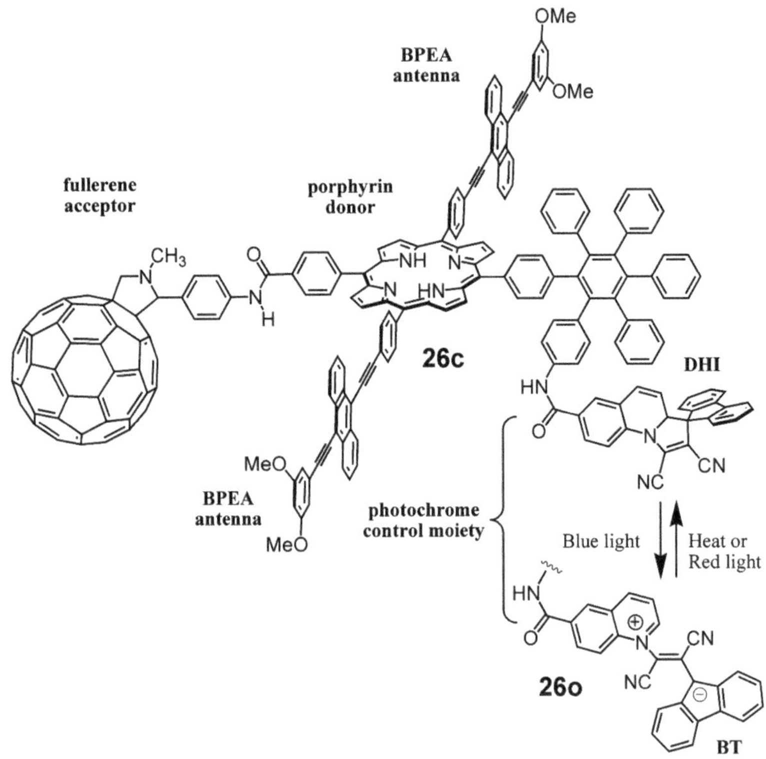

Figure 7.16. Artificial photosynthetic antenna-reaction center with built-in photosensitive control unit. The photochromic moiety can exist in the dihydroindolizine (DHI, **26c**) or betaine (BT, **26o**) forms.

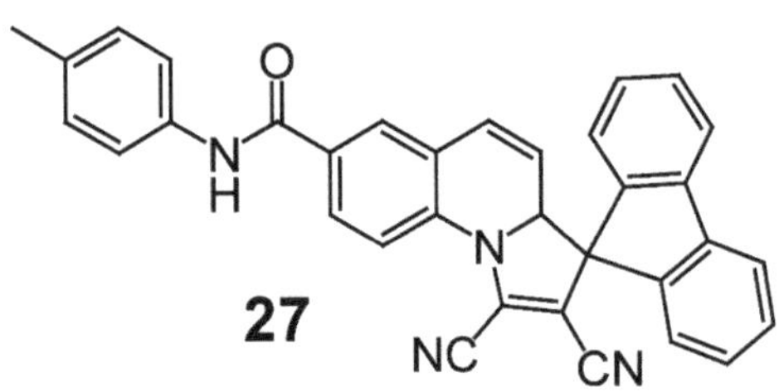

Figure 7.17. Structure of model photochrome **27**.

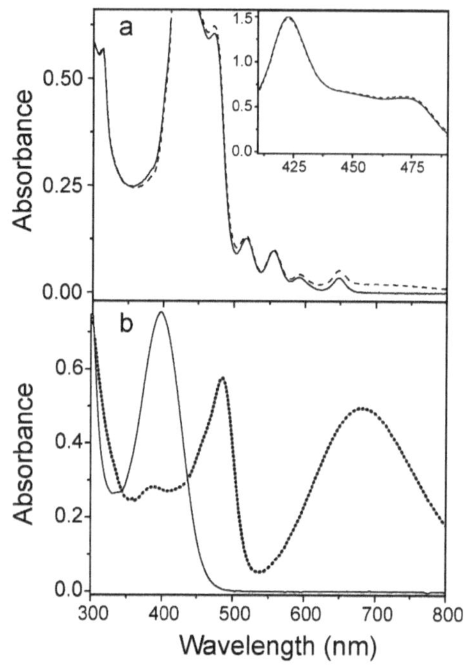

Figure 7.18. (a) Absorption spectra of **26** in 2-methyltetrahydrofuran with the photochromic moiety in the closed (**26c**, ——) and open (**26o**, - - -) forms. (b) Absorption spectra of model photochrome **27** in the DHI (——) and BT (• • • •) forms.

2-methyltetrahydrofuran solution are shown in Fig. 7.18b. In the dihydroindolizine form **27** has an absorption maximum at 395 nm, and no absorbance at wavelengths longer than about 475 nm. Irradiation of the solution with light absorbed by this transition leads to photoisomerization to the betaine, which has absorption bands at 485 nm and 685 nm. Transient absorption measurements show that the excited state of the betaine is very short lived, decaying with a time constant of 0.9 ps.

Figure 7.18a shows the absorption spectra in 2-methyltetrahydrofuran of **26c** and a sample highly enriched in **26o**. The spectrum of **26c** features bands due to the porphyrin at 424, 519, 556, 593, and 650 nm, the BPEA at ~450 (sh) and 474 nm, and the DHI at 395 nm, overlapping the porphyrin Soret. In **26o**, the porphyrin and BPEA spectra are unchanged, but the bands of the betaine are now apparent at wavelengths similar to those of the model compound. The porphyrin moiety of **26** has fluorescence emission maxima at about 655 nm and 720 nm. Thus, there is excellent overlap between the porphyrin emission spectrum and the BT absorption spectrum. This overlap is a requirement for singlet-singlet energy transfer from the porphyrin first excited singlet state to the betaine, which is in turn key to the function of the molecule. Molecule **26o** thermally reverts to **26c** with a time constant of 37 s at 25°C.

The mode of operation of pentad **26** is as follows. When **26c** is exposed to white light in the visible region, light absorbed by any of the porphyrin bands is rapidly converted to excitation in the porphyrin first excited singlet state. Light absorbed by the BPEA antennas is transferred to the porphyrin with a time constant of 4 ps and with a quantum yield of unity, generating the same porphyrin excited state. In the absence of the fullerene, the free-base porphyrin excited state decays with a time constant of 11.4 ns. However, in **26c**, the fullerene quenches the lifetime of the porphyrin first excited singlet state to 2.0 ns. This quenching is due to photoinduced electron transfer to form the $P^{\bullet+}$-$C_{60}^{\bullet-}$ state with a quantum yield of 82%. The charge-separated state decays with a lifetime of 14.0 ns.

In **26c** some of the light in the blue region is absorbed by the DHI moiety, which photoisomerizes to the betaine, yielding **26o**. The photochemistry of this isomer is quite different. The porphyrin excited singlet state is now quenched to a lifetime of only 33 ps by singlet energy transfer to the betaine. This quenching reduces the quantum yield of photoinduced electron transfer to the fullerene to 1%, essentially shutting off the electron transfer process. Thus, **26c** is a successful artificial photosynthetic antenna-reaction center system, whereas **26o** is inactive, and converts the absorbed light energy to heat.

In a solution of **26** at ambient temperatures, the photostationary distribution determines the ratio of **26c/26o**, and this ratio in turn determines the overall quantum yield of formation of the charge-separated state. This is shown in Fig. 7.19, which gives the results of a series of 13 experiments at different white light intensities. At very low light levels, the quantum yield of charge separation is 82%.

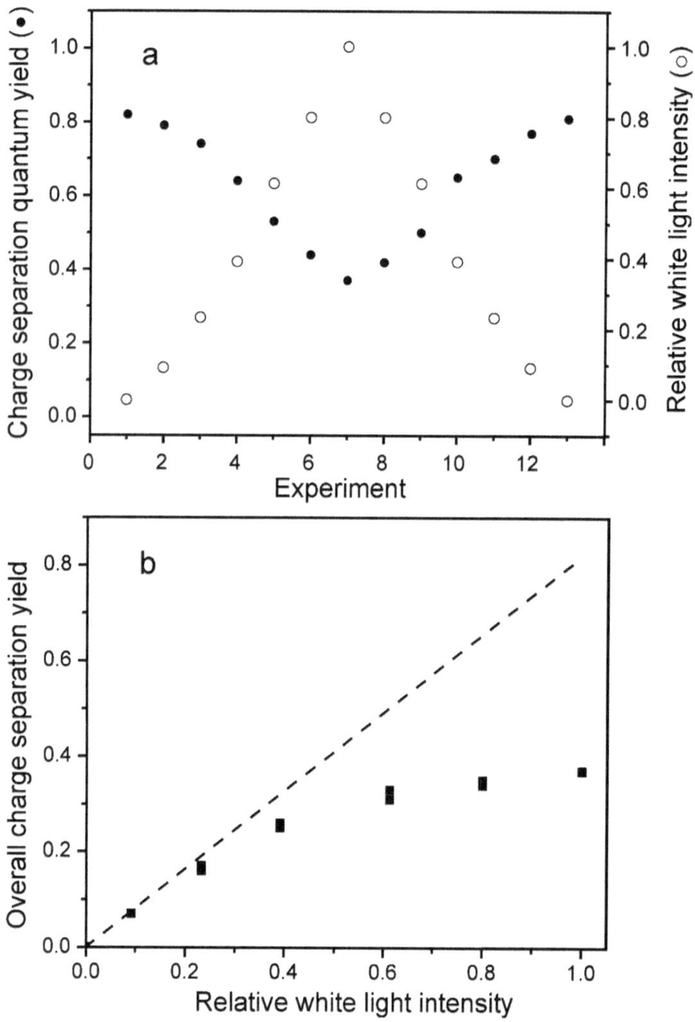

Figure 7.19. (a) Operation of **29** at various light intensities. The quantum yield of charge separation (•) decreases as the white light intensity (o) increases. (b) Relative yield of charge separation as a function of white light intensity.

As the light intensity increases, the DHI in an increasing percentage of the molecules is converted to betaine, which quenches the porphyrin excited states and prevents charge separation. Thus, the quantum yield of charge separation measured for the solution decreases. In this particular study, the lowest quantum yield obtained was 37% for experiment 7, but the yield is even lower at higher light intensities. When the light intensity is reduced, the quantum yield again increases.

Thus, molecule **26** is an artificial photosynthetic antenna-reaction center that internally down regulates the yield of charge separation as the light level increases. This is perhaps more easily appreciated in Fig. 7.19b, which shows the overall yield of charge separation, calculated as the quantum yield times the relative light intensity. Here, the yield of charge-separated states is proportional to the light intensity at low light levels, as would be expected for most photochemical processes. At high light, the yield of charge separation begins to level off because the increased light intensity increases the fraction of **26o** present at equilibrium, and thus reduces the number of molecules that carry out photoinduced electron transfer.

Pentad **26** thus mimics the mechanism by which NPQ is proposed to work in cyanobacteria: increasing light intensity increases the fraction of excited states of the electron donors that are quenched by the control chromophore before they can contribute to photoinduced electron transfer. Of course, the chemical structures in the two systems are very different.

7.6 Conclusions

The molecules discussed in this short review exemplify several of the current themes in artificial photosynthesis research. It is clear that it is now possible to create in the laboratory artificial reaction centers that convert light energy into electrochemical potential via photoinduced electron transfer. The charge-separated states can be formed with quantum yields of essentially unity, can store a significant fraction of the photon energy, and have lifetimes that are long enough to consider using the stored energy to make electricity or drive redox catalysts. It is also possible to enhance the function of these reaction centers by addition of multichromophoric antenna arrays that extend the range of wavelengths that the reaction centers can use effectively. Finally, photoregulatory function can also be incorporated, and may ultimately be necessary to prevent photodamage to these organic systems.

This and related work allows us to enumerate several design principles that are applicable to artificial photosynthesis:

- Using Marcus theory as a guide, electron transfer in artificial reaction centers can be controlled by taking into account thermodynamic driving force, electronic coupling through chemical linkages, and both solvent and internal reorganization energies.
- Long lifetimes coupled with high quantum yields for charge-separated states can be achieved by multistep electron transfer through intermediate donor–acceptor moieties.
- Antenna systems that absorb throughout the visible can be interfaced to artificial reaction centers by optimizing energy transfer rates among the various chromophores and minimizing rates of undesired reactions such as photoinduced electron transfer to and from antenna components.
- Relatively rigid molecular frameworks are necessary for tight control and optimization of electron and energy transfer rates and minimization of populations of molecules with undesired properties. Rigid frameworks also facilitate interpretation of spectroscopic results.
- Self-assembly can be used to advantage in the design of complex multichromophoric systems, but binding constants must be very large to avoid complications from unbound chromophores.
- For both energy and electron transfer, absolute rates are, within limits, irrelevant for achieving high quantum yields. Favorable ratios of rate constants for desired over undesired processes are what matter.
- Artificial photosynthetic systems need not be passive. Photoregulation and other photoprotective mechanisms can be engineered into them in order to limit photodamage.

It must be emphasized that most of the work in the area of artificial photosynthesis, both in our laboratories and elsewhere, has been fundamental in nature, rather than applied. In addition, this review has focused on the parts of the process that involve light absorption and the conversion of the resulting excitation energy to potential energy stored in charge-separated states. Significant work on using this potential energy to carry out useful chemistry at catalytic centers has been performed in a number of laboratories, but this aspect of the problem is not discussed here. The body of work in the artificial photosynthetic area has demonstrated that artificial photosynthetic conversion of light energy to electrical energy or chemical energy stored in fuels is possible [86]. However, the practicality of such a process in terms of questions such as efficiency, cost-effectiveness, durability, etc., is not established, and is being investigated.

Acknowledgment

The research described in this review has been supported by grants from the US Department of Energy and National Science Foundation, as acknowledged in the original publications. YT and DG were supported as part of the Center for Bio-Inspired Solar Fuel Production, an Energy Frontier Research Center funded by the US Department of Energy, Office of Science, Office of Basic Energy Sciences under Award Number DE-SC0001016.

References

1. Ciamician, G. (1912) The photochemistry of the future, *Science*, **36**, 385–394.
2. Blankenship, R. E. (2002) *Molecular Mechanisms of Photosynthesis*, Blackwell Science, Oxford.
3. Förster, T. (1948) Zwischenmolekulare energiewanderung und fluoreszenz, *Annalen der Physik*, **2**, 55–75.
4. Förster, T. (1959) Transfer mechanisms of electronic excitation, *Disc. Faraday Soc.*, **27**, 7–17.

5. Collini, E., Wong, C. Y., Wilk, K. E., Curmi, P. M. G., Brumer, P., and Scholes, G. D. (2010) Coherently wired light-harvesting in photosynthetic marine algae at ambient temperature, *Nature*, **463**, 644–647.
6. Engel, G. S., Calhoun, T. R., Read, E. L., Ahn, T. K., Mancal, T., Cheng, Y. C., Blankenship, R. E., and Fleming, G. R. (2007) Evidence for wavelike energy transfer through quantum coherence in photosynthetic systems, *Nature*, **446**, 782–786.
7. Lee, H., Cheng, Y. C., and Fleming, G. R. (2007) Coherence dynamics in photosynthesis: Protein protection of excitonic coherence, *Science*, **316**, 1462–1465.
8. Scholes, G. D. (2010) Quantum-coherent electronic energy transfer: Did nature think of it first?, *J. Phys. Chem. Lett.*, **1**, 2–8.
9. van Grondelle, R., and Novoderezhkin, V. I. (2010) Quantum design for a light trap, *Nature*, **463**, 614–615.
10. Ahn, T. K., Avenson, T. J., Ballottari, M., Cheng, Y. C., Niyogi, K. K., Bassi, R., and Fleming, G. R. (2008) Architecture of a charge-transfer state regulating light harvesting in a plant antenna protein, *Science*, **320**, 794–797.
11. Berera, R., van Stokkum, I. H. M., d'Haene, S., Kennis, J. T. M., van Grondelle, R., and Dekker, J. P. (2009) A mechanism of energy dissipation in cyanobacteria, *Biophys. J.*, **96**, 2261–2267.
12. Ruban, A. V., Berera, R., Ilioaia, C., van Stokkum, I. H. M., Kennis, J. T. M., Pascal, A. A., van Amerongen, H., Robert, B., Horton, P., and van Grondelle, R. (2007) Identification of a mechanism of photoprotective energy dissipation in higher plants, *Nature*, **450**, 575–578.
13. de Bianchi, S., Ballottari, M., Dall'Osto, L., and Bassi, R. (2010) Regulation of plant light harvesting by thermal dissipation of excess energy, *Biochem. Soc. Trans.*, **38**, 651–660.
14. Durchan, M., Herbstova, M., Fuciman, M., Gardian, Z., Vacha, F., and Polivka, T. (2010) Carotenoids in energy transfer and quenching processes in Pcb and Pcb-PS I complexes from *Prochlorothrix hollandica*, *J. Phys. Chem. B*, **114**, 9275–9282.
15. Marcus, R. A. (1956) The theory of oxidation-reduction reactions involving electron transfer, *J. Chem. Phys.*, **24**, 966–978.
16. Hush, N. S. (1958) Adiabatic rate processes at electrodes. I. Energy-charge relation, *J. Chem. Phys.*, **28**, 962–972.
17. Levich, V. (1966) Present state of the theory of oxidation-reduction in solution (bulk and electrode reactions), *Adv. Electrochem. Electrochem. Eng.*, **4**, 249–371.

18. Dutton, P. L., and Scarpa H. (1978) *Frontiers of Biological Energetics: From Electrons to Tissues*, (eds. Kong, J. L., and Loach, P. A.), Academic Press, New York, pp. 73–82.

19. Tabushi, I., Koga, N., and Yanagita, M. (1979) *Tetrahedron Lett.*, p. 257.

20. Fox, M. A., and Chanon, M. (1988) Intramolecular electron transfer: History and some implications for artificial photosynthesis, in *Photoinduced Electron Transfer, Part D*, (eds. Connolly, J. S., and Bolton, J. R.), Elsevier, Amsterdam, pp. 303–393.

21. Kadish, K. M., Smith, K. M., and Guilard, R. (2000) Intramolecular photoinduced electron transfer reactions of porphyrins, in *The Porphyrin Handbook* (eds. Gust, D., and Moore, T. A.), Academic Press, New York, pp. 153–190.

22. Gust, D., Mathis, P., Moore, A. L., Liddell, P. A., Nemeth, G. A., Lehman, W. R., Moore, T. A., Bensasson, R. V., Land, E. J., and Chachaty, C. (1983) Energy transfer and charge separation in carotenoporphyrins, *Photochem. Photobiol.*, **37S**, S46.

23. Moore, T. A., Gust, D., Mathis, P., Mialocq, J. C., Chachaty, C., Bensasson, R. V., Land, E. J., Doizi, D., Liddell, P. A., Lehman, W. R., Nemeth, G. A., and Moore, A. L. (1984) Photodriven charge separation in a carotenoporphyrin quinone triad, *Nature*, **307**, 630–632.

24. Nishitani, S., Kurata, N., Sakata, Y., Misumi, S., Karen, A., Okada, T., and Mataga, N. (1983) A new model for the study of multistep electron transfer in photosysnthesis, *J. Am. Chem. Soc.*, **105**, 7771–7772.

25. Gust, D., Moore, T. A., and Moore, A. L. (2001) Mimicking photosynthetic solar energy transduction, *Acc. Chem. Res.*, **34**, 40–48.

26. Gust, D., and Moore, T. A. (1991) Mimicking photosynthetic electron and energy transfer, *Adv. Photochem.*, **16**, 1–65.

27. Gust, D., Moore, T. A., and Moore, A. L. (1993) Molecular mimicry of photosynthetic energy and electron transfer, *Acc. Chem. Res.*, **26**, 198–205.

28. Gust, D., Moore, T. A., and Moore, A. L. (2009) Solar fuels via artificial photosynthesis, *Acc. Chem. Res.*, **42**, 1890–1898.

29. Guldi, D. M. (2002) Fullerene-porphyrin architectures; photosynthetic antenna and reaction center models, *Chem. Soc. Rev.*, **31**, 22–36.

30. Imahori, H., and Sakata, Y. (1999) Fullerenes as novel acceptors in photosynthetic electron transfer, *Eur. J. Org. Chem.*, 2445–2457.

31. Wasielewski, M. R. (1992) Photoinduced electron transfer in supramolecular systems for artificial photosynthesis, *Chem. Rev.*, **92**, 435–461.

32. Gust, D., Moore, T. A., Moore, A. L., Lee, S.-J., Bittersmann, E., Luttrull, D. K., Rehms, A. A., DeGraziano, J. M., Ma, X. C., Gao, F., Belford, R. E., and Trier, T. T. (1990) Efficient multistep photoinduced electron transfer in a molecular pentad, *Science*, **248**, 199–201.

33. Liddell, P. A., Sumida, J. P., Macpherson, A. N., Noss, L., Seely, G. R., Clark, K. N., Moore, A. L., Moore, T. A., and Gust, D. (1994) Preparation and photophysical studies of porphyrin-C_{60} dyads, *Photochem. Photobiol.*, **60**, 537–541.

34. Kuciauskas, D., Liddell, P. A., Lin, S., Stone, S., Moore, A. L., Moore, T. A., and Gust, D. (2000) Photoinduced electron transfer in carotenoporphyrin-fullerene triads: Temperature and solvent effects, *J. Phys. Chem. B*, **104**, 4307–4321.

35. Liddell, P. A., Kuciauskas, D., Sumida, J. P., Nash, B., Nguyen, D., Moore, A. L., Moore, T. A., and Gust, D. (1997) Photoinduced charge separation and charge recombination to a triplet state in a carotene-porphyrin-fullerene triad, *J. Am. Chem. Soc.*, **119**, 1400–1405.

36. Terazono, Y., Kodis, G., Liddell, P. A., Garg, V., Moore, T. A., Moore, A. L., and Gust, D. (2009) Multiantenna artificial photosynthetic reaction center complex, *J. Phys. Chem. B*, **113**, 7147–7155.

37. Imahori, H. (2004) Porphyrin-fullerene linked systems as artificial photosynthetic mimics, *Org. Biomol. Chem.*, **2**, 1425–1433.

38. Boyd, P. D. W., and Reed, C. A. (2005) Fullerene-porphyrin constructs, *Acc. Chem. Res.*, **38**, 235–242.

39. Gust, D., Moore, T. A., and Moore, A. L. (2000) Photochemistry of supramolecular systems containing C_{60}, *Photochem. Photobiol. B*, **58**, 63–71.

40. Kodis, G., Terazono, Y., Liddell, P. A., Andréasson, J., Garg, V., Hambourger, M., Moore, T. A., Moore, A. L., and Gust, D. (2006) Energy and photoinduced electron transfer in a wheel-shaped artificial photosynthetic antenna-reaction center complex, *J. Am. Chem. Soc.*, **128**, 1818–1827.

41. D'Souza, F., and Ito, O. (2009) Supramolecular donor-acceptor hybrids of porphyrins/phthalocyanines with fullerenes/carbon nanotubes: electron transfer, sensing, switching, and catalytic applications, *Chem. Commun.*, 4913–4928.

42. Maligaspe, E., Kumpulainen, T., Subbaiyan, N. K., Zandler, M. E., Lemmetyinen, H., Tkachenko, N. V., and D'Souza, F. (2010) Electronic energy harvesting multi BODIPY-zinc porphyrin dyads accommodating fullerene as photosynthetic composite of antenna-reaction center, *Phys. Chem. Chem. Phys.*, **12**, 7434–7444.

43. Helaja, J., Tauber, A. Y., Abel, Y., Tkachenko, N. V., Lemmetyinen, H., Kilpelainen, I., and Hynninen, P. H. (1999) Chlorophylls. IX. the first phyrochlorin–fullerene dyads: Synthesis and conformational studies, *J. Chem. Soc., Perkin Trans.*, **1**, 2403–2408.
44. Bahr, J. L., Kuciauskas, D., Liddell, P. A., Moore, A. L., Moore, T. A., and Gust, D. (2000) Driving force and electronic coupling effects on photoinduced electron transfer in a fullerene-based molecular triad, *Photochem. Photobiol.*, **72**, 598–611.
45. Kuciauskas, D., Liddell, P. A., Moore, A. L., Moore, T. A., and Gust, D. (1998) Magnetic switching of charge separation lifetimes in artificial photosynthetic reaction centers, *J. Am. Chem. Soc.*, **120**, 10880–10886.
46. Maeda, K., Henbest, K. B., Cintolelsi, F., Kuprov, I., Rodgers, C. T., Liddell, P. A., Gust, D., Timmel, C. R., and Hore, P. J. (2008) Chemical compass model of avian magnetoreception, *Nature*, **453**, 387–390.
47. Kodis, G., Liddell, P. A., de la Garza, L., Clausen, P. C., Lindsey, J. S., Moore, A. L., Moore, T. A., and Gust, D. (2002) Efficient energy transfer and electron transfer in an artificial photosynthetic antenna-reaction center complex, *J. Phys. Chem. A*, **106**, 2036–2048.
48. Kuciauskas, D., Liddell, P. A., Lin, S., Johnson, T. E., Weghorn, S. J., Lindsey, J. S., Moore, A. L., Moore, T. A., and Gust, D. (1999) An artificial photosynthetic antenna-reaction center complex, *J. Am. Chem. Soc.*, **121**, 8604–8614.
49. Aratani, N., Osuka, A., Kim, Y. H., Jeong, D. H., and Kim, D. (2000) Extremely long, discrete meso-meso-coupled porphyrin arrays, *Angew. Chem. Int. Ed. Engl.*, **39**, 1458–1462.
50. Bothner-By, A. A., Dadok, J., Johnson, T. E., and Lindsey, J. S. (1996) Molecular dynamics of covalently-linked multi-porphyrin arrays, *J. Phys. Chem.*, **100**, 17,551–17,557.
51. Burrell, A. K., Officer, D. L., Plieger, P. G., and Reid, D. C. W. (2001) Synthetic routes to multiporphyrin arrays, *Chem. Rev.*, **101**, 2751–2796.
52. Balzani, V. (1987) Energy transfer in synthetic porphyrin arrays, in *Supramolecular Photochemistry* (ed. Harriman, A.), D. Reidel Publishing Company, pp. 207–223.
53. Li, J., Ambroise, A., Yang, S. I., Diers, J. R., Seth, J., Wack, C. R., Bocian, D. F., Holten, D., and Lindsey, J. S. (1999) Template-directed synthesis, excited state photodynamics, and electronic communication in a hexameric wheel of porphyrins, *J. Am. Chem. Soc.*, **121**, 8927–8940.
54. Li, J., Diers, J. R., Seth, J., Yang, S. I., Bocian, D. F., Holten, D., and Lindsey, J. S. (1999) Synthesis and properties of star-shaped multiporphyrin-phthalocyanine light-harvesting arrays, *J. Org. Chem.*, **64**, 9090–9100.

55. Lin, V. S. Y., Dimagno, S. G., and Therien, M. J. (1994) Highly conjugated, acetylenyl bridged porphyrins: New models foe light-harvesting antenna systems, *Science*, **264**, 1105–1111.
56. Nakano, A., Osuka, A., Yamazaki, I., Yamazaki, T., and Nishimura, Y. (1998) Windmill-like porphyrin arrays as potent light-harvesting antenna complexes, *Angew. Chem. Int. Ed. Engl.*, **37**, 3023–3027.
57. Paolesse, R., Jaquinod, L., Della Sala, F., Nurco, D. J., Prodi, L., Montalti, M., Di Natale, C., D'Amico, A., Di Carlo, A., Lugli, P., and Smith, K. M. (2000) Beta-fused oligoporphyrins: A novel approach to a new type of extended aromatic system, *J. Am. Chem. Soc.*, **122**, 11295–11302.
58. Rucareanu, S., Mongin, O., Schuwey, A., Hoyler, N., Gossauer, A., Amrein, W., and Hediger, H.-U. (2001) Supramolecular assemblies between macrocyclic porphyrin hexamers and star-shaped porphyrin arrays, *J. Org. Chem.*, **66**, 4973–4988.
59. Cho, H. S., Rhee, H., Song, J. K., Min, C.-K., Takase, M., Aratani, N., Cho, S., Osuka, A., Joo, T., and Kim, D. (2003) Excitation energy transport processes of porphyrin monomer, dimer, cyclic trimer, and hexamer probed by ultrafast fluorescence anisotropy decay, *J. Am. Chem. Soc.*, **125**, 5849–5860.
60. Brodard, P., Matzinger, S., Vauthey, E., Mongin, O., Papamicaël, C., and Gossauer, A. (1999) Investigations of electronic energy transfer dynamics in multiporphyrin arrays, *J. Phys. Chem. A*, **103**, 5858–5870.
61. Hwang, I.-W., Kamada, T., Ahn, T. K., Ko, D. M., Nakamura, T., Tsuda, A., Osuka, A., and Kim, D. (2004) Porphyrin boxes constructed by homochiral self-sorting assembly: Optical separation, exciton coupling, and efficient excitation energy migration, *J. Am. Chem. Soc.*, **126**, 16187–16198.
62. Morandeira, A., Vauthey, E., Schuwey, A. and Gossauer, A. (2004) Ultrafast excited state dynamics of tri- and hexaporphyrin arrays, *J. Phys. Chem. A*, **108**, 5741–5751.
63. Nakamura, Y., Hwang, I.-W., Aratani, N., Ahn, T. K., Ko, D. M., Takagi, A., Kawai, T., Matsumoto, T., Kim, D., and Osuka, A. (2005) Directly meso-meso linked porphyrin rings: Synthesis, characterization, and efficient excitation energy hopping, *J. Am. Chem. Soc.*, **127**, 236–246.
64. Aratani, N., Osuka, A., Cho, H. S., and Kim, D. (2002) Photochemistry of covalently-linked multiporphyrinic systems, *J. Photochem. Photobiol. C*, **3**, 25–52.
65. Bensasson, R. V., Land, E. J., Moore, A. L., Crouch, R. L., Dirks, G., Moore, T. A., and Gust, D. (1981) Mimicry of antenna and photoprotective carotenoid functions by a synthetic carotenoporphyrin, *Nature*, **290**, 329–332.

66. Gust, D., Moore, T. A., Moore, A. L., Devadoss, C., Liddell, P. A., Hermant, R. M., Nieman, R. A., Demanche, L. J., DeGraziano, J. M., and Gouni, I. (1992) Triplet and singlet energy transfer in carotene-porphyrin dyads: role of the linkage bonds, *J. Am. Chem. Soc.*, **114**, 3590–3603.
67. Kodis, G., Herrero, C., Palacios, R., Mariño-Ochoa, E., Gould, S. L., de la Garza, L., van Grondelle, R., Gust, D., Moore, T. A., Moore, A. L., and Kennis, J. T. M. (2004) Light harvesting and photoprotective functions of carotenoids in compact artificial photosynthetic antenna designs, *J. Phys. Chem. B*, **108**, 414–425.
68. Liddell, P. A., Nemeth, G. A., Lehman, W. R., Joy, A. M., Moore, A. L., Bensasson, R. V., Moore, T. A., and Gust, D. (1982) Mimicry of carotenoid function in photosynthesis: Synthesis and photophysical properties of a carotenopyropheophorbide, *Photochem. Photobiol.*, **36**, 641–645.
69. Macpherson, A. N., Gillbro, T., Liddell, P. A., Tatman, D., Durantini, E. N., Gust, D., Moore, T. A., and Moore, A. L. (1998) Singlet energy transfer from carotenoids to cyclic tetrapyrroles in artificial photosynthetic antennas, *Photochem. Photobiol.*, **67S**, 30.
70. Macpherson, A. N., Liddell, P. A., Kuciauskas, D., Tatman, D., Gillbro, T., Gust, D., Moore, T. A., and Moore, A. L. (2002) Ultrafast energy transfer from a carotenoid to a chlorin in a simple artificial photosynthetic antenna, *J. Phys. Chem. B*, **106**, 9424–9433.
71. Mariño-Ochoa, E., Zarate-Kalfopulos, X., Macpherson, A. N., Gust, D., Moore, T. A., and Moore, A. L. (26–30 August 2001) Ultrafast energy transfer from carotenoids to phthalocyanines through the central silicon ligand in artificial photosynthetic antennas. In *Abstracts of Papers, 222nd ACS National Meeting, Chicago, IL, United States.*
72. Berera, R., van Stokkum, I. H. M., Kodis, G., Keirstead, A. E., Pillai, S., Herrero, C., Palacios, R. E., Vengris, M., van Grondelle, R., Gust, D., Moore, T. A., Moore, A. L., and Kennis, J. T. M. (2007) Energy transfer, excited-state deactivation, and exciplex formation in artificial caroteno-phthalocyanine light-harvesting antennas, *J. Phys. Chem. B*, **111**, 6868–6877.
73. Terazono, Y., Liddell, P. A., Garg, V., Kodis, G., Brune, A., Hambourger, M., Moore, T. A., Moore, A. L. and Gust, D. (2005) Artificial photosynthetic antenna-reaction center complexes based on a hexaphenylbenzene core, *J. Porphyrins Phthalocyanines*, **9**, 706–723.
74. Bai, X. and Holdcroft, S. (1993) Molecular control of luminescence from poly(3-hexylthiophenes), *Macromolecules*, **26**, 4457–4460.
75. Terazono, Y., Kodis, G., Liddell, P. A., Garg, V., Gervaldo, M., Moore, T. A., Moore, A. L. and Gust, D. (2007) Photoinduced electron transfer in

a hexaphenylbenzene-based self-assembled porphyrin-fullerene triad, *Photochem. Photobiol.*, **83**, 464–469.

76. Kadish, K. M., Smith, K. M., and Guilard, R. (2000) Equilibrium data of porphyrins and metalloporphyrins, in *The Porphyrin Handbook* (eds. Tabata, M. and Nishimoto, J.), Academic Press, New York, pp. 221–419.
77. D'Souza, F., Deviprasad, G. R., Zandler, M. E., Hoang, V. T., Klykov, A., VanStipdonk, M., Perera, A., El Khouly, M. E., Fujitsuka, M., and Ito, O. (2002) Spectroscopic, electrochemical, and photochemical studies of self-assembled via axial coordination zinc porphyrin-fulleropyrrolidine dyads, *J. Phys. Chem. A*, **106**, 3243–3252.
78. D'Souza, F., Deviprasad, G. R., Zandler, M. E., El Khouly, M. E., Fujitsuka, M., and Ito, O. (2003) Photoinduced electron transfer in "two-point" bound supramolecular triads composed of N,N-dimethylaminophenyl-fullerene-pyridine coordinated to zinc porphyrin, *J. Phys. Chem. A*, **107**, 4801–4807.
79. D'Souza, F., Smith, P. M., Zandler, M. E., McCarty, A. L., Itou, M., Araki, Y., and Ito, O. (2004) Energy transfer followed by electron transfer in a supramolecular triad composed of boron dipyrrin, zinc porphyrin, and fullerene: A model for the photosynthetic antenna-reaction center complex, *J. Am. Chem. Soc.*, **126**, 7898–7907.
80. D'Souza, F., and Ito, O. (2005) Photoinduced electron transfer in supramolecular systems of fullerenes functionalized with ligands capable of binding to zinc porphyrins and zinc phthalocyanines, *Coord. Chem. Rev.*, **249**, 1410–1422.
81. D'Souza, F., Chitta, R., Gadde, S., Islam, D.-M. S., Schumacher, A. L., Zandler, M. E., Araki, Y., and Ito, O. (2006) Design and studies on supramolecular ferrocene-porphyrin-fullerene constructs for generating long-lived charge separated states, *J. Phys. Chem. B*, **110**, 25240–25250.
82. Demmig-Adams, B., and Adams III, W. W. (2002) Antioxidants in photosynthesis and human nutrition, *Science*, **298**, 2149–2153.
83. Li, X. P., Gilmore, A. M., Caffarri, S., Bassi, R., Golan, T., Kramer, D., and Niyogi, K. K. (2004) Regulation of photosynthetic light harvesting involves intrathylakoid lumen pH sensing by the PsbS protein, *J. Biol. Chem.*, **279**, 22866–22874.
84. Wilson, A., Punginelli, C., Gall, A., Bonetti, C., Alexandre, M., Routaboul, J. M., Kerfeld, C. A., van Grondelle, R., Robert, B., Kennis, J. T. M., and Kirilovsky, D. (2008) A photoactive carotenoid protein acting as light intensity sensor, *Proc. Natl. Acad. Sci. U. S. A.*, **105**, 12075–12080.

85. Straight, S. D., Kodis, G., Terazono, Y., Hambourger, M., Moore, T. A., Moore, A. L., and Gust, D. (2008) Self-regulation of photoinduced electron transfer by a molecular nonlinear transducer, *Nat. Nanotechnol.*, **3**, 280–283.

86. Youngblood, W. J., Lee, S.-H. A., Kobayashi, Y., Hernandez-Pagan, E. A., Hoertz, P. G., Moore, T. A., Moore, A. L., Gust, D., and Mallouk, T. E. (2009) Photoassisted overall water splitting in a visible light-absorbing dye-sensitized photoelectrochemical cell, *J. Am. Chem. Soc.*, **131**, 926–927.

Chapter 8

Multiporphyrins-Fullerenes and Multiporphyrins-SWCNTs Mimicking Photosynthetic Antenna-Reaction Center

Francis D'Souza[a] and Osamu Ito[b]

[a] *Department of Chemistry, University of North Texas, 1155 Union Circle, #305070, Denton, TX 76203-5017, USA*

[b] *Fullerene Group, National Institute of Material Science, Tsukuba, and CarbonPhotoScience Laboratory, Kita-Nakayama 2-1-6, Izumi-Ku, Sendai, 981-3215 Japan*

Francis.DSouza@UNT.edu; ito@tagen.tohoku.ac.jp

Recent progress on the design, construction, and photoinduced energy- and electron-transfer applications of self-assembled donor–acceptor arrays has been reviewed, focusing mainly on the multiporphyrin-fullerene and multiporphyrin-SWCNT supramolecular arrays to construct photosynthetic antenna and reaction center models. In addition to the covalently bonded entities, self-assemblies through intermolecular interactions including π–π attraction, ion-pairing, cation-dipole, hydrogen bonding, utilized in the construction of multiporphyrin-fullerene or multiporphyrin-SWCNT are discussed. The control over the energy- and electron-transfer events upon excitation of the multiporphyrins in such supramolecular arrays is accomplished. These studies demonstrate

Multiporphyrin Arrays: Fundamentals and Applications
Edited by Dongho Kim

www.panstanford.com

that it is possible to develop self-assembled multiporphyrin conjugates capable of generating considerably high light converting and long-lived charge-separated species useful for light-energy-harvesting applications.

8.1 Introduction

A deep understanding of the light-induced energy- and electron-transfer events occurring among the well-organized pigments in natural photosynthesis is essential for the construction of artificial molecular and supramolecular systems capable of efficient conversion of light energy into chemical or electrical energy [1–3]. The sophisticated structural studies and a number of spectroscopic studies of natural photosynthetic antenna and reaction centers have provided a wealth of information [4–10]. For example, in the bacterial photosynthetic reaction centers, it has been revealed that the various electron donor and acceptor entities are arranged via non-covalent incorporation into a well-defined protein matrix. The conversion of solar energy in photosynthesis depends on the ability of efficient energy transfer of light-harvesting pigments, and finally, on the formation of long-lived charge-separated states (Fig. 8.1).

The mechanism for lifetime extension of the charge-separated states in natural photosynthetic reaction centers involves distant isolation of the radical cation from the anion species in the charge-separated states, which involves sequential charge hopping using electron-mediator molecules that are positioned appropriately. This evolving attainment eliminates energy-wasting back electron transfer. Additionally, the small reorganization energy ($\lambda_{reorg} \sim$ 0.2 eV) emerged in the photosynthetic reaction center and the well-balanced electronic coupling between the electron donor and acceptor entities have been attributed to the high efficiency of charge separation and slow charge recombination [11, 12].

It has been realized that the construction of molecular devices capable of efficient photo-energy conversion is possible by mimicking the strategy adopted by nature in photosynthesis [13, 14]. Hence, development of covalently bonded molecular triads composed of antenna, electron-donor, and electron-acceptor (Fig. 8.2) has been

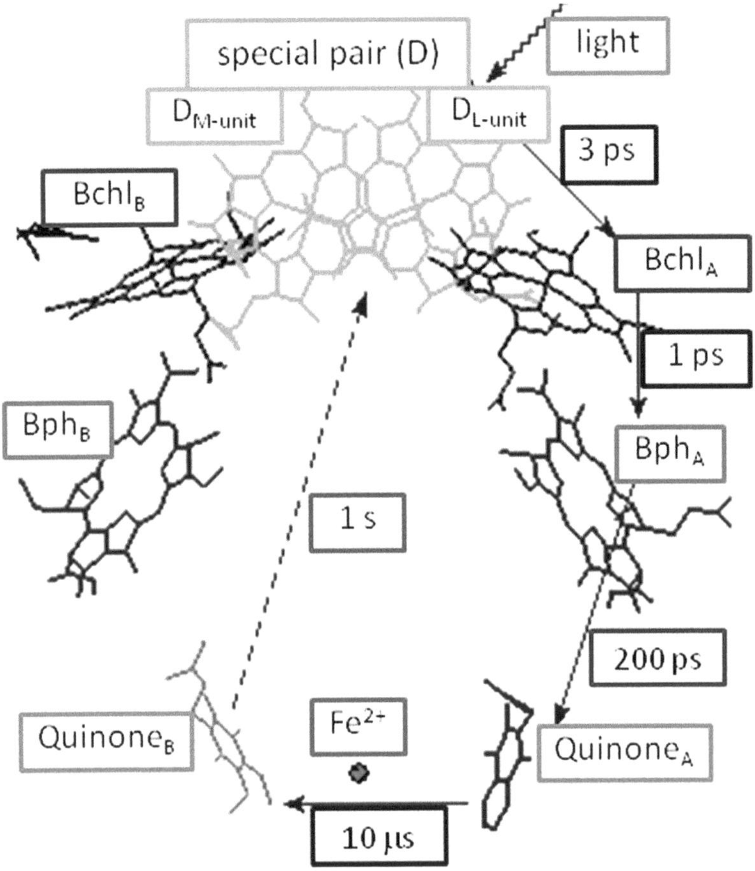

Figure 8.1. Photosynthetic reaction centers of purple bacteria. Adapted from Ref. [4]. See also Color Insert.

one of the important goals during the past two decades [15–20]. In addition to the conversion of solar energy into electric energy as a low-cost alternative for silicon-based photovoltaic materials, utilization of these molecular systems to build new type photovoltaic and optoelectronic devices offers wide applications, including hydrogen production from water [21–24]. Accordingly, a number of molecular systems which satisfy the above prerequisites have been developed [15–20]. In order to increase the rate of forward electron transfer and to slow down the rate of charge recombination,

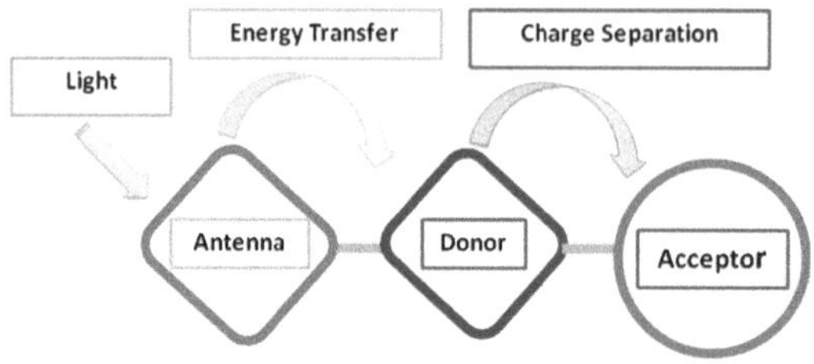

Figure 8.2. Schematic of a covalently linked triad for mimicking photosynthetic antenna-reaction center functionality. See also Color Insert.

improved model compounds with well-adjusted energies of the donor and acceptor entities and finely tuned electronic coupling between them have been elegantly designed and studied. However, one of the draw backs of such covalently bonded molecular arrays is their poor photo-stability considering the tremendous efforts involved in synthesizing such molecular arrays.

In recent years, utilizations of biomimetic supramolecular principles to self-assemble donor–acceptor entities have received considerable attention to overcome the limitations of the covalently linked molecular arrays [25–27]. The combination of weak interactions among the functional molecules makes it possible to define the size and shape of the resulting supramolecular ensembles, achieving targeted supramolecular arrays (Fig. 8.3). For these purposes, different types of weak forces, such as hydrogen bonding, π–π stacking, metal-mediated complexation, complementary

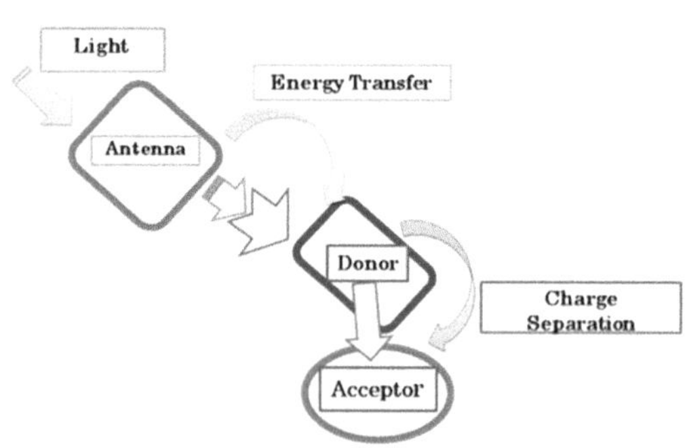

Figure 8.3. Schematic example of a non-covalently linked supramolecular triad system. See also Color Insert.

base-pairing, and electrostatic interactions, have been used. The assemblies constructed using the biomimetic motifs are reversible due to weak-binding interactions, and can be easily tuned by means of the chemical environment and temperature. Consequently, the highly selective and directional intermolecular interactions have afforded a wide variety of new architectures, in which electron-transfer processes between different electron donors and the electron acceptor are shown to take place as efficiently as in the related covalently connected molecular systems.

Several reviews and monographs have been published summarizing the significant findings of this fascinating research area, which have provided us with fortune of information with regard to the nature of the photoactive systems, including the donor–acceptor molecular topology, and nature and distance of the spacer units connecting them. The purpose of this chapter is to summarize key self-assembly approaches with porphyrins [28–31] and phthalocyanines [33–35] developed over the years to form donor–acceptor hybrids in self-assembled systems with photo-stable acceptors such as functionalized nanocarbons. Porphyrins and phthalocyanines are widely utilized donors for electron-transfer studies because of their resemblance to naturally occurring tetrapyrroles and also due to their well-characterized photochemical/photophysical properties such as relatively high absorbance in the visible region and high electron-donor abilities, so that they are suitable for efficient electron transfer in the ground and excited states. Importantly, the porphyrin/phthalocyanine macrocycles are capable of binding a variety of transition metals within their central cavity in which some metal ions such as Zn, Mg, and Ru are coordinately unsaturated. By utilizing the positions axial to the plane of the metal macrocycle ring, multiporphyrin arrays are designed to afford the light-harvesting energy-transfer arrays coupled with charge-separation centers with functionalized nanocarbons such as fullerenes, SWCNT, etc.

For the selection of an ideal electron acceptor, one of the criteria is to employ molecules with small reorganization energy such as fullerenes having large spherical shapes [36–44]. Similarly, 3D carbons with molecular characters such as carbon nanotubes [45–53], and nanohorns [54–63], may also have small reorganization energies. Interestingly, the 2D graphenes having large size π-electron systems may also provide features of electron acceptor [64]. Thus, the all-carbon nanostructures with two- and three-dimensional structures represent exceptional scaffolds for the development of distinctive 2D and 3D architectures. Thus, the remarkable electronic properties of nanocarbons upon incorporation into artificial reaction centers by a combination of the visible-light harvesting and photosensitizing porphyrins and phthalocyanines would yield a range of novel light energy harvesting architectures [36–44]. Their relatively low-reduction potentials, low solvent and internal reorganization energies, low susceptibility to solvent stabilization of the radical anion, and potential ability to act as electron accumulators are additional qualities that would enhance their performance [62, 63]. As a consequence, nanocarbons in molecular arrays is expected to accelerate forward electron transfer (charge-separation; k_{CS}) and slow down the backward electron transfer (charge-recombination; k_{CR}), thus, generating the much desired long-lived charge-separated states. Hence, utilization of these nanocarbons seems to be ideal choice for enhancing the light-harvesting efficiency throughout the solar spectrum, and the conversion of the harvested light into the high-energy state of the charge separation.

The chapter is divided into two parts; the first part is focused on studies involving fullerene as electron acceptor systems, while the second part is focused on studies involving carbon nanotubes systems with multi-tetrapyrrole donors.

8.2 Porphyrin-Fullerene-Porphyrin Supramolecular Systems

In order to construct multiporphyrin-fullerene molecular arrays, a combination of covalent and non-covalent approaches has been widely employed by taking advantage of their inherent properties. Non-covalent assemblies constructed using the biomimetic self-assembly motifs include the merit of reversible binding that can afford self-repairing function, in addition to relatively easier construction saving on laborious synthetic efforts. The fundamental studies to establish self-assembly protocols and photoinduced electron transfer to achieve long-lived charge-separated states have been reviewed in this section including applications of photocatalysis. The subsections are organized according to the different self-assembly methods adopted to form the antenna–donor–acceptor hybrids and multi-donor–acceptor hybrids composed of porphyrin-porphyrin-fullerene and phthalocyanine analogies followed by the studied photoinduced energy- and electron-transfer processes.

8.2.1 *Photoinduced Successive Energy- and Electron Transfer*

A supramolecular triad constructed by the combination of covalent and non-covalent approaches is shown in Fig. 8.4 [65]. A covalently linked free-base porphyrin-fullerene with a pyridine entity, H_2P-C_{60}Py, is coordinated to zinc porphyrin (ZnTPP) to form H_2P-C_{60}Py$\rightarrow$ZnTPP in which $\rightarrow$ is used for coordination bond (Fig. 8.4). The optimized structure is obtained as shown in Fig. 8.4, in which C_{60} moiety is placed between the H_2P and ZnTPP moieties, suitable for energy- and electron-transfer processes toward the central C_{60} unit. That is, ZnTPP is expected to be light-absorbing antenna and energy donor, whereas the H_2P moiety in H_2P-C_{60}Py receives the excited energy and donates its electron to the C_{60} unit resulting in the charge-separation process [65].

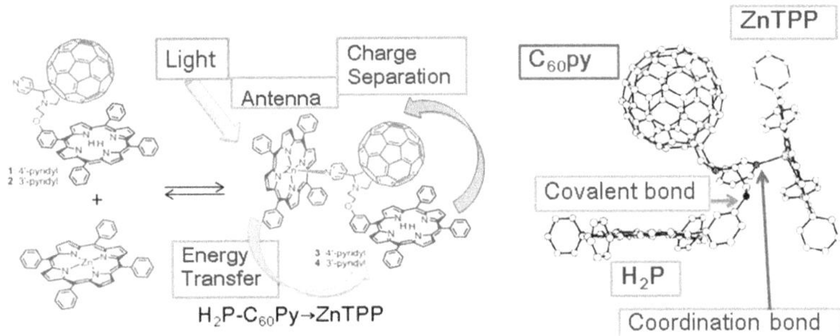

Figure 8.4. (Left) Self-assembly of H_2P-C_{60}Py$\rightarrow$ZnTPP to result in a triad, and the expected roles of the components. (Right) Optimized structure of H_2P-C_{60}Py$\rightarrow$ZnTPP. Modified from Ref. [65]. See also Color Insert.

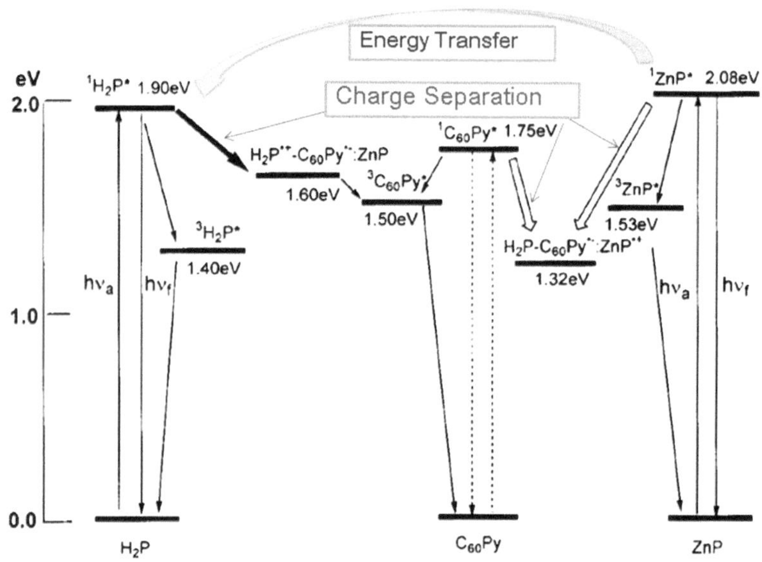

Figure 8.5. Energy level diagram showing the photochemical events of the supramolecular H_2P-C_{60}Py→ZnTPP triad in *o*-DCB. Modified from Ref. [65].

The energy level diagram constructed on the basis of the electrochemical data and the excited states of the components is shown in Fig. 8.5. The higher energy level of ^{1}ZnTPP* than that of $^1H_2P^*$ makes the energy transfer from ^{1}ZnTPP* to H_2P possible. The energy levels of the radical ion pair, $H_2P^{\bullet+}$-$C_{60}^{\bullet-}$Py→ZnTPP and H_2P-$C_{60}^{\bullet-}$Py→ZnTPP$^{\bullet+}$ are lower than those of ^{1}ZnTPP* and $^1H_2P^*$, supporting that the exothermic charge-separation processes are taking place via ^{1}ZnTPP* and $^1H_2P^*$. The charge separation via ^{1}ZnTPP* to generate H_2P-$C_{60}^{\bullet-}$Py→ZnTPP$^{\bullet+}$is more exothermic than that via $^1H_2P^*$ to generate H_2P-$C_{60}^{\bullet-}$Py→ZnTPP$^{\bullet+}$. Since the energy level of H_2P-$C_{60}^{\bullet-}$Py→ZnTPP$^{\bullet+}$ is lower than the triplet states of the components, the charge recombination is possibly taking place to the ground state. On the other hand, the energy level of $H_2P^{\bullet+}$-$C_{60}^{\bullet-}$Py→ZnTPP is higher than these triplet states, thus the charge recombination to the triplet states is possible.

Spectroscopic studies on H_2P-C_{60}Py dyad revealed occurrence of photoinduced charge separation between $^1H_2P^*$ and C_{60}Py producing the radical ion pair, $H_2P^{\bullet+}C_{60}Py^{\bullet-}$. Steady-state fluorescence

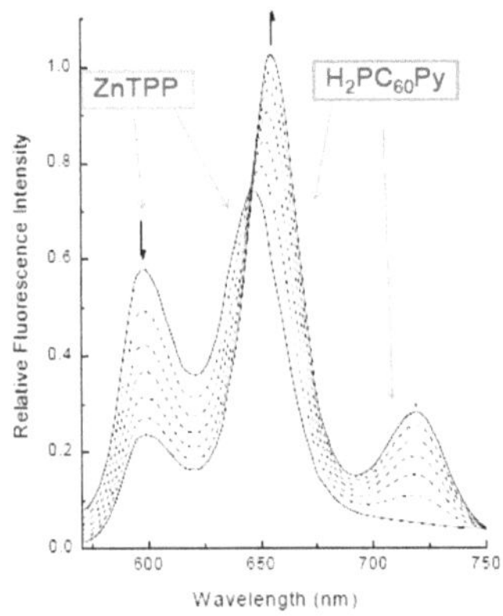

Figure 8.6. Steady-state fluorescence spectral changes of ZnTPP on addition of H_2P-C_{60}Py by predominant excitation of ZnTPP in *o*-dichlorobenzene (*o*-DCB). Modified from Ref. [65].

spectral studies performed on the mixture of ZnTPP and H_2P-C_{60}Py with predominant excitation of ZnTPP resulted in appreciable amounts of ZnTPP-fluorescence quenching as shown in Fig. 8.6, giving an evidence of the supramolecular formation. With concomitant decrease of the ZnTPP fluorescence upon addition of H_2P-C_{60}Py, the appearance of H_2P-fluorescence was observed, suggesting occurrence of energy-transfer to the H_2P moiety, in addition to the increase in its light absorption. Both fluorescence intensities are extremely small, supporting that predominant fluorescence quenching is attributed to the electron transfer to the C_{60}Py moiety. Picosecond time-resolved fluorescence spectral studies also support slow-energy transfer from the ^{1}ZnTPP* to the H_2P moiety, whereas rapid charge separation takes place toward the central C_{60}Py moiety from ^{1}ZnTPP* and $^1H_2P^*$ in H_2P-C_{60}Py→ZnTPP triad.

Nanosecond transient absorption spectral studies were performed to determine the nature of the photo-processes via the excited state in the supramolecular triad systems (Fig. 8.7). The efficiency of intermolecular electron transfer was evaluated by monitoring the absorbance of the transient absorption bands at 1000 nm due to the $C_{60}^{\bullet-}$ moiety; the initial absorbance at 1000 nm for H_2P-C_{60}Py→ZnTPP is slightly higher than that of H_2P-C_{60}Py. This observation indicates that the photoinduced charge separation

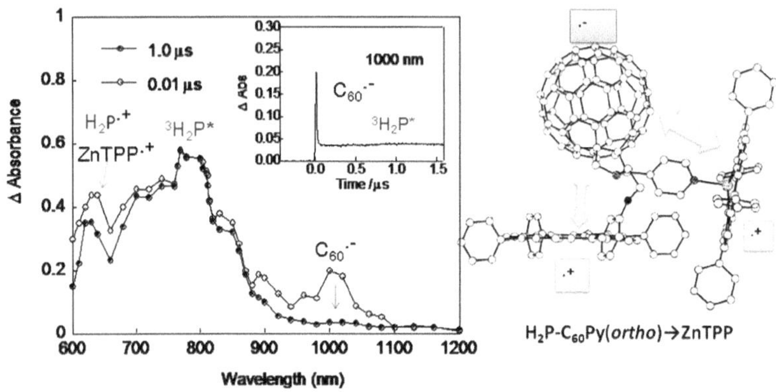

Figure 8.7. (Left) Nanosecond transient absorption spectra of H_2P-$C_{60}Py\rightarrow ZnTPP$ triad by the predominant excitation of ZnTPP in *o*-DCB. (Right) Close distances between electron and hole in H_2P-$C_{60}Py$(*para*)$\rightarrow$ZnTPP. Modified from Ref. [65].

via $^1ZnTPP^*$, generating the radical ion pair (RIP) like as H_2P-$C_{60}^{\bullet -}Py\rightarrow ZnTPP^{\bullet +}$ takes place in addition to the photosensitized charge separation via $^1H_2P^*$, generating $H_2P^{\bullet +}$-$C_{60}^{\bullet -}Py\rightarrow ZnTPP$. From the time profile at 1000 nm, the charge recombination rate constant (k_{CR}) is evaluated to be ca. 3×10^7 s^{-1}, corresponding to the lifetime of the RIP (τ_{RIP}) to be ca. 30 ns in *o*-DCB [65].

On the other hand, boron dipyrrin (BDP), which is well known as an energy funneling antenna is employed to observe clear energy transfer followed by the charge separation process. Thus, a new supramolecular triad, $C_{60}Im\rightarrow ZnP$-BDP was designed and constructed (Fig. 8.8), in which the ZnP moiety in $C_{60}Im\rightarrow ZnP$ unit is energy receiver from $^1BDP^*$ and the charge separation takes place through the coordination bond via the resultant $^1ZnP^*$ [66]. The optimized structure in Fig. 8.8 shows arrangement of antenna-donor–acceptor in this order. The HOMO of the triad is localized on the ZnP unit, whereas the LUMO is localized on the C_{60} moiety, predicting the most stable RIP as $C_{60}^{\bullet -}Im\rightarrow ZnP^{\bullet +}$-BDP.

Selective excitation of the BDP moiety is possible with 550 nm light resulting in an efficient energy transfer ($k_{EnT}^{S} = 9.2\times 10^9$ s^{-1}) from $^1BDP^*$ to ZnP, creating $^1ZnP^*$ with quantum yield of ~0.8, as evaluated from the fluorescence decay of BDP at 522 nm and

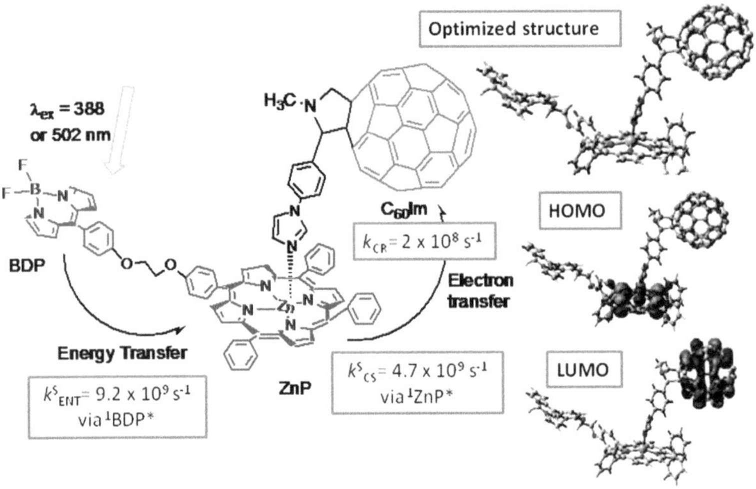

Figure 8.8. Structure of C_{60}Im→ZnP-BDP triad, its optimized structure, the HOMO and the LUMO. Modified from Ref. [66]. See also Color Insert.

the corresponding rise of ZnP fluorescence at 600 nm (Fig. 8.9; inset). Then, the ZnP-fluorescence lifetime of ZnP-BDP was shorter than that of the pristine ZnP, suggesting that the slow charge separation takes place to generate $ZnP^{\bullet+}$-$BDP^{\bullet-}$ with $k^S_{CS} = 3.1 \times 10^8\ s^{-1}$. Upon forming the supramolecular C_{60}Im→ZnP-BDP triad, the ZnP-fluorescence lifetime became further shorten due to the fast charge separation to the C_{60}Im moiety via $^1ZnP^*$, generating $C_{60}^{\bullet-}$Im→$ZnP^{\bullet+}$-BDP with $k^S_{CS} = 4.7 \times 10^9\ s^{-1}$ [66].

The confirmation of the charge separation was obtained by the transient absorption spectra of C_{60}Im→ZnP-BDP (Fig. 8.10), which showed the $C_{60}^{\bullet-}$-absorption band at 1000 nm, although the $ZnP^{\bullet+}$-absorption band expected to appear in the 600–650 nm was buried in the broad bands of the triplet states. These processes are well interpreted by the energy level diagram of C_{60}Im→ZnP-BDP showing the photochemical events. The thick arrows represent fast and efficient processes, whereas thin and dashed arrows represent less slow and minor processes [66]. From the decay of the $C_{60}^{\bullet-}$-transient band, the k_{CR1} for $C_{60}^{\bullet-}$Im→$ZnP^{\bullet+}$-BDP was evaluated to be $2 \times 10^8\ s^{-1}$, corresponding to the lifetime of the RIP (τ_{RIP}) to be ca. 5 ns, which is not so long because of the close structure between

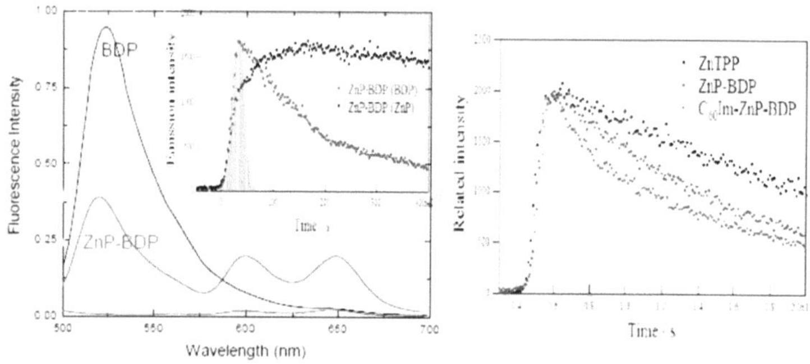

Figure 8.9. (Left) Steady-state fluorescence spectra in *o*-DCB; (black) BDP, (red) ZnP-BDP, and (purple) C_{60}Im→ZnP-BDP. Inset: Fluorescence time profiles of ZnP-BDP monitored at ZnP and BDP fluorescence wavelengths. (Right) Fluorescence time-profiles at 600 nm of ZnP; (black) ZnTPP, (blue) ZnP-BDP, and (red) C_{60}Im→ZnP-BDP in *o*-DCB. Hatched area shows the laser pulse profile of the excited light at 550 nm. Modified from Ref. [66]. See also Color Insert.

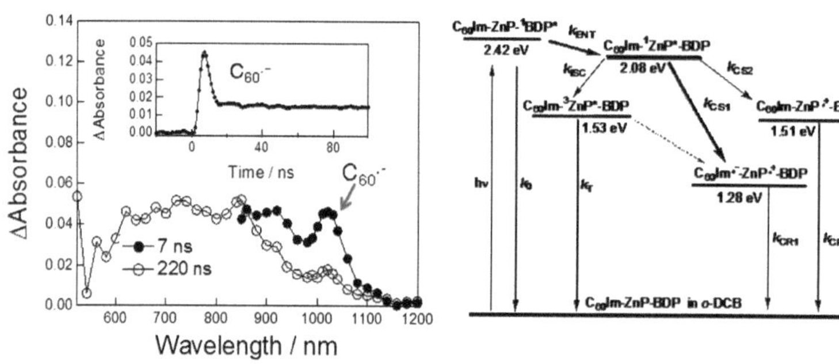

Figure 8.10. (Left) Transient absorption spectra of the supramolecular triad formed by coordinating C_{60}Im (0.25 mM) to ZnP-BDP in *o*-DCB at 7 ns (•) and 160 ns (○) after the 500 nm laser light irradiation. Inset: Absorption-time profile at 1020 nm. (Right) Energy level diagram showing the different photochemical events of the supramolecular C_{60}Im-ZnP-BDP triad after excitation of the BDP moiety. Modified from Ref. [66].

$C_{60}^{\bullet-}$ and $ZnP^{\bullet+}$ connected with coordination bond (Fig. 8.8). The k_{CR2} for C_{60}Im→$ZnP^{\bullet+}$-$BDP^{\bullet-}$ was faster, because of the intramolecular event through the short covalent bond.

These to build antenna-reaction center was further extended to crown ether—ammonium cation interacting supramolecular

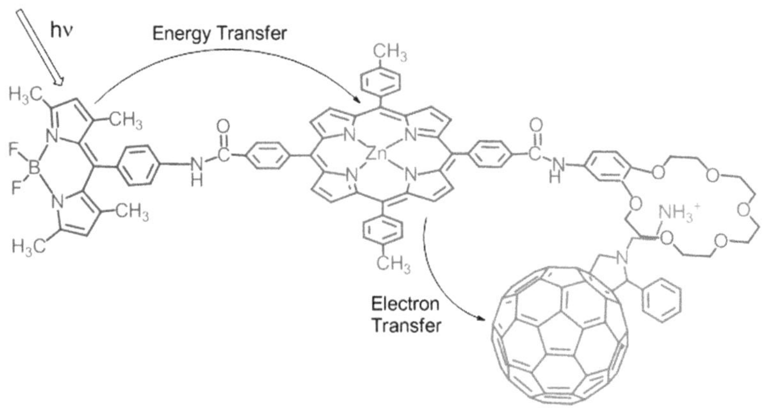

Figure 8.11. Structure of the C_{60}-NH_3^+; (Crown)ZnP-BDP) supramolecular triad. Modified from Ref. [67]. See also Color Insert.

triad as shown in Figure 8.11. For the construction, first, BDP was covalently attached to a ZnP entity bearing a benzo-18-crown-6 host segment at the opposite end of the porphyrin ring, (Crown)ZnP-BDP (Fig. 8.11). Next, an alkyl ammonium-functionalized fullerene (C_{60}-NH_3^+) is used to self-assemble by inserting the –NH_3^+ moiety to crown ether entity via ion–dipole interactions (NH_3^+; Crown). The newly formed supramolecular triad (C_{60}-NH_3^+; (Crown)ZnP-BDP) was fully characterized by spectroscopic and computational methods [67]. Selective excitation of the BDP moiety in the dyad resulted in efficient energy transfer over 97% efficiency creating ^{1}ZnP*. Upon forming the supramolecular C_{60}-NH_3^+; (Crown)ZnP-BDP triad, the ^{1}ZnP* results in electron transfer to the C_{60}-NH_3^+ moiety, yielding a charge-separated RIP. Nanosecond transient absorption studies confirm the long-living charge-separated state ($C_{60}^{\bullet-}$-NH_3^+; (Crown)$ZnP^{\bullet+}$-BDP). This supramolecular system represents a successful model to mimic the "combined antenna-reaction center" events of photosynthesis.

The third example was to utilize ferric porphyrin (Fe(P)Cl) as a final electron acceptor by constructing H_2P-C_{60}Py→Fe(P)Cl triads as shown in Fig. 8.12, in which the supramolecular triads were constructed by using "covalent-coordinate" binding strategy. For this aim, a H_2P was covalently linked to a C_{60}Py, and then the Py was

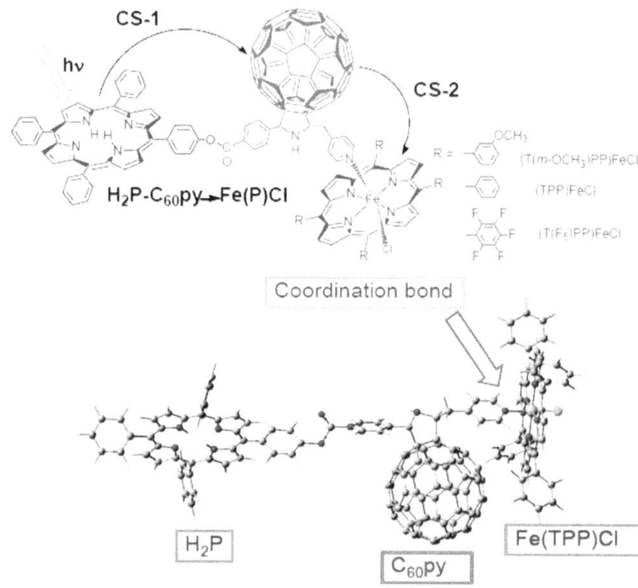

Figure 8.12. Structures of H_2P-C_{60}Py→Fe(P)Cl triads and the optimized structure of H_2P-C_{60}py→Fe(TPP)Cl triad. Modified from Ref. [68]. See also Color Insert.

subsequently utilized to coordinate to Fe(P)Cl) [68]. The newly formed triads were characterized by spectral methods and the structures were deduced from DFT calculations using DFT-B3LYP/3-21G(*) methods.

Cyclic voltammetry studies revealed that the potential of the first reduction process depends upon the substituents on the porphyrin macrocycle and followed the order: Fe(T(*m*-OCH_3)PP)Cl > Fe(TPP)Cl > Fe(T(F_5)PP)Cl. The first reduction process corresponds to the reduction from Fe^{III}(P)Cl to Fe^{II}(P)Cl, in which the latter is equivalent to $Fe^{III}(P^{\bullet-})$Cl. Importantly, these reduction potentials are smaller than that of the first reduction of C_{60}Py, suggesting higher electron accepting property of Fe(P)Cl than C_{60}Py (Fig. 8.13).

The energy-level diagram was constructed with the use of the redox potential and spectral data suggesting the occurrence of sequential electron transfer from $^1H_2P^*$ to the covalently linked C_{60}Py; then electron migration from $C_{60}^{\bullet-}$Py to Fe^{III}(P), generating

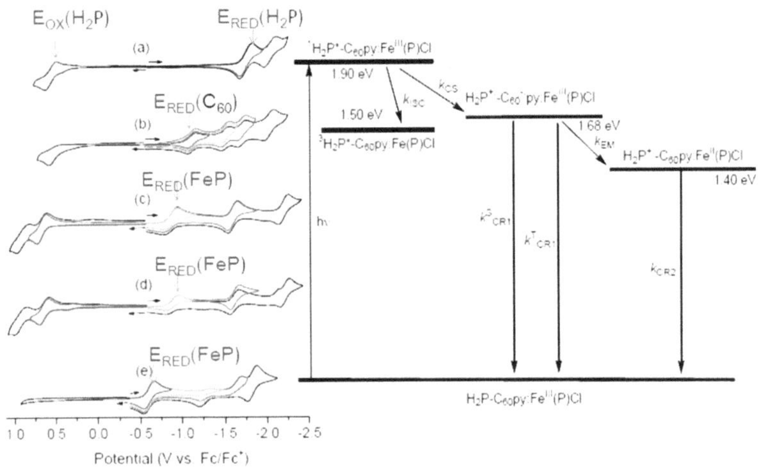

Figure 8.13. Cyclic voltammograms of (a) H_2P, (b) H_2P-C_{60}Py, (c) Fe(T(*m*-OCH_3)PP)Cl, (d) Fe(TPP)Cl, and (e) Fe(T(F_5)PP)Cl in *o*-DCB. Right: Energy level diagram. Modified from Ref. [68].

Fe^{II}(P) (equilibrating Fe($P^{\bullet -}$)Cl). Consequently, distant separation of the hole of $H_2P^{\bullet +}$ and the electron of Fe^{II}(P) anticipates the long-lived charge-separated RIPs, $H_2P^{\bullet +}$-C_{60}Py→Fe($P^{\bullet -}$)Cl.

Although H_2P-fluorescence intensity quenching and an increase in the fluorescence decay rate of the H_2P moiety were observed by the covalent connection of the C_{60}Py moiety, further quenching was observed on coordination to Fe(P)Cl as shown in Fig. 8.14 (left) and inset, respectively. Then, an increase of the k_{CS}^{S} values was observed from 1×10^9 s^{-1} for H_2P-C_{60}Py to (3–5) $\times$ 10^9 s^{-1} for H_2P-C_{60}Py→Fe(P)Cl. The substituents on the Fe(P)Cl also affected the H_2P-fluorescence decay rates through the C_{60}Py. The nanosecond transient absorption spectra showed that the $C_{60}^{\bullet -}$Py moiety at 1000 nm is generated at the initial charge separation via $^1H_2P^*$, producing $H_2P^{\bullet +}$-$C_{60}Py^{\bullet -}$ →Fe(P)Cl. Then, the $C_{60}^{\bullet -}$Py moiety decayed within 100 ns with an increase of the 620 nm band due to Fe($P^{\bullet -}$)Cl (Fig. 8.14 (right)), giving kinetic and spectral evidence of electron migration from the central $C_{60}Py^{\bullet -}$ to the terminal acceptor Fe(P)Cl, thus, producing the final RIP, $H_2P^{\bullet +}$-C_{60}Py→Fe^{III}($P^{\bullet -}$)Cl (equiv. $H_2P^{\bullet +}$-C_{60}Py→Fe^{II}(P)Cl). Attempts were made to obtain the lifetime of the final charge-separated species by monitoring the

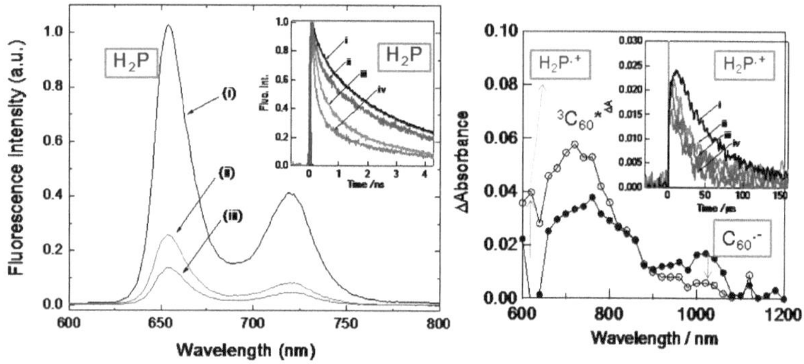

Figure 8.14. (Left) Steady-state fluorescence spectrum of (i) H_2TPP, (ii) H_2P-$C_{60}Py$ dyad, and (iii) H_2P-$C_{60}Py \rightarrow Fe(TPP)Cl$ triad in *o*-DCB; $\lambda_{ex} = 516$ nm. Inset. Fluorescence decays (600–750 nm) of $H_2PC_{60}Py \rightarrow Fe(P)Cl$; $\lambda_{ex} = 400$ nm. (i) $H_2PC_{60}Py$, (ii) $H_2PC_{60}Py \rightarrow Fe(T(m\text{-}OCH_3)PP)Cl$, (iii) $H_2PC_{60}Py \rightarrow Fe(TPP)Cl$, and (iv) $H_2PC_{60}Py \rightarrow Fe((F_5)TPP)Cl$. (Right): Nanosecond transient absorption spectra of H_2P-$C_{60}Py \rightarrow Fe(T(m\text{-}OCH_3)PP)Cl$ observed by 532 nm laser irradiation in at 0.01 μs (•) and 0.10 μs (○) in Ar-saturated *o*-DCB. Inset: Absorption-time profiles at 620 nm. Modified from Ref. [68].

decay of $H_2P^{\bullet +}$ at 620 nm. Lifetimes of the order of 20 μs were obtained, although this value is affected by the nearby long-living triplet states of porphyrin [68].

8.2.2 *Photoinduced Direct Charge Separation Between Multiporphyrin-Fullerene Systems*

Self-assembled bis(porphyrin)-fullerene conjugates were formed by utilizing fullerene derivative with two axially coordinating pyridine ligands. For simple ZnP, the bis-pyridine functionalized fullerene ($PyC_{60}Py$) was axially coordinated to two ZnP molecules independently in a Y-shape (Fig. 8.15) [69]. For ZnP dimer, the moderately stable supramolecular complex is formed as shown in Fig. 8.15 [70]. In both cases, the HOMO is localized on one of the ZnP moiety, whereas the LUMO is localized on the C_{60} moiety, suggesting the RIP like $ZnP^{\bullet +} \leftarrow PyC_{60}^{\bullet -}Py \rightarrow ZnP$ is formed. The time-resolved fluorescence measurements reveal the slow charge separation ($k_{CS} \approx 10^8$ s^{-1}) with quantum yield as low as 0.14.

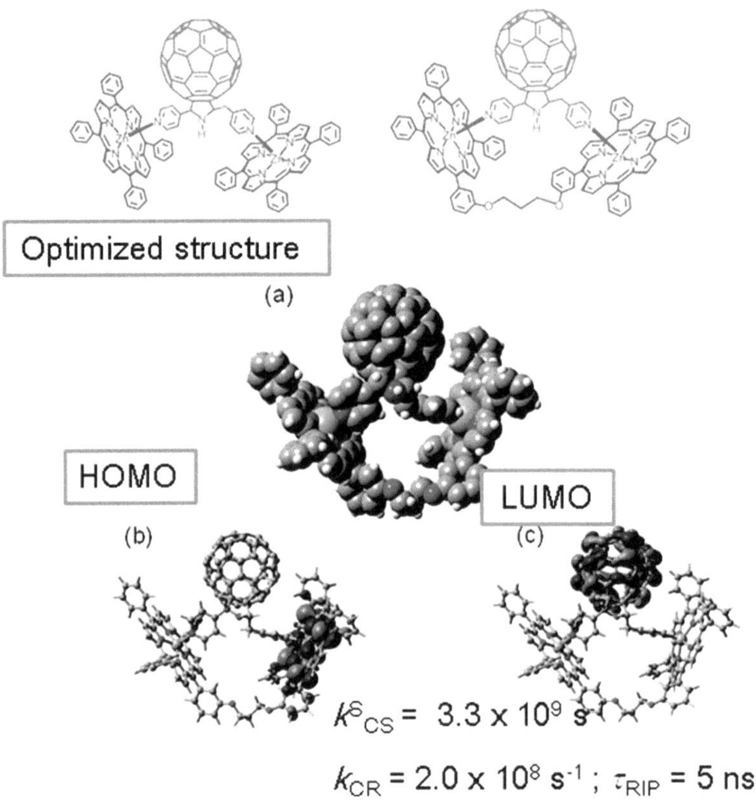

Figure 8.15. Two bisporphyrins-fullerene supramolecular complexes; (a) optimized structure, (b) the HOMO, and (c) LUMO of the ZnP dimer complex with PyC_{60} Py. Modified from Refs. [69, 70]. See also Color Insert.

Furthermore, ZnNc was utilized to self-assemble with PyC_{60}Py [71]. Although the optimized structure of the resulting complex was close to that of the ZnP coordinated complex, association constant (K_{asso}) is higher than the ZnP analog due to higher donor ability of ZnNc (Fig. 8.16 (right)). Little or no π–π interaction between two ZnNc wings was observed experimentally or via computational MOs. The main ZnNc-fluorescence quenching pathway involved charge separation from ^{1}ZnNc* to the C_{60} moiety with $k^S_{CS} = 5.7 \times 10^9$ s^{-1}, indicating an efficient photochemical electron-transfer process [71].

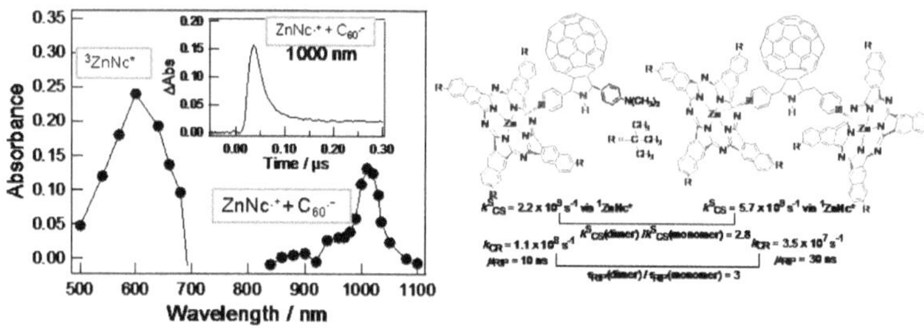

Figure 8.16. Nanosecond transient spectrum at 50 ns after 532-nm laser pulsed light irradiation of ZnNc (0.2 mM) in ZnNc←PyC$_{60}$Py→ZnN in Ar-saturated toluene. Kinetic parameters for ZnNc←PyC$_{60}$ and ZnNc←PyC$_{60}$Py→ZnNc in *o*-DCB. Modified from Refs. [71, 72].

Transient absorption spectrum observed at 50 ns exhibits the bands in the 980–1020 nm region corresponding to the formation of ZnNc$^{\bullet+}$ and $C_{60}^{\bullet-}$, indicating the formation of the RIP [72]. The time profile for the 1000 nm band showed a quick rise-and-decay behavior, corresponding to the fast charge separation via ^{1}ZnNc* within the laser pulse and the subsequent charge recombination, respectively (Fig. 8.16, inset). From the decay time-profile, the k_{CR} of the supramolecular RIP was estimated to be 3.5×10^7 s^{-1}, corresponding to $\tau_{RIP} = 30$ ns. The τ_{RIP} value for ZnNc←PyC$_{60}$Py→ZnN is slightly longer compared to ZnNc←PyC$_{60}$Py and the ZnP analogy [69].

Hydrogen bonding motif has been developed to construct multiple porphyrin-fullerene conjugates. Bis porphyrin-substituted diamidopyridine affords an example for multiple porphyrin–fullerene conjugate via the "three-point" hydrogen bonding as shown in Fig. 8.17 [73]. The steady-state absorption peak shift and fluorescence quenching support the conjugate formation. A more structural information can be obtained by the optimized geometry calculated by the B3LYP/3-21G$^{(*)}$ method; that is, one of the porphyrin units of the dimer is closer to the fullerene entity via the "three-point" hydrogen bonding (Fig. 8.17). The HOMO is localized to the porphyrin closer to the C$_{60}$ sphere, to which the LUMO is localized. The second HOMO is localized to the other porphyrin at far position from the C$_{60}$ sphere. The positioning of the porphyrin entity with respect to the fullerene entity (near or far) seems to

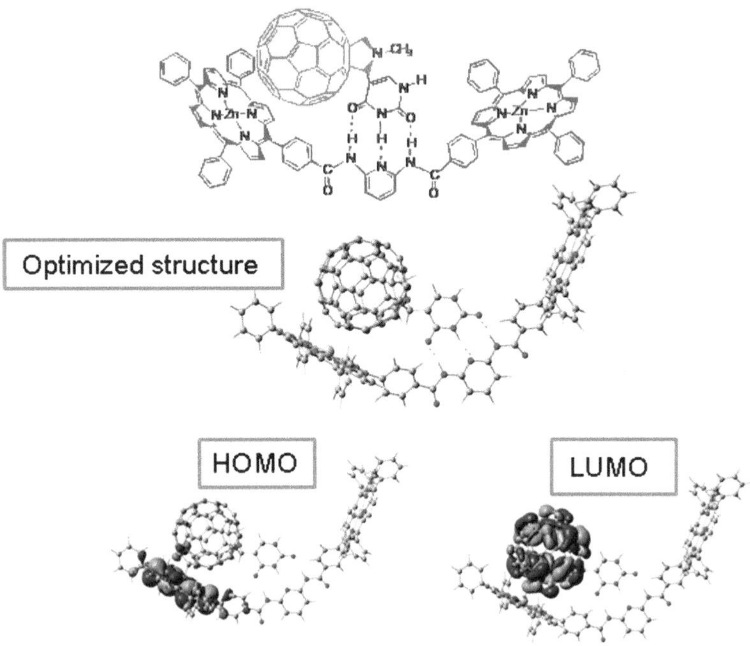

Figure 8.17. The optimized structure of bisporphyrin-fullerene assembled via hydrogen bonding; calculated structure by B3LYP/3-21G(*) method (red = O, blue = N, gray = C, and white = H) and the location of HOMO and LUMO. Modified from Ref. [73].

influence kinetics of charge-separation and charge-recombination events, thus, delineating the structural importance in governing the electron-transfer rates.

Picosecond time-resolved emission and nanosecond-transient absorption techniques were employed to evaluate kinetics of electron transfer and to characterize the electron-transfer products. The k_{CS}^{S} values via the $^{1}ZnP^{*}$ moiety in the complex were evaluated to be 1.4×10^{10} s^{-1} and 2.8×10^{9} s^{-1} corresponding to the near-side and far-side located ZnP units with respect to the C_{60} entity, respectively. By monitoring the decay of the $C_{60}^{\bullet-}$ peak at 1020 nm, two rates were evaluated to be 2.2×10^{7} and 5.8×10^{6} s^{-1} corresponding to charge recombination from the near-side and far-side positioned $ZnP^{\bullet+}$ entities (Fig. 8.18). This resulted in $\tau_{RIP} = 45$ ns and 172 ns, respectively. It is assumed that the fast electron-transfer processes for the

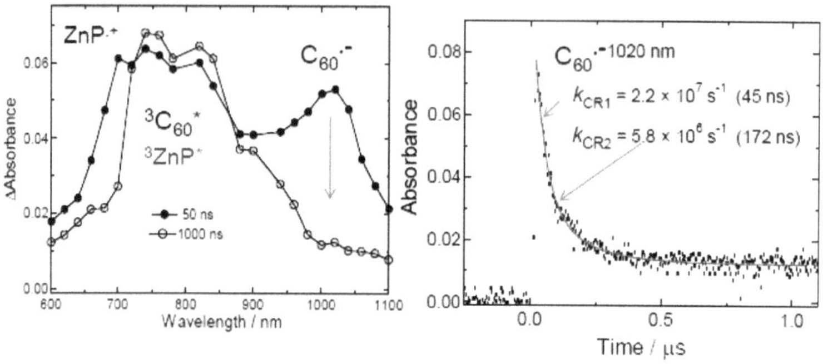

Figure 8.18. (Left) Transient absorption spectra observed by the excitation of bisporphyrin-fullerene assembly with 532-nm laser light in acetonitrile:dichlorobenzene (6:4 v/v). (Right) Time-profile of the 1020 nm peak; curve-fitting gave two rate constants. Modified from Ref. [73].

closer ZnP-C_{60} pair occur via through-bond mechanism, whereas the slower processes for distant pair occur via through-space mechanism. These k_{CR} values are smaller than the respective k_{CS}^{S} values by a factor of 1000, suggesting prolongation of the charge-separation state, affording possible application to an artificial photosynthetic model.

Recently, molecular systems in the rotaxane interlocked mechanisms have emerged as one of the important class of supramolecular compounds [74–78]. One of the most striking features of these interlocked systems is that the constituting components are arranged spatially within a certain distance without the aid of covalent bonds. The photoinduced electron-transfer processes between the spatially positioned donor and acceptor in rotaxanes are expected to be much different from the covalent bonded and coordinated multiporphyrin molecular arrays. Rotaxanes possessing a macrocyclic ring which is threaded on a rod with stoppers that prevent dissociation have been reported [78–82]. Watanabe et al. attempted to synthesize rotaxanes comprising a necklace pendent C_{60} moiety and zinc porphyrin dimer with varying axle lengths as shown in Fig. 8.19 [83, 84]. The intra-rotaxane photoinduced electron-transfer processes between the spatially positioned C_{60} and ZnP in rotaxanes have been investigated using time-resolved spectra; measurements with the aid of the molecular dynamics calculations.

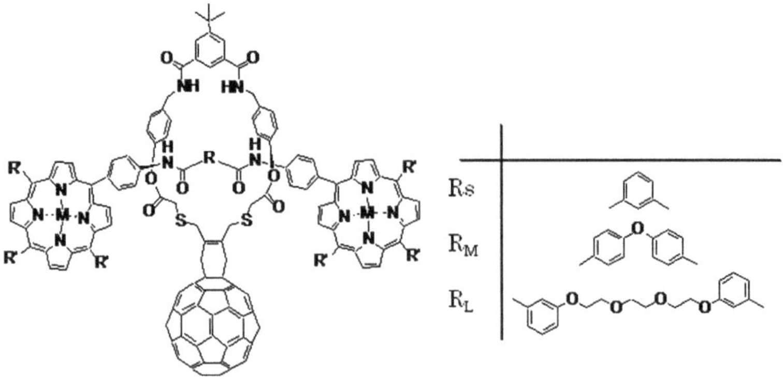

Figure 8.19. Rotaxanes comprising a necklace pended C_{60} moiety and zinc porphyrin dimer ((ZnP; C_{60}; ZnP)$_{Rot}$) with varying axle lengths, R_S, R_M, and R_L. Modified from Refs. [83, 84].

Optimized molecular structures for (ZnP; C_{60}; ZnP)$_{Rot}$ R_S, R_M, and R_L calculated by the MM3 force field show the dynamic fluctuations in molecular structures, which is one of the characteristics of the rotaxanes (Fig. 8.20 (left)). The center-to-center distances (R_{CC}) between the C_{60} and ZnP moieties show some distributions (Fig. 8.20 (left)), from which average R_{CC} values were evaluated, in addition to the minimal and maximal values. The most probable

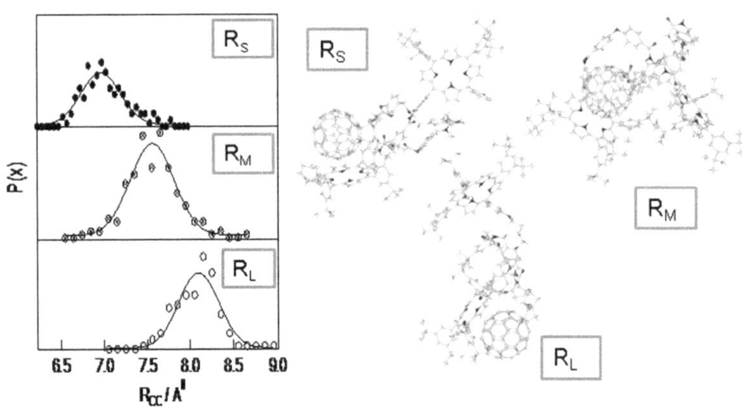

Figure 8.20. (a) Distribution of (P(x)) vs. R_{CC} of (ZnP; C_{60}; ZnP)$_{Rot}$. Optimized structures calculated with probable R_{CC} values. Modified from Refs. [83, 84].

structures are shown in Fig. 8.20 (right), in which one of the ZnP moieties approaches the C_{60} moiety.

The τ_F values of (ZnP; C_{60}; ZnP)$_{Rot}$ were evaluated from the fluorescence time profiles at 625 nm of the ZnP moieties. The τ_F values become longer with increasing in the distance between the C_{60} and ZnP at near side. From these τ_F values, the k^S_{CS} values via $^1ZnP^*$ were evaluated to be 1.1×10^{10} s^{-1} for R_S, 1.0×10^9 s^{-1} for R_M, and 4.2 $\times 10^8$ s^{-1} for R_L in PhCN at RT. With increase in the axle length from R_S to R_L, both k^S_{CS} values tend to decrease. For R_S, the Φ^S_{CS} value was larger than 0.9 indicating that the CS process via $^1ZnP^*$ is predominant, whereas for R_M and R_L, the Φ^S_{CS} values were less than 0.6 and 0.45, respectively, indicating that the CS process via the $^3ZnP^*$ moiety is also possible [83, 84].

The transient absorption spectra of (ZnP; C_{60}; ZnP)$_{Rot}$ in PhCN are shown in Fig. 8.21, in which the 1000 nm band was attributed to the $C_{60}^{\bullet-}$ moiety, whereas the $ZnP^{\bullet+}$ moiety absorption at 620 nm may be hidden in the nearby triplet-state absorption bands of the $^3C^*_{60}$ moiety (700 nm). In the time profiles at 1000 nm for short axle rotaxane, R_S, a quick rise within the 6-ns laser pulse was observed, suggesting that this quick rise corresponds to the charge-separation process via the $^1ZnP^*$ moiety. On the other hand, a slow rise

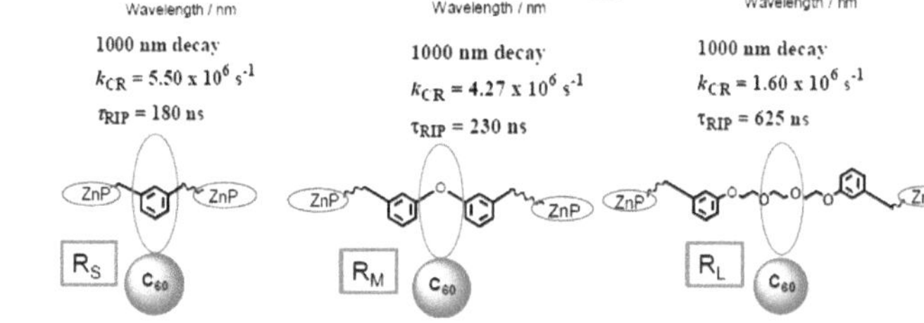

Figure 8.21. Transient absorption spectra at (•)100 ns and (○) 1000 ns; (inset) time profiles of C_{60} radical anion at 1000 nm of (ZnP; C_{60}; ZnP)$_{Rot}$. (Lower part) k_{CR} and τ_{RIP}. Modified from Refs. [83, 84].

(50–300 ns) after quick rise of the 1000 nm band was observed for long axle rotaxane, R_L, suggesting that the charge-separation process via $^3ZnP^*$ (or $^3C_{60}^*$) is included in addition to that via $^1ZnP^*$. For intermediate axle (R_M), the slow rise part of the 1000 nm band was narrow (within 100 ns), suggesting that the charge-separation rate via $^3ZnP^*$(or $^3C_{60}^*$) is fast. It is notable that the partition of the slow charge separation via the triplet states increases with lowering the temperature, probably, because the charge-separation rate via the $^1ZnP^*$ moiety is slower than the intersystem crossing rate at lower temperature.

The charge recombination rate monitored with the decay of the $C_{60}^{\bullet-}$ moiety was found to be 5.5×10^6 s^{-1} in PhCN for R_S-type $(ZnP^{\bullet+}; C_{60}^{\bullet-}; ZnP)_{Rot}$, corresponding to τ_{RIP} of 180 ns. This relatively short τ_{RIP} value also supports that R_S-type $(ZnP^{\bullet+}; C_{60}^{\bullet-}; ZnP)_{Rot}$ possesses the singlet spin-character corresponding to the charge separation via the $^1ZnP^*$ moiety. In the case of R_L-type $(ZnP;C_{60};ZnP)_{Rot}$, the τ_{RIP} value was as long as 625 ns due to triplet spin- character, corresponding to the charge separation via the $^3ZnP^*$ moiety. In the case of R_M-type $(ZnP^{\bullet+}; C_{60}^{\bullet-}; ZnP)_{Rot}$, the τ_{RIP} value was evaluated to be 230 ns in PhCN at RT; with decreasing temperature, the decay at 1000 nm slowed down, due to the participation of the charge separation via the $^3ZnP^*$ moiety that increases with lowering the temperature [83, 84].

The τ_{RIP} values become long with decreasing solvent polarity in each $(ZnP^{\bullet+}; C_{60}^{\bullet-}; ZnP)_{Rot}$, suggesting that the charge-separation process locates in the Marcus-inverted region, since the $-\Delta G_{CR}$ value is larger than the reorganization energy. This principle may be operative for both RIPs with singlet and triplet spin characters. Therefore, even in R_S-type $(ZnP^{\bullet+}; C_{60}^{\bullet-}; ZnP)_{Rot}$, the τ_{RIP} value of 180 ns is quite long compared with the RIPs that do not contain the $C_{60}^{\bullet-}$ moiety.

In summary, it is found that the charge-separation routes in rotaxanes easily change from the singlet route to the triplet route with varying the axle length, the temperature, and solvent polarity. This is because of the fluctuating components that are between the fixed donor–acceptor conjugates with chemical bonds benefiting the singlet route and freely-moving mixture of the donor and acceptor in solution preferring to the triplet route. On the basis of these results, one can control the spin characters of the RIPs and the τ_{RIP} values. However, one of our aims to introduce two ZnP units in a rotaxane is to observe a hole-blitzing through the two ZnP moieties in each RIP, a phenomenon that is yet to be confirmed.

8.2.3 *Photoinduced Direct Charge Separation via Co-Facial Dimer Models*

Owing to the structural similarity of the special pair of bacterial photosynthesis, supramolecular architectures involving co-facial porphyrin dimers are anticipated to be better biomimetic models of reaction center complexes. By using well-defined supramolecular approaches, a co-facial porphyrin dimer was formed from tetrakis *meso*-(benzo-15-crown-5)porphyrinatozinc (ZnTCP). Addition of K^+ induces dimerization of ZnTCP, generating K_4^+(ZnTCP)$_2$ [85–88]. The dimer was subsequently self-assembled with functionalized fullerene with pyridine and — NH_3^+ (PyC$_{60}$-NH_3^+) *via* axial coordination and crown ether — NH_3^+ complexation to form the bis((donor)$_2$-acceptor) tetrad (K_4^+(ZnTCP)$_2$:(PyC$_{60}$-NH_3^+)$_2$) as shown in Fig. 8.22, mimicking the non-covalently bound entities of the photosynthetic reaction center [89]. The optimized structure obtained by B3LYP/3-21G(*) method supports a stable complex formation; that is, the adopted self-assembly methodology yielded supramolecular complexes of higher stability with defined geometry and orientation (Fig. 8.22 (left)). Further, the steady-state absorption spectral changes indicated K^+-induced dimerization of ZnTCP, generating K_4^+(ZnTCP)$_2$ by the large blue shift of the ZnTCP band at 435 nm.

Further addition of PyC$_{60}$-NH_3^+ confirmed the formation of (K_4^+(ZnTCP)$_2$:(PyC$_{60}$-NH_3^+)$_2$ by the decrease of the 415 nm band of K_4^+(ZnTCP)$_2$ as shown in Fig. 8.22 (right). This decrease in the 415 nm band is characteristic of the coordination of the Py moiety to Zn atom, suggesting that the PyC$_{60}$-NH_3^+ is fixed near to the porphyrin plane with coordination of N-atom lone pair to the central zinc ions of (ZnTCP)$_2$. Although the K^+-included dimer, K_4^+(ZnTCP)$_2$, would be weakened by the -NH_3^+ complexation, any dissociation of the dimer with repulsion of K^+ with the -NH_3^+ does not occur

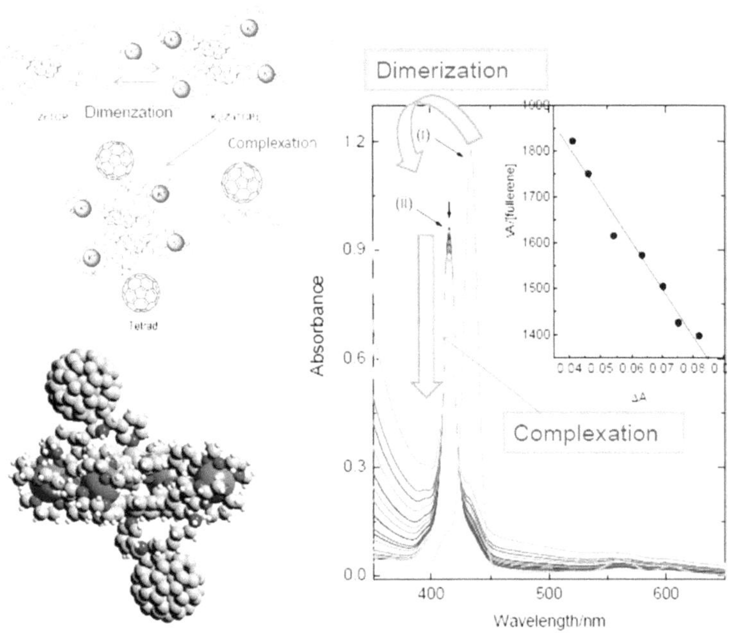

Figure 8.22. (Left) Construction of the supramolecular 'special pair' porphyrin dimer-fullerene tetrad, $(K_4^+(ZnTCP)_2{:}(PyC_{60}\text{-}NH_3^+)_2$, and its optimized structure obtained by B3LYP/3-21G(*) method (purple = K^+, red = O, blue = N, gray = C, and white = H). (Right) UV-visible spectrum of (i) zinc tetrakis(benzo-15-crown-5)porphyrin (ZnTCP = 2.0 μM), (ii) K^+ induced $K_4^+(ZnTCP)_2$ and then on increasing addition of PyC_{60} NH_3^+ (1.3 μM each addition) in PhCN. The inset shows Scatchard plot constructed for evaluating the binding constant. Modified from Ref. [89].

primarily owing to multi-binding modes, as confirmed by keeping the 415 nm band [89].

The ZnTCP-fluorescence intensity was quenched on addition of K^+ to form the dimer and addition of PyC_{60} NH_3^+ to form the supramolecular conjugate caused further quenching, accompanying with acceleration of the ZnTCP-fluorescence decay rates on addition of K^+ and further addition of $PyC_{60}\text{-}NH_3^+$ as shown in Fig. 8.23 (left). From the τ_F values, the k_{CS}^S value for charge separation from $^1(ZnTCP)_2^*$ to $PyC_{60}\text{-}NH_3^+$ was found to be relatively large (4.3×10^9 s^{-1}). The nanosecond transient absorption spectra of

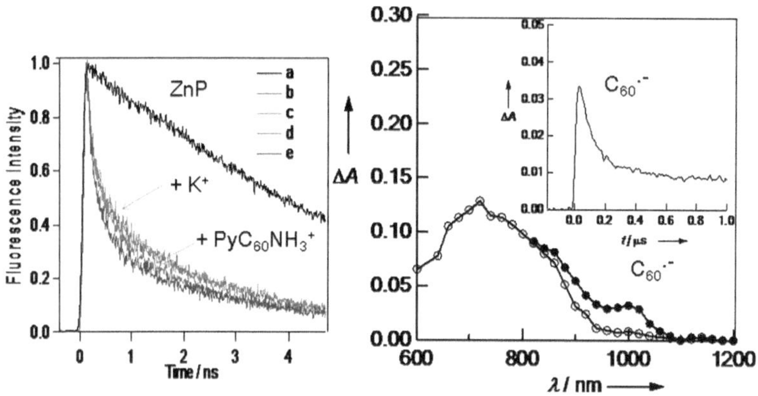

Figure 8.23. (Left) Fluorescence time-profile of ZnP at 600 nm. (Right) Nanosecond transient absorption spectra of $K_4^+(ZnTCP)_2$ (0.1 mM) with $3PyC_{60}\ NH_3^+$ (0.11 mM) in benzonitrile observed by 550 nm laser irradiation in at 30 ns (•) and 500 ns (○). The inset shows the time profile at 1000 nm. Modified from Ref. [89].

$(_4^+(ZnTCP)_2{:}(PyC_{60}\text{-}NH_3^+)_2$ shown in Fig. 8.23 (right) exhibited the 1000 nm band characteristic of $C_{60}^{\bullet-}$, supporting the RIP formation. From the decay of the $C_{60}^{\bullet-}$ moiety at 1000 nm until 200 ns, the k_{CR} value was found to be $1.0 \times 10^7\ s^{-1}$ ($\tau_{RIP} = 100$ ns); slow decay part after 200 ns may be the tail of the 700 nm band due to the $^3C_{60}^*$ moiety. Efficient charge separation and relatively slow reverse electron transfer can be measured with the ratio (k_{CS}^S/k_{CR}), which is as large as 430, supporting that the generated electron and hole are kept for duration capable to utilize for subsequent electron mediating systems. This is very important in the photosynthetic light energy conversion, achieving a novel biomimetic model [89].

Furthermore, this methodology was extended to Mg central atom forming $K_4^+(MgTCP)_2{:}(PyC_{60}\text{-}NH_3^+)_2$; $k_{CS}^S = 4.6 \times 10^9\ s^{-1}$, $k_{CR} = 1.0\times 10^7\ s^{-1}$, and $k_{CS}^S/k_{CR} = 470$. These evaluated values indicate that the Mg analogy is slightly better photosynthetic model than the Zn analogy [90].

More recently, this methodology was further extended to zinc phthalocyanine (ZnPc) analog [91]. A co-facial ZnPc dimer was formed via K^+-induced dimerization of zinc tetrakis(penta-oxatrideca-methylene)phthalocyanine (ZnTCPc) [91]. This dimer,

$K_4^+(ZnTCPc)_2$, was subsequently self-assembled with PyC_{60}-NH_3^+ to form $K_4^+(ZnTCPc)_2$:$(PyC_{60}$-$NH_3^+)_2$, mimicking the bacterial photosynthetic reaction center. The phthalocyanine macrocycle-based model system exhibits distinct evidence of the RIP in the transient absorption spectra ($ZnTCPc^{\bullet+}$), which is superior to the porphyrin analog, further supporting longer-lived charge-separated state for the ZnPc analog, a key step of photosynthetic light energy conversion process. The importance of co-facial dimer in stabilizing the charge-separated state was borne out in these three supramolecular 'reaction center' donor–acceptor mimics.

8.2.4 *Photoinduced Direct Charge Separation in Bridged Dimer Models*

As a second set of examples for photosynthetic models, oxoporphyrinogen (OxP) entity is employed as a base of building block of the 3D architectures including multiple ZnP and C_{60} moieties, due to the presence of four hemi-quinonoids on the macrocycle periphery [92, 93]. At first, the ZnP units are covalently bonded to the internal N-atoms inside of OxP, forming $(ZnP)_2$-OxP. The OxP with quinone moieties acts as an electron acceptor, generating ZnP($ZnP^{\bullet+}$)-$OxP^{\bullet-}$ ($k_{CS}^{S} = 9.3 \times 10^9$ s^{-1} in *o*-DCB) on photoexcitation of ZnP [94]. By further addition of C_{60}Im, the supramolecular pentad (C_{60}Im$\rightarrow$ZnP$)_2$-OxP was constructed by adopting a "coordinate" bonding approach.

The molecular structure and its optimized structure of (C_{60}Im$\rightarrow$ZnP$)_2$-OxP obtained by B3LYP/3-21G$^{(*)}$ calculations are also shown in Fig. 8.24 [95]. In agreement with the X-ray and computational analyses, the OxP in (C_{60}Im$\rightarrow$ZnP$)_2$-OxP revealed a highly ruffled structure. Also, based on the ^{1}H NMR spectral data, two ZnP units of (C_{60}Im$\rightarrow$ZnP$)_2$-OxP are present as cis-form, pointing in the same direction with respect to the OxP plane [96–98].

In the optimized structure of (C_{60}Im$\rightarrow$ZnP$)_2$-OxP, the Zn-Zn distance (tR_{CC}) between two ZnP moieties was found to be ~19 Å; thus, the two fullerene entities are separated by a distance of 29 Å, while the closer R_{CC} between the ZnP and C_{60} is ~13 Å. The formation of such complex was characterized by spectral and electrochemical techniques [95–98].

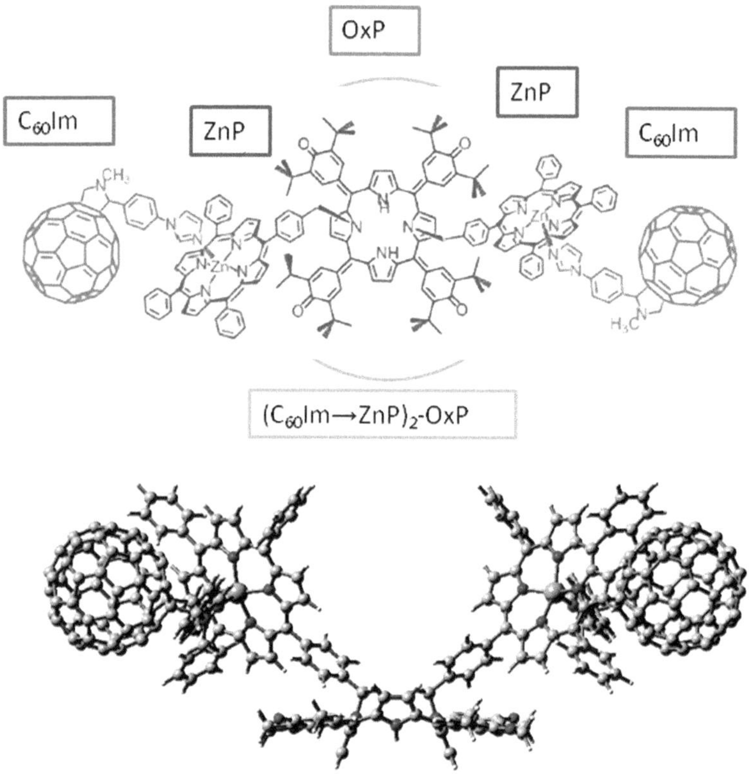

Figure 8.24. Supramolecular formation of $(C_{60}Im{\rightarrow}ZnP)_2$-OxP with covalent and coordinate bonds and its optimized structure. Modified from Ref. [95].

Free-energy calculations suggest that the photoinduced electron/energy transfer from $^1ZnP^*$ to $C_{60}Im$ and OxP is feasible in $(C_{60}Im{\rightarrow}ZnP)_2$-OxP. The k_{CS}^{S} value via $^1ZnP^*$ was determined from theτ_F value to be 1.2×10^{10} s^{-1} in *o*-DCB. This value is larger than that $(ZnP)_2$-OxP suggesting that the employed fullerene acts as a primary electron acceptor vs. $^1ZnP^*$. Indeed, the nanosecond transient absorption studies provided an evidence for the occurrence of the charge separation, generating $(C_{60}Im{\rightarrow}ZnP)(C_{60}^{\bullet-}Im{\rightarrow}ZnP^{\bullet+})_2$-OxP. The τ_{RIP} value was found to be ca. 100 ns indicating charge stabilization to some extent in this supramolecular system [95].

When two ZnP units were covalently linked to the N-atoms of OxP by the acetylene linkages and by replacing C_{60}Im with $C_{60}(Py)_2$ entity, supramolecular $C_{60}(Py)_2 \rightarrow (ZnP)_2$-OxP complexes with defined distance and orientation are obtained. Its optimized structure of $C_{60}(Py)_2 \rightarrow (ZnP)_2$-OxP obtained by B3LYP/3-21(G(*) method is as shown in Fig. 8.25. The MO calculation also supports its higher stability. The HOMO and LUMO were localized on the ZnP and C_{60}Im, respectively, suggesting that the most stable RIP is like as $(C_{60}^{\bullet-}(Py)_2 \rightarrow (ZnP^{\bullet+})ZnP)_2$-OxP [99].

When four ZnP units were covalently linked to the N-atoms of OxP by the acetylene linkages forming $(ZnP)_4$-OxP, further addition of $C_{60}Py_2$ was capable to accommodate two entities of $C_{60}(Py)_2$, forming $(C_{60}(Py)_2 \rightarrow (ZnP)_2)_2$-OxP (Fig. 8.26) [99].

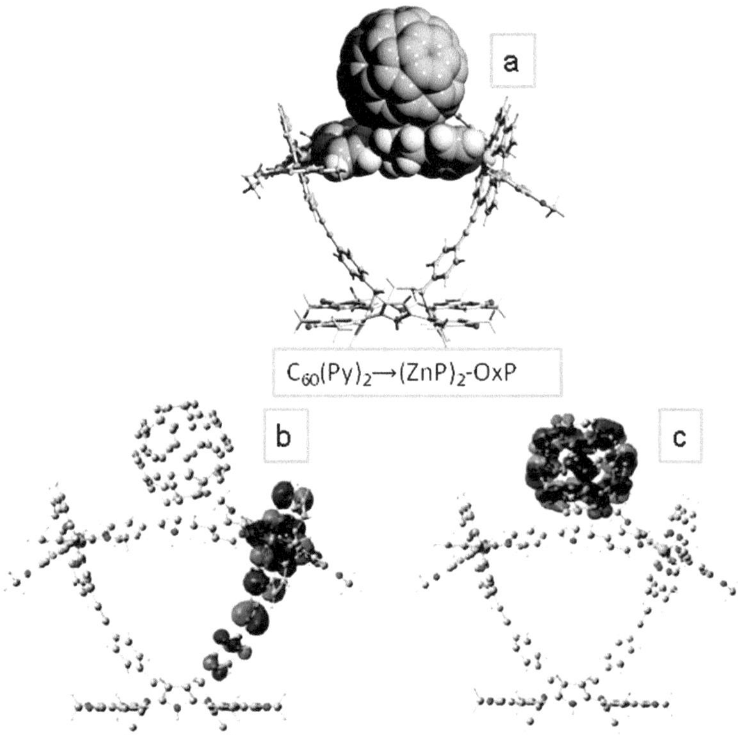

Figure 8.25. B3LYP/3-21(G(*) optimized structure of (a) $C_{60}(Py)_2 \rightarrow (ZnP)_2$-OxP, (b) HOMO and (c) LUMO. Modified from Ref. [99].

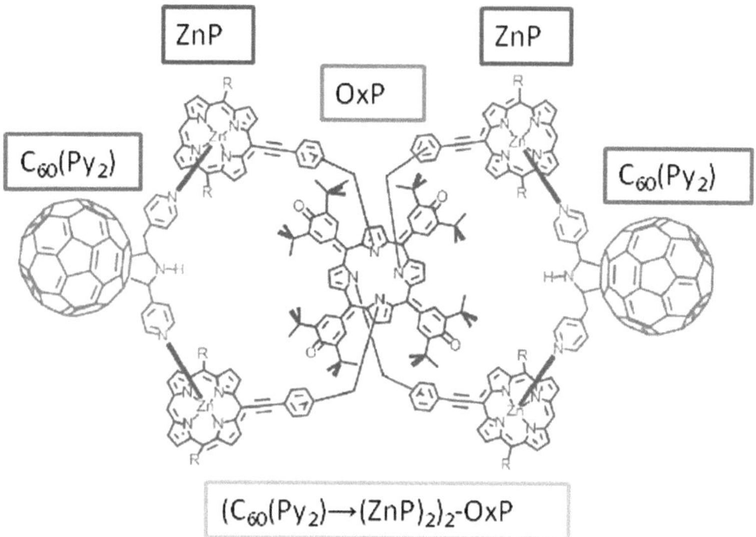

Figure 8.26. Structure of $(C_{60}(Py)_2 \rightarrow (ZnP)_2)_2$-OxP supramolecular donor–acceptor hybrid. Modified from Ref. [99].

The optimized structure of $(C_{60}(Py)_2 \rightarrow (ZnP)_2)_2$-OxP is obtained by B3LYP/3-21(G(*) method as shown in Fig. 8.27 [99]. The composite molecular array has symmetric 3D-architecture with relatively high stability. The positions of the two C_{60} spheres are on the opposite sides of the OxP plane and between two ZnP moieties (Fig. 8.27).

The photophysical parameters are evaluated from the time-resolved spectroscopic studies. Efficient charge separation is observed ($k_{CS}^{S} = 2.4 \times 10^{10}$ s^{-1}) for $(C_{60}(Py)_2 \rightarrow (ZnP)_2)_2$-OxP, which is slightly larger than that of $C_{60}(Py)_2 \rightarrow (ZnP)_2$-OxP. The charge recombination rate is evaluated from the decay of the transient absorption band of $C_{60}^{\bullet -}$ at 1000 nm as shown in Fig. 8.28, giving $k_{CR} = 1.0 \times 10^{7}$ s^{-1} in *o*-DCB. The k_{CS}^{S}/k_{CR} values of 2000–3000 nicely demonstrate kinetically well-stabilization of the 3D-supramolecular architecture. The well-defined supramolecular systems having up to seven photo- and redox-active entities are capable to absorb the light in the wide-range wavelength; then, the light absorption surely make it possible to create the charge-separated state, resulting in the high light-harvesting capability.

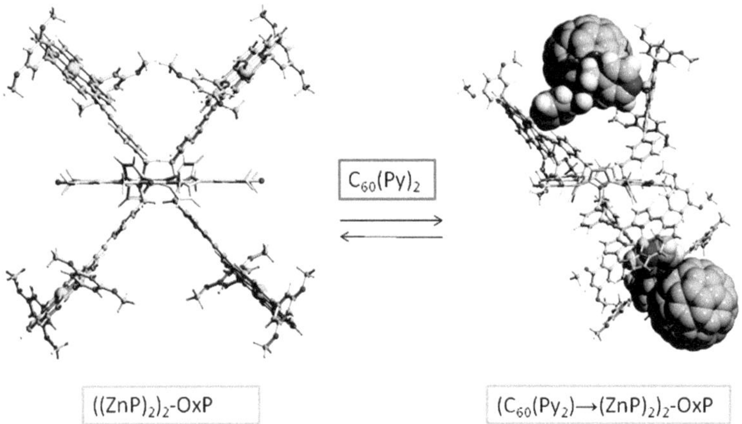

Figure 8.27. Supramolecular formation of $(C_{60}(Py)_2\rightarrow(ZnP)_2)_2$-OxP from $((ZnP)_2)_2$-OxP and $C_{60}(Py_2)$ in *o*-DCB. Modified from Ref. [99].

Furthermore, the self-assembly strategies with multiple binding sites of metalloporphyrins (or metallophthalocyanines) and fullerenes could produce supramolecular conjugates resulting in modulation of electron-transfer processes, a property that could be easily employed as a switching mechanism in molecular electronics.

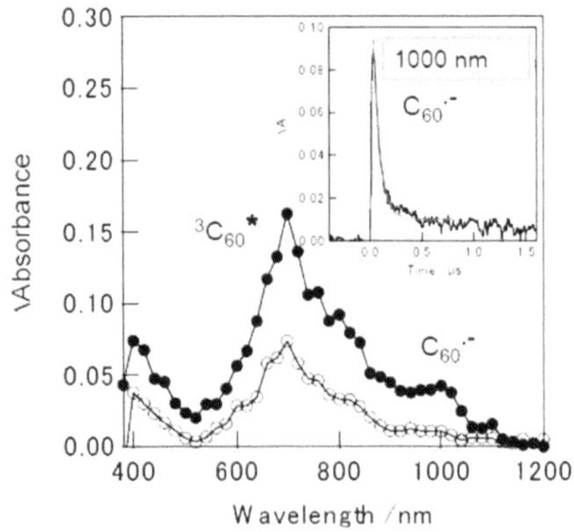

Figure 8.28. Nanosecond transient absorption spectra of $(ZnP_p)_4$-OxP with $C_{60}(Py)_2$ observed by 532 nm laser irradiation in at 0.1 μs (•) and 1.0 μs (○) in *o*-DCB. Inset: Absorption-time profile. Modified from Ref. [99].

8.3 Multiporphyrin—SWCNT Donor–Acceptor Hybrids

Development of nanomaterials and their utilization in building composite materials and photovoltaic devices are topics of current interests in the field of nanotechnology, because these materials provide a systematic platform to understand the basic concepts of the dimensionality and confinement in materials [45–53]. Among many nanosized materials available to-date, single-walled carbon nanotubes (SWCNT) stand out to be unique materials due to their remarkable mechanical and electronic properties.

SWCNTs are usually a mixture of semiconducting, conducting, or metallic tubes. The proper justification for the electrical conductivity of these SWCNTs has been explained by their chirality index. Each SWCNT has distinct π and π^*-orbital levels with small HOMO–LUMO gap corresponding to a narrow band gap of the electronic band structures. The high-lying π-orbital is capable of donating an electron to the nearby electron acceptor, whereas the low-lying π^*-orbital can accept an electron from the nearby electron donor. Upon receiving the electron, the electron transport along the axis of the micrometers-long carbon structure can contribute to a decrease of probability for back electron transfer to the cation radical at the donor site, and this is a main difference of nanotubes from the earlier discussed fullerene electron acceptors.

The donor–acceptor hybrids appending multiple ZnP or ZnPc on SWCNTs, $(ZnP)_n$/SWCNT, or $(ZnPc)_n$/SWCNT can relative easily be prepared by the covalent and non-covalent methodologies. In order to exploit the advantages of self-assembly and self-repairing mechanism, π-systems capable of attaching to the SWCNT are often utilized. In this sense, direct attachments of ZnP and ZnPc are possible via the π–π nteraction using π-surfaces of SWCNT. However, to control the distance between the ZnP or ZnPc and the surface of SWCNT, intermediate agents capable to connect the SWCNT with ZnP or ZnPc are sought. This approach also increases the solubility (or dispensability) of the $(ZnP)_n$/SWCNT or $(ZnPc)_n$/SWCNT nanoconjugates into various solvents, making device fabrication relatively easier.

Hence, from the above discussion it can be inferred that non-covalent functionalization of SWCNT using π–π interactions of large molecular π-systems is a viable approach. Accordingly, pyrene (Pyr) derivatives bearing nitrogenous bases such as pyridine or imidazole were prepared, in which pyrene π-system stacked on nanotubes, and pyridine or imidazole on the other terminals coordinates to ZnP and ZnPc (or ZnNc), forming "double-decker" architectures. One example is shown in Fig. 8.29 for $(ZnNc{\leftarrow}ImPyr)_n$/SWCNT, in which the ZnNc moiety absorbs the visible light and donates an electron to SWCNT as acceptor over ImPyr as medium. This approach in turn provided better photoinduced charge-separated states with sufficiently long lifetimes [100].

The structures and photophysical behavior of $(ZnNc{\leftarrow}ImPyr)_n$/SWCNT nanohybrids are characterized by using TEM, Raman spectra, UV-visible-near-IR absorption and fluorescence spectra, and electrochemical methods. UV-visible-near-IR absorption spectra of the dispersed solutions reveal the characteristic peaks of the SWCNT in the near-IR region (700–1600 nm), which provides further evidence for the preservation of the π–π-conjugated framework of SWCNT as shown in Fig. 8.30 (left).

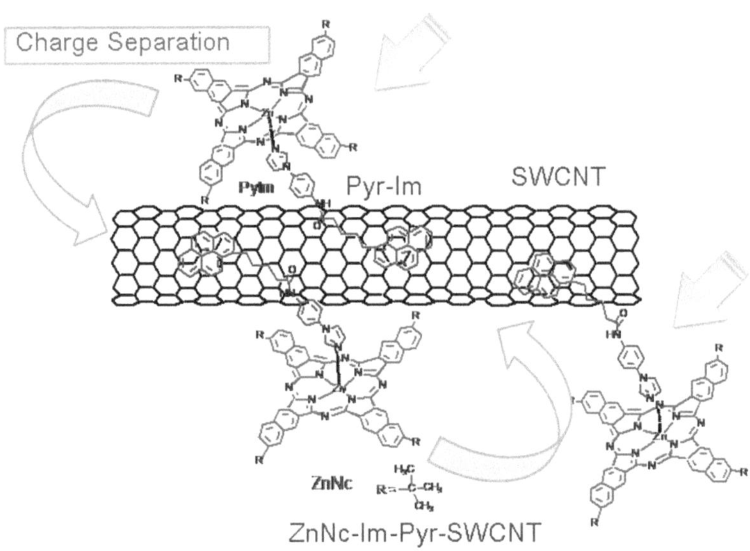

Figure 8.29. Supramolecular architecture of $(ZnNc{\leftarrow}ImPyr)_n$/SWCNT by π–π stacking and metal-ligand axial coordination. Modified from Reference [100] in which $(ZnNc{\leftarrow}ImPyr)_n$/SWCNT is simply written as ZnNc-Im-Pyr-SWCNT.

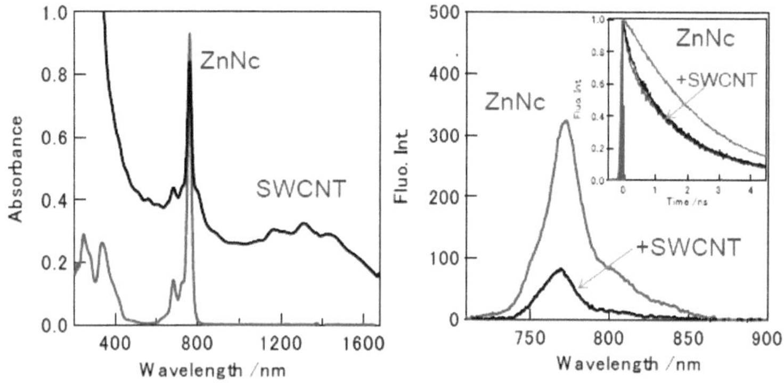

Figure 8.30. (Left) Steady-state absorption spectra of (ZnP←ImPyr)$_n$/SWCNT and ZnNc-fluorescence quenching of (ZnP←ImPyr)$_n$/SWCNT in DMF. Modified from Ref. [100]. See also Color Insert.

Excited state dynamics of (ZnNc←ImPyr)$_n$/SWCNT and (ZnP←ImPyr)$_n$/SWCNT nanohybrids are elucidated by monitoring the ZnNc (or ZnP) fluorescence intensity, which is quenched in the presence of SWCNT (Fig. 8.30 (right)). The ZnNc-fluorescence exhibits quick decay in the nanohybrids as compared to the slow decay of the unbound ZnNc-fluorescence, in agreement with the steady-state fluorescence quenching measurements. From the initial fluorescence decays, the ZnNc fluorescence lifetimes (τ_F) were evaluated to be 415 ps (fraction = 41%) in DMF and 450 ps (fraction = 38%) in THF for (ZnNc←ImPyr)$_n$/SWCNT, which are shorter than unbound donor entity (~2000 ps for unbound ZnNc and ZnP), implying rapid deactivation of ^{1}ZnNc* by the SWCNT. Since the τ_F values become short with solvent polarity and based on the free-energy considerations, it was reasonable to assume occurrence of charge-separation process within the nanohybrids; namely, one-electron oxidation of the ZnNc unit and one-electron reduction of nanotubes occur as represented forming [(ZnNc$^{\bullet+}$←ImPyr)$_n$/SWCNT$^{n(\bullet-)}$. The k_{CS}^S values via ^{1}ZnNc* are evaluated to be (2–3) × 10^9 s^{-1} from the shortening of the τ_f values, on assuming that the energy-transfer process to SWCNT is not a major deactivation process. The magnitudes of k_{CS}^S via ^{1}ZnNc* suggested that the charge-separation process was efficient within the nanohybrids and such processes are much facile in more polar DMF solution. Similar observations were also made for the (ZnP←ImPyr)$_n$/SWCNT hybrids [101].

By photoexciting the ZnNc or ZnP entities, it was possible to identify the transient photoproducts and to follow kinetics of their decay rates corresponding to the charge-recombination process. In the absence of oxygen, transient absorption bands in the wide wavelength region from the UV-vis region to near-IR region were observed for the nanohybrids. For (ZnNc←ImPyr)$_n$/SWCNT, the transient absorption of ZnNc$^{\bullet+}$ appeared as a characteristic band in the 900–1000 nm region as shown in Fig. 8.31, although the observed band of ZnNc$^{\bullet+}$ in the nanohybrids was broader than the isolated one. In addition, new band appeared at 1400 nm, which was not found for fullerene radical anion conjugates; thus, this band is tentatively attributed to the excess electron trapped in SWCNT (SWCNT$^{\bullet-}$), because the decay rate is the same as that of ZnNc$^{\bullet+}$ due to the electron–hole recombination. Absence of the characteristic feature of the triplet-triplet absorption of the ZnNc suggested efficient quenching of ^{1}ZnNc* by the SWCNT, interrupting the ISC process. The k_{CR} and τ_{RIP} values evaluated from the decay of ZnNc$^{\bullet+}$-absorption are found to be $1.4 \times 10^7 s^{-1}$ and 60 ns in DMF, respectively. These values indicate charge stabilization to some extent in these novel donor–acceptor hybrids [100].

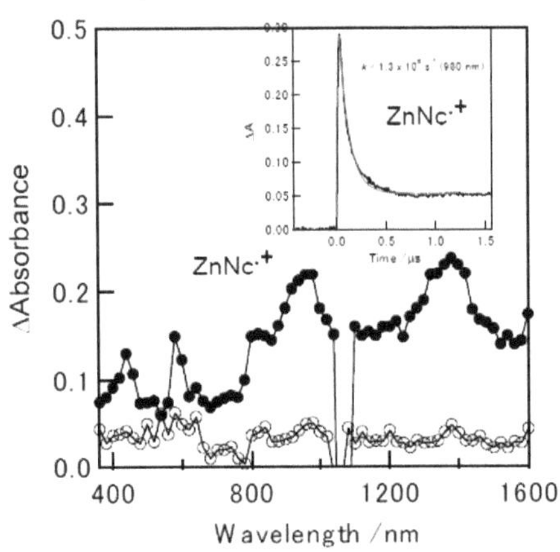

Figure 8.31. Nanosecond transient absorption spectra of (ZnNc←ImPy)$_n$/SWNT in DMF observed by 532-nm laser irradiation in at 0.1 μs (•) and 1.0 μs (○). Inset: Absorption-time profile. Modified from Ref. [100].

Photoinduced electron-transfer based catalytic processes to accumulate the oxidized and reduced products of a secondary electron donor and a secondary electron acceptor provide practical applications in addition to offering overwhelming evidence of charge separation of the original donor–acceptor hybrids In our investigations, although the chargeseparation process yielding $(ZnNc^{\bullet+} \leftarrow ImPyr)_n/SWCNT^{n(\bullet-)}$ and $(ZnP^{\bullet+} \leftarrow ImPyr)_n/SWCNT^{n(\bullet-)}$ was confirmed by the time-resolved spectral methods, a further evidence was obtained with the aid of an electron mediator, hexyl viologen dication (HV^{2+}) and an electron-hole shifting reagent, 1-benzyl-1,4-dihydronicotinamide (BNAH). Figure 8.32 shows steady-state absorption spectral changes observed by the repeated 532 nm laser light irradiation of $(ZnNc \leftarrow ImPyr)_n/SWCNT$ in the presence of HV^{2+} and BNAH. Accumulation of $HV^{\bullet+}$ was observed around 620 nm indicating electron pooling in solution, which occurs by the electron mediating process from $SWCNT^{\bullet-}$ to HV^{2+}; indeed,

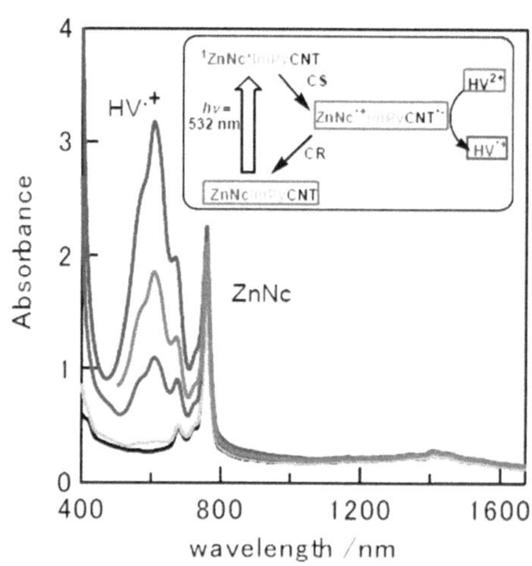

Figure 8.32. Steady-state absorption spectral changes of $(ZnNc \leftarrow ImPyr)_n/SWCNT$ (i) in the presence of 0.5 mM HV^{2+} and BNAH (in 0.5 cm cell length) before and after repeated 532 nm laser light irradiation (ca. 3 mJ/pulse), [BNAH] = (ii) 0, (black) (iii) 0.5, (pink) (iv) 1.0, (red) and (v) 1.5, (blue) in deaerated DMF. Inset: Photoinduced electron-pooling mechanism. $(ZnNc \leftarrow ImPyr)_n/SWCNT$ is simply written as ZnNcImPyCNT; SWCNT, CS; charge separation, CR; charge recombination. Modified from Ref. [100].

the HV^{2+} concentration increased the maximal $HV^{\bullet+}$ absorbance (Abs($HV^{\bullet+}$)) due to bimolecular process. Further, the resulting $HV^{\bullet+}$ concentration increased with the BNAH concentration, because the hole-shift process from $ZnNc^{\bullet+}$to BNAH was also a bimolecular process, resulting in $BNAH^{\bullet+}$, which further irreversibly changes to 1-benzyl-4-nicotinamide cation (BNA^{+}) as shown in inserted scheme in Fig. 8.32. Photocatalytic generation of $HV^{\bullet+}$(60–70% maximal yields) was supported by the invariant absorption intensity of ZnNc at 790 nm during these photochemical processes (Fig. 8.32) [100].

In the case of the covalently bonded $(ZnP)_n$-SWCNT with short covalent bond linkage (Fig. 8.32 (left)), $k_{CS}^{S} = 4.2 \times 10^9\ s^{-1}$ via $^1ZnP^*$, $k_{CR} = 4.0 \times 10^6\ s^{-1}$, $\tau_{RIP} = 250$ ns, and $MV^{\bullet+}$(60 % maximal yield) were obtained [102]. Furthermore, the photosensitized H_2-generation was observed for $(ZnP)_n$-SWCNT in the presence of excess BNAH and a colloidal Pt protected with polyvinylpyrrolidone (Pt-PVP) in EtOH-aqueous buffer solution under visible light (>500 nm) irradiation. The H_2 yields were pH dependent as seen in Fig. 8.33 (right), decreasing with increase in pH. This implies

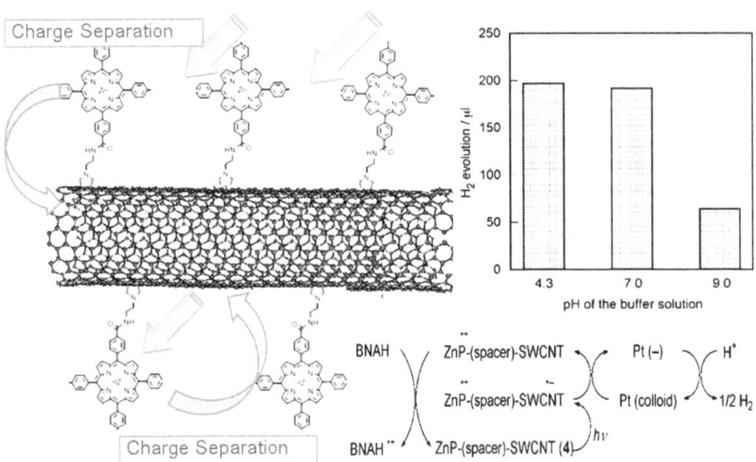

Figure 8.33. Photocatalytic H_2 evolution in EtOH aqueous buffer solution with the 60 min Xe-lump light (>500 nm) illumination of $(ZnP)_n$-SWCNT in the presence of BNAH, and colloidal Pt-PVP at RT. Modified from Ref. [102].

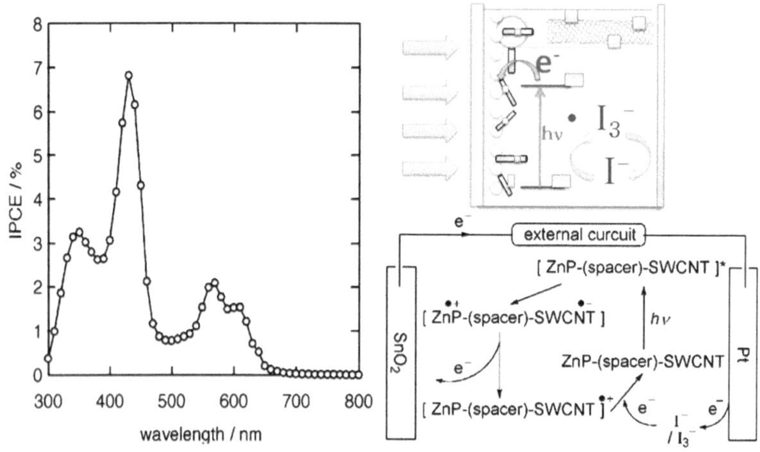

Figure 8.34. The incident photon-to-photocurrent efficiency (IPCE) vs. wavelength of the illumination light of the FTO/SnO_2 electrode coated with $(ZnP)_n$-SWCNT. Counter electrode, Pt; electrolyte, NaI and I_2 in CH_3CN. Modified from Ref. [102].

that H^+ accepts an electron from Pt catalysis, to which $(ZnP^{\bullet +})_n$-SWCNT$)^{n(\bullet -)}$ supplies excess electron (scheme in Fig. 8.33). Added BNAH traps the hole of $ZnP^{\bullet +}$ to prevent charge recombination with $SWCNT^{\bullet -}$, resulting in an increase in H_2 evolution [102].

When a solar cell was constructed with the FTO/SnO_2 electrode modified by $(ZnP)_n$-SWCNT coating, the photon-to-current responses were observed according to the on–off cycles of the Xe-light (>500 nm) illumination. The incident photon-to-current efficiency (IPCE) vs. wavelength of the illumination light was measured using monochromatic light. The IPCE spectrum is shown in Fig. 8.34, in which three peaks appeared at 430, 570, and 610 nm well resembled the absorption peaks of $(ZnP)_n$-SWCNT. This IPCE spectrum strongly suggests that the photocurrent was derived from the light absorption of the porphyrin unit in $(ZnP)_n$-SWCNT. The maximal IPCE value was 6.8% at 430 nm, which is almost the same level as the reported values of other similar multiporphyrins and SWCNT supramolecular electrodes [103, 104].

8.4 Summary

Recent progress in the design and electron transfer of self-assembled multiple porphyrin-fullerene and multiple porphyrin-SWCNT has been summarized, focusing on the photosynthetic antenna-reaction center models. The described works in this chapter demonstrate that combination of the covalent bonding constructions with the non-covalent self-assembly approaches could be successfully utilized to build multiple porphyrin supramolecular systems undergoing the sequential energy- and electron-transfer events. As self-assmeblies, π–π interactions, ion-pairing, cation-diploe, and hydrogen bonding have been rationally used to construct such multiple porphyrin reaction center model with light-harvesting units. Thus, control over the energy- and electron-transfer events is accomplished in these conjugates held by multiple binding modes. The works reviewed here also shows that it is possible to develop self-assembly methodologies ultimately generating considerably long-lived charge-separated radical ion pairs useful for light energy harvesting applications. Additionally, they could be employed as photocatalysts to generate redox active products and production of hydrogen.

For future studies, it is highly desirable to continue the fundamental studies on photoinduced energy- and elecron transfer in more complex supramolecular systems involving multiple photosensitizing donor–acceptor pairs, since such molecular assemblies are advantageous to absorb the solar light in the wide wavelength region. It must be noted, however, that each supramolecule consisting of multiple porphyrin and fullerene would absorb only one photon at a time inducing only one electron-transfer event. Therefore, further fabrication of these supramolecular assemblies in such a fashion that they would accelerate the successive energy-transfer and electron-transfer/electron-mediating processes are needed.

In the case of photosensitized studies involving multiple porphyrin-SWCNT assemblies, it is also desired to extend the studies to sorted semi-conducting SWCNTs of appropriate sizes to match the energetic location of the conduction band, instead of the samples containing both metallic and semi-conducting nanotubes [105].

In addition to fundamental studies involving photoinduced electron transfer originating from these supramolecular multiple porphyrin-nanocarbon conjugates, practical applications of supramolecular conjugates, especially in the areas of photovoltaic, sensing and photocatalysis including hydrogen production, are promising. As for the multiple poprhyrin photosenseitized nanocarbon hybrids, uses of other forms of carbon nanostructures, such as graphenes have just started emerging out [106]. More can be expected from these novel materials for electron-transfer applications in the coming years [107, 101].

Acknowledgement

The work described from the authors group was financially supported by the National Science Foundation and National Institute for Materials Science of Japan.

References

1. Deisenhofer, J., and Norris. J. R. (1993) *The Photosynthetic Reaction Center*, Academic Press, San Diego.
2. Blankenship, R. E., Madigan, M. T., and Bauer C. E. (1995) *Anoxygenic Photosynthetic Bacteria*, Kluwer Academic Publishers, Dordrecht.
3. Renger, G. (2008) *Primary Processes of Photosynthesis: Principles and Apparatus*, RSC Publishing, Cambridge.
4. Deisenhofer, J., Epp, O., Miki, K., Huber, R., and Michel, H. (1986) Structure of the protein subunits in the photosynthetic reaction center of Rhodopseudomonas viridis at 3Å resolution, *Nature*, **318**, 618–624.
5. Allen, J. P., Feher, G., Yeates, T. O., Rees, D. C., Deisenhofer, J., Michel, H., and Huber, R. (1986) Structural homology of reaction centers from Rhodopseudomonas sphaeroides and Rhodopseudomonas viridis as determined by X- ray diffraction, *Proc. Natl. Acad. Sci. USA*, **83**, 8589–8593.
6. Deisenhofer, J., and Michel, H. (1989) The photosynthetic reaction center from the purple bacterium Rhodopseudomonas viridis, *Science*, **245**, 1463–1673.

7. Kirmaier, C., and Holton, D. (1993) *The Photosynthetic Reaction Center* (ed. Deisenhofer, J., and Norris, J. R.), vol. 2, Academic Press, San Diego, p. 49.
8. Jordan, P., Fromme, P., Witt, H.-T., Klukas, O., Seanger, W., and Krauss. N. (2001) Three-dimensional structure of cyanobacterial photosystem I at 2.5 Å resolution, *Nature*, **411**, 909–917.
9. Demmig-Adams, B. (2006) *Photoprotection, Photoinhibition, Gene Regulation and Environment*, Kluwer Academic.
10. Kohler, J. (2006) *Chlorophylls and Bacteriochlorophylls: Biochemistry, Biophysics, Functions and Applications*, Springer Verlag.
11. Laporte, L., McDowell, L. M., Kirmaier, C., Schenck, C. C., and Holten, D. (1993) Insights into the factors controlling the rates of the deactivation processes that compete with charge separation in photosynthetic reaction centers, *Chem. Phys.*, **176**, 615–629.
12. Rabenstein, B., Ullmann, G. M., and Knapp, E. W. (1998) Energetics of electron- transfer and protonation reactions of the quinones in the photosynthetic reaction center of Rhodopseudomonas viridis, *Biochem.*, **37**, 2488–2495.
13. Connolly, J. S. (ed.) (1981) *Photochemical Conversion and Storage of Solar Energy* Academic Press, New York.
14. Bard, A. J., and Fox, M. A. (1995) Artificial photosynthesis: Solar splitting of water to hydrogen and oxygen, *Acc. Chem. Res.*, **28**, 141–145.
15. Friesner, R. A., and Won, Y. (1989) Spectroscopy and electron transfer dynamics of the bacterial photosynthetic reaction center, *Biochim. Biophys. Acta, Bioenergetics*, **977**, 99–112.
16. Paddon-Row, M. N. (1994) Investigating long-range electron-transfer processes with rigid, covalently linked donor-(norbornylogous bridge)-acceptor systems, *Acc. Chem. Res.*, **27**, 18–25.
17. Maruyama, K., Osuka, A., and Mataga, N. (1994) A chemical approach towards photosynthetic reaction center, *Pure Appl. Chem.*, **66**, 867–872.
18. Lehn, J.-M. (1995) *Supramolecular Chemistry: Concepts and Perspectives*, VCH Weinheim, Germany.
19. Osuka, A., Mataga, N., and Okada, T. (1997) A chemical approach towards the photosynthetic reaction center , *Pure Appl. Chem.*, **69**, 797–802.
20. Gust, D., Moore, T. A., and Moore, A. L. (2009) Solar fuels via artificial photosynthesis, *Acc. Chem. Res.*, **42**, 1890–1898.

21. Sheats, J. R. (2004) Manufacturing and commercialization issues in organic electronics, *J. Mater. Res.*, **19**, 1974–1989.
22. Grätzel, M. (2006) The advent of mesoscopic injection solar cells, *Prog. Photovolt Res. Appl.*, **14**, 429–442.
23. Yum, J.-H., Chen, P., Grätzel, M., and Nazeeruddin, M. K. (2008) Recent developments in solid-state dye-sensitized solar cells, *ChemSusChem*, **1**, 699–707.
24. Balzani, V., Credi, A., and Venturi, M. (2008) Photochemical conversion of solar energy, *ChemSusChem*, **1**, 26–58.
25. Nelson, J. (2001) Perspectives: Solar energy: Solar cells by self-assembly, *Science*, **293**, 1059–1960.
26. Sun, S.-S. and Lees, A. J. (2002) Transition metal based supramolecular systems: synthesis, photophysics, photochemistry and their potential applications as luminescent anion chemosensors, *Coord. Chem. Rev.*, **230**, 171–192.
27. Wasielewski, M. R. (2009) Self-assembly strategies for integrating light harvesting and charge separation in artificial photosynthetic systems. *Acc. Chem. Res.*, **42**, 1910–1921.
28. Kadish, K. M., Smith, K. M., and Guilard, R. (eds) (2000) *The Porphyrin Handbook*, vol. 1–2, Academic Press, San Diego.
29. Drain, C. M., Varotto, A., and Radivojevic, I. (2009) Self-organized porphyrinic materials, *Chem. Rev.*, **109**, 1630–1658.
30. Beletskaya, I., Tyurin, V. S., Tsivadze, A. Yu., Guilard, R., and Stern, C. (2009) Supramolecular chemistry of metalloporphyrins, *Chem. Rev.*, **109**, 1659–1713.
31. D'Souza, F., and Ito, O. (2010) Tetrapyrrole-nanocarbon hybrids: Self-assembly and photoinduced electron transfer, in *Handbook of Porphyrin Science*, (ed. Kadish, K. M., Smith, K. M., and Guilard R.), vol. 1, World Scientific Publishing, pp. 307–438.
32. McKeown, N. B. (ed.) (1998) *Phthalocyanine Materials: Structure, Synthesis and Function*, Cambridge University Press, Cambridge.
33. Kobayashi, N. (2002) Dimers, trimers and oligomers of phthalocyanines and related compounds, *Coord. Chem. Rev.*, **227**, 129–152.
34. de la Torre, G., Claessens, C. G., and Torres, T. (2007) Phthalocyanines: Old dyes, new materials. Putting color in nanotechnology, *Chem. Commun.*, **43**, 2000–2015.
35. Bottair, G., de la Torre, G., Guldi, D. M., and Torres, T. (2010) Covalent and noncovalent phthalocyanine-carbon nanostructure systems: Syn-

thesis, photoinduced electron transfer, and application to molecular photovoltaics, *Chem. Rev.*, **110**, 6768–6816.

36. Sariciftci, N. S., Smilowitz, L., Heeger, A. J., and Wudl, F. (1992) Photoinduced electron transfer from a conducting polymer to buckminsterfullerene, *Science*, **258**, 1474–1476.
37. Yoshino, K., Xiao, H. Y., Morita, S., and Zakhidov, A. A. (1993) Difference in doping effects of C_{60} and C_{70} in poly(3-hexylthiophene), *Jpn. J. Appl. Phys.*, **32**, L140–L143.
38. Imahori, H., and Sakata, Y. (1997) Donor-linked fullerenes. Photoinduced electron transfer and its potential application, *Adv. Mater.*, **9**, 537–546.
39. Kadish, K. M., and Ruoff, R. S. (eds) (2000) *Fullerenes, Chemistry, Physics and Technology*, Wiley-Interscience, New York.
40. Fukuzumi, S., and Guldi, D. M. (2001) *Electron Transfer in Chemistry* (ed. Balzani, V.), vol. 2, Wiley-VCH, Weinheim, pp. 270–337.
41. El-Khouly, M. E., Ito, O., Smith, P. M., and D'Souza, F. (2004) Intermolecular and supramolecular photoinduced electron transfer processes of fullerene-porphyrin /phthalocyanine systems, *J. Photochem. Photobiol. C: Photochem. Rev.*, **5**, 79–104.
42. Fukuzumi, S. (2008) Development of bioinspired artificial photosynthetic systems, *Phys. Chem. Chem. Phys.*, **10**, 2283–2297.
43. Chitta, R., and D'Souza, F. (2008) Self-assembled tetrapyrrole-fullerene and tetrapyrrole-carbon nanotube donor-acceptor hybrids for light induced electron transfer applications, *J. Mater. Chem.*, **18**, 1440–1471.
44. Araki, Y., and Ito, O. (2008) Factors controlling lifetimes of photoinduced charge-separated states of fullerene-donor molecular systems, *J. Photochem. Photobiol. C: Photochem. Rev.*, **9**, 93–110.
45. Dresselhaus, M. S., Dresselhaus, G., and Avouris, P. (2001) *Carbon Nanotubes: Synthesis, Structure, Properties and Applications*, Springer, Berlin.
46. Sun, Y.-P., Fu, K., and Hung, W. (2002) Functionalized carbon nanotubes: Properties and applications, *Acc. Chem. Res.*, **35**, 1096–1104.
47. Poole, C. P., and Owens. F. J. (eds) (2003) *Introduction to Nanotechnology*, Wiley-Interscience, Weinheim.
48. Wolf, E. L. (ed.) (2004) *An Introduction to Modern Concepts in Nanoscience*, John Wiley and Sons, New York.
49. Reich, S., Thomsen C., and Maultzsch, J. (2004) *Carbon Nanotubes: Basic Concepts and Physical Properties*, Wiley-VCH, Weinheim.

50. Guldi, D. M., Rahman, G. M. A., Zerbetto, F., and Prato, M. (2005) Carbon nanotubes in electron donor-acceptor nanocomposites, *Acc. Chem. Res.*, **38**, 871–878.

51. Meyyappan, M. (2006) *Carbon Nanotubes, Science and Application*, Wiley-VCH, Weinheim.

52. Sgobba, V., Rahman, G. M. A., Ehli, C., and Guldi, D. M. (2007) Covalent and non-covalent approaches toward multifunctional carbon nanotubes materials, in *Fullerenes-Principles and Applications* (ed. Langa, F., and Nierengarten, J.-F.), Royal Society Chemistry, Cambridge.

53. Sgobba V., and Guldi, D. M. (2009). Carbon nanotubes-electronic/electrochemical properties and application for nanoelectronics and photonics, *Chem. Soc. Rev.*, **38**, 165–184.

54. Pagona, G., Sandanayaka, A. S. D., Araki, Y., Fan, J., Tagmatarchis, N., Yudasaka, M., Ijima, S., and Ito, O. (2006) Electronic interplay on illuminated aqueous carbon nanohorn-porphyrin ensembles *J. Phys. Chem. B*, **110**, 20729–20732.

55. Pagona, G., Sandanayaka, A. S. D., Araki, Y., Fan, J., Tagmatarchis, N., Charalambidis, G., Trohopoulos, P., Coutsolelos, A. G., Boitrel, B., Yudasaka, M., Iijima, S., and Ito, O. (2007) Covalent connection of carbon nanohorns with porphyrin: Photo-induced electron transfer and energy transfer, *Adv. Funct. Mater.*, **17**, 1705–1711.

56. Cioffi, C., Campidelli, S., Sooambar, C., Marcaccio, M., Marcolongo, G., Meneghetti, M., Paolucci, D., Paolucci, F., Ehli, C., Rahman, G. M. A., Sgobba, V., Guldi, D. M., and Prato, M. (2007) Synthesis, characterization, and photoinduced electron transfer in functionalized single wall carbon nanohorns, *J. Am. Chem. Soc.*, **129**, 3938–3945.

57. Sandanayaka, A. S. D., Pagona, G., Tagmatarchis, N., Yudasaka, M., Iijima, S., Araki, Y., and Ito, O. (2007) Photoinduced electron-transfer processes of carbon nanohorns covalently linked with pyrene chromophores: Towards intramolecular charge-separation and electron-migration systems, *J. Mater. Chem.*, **17**, 2540–2546.

58. Pagona, G., Sandanayaka, A. S. D., Maigné, A., Fan, J., Papavassiliou, G. C., Petsalakis, I. D., Steele, B. R., Tagmatarchis, N., Yudasaka, M., Iijima, S., and Ito, O. (2007) Photo-induced electron-transfer on aqueous carbon nanohorn-pyrene-tetrathiafulvalene architectures, *Chem. Eur. J.* **13** 7600–7607.

59. Rotas, G., Sandanayaka, A .S. D., Tagmatarchis, N., Ichihashi, T., Yudasaka, M., Iijima, S., and Ito, O. (2008) Terpyridine Cu^{II}-carbon nanohorns: Metallo-nanocomplexes for photoinduced charge-separation, *J. Am. Chem. Soc.* **130** 4725–4747

60. Zhang, M., Yudasaka, M., Murakami, T., Ajima, K., Sandanayaka, A. D. S., Ito, O. Tsuchita, K., and Iijima, S. (2008) Fabrication of ZnPc-nanohorn-protein for photodynamic and hyperthermia double cancer-therapy, *Proc. Natl. Acad. Sci.* (USA), **105**, 14773–14778.
61. Sandanayaka, A. S. D., Ito, O., Zhang, M., Ajima, K., Iijima, S., Yudsasaka, M., Murakami, T., and Tsuchita, K. (2009) Photoinduced electron-transfer in zinc phthalocyanine loaded on single walled carbon nanohorns in aqueous solution, *Adv. Mater.*, **21**, 4366–4371.
62. Echegoyen, L., and Echegoyen, L. E. (1998) Electrochemistry of fullerenes and their derivatives, *Acc. Chem. Res.*, **31**, 593–601.
63. Imahori, H., Hagiwara, K., Akiyama, T., Akoi, M., Taniguchi, S., Okada, T., Shirakawa, M., and Sakata, Y. (1996) The small reorganization energy of C_{60} in electron transfer, *Chem. Phys. Lett.*, **263**, 545–550.
64. Karousis, N., Sandanayaka, A. S. D., Hasobe, T., Economopoulos, S. P., Sarantopoulous, E., and Tagmatrarchis, N. (2011) Graphene oxide with covalently linked porphyrin antennae: Synthesis, characterization and photophysical properties, *J. Mater. Chem.* **21**, 109–117.
65. D'Souza, F., Deviprasad, G. R., Zandler, M. E., El-Khouly, M. E., Fujitsuka, M., and Ito, O. (2002) Electronic interactions and photoinduced electron transfer in covalently linked porphyrin-C_{60}(pyridine) dyads and supramolecular triads formed by self-assembling the dyads and zinc porphyrin, *J. Phys. Chem. B*, **106**, 4952–4962.
66. D'Souza, F., Smith, P. M., Zandler, M. E., McCarty, A. L., Itou, M., Araki, Y., and Ito, O. (2004). Energy transfer followed by electron transfer in a supramolecular triad composed of boron dipyrrin, zinc porphyrin, and fullerene: A model for the photosynthetic antenna-reaction center complex, *J. Am. Chem. Soc.*, **126**, 7898–7907.
67. Maligaspe, E., Tkachenko, N. V., Subbaiyan, N. K., Chitta, R., Zandler, M. E., Lemmetyinen, H., and D'Souza, F. (2009) Photosynthetic antenna-reaction center mimicry: Sequential energy- and electron transfer in a self-assembled spramolecular triad composed of boron dipyrrin, zinc porphyrin and fullerene, *J. Phys. Chem. A*, **113**, 8478–8489.
68. D'Souza, F., Gadde, S., Schumacher, A. L., Zandler, M. E., Sandanayaka, A. S. D., Araki, Y., and Ito, O. (2007) Supramolecular triads of free-base porphyrin, fullerene and ferric porphyrins *via* 'covalent-coordinate' binding approach: Formation, sequential electron transfer and charge stabilization, *J. Phys. Chem. C*, **111**, 11123–11130.
69. El-Khouly, M. E., Gadde, S., Deviprasad, G. R., Fujitsuka, M., Ito, O., and D'Souza, F. (2003) Self-assembled supramolecular triad composed of

fulleropyrrolidine bearing two pyridine moieties axially coordinated to two zinc porphyrins, *J. Porphyrins Phthalocyanines*, **7**, 1–7.

70. D'Souza, F., Gadde, S., Zandler, M. E., Itou, M., Araki, Y., and Ito, O. (2004) Supramolecular complex composed of a covalently linked zinc porphyrin dimer and fulleropyrrolidine bearing two axially coordinating pyridine entities, *Chem. Commun.*, **40**, 2276–2277.
71. D'Souza, F., Gadde, S., El-Khouly, M. E., Zandler, M. E., Araki, Y., and Ito, O. (2005) A supramolecular Star Wars Tie Fighter Ship: electron transfer in a self-assembled triad composed of two zinc naphthalocyanines and a fullerene, *J. Porphyrins Phthalocyanines*, **9**, 698–705.
72. El-Khouly, M. E., Rogers, L. M., Zandler, M. E., Gadde, S., Fujitsuka, M., Ito, O., and D'Souza, F. (2003) Studies on intra-supramolecular and intermolecular electron-transfer processes between zinc naphthalocyanine and imidazole-appended fullerene,*ChemPhysChem*, **4**, 474–481.
73. Gadde, S., Islam, D. M. S., Wijesinghe, C. A., Subbaiyan, N. K., Zandler, M. E., Araki, Y., Ito, O., and D'Souza, F. (2007) Light induced electron transfer of a supramolecular bis zinc porphyrin-fullerene triad constructed *via* diacetylamidopyridine:uracil hydrogen bonding motif, *J. Phys. Chem. C*, **111**, 12500–12503.
74. Dietrich-Buchecker, C. O., and Sauvage, J. P. (1987) Interlocking of molecular threads: from the statistical approach to the templated synthesis of catenands, *Chem. Rev.*, **87**, 795–810.
75. Jager, R., and Vogtle, F. (1997) A new synthetic strategy towards molecules with mechanical bonds: nonionic template synthesis of amide-linked catenanes and rotaxanes, *Angew. Chem., Int. Ed.*, **36**, 930–944.
76. Schliwa, M. (2003) *Molecular Motors*, Wiley-VCH, Weinheim.
77. Yashima, E., Maeda, K., and Furusho, Y. (2008) Single- and double-stranded helical polymers: Synthesis, structures, and functions, *Acc. Chem. Res.*, **41**, 1166–1180.
78. Crowley, J. D., Goldup, S. M., Lee, A.-L., Leigh, D. A., and McBurney, R. T. (2009) Active metal template synthesis of rotaxanes, catenanes and molecular shuttles, *Chem. Soc. Rev.*, **38**, 1530–1541.
79. Li, K., Schuster, D. I., Guldi, D. M., Herranz, M. A., and Echegoyen, L. (2004) Convergent synthesis and photophysics of [60]fullerene/porphyrin-based rotaxanes, *J. Am. Chem. Soc.*, **126**, 3388–3389.
80. Li, K., Bracher, P. J., Guldi, D. M., Herranz, M. A., Echegoyen, L., and Schuster, D. I. (2004) [60]Fullerene-stoppered porphyrinorotaxanes: Pro-

nounced elongation of charge-separated-state lifetimes, *J. Am. Chem. Soc.*, **126**, 9156–9157.

81. Schuster, D. I., Li, K., and Guldi, D. M. (2006) Porphyrin-fullerene photosynthetic model systems with rotaxane and catenane architectures, *C. R. Chimie*, **9**, 892–908.
82. Jakob, M., Berg, A., Rubin, R., Levanon, H., Li, K., and Schuster, D. I. (2009) Photoinduced electron transfer in porphyrin- and fullerene/porphyrin-based rotaxanes as studied by time-resolved EPR spectroscopy, *J. Phys. Chem. A*, **113**, 5846–5854.
83. Watanabe, N., Kihara, N., Furusho, Y., Takata, T., Araki, Y., and Ito, O. (2003) Photoinduced intrarotaxane electron transfer between zinc porphyrin and [60]fullerene in benzonitrile, *Angew. Chem., Int. Ed.*, **42**, 681–683.
84. Sandanayaka, A. S. D., Watanabe, N., Ikeshita, K. I., Araki, Y., Kihara, N., Furusho, Y., Ito, O., and Takata, T. (2005) Synthesis and photoinduced electron transfer processes of rotaxanes bearing [60]fullerene and zinc porphyrin: Effects of interlocked structure and length of axle with porphyrins, *J. Phys. Chem. B*, **109**, 2516–2525.
85. Thanabal, V., and Krishnan, V. (1982) Porphyrins with multiple crown ether voids: novel systems for cation complexation studies, *J. Am. Chem. Soc.*, **104**, 3643–3650.
86. Thanabal, V., and Krishnan, V. (1982) Cation-induced crown porphyrin dimers of oxovanadium(IV), *Inorg. Chem.*, **21**, 3606–3613.
87. van Willigen, H., and Chandrashekar, T. K. (1986) ENDOR study of copper(II) crown porphyrin dimerization, *J. Am. Chem. Soc.*, **108**, 709–713.
88. Chitta, R., Rogers, L. M., Wanklyn, A., Karr, P. A., Kahol, P. K., Zandler, M. E., and D'Souza, F. (2004) Electrochemical, spectral, and computational studies of metalloporphyrin dimers Formed by cation complexation of crown ether cavities, *Inorg. Chem.*, **43**, 6969–6978.
89. D'Souza, F., Chitta, R., Gadde, S., Rogers, L. M., Karr, P. A., Zandler, M. E. Sandanayaka, A. S. D., Araki, Y., and Ito, O. (2007) Photosynthetic reaction center mimicry of special Pair' dimer linked to electron acceptors by supramolecular approach: Self-assembled cofacial zinc Porphyrin dimer complexed with fullerene(s), *Chem.–Eur. J.*, **13**, 916–922.
90. Sandanayaka, A. S. D., Subbaiyan, N. K., Chitta, R., Araki, Y., Ito, O., and D'Souza, F. (2008) Co-facial magnesium porphyrin dimer complexed with fullerene: Photosynthetic reaction center model of 'special pair' self-assembled to electron acceptor, *J. Porphyrins Phthalocyanines*, **12**, 857–865.

91. D'Souza, F., Maligaspe, E., Okubo, K., Zandler, M. E., Subbaiyan, N. K., and Fukuzumi, S. (2009) Photosynthetic reaction center mimicry: Low reorganization energy driven charge stabilization in self-assembled cofacial zinc phthalocyanine dimer-fullerene conjugate, *J. Am. Chem. Soc.*, **131**, 8787–8797.

92. Hill, J. P., Hewitt, I. J., Anson, C. E., Powell, A. K., McCarty, A. L., Karr, P. A., Zandler, M. E., and D'Souza, F. (2004) Highly nonplanar, electron deficient, N-substituted tetra-oxocyclohexadienylidene porphyrinogens: Structural, computational, and electrochemical investigations, *J. Org. Chem.*, **69**, 5861–5869.

93. Hill, J. P., Schmitt, W., McCarty, A. L., Ariga, K., and D'Souza, F. (2005) Structures, spectral and electrochemical properties of N-(naphth-2-ylmethyl)-appended porphyrinogens, *Eur. J. Org. Chem.*, **14**, 2893–2902.

94. Hill, J. P., Sandanayaka, A. S. D., McCarty, A. L., Karr, P. A., Zandler, M. E., Charvet, R., Ariga, K., Araki, Y., Ito, O., and D'Souza, F. (2006) A novel bis(zinc-porphyrin)-oxoporphyrinogen donor-acceptor triad: Synthesis, electrochemical, computational and photochemical studies, *Eur. J. Org. Chem.*, 595–603.

95. Schumacher, A. L., Sandanayaka, A. S. D., Hill, J. P., Ariga, K., Karr, P. A., Araki, Y., Ito, O., and D'Souza, F. (2007) Supramolecular triad and pentad composed of zinc porphyrin(s), oxoporphyrinogen, and fullerene(s): Design and electron transfer studies, *Chem.–Eur. J.*, **13**, 4628–4635.

96. Hill, J. P., Ariga, K., Schumacher, A. L., Karr, P. A., and D'Souza, F. (2007) Pyren-1-ylmethyl N-substituted oxoporphyrinogens, *J. Porphyrins Phthalocyanines*, **11**, 390–396.

97. Schumacher, A. L., Hill, J. P., Ariga, K., and D'Souza, F. (2007) Highly effective electrochemical anion sensing based on oxoporphyrinogen, *Electrochem. Commun.*, **9**, 2751–2754.

98. Hill, J. P., Ariga, K., and D'Souza, F. (2009) Structures and properties of hemiquinone- substituted oxoporphyrinogens, *J. Porphyrins Phthalocyanines,* **13**, 60–69.

99. Xie, Y. Hill, J. P. Schumacher, A. L., Sandanayaka, A. S. D. Araki, Y. Karr, P. A. Labuta, J. D'Souza, F. Ito, O. Anson, C. E. Powell A. K. and Ariga, K. (2008) Twisted, two-faced porphyrins as hosts for bispyridyl fullerenes: Construction and photophysical properties, *J. Phys. Chem. C*, **112**, 10559–10579

100. Chitta, R., Sandanayaka, A. S. D., Schumacher, A. L., D'Souza, L., Araki, Y., Ito, O., and D'Souza, F. (2007) Donor-acceptor nanohybrids of zinc

naphthalocyanine or Zinc porphyrin non-covalently linked to single-wall carbon nanotubes for photoinduced electron transfer, *J. Phys. Chem. C*, **111**, 6947–6955.

101. D'Souza, F., Sandanayaka, A. S. D., and Ito, O. (2010) SWNT-based supramolecular nanoarchitectures with photosensitizing donor and acceptor molecules, *J. Phys. Chem. Letters (Perspective)*, **1**, 2586–2593.
102. Arai, T., Nobukuni, S., Sandanayaka, A. D. S., and Ito, O. (2009) Zinc porphyrin covalently bound to the side wall of single-walled carbon nanotubes via flexible bonds: Photoinduced electron transfer in polar solvent, *J. Phys. Chem. C.*, **113**, 14493–14499.
103. Guldi, D. M., Rahman, G. M. A., Qin, S. H., Tchoul, M., Ford, W. T., Marcaccio, M., Paolucci, D., Paolucci, F., Campidelli, S., and Prato, M. (2006) Versatile coordination chemistry towards multifunctional carbon nanotube nanohybrids, *Chem.-Eur. J.*, **12**, 2152–2161.
104. Boul, P. J., Cho, D.-G., Rahman, G. M. A., Marquez, M., Ou, Z., Kadish, K. M., Guldi, D. M., and Sessler, J. L. (2007) Sapphyrin-nanotube assemblies, *J. Am. Chem. Soc.*, **129**, 5683–5687.
105. Maligaspe, E., Sandanayaka, A. S. D., Hasobe, T., Ito, O., and D'Souza, F. (2010) Sensitive efficiency of photovoltaic cell to band gaps of semiconductive SWNTs with the supramolecularly attached zinc porphyrin bearing pyrene glues, *J. Am. Chem. Soc.*, **113**, 8158–8164.
106. Imahori, H., and Umeyama, T. (2009) Donor-acceptor nanoarchitecture on semiconducting electrodes for solar energy conversion, *J. Phys. Chem. C*, **113**, 9029–9039.
107. D'Souza, F., and Ito, O. (2009) Supramolecular donor-acceptor hybrids of porphyrins/phthalocyanines with fullerenes/carbon nanotubes: Electron transfer, sensing, switching, and catalytic applications, *Chem. Commun. Feature Article*, 4913–4928.

Chapter 9

Application of Multiporphyrin Arrays to Solar Energy Conversion

Simon Mathew[a] and Hiroshi Imahori[a,b,c]

[a] *Institute for Cell-Material Sciences (iCeMS), Kyoto University, Nishikyo-ku, Kyoto, 615-8510, Japan*
[b] *Department of Molecular Engineering, Graduate School of Engineering, Kyoto University, Nishikyo-ku, Kyoto, 6615-8510, Japan*
[c] *Fukui Institute for Fundamental Chemistry, Kyoto University, Sakyo-ku, Kyoto 606-8103, Japan*
Imahori@scl.kyoto-u.ac.jp

9.1 Introduction

9.1.1 *General Introduction*

One of the greatest challenges facing society today is the quest to find an alternate means of energy generation, in order to eliminate the use of finite feedstocks such as fossil fuels. Additionally, the production of electricity without harmful byproducts or undesirable outcomes (CO_2 emission, nuclear waste, food vs. fuel, etc.) is essential to ensure no negative effects Arising from the use of alternate sources of energy. As it stands currently, yearly global consumption has reached approximately 15 TW, and is provided by a range of sources that are predominantly non-renewable by nature [1]. However, the bulk of this energy arrives, indirectly, through the

Multiporphyrin Arrays: Fundamentals and Applications
Edited by Dongho Kim

www.panstanford.com

absorption of light, for example the fossilization of plant matter gives rise to coal [1]. Direct use of sunlight to effect electrical current generation is an attractive means of energy production, as the sun provides a staggering 120,000 TW to the earth's surface [1]. The use of solar energy conversion to produce energy is possible through the exploitation of the photovoltaic effect using inorganic substrates (i.e., silicon-based solar cells), however the performance of these cells is limited by the purity (i.e., monocrystallinity) of the materials used, making cost a serious factor when considering the economic viability of solar energy conversion.

Nature performs its own solar energy conversion process, namely photosynthesis, which occurs in green plants and photosynthetic bacteria. The color of these organisms is due to the presence of chlorophyll dyes within them and extends the ability to perform work upon the absorption of light. The mechanism of energy conversion in the photosynthetic apparatus is intrinsically different from that occurring in inorganic solar cells. Instead of the photovoltaic effect, the photosynthetic apparatus uses photoinduced energy and electron transfer to enable the funneling of energy of different wavelengths to a "reaction center". The reaction center is an elaborate array of dyes and other electroactive cofactors that function as light absorbers and/or electron donors/acceptors, and it is this array that allows the process of charge separation to occur. This formation of separated charges in this system is the source of electrical potential that allows the organism to perform useful work.

Solar energy conversion methods using photoinduced energy and electron transfer processes provide a cheaper route toward solar energy conversion devices, as they utilize organic light absorbing molecules and polymers (bulk heterojunction solar cells, BHJSCs) which are sometimes adsorbed onto a cheap, semiconducting substrate (dye-sensitized solar cells, DSSCs). These two types of solar cell are regarded as the most promising solar solution for meeting energy demands not only because of the cost reduction associated with the use of organic materials, but also because of the unique features made available from their organic form (i.e., flexibility, lightweight, and colorfulness) [2, 3]. Both BHJSCs and DSSCs will be discussed in greater detail.

9.1.2 *Bulk Heterojunction Solar Cells*

In a BHJSC, electron donor and acceptor compounds (usually a donor polymer and a fullerene acceptor) are mixed together in solution and cast into a film, which maintains contact between two electrodes. One of the electrodes is typically a conductive, glass electrode, either tin-doped indium oxide (ITO) or fluorine-doped tin oxide (FTO). As shown in Fig. 9.1, as the photoactive film of donor and acceptor is essentially symmetrically distributed, the anode and the cathode must be defined by the metals at the respective electrodes [4]. Additionally, a layer of conductive polymer, typically poly(3,4-ethylenedioxythiophene) poly(styrenesulfonate), (PEDOT:PSS) is applied to the ITO glass to help smoothen the surface.

The process of charge generation in this cell is unique as it relies on the formation of excitons at the donor–acceptor heterojunction upon light absorption by the donor polymer component of the cell [5]. Excitons—electron–hole pairs bound strongly by coulombic attraction—at the interface between donor and acceptor, are able to undergo a photoinduced electron transfer to the fullerene. This results in the formation of mobile charge carriers, which are the separated electrons and holes that are able to migrate toward the respective electrodes [6]. There is, however, the possibility that the separated charges recombine at another heterojunction. Therefore, it is highly desirable to produce organic thin films with a bicontinuous interpenetrating network of donor and acceptor domains to promote efficient migration of charges to the electrodes [4, 7]. With

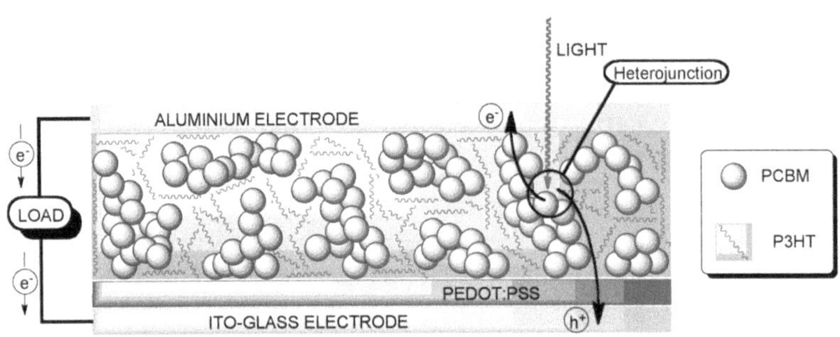

Figure 9.1. Construction and operation of a BHJSC.

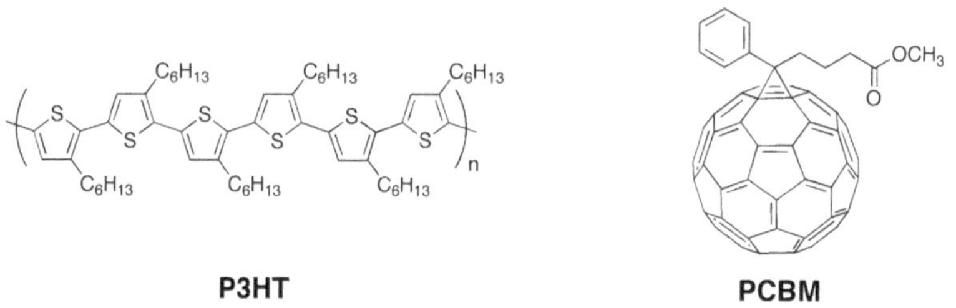

Figure 9.2. Structures of P3HT and PCBM.

regard to the molecular components of the photoactive layer, Fig. 9.2 depicts the classic components used in the BHJSC in Fig. 9.1, composed of a donor polymer, poly(3-hexylthiophene) (P3HT) and an acceptor molecule. In this case the acceptor is the fullerene [6,6]-phenyl C_{61}-butyric acid methyl ester (PCBM), considered the golden standard since its inception in 1995 [8].

9.1.3 *Dye-Sensitized Solar Cells*

The conception of DSSCs in 1985 saw an explosion in literature reports concerning energy generation with these unique constructs [9–13]. The general construct of the device, presented in Fig. 9.3, involves a photoanode that is composed of conductive glass (in this case FTO) to which a thin film of nanocrystalline oxide (2–12 μm) is applied—usually a wide bandgap *n*-type semiconducting metal oxide such as TiO_2, SnO_2, or ZnO. Dye molecules, organic or organometallic in nature (or in some cases a semiconducting quantum dot) are adsorbed onto the nanocrystalline semiconducting oxide. The modified electrode is combined with a hole conducting material (redox electrolyte such as I^-/I_3^- or a *p*-type semiconductor) and a counter electrode (in this case Pt coated FTO glass serves as the photoinert cathode) to complete the fabrication of a photoelectrochemical cell. When the photosensitizer (S) absorbs light, the molecule is able to inject an electron from its excited state (S^*) into the conduction band (CB) of the semiconductor, and this cationic dye radical ($S^{\bullet+}$) can then be regenerated by the redox electrolyte.

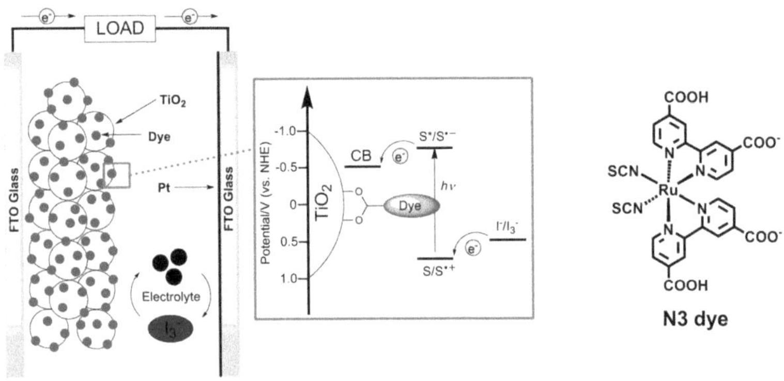

Figure 9.3. Construction and operation of a DSSC and the N3 ruthenium dye.

The role of the dye in the energy generation process within the DSSC is critical, with the use of ruthenium complexes (Fig. 9.3) displaying superior performance thus far [14–21]. Such dyes possess a broad absorption spectrum (in order to absorb the most amount of available light) and a three-dimensional (3D) geometry due to the metal–ligand binding structure. Such dyes are at a disadvantage however due to the limited natural abundance (0.001 ppm in earth's core) of the rare earth metal ruthenium, which makes the resulting dye expensive [22]. The recent focus on tetrapyrrolic macrocycles, similar to naturally occurring chlorophylls, have demonstrated that this family of molecules can provide an alternative to the expensive ruthenium dyes in the present day industry standard [23].

9.1.4 *Parameters Used in Photovoltaic Cell Characterization*

As the performance of multiporphyrin constructs as part of photoelectrochemical cells will be discussed, it is prudent to give a brief introduction to the parameters obtained when characterizing solar cell devices. The key parameters mainly referred to are the incident photon-to-current conversion efficiency (IPCE), photocurrent/voltage curve (J-V curve), open circuit voltage (V_{OC}), short-circuit photocurrent density (J_{SC}), fill factor (ff), and the overall solar energy conversion efficiency (η).

The IPCE of the solar cell is defined as the amount of current (i.e., number of electrons) flowing through the external circuit divided by the number of incident photons, expressed as a percentage. It is related to the wavelength of dye absorption by the following equation:

$$IPCE = LHE \times \phi_{\mathrm{inj}} \times \eta_{\mathrm{col}}$$

where *LHE* is light-harvesting efficiency, ϕ_{inj} is electron injection efficiency, and η_{col} is electron collection efficiency.

The remaining information can be obtained from the acquisition of a J-V curve. The V_{oc} is the potential between the electrodes when the device is exposed to light but there is no current flowing. This value corresponds to the difference between the Fermi level of the conduction band of the oxide (e.g., TiO_2) and the redox potential of the electrolyte (i.e., I^-/I_3^-). The J_{sc} represents the maximum photocurrent per unit area when there is zero resistance between the electrodes and is related to the interaction between the semiconducting oxide and the sensitizer. The *ff* is the maximum theoretical power output ($J_{\mathrm{max}} \times V_{\mathrm{max}}$) divided by the product of J_{sc} and V_{oc} (observed power output) as presented below. It is a measure of the actual power output in comparison to the theoretical maximum:

$$ff = (J_{\mathrm{max}} \times V_{\mathrm{max}})/(J_{\mathrm{sc}} \times V_{\mathrm{oc}})$$

The η value is a practical measurement relating the maximum output of electrical energy proportional to the energy of the incident sunlight (I_0). It is given in equation below:

$$\eta = (J_{\mathrm{sc}} \times V_{\mathrm{oc}} \times ff)/I_0$$

In order to have high-performance organic-based solar cells, we must consider the following criteria in their design: (1) excellent light-harvesting ability in the visible and near infrared (NIR) region, (2) efficient energy transfer to the heterojunction (if applicable) and excellent charge separation at the heterojunction, and (3) effective transport of the generated holes and electrons to their respective electrodes in order to minimize undesirable charge recombination processes [24]. In order to improve light-harvesting ability, multiporphyrin arrays have been utilized as well as the corresponding complexes and composites with C_{60} fullerene (and derivatives).

The use of porphyrins, synthetic analogues of naturally occurring chlorophyll used in photosynthesis for light harvesting have proven to be an excellent choice, due to their light-harvesting properties, stability and synthetic accessibility. Furthermore, the naturally occurring photosynthetic apparatus utilizes many dye molecules arranged in a specific manner in order to achieve a high quantum yield for the process of charge separation. The use of porphyrins in conjunction with C_{60} is also advantageous as these two families of molecules exhibit low reorganization energies upon electron transfer, leading to the possibility of obtaining long-lived charge separated states with high quantum yields [25–29].

In addition, electrophoretic deposition has been extensively utilized in the fabrication of photovoltaic devices because it is fast and economical. This deposition technique makes it possible to control the nano- and micro-structure of the organic molecules on the electrode surface, which is a critical feature to control within an organic film [2].

9.2 Giant Multiporphyrin Arrays Assembled on Inorganic Substrates

9.2.1 *Porphyrin-Modified Gold Nanoparticles*

Self-assembled monolayers (SAMs) of thiol-appended porphyrins on gold substrates have been previously studied with the aim of using such constructs for light harvesting. The inherent two-dimensional (2D) character of these constructs intrinsically limits the light-harvesting efficiency as the porphyrin monolayer can only absorb a small fraction of light [30–35]. The inclusion of 3D architecture into any light-harvesting construct in order to maximize light-harvesting ability is advantageous, as the physical properties of nanosized materials are quite different from that of the bulk material.

By moving from porphyrin-thiol SAMs on gold surfaces to monolayer protected gold nanoparticles (MPPs), the light-harvesting efficiency is drastically improved as a result of the increased surface area due to the adoption a 3D architecture drastically improved the light-harvesting efficiency (Fig. 9.4) [36]. Furthermore, secondary

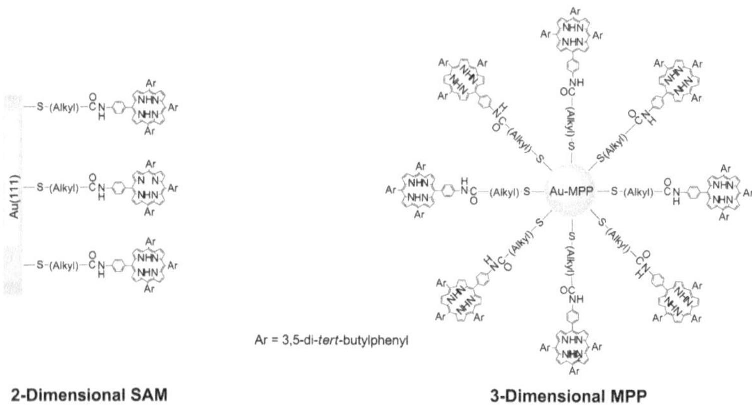

Figure 9.4. From porphyrin SAMs to porphyrin MPPs [36].

organization of the free-base porphyrin (H_2P) thiolate onto the gold nanoparticle did not strongly quench the fluorescence of the porphyrin within the MPP, in contrast to the bulk material (porphyrin-SAM) [36].

The resulting porphyrin-MPPs possess a well defined size (8–9 nm) and geometry (spherical) and could be subjected to tertiary and quaternary organization through the selective complexation of C_{60} molecules and subsequent clusterization (Fig. 9.5) [37].

These organizational processes were achieved in one synthetic step, thus allowing the formation of porphyrin-MPP+C_{60} clusters, as well as the reference clusters H_2P-C_{11}-MPP (without C_{60}) and H_2P+C_{60} (without gold nanoparticles). Both of the porphyrin-functionalized MPP clusters displayed broader absorption spectra, implying successful clusterization. Additionally, the absorption spectrum of H_2P-C_{11}-MPP+C_{60} showed a broad wavelength absorption in the 700–800 nm region, indicative of the diagnostic charge-transfer (CT) absorption band arising from the π-complex formed between the porphyrin and C_{60} [37].

Furthermore, it was found that moving from an alkanethiol with 11 methylene groups to one with 24, the Soret band experienced less broadening, which is indicative of decreased interaction between porphyrin units. This was directly attributed to the use of a longer alkyl spacer between the porphyrin and the gold nanoparticle

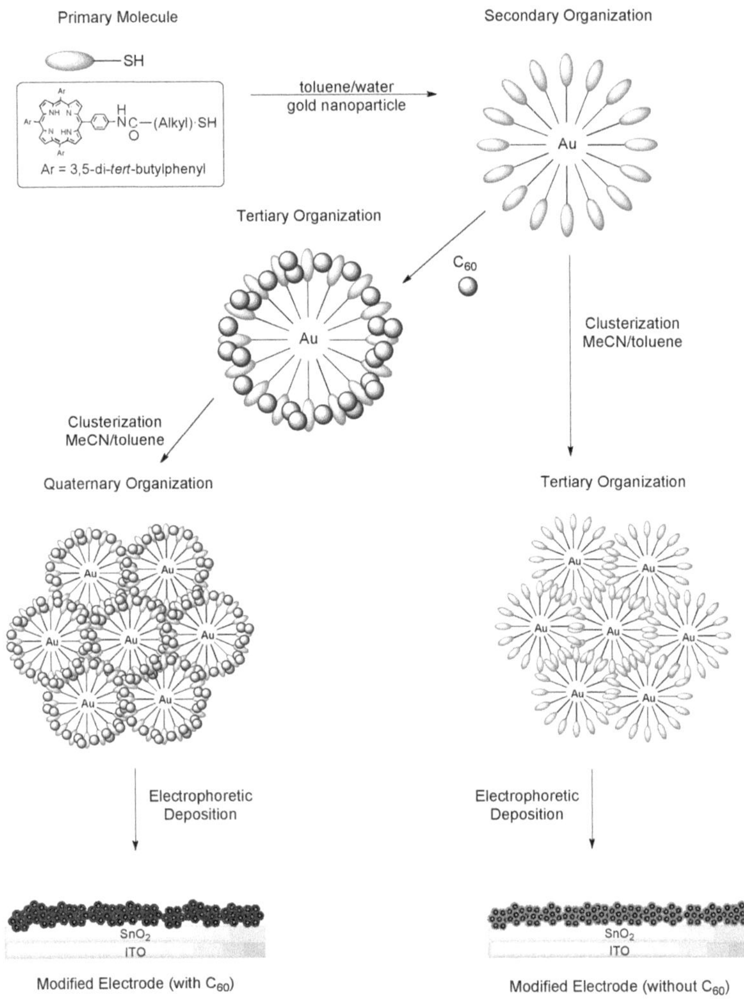

Figure 9.5. Illustration of higher order organizational processes with porphyrin-MPP-C_{60} complexes [37].

(Fig. 9.6) [38]. In the visible-NIR absorption spectrum, this distance dependence causes the charge-transfer absorption of the H_2P-C_{24}-MPP+C_{60} composite to be much weaker in intensity.

Electrodeposition of the clusters onto an optically transparent electrode (OTE) of nanostructured SnO_2 on ITO was achieved by

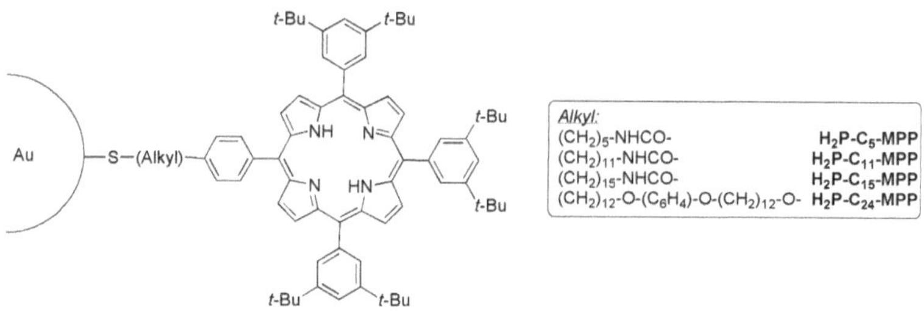

Figure 9.6. Comparison of porphyrin MPPs with different alkyl chain lengths [38, 39].

subjecting suspensions of the clusters to a high electric (DC) field (200 V) [37]. The resulting electrode possessed quaternary organization of the clusters and for the ITO/SnO_2/(H_2P-C_{11}-MPP+C_{60}) the formation of an interpenetrating network of porphyrin and C_{60} molecules [37, 39].

9.2.1.1 Effect of alkyl chain

The modified electrodes were subjected to photoelectrochemical measurements by using a standard two-electrode system, with a working electrode and a Pt-wire gauze electrode in 0.5 M NaI and 0.01 M I_2 in acetonitrile. The IPCE value of the ITO/SnO_2/(H_2P-MPP+C_{60}) system was monitored as a function of increasing C_{60} concentration ([C_{60}] in the initial toluene solution used in the complexation/clusterization stage). The system exhibited an outstanding increase in the IPCE with increasing C_{60} content, reaching a maximum IPCE value of 28% at 490 nm when using a ratio of $H_2P/C_{60} = 37 : 63$ [37, 39].

By increasing the number of methylene groups on the alkyl chain to 15, the ability of the H_2P-C_{15}-MPP to accommodate C_{60} molecules is greatly enhanced, resulting in a maximum IPCE of 54% at ~600 nm when using a ratio of $H_2P/C_{60} = 1{:}2$ [39]. Also, the device exhibited a very broad photoresponse (~1000 nm) that extended into the NIR region of the spectrum [39].

Further extension of the alkyl linker to 24 methylene carbons resulted in the IPCE of the photoelectrochemical device, dropping off to 8.5% at 445 nm when using a ratio of $H_2P/C_{60} = 1{:}8$ [38].

It was surmised that the spacer between the porphyrin and the gold nanoparticle core was much too long and flexible to enable efficient supramolecular complexation. As a result the complexation event between porphyrins on the H_2P-C_{24}-MPP and C_{60} is somewhat inhibited.

It was concluded that the observed quenching of the porphyrin-excited singlet state occurred via an electron transfer to the C_{60} within the construct rather than energy transfer to the gold nanoparticle. In contrast to this, an analogous system possessing the shorter alkanethiol linkage ITO/SnO_2/(H_2P-C_5-MPP+C_{60}) showed an IPCE that was reduced sevenfold because the length of the alkyl chain inhibits the accommodation of C_{60} between porphyrin rings [37, 39]. Also, the closer proximity of the porphyrin to the gold nanoparticle enables energy transfer to the gold nanoparticle to dominate over electron transfer to the C_{60}, effectively quenching the excited singlet state of the porphyrin [37, 39]. This results in an IPCE value reduced by one order of magnitude relative to the ITO/SnO_2/(H_2P-C_{11}+C_{60}) system, clearly demonstrating that the large improvements in the photoelectrochemical properties are due to the 3D structure between the porphyrin and the fullerene.

9.2.1.2 Effect of porphyrin

To examine the effect of different types of porphyrins within the MPP construct, a comparison between free-base and zinc porphyrin (ZnP) analogues was undertaken [39]. Thus the photoelectrochemical devices ITO/SnO_2/(ZnP-C_{11}-MPP+C_{60}) and ITO/SnO_2/(ZnP-C_{15}-MPP+C_{60}) were constructed (Fig. 9.7). The photocurrent action spectra of the zinc analogues gave IPCE values that were over 50% decreased from that of the free-base porphyrin containing constructs.

This significant drop in performance was attributed to the changes in the driving force of electron transfer process that arise upon metalation of the porphyrin core. For instance, in the case of a zinc porphyrin, the driving force for the formation of the charge separation product from excited zinc porphyrin and C_{60} becomes larger when compared to the free-base porphyrin. This is because of the reduction potential of excited singlet state ZnP (−1.0 V vs. NHE)

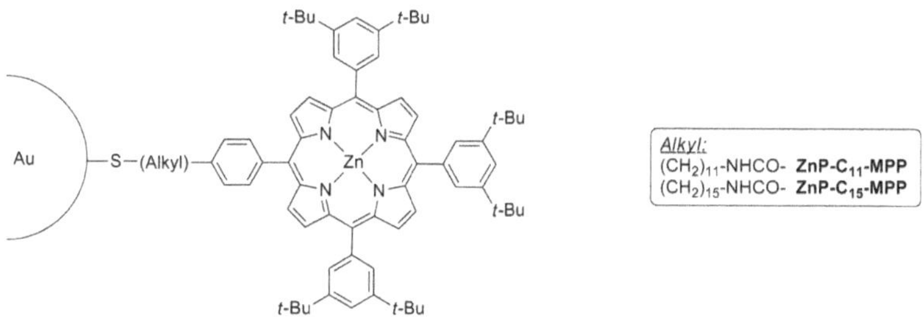

Figure 9.7. Schematic of zinc porphyrin-MPP constructs [39].

is lower than that of H_2P. In a consistent manner, the driving force for electron transfer from the I^-/I_3^- redox couple to the oxidized zinc porphyrin $ZnP^{\bullet+}$ is smaller than that for $H_2P^{\bullet+}$ as a result of the lower oxidation potential of ZnP (+1.0 V vs. NHE). Therefore, the reduced IPCE value is presumably due to slower electron transfer from I^-/I_3^- to $ZnP^{\bullet+}$ as a result of a decreased driving force for electron transfer.

9.2.1.3 Effect of fullerene: C_{60} vs. C_{70}

Substitution of the electron acceptor molecule C_{60} for C_{70} in this system (ITO/SnO_2/(ZnP-C_{15}-MPP+fullerene)) was attempted in order to probe the relationship between the shape of the fullerene and the performance of the corresponding device [39]. While C_{60} possesses a spherical geometry, C_{70} has a rugby ball shape (prolate spheroid), therefore the steric requirements of each are anticipated to be different upon complexation between two porphyrin rings on the functionalized MPP (Fig. 9.8).

Upon recording photocurrent action spectra it was shown that the composite electrode that utilized C_{70} experienced fourfold lower IPCE at the maximum value [39]. The 3D steric control between porphyrins and fullerenes on the MPP is essential in efficient charge separation and (hole and electron) transport. Although there is virtually no difference in the electron transfer properties of C_{60} and C_{70} [40–42], the uniform, spherical shape of C_{60} is much better than the rugby ball shape of C_{70} and it is this larger degree of uniformity gained with C_{60} that promotes effective formation of supramolecular clusters. More importantly, during the clusterization process

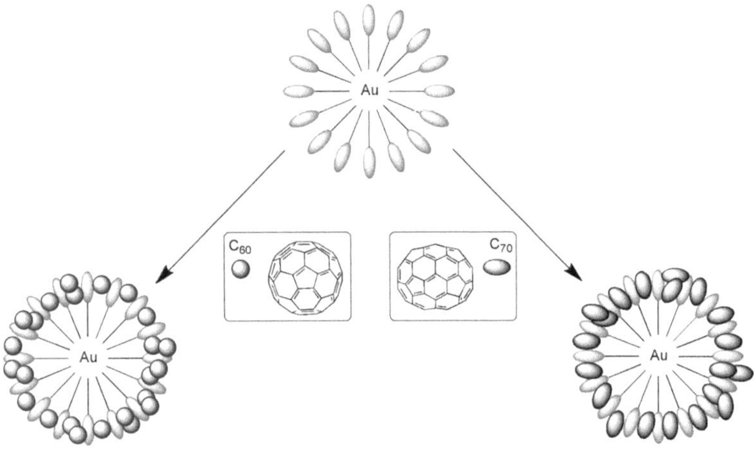

Figure 9.8. Porphyrin-MPP constructs using C_{60} and C_{70}.

the formation of an interpenetrating network of fullerene molecules enabled efficient electron hopping to the anode, resulting in a larger IPCE value [39].

9.2.1.4 Introducing tailored holes into the MPP surface

Modification of the H_2P-C_{11}-MPP particle was performed to create tailored holes on the porphyrin-modified gold nanoparticle surface, in order to assess the effect of this change in surface on the binding of C_{60} molecules [43, 44]. Surface hole formation was achieved using an exchange reaction between porphyrin alkylthiolates on the gold nanoparticle surface with short alkylthiols, resulting in the increase in free volume between porphyrin moieties on the MPP exterior (Fig. 9.9). These alkylthiols also possessed a terminal carboxyl group to induce amphiphilicity on the surface of the porphyrin-MPP, thereby suppressing aggregation of the construct. It was determined that there are 90 porphyrins and 60 carboxyl-functionalized alkyl thiolates on the surface of the MPP, equating to a 40% reduction of porphyrin moieties compared to the reference system H_2P-C_{11}-MPP.

In a similar manner to the well studied reference system (H_2P-C_{11}-MPP), the modified construct H_2P-C_{11}($C_{15}CO_2H$)-MPP can also form supramolecular complexes with C_{60} in toluene, and can be

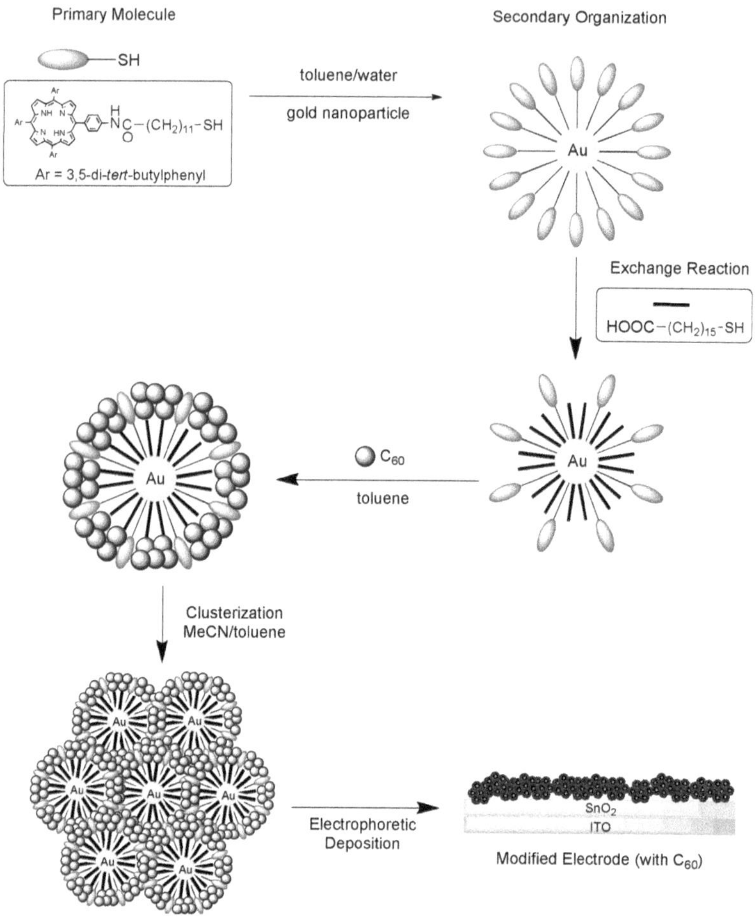

Figure 9.9. Porphyrin-MPP constructs with tailored holes [43, 44].

further clusterized using a mixed acetonitrile/toluene solvent. The absorption spectrum of (H_2P-C_{11}($C_{15}CO_2H$)-MPP+C_{60}) in acetonitrile/toluene (1:2) was broader than that of the toluene solution, indicating the formation of a π-complex formed between the porphyrin and C_{60} in the mixed solvent system. This is similar to the previously reported reference compound (H_2P-C_{11}-MPP+C_{60}) [37, 39].

Dynamic light scattering (DLS) analysis in the mixed acetonitrile/toluene (1:2) solvent shows that H_2P-C_{11}($C_{15}CO_2H$)-MPP and (H_2P-C_{11}($C_{15}CO_2H$)-MPP+C_{60}) clusters have sizes of 190 nm and 180 nm respectively, which increase after incubation periods. This is in stark contrast to the individual solutions of H_2P-C_{11}($C_{15}CO_2H$)-MPP and C_{60}, whose mean diameter (as observed by DLS) remains constant upon incubation. This reinforces the notion that the C_{60} molecules are incorporated into the tailored void spaces between porphyrin moieties on the MPP. Transmission electron microscopy (TEM) studies also show that (H_2P-C_{11}($C_{15}CO_2H$)-MPP+C_{60}) forms clusters of well defined size (2–3 μm) and shape (rod-like), further supporting the idea that H_2P-C_{11}($C_{15}CO_2H$)-MPP molecules self-assemble with the C_{60} in the mixed solution to give large donor–acceptor nanoclusters with an interpenetrating network.

Electrode modification was also performed in an analogous manner by electrophoretic deposition (200 V, 1 min) onto an ITO electrode with nanostructured SnO_2. Photoelectrochemical properties were assessed as a function of C_{60} concentration within the ITO/SnO_2/(H_2P-C_{11}($C_{15}CO_2H$)-MPP+C_{60}) by keeping [H_2P] within H_2P-C_{11}($C_{15}CO_2H$)-MPP and H_2P-C_{11}-MPP constant at 0.17 mM and increasing the concentration of C_{60} in the acetonitrile/toluene solution from 0 to 1.0 mM in the simultaneous complexation/clusterization step (prior to electrophoretic deposition).

For the ITO/SnO_2/(H_2P-C_{11}($C_{15}CO_2H$)-MPP+C_{60}) device a maximum IPCE value of 42% was achieved at 475 nm ([H_2P]/[C_{60}] = 1:6), whereas the reference device ITO/SnO_2/(H_2P-C_{11}-MPP+C_{60}) managed an IPCE maximum of 16% at 440 nm ([H_2P]:[C_{60}] = 1:3). This near threefold improvement in the IPCE maximum demonstrated that the presence of larger bucket-shaped holes in H_2P-C_{11}($C_{15}CO_2H$)-MPP improved the photoelectrochemical properties of the resulting device relative to the small, wedge-shaped holes in H_2P-C_{11}-MPP. The reasoning behind this improvement is the increased amount of C_{60} in the ITO/SnO_2/(H_2P-C_{11}($C_{15}CO_2H$)-MPP+C_{60}) device, enabling the formation of a more effective network of C_{60} in which electrons produced by photoinduced charge separation can be relayed to the conduction band of the SnO_2 electrode via an electron-hopping mechanism [39].

9.2.1.5 Current–voltage ($J-V$) characterization of cells

The current–voltage characteristics of the ITO/SnO_2/(H_2P-C_{11}-MPP+C_{60}), ITO/SnO_2/H_2P-C_{11}-MPP and ITO/SnO_2/(H_2P-C_{11}+C_{60}) devices were recorded in the dark and under visible light irradiation (>370 nm) [37]. Using an input power of 3.4 mW.cm^{-2}, the ITO/SnO_2/(H_2P-C_{11}-MPP+C_{60}) system possessed a much higher fill factor (0.35), open circuit voltage (V_{oc}, 420 mV), short circuit current density (J_{sc}, 0.14 mA.cm^{-2}), and overall power conversion efficiency (η, 0.61%) when compared to the reference systems. Moving to the ITO/SnO_2/(H_2P-C_{15}-MPP+C_{60}) the performance improved further. Using an input power of 11.2 mW.cm^{-2}, an increased *ff* (0.43) was observed along with an improvement in J_{sc} (1.0 mA.cm^{-2}), giving an overall power conversion efficiency of 1.5% [37].

The photocurrent generation observed in the system is initiated by a photoinduced electron transfer from the porphyrin-excited singlet state ($^1H_2P^*/H_2P^{\bullet+} = -0.7$ V vs. NHE) within the construct to C_{60} ($C_{60}/C_{60}^{\bullet-} = -0.2$ V vs. NHE) in the porphyrin/C_{60} complex. This is then followed by electron injection from the reduced fullerene into the conduction band of SnO_2 (0 V vs. NHE) nanocrystallite. Meanwhile, the oxidized porphyrin ($H_2P/H_2P^{\bullet+} = 1.2$ V vs. NHE) can undergo electron transfer reduction with the iodide ($I_3^-/I^- = 0.5$ V vs. NHE) in the electrolyte to regenerate the ground state dye [37].

9.2.2 *Porphyrin-Modified Silica Particles*

In order to improve the efficiency of the corresponding photoelectrochemical device, the metal nanoparticle-porphyrin conjugates were modified to address two outstanding design flaws. Firstly, the need to eliminate the use of expensive noble metals must be addressed in order to reduce the cost of the device. Secondly, the elimination of undesirable quenching of the porphyrin-excited singlet state as a result of a competing energy transfer processes to the gold surface must be addressed, a factor that is thought to be the most viable means of improving the photoelectrochemical properties of the corresponding device. Thus the synthesis, characterization, and photovoltaic properties of porphyrin-modified silica particles are discussed.

9.2.2.1 Porphyrin-modified silica microparticles

Porphyrin-functionalized silica microparticles (H_2P-mSiO_2) were synthesized via the coupling of an activated porphyrin ester with (aminopropyl)silylated silica microparticles (mSiO_2), (Fig. 9.10) by heating a toluene mixture at reflux for 24 hours [45]. In addition to this, a reference compound was synthesized via the reaction of the activated ester with butylamine.

The porphyrin-modified silica microparticles were characterized by elemental analysis, optical microscopy, diffuse reflectance absorbance and IR spectroscopy. Porphyrin attachment to silica microparticles was demonstrated using optical microscopy, with the resulting images showing purple-red spheres with an average size of 5 μm. On the other hand, the (aminopropyl)silylated microparticle exhibited spheres of similar size, but bereft of

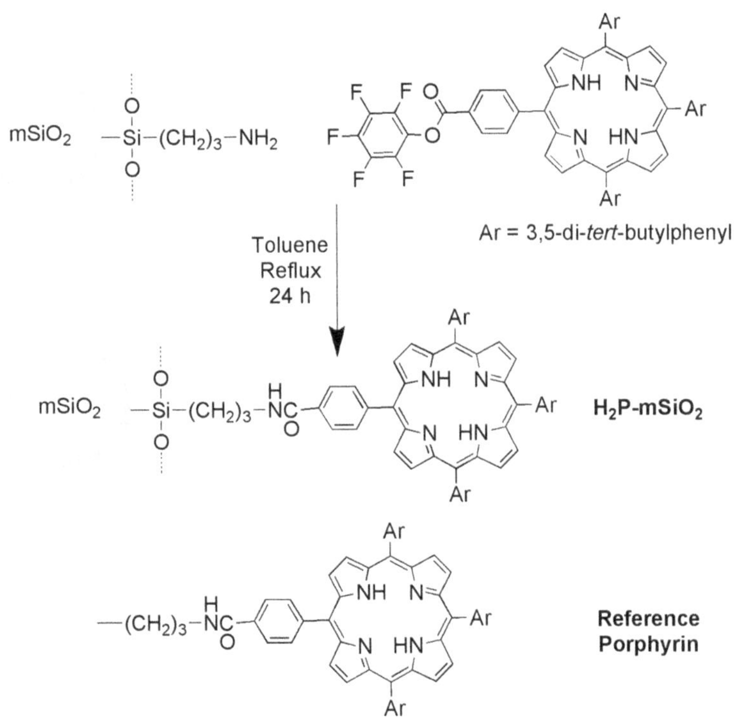

Figure 9.10. Synthesis of porphyrin-modified silica microparticles [45].

pigmentation. IR spectroscopy revealed the presence of the amide linkage between the silica microparticle and the porphyrin at the characteristic value of 1670 cm^{-1}, confirming the covalent immobilization of the dye onto the silica surface. Diffuse reflectance absorbance spectroscopy showed a typical free-base porphyrin absorption spectrum with the Soret absorption experiencing a bathochromic shift of 2 nm, relative to a chloroform solution of the butylamido-porphyrin model compound. This indicated a degree of perturbation upon linkage to the silica surface [45]. Based on the elemental analysis of the porphyrin-modified silica microparticles (C:13.9%; H:1.78%; N:1.64%), the surface area of the silica microparticles (224 $m^2.g^{-1}$) and the density of amino groups (3 $\mu mol.m^{-2}$) on the silica, it was determined that the average distance between porphyrin groups on the silica particles was approximately 1.6 nm, which is enough to incorporate a molecule of C_{60} ($\sim$1 nm) [39].

In a manner akin to the gold nanoparticles discussed above, the porphyrin-modified silica microparticles were suspended in a solution of toluene/C_{60} to enable complexation of fullerenes and subsequent clusterization upon the introduction of the mixed acetonitrile/toluene solvent. The clusterization process was performed at constant porphyrin concentration ($[H_2P] = 0.17$ mM) and the concentration of C_{60} was varied from 0 to 0.5 mM, in order to study the concentration dependence of C_{60} on the photoelectrochemical properties of the corresponding working electrodes. The composite clusters were electrophoretically deposited onto nanostructured SnO_2 electrodes by exposure to DC current (300 V, 2 min) to yield a modified working electrode. When compared to ITO/SnO_2/H_2P-mSiO_2, the SEM of the ITO/SnO_2/(H_2P-mSiO_2+C_{60}) showed a rougher surface upon the inclusion of the fullerene, demonstrating the formation of composite clusters with C_{60}.

The photoelectrochemical properties of the device were assessed using a three-electrode system in a similar manner to the work presented in the above section. An investigation of the effects of C_{60} concentration on the photocurrent action spectra revealed that the best performance was achieved with a porphyrin/fullerene ratio of 1:2. An IPCE maximum of 17% was reached at 420 nm using 125 $\mu W.cm^{-2}$ of light. This was a fourfold increase on the photoelectrochemical device composed of a porphyrin-modified gold nanoparticles featuring a similar spacer length, which gave an IPCE maximum of 4% under similar conditions [39, 43]. This clearly demonstrates that the substitution of the gold core of the construct for a silica microparticle is responsible for the improvement in the photocurrent generation efficiency due to the inhibition of energy transfer quenching processes of the excited singlet porphyrin component of the system upon light absorption.

9.2.2.2 Porphyrin-modified silica nanoparticles

In an effort to further improve the light-harvesting ability of the porphyrin-modified silica microparticle construct described above, a similar multiporphyrin array utilizing a silica nanoparticle was synthesized [46]. This was accomplished via the reaction of the same porphyrin activated ester with an (aminopropyl)silylated silica nanoparticle as shown in Fig. 9.11.

Silica nanoparticles were synthesized according to literature procedures to yield (aminopropyl)silyloxy nanoparticles with a diameter of 54 nm, as determined from TEM measurements. In a similar manner to the preceding investigation, the optical microscopic characterization of the modified nanoparticle revealed purple-red colored spheres which contrasts with the uncolored starting material. IR spectroscopy confirmed the presence of the amide linkage between porphyrin and silica nanoparticle with the characteristic amide carbonyl vibration at 1657 cm^{-1}. Elemental

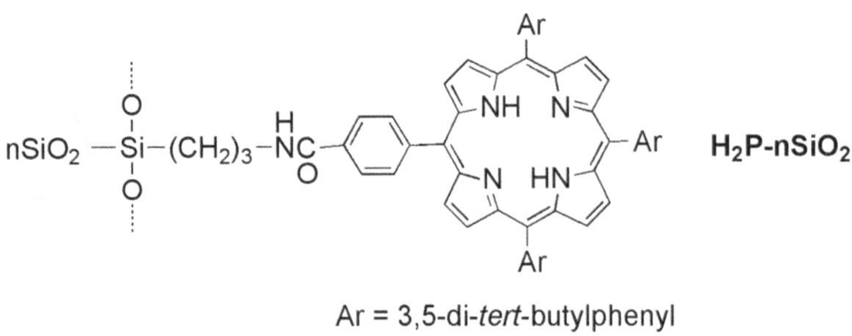

Figure 9.11. Porphyrin-modified silica nanoparticle [46].

analysis performed on the porphyrin-modified silica nanoparticles allowed the extent of porphyrin attachment to be determined, and estimated to be 25%. This value was consistent with the value determined by UV-vis spectroscopy (24%), achieved by measuring the initial and final concentration of porphyrin starting material in the reaction mixture used in the coupling reaction. Furthermore, the separation distance between porphyrin residues on the nanoparticle was calculated to be ~1.7 nm, similar to the value obtained for the porphyrin-modified microparticle (~1.6 nm), thus allowing the porphyrin-modified nanoparticle to complex C_{60} molecules between porphyrin rings [47–51].

UV-vis spectroscopy revealed that there was no shift of the Soret band upon porphyrin attachment, thus it can be assumed that the environment of the porphyrin is similar to that experienced in solution. However, fluorescence spectra showed that the emission relative to the butylamido-porphyrin model compound was lower upon covalent attachment to the silica nanoparticle. Time-resolved fluorescence spectroscopy in toluene (excitation at 400 nm, monitoring emission at 650 nm) revealed that the porphyrin-modified silica nanoparticle experienced a ~30% shorter fluorescence lifetime (6.2 ns) than that of the model butylamido-porphyrin compound (9.6 ns). Therefore, the decrease in steady-state emission cannot be attributed to energy transfer from the excited singlet porphyrin to the silica nanoparticle as a greater decrease would be expected, as with the porphyrin-modified gold nanoparticles described above, which experience a decrease in the observed fluorescence lifetime of two orders of magnitude when energy transfer occurs [36, 52–54]. The decrease in the fluorescence lifetime was attributed to self-quenching of the porphyrin-excited singlet state, a phenomenon that had been observed previously for SAMs of porphyrins on ITO electrodes.

As with the constructs already discussed, composite clusters with C_{60} were formed by maintaining a constant concentration of porphyrin units ($[H_2P] = 0.08$ mM). The concentration of the fullerene ($[C_{60}] = 0$ to 0.5 mM) in the mixed solvent system (acetonitrile/toluene, 2:1) was varied to enable the effect of fullerene concentration within the composite clusters and the corresponding efficiency of the photoelectrochemical device to be studied.

Having incorporated C_{60} into the modified nanoparticle and associated these into larger clusters, electrophoretic deposition (500 V, 2 min) was undertaken to attach these clusters onto nanostructured SnO_2 electrodes, yielding a modified electrode. SEM analysis revealed the rough nature of the nanoparticles in the ITO/SnO_2/(H_2P-nSiO_2+C_{60}) modified electrodes, and when compared to the ITO/SnO_2/H_2P-nSiO_2 electrode, was considered to be characteristic evidence of the formation of interpenetrating composite clusters of H_2P-nSiO_2 and C_{60} molecules [46].

Upon recording photocurrent action spectra for modified electrodes with increasing levels of C_{60} used in the clusterization process, it was revealed that using a 1:4 ratio of porphyrin/fullerene resulted in a maximum IPCE of 10% at 400 nm. This value is 2.5 times higher than the maximal IPCE value (4%) for the analogous porphyrin-modified gold nanoparticle featuring a similar chain length, thus demonstrating the effect of substitution of the metal core for a silica one [37–39, 43, 44].

Interestingly, this maximum IPCE value of 10% for the ITO/SnO_2/(H_2P-nSiO_2+C_{60}) is much smaller than the value reported for the analogous porphyrin-modified microparticle (17%). This was attributed to the difference in the amount of excess C_{60} present in the working electrode and results in a poor connection between nanoparticles [45]. This leads to poor packing of the larger clusters in the mixed solvent, and the resulting clusterized particles of the modified working electrode exhibit irregular sizes and shapes. Thus the electron and hole mobilities in the ITO/SnO_2/(H_2P-nSiO_2+C_{60}) construct are not as efficient as that present in the ITO/SnO_2/(H_2P-mSiO_2+C_{60}).

9.2.3 *Porphyrin-Modified CdSe Nanoparticles*

Having overcome the unfavorable, competing energy transfer process by substitution of the metal core with a silica particle, the next step in improving the performance of the multiporphyrin array was to restore light-harvesting ability to the core. Previous investigations show that, the use of metal nanoparticles will always result in some degree of unfavorable quenching [36, 54], thus the use of a semiconducting nanoparticle is warranted. Like

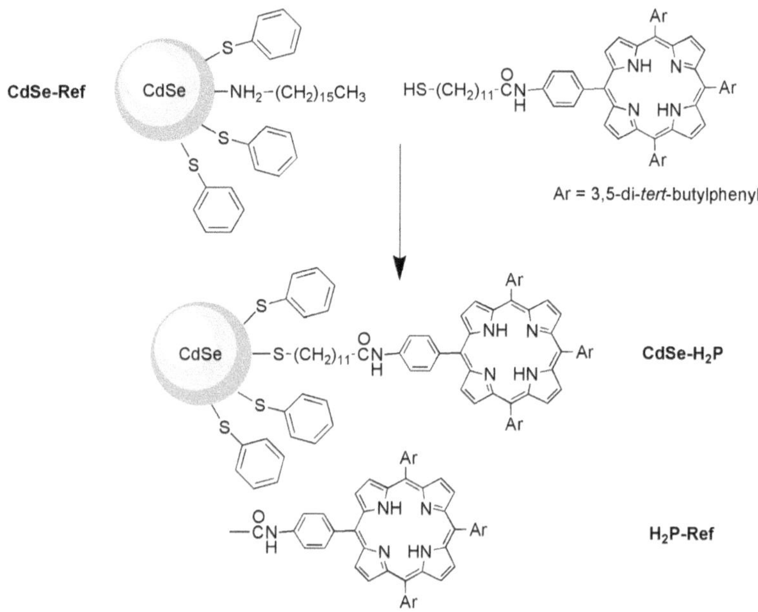

Figure 9.12. Synthesis of porphyrin-modified CdSe nanoparticles [60].

their metal nanoparticle counterparts, semiconducting nanoparticles exhibit light-harvesting properties that can lead to a subsequent energy transfer to the immobilized porphyrin on the nanoparticle, if the bandgap is larger than that of the porphyrin [55–59]. Therefore, not only can the unfavorable quenching process of the excited porphyrin singlet state be suppressed, but the light harvesting from the semiconducting nanoparticle can contribute favorably to photoinduced processes.

This lead to the synthesis of a porphyrin-modified CdSe nanoparticle [60]. Figure 9.12 outlines the synthesis of the construct, achieved by the place-exchange reactions occurring upon treatment of a hexadecylamine-linked CdSe nanoparticle (CdSe-ref, core size 3.6 nm) with a porphyrin alkylthiol.

The bandgap of the CdSe-nanoparticle was calculated from the onset of absorption in toluene, yielding a value of 2.17 eV, whereas the porphyrin reference compound (H_2P-Ref, Fig. 9.12) gave a value of 1.90 eV, as determined by absorption and emission data. As the bandgap of the CdSe-ref is larger than the H_2P-ref, the occurrence of photoinduced energy transfer from the nanoparticle to the porphyrin within the assembly is possible.

UV-vis absorption spectra of the porphyrin-modified nanoparticle revealed that the number of porphyrin residues on the surface of a single CdSe nanoparticle could be modulated as a function of reaction time, through estimation by using the extinction coefficient of the porphyrin and the nanoparticle. After a reaction time of 24 hours the number of porphyrin units on each nanoparticle reached a maximum of 21.

Further evidence for the structure of the porphyrin-modified CdSe nanoparticle was provided by X-ray photoelectron spectroscopy (XPS), elemental analyses and to a lesser extent ^{1}H NMR spectroscopy. Calculations performed on a model nanoparticle (assuming a spherical particle and an internal Cd/Se ratio of 1:1, with 50% of the outermost atoms of the nanoparticle being Cd) demonstrated that there are 140 ligands on the CdSe-ref nanoparticle. Elemental analysis of the CdSe-ref gives a ligand ratio of 1:1.2 for thiophenol/hexadecylamine, which corresponds to 63 thiophenol and 77 hexadecylamine moieties on the surface of the nanoparticle. Using the elemental analysis results for the H_2P-CdSe the ratio of ligands changes to 3:1:1 for thiophenol/hexadecylamine/porphyrin alkylthiol. This corresponds to 63 thiophenol, 21 hexadecylamine and 21 porphyrin alkylthiol ligands, which supports the UV-vis analysis of CdSe-H_2P. However the calculated ligand occupancy for CdSe-H_2P revealed a loss of 35 unoccupied hexadecylamine residues upon ligation of the porphyrin to the nanoparticle. This discrepancy can be explained by the sheer increase in molecular size of the porphyrin ligand (1.1 nm^2) compared to that of hexadecylamine (0.2 nm^2), resulting in a moderate density of porphyrin moieties on each nanoparticle. This lower density of porphyrin moieties on the nanoparticle is also consistent with the lack of red-shifting experienced by the Soret band upon complexation to the CdSe [60].

The emission properties of the porphyrin-modified CdSe nanoparticle were initially investigated by recording steady-state fluorescence spectra. Using selective excitation of the CdSe nanoparticle at 380 nm, it was found that the emission from the CdSe (at $\sim$550 nm) decreased with the increasing number of porphyrin ligands on the nanoparticle. This decrease was accompanied by an increase in the porphyrin-derived emission at 650 nm and 720 nm, implying that photoinduced energy transfer from the CdSe core to the peripheral porphyrin moiety occurred. Recording excitation spectra by monitoring the porphyrin-based emission at 651 nm corroborated the energy transfer process. By removing the porphyrin-based absorbance in the excitation spectrum, the result was an excitation spectrum matching that of the CdSe-ref, with a characteristic peak at 538 nm. Emission spectra were also recorded using excitation at 421 nm where the absorbance from the porphyrin component is the same for each sample. The emission from the CdSe nanoparticle as well as the energy transfer from CdSe to H_2P within the porphyrin-modified nanoparticle were subtracted and the resulting spectra revealed that the intensity of porphyrin emission (derived directly from porphyrin excitation) remained proportional to the number of attached porphyrins. Thus it could be concluded that unfavorable deactivation processes such as energy and electron transfer or self-quenching were suppressed. This is in stark contrast to the other porphyrin-modified nanoparticles discussed in this chapter, and agrees with the energy diagram presented in Fig. 9.13.

Differential pulse voltammetry (in CH_2Cl_2, 0.1M n-Bu_4NPF_6) of the CdSe-ref was performed in order to determine the oxidation potential of the nanoparticle. The oxidation peak appeared at 1.47 V (vs. Ag/0.01 M $AgNO_3$) and had a broad shape due to the size distribution of the CdSe nanoparticles. Using this value and the value of the bandgap obtained spectroscopically, the energy levels of the CB and the valence band (VB) were determined to be -1.02 V and $+1.15$ V (vs. NHE) respectively. In a similar manner the energy levels for the H_2P-ref were resolved and an energy diagram was constructed (Fig. 9.13).

From the energy diagram it can be seen that competing photoinduced electron transfers from the CdSe-excited state to the ground state porphyrin ($\Delta G = 0.02$ eV), from the excited singlet state of the porphyrin to the CB of CdSe ($\Delta G = 0.22$ eV) and from the VB of CdSe to the excited singlet state of the porphyrin ($\Delta G = 0.29$ eV) are all positive with regard to the free energy released, thus making these processes forbidden. Although the photoinduced electron

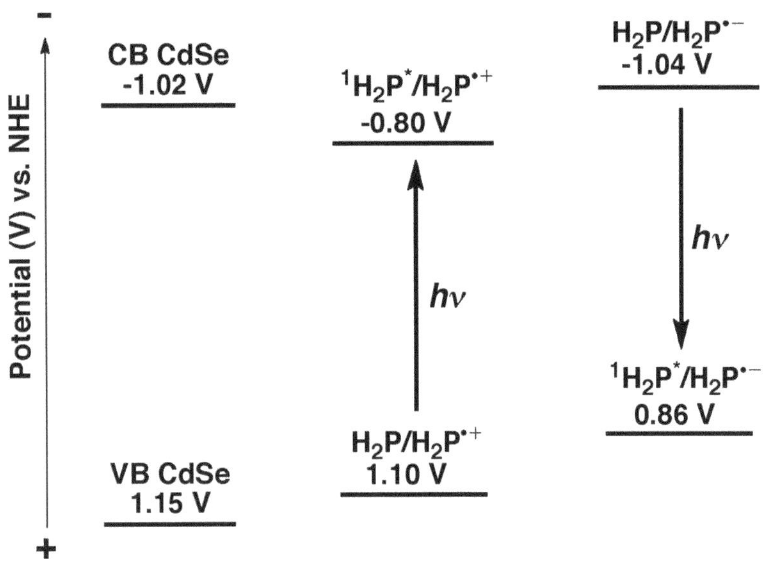

Figure 9.13. Energy diagram for the CdSe-H_2P nanoparticle.

transfer process from the H_2P to the VB of the CdSe-excited state ($\Delta G = -0.05$ eV) is slightly allowed, the electron transfer did not take place. Therefore, energy transfer from the CdSe-excited state to the porphyrin is the sole process occurring in the CdSe-excited state of the porphyrin-modified nanoparticle. It is forbidden for the porphyrin-excited state to be deactivated by either energy or electron transfer to the CdSe core.

9.3 Multiporphyrin Assemblies Formed Through Organic Molecule Clusterization

9.3.1 *Porphyrin-Carbon Nanotube Constructs*

A recurring theme in the realization of high performance bulk heterojunction solar cells is the formation of a "nanohighway" for efficient hole and electron transport [61–69]. Preorganization of the donor and acceptor components (in D-A multilayers mounted on electrodes) plays a large part in achieving efficient photocurrent

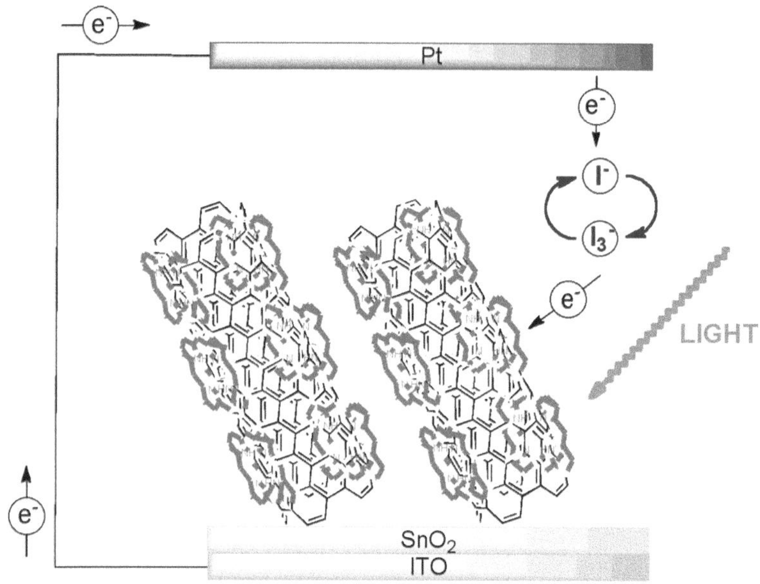

Figure 9.14. Supramolecular assembly of a protonated porphyrin and SWCNT [69].

generation, with the importance of this highlighted in the previous section.

The one-dimensional (1D), nanowire-like structure of carbon nanotubes (CNTs) is an attractive option to provide the nanohighway-type architectures on electrodes. However, the dispersability of the CNTs is poor because the strong π–π interactions between individual CNTs in a bundle make them difficult to process in solution [61, 62, 70]. The bundles of CNTs remain intact when used in conjunction with polymers and the like for the purposes of making photovoltaic devices, thus precluding the formation of the important interpenetrating D-A bicontinuous nanohighway and promoting the self-quenching of the excited state CNTs.

Previously, protonated free-base porphyrins were used to create D-A assemblies wherein the nanotube acts as a template for porphyrin complexation to create highly ordered supramolecular assemblies of single-walled CNT (SWCNT)-porphyrin composites, as shown in Fig. 9.14 [69].

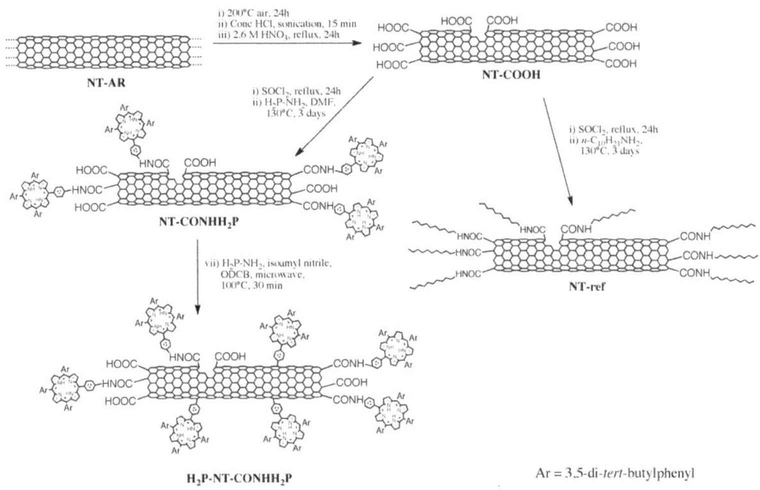

Figure 9.15. Synthesis of covalently linked porphyrin-SWCNT assemblies [77].

These composites were electrophoretically deposited onto a nanostructured SnO_2 electrode and subsequently fabricated into a photoelectrochemical device. Using a redox electrolyte of NaI (0.5 M) and I_2 (0.01 M) and a Pt counter electrode, this modified electrode showed a maximum IPCE of 13% using an applied potential of 0.2 V. Under a light input power of 12.4 mW. cm^{-2}, $\eta = 0.012\%$ was achieved, with $J_{SC} = 0.10$ mA. cm^{-2}, $V_{OC} = 60$ mV, and $ff = 0.24$.

The ability to covalently attach additional functionality to the termini, defect sites and sidewalls of SWCNTs provides an excellent means of dispersing the bundled nanotubes [71–76]. In particular, bulky porphyrin units bearing *tert*-butyl groups at the *meta*-position of the phenyl rings were employed as large chromophores, with these also acting as exfoliating agents for the SWCNTs [77].

This was realized through covalent linkage of the acid chloride derivative of the acid-treated SWCNT (NT-COOH) with amine-functionalized porphyrins (Fig. 9.15) [78, 79]. In this study, a sequential functionalization of the SWCNT was performed via reaction of the amido-linked porphyrin-SWCNTs (NT-CONHH$_2$P) with the diazonium salt of an amino porphyrin synthon to introduce more bulky porphyrin moieties at defect sites and onto the sidewall to give H_2P-NT-CONHH$_2$P [73, 80]. The result is effective suppression of nanotube bundle formation and improved solubility in organic solvents. More importantly, the self-quenching of excited SWCNTs as well as excited porphyrins would be suppressed as a result of the exfoliation.

The maximum concentration in DMF for each of the SWCNT-based compounds was determined after sonication for 30 minutes, with a solubility order of H_2P-NT-CONHH$_2$P (0.5 g.L^{-1}) > NT-CONHH$_2$P (0.3 g.L^{-1}) > NT-COOH (0.15 g.L^{-1}) > NT-ref (0.1 g.L^{-1}) $\gg$ NT-AR (0.001 g.L^{-1}). The trend shows that the extent of functionalization and indeed double functionalization (terminals, defect sites, and sidewalls) with porphyrins is highly effective in exfoliating the SWCNT bundles.

The porphyrin-modified SWCNTs were additionally characterized through the use of AFM, TEM, FT-IR, resonant Raman spectroscopy, TGA, UV-vis-NIR absorption, and fluorescence spectroscopy. AFM images were taken after spin coating pre-sonicated DMF solutions/suspensions onto freshly cleaved mica. The presence of fibrous structures is observed in each of the images due to the tendency of the individual nanotubes to bundle. However, the effectiveness of the nanotubes exfoliation was consistent with the trend in the solubility data. Thus the sizes of the bundles are as follows: NT-AR > NT-COOH (1.5–8 nm) > NT-ref (3–12 nm) > NT-CONHH$_2$P (1–8 nm) > H_2P-NT-CONHH$_2$P (1–3 nm).

The resonant Raman spectroscopic analysis is useful in providing hybridization information for the carbon atoms of the nanotube. The ratio of the tangential mode (G band) around 1600 cm^{-1} and the disorder mode (D band) around 1350 cm^{-1}, known as the G/D ratio, reflects the relative number of sp^3 carbon in the nanotube and thus can be used to estimate the degree of functionalization [81]. Furthermore, the Raman signals arising from the porphyrin and alkyl moieties possess negligible intensity compared to those originating from the nanotube, adding to the convenience of the analysis. As expected, the G/D ratios of NT-ref and NT-CONHH$_2$P are very close, with values of 5.8 and 5.6 respectively. Both of these are much smaller than that of the NT-AR (G/D = 14) starting material because the acid treatment of the nanotubes increases the number of terminal and defect sites within the sample. The G/D ratio is further decreased to 2.8 in the highly substituted H_2P-NT-CONHH_2P and occurs as a result of sidewall functionalization, causing more nanotube carbons to be converted from sp^2 to sp^3 hybridization.

Analysis of the porphyrin-functionalized SWCNTs reveals the presence of a Soret band around 420 nm and Q-bands from 500 to 650 nm, as well as a broad absorption extending into the NIR region. Also, the spectra match the superposition of the UV-vis-NIR spectra of NT-ref and H_2P-ref, giving unequivocal evidence of porphyrin functionalization on the nanotube. The Soret absorptions of chromophores in both of the porphyrin-modified SWCNTs are shifted bathochromically by 2 nm upon covalent attachment of the porphyrin to the nanotube. In addition to this, the Soret absorption in H_2P-NT-CONHH_2P is broader than that of NT-CONHH_2P, implying that there is a larger degree of ground-state electronic communication between the SWCNTs and porphyrin units attached to the sidewall [82]. It was also determined spectroscopically that the degree of sidewall functionalization was twice as large as that of terminal and defect functionalization. Furthermore, it was determined that porphyrin residues on the sidewall of the nanotube occur every 360 carbon atoms. These figures can be extrapolated to give an interporphyrin distance of 3.2 nm on a SWCNT possessing a diameter of 1.0 nm.

The acquisition of steady-state fluorescence data was achieved by excitation of the porphyrinic Soret band. An emission spectrum reminiscent of a free-base porphyrin was recorded and no emission from the SWCNT was observed. The fluorescence intensities of H_2P-NT-CONHH_2P (7%) and NT-CONHH_2P (12%) were heavily reduced compared to a sample of H_2P-ref, implying the occurrence of photoinduced energy/electron transfer or charge transfer. Also the stronger quenching experienced by H_2P-NT-CONHH_2P is in good agreement with the larger degree of ground state electronic communication, as observed by Soret broadening in the UV-vis-NIR spectra.

Modified electrodes were fabricated through electrophoretic deposition (80 or 100 V, 120 s) to create thin films of the porphyrin-modified nanotubes on a nanostructured SnO_2 electrode on FTO. The resulting modified electrodes were subjected to analysis by optical microscopy, which was able to detect the presence of SWCNT bundles that were the least pronounced in the H_2P-NT-CONHH$_2$P sample the result of efficient exfoliation. AFM analysis of the bundle heights on the modified electrodes was also performed, giving a similar pattern to the solubility experiments and AFM experiments described above. The measured bundle sizes are as follows: NT-ref (100 ± 20 nm) > NT-COOH (60 ± 20 nm) > NT-CONHH$_2$P (30 ± 10 nm) > H_2P-NT-CONHH$_2$P (23 ± 7 nm).

The modified electrodes were subjected to photoelectrochemical measurements in acetonitrile containing 0.5 M LiI and 0.01 M I_2, a Pt wire counter electrode and a reference electrode. By excitation at 400 nm, photocurrent action spectra were recorded and the maximum IPCE values obtained as follows: FTO/SnO_2/NT-ref (2.3%) < FTO/SnO_2/NT-COOH (2.6%) < FTO/SnO_2/NT-CONHH$_2$P (4.0%) < FTO/SnO_2/H_2P-NT-CONHH$_2$P (4.9%). Once again, these values are consistent with the AFM results for the analysis of the SWCNT bundle sizes (or alternatively, exfoliation efficiency). This suggests that the suppression of nanotube bundle formation decreases the degree of self-quenching experienced by SWCNTs through intertube interactions, resulting in more efficient photocurrent generation [77].

An analogous device employing TiO_2 in place of SnO_2 was fabricated, and under the same conditions a maximum IPCE of 5.6% at 400 nm was recorded. However, in comparison to the FTO/TiO_2 device (12% at 400 nm) the photocurrent was reduced. This is primarily due to the difference in the level of the conduction band of the two semiconductors ($E_{CB}(SnO_2) = 0$ V vs. NHE and $E_{CB}(TiO_2) = -0.5$ V vs. NHE) (Fig. 9.16) [83, 84]. The result is an exothermic character for the electron injection from the excited state SWCNT to the CB of SnO_2, with the same scenario being endothermic if the semiconductor is TiO_2.

We can therefore formulate an energy diagram using (8, 6) semiconducting SWCNTs, which are the major species present in SWCNTs. This is calculated by the experimentally determined, relational expression of the Fermi level with the diameter of SWCNTs and the diameter and bandgap energy of semiconducting SWCNTs. From this, we can see that photocurrent generation is initiated by electron injection from the CB of SWCNT-excited states of diameter 0.9–1.1 nm (–0.5 to 0.0 V vs. NHE) into the CB of the nanostructured

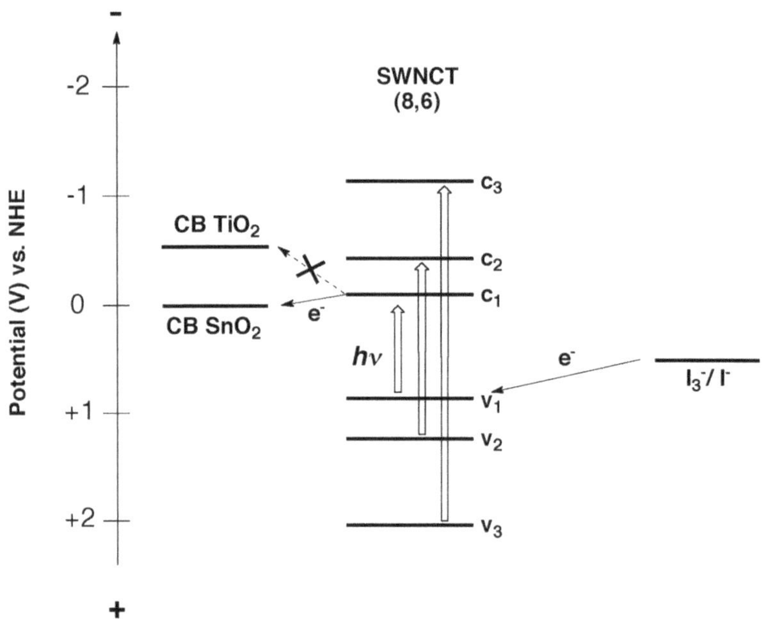

Figure 9.16. Energy diagram of photocurrent generation for (8, 6) SWCNTs [77].

SnO_2 ($E_{CB} = 0$ vs. NHE). Likewise, an electron transfer can take place from the redox couple ($I^-/I_3^- = 0.5$ V vs. NHE) in the electrolyte to the VB of the s-SWCNT (0.5–1.2 V vs. NHE).

It should also be noted that there is no photocurrent response from the porphyrin moieties and that this lack of contribution is in disagreement with the quenching of the porphyrin-excited singlet state upon attachment to SWCNTs. Photoinduced energy transfer can be excluded as the cause of the quenching process, as production of the SWCNT-excited state due to energy transfer from the porphyrin would lead to photocurrent response derived from the porphyrin absorption. Photoinduced electron transfer from the porphyrin to the SWCNT is also plausible mechanism, but the reports of fast charge separation and slow charge recombination seen in porphyrin-SWCNT systems in an analogous manner to porphyrin-fullerene systems tend to suggest that this would lead to highly efficient photocurrent generation [26, 27, 29, 69, 85–88].

Finally, the utilization of a CT quenching mechanism was invoked to explain the photoinduced phenomena of the system. The CT quenching mechanism is based on similar observations of porphyrin-fullerene dyads possessing short spacers [89–91]. Therefore, in this system, the porphyrin-excited singlet state and the SWCNT form an exciplex through a CT interaction, which decays to the ground state without forming the charge-separated state, thus there is no contribution from porphyrin absorption to photocurrent generation. Additionally, covalent attachment of the porphyrin units to the terminals, the defect sites, and the sidewalls of the SWNTs influences the electronic structures and energy levels of the SWNTs significantly, leading to alteration of the photophysics.

9.3.2 *Porphyrin-Modified Dendrimers*

The use of peripherally functionalized porphyrin dendrimers to promote complexation with C_{60} and subsequent clusterization was also investigated by preparing a series of multiporphyrin dendrimers (Fig. 9.17) [92, 93]. To enable comparisons of the effects of the dendritic/multiporphyrinic structure, H_2P-ref was employed as a model compound. Similarly, the photoelectrochemical properties of a covalent-linked porphyrin-fullerene dimer, H_2P-C_{60}, were also investigated in order to determine the role of C_{60} and its supramolecular assembly within the construct in promoting efficient photocurrent generation.

The compounds were subjected to complexation of C_{60} and subsequent clusterization in acetonitrile/toluene (3:1), keeping the concentration of porphyrin units constant at 0.19 mM and an optimized concentration of $[C_{60}] = 0.31$ mM (ascertained by assessing the photoelectrochemical properties of (H_2P-ref+C_{60}). These clusters were further subjected to a high DC current (500 V, 1 min) to promote electrophoretic deposition onto an OTE of nanostructured SnO_2.

The photoelectrochemical properties of the modified electrodes were assessed in acetonitrile with 0.5 M NaI, 0.01 M I_2, and a Pt wire counter electrode. The modified electrode ITO/SnO_2/(H_2P-ref+C_{60}) gave a maximum IPCE value of ~4%. The IPCE values of the ITO/SnO_2/(D_8P_8+C_{60}) and the ITO/SnO_2/($D_{16}P_{16}$+C_{60})

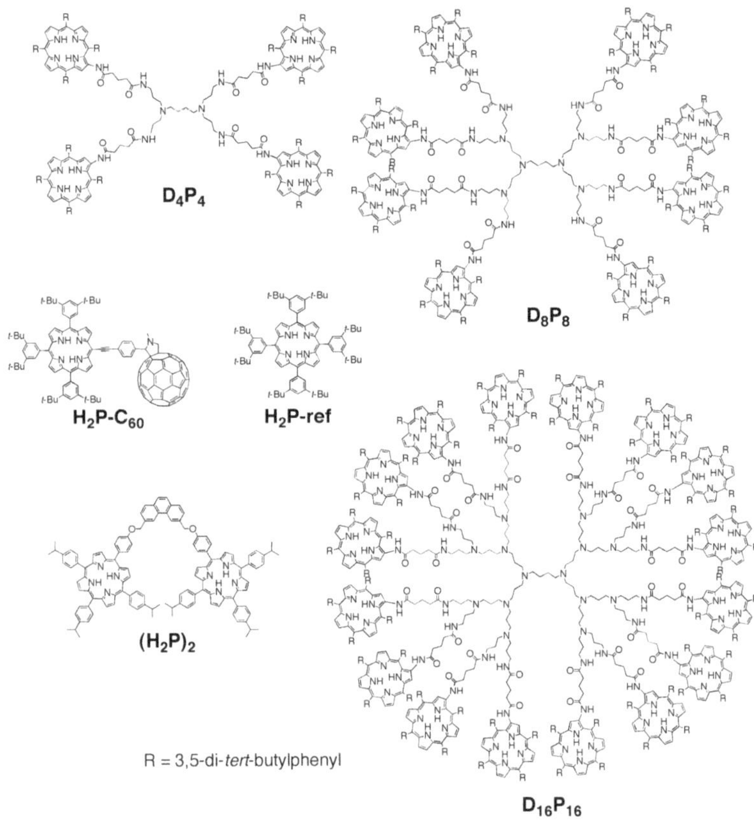

Figure 9.17. Dendritic porphyrins and reference compounds from [93].

modified electrodes were determined to be approximately 10% and 2%. This value is in stark contrast to that exhibited by the ITO/SnO_2/(D_4P_4+C_{60}) system, where the IPCE rose to 15%, in addition to possessing a broader photoresponse to that of the reference systems. Additionally, the ITO/SnO_2/($(H_2P)_2$+C_{60}) system exhibited a relatively high IPCE with a maximum of ~14% recorded. These values are consistent with the TEM data of the cluster sizes and shapes and also with the associative strength between the dendrimers and the C_{60}.

Upon recording the current-voltage characteristics of the cells it was revealed that the ITO/SnO_2/(D_4P_4+C_{60}) system has a *ff* =

0.31, V_{oc} = 220 mV, J_{sc} = 0.29 mA.cm^{-2} and an η = 0.32%. Under similar conditions the ITO/SnO_2/($(H_2P)_2$+C_{60}) system exhibited an overall η = 0.15%. Thus, the J–V characteristics of both systems are greatly improved, especially the ITO/SnO_2/(D_4P_4+C_{60}) system, giving an approximately tenfold improvement on the reference ITO/SnO_2/(H_2P-ref+C_{60}) system (η = 0.035%) under identical conditions.

9.3.3 *Porphyrin-Functionalized α-Polypeptides*

Porphyrin-functionalized α-polypeptides have also been used to provide multiporphyrin constructs that are able to complex C_{60} fullerene and undergo clusterization to form giant multiporphyrin arrays. Figure 9.18 shows the porphyrin-peptide oligomers used in this study [94, 95].

In the clusterization process, the concentration of the porphyrin unit was kept constant at 0.19 mM for each sample and the concentration of C_{60} kept constant (but in excess) at 0.31 mM. A fast

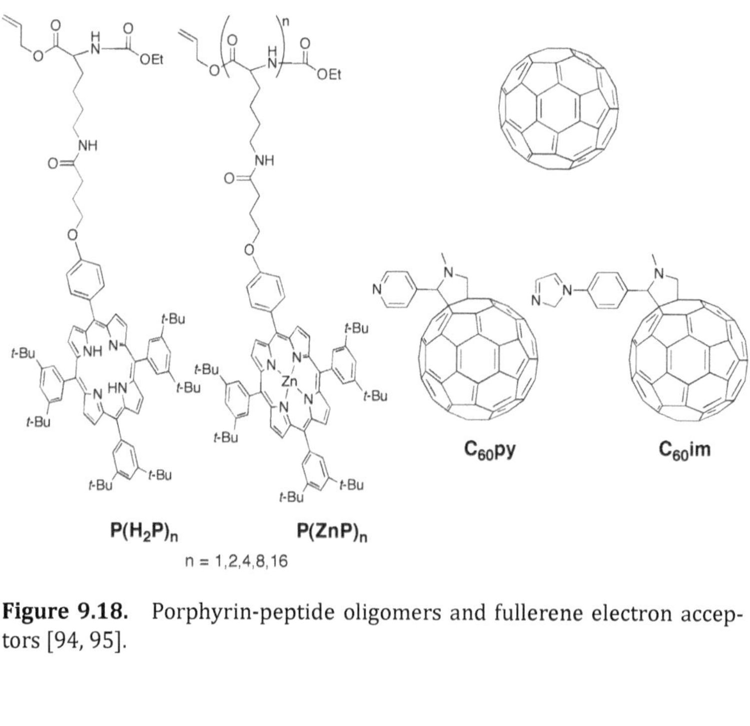

Figure 9.18. Porphyrin-peptide oligomers and fullerene electron acceptors [94, 95].

injection process formed composite clusters where a toluene solution of the porphyrin-peptide and C_{60} was mixed with acetonitrile to yield a mixed solution of acetonitrile/toluene (3:1). This allowed both supramolecular complexation of the fullerene and subsequent clusterization to be achieved in the same synthetic step. The composites were electrophoretically deposited onto and optically transparent ITO/SnO_2 electrodes by exposing the cluster suspension to a high DC voltage (500 V, 1 min). The resulting modified electrode was then subjected to photoelectrochemical measurements in a standard two-electrode system consisting of a working electrode, a Pt wire gauze electrode and electrolyte composed of 0.5 M NaI and 0.01 M I_2 in acetonitrile. Photocurrent action spectra and the current–voltage characteristics of the cells were obtained, the results of which are summarized in Table 9.1.

IPCE measurements revealed a number of interesting observations. Firstly, the IPCE increases dramatically with the number of porphyrins on the polypeptide chain. This is indicative of efficient photoinduced electron transfer from the porphyrin-excited singlet state to the fullerene in the supramolecular complex. This result is also reflected in the association constants and quenching data obtained from steady-state fluorescence analysis of the ((H_2P or ZnP)+C_{60}) complex.

Secondly, the ITO/SnO_2/(P(H_2P)$_{16}$+C_{60}) system gave a maximum IPCE of 48% at 600 nm, in addition to a broad photoresponse

Table 9.1. IPCE and solar cell efficiencies from (porphyrin-peptide+C_{60}) modified electrodes [95]

System	IPCE (%)	η (%)	System	IPCE (%)	η (%)
(P(ZnP)$_1$+C_{60})	4	0.047	(P(H_2P)$_1$+C_{60})	4	0.043
(P(ZnP)$_2$+C_{60})	12		(P(H_2P)$_2$+C_{60})	26	
(P(ZnP)$_4$+C_{60})	22		(P(H_2P)$_4$+C_{60})	36	
(P(ZnP)$_8$+C_{60})	33		(P(H_2P)$_8$+C_{60})	42	1.3
(P(ZnP)$_{16}$+C_{60})	56	1.4	(P(H_2P)$_{16}$+C_{60})	48	1.6
(P(ZnP)$_{16}$+C_{60}im)	15				
(P(ZnP)$_{16}$+C_{60}py)	20				

(up to 950 nm). The high degree of light absorption in the composite was ascribed to the ability of the polypeptide structure to control the 3D structure of the porphyrin and C_{60} within the clusters. Looking at the zinc porphyrin-polypeptide oligomers, the same trend with regard to IPCE and oligomer size was observed. In a similar manner to the free-base analogue, the maximum IPCE was also obtained for the ITO/SnO_2/(P(ZnP)$_{16}$)+C_{60}) system, where a much higher value of 56% at 480 nm was observed. This change can be attributed to the difference in driving force for electron transfer being more favorable when using zinc porphyrins within the oligomer rather than the free-base analogue. Also, the photocurrent action spectra of the ITO/SnO_2/(P(H_2P)$_n$+C_{60}) electrode possess a larger and broader response than the zinc analogue, which also supports the presence of a stronger supramolecular interaction between porphyrin and fullerene in the polypeptide structure of the free-base analogues.

In addition to the photoelectrochemical study of modified electrodes comprised of C_{60} composites, the use of fullerenes able to ligate with the zinc porphyrin was investigated. Thus, the ITO/SnO_2/(P(ZnP)$_n$+C_{60}py) and ITO/SnO_2/(P(ZnP)$_n$+C_{60}im) electrodes were fabricated. As presented in Table 9.1, the introduction of a ligand onto the fullerene moiety decreases the IPCE maximum to 20% and 15%, respectively, for the ITO/SnO_2/(P(ZnP)$_{16}$+C_{60}py) and ITO/SnO_2/(P(ZnP)$_{16}$+C_{60}im) systems. An important conclusion to be drawn from this is that the performance of these two modified electrodes was severely decreased, although the introduction of a ligand markedly increased the binding constant between metalloporphyrin and fullerene moieties, suggesting that the strength of the association between donor and acceptor is not necessarily responsible for improving the IPCE values. The other difference is the quality of electron and hole transport in the thin film of the modified electrode, as this has continually been shown to have a drastic effect on the magnitude of the photocurrent generation. Thus, it was surmised that the steric hindrance of the ligand moiety was responsible for the decrease in the maximum IPCE values obtained for the ITO/SnO_2/(P(ZnP)$_n$+C_{60}py) and ITO/SnO_2/(P(ZnP)$_n$+C_{60}im) modified electrodes.

9.4 Multiporphyrin Arrays Composed of Donor–Acceptor Molecules and Composites

9.4.1 *Porphyrin Clusters*

Porphyrin clusters comprised of 5,15-bis (3,5-di-*tert*-butylphenyl)porphyrin (Fig. 9.19) were made by a clusterization process in the acetonitrile/toluene = 9:1 mixed solvent [96]. Using this fast injection method, a transparent suspension of porphyrin cluster was formed. The UV-vis spectroscopy of the porphyrin clusters revealed that although the structure of the Q-bands was preserved, the Soret band experienced slight blue-shift (<5 nm) and noticeable broadening when compared to toluene solutions of the monomeric porphyrin. Such changes are indicative of the presence of intermolecular interactions that influence the electronic transitions of the Soret and Q-bands in different ways [97–100].

Electrophoretic deposition of the porphyrin clusters onto an ITO/TiO_2 electrode was achieved by exposure to DC current (500 V, 10 min). The absorption spectrum of the modified electrode showed a slight (3 nm) red shift in the absorption maximum (which originates from Soret absorption by the porphyrin). AFM analysis of the modified electrode revealed the presence of uniform clusters with size 50–100 nm arranged on the electrode to give a nanoporous morphology, imparting a very high surface area to the ITO/TiO_2/H_2P thin film. A photoelectrochemical cell was constructed using an electrolyte of NaI (0.5 M), I_2 (0.01 M) in acetonitrile. The maximum IPCE of the device was 2%, measured at an applied potential of 0.2 V vs. SCE.

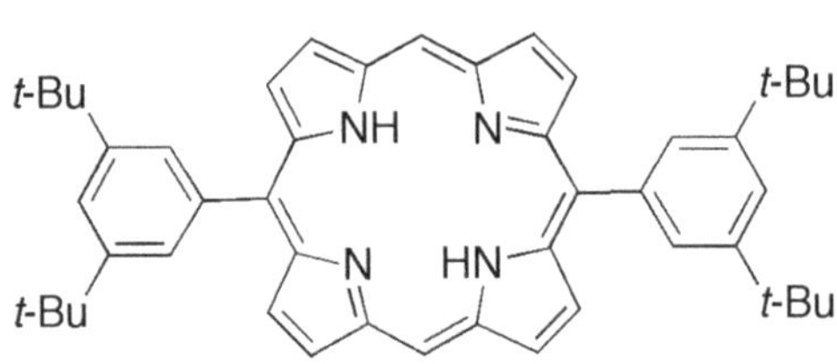

Figure 9.19. 5,15-Bis(3,5-di-*tert*-butylphenyl)porphyrin [96].

9.4.2 *Porphyrin-Fullerene Cluster Composites*

The clusterization of monomeric porphyrins to yield supramolecular synthons for use in light harvesting was extended to incorporate the use of a monomeric acceptor molecule, C_{60} fullerene [101]. The assembly of equimolar amounts of C_{60} and 5,15-bis(3,5-di-*tert*-butylphenyl)porphyrin (0.19 mM each) into molecular clusters was achieved in the mixed solvent acetonitrile/toluene (3:1). Subsequent electrophoretic deposition (500 V) of the clusters onto an optically transparent ITO/SnO_2 was performed to yield modified electrodes of ITO/SnO_2/(H_2P+C_{60}) as well as reference electrodes composed of respective individual porphyrin and fullerene clusters, ITO/SnO_2/H_2P and ITO/SnO_2/C_{60}.

The construction of a photoelectrochemical cell was performed using the electrolyte NaI (0.5 M) and I_2 (0.01 M) in acetonitrile, a Pt gauze counter electrode and a saturated calomel electrode reference electrode (SCE). Using an applied potential of 0.2 V vs. SCE, photocurrent action spectra were recorded for the three modified electrodes. A maximum IPCE value of 17% was achieved with the ITO/SnO_2/(H_2P+C_{60}) modified electrode, with much smaller IPCE values being obtained for the ITO/SnO_2/H_2P and ITO/SnO_2/C_{60} photoelectrochemical devices. Furthermore, the sum of the IPCE maxima of the single component cluster electrodes (~6%) was significantly smaller than that of the mixed component cluster. This clearly demonstrates that interplay between the excited states of each component on the composite cluster is responsible for the larger photocurrent generation.

9.4.2.1 The effect of the porphyrin

Work on the system formed from composite clusters of monomeric porphyrins and fullerenes was extended to investigate the substitution effects of the porphyrin utilized in the clusterization process [102]. Clusters of five different porphyrins possessing different steric and electronic qualities (Fig. 9.20) were studied.

UV-vis absorption spectra of the molecular complexes between the porphyrins and C_{60} in toluene were compared with the summative absorption spectra of respective monomers under the same

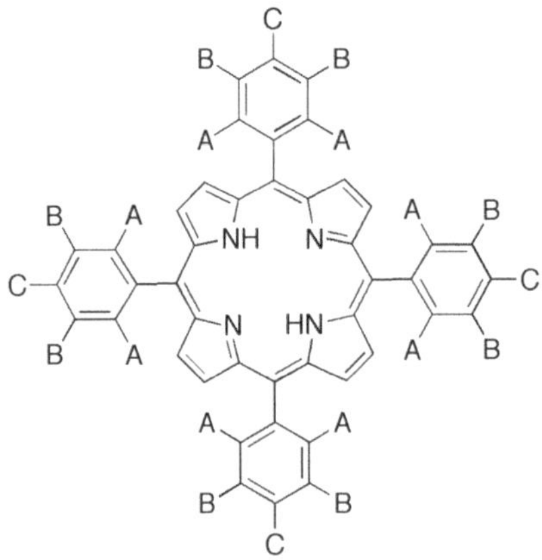

3,5-OMeTPP: A,C=H, B=OMe
3,4,5-OMeTPP: A=H, B,C=OMe
2,6-OMeTPP: B,C=H, A=OMe
TPP: A,B,C =H
3,5-CF_3TPP: A,C=H, B=CF_3

Figure 9.20. Porphyrins used in this study [102].

conditions. The appearance of pronounced CT absorption from 650 to 800 nm was observed for the (3,5-OMeTPP+C_{60}) solution, with a similar absorption appearing in the spectrum of the (3,4,5-OMeTPP+C_{60}) solution. However, the solutions of 2,6-OMeTPP and 3,5-CF_3TPP with C_{60} afforded no CT absorption in their respective UV-vis absorption spectra. This comes as a result of the large steric hindrance and the poor electron donating ability of the respective porphyrins, which inhibits supramolecular complexation. The association constants of the molecular complexes were obtained via a UV-vis titration experiment, the result of which is summarized in Table 9.2.

Self-organization of the porphyrins and C_{60} into composite clusters was achieved by fast injection of a toluene solution of the respective porphyrin/C_{60} into acetonitrile. The result is the formation of

Table 9.2. Association constants and IPCE data for the (porphyrin+C_{60}) complexes [102]

Porphyrin	Association Constant K (M^{-1})	IPCE (%)
3,5-OMeTPP	1.3×10^3	59
3,4,5-OMeTPP	1.4×10^3	10
2,6-OMeTPP	Not observed	5.7
3,5-CF_3TPP	Not observed	4.0
TPP	12.2×10^2	0.5

composite clusters in a mixed solvent (acetonitrile/toluene = 3:1) containing [H_2P] = 0.017 mM and [C_{60}] = 0.1 mM. The absorption spectra of the (H_2P+C_{60}) clusters in the mixed solvent afford much broader spectra than those in toluene alone, implying successful formation of the composite clusters. Furthermore, the broad absorption of the cluster suspensions in the 700–800 nm region is diagnostic of the CT absorption as a result of the π-complex formed between the porphyrin and C_{60} [50, 103–106]. Also, the UV-vis spectrum of (3,5-OMeTPP+C_{60}) possesses two Soret absorptions at 418 and 443 nm. The latter Soret absorption can be assigned to *J*-type aggregation of the porphyrins with C_{60} molecules, with the former originating from porphyrin units that are poorly organized within the cluster solution [30, 107–109]. DLS analysis of the cluster solutions revealed narrow size distributions for all the porphyrin/C_{60} composites, from 110 nm to 130 nm, with the (3,5-OMeTPP+C_{60}) cluster solution showing another cluster size centered at 2 μm.

Analysis of the (H_2P+C_{60}) clusters by cross-polarization magic angle spinning (CP-MAS) afforded solid-state ^{13}C NMR spectra. The spectra revealed new resonances between 143.9 and 138.9 ppm, upfield from that of pristine C_{60} (144.1 ppm) in the spectra obtained for (3,5-OMeTPP+C_{60}) and (3,4,5-OMeTPP+C_{60}). The upfield shifting of the fullerene peaks gave unequivocal evidence of the presence of $C_{60}^{\delta-}$ within the complex, consistent with the formation of a CT complex [102].

Thin films of the (H_2P+C_{60}) composites on nanostructured SnO_2 electrodes were constructed by electrophoretic deposition of the cluster solutions in the mixed solvent through the application of a high electric (DC) field (200 V, 2 min). The clusters were deposited onto the ITO/SnO_2 electrode to give modified working electrodes, denoted as ITO/SnO_2/(H_2P+C_{60}). Interestingly, the absorption spectrum for the (3,5-OMeTPP+C_{60}) modified electrode only possessed a singular Soret absorption at 440 nm as a result of *J*-aggregation of the 3,5-OMeTPP in the cluster solution.

SEM analysis of the ITO/SnO_2/(3,5-OMeTPP+C_{60}) revealed the presence of well-defined cubic clusters with dimensions of 1.0 × 3.0 μm, which is in good agreement with the observed 2 μm clusters seen in the DLS analysis. However, SEM analysis of clusters made from the other porphyrins displayed either no cluster formation or poorly defined clusters. Optical microscopy of the ITO/SnO_2/(3,5-OMeTPP+C_{60}) electrode using polarized light found that the large cubic clusters were in fact microcrystalline. AFM analysis of the porphyrin/fullerene films that lie below the cubic clusters showed close packing of clusters that are somewhat planar in shape. A similar shape of cluster was also observed for the ITO/SnO_2/(3,5-CF_3TPP+C_{60}). In contrast to this, AFM analysis of the ITO/SnO_2/(3,4,5-OMeTPP+C_{60}), ITO/SnO_2/(2,6-OMeTPP+C_{60}) and ITO/SnO_2/(TPP+C_{60}) electrodes showed the presence of small spherical clusters. The sizes of the clusters measured by AFM were consistent with the results obtained from the DLS analysis.

The photoelectrochemical properties of the modified electrodes were examined in acetonitrile with LiI (0.5 M) and I_2 (0.01 M), using a Pt counter electrode and a SCE reference. Photocurrent action spectra were recorded to gain an insight into the substituent effects on the photoelectrochemical properties, with the results presented in Table 9.2. For the ITO/SnO_2/(3,5-OMeTPP+C_{60}) system, an increasing relative ratio of C_{60} in the clusters resulted in a remarkable increase in the maximum IPCE to 59% at 425 nm ([H_2P]/[C_{60}] = 1:6). Similar behavior was found for the other electrodes under the same conditions, albeit with much lower IPCE values. The huge maximum IPCE value of the ITO/SnO_2/(3,5-OMeTPP+C_{60}) system (59%) compared to the other systems in this study (0.5–10%) clearly shows that the unique molecular arrangement in the ITO/SnO_2/(3,5-OMeTPP+C_{60}) device is responsible for the resulting photocurrent generation efficiency.

To analyze the stoichiometry of the clusters in more detail, desorption was performed on the modified electrodes using toluene and the ratio of [H_2P]/[C_{60}] was determined. The complexation ratios ([H_2P]/[C_{60}]) in the ITO/SnO_2/(2,6-OMeTPP+C_{60}) and ITO/SnO_2/(3,5-OMeTPP+C_{60}) electrodes were determined to be between 1:10 (for 3,5-OMeTPP) and 1:350 (for 2,6-OMeTPP). The order of the ratios of [H_2P]/[C_{60}] closely matched the complexation strength of the porphyrins for C_{60} and bears some resemblance to the photocurrent generation ability for the films, with some exceptions. Of these exceptions are the ITO/SnO_2/(2,6-OMeTPP+C_{60}), the ITO/SnO_2/(3,5-CF_3TPP+C_{60}), and the ITO/SnO_2/(TPP+C_{60}) devices. The relatively high IPCE of the ITO/SnO_2/(2,6-OMeTPP+ C_{60}) device may occur as a result of the strong electron donating ability of the particular porphyrin used (2,6-OMeTPP), in comparison to the poor electron donating ability of 3,5-CF_3TPP and TPP. In comparing the higher maximum IPCE of the ITO/SnO_2/(3,5-CF_3TPP+C_{60}) and ITO/SnO_2/(TPP+C_{60}) modified electrodes, the better molecular packing of the 3,5-CF_3TPP is said to be responsible for the improved photocurrent generation of the device.

To investigate the influence of molecular arrangement on the photoelectrochemical properties of the porphyrin/C_{60} clusters, single-crystal X-ray analysis was undertaken. However under the conditions used, 3,5-OMeTPP.2C_{60} was the only co-crystallized solid obtained. X-ray analyses revealed the presence of CH–π interactions between the methoxy groups of the porphyrin and the π-system of the fullerene, assisting the complexation between the two molecules. In addition to this, the presence of a weak hydrogen bonding interaction between methoxy groups was also suggested. Thus, within the single crystal, the porphyrin and fullerene molecules form an alternating layer structure, with the porphyrins arranged in a 1D chain and the C_{60} molecules arranged in a 2D sheet.

By performing a diffuse reflectance spectroscopy of a ground sample of the single crystal, a single Soret band at 439 nm was detected, along with the CT absorption at 700–800 nm, which is highly consistent with the observed absorption spectrum of the ITO/SnO_2/(3,5-CF_3TPP+C_{60}) modified electrode. This, combined with the observation that the ITO/SnO_2/(3,5-CF_3TPP+C_{60}) electrode exhibits strong quenching of porphyrinic emission with a concurrent evolution of CT emission at 825 nm, suggests that such specific supramolecular interactions as observed in the single crystal are responsible for the high degree of order within the composite, resulting in efficient photocurrent generation.

9.4.2.2 Porphyrin-fullerene dyads

The use of a porphyrin-fullerene dyad, shown in Fig. 9.21, was also investigated as the monomeric molecule exhibited excellent ability to undergo photoinduced charge separation abilities. A cluster solution in mixed solvent (toluene/acetonitrile = 1:19) was achieved by injection of a toluene solution of the dyad into acetonitrile [110].

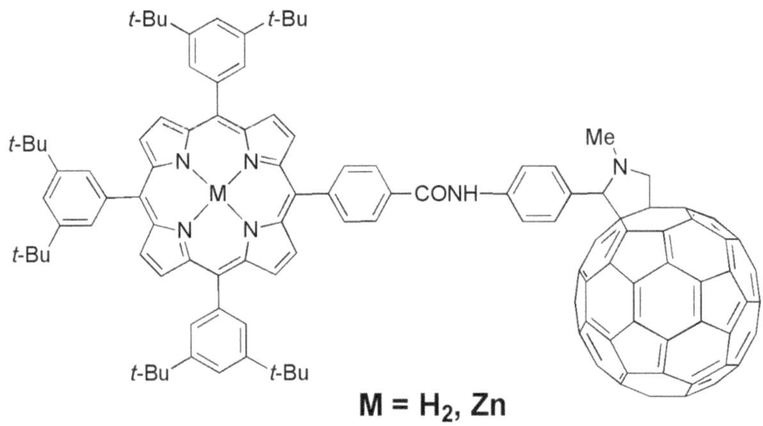

Figure 9.21. Porphyrin-fullerene dyad [110].

UV-vis spectra of the clusters in the mixed solvent revealed a red shift of 16 nm in the porphyrin-derived Soret absorption when compared to the dyad in toluene. Steady-state emission spectra demonstrated the ability of the clusters to undergo photoinduced charge separation, and this observation was confirmed through the use of time-resolved transient absorption. The porphyrin-fullerene dyad clusters were cast onto nanostructured SnO_2 films and when fabricated into a cell, maximum IPCE values of 0.42% and 0.36% were achieved for the free-base and zinc analogues, respectively. Comparison of these results to the analogous monolayer devices showed dramatic increases in the light-harvesting properties of the electrodes of 40-fold and 20-fold, respectively, for the free-base and zinc-porphyrin dyad clusters. However, the covalent inclusion of the C_{60} moiety resulted in a lower IPCE value, as the formation of an interpenetrating network of fullerenes cannot be formed upon clusterization.

9.4.3 *Solution-Processed Donor–Acceptor Composites and Complexes*

9.4.3.1 ZnO nanorod

The unique photocurrent generating ability of the C_{60} fullerene and zinc tetrakis(3,5-di-*tert*-butylphenyl)porphyrin supramolecular complex was extended through the use of semiconducting ZnO nanowires and nanoparticles [111]. Additionally, a fullerene monolayer was employed to further improve photocurrent generation and provide a better methodology for achieving bottom-up self-organization.

The use of ZnO nanowires was investigated in order to improve performance of the cell by increasing the interfacial area of the electrode. Additionally, the efficiency of the device may be increased due to the fact that the 1D structure of ZnO nanorods (ZnO-r) results in high electron mobility as well as facile, low temperature syntheses, and the ability to control the morphology via solution processing [112, 113].

Initially, ZnO nanocrystals were formed on the FTO electrode by immersion in an ethanolic $Zn(OAc)_2$ solution, with the nanotubes then grown by subsequent immersion in an ethanolic $Zn(NO_3)_2$ solution. The density and size of the ZnO nanocrystal seeds were controlled by the concentration and immersion time of the zinc acetate solution, whereas the length of the ZnO nanorods was determined by the immersion time in the zinc acetate solution. These ZnO-nanorod electrodes were also compared to the ZnO nanoparticle (ZnO-p) electrodes.

Initially, C_{60} monolayers were formed on the ZnO nanorod and nanoparticle electrodes by immersing the FTO/ZnO-r and FTO/ZnO-p into a 0.2 mM solution of C_{60}-acid (Fig. 9.22). The electrodes were further modified by spin coating a DMF solution of ZnP and C_{60}-acid composite to yield the FTO/ZnO-r/C_{60}-acid/(ZnP+C_{60}-acid) and FTO/ZnO-p/C_{60}-acid/(ZnP+C_{60}-acid) electrodes.

The modified electrodes were incorporated into a photoelectrochemical device using a platinum counter electrode and an electrolyte made from 0.5 M LiI and 0.01 M I_2 in acetonitrile. Photocurrent action spectra of the modified electrodes revealed that the maximum IPCE of the FTO/ZnO-r/C_{60}-acid/(ZnP-C_{60}-acid) device was 30% at 400 nm, using the optimized ratio of [ZnP]/[C_{60}-acid] as 1:6. The presence of the C_{60} monolayer was revealed by the fabrication of FTO/ZnO-r/ZnP, which exhibited virtually the same IPCE as that of the FTO/ZnO-r electrode, demonstrating that electron injection from the porphyrin-excited singlet state into the CB of ZnO is not possible. The maximum IPCE of the FTO/ZnO-r/C_{60}-acid device is

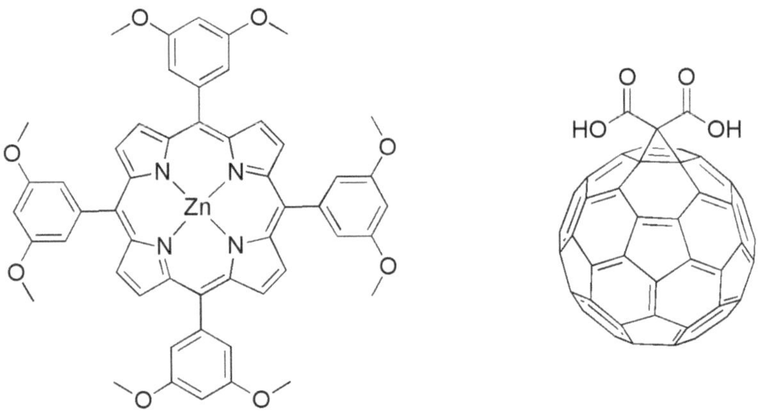

ZnP C_{60}-acid

Figure 9.22. Structures of C_{60} acid and porphyrin used in this study [111].

much higher than that of FTO/ZnO-r, meaning that the excited state of C_{60} in the electrode can undergo an electron transfer with the I^- in the electrolyte to produce $C_{60}^{\bullet-}$. This can then inject an electron into the CB of ZnO, hence the noticeable increase in photocurrent generation. Introducing a porphyrin or fullerene into the modified electrode to give FTO/ZnO-r/C_{60}-acid/ZnP or FTO/ZnO-r/C_{60}-acid/C_{60}-acid leads to a small increase in photocurrent generation, showing that electron injection from the LUMO of the C_{60}-acid monolayer is a poor pathway to effect electron injection from the excited state of ZnP or C_{60}-acid. However, the FTO/ZnO-r/C_{60}-acid/(ZnP+C_{60}-acid) modified electrode displays superior photocurrent generation ability when compared to the other electrodes, effectively demonstrating that an initial charge separation between ZnP and C_{60}-acid in the presence of the C_{60}-acid monolayer has a profound consequence on cell performance. The FTO/ZnO-r/C_{60}-acid/(ZnP+C_{60}-acid) device showed $J_{sc} = 1.37$ mA.cm^{-2}, $V_{oc} = 0.26$, and $ff = 0.4$, yielding a η value of 0.14% under standard AM 1.5 conditions.

Comparing this result to that of the analogous device using ZnO nanoparticles, FTO/ZnO-p/C_{60}-acid/ZnP, a similar IPCE value was obtained. This was a surprising result given that the nanoparticle possessed a threefold greater surface area (and threefold greater concentration of C_{60}-acid) than the nanorod electrode. Cross-sectional SEM revealed that the ability of the (ZnP-C_{60}-acid) composite to successfully infiltrate (and as a result completely cover) the nanorod array during spin coating was the key process resulting in the differences observed [111].

9.4.4 *Bicontinuous Donor–Acceptor Arrays*

Bicontinuous donor–acceptor (D-A) arrays on a molecular level have been shown to enable efficient charge separation and facilitate the transportation of these separated charges to their respective electrodes. In an effort to realize a vertical alignment of bicontinuous D-A arrays on a flat FTO/SnO_2 electrode, a Pd-mediated stepwise growth of porphyrin brushes was performed, as shown in Fig. 9.23 [114]. Additionally, the macrocycle used was a zinc porphyrin, enabling the infiltration of pyridyl-fullerene molecules (Py-C_{60}, Fig. 9.24) using a combination of axial coordination of the pyridyl group of the derivatized fullerene to the metalloporphyrin and π–π interactions between fullerenes within the fullerene moieties.

This elaborate architecture was constructed by initial adsorption of a pyridyl Zn porphyrin acid (Py-ZnP-Acid) (Fig. 9.23) on a smooth FTO/SnO_2 electrode (roughness factor 1.1) via the car-

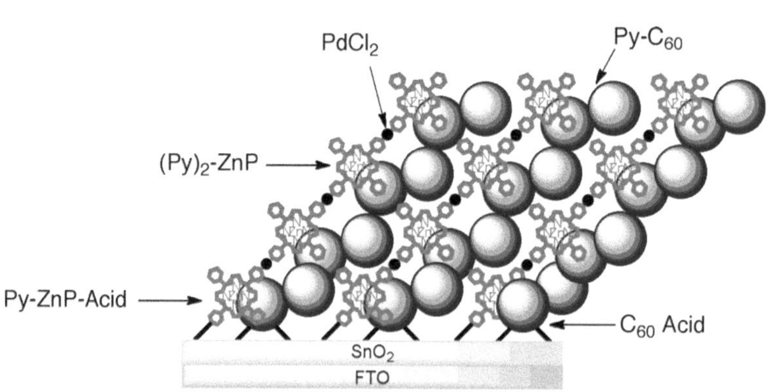

Figure 9.23. Schematic of a porphyrin-fullerene array on FTO/SnO_2 [114].

boxylic acid residue on the macrocycle. Following this, subsequent treatments with Pd(II) and coordination of a bis(pyridyl)porphyrin ($(Py)_2$-ZnP) enabled controllable, stepwise growth of porphyrin layers, easily modulated by cycling the deposition process. In this manner, electrodes with increasing porphyrin layers were constructed, denoted as FTO/SnO_2/$(ZnP)_n$, where n = 1–8. UV-vis absorption measurements performed after the growth of each successive layer showed the amount of assembled porphyrin increased linearly with the number of layers, and leveled off at n = 8.

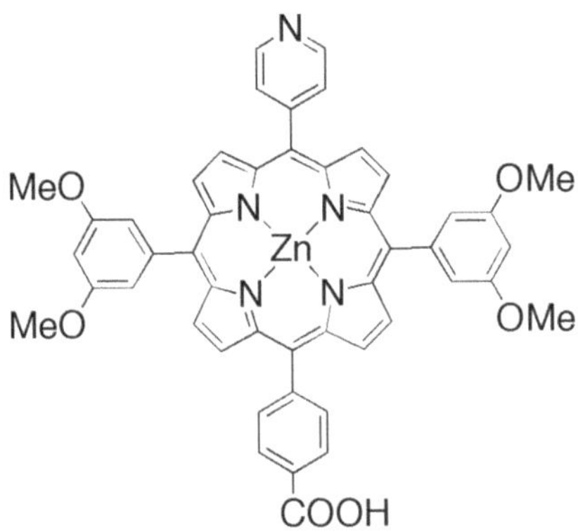

Py-ZnP-Acid

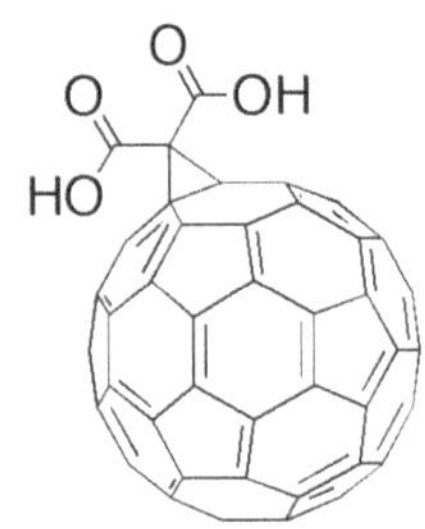

C_{60} Acid

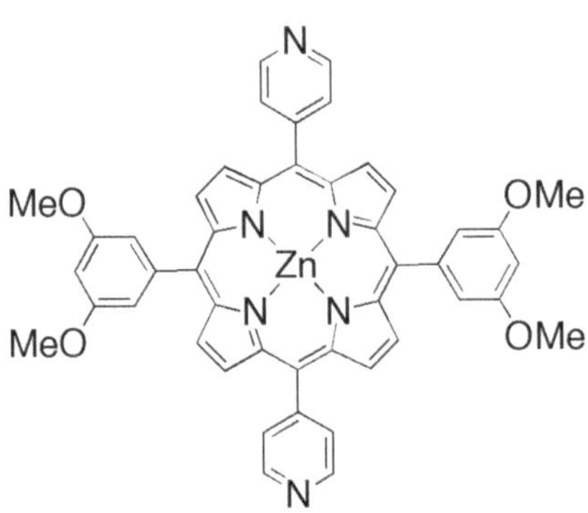

$(Py)_2$-ZnP

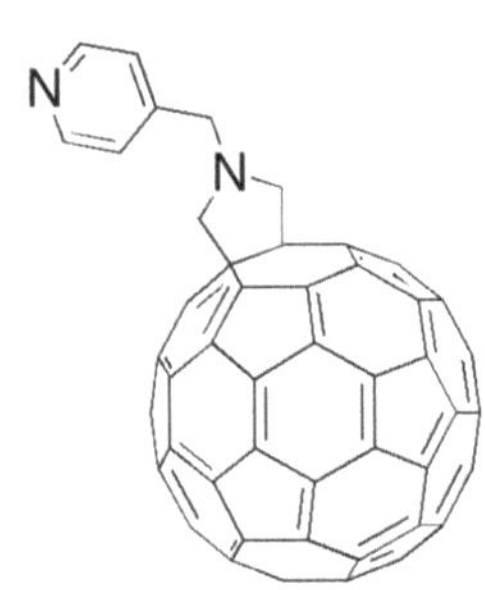

Py-C_{60}

Figure 9.24. Molecular synthons used to form a bicontinuous D-A array on FTO/SnO_2 [114].

Within the layered porphyrin surface, the production of ZnP-C_{60} composites was performed in a similar manner, the only difference being the initial adsorption of both Py-ZnP-Acid and C_{60}-Acid. This electrode, FTO/SnO_2/(C_{60}-Acid+$(ZnP)_n$) was then immersed in an ODCB solution of Py-C_{60}, which has the ability to coordinate to the zinc porphyrins, infiltrating the linear porphyrin brushes. Steady-state fluorescence of the FTO/SnO_2/(C_{60}-Acid+$(ZnP)_n$+Py-C_{60}) electrode revealed strong quenching of porphyrin emission compared to that of the reference, suggesting that the singlet-excited state of the porphyrin ($^1ZnP^*$) is quenched by electron transfer to the coordinated fullerene moieties.

AFM of the modified electrodes showed that for the FTO/SnO_2/(C_{60}-Acid+$(ZnP)_n$+Py-C_{60}) where $n = 2$ to 5, a smooth morphology was observed suggesting an intimate mixing of components. The incorporation of Py-C_{60} suppresses porphyrin aggregation, resulting in a D-A bicontinuous array on the SnO_2 electrode.

Plotting the IPCE of the modified electrodes as a function of the number of porphyrin layers revealed that for the FTO/SnO_2/(C_{60}-Acid+$(ZnP)_n$) an increase was observed up to $n = 3$ (~10%) and then it decreased gradually. The origin of the observed current generation was an electron injection from the porphyrin singlet-excited state into the CB of SnO_2. The maximum at $n = 3$, which roughly corresponds to an exciton diffusion length of ~6 nm, is reminiscent of that observed from copper phthalocyanine (8 ± 3 nm in BHJSCs). The infiltration of Py-C_{60} into this construct results in the IPCE maximum in the FTO/SnO_2/C_{60}-Acid+($(ZnP)_n$+Py-C_{60}) to appear at $n = 5$, with an increased IPCE of 21%. Thus, exciton and carrier losses within the porphyrin arrays are reduced upon the inclusion of Py-C_{60} as a result of improved exciton capture and charge dissociation. Photocurrent generation in the FTO/SnO_2/(C_{60}-Acid+$(ZnP)_n$+Py-C_{60}) system is presented below in Fig. 9.25.

From the excited singlet state of the porphyrin ($^1ZnP^*$) the process of charge separation between the porphyrin and fullerene occurs as well as the observed direct electron injection to the CB of SnO_2. In the case of charge separation, the generated C_{60} radical anion injects an electron into the SnO_2 via an electron-hopping

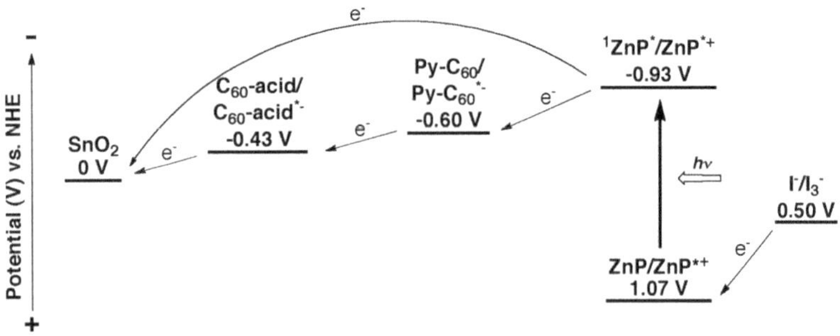

Figure 9.25. Schematic illustration of photocurrent generation in the FTO/SnO_2/(C_{60}-Acid+(ZnP)$_n$+Py-C_{60}) modified electrode [114].

mechanism through the infiltrated C_{60} molecules coordinated to the porphyrin brush. Thus the presence of the bicontinuous D-A array is responsible for enhanced photocurrent generation.

References

1. Gräetzel, M. (2009) Recent advances in sensitized mesoscopic solar cells, *Acc. Chem. Res.*, **42**, 1788–1798.
2. Imahori, H. (2007) Electrophoretic deposition of donor-acceptor nanostructures on electrodes for molecular photovoltaics, *J. Mater. Chem.*, **17**, 31–41.
3. Krebs, F. C. (2009) Fabrication and processing of polymer solar cells: A review of printing and coating techniques, *Sol. Energy Mater. Sol. Cells*, **93**, 394–412.
4. Yu, G., Gao, J., Hummelen, J. C., Wudl, F. and Heeger, A. J. (1995) Polymer photovoltaic cells: Enhanced efficiencies via a network of internal donor-acceptor heterojunctions, *Science*, **270**, 1789–1791.
5. Benanti T. and Venkataraman, D. (2006) Organic solar cells: An overview focusing on active layer morphology, *Photosynth. Res.*, **87**, 73–81.
6. Kippelen B. and Brédas J. L. (2009) Organic photovoltaics, *Energy Environ. Sci.*, **2**, 251–261.
7. Sagawa, T. Yoshikawa, S. and Imahori, H. One-Dimensional Nanostructured Semiconducting Materials for Organic Photovoltaics, *J. Phys. Chem. Lett.*, **1**, 1020–1025.

8. Hummelen, J. C., Knight, B. W., LePeq, F., Wudl, F., Yao, J. and Wilkins C. L. (1995) Preparation and characterization of fulleroid and methanofullerene derivatives, *J. Org. Chem.*, **60**, 532–538.
9. Desilvestro, J., Grätzel, M., Kavan, L., Moser, J. and Augustynski, J. (1985) Highly efficient sensitization of titanium dioxide, *J. Am. Chem. Soc.*, **107**, 2988–2990.
10. Vlachopoulos, N., Liska, P., Augustynski, J. and Grätzel, M. (1988) Very efficient visible light energy harvesting and conversion by spectral sensitization of high surface area polycrystalline titanium dioxide films, *J. Am. Chem. Soc.*, **110**, 1216–1220.
11. O'Regan, B. and Grätzel, M. (1991) A low-cost, high-efficiency solar cell based on dye-sensitized colloidal TiO_2 films, *Nature*, **353**, 737–740.
12. Nazeeruddin, M. K., Kay, A., Rodicio, I., Humphry-Baker, R., Mueller, E., Liska, P., Vlachopoulos, N. and Grätzel M. (1993) Conversion of light to electricity by *cis*-X_2bis(2,2'-bipyridyl-4,4'-dicarboxylate)ruthenium(II) charge-transfer sensitizers ($X = Cl^-$, Br^-, I^-, CN^-, and SCN^-) on nanocrystalline titanium dioxide electrodes, *J. Am. Chem. Soc.*, **115**, 6382–6390.
13. Grätzel, M. (2001) Photoelectrochemical cells, *Nature*, **414**, 338–344.
14. Chen, C. Y., Wu, S. J., Wu, C. G., Chen, J. G. and Ho, K. C. (2006) A Ruthenium Complex with Superhigh Light-Harvesting Capacity for Dye-Sensitized Solar Cells, *Angew. Chem. Int. Ed.*, **45**, 5822–5825.
15. Chen, C. Y., Wu, S. J., Wu, C. G., Chen, J. G. and Ho, K. C. (2007) A new route to enhance the light-harvesting capability of ruthenium complexes for dye-sensitized solar cells, *Adv. Mater.*, **19**, 3888–3891.
16. Chen, C. Y., Wu, S. J., Wu, C. G., Chen, J. G. and Ho, K. C. (2007) New ruthenium complexes containing oligoalkylthiophene-substituted 1,10-phenanthroline for nanocrystalline dye-sensitized solar cells, *Adv. Funct. Mater.*, **17**, 29–36.
17. Gao, F., Wang, Y., Zhang, J., Shi, D., Wang, M., Humphry-Baker, R., Wang, P., Zakeeruddin, S. M. and Grätzel, M. (2008) A new heteroleptic ruthenium sensitizer enhances the absorptivity of mesoporous titania film for a high efficiency dye-sensitized solar cell, *Chem. Commun.*, 2635–2637.
18. Qin, H., Wenger, S., Xu, M., Gao, F., Jing, X., Wang, P., Zakeeruddin, S. M. and Grätzel, M. (2008) An organic sensitizer with a fused dithienothiophene unit for efficient and stable dye-sensitized solar cells, *J. Am. Chem. Soc.*, **130**, 9202–9203.

19. Gao, F., Wang, Y., Shi, D., Zhang, J., Wang, M., Jing, X., Humphry-Baker, R., Wang, P., Zakeeruddin, S. M. and Grätzel, M. (2008) Enhance the optical absorptivity of nanocrystalline TiO_2 film with high molar extinction coefficient ruthenium sensitizers for high performance dye-sensitized solar cells, *J. Am. Chem. Soc.,* **130**, 10720–10728.

20. Cao, Y., Bai, Y., Yu, Q., Cheng, Y., Liu, S., Shi, D., Gao, F. and Wang P. (2009) Dye-sensitized solar cells with a high absorptivity ruthenium sensitizer featuring a 2-(Hexylthio)thiophene conjugated bipyridine, *J. Phys. Chem. C,* **113**, 6290–6297.

21. Choi, H., Baik, C., Kim, S., Kang, M.-S., Xu, X., Kang, H., S., Kang, S., O., Ko, J., Nazeeruddin, M. K. and Grätzel, M. (2008) Molecular engineering of hybrid sensitizers incorporating an organic antenna into ruthenium complex and their application in solar cells, *New J. Chem.,* **32**, 2233–2237.

22. Mishra, A., Fischer, M. and Bäuerle, P. (2009) Metal-free organic dyes for dye-sensitized solar cells: from structure: property relationships to design rules, *Angew. Chem. Int. Ed.,* **48**, 2474–2499.

23. Campbell, W. M., Burrell, A. K., Officer, D. L. and Jolley, K.W. (2004) Porphyrins as light harvesters in the dye-sensitised TiO_2 solar cell, *Coord. Chem. Rev.,* **248**, 1363–1379.

24. Ooyama, Y. and Harima Y. (2009) Molecular designs and syntheses of organic dyes for dye-sensitized solar cells, *Eur. J. Org. Chem.,* **2009**, 2903–2934.

25. Imahori, H., Hagiwara, K., Akiyama, T., Aoki, M., Taniguchi, S., Okada, T., Shirakawa, M. and Sakata, Y. (1996) The small reorganization energy of C_{60} in electron transfer, *Chem. Phys. Lett.,* **263**, 545–550.

26. Imahori, H. and Sakata, Y. (1997) Donor-linked fullerenes: photoinduced electron transfer and its potential application, *Adv. Mater.,* **9**, 537–546.

27. Imahori, H. and Sakata, Y. (1999) Fullerenes as novel acceptors in photosynthetic electron transfer, *Eur. J. Org. Chem.,* **1999**, 2445–2457.

28. Imahori, H., Mori, Y. and Matano Y. (2003) Nanostructured artificial photosynthesis, *J. Photochem. Photobiol. C,* **4**, 51–83.

29. Imahori, H. (2004) Porphyrin-fullerene linked systems as artificial photosynthetic mimics, *Org. Biomol. Chem.,* **2**, 1425–1433.

30. Imahori, H., Norieda, H., Nishimura, Y., Yamazaki, I., Higuchi, K., Kato, N., Motohiro, T., Yamada, H., Tamaki, K., Arimura, M. and Sakata, Y. (2000) Chain length effect on the structure and photoelectrochemical

properties of self-assembled monolayers of porphyrins on gold electrodes, *J. Phys. Chem. B*, **104**, 1253–1260.

31. Imahori, H., Yamada, H., Nishimura, Y., Yamazaki, I. and Sakata, Y. (2000) Vectorial multistep electron transfer at the gold electrodes modified with self-assembled monolayers of ferrocene-porphyrin-fullerene triads, *J. Phys. Chem. B*, **104**, 2099–2108.
32. Imahori, H., Norieda, H., Yamada, H., Nishimura, Y., Yamazaki, I., Sakata, Y. and Fukuzumi, S. (2000) Light-harvesting and photocurrent generation by gold electrodes modified with mixed self-assembled monolayers of boron-dipyrrin and ferrocene-porphyrin-fullerene triad, *J. Am. Chem. Soc.*, **123**, 100–110.
33. Yamada, H., Imahori, H., Nishimura, Y., Yamazaki, I. and Fukuzumi, S. (2002) Enhancement of photocurrent generation by ITO electrodes modified chemically with self-assembled monolayers of porphyrin-fullerene dyads, *Adv. Mater.*, **14**, 892–895.
34. Yamada, H., Imahori, H., Nishimura, Y., Yamazaki, I., Ahn, T. K., Kim, S. K. Kim, D. and Fukuzumi, S. (2003) Photovoltaic properties of self-assembled monolayers of porphyrins and porphyrin-fullerene dyads on ITO and gold surfaces, *J. Am. Chem. Soc.*, **125**, 9129–9139.
35. Uosaki, K., Kondo, T., Zhang, X. Q. and Yanagida, M. (1997) Very efficient visible-light-induced uphill electron transfer at a self-assembled monolayer with a porphyrin-ferrocene-thiol linked molecule, *J. Am. Chem. Soc.*, **119**, 8367–8368.
36. Imahori, H., Arimura, M., Hanada, T., Nishimura, Y., Yamazaki, I., Sakata, Y. and Fukuzumi, S. (2001) Photoactive three-dimensional monolayers: porphyrin-alkanethiolate-stabilized gold clusters, *J. Am. Chem. Soc.*, **123**, 335–336.
37. Hasobe, T., Imahori, H., Kamat, P. V. and Fukuzumi, S. (2003) Quaternary self-organization of porphyrin and fullerene units by clusterization with gold nanoparticles on SnO_2 electrodes for organic solar cells, *J. Am. Chem. Soc.*, **125**, 14962–14963.
38. Imahori, H., Fujimoto, A., Kang, S., Hotta, H., Yoshida, K., Umeyama, T., Matano, Y. and Isoda, S. (2006) Structure and photoelectrochemical properties of nanostructured SnO_2 electrodes deposited electrophoretically with the composite clusters of porphyrin-modified gold nanoparticle with a long spacer and fullerene, *Tetrahedron*, **62**, 1955–1966.
39. Hasobe, T., Imahori, H., Kamat, P. V., Ahn, T. K., Kim, S. K., Kim, D., Fujimoto, A., Hirakawa, T. and Fukuzumi, S. (2005) Photovoltaic cells

using composite nanoclusters of porphyrins and fullerenes with gold nanoparticles, *J. Am. Chem. Soc.*, **127**, 1216–1228.

40. Dubois, D., Kadish, K. M., Flanagan, S., Haufler, R. E., Chibante, L. P. F. and Wilson L. J. (1991) Spectroelectrochemical study of the C_{60} and C_{70} fullerenes and their mono-, di-, tri- and tetraanions, *J. Am. Chem. Soc.*, **113**, 4364–4366.

41. Fujitsuka, M., Luo, C. and Ito, O. (1998) Electron-transfer reactions between fullerenes (C_{60} and C_{70}) and tetrakis(dimethylamino) ethylene in the ground and excited states, *J. Phys. Chem. B*, **103**, 445–449.

42. Park, J., Kim, D., Suh, Y. D. and Kim, S. K. (1994) Comparative study on photoinduced electron transfer from *N*, *N*-Dimethylaniline (DMA) and 4,4'-Methylenebis(*N*, *N*-dimethylaniline) (BDMA) to C_{60} and C_{70} in toluene, *J. Phys. Chem.*, **98**, 12715–12719.

43. Imahori, H., Fujimoto, A., Kang, S., Hotta, H., Yoshida, K., Umeyama, T., Matano, Y. and Isoda, S. (2005) Molecular photoelectrochemical devices: supramolecular incorporation of C_{60} molecules into tailored holes on porphyrin-modified gold nanoclusters, *Adv. Mater.*, **17**, 1727–1730.

44. Imahori, H., Fujimoto, A., Kang, S., Hotta, H., Yoshida, K., Umeyama, T., Matano, Y., Isoda, S., Isosomppi, M., Tkachenko, N. V. and Lemmetyinen H. (2005) Host–guest interactions in the supramolecular incorporation of fullerenes into tailored holes on porphyrin-modified gold nanoparticles in molecular photovoltaics, *Chem. Eur. J.*, **11**, 7265–7275.

45. Imahori, H., Mitamura, K., Umeyama, T., Hosomizu, K., Matano, Y., Yoshida, K. and Isoda, S. (2006) Efficient photocurrent generation by SnO_2 electrode modified electrophoretically with composite clusters of porphyrin-modified silica microparticle and fullerene, *Chem. Commun.*, 406–408.

46. Imahori, H., Mitamura, K., Shibano, Y., Umeyama, T., Matano, Y., Yoshida, K., Isoda, S., Araki, Y. and Ito O. (2006) A photoelectrochemical device with a nanostructured SnO_2 electrode modified with composite clusters of porphyrin–modified silica nanoparticle and fullerene, *J. Phys. Chem. B*, **110**, 11399–11405.

47. Evans, D. R., Fackler, N. L. P., Xie, Z., Rickard, C. E. F., Boyd, P. D. W. and Reed, C. A. (1999) π-Arene/cation structure and bonding. Solvation versus ligand binding in iron(III) tetraphenylporphyrin complexes of benzene, toluene, p-xylene, and [60]fullerene, *J. Am. Chem. Soc.*, **121**, 8466–8474.

48. Sun, D., Tham, F. S., Reed, C. A., Chaker, L. and Boyd P. D. W. (2002) Supramolecular fullerene-porphyrin chemistry. Fullerene complexation by metalated "Jaws Porphyrin" hosts, *J. Am. Chem. Soc.*, **124**, 6604–6612.
49. Tashiro, K., Aida, T., Zheng, J. Y., Kinbara, K., Saigo, K., Sakamoto, S. and Yamaguchi, K. (1999) A cyclic dimer of metalloporphyrin forms a highly stable inclusion complex with C_{60}, *J. Am. Chem. Soc.*, **121**, 9477–9478.
50. Imahori, H., Hagiwara, K., Aoki, M., Akiyama, T., Taniguchi, S., Okada, T., Shirakawa, M. and Sakata Y. (1996) Linkage and solvent dependence of photoinduced electron transfer in zinc porphyrin-C_{60} dyads, *J. Am. Chem. Soc.*, **118**, 11771–11782.
51. Shirakawa, M., Fujita, N. and Shinkai, S. (2003) [60]Fullerene-motivated organogel formation in a porphyrin derivative bearing programmed hydrogen-bonding sites, *J. Am. Chem. Soc.*, **125**, 9902–9903.
52. Imahori, H., Kashiwagi, Y., Hanada, T., Endo, Y., Nishimura, Y., Yamazaki, I. and Fukuzumi, S. (2003) Metal and size effects on structures and photophysical properties of porphyrin-modified metal nanoclusters, *J. Mater. Chem.*, **13**, 2890–2898.
53. Fukuzumi, S., Endo, Y., Kashiwagi, Y., Araki, Y., Ito, O. and Imahori, H. (2003) Novel photocatalytic function of porphyrin-modified gold nanoclusters in comparison with the reference porphyrin compound, *J. Phys. Chem. B*, **107**, 11979–11986.
54. Imahori, H., Kashiwagi, Y., Endo, Y., Hanada, T., Nishimura, Y., Yamazaki, I., Araki, Y., Ito, O. and Fukuzumi, S. (2003) Structure and photophysical properties of porphyrin-modified metal nanoclusters with different chain lengths, *Langmuir*, **20**, 73–81.
55. Zenkevich, E., Cichos, F., Shulga, F., Petrov, E., P., Blaudeck, T. and von Borczyskowski, C. (2005) Nanoassemblies designed from semiconductor quantum dots and molecular arrays, *J. Phys. Chem. B*, **109**, 8679–8692.
56. Schmelz, O., Mews, A., Basché, T., Herrmann, A. and Müllen, K. (2001) Supramolecular complexes from CdSe nanocrystals and organic fluorophors, *Langmuir*, **17**, 2861–2865.
57. Robel, I., Subramanian, V., Kuno, M. and Kamat, P. V. (2006) Quantum dot solar cells. Harvesting light energy with CdSe nanocrystals molecularly linked to mesoscopic TiO_2 films, *J. Am. Chem. Soc.*, **128**, 2385–2393.

58. Sheeney-Haj-Ichia, L., Basnar, B. and Willner, I. (2005) Efficient generation of photocurrents by using CdS/Carbon nanotube assemblies on electrodes, *Angew. Chem. Int. Ed.*, **117**, 80–85.
59. Dayal, S., Lou, Y., Samia, A. C. S., Berlin, J. C., Kenney, M. E. and Burda, C. (2006) Observation of non-Förster-type energy-transfer behavior in quantum dot-phthalocyanine conjugates, *J. Am. Chem. Soc.*, **128**, 13974–13975.
60. Kang, S., Yasuda, M., Miyasaka, H., Hayashi, H., Kawasaki, M., Umeyama, T., Matano, Y., Yoshida, K., Isoda, S. and Imahori, H. (2008) Light harvesting and energy transfer in multiporphyrin-modified CdSe nanoparticles, *ChemSusChem*, **1**, 254–261.
61. Romero, D. B., Carrard, M., De Heer, W. and Zuppiroli, L. (1996) A carbon nanotube/organic semiconducting polymer heterojunction, *Adv. Mater.*, **8**, 899–902.
62. Ago, H., Petritsch, K., Shaffer, M. S. P., Windle, A. H. and Friend, R. H. (1999) Composites of carbon nanotubes and conjugated polymers for photovoltaic devices, *Adv. Mater.*, **11**, 1281–1285.
63. Bhattacharyya, S., Kymakis, E. and Amaratunga, G. A. J. (2004) Photovoltaic properties of dye functionalized single-wall carbon nanotube/conjugated polymer devices, *Chem. Mater.*, **16**, 4819–4823.
64. Landi, B. J., Raffaelle, R. P., Castro, S. L. and Bailey, S. G. (2005) Single-wall carbon nanotube–polymer solar cells, *Prog. Photovolt Res. Appl.*, **13**, 165–172.
65. Jang, S. R., Vittal, R. and Kim, K. J. (2004) Incorporation of functionalized single-wall carbon nanotubes in dye-sensitized TiO_2 solar cells, *Langmuir*, **20**, 9807-9810.
66. Rahman, G. M. A., Guldi, D. M., Cagnoli, R., Mucci, A., Schenetti, L., Vaccari, L. and Prato, M. (2005) Combining single wall carbon nanotubes and photoactive polymers for photoconversion, *J. Am. Chem. Soc.*, **127**, 10051–10057.
67. Sgobba, V., Rahman, G., Guldi, D., Jux, N., Campidelli, S. and Prato, M. (2006) Supramolecular assemblies of different carbon nanotubes for photoconversion processes, *Adv. Mater.*, **18**, 2264–2269.
68. Barazzouk, S., Hotchandani, S., Vinodgopal, K. and Kamat, P. V. (2004) Single-wall carbon nanotube films for photocurrent generation. A prompt response to visible-light irradiation, *J. Phys. Chem. B*, **108**, 17015–17018.

69. Hasobe, T., Fukuzumi, S. and Kamat, P. V. (2006) Organized assemblies of single wall carbon nanotubes and porphyrin for photochemical solar cells: Charge injection from excited porphyrin into single-walled carbon nanotubes, *J. Phys. Chem. B*, **110**, 25477–25484.
70. Star, A., Stoddart, J., F., Steuerman, D., Diehl, M., Boukai, M., Wong, E. W., Yang, X., Chung, S. W., Choi, H. and Heath, J. R. (2001) Preparation and properties of polymer-wrapped single-walled carbon nanotubes, *Angew. Chem. Int. Ed.*, **40**, 1721–1725.
71. Stephenson, J. J., Hudson, J. L., Azad, S. and Tour, J. M. (2005) Individualized single walled carbon nanotubes from bulk material using 96% sulfuric acid as solvent, *Chem. Mater.*, **18**, 374–377.
72. Dyke C. A. and Tour, J. M. (2004) Covalent functionalization of single-walled carbon nanotubes for materials applications, *J. Phys. Chem. A*, **108**, 11151–11159.
73. Bahr, J. L. and Tour, J. M. (2001) Highly functionalized carbon nanotubes using *in situ* generated diazonium compounds, *Chem. Mater.*, **13**, 3823–3824.
74. Georgakilas, V., Voulgaris, D., Vázquez, E., Prato, M., Guldi, D. M., Kukovecz, A. and Kuzmany, H. (2002) Purification of HiPCO carbon nanotubes via organic functionalization, *J. Am. Chem. Soc.*, **124**, 14318–14319.
75. Coleman, K. S., Bailey, S. R., Fogden, S. and Green, M. L. H. (2003) Functionalization of single-walled carbon nanotubes via the bingel reaction, *J. Am. Chem. Soc.*, **125**, 8722–8723.
76. Umeyama, T., Tezuka, N., Fujita, M., Matano, Y., Takeda, N., Murakoshi, K., Yoshida, K., Isoda, S. and Imahori, H. (2007) Retention of intrinsic electronic properties of soluble single-walled carbon nanotubes after a significant degree of sidewall functionalization by the bingel reaction, *J. Phys. Chem. C*, **111**, 9734–9741.
77. Umeyama, T., Fujita, M., Tezuka, N., Kadota, N., Matano, Y., Yoshida, K., Isoda, S. and Imahori, H. (2007) Electrophoretic deposition of single-walled carbon nanotubes covalently modified with bulky porphyrins on nanostructured SnO_2 electrodes for photoelectrochemical devices, *J. Phys. Chem. C*, **111**, 11484–11493.
78. Sun, Y. P., Huang, W., Lin, Y., Fu, K., Kitaygorodskiy, A., Riddle, L. A., Yu, Y. J. and Carroll, D. L. (2001) Soluble dendron-functionalized carbon nanotubes: preparation, characterization, and properties, *Chem. Mater.*, **13**, 2864–2869.

79. Herranz, M. Á., Martín, N., Campidelli, S., Prato, M., Brehm, G. and Guldi, D. M. (2006) Control over electron transfer in tetrathiafulvalene-modified single-walled carbon nanotubes, *Angew. Chem. Int. Ed.*, **45**, 4478–4482.

80. Bahr, J. L. and Tour, J. M. (2002) Covalent chemistry of single-wall carbon nanotubes, *J. Mater. Chem.*, **12**, 1952–1958.

81. Dresselhaus, M. S., Dresselhaus, G., Saito, R. and Jorio, A. (2005) Raman spectroscopy of carbon nanotubes, *Phys. Rep.*, **409**, 47–99.

82. Saito, R., Dresselhaus, G. and Dresselhaus, M. S. (eds.), (1998) *Physics of Carbon Nanotube*, Imperial College Press, London

83. Kamat, P. V., Barazzouk, S., Thomas, K. G. and Hotchandani, S. (2000) Electrodeposition of C_{60} cluster aggregates on nanostructured SnO_2 films for enhanced photocurrent generation, *J. Phys. Chem. B*, **104**, 4014–4017.

84. Kamat, P. Barazzouk, S., Hotchandani, S. and Thomas, K. (2000) Nanostructured thin films of C_{60}–Aniline dyad clusters: electrodeposition, charge separation, and photoelectrochemistry, *Chem. Eur. J.*, **6**, 3914–3921.

85. Guldi, D. M., Rahman, G. M. A., Zerbetto, F. and Prato, M. (2005) Carbon nanotubes in electron donor-acceptor nanocomposites, *Acc. Chem. Res.*, **38**, 871–878.

86. Imahori, H. (2004) Giant multiporphyrin arrays as artificial light-harvesting antennas, *J. Phys. Chem. B*, **108**, 6130–6143.

87. Imahori, H. and Fukuzumi, S. (2004) Porphyrin- and fullerene-based molecular photovoltaic devices, *Adv. Funct. Mater.*, **14**, 525–536.

88. Imahori, H. (2007) Creation of fullerene-based artificial photosynthetic systems, *Bull. Chem. Soc. Jpn.*, **80**, 621–636.

89. Tkachenko, N. V., Rantala, L., Tauber, A. Y., Helaja, J., Hynninen, P. H. and Lemmetyinen, H. (1999) Photoinduced electron transfer in phytochlorin-[60]fullerene dyads, *J. Am. Chem. Soc.*, **121**, 9378–9387.

90. Kesti, T. J., Tkachenko, N. V., Vehmanen, V., Yamada, H., Imahori, H., Fukuzumi, S. and Lemmetyinen, H. (2002) Exciplex intermediates in photoinduced electron transfer of porphyrin-fullerene dyads, *J. Am. Chem. Soc.*, **124**, 8067–8077.

91. Tkachenko, N. V., Lemmetyinen, H., Sonoda, J., Ohkubo, K., Sato, T., Imahori, H. and Fukuzumi, S. (2003) Ultrafast photodynamics of exciplex formation and photoinduced electron transfer in porphyrin-fullerene dyads linked at close proximity, *J. Phys. Chem. A*, **107**, 8834–8844.

92. Hasobe, T., Kashiwagi, Y., Absalom, M., Sly, J., Hosomizu, K., Crossley, M., Imahori, H., Kamat, P. and Fukuzumi, S. (2004) Supramolecular photovoltaic cells using porphyrin dendrimers and fullerene, *Adv. Mater.*, **16**, 975–979.
93. Hasobe, T., Kamat, P. V., Absalom, M. A., Kashiwagi, Y., Sly, J., Crossley, M. J., Hosomizu, K., Imahori, H. and Fukuzumi, S. (2004) Supramolecular photovoltaic cells based on composite molecular nanoclusters: Dendritic porphyrin and C_{60}, porphyrin dimer and C_{60}, and porphyrin-C_{60} dyad, *J. Phys. Chem. B*, **108**, 12865–12872.
94. Hasobe, T., Kamat, P. V., Troiani, V., Solladié, N., Ahn, T. K., Kim, S. K., Kim, D., Kongkanand, A., Kuwabata, S. and Fukuzumi, S. (2005) Enhancement of light-energy conversion efficiency by multi-porphyrin arrays of porphyrin-peptide oligomers with fullerene clusters, *J. Phys. Chem. B*, **109**, 19–23.
95. Hasobe, T., Saito, K., Kamat, P. V., Troiani, V., Qiu, H., Solladié, N., Kim, K. S., Park, J. K., Kim, D., D'Souza, F. and Fukuzumi, S. (2007) Organic solar cells. Supramolecular composites of porphyrins and fullerenes organized by polypeptide structures as light harvesters, *J. Mater. Chem.*, **17**, 4160–4170.
96. Hasobe, T., Imahori, H., Fukuzumi, S. and Kamat, P. V. (2003) Nanostructured assembly of porphyrin clusters for light energy conversion, *J. Mater. Chem.*, **13**, 2515–2520.
97. Kim, Y., S., Liang, K., Law, K. Y. and Whitten D. G. (1994) An investigation of photocurrent generation by squaraine aggregates in monolayer-modified tin oxide (SnO_2) electrodes, *J. Phys. Chem.*, **98**, 984–988.
98. Yang, J. H., Chen, Y. M., Ren, Y. L., Bai, Y. B., Wu, Y., Jang, Y. S., Su, Z. M., Yang, W. S., Wang, Y., Q., Zao, B. and Li, T., J. (2000) Identification of *H*-aggregate in a monolayer amphiphilic porphyrin-TiO_2 nanoparticle heterostructure assembly and its influence on the photoinduced charge transfer, *J. Photochem. Photobiol. A*, **134**, 1–7.
99. Wu, D., G., Huang, C. H., Gan, L. B. and Huang, Y. Y. (1998) Photoelectric response from two amphiphilic nitrobenzene dyes with alkoxyl on a monolayer-modified ITO electrode, *Langmuir*, **14**, 3783–3787.
100. Wu, D. G., Huang, Y. Y., Huang, C. H. and Gan, L. B. (1998) Photoelectric conversion from a nitrobenzene dye monolayer modified ITO electrode, *J. Chem. Soc. Faraday Trans.*, **94**, 1411–1415.
101. Hasobe, T., Imahori, H., Fukuzumi, S. and Kamat, P. V. (2003) Light energy conversion using mixed molecular nanoclusters. Porphyrin and

C_{60} cluster films for efficient photocurrent generation, *J. Phys. Chem. B,* **107**, 12105–12112.

102. Imahori, H., Ueda, M., Kang, S., Hayashi, H., Hayashi, S., Kaji, H., Seki, S., Saeki, A., Tagawa, S., Umeyama, T., Matano, Y., Yoshida, K., Isoda, S., Shiro, M., Tkachenko, N. and Lemmetyinen, H. (2007) Effects of porphyrin substituents on film structure and photoelectrochemical properties of porphyrin/fullerene composite clusters electrophoretically deposited on nanostructured SnO_2 electrodes, *Chem. Eur. J.,* **13**, 10182–10193.

103. Imahori, H., Tkachenko, N. V., Vehmanen, V., Tamaki, K., Lemmetyinen, H., Sakata, Y. and Fukuzumi, S. (2001) An extremely small reorganization energy of electron transfer in porphyrin-fullerene dyad, *J. Phys. Chem. A,* **105**, 1750–1756.

104. Tkachenko, N. V., Guenther, C., Imahori, H., Tamaki, K., Sakata, Y., Fukuzumi, S. and Lemmetyinen, H. (2000) Near infra-red emission of charge-transfer complexes of porphyrin-fullerene films, *Chem. Phys. Lett.,* **326**, 344–350.

105. Vehmanen, V., Tkachenko, N. V., Imahori, H., Fukuzumi, S. and Lemmetyinen, H. (2001) Charge-transfer emission of compact porphyrin-fullerene dyad analyzed by Marcus theory of electron-transfer, *Spectrochim. Acta A,* **57**, 2229–2244.

106. Wahadoszamen, M., Nakabayashi, T., Kang, S., Imahori, H. and Ohta, N. (2006) external electric field effects on absorption and fluorescence spectra of a fullerene derivative and its mixture with zinc-tetraphenylporphyrin doped in a PMMA film, *J. Phys. Chem. B,* **110**, 20354–20361.

107. Maiti, N. C., Mazumdar, S. and Periasamy, N. (1998) *J*- and *H*-aggregates of porphyrin-surfactant complexes: time-resolved fluorescence and other spectroscopic studies, *J. Phys. Chem. B,* **102**, 1528–1538.

108. Kano, K., Fukuda, K., Wakami, H., Nishiyabu, R. and Pasternack, R. F. (2000) Factors influencing self-aggregation tendencies of cationic porphyrins in aqueous solution, *J. Am. Chem. Soc.,* **122**, 7494–7502.

109. Okada, S. and Segawa, H. (2003) Substituent-control exciton in *J*-aggregates of protonated water-insoluble porphyrins, *J. Am. Chem. Soc.,* **125**, 2792–2796.

110. Imahori, H., Hasobe, T., Yamada, H., Kamat, P. V., Barazzouk, S., Fujitsuka, M., Ito, O. and Fukuzumi, S. (2001) Spectroscopy and

photocurrent generation in nanostructured thin films of porphyrin-fullerene dyad clusters, *Chem. Lett.*, **30**, 784–785.

111. Hayashi, H., Kira, A., Umeyama, T., Matano, Y., Charoensirithavorn, P., Sagawa, T., Yoshikawa, S., Tkachenko, N. V., Lemmetyinen, H. and Imahori, H. (2009) Effects of electrode structure on photoelectrochemical properties of ZnO electrodes modified with porphyrin-fullerene composite layers with an intervening fullerene monolayer *J. Phys. Chem. C*, **113**, 10819–10828.
112. Kaidashev, E. M., Lorenz, M., von Wenckstern, H., Rahm, A., Semmelhack, H. C., Han, K. H., Benndorf, G., Bundesmann, C., Hochmuth, H. and Grundmann, M. (2003) High electron mobility of epitaxial ZnO thin films on c-plane sapphire grown by multistep pulsed-laser deposition, *Appl. Phys. Lett.*, **82**, 3901–3903.
113. Greene, L. E., Law, M., Goldberger, J., Kim, F., Johnson, J.C., Zhang, Y., Saykally, R. J. and Yang, P. (2003) Low-temperature wafer-scale production of ZnO nanowire arrays, *Angew. Chem. Int. Ed.*, **42**, 3031–3034.
114. Kira, A., Umeyama, T., Matano, Y., Yoshida, K., Isoda, S., Park, J. K., Kim, D. and Imahori, H. (2009) Supramolecular donor-acceptor heterojunctions by vectorial stepwise assembly of porphyrins and coordination-bonded fullerene arrays for photocurrent generation, *J. Am. Chem. Soc.*, **131**, 3198–3200.

Chapter 10

Self-Assembled Composite Materials of Porphyrins for Optoelectronics

Taku Hasobe
Department of Chemistry, Faculty of Science and Technology,
Keio University, 3-14-1 Hiyoshi, Kohoku-ku, Yokohama, Kanagawa 223-8522, Japan
PRESTO, Japan Science and Technology Agency (JST),
4-1-8 Honcho, Kawaguchi, Saitama, 332-0012, Japan
hasobe@chem.keio.ac.jp

10.1 Introduction

Construction of molecular architectures based on aggregation or self-association for controlled size and shape is a simple and convenient method [1–5]. Of particular interest are efficient energy- and electron-transfer processes [6–8], which are very related to the molecular assembly in natural photosynthetic system, providing better understanding of photofunctional molecular architectures. These fundamental researches of such organic materials also entail synthetic and photophysical processes of molecular aggregates with nanometer scales since their optical properties significantly differ from those of monomeric species [1, 2].

One of the promising candidates for photofunctional materials is a porphyrin dye that is involved as a chromophore and an

Multiporphyrin Arrays: Fundamentals and Applications
Edited by Dongho Kim

www.panstanford.com

electron donor in a number of important biological systems including the primary photochemical reactions of chlorophylls (porphyrin derivatives) in the photosynthetic reaction centers [9–13]. Therefore, porphyrins are also suitable for efficient electron transfer because of small reorganization energies [14, 15]. In addition, rich and extensive absorption features of porphyrinoid systems guarantees increased absorption cross-sections and an efficient use of the solar spectrum [16]. On the other hand, porphyrins are also promising building blocks for superstructures. So far, various nanoassemblies composed of porphyrins such as particles, tubes sheets, wires and rods are widely reported [17–41]. The promising applications of these molecular architectures are photonics, electronics, sensor, catalyses and light energy conversion etc.

Moreover, a combination of porphyrins and other acceptor molecules/materials (e.g., fullerenes, carbon nanotubes, other dyes, and metal and semiconductor nanoparticles, etc.) seems ideal for fulfilling an enhanced light-harvesting efficiency of chromophores throughout the solar spectrum and a highly efficient conversion of the absorbed light into the high energy state of the charge separation by photoinduced electron transfer [42–48]. Especially, porphyrins and fullerenes are also known to form supramolecular complexes, which contain closest contacts between one of the electron-rich 6:6 bonds of the guest fullerene and the geometric center of the host porphyrin. Such a strong interaction between porphyrins and fullerenes is likely to be a good driving force for the formation of supramolecular complexes [49, 50].

One of the representative applications of porphyrin composite materials is an optoelectronic application such as photovoltaics. In photovoltaic systems, sunlight is directly converted into electricity by the photovoltaic effect. In the organic film, the photovoltaic processes are performed by the following three ones: (i) light-harvest and exciton diffusion, (ii) charge separation, and (iii) carrier transport [5]. In such organic blend films, the photons are absorbed by the donor or the acceptor moieties, by both. The sun produces a broad solar spectrum output from ultraviolet to infrared region, whereas absorption region of organic molecules (monomer) is generally very limited. Therefore, the cooperative association or aggregation of organic dyes effectively contributes to broadening of the absorption region. This is the first reason for requirement of molecular assembly for photovoltaics. Then, if the donor moieties absorb most of incident light energy, excitons (electron–hole pairs) are generated on the donor or acceptor domain following promotion of a π-electron from HOMO level to LUMO level. To dissociate into charge carriers, the excitons need to migrate toward a donor/acceptor interface (exciton diffusion). However, exciton diffusion range is generally limited in case of organic molecules ($\sim$a few nm) [5]. Consequently, close organization between donor and acceptor moieties for the efficient charge separation is required for the utilization of excitation energy (second reason). Finally, the created charges have to escape from their mutual Columb attraction before migration toward the electrodes takes place (typically via hopping mechanism). In this process, mobilities of molecular aggregated structures largely dominate the carrier transport proeprties. As reported previously [51, 52], mobility is largely dependent on the grain structures (i.e., grain-size/boundary). This means that the size and morphological control of molecular assemblies is also a critical factor to efficient transport charge carriers (third reason).

In this chapter, based on these ideas, the focus is on the recent advances in the photofunctional supramolecular materials of porphyrins for optoelectronics such as photovoltaics. The synthesis, photophysical and photovoltaic properties are widely discussed.

10.2 Porphyrin Self-Assembled Nanoparticles

10.2.1 *Synthesis of Porphyrin Nanoparticles by Mixing Solvent Techniques*

(Polyetheylene)glycol (PEG) derivatives have been widely used in nanoparticle preparations to prevent agglomeration and precipitation of amorphous solids, such as for CdS. In 2002, Drain and coworkers report the synthesis and characterization of porphyrin nanoparticles prepared using mixing solvent techniques (Fig. 10.1) [18]. PEG was used in experiments to prepare nanoparticles composed of *meso*-substituted tetraphenylporphyrins (Table 10.1).

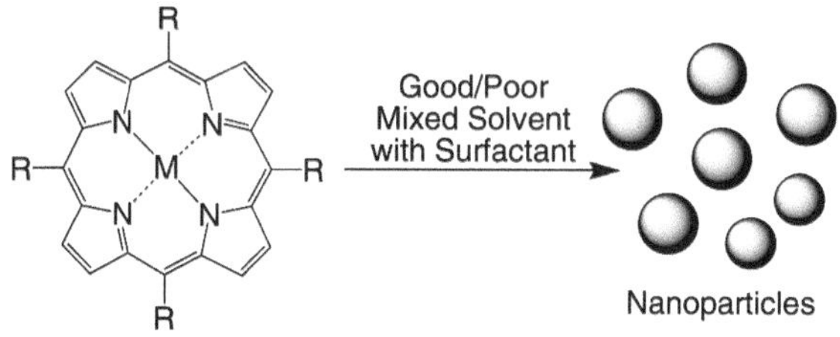

Figure 10.1. Schematic illustration of preparation of porphyrin nanoparticles using mixing solvent techniques in this study.

These organic nanoparticles (ONPs) are formed by adding a guest solvent, in which the porphyrinoid is insoluble, to a solution of the macrocycle in a host solvent with a small percentage of PEG as a stabilizer and vigorous mixing. Particle size and chromophore organization depend on the molecule and the conditions used to form the nanoparticles. Control experiments without PEG yielded no stable porphyrin nanoparticles. Consistent with other nanoparticle preparations, the stabilizer prevents agglomeration and is a factor in determining nanoparticle size. Dynamic light scattering (DLS) was used for initial characterization of the porphyrin nanoparticles. The stable porphyrin nanoparticles with ca. 20–200 nm diameter are successfully observed. Moreover, it was found that nanoparticles of several metalloporphyrins have

Table 10.1. Absorption maximum (λ_{max}), triplet quenching rate constant (k_q), triplet lifetime (τ_T^0), quantum yield of triplet quenching (ϕ_q^T), and T-T annihilation rate constants(γ_T) of porphyrin nanoassemblies at room temperature

Sample	λ_{max}/nm	k_q /s^{-1}	τ_T^0/ns	ϕ_q^T a	γ_T/mol^{-1} dm^3 s^{-1} b
$H_2P(CO_2H)_4$ /TriEG	460	3.9×10^7	26	0.35	6.5×10^{10}
$H_2P(CO_2H)_4$ /TetraEG	460	2.8×10^7	36	0.51	5.8×10^{10}
$H_2P(CO_2H)_4$ /HexaEG	460	1.8×10^7	55	0.77	4.7×10^{10}
$H_2P(CO_2H)_4$ /HeptaEG	440	1.5×10^7	67	0.83	3.9×10^{10}

Source: reprinted with permission from [53].
[a]Calculated by equation $F_q^T = [0.96 - F_q^S(\text{ave})]$ assuming Φ^s of $^1H_2P^*$ $(CO_2H)_4$ of monomer is 0.96.
[b]Calculated by Eq. 10. 4.

significantly better catalytic properties in terms of turnovers than the same solvated metalloporphyrin [18, 19].

10.2.2 *Size-Control of Porphyrin Nanoparticles and Their Quenching Properties of Excited States*

As stated above, preparation of porphyrin nanoparticles has been reported previously. However, little attention has been drawn toward the photophysical properties, so far. A systematic examination of the structural and photophysical properties provide us fundamental informations for the future light-driven applications. In this work, an ability to control the structures and photodynamics of porphyrin nanoparticles by using surfactants such as ethylene glycols (EG) with different spacer lengths is reported [53].

Stabilizers such as ethylene glycol (EG) are useful for the formation of stable colloidal systems by host–guest solvent method. In this systems, the four types of EG derivatives with different chain lengths such as triethylene glycol (TriEG), tetraethylene glycol (TetraEG), hexaethylene glycol (HexaEG) and heptaethylene glycol (HeptaEG) are shown in Fig. 10.2. Porphyrin nanoassemblies are prepared by adding an excess volume of poor solvent (H_2O) to a $H_2P(CO_2H)_4$ solution in good solvent (THF) mixed with small volume of EG

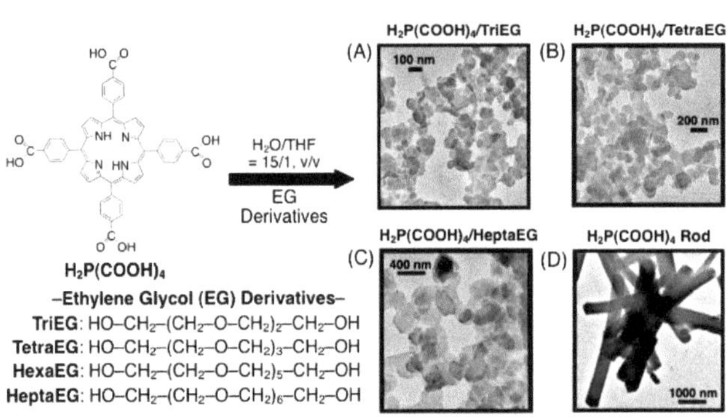

Figure 10.2. Size-control of porphyrin nanoparticles using EG derivatives. (A) $H_2P(CO_2H)_4$/TriEG, (B) $H_2P(CO_2H)_4$/TetraEG, (C) $H_2P(CO_2H)_4$/HeptaEG, and (D) $H_2P(CO_2H)_4$ pristine rods. Reprinted with permission from [5, 53].

surfactants while vigorously stirring. In all systems, the final concentrations of $H_2P(CO_2H)_4$ are fixed as 0.05 mM in H_2O/THF (15/1, v/v) containing 1.5% volume of EG derivative. The reference system without EG is also performed under the same condition (H_2O/THF = 15/1, v/v).

Transmission electron micrographs (TEM) are employed to examine the sizes and shapes of these porphyrin nanoassemblies as shown in Fig. 10.2. In the case of $H_2P(CO_2H)_4$/EG composite nanoassemblies (A–C in Fig. 10.2), we can see a lot of nanoparticles, which is in sharp contrast with much larger size of pristine $H_2P(CO_2H)_4$ rods without EG (D in Fig. 10.2) [53]. This indicates that $H_2P(CO_2H)_4$ moieties are effectively organized by EG derivatives to form spherical assemblies. The main driving force is likely due to the hydrogen bonding interaction between the -COOH group in $H_2P(CO_2H)_4$ and the -OH group in EG surfactants. Additionally, with increasing chain lengths of EG surfactants, the average diameters of nanoparticles also increase from ca. 100 to 400 nm [53]. This trend is in good agreement with that measured by dynamic light scattering (DLS). These indicate that porphyrin nanoparticles are formed in a mixed H_2O/THF solution.

X-ray diffraction (XRD) measurement further demonstrates the following two important structural aspects. One is that all $H_2P(CO_2H)_4$/EG nanoparticles have the similar internal crystal alignments. The other is that $H_2P(CO_2H)_4$ assemblies surrounded by EG surfactants possess enhanced crystallinities and organized internal structures as compared to the pristine $H_2P(CO_2H)_4$ rods [53]. Additionally, the porphyrin nanoparticles are also stable in solution without precipitation for several days.

To examine the excited species of $H_2P(CO_2H)_4$ nanoparticles, transient absorption spectra were measured following 532 nm laser light pulse irradiation of $H_2P(CO_2H)_4$ moiety. Upon the excitation of $H_2P(CO_2H)_4$/TriEG nanoparticles in H_2O/THF (15/1, v/v), transient absorption spectra are observed as shown in Fig. 10.3, where the broad bands appear in 460 and 880 nm regions. The observed broad bands in 460 and 880 regions are ascribed to the triplet state of porphyrin [$^3H_2P^*(CO_2H)_4$]. An inset of the decay obeys first-order kinetics with a certain range of rate constant 1.5×10^7 s^{-1} to 3.9×10^7 s^{-1} (Table 10.1). The triplet lifetimes of the

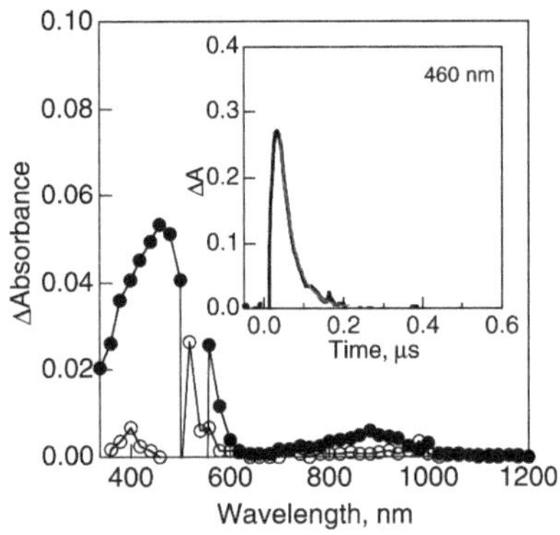

Figure 10.3. Nanosecond transient absorption spectra of $H_2P(CO_2H)_4$/TriEG nanoassemblies (H_2O/THF, 15/1, v/v) observed by 532 nm laser light (ca. 3 mJ/pulse) irradiation at 0.1 μs (λ) and 1.0 μs (○) in Ar-saturated solution. Inset: time profile at 460 nm. $[H_2P(CO_2H)_4] = 0.05$ mM in H_2O/THF (15/1, v/v). Reprinted with permission from [53].

transient species are namely calculated to be range from 67 to 26 ns (Table 10.1), which is infinitely shorter than that of the corresponding monomer $H_2P(CO_2H)_4$ condition (40 μs). The efficient triplet quenching of $H_2P(CO_2H)_4$ nanoparticle is due to the aggregation effects of $H_2P(CO_2H)_4$ moieties such as triplet-triplet (T-T) annihilation (*vide infra*). Moreover, the lifetime of triplet state decreases with decreasing particle-sizes (Table 10.1). This trend is also very similar to that of fluorescence lifetime measurement (i.e., excited singlet state).

10.2.3 *Particle Size-Dependent Annihilation Processes*

In quenching process of the excited states in these aggregated modes, the exciton-exciton annihilation process largely controls the actual lifetimes since the exciton diffusion and exciton-exciton annihilation alternately occur within nanoparticles [54, 55]. Therefore, to observe the triplet-triplet (T-T) annihilation processes within nanoassemblies, further experiments are performed. When the laser power is increased at the same concentration of $H_2P(CO_2H)_4$ nanoassemblies, the initial decay rate increased with changing decay kinetics from first order to a mixed order (first and second order) because of the triplet-triplet annihilation (Eq. 10.1) [53, 56].

$$[^3(H_2P)^*] + [^3(H_2P)^*]\Diamond[^1(H_2P)^*] + [H_2P] \tag{10.1}$$

In order to observe the T-T annihilation rate within nanoassemblies, following mathematical equation can be expressed as

$$\frac{dn}{dt} = -kn(t) - \gamma(t)n(t)^2 \tag{10.2}$$

where n is the population of the excited state, $n(t = 0) = n_0$, and k is the decay rate constant of the excited state. In large aggregates with a high mobility of excitons, the annihilation rate constant, $\gamma\ (t)$, can be assumed to be time-independent and the solution to Eq. 10.2 yields

$$n(t) = \frac{n_0 \exp(-kt)}{1 + n_0\gamma k^{-1}\,[1 - \exp(-kt)]} \tag{10.3}$$

At low initial excitation concentrations for excited singlet state, Eq. 10.3 can be rewritten thus:

$$\frac{1}{c(t)} = \left(\frac{1}{c_0} + \frac{\gamma_T}{k_T}\right)\exp(k_T t) - \frac{\gamma_T}{k_T} \tag{10.4}$$

where c is the population of the excited triplet state, $c(t = 0) = c_0$, and k_T is the decay rate of the excited triplet state of $H_2P(CO_2H)_4$ monomer. The γ_T is T-T annihilation rate constant. The observed T-T annihilation rate constants (γ_T) for nanoassemblies are shown in Table 10.1. The γ_T values are dependent on nanoparticle sizes, and evaluated to be 3.9–6.5 $\times 10^{10}$ mol^{-1} dm^3 s^{-1} [53]. Thus, we observed quite fast and efficient γ_T values in the form of aggregates. The observed S-S annihilation rate constants (γ_S) for $H_2P(CO_2H)_4$/TriEG and $H_2P(CO_2H)_4$/TetraEG nanoparticles (8.8 $\times 10^8$ and 8.3 $\times$ 10^8 mol^{-1}dm^3 s^{-1}) are much smaller than those of γ_T. The difference between γ_S and γ_T may be dependent on the freedom degree of exciton motion of excited singlet or triplet state in the crystalline structures [54]. Thus, based on the above discussion, we can conclude that size-control of organic nanoparticles plays an important role for the exciton diffusion property.

10.3 Photovoltaic Application of Porphyrin-Based Nanoparticles

10.3.1 *Porphyrin Pristine Nanoparticles*

These self-assembled nanoparticles can be deposited as thin films on nanostructured electrode using an electrophoretic technique for fabrication of photoelectrochemical cells (Fig. 10.4). The fabrication method is as follows [57]. A known amount (~2 mL) of 5,15-bis(3,5-di-*tert*-butylphenyl)porphyrin [$H_2P(tBu)_2$] nanoparticle, which is clusterized in acetonitrile/toluene (9/1, v/v), is transferred to a cuvette in which optically transparent electrode (OTE) and TiO_2-coated film electrode (OTE/TiO_2) were kept at a distance of ~6 mm using a Teflon spacer. By applying electric field between two electrodes, the deposition of the film can be visibly seen as the solution becomes colorless with simultaneous brown coloration of the OTE/TiO_2 electrode [denoted as OTE/TiO_2/$(H_2P(tBu)_2)_n$] [57]. OTE/TiO_2/$(H_2P(tBu)_2)_n$ is highly photoactive and capable of undergoing charge separation under visible light excitation.

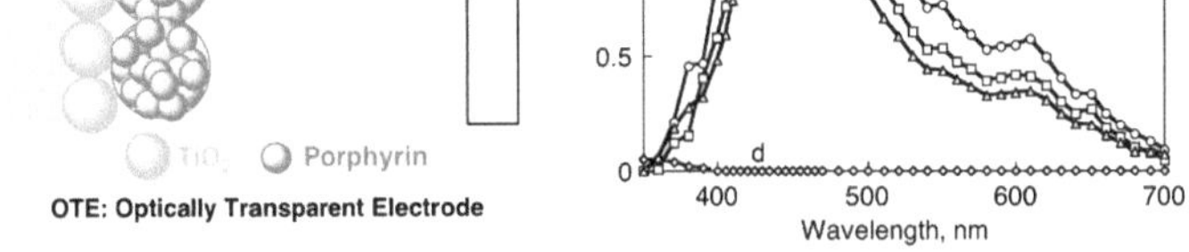

Figure 10.4. (A) Illustrations of the porphyrin nanoparticles-deposited photoelectrochemical cell (left) and porphyrin–fullerene composite nanoparticles (right). (B) Comparison of photocurrent response (IPCE values) of (a) OTE/SnO_2/$(H_2P(tBu)_2+C_{60})_n$, (b) OTE/SnO_2/$(C_{60})_n$, (c) OTE/SnO_2/$(H_2P(tBu)_2)_n$, and (d) the sum of the IPCE response of OTE/SnO_2/$(C_{60})_n$ (b) and OTE/SnO_2/$(H_2P(tBu)_2)_n$ (c); an applied bias potential: 0.2 V vs. SCE, electrolyte: 0.5 M NaI and 0.01 M I_2 in acetonitrile. Reprinted with permission from [57].

Photoelectrochemical performance of $OTE/TiO_2/(H_2P(tBu)_2)_n$ was measured in standard two-compartment electrolyte cell [57]. Photoexcitation of the porphyrin film electrode assembly in a photoelectrochemical cell with visible light produces relatively high photocurrent generation. A maximum photocurrent of 0.15 mA cm^{-2} and a photovoltage of 250 mV were attained using I_3^-/I^- redox couple under white light illuminatiom. The incident photon-to-photocurrent generation efficiency (IPCE) was calculated by normalizing the photocurrent values for incident light energy and intensity using Eq. 10.5 [57].

$$\text{IPCE}(\%) = 100 \times 1240 \times I_{sc}/(W_{in} \times \lambda) \qquad (10.5)$$

where I_{sc} is the short circuit photocurrent (A/cm^2), W_{in} is the incident light intensity (W/cm^2), and λ is the wavelength (nm). An IPCE of 2.0% has been achieved at an applied bias potential of 0.2 V *vs.* SCE [57].

10.3.2 *Composite Nanoparticles Composed of Porphyrins and Fullerenes by Simple Blend*

By using the above method, organic nanoparticles composed of two different moieties (e.g., donor and acceptor moieties) in suspension were easily deposited onto nanostructured SnO_2 [OTE/SnO_2] [58]. These composite molecular nanoparticles of C_{60} (acceptor) and H_2P (donor) prepared in acetonitrile/toluene mixed solvent absorb light over entire spectrum of visible light.

The highly colored composite nanoparticles can be assembled as three-dimensional array onto nanostructured SnO_2 films using the same electrophoretic deposition approach [$OTE/SnO_2/(H_2P(tBu)_2+C_{60})_n$]. $OTE/SnO_2/(H_2P(tBu)_2+C_{60})_n$ film exhibits an IPCE as high as 17% at an applied potential of 0.2 V *vs.* SCE which is significantly higher than the additive effect observed from either porphyrin (IPCE = 1.6%) or fullerene clusters (IPCE = 5.0%) under same photoelectrochemical conditions (Fig. 10.5) [58]. The high IPCE observed with $H_2P(tBu)_2$ and C_{60} composites demonstrate the synergy of these systems towards yielding efficient photoinduced charge separation within these composite nanoassemblies.

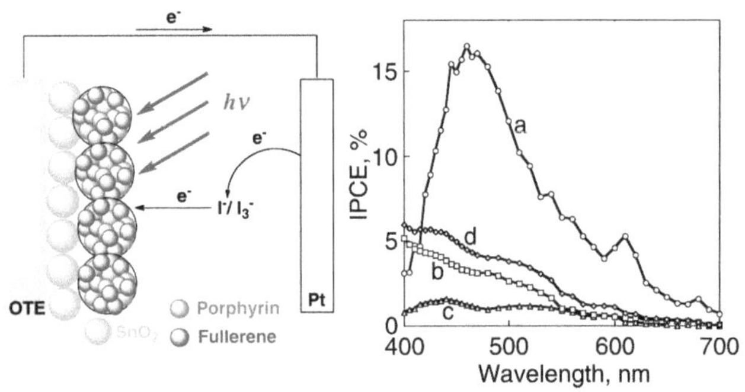

Figure 10.5. Illustrations of of the porphyrin–fullerene composite nanoparticles deposited-photoelectrochemical cell (left). Comparison of photocurrent response (IPCE values) of (a) OTE/SnO_2/($H_2P(tBu)_2+C_{60})_n$, (b) OTE/SnO_2/$(C_{60})_n$, (c) OTE/SnO_2/$(H_2P)_n$, and (d) the sum of the IPCE response of OTE/SnO_2/$(C_{60})_n$ (b) and OTE/SnO_2/$(H_2P)_n$ (c); an applied bias potential: 0.2 V *vs.* SCE, electrolyte: 0.5 M NaI and 0.01 M I_2 in acetonitrile (right). Reprinted with permission from [58].

Thus, composite nanoparticles of porphyrins (donor) and fullerenes (acceptor) effectively work as photoactive materials in photovoltaic cells.

10.3.3 *Supramolecular Nanoparticles Composed of Multiporphyrin Arrays and Fullerenes for Efficient Charge Separation*

As shown above, molecular nanoassembly composed of porphyrins and fullerenes by simple blend is a useful way for construction of photovoltaic systems. However, in such a system, the serious problem is the limited IPCE value (∼17%). To get the high photovoltaic properties, organization of donor and acceptor moieties for efficient charge separation and carrier transport is required. Therefore, novel strategies for self-organization of porphyrin and fullerene moieties using multiporphyrin arrays are performed.

Figure 10.6 (left) shows chemical structures of porphyrin-peptide oligomers with different number of porphyrins in a polypeptide unit (porphyrin functionalized α-polypeptides) [59, 60]. Such porphyrin oligomers with a polypeptidic backbone are flexi-

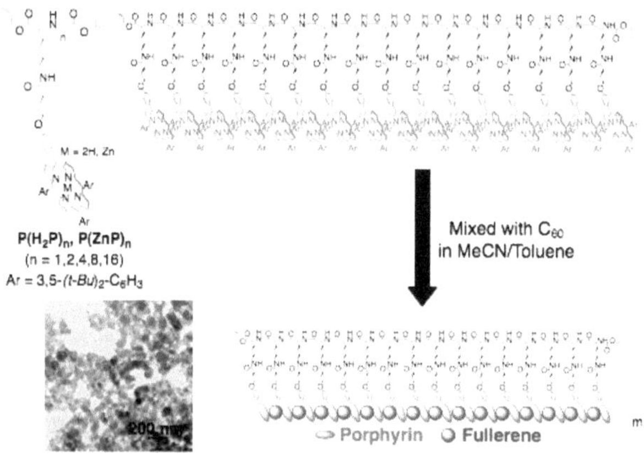

Figure 10.6. Chemical structures of porphyrin-peptide oligomers employed in this study (left). The right figure shows an organization process of $P(H_2P)_{16}$ with C_{60} [denoted as $(P(H_2P)_{16} + C_{60})_m$]. TEM image is $(P(H_2P)_{16}+C_{60})_m$. Reprinted with permission from [59].

ble enough to accommodate C_{60} between the porphyrin units. Figure 10.6 (right) also shows the preparation procedure and TEM image of $(P(H_2P)_{16}+C_{60})_m$. The sizes and shapes of $(P(H_2P)_{16}+C_{60})_m$ are much controlled as compared with those of $(P(ZnP)_1+C_{60})_m$ [59].

Figure 10.7A shows the IPCE of the $(P(H_2P)_n+C_{60})_m$ ($n = 1, 2, 4, 8, 16$) modified electrode at a constant concentration ratio of porphyrin to C_{60}. The IPCE value of $(P(H_2P)_n+C_{60})_m$ system exhibits a remarkable increase with increasing the number of porphyrins in a polypeptide unit. In particular, the $(P(H_2P)_{16}+C_{60})_m$ system has the maximum IPCE value of 48% at 600 nm as well as the broad photoresponse, extending into the infrared region up to 1000 nm [59]. Such an effective light energy conversion is ascribed to the polypeptide structure to control three-dimensional organization between porphyrins and C_{60}. The maximum IPCE of $(P(ZnP)_n+C_{60})_m$ system also exhibits a remarkable increase with increasing the number of porphyrins in a polypeptide unit. The maximum IPCE of $(P(ZnP)_{16}+C_{60})_m$ is determined to be 56% at 480 nm (Fig. 10.7B) [59].

The major reason for increased IPCE values with an increase of the number porphyrins in a poly-peptide unit is the charge-

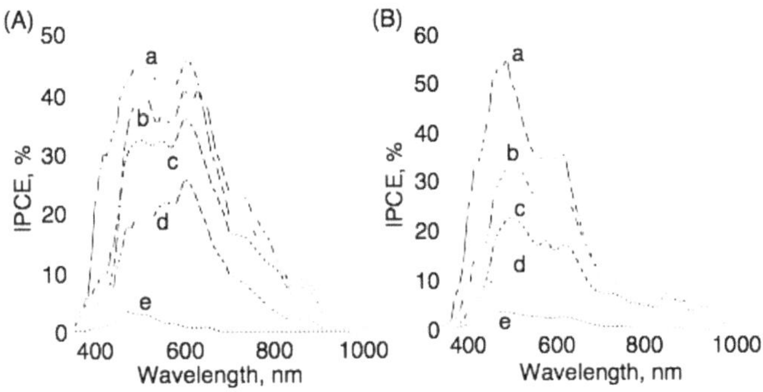

Figure 10.7. (A) The photocurrent action spectra (IPCE *vs.* wavelength) of (a) $(P(H_2P)_{16}+C_{60})_m$, (b) $(P(H_2P)_8+C_{60})_m$, (c) $(P(H_2P)_4+C_{60})_m$, (d) $P(H_2P)_2+C_{60})_m$, and (e) $(P(H_2P)_1+C_{60})_m$ on OTE/SnO_2 electrodes. (B) The photocurrent action spectra of (a) $(P(ZnP)_{16}+C_{60})_m$, (b) $(P(ZnP)_8+C_{60})_m$, (c) $(P(ZnP)_4+C_{60})_m$, (d) $(P(ZnP)_2+C_{60})$, and (e) $(P(ZnP)_1+C_{60})_m$ on OTE/SnO_2 electrodes. Reprinted with permission from [59].

separation process between porphyrins and fullerenes. The UV-vis absorption and fluorescence spectral changes give us useful information of supramolecular formation between porphyrins and fullerenes. The binding constants between porphyrins and fullerenes are thereby evaluated quantitatively. The spectral changes indicate that C_{60} forms 1 : 1 complex with the porphyrin moiety. The apparent formation constant (K) between $P(H_2P)_{16}$ and C_{60} determined from the fluorescence quenching (8.2×10^4 M^{-1}) is significantly larger than that determined from the UV-vis spectral change (1.4×10^4 M^{-1}). This indicates that the excited energy migration between the porphyrin units occurs efficiently prior to the electron transfer. The K values of various types of porphyrin-peptide oligomers [$P(H_2P)_n$or $P(ZnP)_n$] and C_{60} derivatives are listed in Table 10.2 [59]. The association constant between porphyrins and C_{60} increases with increasing number of porphyrins in a polypeptide unit. This indicates that the interaction between porphyrin and C_{60} becomes stronger with increasing the number of porphyrins in the porphyrinic polypeptide. Such differences in the association constants are in good agreement with the IPCE results (Fig. 10.7) [59].

The photodynamics of the composite molecular assemblies was further measured by femtosecond time-resolved transient

Table 10.2. Comparison of binding constants between porphyrins and C_{60}

	K, M^{-1}			K, M^{-1}	
Porphyrin	UV -vis	Fluorescence	Porphyrin	UV -vis	Fluorescence
$P(ZnP)_2$	—	—	$P(H_2P)_2$	—	—
$P(ZnP)_4$	5.9×10^2	5.0×10^3	$P(H_2P)_4$	2.7×10^3	9.5×10^3
$P(ZnP)_8$	4.1×10^3	1.4×10^4	$P(H_2P)_8$	5.3×10^3	2.2×10^4
$P(ZnP)_{16}$	6.7×10^3	3.9×10^4	$P(H_2P)_{16}$	1.4×10^4	8.2×10^4

Source: Reprinted with permission from [59].

absorption spectra to examine the charge-separation process (Fig. 10.8). Upon 387 nm laser pulse excitation we observe a broad absorption in the 450–510 nm as we populate the porphyrin singlet excited state. This transient quickly decays to form triplet excited state. In addition, a strong transient bleaching (around 610 nm) arises from the fluorescence. In contrast to $(P(ZnP)_1+C_{60})_m$ system (Fig. 10.8A), the strong absorption arising from singlet and triplet excited states of porphyrins (450–510 nm) is missing in the case of $(P(ZnP)_8+C_{60})_m$ (Fig. 10.8B). Instead a broad absorption at around 650 nm appears after the laser pulse excitation as shown in Fig. 10.8B. Such broad absorption spectra are clear indication of formation of the porphyrin radical cation. It is also interesting

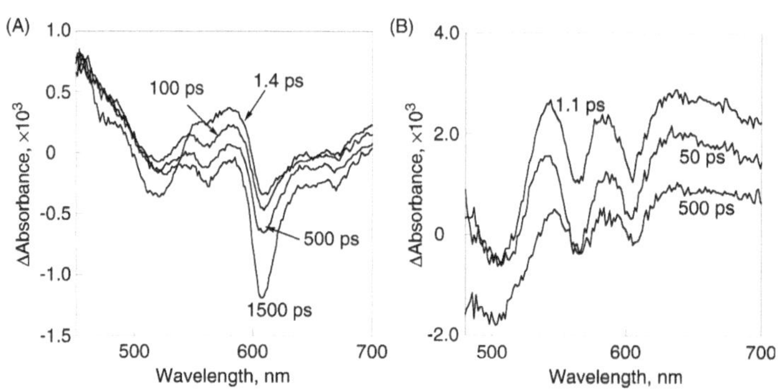

Figure 10.8. Time-resolved absorption spectra recorded following 387 nm laser pulse excitation of (A) $(P(ZnP)_1+C_{60})_m$ and (B) $(P(ZnP)_8+C_{60})_m$. All samples were prepared in argon-saturated acetonitrile/toluene (3/1, v/v) and measurements were made at 298 K. Reprinted with permission from [59].

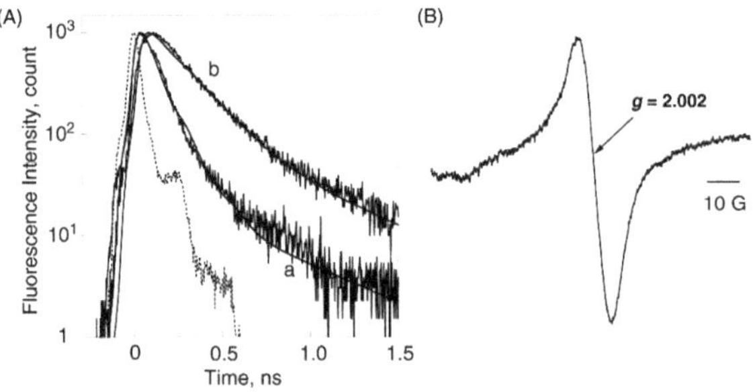

Figure 10.9. (A) Fluorescence decay curves of (a) $(P(H_2P)_{16}+C_{60})_m$ and (b) $(P(H_2P)_{16})_m$ modified electrodes. The fluorescence decays are observed at 650 nm by the time-correlated single-photon counting method. The excitation wavelength is 420 nm and IRF (instrument response function) curves are presented for each decay. (B) ESR spectrum of photoirradiated $(P(ZnP)_{16}+C_{60})_m$ in acetonitrile/toluene (3/1, v/v) under photoirradiation of a high-pressure mercury lamp, measured at 123 K. Reprinted with permission from [59].

to note that the bleaching originating from fluorescence (610 nm region) is absent in the transient spectra (Fig. 10.8B). Thus, different supramolecular formations between porphyrins and fullerens qualitatively have an effect on the photoinduced electron-transfer process.

The fluorescence lifetime and electron spin resonance (ESR) results also support the investigation of photoinduced electron-transfer process. The fluorescence decay curves are shown in Fig. 10.9A. The fluorescence decay of of $(P(H_2P)_{16}+C_{60})_m$ is strongly quebched as compared to that of $(P(H_2P)_{16})_m$ without C_{60}. This indicates that efficient photoinduced electron tranfer takes place from $^1P(H_2P)_{16}^*$ to C_{60}. The fluorescence lifetimes of first major component decrease with increasing the number of porphyrins in a polypeptide unit because of acceleration of the forward electron tranfer [59].

The formation of the radical cation of $P(ZnP)_{16}$ and the radical anion of C_{60} in $(P(ZnP)_{16}+C_{60})_m$ composite assemblies upon photoexcitation is confirmed by ESR measurements performed in frozen acetonitrile/toluene under photoirradiation. The resulting ESR spectrum of $(P(ZnP)_{16}+C_{60})_m$ upon photoirradiation in acetonitrile/toluene at 123 K is shown in Fig. 10.9B. The ESR signal

(g = 2.002) is assigned to the overlap of two signals due to porphyrin radical cation and $C_{60}^{\bullet-}$ [59]. Thus, our supramolecular strategy using multiporphyrin arrays and fullerenes for photovoltaics successfully contributes to the drastic improvement of light energy conversion properties [59]. In addition to this system, we have also obseved the similar drastic improvements in dffirent systems such as porphyrin-alkanethiolate monolayer-protected gold nanoparticles [48, 61], porphyrin dendrimer [62, 63], TiO_2 nanoparticle/nanotubes, and other dye molecules [47].

Photocurrent generation in the present systems is initiated by photoinduced charge separation from the singlet excited state of free base porphyrin ($^1H_2P^*/H_2P^{\bullet+} = -0.7$ V vs NHE) [59] or zinc porphyrin ($^1ZnP^*/ZnP^{\bullet+} = -1.0$ V vs NHE) [59] to C_{60} ($C_{60}/C_{60}^{\bullet-} = -0.2$ V vs NHE) [59] in the porphyrin-C_{60} complex rather than direct electron injection to conduction band of SnO_2 (0 V vs NHE) system [59]. The reduced C_{60} injects electrons into the SnO_2nanocrystallites, whereas the oxidized porphyrin ($H_2P/H_2P^{\bullet+} = 1.2$ V or $ZnP/ZnP^{\bullet+} = 1.0$ V vs NHE) [59] undergoes the electron-transfer reduction with the iodide ($I_3^-/I^- = 0.5$ V vs NHE) [59] in the electrolyte system. In such a case, electron injection following the charge separation occurs more effectively to a nanostructured SnO_2 electrode.

10.4 Bar-Shaped Molecular Assemblies of Porphyrins

10.4.1 *Sonication-Assisted Porphyrin Nanorods and Their Photocurrent Generation*

Ultrasonic wave is often utilized for formation of crystal nucleus and growth in the preparation of molecular crystals but seldom favors the formation of an ordered assembly. We have constructed novel supramolecular nanorods and fibers controlled by intermolecular interactions such as π–π stacking interaction of porphyrins without the other bonds/interactions or aiding surfactants, which are prepared in sonicated solution medium [64].

The porphyrin nanorod was prepared by the following procedure. First, 3.5 mmol dm^{-3} 5,15-bis(3,5-di-*tert*-butylphenyl) porphyrin [$H_2P(tBu)_2$] solution was prepared in toluene. Then,

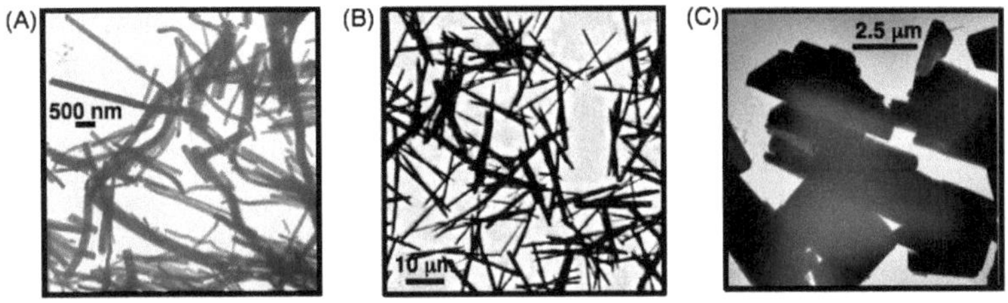

Figure 10.10. (A) TEM images of $H_2P(tBu)_2$ nanorod, (B) optical microscope images of $H_2P(tBu)_2$ fiber, and (C) reference $H_2P(tBu)_4$ assembly. Reprinted with permission from [64].

the toluene solution is simply mixed with 9 times volume of acetonitrile (final concentration: 0.35 mmol dm^{-3} in a poor/good solvent = 9/1, v/v). The solution was sonicated (45 kHz) for 30 min at 15°C to form macroscopic self-assembled porphyrins such as rod-like structures by using our original set-up. To check the substituent effect of the porphyrin ring on the macroscpic aggregate structures, a reference porphyrin: 5,10,15,20-tetra(3,5-di-*tert*-butylphenyl)porphyrin [$H_2P(tBu)_4$] is also employed.

Figure 10.10A, B, and C shows TEM images of $H_2P(tBu)_2$ nanorod, $H_2P(tBu)_2$ nanorod and $H_2P(tBu)_4$ assembly, respectively. In contrast with large square crystal structures (2–3 μm scale) of the $H_2P(tBu)_4$ assembly (Fig. 10.10C), $H_2P(tBu)_2$ nanorods (Fig. 10.10A) form a lot of long bar shaped-structures. The structure of $H_2P(tBu)_2$ nanorods are 360 ± 130 nm in diameter and 5.02 ± 1.94 μm in length. To construct much longer fiber structures, the nanorod suspended solution is further allowed to stand for 6 days. Then, the crystallized porphyrins were stirred for 5 min at room temperature. The nanorod crystals finally grow up to be longer fiber structures. In Fig. 10.10B, we can clearly see much longer fiber structures analyzed as 890 ± 270 nm in diameter and 27.2 ± 6.9 μm in length. In the formation of the fiber structures, a large increase of length from 5.0 to 27.2 μm (~6 times) relative to that of diameter (~2 times) indicates that anisotropic crystal growth largely occurs toward the length direction. Thus, we have successfully controlled the structures of pophyrin fibrous assemblies within a certain definite range. This trend is also observed between 5,15-diphenyl-porphyrin and 5,10,15,20-tetraphenyl-porphyrin. Additionally, the internal structures of the self-assembled porphyrins are investigated by XRD analysis. We can observe strong diffraction peaks of *a* and *c* axes such as (100) and (002) in the unit cell, whereas the diffraction intensity based on *b* axis is very weak. Considering the unit cell structure (monoclinic structure: dihedral angles 90° in axes *a*–*b* and *b*–*c*), growth direction of the rod-crystals may be *b* axis direction [64].

In Fig. 10.10C, the absorption spectrum of $H_2P(tBu)_2$ nanorod (spectrum *a*) exhibits much broader and more intense absorption in the visible and near infrared regions than those of the corresponding simple drop-cast film prepared by $H_2P(tBu)_2$ in toluene (spectrum *b*) and toluene solution (spectrum *c*). This change of absorption property is likely because of strong supramolecular π–π interaction between a porphyrin and its nearest neighbor in nanorod structures. The IPCE spectrum (spectrum *a* in Fig. 10.11B) also shows a broad photoresponse in the visible region (maximum IPCE: ~5.5% at 460 nm) [64], which parallels the corresponding absorption (spectrum *a* in Fig. 10.11C) and excitation spectra (spectrum *b* in Fig. 10.11B). These experiments confirmed the role of $H_2P(tBu)_2$ nanorod towards harvesting light energy and generating photocurrent during the operation of a photoelectrochemical cell.

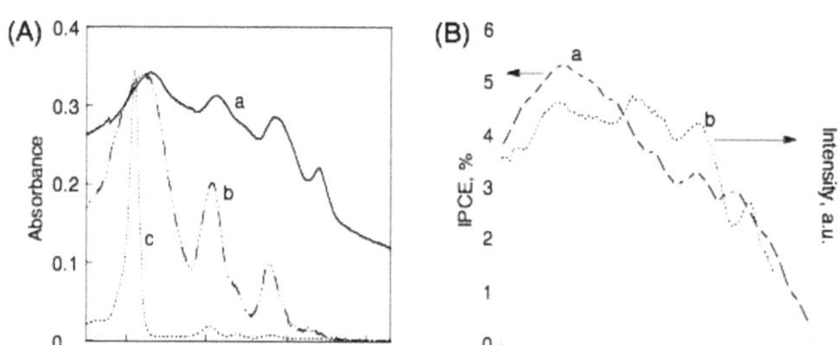

Figure 10.11. (A) Absorption spectra of (a) $H_2P(tBu)_2$ nanorod film, (b) $H_2P(tBu)_2$ drop-cast film on quartz plates, and (c) $H_2P(tBu)_2$ (4×10^{-6} mol dm^{-3}) in toluene. (B) (a) Photocurrent (IPCE) action spectrum of $H_2P(tBu)_2$ nanorod on an OTE/SnO_2. Electrolyte: 0.5 mol dm^{-3} LiI and 0.01 mol dm^{-3} I_2 in acetonitrile. (b) Excitation spectrum of $H_2P(tBu)_2$ nanorod film; observed at 710 nm. Reprinted with permission from [64].

10.4.2 *Fullerenes-Encapsulated Porphyrin Hexagonal Nanorods for Photovoltaics*

Considering the photovoltaic mechanism (*vide supra*), a bar-shaped structure composed of two different molecules with separated inside and outside layers is a good candidate for the efficient carrier generation (hole and electron). Therefore, we have succeeded in construction of new molecular composites: fullerenes-encapsulated porphyrin hexagonal nanorods composed of zinc *meso*-tetra (4-pyridyl) porphyrin [ZnP(Py)$_4$] and fullerene derivatives [denoted as fullerene-ZnP(Py)$_4$ nanorod], which are prepared by aiding surfactant: cetyltrimethylammonium bromide (CTAB) in a DMF/acetonitrile mixed solvent (Fig. 10.12) [65, 66].

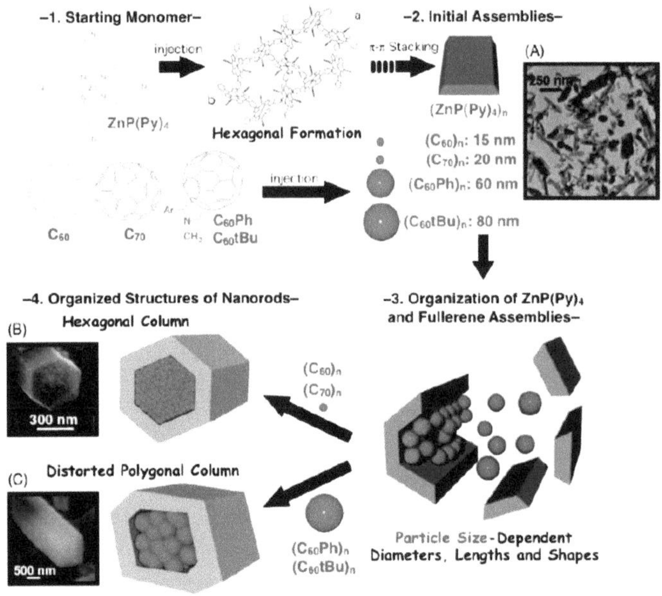

Figure 10.12. An illustration of fullerene nanoparticle-promoted organization of ZnP(Py)$_4$ nanorods. CTAB is omitted for clarity. The electron micrographs show (A) immediate formations of C_{60}*t*Bu and ZnP(Py)$_4$ composites (1 min) after injection, (B) hexagonal C_{60}–ZnP(Py)$_4$ rod, and (C) distorted polygonal C_{60}*t*Bu–ZnP(Py)$_4$ rod. Reprinted with permission from [66].

Table 10.3. Analyzed structures of $ZnP(Py)_4$ tube, and C_{60}–$ZnP(Py)_4$ rod, C_{70}–$ZnP(Py)_4$ rod, $C_{60}Ph$–$ZnP(Py)_4$ rod, $C_{60}tBu$–$ZnP(Py)_4$ rod and the corresponding fullerene nanoparticles

Nanorod/Nanotube	Rod length (μm)[a]	Rod outside Diameter (nm)[a]	Fullerene particle diameter (nm)[b]
$ZnP(Py)_4$ tube	2.13 ± 0.27	540 ± 30	-
C_{60}–$ZnP(Py)_4$ rod	4.12 ± 0.94	490 ± 90	15
C_{70}–$ZnP(Py)_4$ rod	4.50 ± 0.71	480 ± 40	20
$C_{60}Ph$–$ZnP(Py)_4$ rod	6.45 ± 1.71	700 ± 160	60
$C_{60}tBu$–$ZnP(Py)_4$ rod	7.82 ± 2.07	940 ± 240	80

Source: Reprinted with permission from [66].
[a]Analyzed by SEM.
[b]Average values analyzed by DLS

The preparation method is as follows. A mixed proper ratio of $ZnP(Py)_4$ and fullerene derivatives in DMF solution (*1. Starting Monomer*) was injected into 7.5 times volume of continuously stirred 0.20 mM CTAB acetonitrile solution at room temperature. The final concentrations of $ZnP(Py)_4$ and fullerene derivatives are 0.03 and 0.02 mM in DMF/acetonitrile (2/15, v/v), respectively. On injecting, they largely form $ZnP(Py)_4$ flake assemblies and fullerene-based nanoparticles, separately (Fig. 10.12A). With an increase in the bulkiness of substituent groups in fullerene derivatives, the average diameters increased from 15 to 80 nm (Table 10.3). In $ZnP(Py)_4$ assemblies, with the diffusion of DMF into acetonitrile, the Zn-N axial coordination of pyridyl N-atoms to zinc atoms of $ZnP(Py)_4$ promotes the growth of aggregates, which continue to grow into a flake structure. In this case, the organization process of $ZnP(Py)_4$ moieties is initially derived from coordination bond in contrast with C_{60} assemblies based on relatively weak π–π interactions. Then, fullerene-$ZnP(Py)_4$ nanorods are finally formed (Fig. 10.12B and C). In scanning electron microscopy (SEM) measurement, $ZnP(Py)_4$ pristine hexagonal nanotubes with a large hollow structure [denoted as $ZnP(Py)_4$ tube] are observed, whereas the hollow hole is completely closed in the case of nanorods composed of fullerenes (C_{60} and C_{70}) and $ZnP(Py)_4$ [fullerene-$ZnP(Py)_4$ rod]. In C_{60} derivative-$ZnP(Py)_4$ rods, the distorted polygonal columnar structures with large diameter and length are formed, as compared to the hexagonal structures of C_{60}-ZnP(Py)$_4$ and C_{70}-ZnP(Py)$_4$ rods. The reference ZnP(Py)$_4$ hexagonal nanotube without C_{60} was also prepared in the same manner for comparison [denoted as ZnP(Py)$_4$ nanotube].

ZnP(Py)$_4$ tube shows a bar-like structure with a large hollow hole, whereas the hole is completely closed in C_{60}-ZnP(Py)$_4$ rods (Fig. 10.12B). C_{60}-ZnP(Py)$_4$ nanorods are analyzed as 4.12 ± 0.94 μm in length and 490 ± 90 nm in outside diameter by SEM images. As compared to ZnP(Py)$_4$ tubes analyzed as 2.13 ± 0.27 μm in length and 540 ± 30 nm in outside diameter, a large increase of length direction from 2.13 to 4.12 μm relative to unchanged diameters (~500 nm) indicates that anisotropic crystal growth largely occurs toward the length direction due to π–π interaction of encapsulated C_{60} moieties within ZnP(Py)$_4$ assembly [65, 66]. Moreover, the increased sizes of bar-shaped structures are in good agreement with the trend of fullerene nanoparticles (Table 10.3). This indicates that an encapsulated process of fullerene assemblies by ZnP(Py)$_4$ flake assemblies plays an important role in the final structural organization.

The absorption spectrum of fullerene-ZnP(Py)$_4$ rod exhibits much broader and more intense absorption in the visible and near infrared regions than those of the corresponding monomers: ZnP(Py)$_4$ or C_{60} in DMF because of aggregated interactions (Fig. 10.13A). Such a broad absorption property is useful for solar energy conversion. The picosecond fluorescence lifetime and nanosecond transient absorption measurements of fullerene-ZnP(Py)$_4$ rod suggest the efficient quenching of the singlet excited state by C_{60} moiety because of photoinduced electron transfer (PET), which is in sharp contrast with PET *via* ^{3}ZnP(Py)$_4^*$ in reference non-organized system prepared without CTAB. In the transient absorption spectrum of C_{70}–ZnP(Py)$_4$ rod at 0.1 μs, the broad peak observed at 500–780 nm can be attributed to the radical cation of the ZnP(Py)$_4$, [ZnP$^{\bullet+}$(Py)$_4$] moieties. The inset time profile in Fig. 10.13B shows the rapid decay of the $C_{70}^{\bullet-}$ at 1380 nm. This indicates that the chargeseparation process takes place via the ^{1}ZnP*(Py)$_4$ moiety. Thus, the formation of a charge-separation state, $C_{70}^{\bullet-}$–ZnP$^{\bullet+}$(Py)$_4$, in the nanorods was suggested by the transient bands. The decay rates of the transient absorption bands

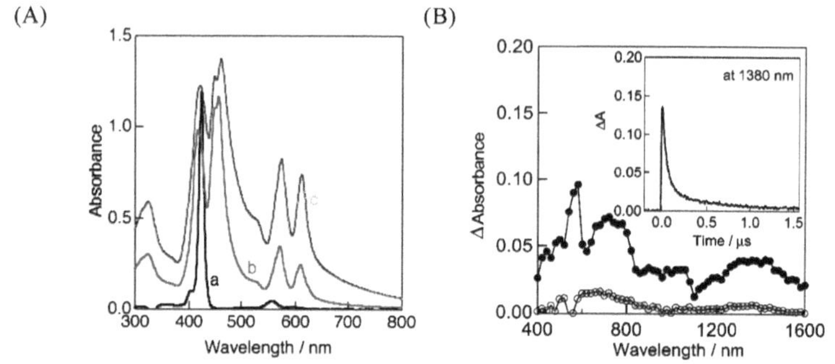

Figure 10.13. (A) Steady-state absorption spectra of (a) 6 μM ZnP $(Py)_4$ monomer, (b) $ZnP(Py)_4$ tube, and (c) $C_{60}t$Bu–$ZnP(Py)_4$ rod in DMF/acetonitrile (15/2, v/v). (B) Nanosecond transient absorption spectra of C_{70}–$ZnP(Py)_4$ rod in Ar-saturated DMF/acetonitrile (2/15, v/v) after 532 nm laser irradiation at 0.1 1s (•) and 1.0 μs (○). Inset: the time profiles of the anion radical of C_{70} monitored at 1380 nm. Reprinted with permission from [66].

can be attributed to charge recombination, after the formation of the chargeseparation state, $C_{70}^{\bullet-}$–$ZnP^{\bullet+}(Py)_4$. In the case of C_{60}–$ZnP(Py)_4$ rod, C_{60}Ph-$ZnP(Py)_4$ rod, and $C_{60}t$Bu–$ZnP(Py)_4$ rod, the radical anion appears at 1080 or 1020 nm, respectively. The forward and back electron-transfer rate constants are largely dependent on the structures of the nanorods.

The photocurrent action spectrum of OTE/fullerene-$ZnP(Py)_4$-nanorod show a broad photoresponse in the visible region, and the maximum IPCE (~35%) is much larger than the sum of two individual IPCE values (~6.5%) of OTE/$ZnP(Py)_4$-nanotube and OTE/C_{60}-assembly under the same condition [65, 66]. These results clearly indicate that an organized structure between C_{60} and $ZnP(Py)_4$ as well as an excellent electron acceptor property of C_{60} play a role for the improvement of light energy conversion property.

10.4.3 *Supramolecular Patterning on Surfaces*

The construction of complicated submicrometer-scaled patterns on surfaces serves as basis for the fabrication of integrated circuits and MEMS etc [67]. Aggregation or self-assembly of molecules on a surface can be a simple and versatile approach, if the simple physical processes such as dewetting are effectively combined with supramolecular techniques [38, 67–69]. Especially, disc-like molecules (polycyclic aromatic hydrocarbons: PAH) functionalized with alkyl groups such as triphenylene derivatives self-assemble into supramolecular columnar structures that organize into hexagonal and nematic phases because of the stacking π–π interactions [70–73]. Therefore, such stacking interaction of PAH connected by dyes enables us to construct organized assemblies of dyes. Therefore, a combination of triphenylenes and porphyrins has promising potentials for such aggregate formations and the light-driven functionalities (charge/energy transfer etc).

First, a new series of triphenylene core-centered porphyrin hexamers with different alkyl chain lengths and the respective reference compounds is synthesized as shown in Fig. 10.14 [74]. Then, by casting the toluene solutions of these porphyrin hexamers onto carbon-coated films, monolayered patterning (70–80 Å) is

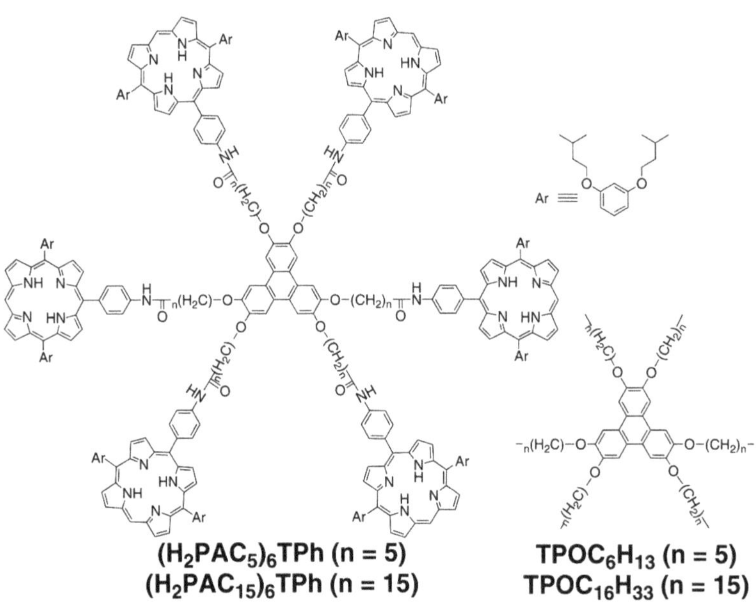

Figure 10.14. Triphenylene core-centered porphyrin hexamers and reference compounds. Reprinted with permission from [74].

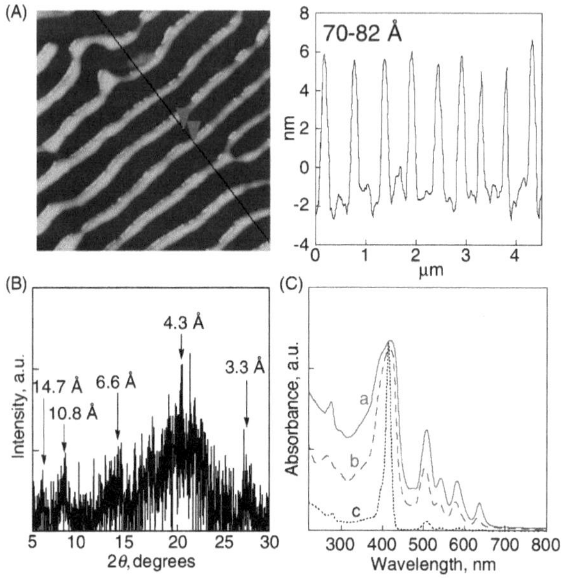

Figure 10.15. (A) AFM image of $(H_2PAC_{15})_6TPh$ patterns on a carbon-coated film, (B) XRD data of $(H_2PAC_{15})_6TPh$ patterns, (C) UV spectra of (a) $(H_2PAC_{15})_6TPh$ patterns (film) and (b) $(H_2PE)_6TPh$ patterns (film), (c) 3 μM $(H_2PAC_{15})_6TPh$ in THF. Reprinted with permission from [74].

systematically observed (Fig. 10.15A). These aggregate structures are in sharp contrast with the macroscopic bundle assemblies of the corresponding triphenylene references. In particular, with increasing alkyl chain lengths between porphyrin and triphenylene units, highly aligned and ordered molecular patterns are formed.

To further discuss the internal structures, we also measured X-ray diffraction (XRD) patterns of the aggregate formations (Fig. 10.15B). In the wide-angle region, two diffuse reflections are observed. The broad one centered at about 4.3 Å corresponds to the aliphatic chains [75]. The relatively sharper one observed at higher 2θ-values and separated from the broad one (3.3 Å) is due to the stacking of the triphenylene core one on top of another [75]. Figure 10.15C also shows the absorption spectrum of the aggregate structures. In contrast to the porphyrin band, the triphenylene band in $(H_2PAC_{15})_6TPh$ assembly (274 nm) becomes blue-shifted, which suggests the columnar stacked-formation [76]. Based

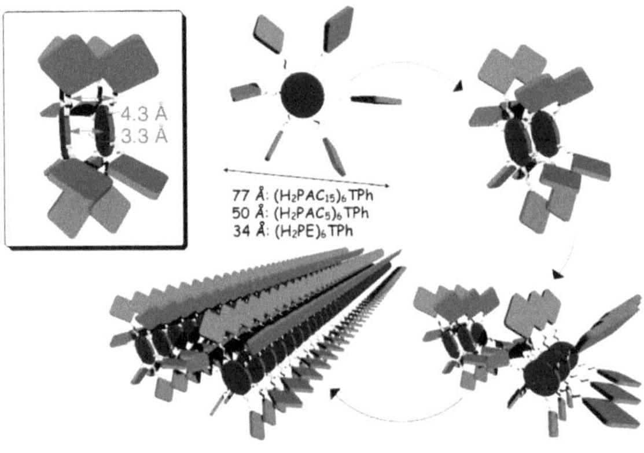

Figure 10.16. A schematic illustration of organization process. The purple square, yellow wavy line and blue round shapes stand for porphyrin, alkyl chain, and triphenylene units, respectively. The molecular sizes were estimated by DFT calculation. The insertion: proposed stacking structure of $(H_2PAC_{15})_6TPh$ units. Reprinted with permission from [74]. See also Color Insert.

on these results, the molecular stacking formation in $(H_2PAC_{15})_6TPh$ fiber is proposed (Fig. 10.16). The stacking distances of triphenylenes (3.3 Å) and alkyl chains (4.3 Å) are evaluated by XRD result (Fig. 10.15B). NMR, FTIR, and absorption data qualitatively support the proposed structure. Further, dewetting process should be emphasized as stated above.

10.5 Supramolecular Formation of Carbon Nanotubes with Porphyrins

10.5.1 *Composite Nanorods Composed of Protonated Porphyrins and Single-Wall Carbon Nanotubes*

Single-wall carbon nanotube (SWCNT) has the characteristic π-conjugated cylinder structure. Therefore, SWCNT can provide an ideal network to promote charge transfer with dye molecules and transport electrons to the collecting surface (e.g., photovoltaic cells). On the other hand, protonated porphyrins provide a convenient way to construct ordered molecular assemblies. Therefore, we have prepared novel SWCNTs driven aggregation of protonated porphyrins

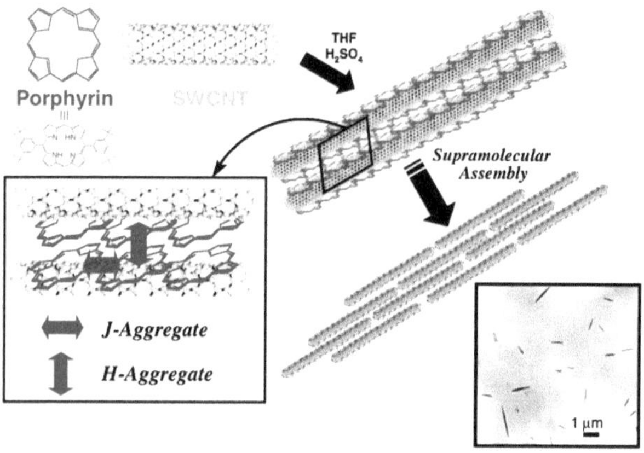

Figure 10.17. Illustration of supramolecular assembly between protonated porphyrins and SWCNT in this study. Reprinted with permission from [77].

to produce supramolecular assemblies in the form of macroscopic bundles (Fig. 10.17).

The general strategy and procedure of constructing supramolecular assembly between 5,15-bis(3,5-di-*tert*-butylphenyl)porphyrin [$H_2P(tBu)_2$] and SWCNTs is shown in Fig. 10.11. One milligram purified SWCNTs were added to 0.2 mM solution of porphyrin in THF containing 1.0% H_2SO_4 (v/v). The suspension was sonicated for 1 h at room temperature, followed by centrifugation of the suspension for 30 min at 10,000 rpm to separate the SWCNTs-$H_4P(tBu)_2^{2+}$ composite. The solid was resuspended in acidified THF with sonication (1 h) and centrifuged again to wash out any unbound porphyrin. The first step involves binding of $H_2P(tBu)_2$ on the surface of SWCNTs. This is followed by Van der Waal's interaction to form aligned macroscopic rods. Protonated form of $H_2P(tBu)_2$ [$H_4P(tBu)_2^{2+}$] and SWCNT composites form 0.5–3.0 μm sized rod-like structures [77].

These rod-assemblies can be assembled onto nanostructured SnO_2 films (OTE/SnO_2) using similar electrophoretic deposition method (Fig. 10.18). These organized assemblies are photoactive and absorb strongly in the entire visible region. The IPCE of OTE/SnO_2/SWCNTs-$H_4P(tBu)_2^{2+}$ is ~13% under an applied potential of 0.2 V vs. SCE [78]. This value is much larger than those

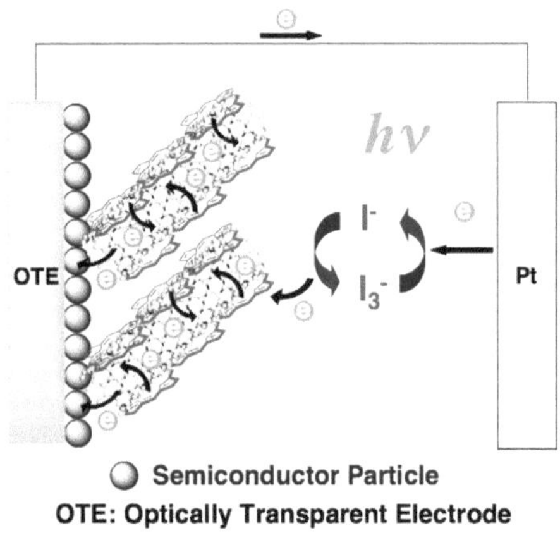

Figure 10.18. Illustration of the photoelectrochemical solar cell of supramolecular assemblies composed of protonated porphyrins and single-wall carbon nanotubes (SWCNT). Reprinted with permission from [78].

of the respctive single-component systems. Thus, the dual role of SWCNT in promoting photoinduced charge separation and facilitating charge transport is presented.

10.5.2 *Stacked-Cup Carbon Nanotubes*

Where as conventional carbon nanotubes are made up of seamless cylinders of hexagonal carbon network, the cup-stacked structure provides a hollow tubular morphology (Fig. 10.19) [79, 80]. This stacked-cup morphology provides a large portion of exposed and reactive edges in the outer and inner surfaces of the hollow tubes. The availability of inner and outer edges of these stacked-cups to chemical functionalization or surface modification opens up new avenues to utilize them in electronic and catalytic applications. Upon ball-milling the longer stacked-cups can be broken into smaller stacked-cups and further increase the number of accessible active sites [81]. These stacked-cup carbon nanotubes (SCCNT) films on OTE/SnO_2 undergo charge separation and deliver photocurrent under visible light irradiation. A maximum IPCE of 17% at +0.2 V vs. SCE is much larger than those reported for pristine carbon nanostructures. For example, C_{60} films exhibit a maximum IPCE of ~5% and SWCNT films exhibit a maximum IPCE less than 0.2% [82].

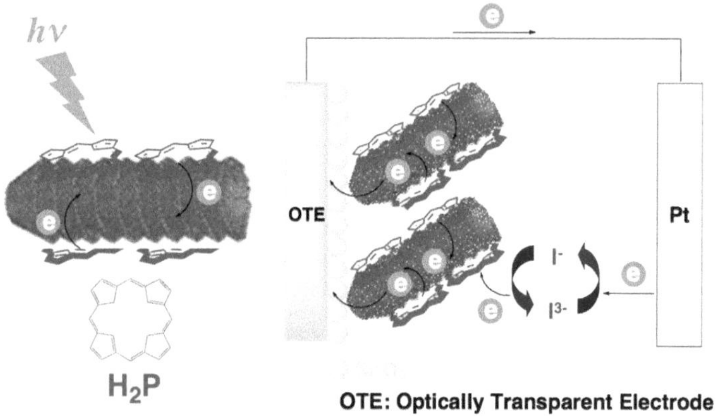

Figure 10.19. Illustration and SEM image of stacked-cup carbon nanotubes (SCCNT) and the application to the photoelectrochemical cell in this study. Reprinted with permission from [46].

Moreover, molecular assemblies composed of SCCNT and 5,15-bis(3,5-di-*tert*-butylphenyl)porphyrin [$H_2P(tBu)_2$] prepared in acetonitrile/toluene (5/1, v/v) were deposited as three-dimensional arrays onto nanostructured SnO_2 films to further improve the photoelectrochemical response in the visible. The composite electrode of SCCNT-$H_2P(tBu)_2$ (OTE/SnO_2/SCCNT-$H_2P(tBu)_2$) exhibits IPCE of 32% under an applied potential of 0.2 V vs. SCE [46]. The observed increase in IPCE is greater than that of the additive effects observed from the single-component systems (viz., OTE/SnO_2/SCCNT or OTE/SnO_2/$H_2P(tBu)_2$).

To check the interaction between $H_2P(tBu)_2$ and SCCNT, the steady-state fluorescence spectra of OTE/SnO_2/SCCNT– $H_2P(tBu)_2$ film (spectrum *a*) and OTE/SnO_2/$H_2P(tBu)_2$ film (spectrum *b*) were measured (Fig. 10.20A). The excitation wavelength was at 410 nm which facilitated the excitation of the soret $H_2P(tBu)_2$ band. The $H_2P(tBu)_2$ assembly film exhibits twin peak with maximum at 655 and 710 nm. The fluorescence intensity of $H_2P(tBu)_2$ film deposited on SCCNT exhibited significantly lower emission yield, thus confirming the quenched excited state.

Taking into consideration efficient photocurrent generation of OTE/SnO_2/SCCNT–$H_2P(tBu)_2$ electrode, this quenching property is largely ascribed to photoinduced electron transfer from the singlet

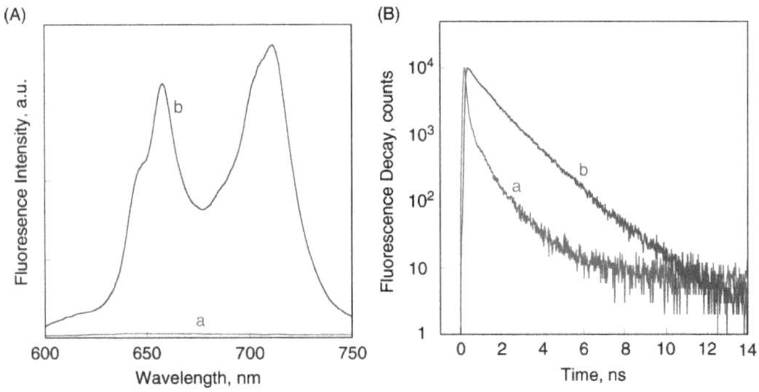

Figure 10.20. (A) Steady-state fluorescence spectra of (a) OTE/SnO_2/SCCNT- $H_2P(tBu)_2$ and (b) OTE/SnO_2/$H_2P(tBu)_2$. Excitation wavelength is 410 nm. (B) Fluorescence decay curves of (a) OTE/SnO_2/SCCNT–$H_2P(tBu)_2$ and (b) OTE/SnO_2/ $H_2P(tBu)_2$ observed at 710 nm by single photon counting method. Excitation wavelength is 373 nm. Reprinted with permission from [46].

excited state of $H_2P(tBu)_2$ to SCCNT (Eqs. 10. 6 and 10. 7).

$$H_2P(tBu)_2 + h\nu \rightarrow {}^1H_2P(tBu)_2* \qquad (10.6)$$

$${}^1H_2P(tBu)_2 * + SCCNT \rightarrow {}^1H_2P(tBu)_2^{\bullet+} + SCCNT(e) \qquad (10.7)$$

The measured fluorescence lifetime of OTE/SNO_2/SCCNT–$H_2P(tBu)_2$ with that of OTE/SnO_2/$H_2P(tBu)_2$ (Fig. 10.20B) provides further insight into the kinetics of electron-transfer process. The emission decay of $H_2P(tBu)_2$ assembly film can be fitted to biexponential decay kinetics. The varying degree of aggregation is likely to influence the decay behavior of the excited porphyrin. The two lifetime components of the OTE/SnO_2/$H_2P(tBu)_2$ film were 1.26

Table 10.4. Comparison of Fluorescence Lifetimes of OTE/SnO_2/$H_2P(tBu)_2$–SCCNT and OTE/SnO_2/$H_2P(tBu)_2$

	τ_1	τ_2	τ_3
$H_2P(tBu)_2$–SCCNT	119 ps (48%)	797 ps (38%)	2.36 ns (14%)
$H_2P(tBu)_2$	—	1.26 ns (48%)	2.35 ns (52%)

Source: Reprinted with permission from [46].
Note: The lifetimes were obtained from the analysis of the emission decay using multiexponential kinetics expression of the form $I(t) = \sum A_n \exp(-t/\tau_n)$.

ns and 2.35 ns, respectively (Table 10.4). The excited $H_2P(tBu)_2$ in $OTE/SnO_2/SCCNT$– $H_2P(tBu)_2$ electrode deactivates faster than in $OTE/SnO_2/H_2P(tBu)_2$ electrode. The emission decay was analyzed using tri-exponential kinetic fit and the lifetimes are summarized in Table 10.4. An additional fast decay component (120 ps) was evident in the $H_2P(tBu)_2$ film coated on SCCNT. Nearly 50% of the decay occurred via this fast deactivation pathway. If we attribute this fast decay component of 0.12 ns to the electron-transfer pathway of excited state deactivation, we obtain the rate constant for electron transfer as 8×10^9 s^{-1}. It should be noted that this value is an upper estimate and other slower components may also contribute to the electron-transfer process.

In addition to the above SWCNT- or SCCNT-based systems, we have also reported carbon nanohorn–porphyrin [45] and CdS-MWCNT composites [83]. Thus, organization of carbon nanotubes with dye molecules such as porphyrins demonstrates efficient light energy conversion properties.

10.6 Concluding Remarks

In this chapter, recent developments of supramolecular assemblies of porphyrins for optoelectronics (esp. photovoltaics) are widely described. In this strategy, utilization of supramolecular techniques enables us to organize molecular aggregates, which possess photovoltaic fuctionalities such as light-harvest (and exciton diffusion), charge separation, and carrier transport. This method is in sharp contrast with the general top–down approaches such as vacuum deposition and spin-coat techniques.

The major problem is a random orientation of molecular assemblies on the electrode surface, which hamper the efficient carrier transport and interfacial control between organic layer and electrode. Therefore, one of the new useful strategies for the improvement of the conversion property may be precise alignment of these molecular assemblies onto the surface electrode using patterning technique. Finally, by overcoming these problems together with development of new molecular architectures, such supramolecular strategies could pave the way for developing future electronics and light energy conversion systems.

Acknowledgement

The author thanks and expresses gratitude to his collaborators and coworkers, whose names appear in the references. Figures, schemes, and tables are reproduced from the related references with permission of the American Chemical Society, the Royal Society of Chemistry, and Wiley-VCH.

References

1. Jacobs, H. O., Tao, A. R., Schwartz, A., Gracias, D. H. and Whitesides, G. M. (2002) Fabrication of a cylindrical display by patterned assembly, *Science*, **296**, 323–325.
2. Whitesides, G. M., and Boncheva, M. (2002) Beyond molecules: Self-assembly of mesoscopic and macroscopic components, *Proc. Natl. Acad. Sci. U. S. A.*, **99**, 4769–4774.
3. Lehn, J.-M., (1990) Perspectives in supramolecular chemistry†-†from molecular recognition towards molecular information processing and self-organization, *Angew. Chem., Int. Ed. Engl.*, **29**, 1304–1319.
4. Philp, D., and Stoddart, J. F. (1996) Self-assembly in natural and unnatural systems, *Angw. Chem. Int. Ed. Engl.*, **35**, 1154–1196.
5. Hasobe, T. (2010) Supramolecular nanoarchitectures for light energy conversion, *Phys. Chem. Chem. Phys.*, **12**, 44–57.
6. Gust, D., and Moore, T. A. (1989) Mimicking photosynthesis, *Science*, **244**, 35–41.
7. Fukuzumi, S. (2006) Bioinspired electron-transfer systems and applications, *Bull. Chem. Soc. Jpn.*, **79**, 177–195.
8. Imahori, H., Tamaki, K., Guldi, D. M., Luo, C., Fujitsuka, M., Ito, O., Sakata, Y., and Fukuzumi, S. (2001) Modulating charge separation and charge recombination dynamics in porphyrin-fullerene linked dyads and triads: marcus-normal versus inverted region, *J. Am. Chem. Soc.*, **123**, 2607–2617.
9. Hu, X., Ritz, T., and Schulten, K. (1998) Architecture and mechanism of the light-harvesting apparatus of purple bacteria, *PNAS*, **95**, 5935–5941.
10. McDermott, G., Prince, S. M., Freer, A. A., Hawthornthwaite-Lawless, A. M., Papiz, M. Z., Cogdell, R. J., and Isaacs, N. W. (1995) Crystal structure of an integral membrane light-harvesting complex from photosynthetic bacteria, *Nature*, **374**, 517–527.

11. Deisenhofer, J., and Michel, H. (1989) The photosynthetic reaction center from the purple bacterium Rhodopseudomonas viridis, *Science*, **245**, 1463–1473.
12. Gust, D., Moore, T. A., and Moore, A. L. (1993) Molecular mimicry of photosynthetic energy and electron transfer, *Acc. Chem. Res.*, **26**, 198–205.
13. Wasielewski, M. R. (2009) Self-assembly strategies for integrating light harvesting and charge separation in artificial photosynthetic systems, *Acc. Chem. Res.*, **42**, 1910–1921.
14. Imahori, H., Yamada, H., Guldi, D. M., Endo, Y., Shimomura, A., Kundu, S., Yamada, K., Okada, T., Sakata, Y., and Fukuzumi, S. (2002) Comparison of reorganization energies for intra- and intermolecular electron transfer, *Angew. Chem. Int. Ed.*, **41**, 2344–2347.
15. Araki, Y., and Ito, O. (2008) Factors controlling lifetimes of photoinduced charge-separated states of fullerene-donor molecular systems, *J. Photochem. Photobiol. C: Photochem. Rev.*, **9**, 93–110.
16. Gust, D., Moore, T. A., and Moore, A. L. (2001) Mimicking photosynthetic solar energy transduction, *Acc. Chem. Res.*, **34**, 40–48.
17. Drain, C. M., Smeureanu, G., Patel, S., Gong, X., Garnod, J., and Arijeloyea, J. (2006) Porphyrin nanoparticles as supramolecular systems, *New J. Chem.*, **30**, 1834–1843.
18. Gong, X., Milic, T., Xu, C., Batteas, J. D., and Drain, C. M. (2002) Preparation and characterization of porphyrin nanoparticles, *J. Am. Chem. Soc.*, **124**, 14290–14291.
19. Drain, C. M., Varotto, A., and Radivojevic, I. (2009) Self-organized porphyrinic materials, *Chem. Rev.*, **109**, 1630–1658.
20. Radivojevic, I., Likhtina, I., Shi, X., Singh, S., and Drain, C. M. (2010) Self-organized nanofibers and nanorods of porphyrins bearing hydrogen bonding motifs, *Chem. Commun.*, **46**, 1643–1645.
21. Medforth, C. J., Wang, Z., Martin, K. E., Song, Y., Jacobsen, J. L., and Shelnutt, J. A. (2009) Self-assembled porphyrin nanostructures, *Chem. Commun.*, 7261–7277.
22. Wang, Z., Ho, K. J., Medforth, C. J., and Shelnutt, J. A. (2006) Porphyrin nanofiber bundles from phase-transfer ionic self-assembly and their photocatalytic self-metallization, *Adv. Mater.*, **18**, 2557–2560.
23. Wang, H., Song, Y., Medforth, C. J., and Shelnutt, J. A. (2006) Interfacial synthesis of dendritic platinum nanoshells templated on benzene nanodroplets stabilized in water by a photocatalytic lipoporphyrin, *J. Am. Chem. Soc.*, **128**, 9284–9285.

24. Wang, Z., Li, Z., Medforth, C. J., and Shelnutt, J. A. (2007) Self-assembly and self-metallization of porphyrin nanosheets, *J. Am. Chem. Soc.*, **129**, 2440–2441.
25. Martin, K. E., Wang, Z., Busani, T., Garcia, R. M., Chen, Z., Jiang, Y., Song, Y., Jacobsen, J. L., Vu, T. T., Schore, N. E., Swartzentruber, B. S., Medforth, C. J., and Shelnutt, J. A. (2010) Donor–acceptor biomorphs from the ionic self-assembly of porphyrins, *J. Am. Chem. Soc.*, **132**, 8194–8201.
26. Wang, Z., Medforth, C. J., and Shelnutt, J. A. (2004) Self-metallization of photocatalytic porphyrin nanotubes, *J. Am. Chem. Soc.*, **126**, 16720–16721.
27. Wang, Z., Medforth, C. J., and Shelnutt, J. A. (2004) Porphyrin nanotubes by ionic self-assembly, *J. Am. Chem. Soc.*, **126**, 15954–15955.
28. Lee, S. J., Hupp, J. T., and Nguyen, S. T. (2008) Growth of narrowly dispersed porphyrin nanowires and their hierarchical assembly into macroscopic columns, *J. Am. Chem. Soc.*, **130**, 9632–9633.
29. Lee, S. J., Mulfort, K. L., Zuo, X., Goshe, A. J., Wesson, P. J., Nguyen, S. T., Hupp, J. T., and Tiede, D. M. (2008) Coordinative self-assembly and solution-phase x-ray structural characterization of cavity-tailored porphyrin boxes, *J. Am. Chem. Soc.*, **130**, 836–838.
30. Lee, S. J., Jensen, R. A., Malliakas, C. D., Kanatzidis, M. G., Hupp, J. T., and Nguyen, S. T. (2008) Effect of secondary substituent on the physical properties, crystal structures, and nanoparticle morphologies of (porphyrin)Sn(OH)2: diversity enabled via synthetic manipulations, *J. Mater. Chem.*, **18**, 3640–3642.
31. Harada, R., Matsuda, Y., Okawa, H., and Kojima, T. (2004) A porphyrin nanotube: size-selective inclusion of tetranuclear molybdenum–oxo clusters, *Angew. Chem., Int. Ed.*, **43**, 1825–1828.
32. Fukuzumi, S., and Kojima, T. (2008) Photofunctional nanomaterials composed of multiporphyrins and carbon-based p-electron acceptors, *J. Mater. Chem.*, **18**, 1427–1439.
33. Hu, J. S., Guo, Y. G., Liang, H. P., Wan, L. J., and Jiang, L. (2005) Three-dimensional self-organization of supramolecular self-assembled porphyrin hollow hexagonal nanoprisms, *J. Am. Chem. Soc.*, **127**, 17090–17095.
34. Wang, Z., Lybarger, L. E., Wang, W., Medforth, C. J., Miller, J. E., and Shelnutt, J. A. (2008) Monodisperse porphyrin nanospheres synthesized by coordination polymerization, *Nanotechnology*, **19**, 395–604.
35. Floris, H., Cameron, C. L., Marko, M. L. N., Jeroen, C. G., Peter, C. M. C., Antje, L., George, F., Philippe, E. L. G. L., Albertus, P. H. J. S., and Meijer,

E. W. (2010) Dilution-induced self-assembly of porphyrin aggregates: a consequence of coupled equilibria, *Angew. Chem. Int. Ed.*, **49**, 3939–3942.

36. Ozawa, H., Tanaka, H., Kawao, M., Uno, S., and Nakazato, K. (2009) Preparation of organic nanoscrews from simple porphyrin derivatives, *Chem. Commun.*, 7411–7413.
37. Elemans, J. A. A., Hameren, W., R. V., Nolte, R. J. M., and Rowan, A. E. (2006) Molecular materials by self-assembly of porphyrins, phthalocyanines, and perylenes, *Adv. Mater.*, **18**, 1251–1266.
38. van Hameren, R., Schon, P., van Buul, A. M., Hoogboom, J., Lazarenko, S. V., Gerritsen, J. W., Engelkamp, H., Christianen, P. C. M., Heus, H. A., Maan, J. C., Rasing, T., Speller, S., Rowan, A. E., Elemans, J. A. A. W., and Nolte, R. J. M. (2006) Macroscopic hierarchical surface patterning of porphyrin trimers via self-assembly and dewetting, *Science*, **314**, 1433–1436.
39. Lensen, M. C., Elemans, J. A. A. W., Dingenen, S. J. T. V., Gerritsen, J. W., Speller, S., Rowan, A. E., and Nolte, R. J. M. (2007) Giant porphyrin disks: control of their self-assembly at liquid-solid interfaces through metal-ligand interactions, *Chem. Eur. J.*, **13**, 7948–7956.
40. Pieter, A., de Witte, J., Castriciano, M., Cornelissen, J. J. L. M., Scolaro, L. M., Nolte, R. J. M., and Rowan, A. E. (2003) Helical polymer-anchored porphyrin nanorods, *Chem. Eur. J.*, **9**, 1775–1781.
41. Lensen, M. C., Takazawa, K., Elemans, J. A. A. W., Jeukens, C. R. L. P. N., Christianen, P. C. M., Maan, J. C., Rowan, A. E., and Nolte, R. J. M. (2004) Aided self-assembly of porphyrin nanoaggregates into ring-shaped architectures, *Chem. Eur. J.*, **10**, 831–839.
42. Kay, A., and Graetzel, M. (1993) Artificial photosynthesis. 1. Photosensitization of titania solar cells with chlorophyll derivatives and related natural porphyrins, *J. Phys. Chem.*, **97**, 6272–6277.
43. Gust, D., Moore, T. A., and Moore, A. L. (2009) Solar fuels via artificial photosynthesis, *Acc. Chem. Res.*, **42**, 1890–1898.
44. Fukuzumi, S., Endo, Y., Kashiwagi, Y., Araki, Y., Ito, O., and Imahori, H. (2003) Novel photocatalytic function of porphyrin-modified gold nanoclusters in comparison with the reference porphyrin compound, *J. Phys. Chem. B*, **107**, 11979–11986.
45. Pagona, G., Sandanayaka, A. S. D., Hasobe, T., Charalambidis, G., Coutsolelos, A. G., Yudasaka, M., Iijima, S., and Tagmatarchis, N. (2008) Characterization and photoelectrochemical properties of nanostructured thin film composed of carbon nanohorns covalently functionalized with porphyrins, *J. Phys. Chem. C*, **112**, 15735–15741.

46. Hasobe, T., Murata, H., and Kamat, P. V. (2007) Photoelectrochemistry of stacked-cup carbon nanotube films. tube-length dependence and charge transfer with excited porphyrin, *J. Phys. Chem. C*, **111**, 16626–16634.

47. Hasobe, T., Fukuzumi, S., Hattori, S., and Kamat, P. V. (2007) Shape- and functionality-controlled organization of tio2-porphyrin-c60 assemblies for improved performance of photochemical solar cells, *Chem. Asian J.*, **2**, 265–272.

48. Hasobe, T., Imahori, H., Kamat, P. V., Ahn, T. K., Kim, S. K., Kim, D., Fujimoto, A., Hirakawa, T., and Fukuzumi, S. (2005) Photovoltaic cells using composite nanoclusters of porphyrins and fullerenes with gold nanoparticles, *J. Am. Chem. Soc.*, **127**, 1216–1228.

49. Boyd, P. D. W., and Reed, C. A. (2005) Fullerene-porphyrin constructs, *Acc. Chem. Res.*, **38**, 235–242.

50. Tashiro, K., and Aida, T. (2007) Metalloporphyrin hosts for supramolecular chemistry of fullerenes, *Chem. Soc. Rev.*, **36**, 189–197.

51. Horowitz, G., and Hajlaoui, M. E. (2000) Mobility in polycrystalline oligothiophene field-effect transistors dependent on grain size, *Adv. Mater.*, **12**, 1046–1050.

52. Di Carlo, A., Piacenza, F., Bolognesi, A., Stadlober, B., and Maresch, H. (2005) Influence of grain sizes on the mobility of organic thin-film transistors, *Appl. Phys. Lett.*, **86**, 263501–263503.

53. Sandanayaka, A. S. D., Araki, Y., Wada, T., and Hasobe, T. (2008) Structural and photophysical properties of self-assembled porphyrin nanoassemblies organized by ethylene glycol derivatives, *J. Phys. Chem. C*, **112**, 19209–19216.

54. Yokoi, K., and Ohba, Y. (1987) Thermal effects on triplet exciton processes in anthracene crystals, *J. Chem. Phys.*, **86**, 3318–3324.

55. Yago, T., Tamaki, Y., Furube, A., and Katoh, R. (2008) Self-trapping limited exciton diffusion in a monomeric perylene crystal as revealed by femtosecond transient absorption microscopy, *Phys. Chem. Chem. Phys.*, **10**, 4435–4441.

56. Valkunas, L., Åkesson, E., Pullerits, T., and Sundström, V. (1996) Energy migration in the light-harvesting antenna of the photosynthetic bacterium Rhodospirillum rubrum studied by time-resolved excitation annihilation at 77 K, *Biophy. J.*, **70**, 2373–2379.

57. Hasobe, T., Imahori, H., Fukuzumi, S., and Kamat, P. V. (2003) Nanostructured assembly of porphyrin clusters for light energy conversion, *J. Mater. Chem.*, **13**, 2515–2520.

58. Hasobe, T., Imahori, H., Fukuzumi, S., and Kamat, P. V. (2003) Light energy harvesting using mixed molecular nanoclusters. porphyrin and C60 cluster films for efficient photocurrent generation, *J. Phys. Chem. B*, **107**, 12105–12112.
59. Hasobe, T., Saito, K., Kamat, P. V., Troiani, V., Qiu, H., Solladié, N., Kim, K. S., Park, J. K., Kim, D., D'Souza, F., and Fukuzumi, S. (2007) Organic solar cells. Supramolecular composites of porphyrins and fullerenes organized by polypeptide structures as light harvesters, *J. Mater. Chem.*, **17**, 4160–4170.
60. Hasobe, T., Kamat, P. V., Troiani, V., Solladie, N., Ahn, T. K., Kim, S. K., Kim, D., Kongkanand, A., Kuwabata, S., and Fukuzumi, S. (2005) Enhancement of light-energy conversion efficiency by multi-porphyrin arrays of porphyrin-peptide oligomers with fullerene clusters, *J. Phys. Chem. B*, **109**, 19–23.
61. Hasobe, T., Imahori, H., Fukuzumi, S., and Kamat, P. V. (2003) Quaternary self-organization of porphyrin and fullerene units by clusterization with gold nanoparticles on sno2 electrodes for organic solar cells, *J. Am. Chem. Soc.*, **125**, 14962–14963.
62. Hasobe, T., Kashiwagi, Y., Absalom, M. A., Hosomizu, K., Crossley, M. J., Imahori, H., Kamat, P. V., and Fukuzumi, S. (2004) Supramolecular photovoltaic cells of porphyrin dendrimers and fullerene, *Adv. Mater.*, **16**, 975–979.
63. Hasobe, T., Kamat, P. V., Absalom, M. A., Kashiwagi, Y., Sly, J., Crossley, M. J., Hosomizu, K., Imahori, H., and Fukuzumi, S. (2004) Supramolecular photovoltaic cells based on composite molecular nanoclusters: dendritic porphyrin and C60, porphyrin dimer and C60, and porphyrin-C60 dyad, *J. Phys. Chem. B*, **108**, 12865–12872.
64. Hasobe, T., Oki, H., Sandanayaka, A. S. D., and Murata, H. (2008) Sonication-assisted supramolecular nanorods of meso-diaryl-substituted porphyrins, *Chem. Commun.*, 724–726.
65. Hasobe, T., Sandanayaka, A. S. D., Wada, T., and Araki, Y. (2008) Fullerene-encapsulated porphyrin hexagonal nanorods. An anisotropic donor-acceptor composite for efficient photoinduced electron transfer and light energy conversion, *Chem. Commun.*, 3372–3374.
66. Sandanayaka, A. S. D., Murakami, T., and Hasobe, T. (2009) Preparation and photophysical and photoelectrochemical properties of supramolecular porphyrin nanorods structurally controlled by encapsulated fullerene derivatives, *J. Phys. Chem. C*, **113**, 18369–18378.
67. Verbiest, T., Elshocht, S. V., Kauranen, M., Hellemans, L., Snauwaert, J., Nuckolls, C., Katz, T. J., and Persoons, A. (1998) Strong enhancement of

nonlinear optical properties through supramolecular chirality, *Science*, **282**, 913–915.

68. Percec, V., Glodde, M., Bera, T. K., Miura, Y., Shiyanovskaya, I., Singer, K. D., Balagurusamy, V. S. K., Heiney, P. A., Schnell, I., Rapp, A., Spiess, H. W., Hudson, S. D., and Duan, H. (2002) Self-organization of supramolecular helical dendrimers into complex electronic materials, *Nature*, **417**, 384–387.
69. Deegan, R. D., Bakajin, O., Dupont, T. F., Huber, G., Nagel, S. R., and Witten, T. A. (1997) Capillary flow as the cause of ring stains from dried liquid drops, *Nature*, **389**, 827–829.
70. Wu, J., Grimsdale, A. C., and Müllen, K. (2005) Combining one-, two- and three-dimensional polyphenylene nanostructures, *J. Mater. Chem.*, **15**, 41–52.
71. Schmidt-Mende, L., Fechtenkötter, A., Müllen, K., Moons, E., Friend, R. H., and MacKenzie, J. D. (2001) Self-organized discotic liquid crystals for high-efficiency organic photovoltaics, *Science*, **293**, 1119–1122.
72. Bushby, R. J., and Lozman, O. R. (2002) Discotic liquid crystals 25 years on, *Current Opinion in Colloid & Interface Science*, **7**, 343–354.
73. Laschat, S., Baro, A., Steinke, N., Giesselmann, F., Hägele, C., Scalia, G., Judele, R., Kapatsina, E., Sauer, S., Schreivogel, A., and Tosoni, M. (2007) Discotic liquid crystals: from tailor-made synthesis to plastic electronics, *Angew. Chem. Int. Ed.*, **46**, 4832–4887.
74. Hasobe, T., Rabbani, M. G., Sandanayaka, A. S. D., Sakai, H., and Murakami, T. (2010) Synthesis and aggregate formation of triphenylene core-centered porphyrin hexamers, *Chem. Commun.*, **46**, 889–891.
75. Kumar, S., Manickam, M., Varshney, S. K., Rao, D. S. S., and Prasad, S. K. (2000) Novel heptasubstituted triphenylene discotic liquid crystals, *J. Mater. Chem.*, **10**, 2483–2489.
76. Xiao-Shuang, F., and Cai-Yuan, P. (2002) A cylindrical polymer: a stack of discotic triphenylene within comblike polymethacrylate, *ChemPhysChem*, **3**, 539–543.
77. Hasobe, T., Fukuzumi, S., and Kamat, P. V. (2005) Ordered assembly of protonated porphyrin driven by single wall carbon nanotubes. J- and H-aggregates to nanorods, *J. Am. Chem. Soc.*, **127**, 11884–11885.
78. Hasobe, T., Fukuzumi, S., and Kamat, P. V. (2006) Organized assemblies of single wall carbon nanotubes and porphyrin for photochemical solar cells: charge injection from excited porphyrin into single-walled carbon nanotubes, *J. Phys. Chem. B*, **110**, 25477–25484.

79. Endo, M., Kim, Y. A., Hayashi, T., Yanagisawa, T., Muramatsu, H., Ezaka, M., Terrones, H., Terrones, M., and Dresselhaus, M. S. (2003) Microstructural changes induced in "stacked cup" carbon nanofibers by heat treatment, *Carbon*, **41**, 1941–1947.
80. Endo, M., Kim, Y. A., Hayashi, T., Fukai, Y., Oshida, K., Terrones, M., Yanagisawa, T., Higaki, S., and Dresselhaus, M. S. (2002) Structural characterization of cup-stacked-type nanofibers with an entirely hollow core, *Appl. Phys. Lett.*, **80**, 1267–1269.
81. Kim, Y. A., Hayashi, T., Fukai, Y., Endo, M., Yanagisawa, T., and Dresselhaus, M. S. (2002) Effect of ball milling on morphology of cup-stacked carbon nanotubes, *Chem. Phys. Lett.*, **355**, 279–284.
82. Hasobe, T., Fukuzumi, S., and Kamat, P. V. (2006) Stacked-cup carbon nanotubes for photoelectrochemical solar cells, *Angew. Chem. Int. Ed.*, **45**, 755–759.
83. Mountrichas, G., Sandanayaka, A. S. D., Economopoulos, S. P., Pispas, S., Ito, O., Hasobe, T., and Tagmatarchis, N. (2009) Photoinduced electron transfer in aqueous carbon nanotube/block copolymer/CdS hybrids: application in the construction of photoelectrochemical cells, *J. Mater. Chem.*, **19**, 8990–8998.

Chapter 11

Vectorial Photoinduced Charge Transfer in Langmuir-Blodgett Films of Porphyrin-Based Donor-Acceptor Systems

Nikolai V. Tkachenko and Helge Lemmetyinen
Department of Chemistry and Bioengineering,
Tampere University of Technology,
P.O. Box 541, FIN-33100 Tampere, Finland
nikolai.tkachenko@tut.fi

Langmuir-Blodgett (LB) method can be used to fabricate molecular films with uniform orientation of molecules. This property can be used to achieve vectorial photoinduced electron transfer (PET) across organic film by employing specially designed donor-acceptor (DA) molecules. A series of porphyrin-based DA compounds was synthesized and studied in different LB film arrangements. The porphyrin-fullerene dyads with a face-to-face orientation of the donor and acceptor and a short inter-chromophore distance have demonstrated a fast (<200 fs) and efficient PET, and a long lifetime of the charge-separated state (>100 ns). The latter was achieved due to tight packing of the dyads, which permits charge migration between the donors and acceptors, thus resulting in formation of intramolecular charge-separated states in the layer. The long-living

Multiporphyrin Arrays: Fundamentals and Applications
Edited by Dongho Kim

www.panstanford.com

charges can be further utilized by adding secondary donor and acceptor layers also deposited by LB method.

11.1 Introduction

Porphyrins are one of the key compounds in mimicking photoinduced electron transfer (PET) reaction in natural photosynthesis, and they were actively involved in synthesis of model donor-acceptor (DA) dyads, triads, and more complex compounds [1–3]. The main benefits of using porphyrins in chemical engineering of organic DA systems are (1) well-developed synthetic methods providing variety of roots for chemical engineering, (2) strong absorption in the visible part of the spectrum, which is the key requirements for the photo-initiation of reactions, and (3) reasonably low oxidation potential compatible to that of chlorophylls, which justify combining porphyrin donors with well-known acceptors such as quinones and fullerenes. A great deal of information is available now on mechanisms of the PET reactions based on studies of such DA compounds.

The donor-acceptor molecules show the principle how the charge transfer can be driven by the light in molecular systems. The next step in utilization of this useful property is assembling the functional compounds into relatively large molecular systems which are capable of unidirectional charge transfer under light illumination. The ultimate goal for this system is to achieve a long distance photo-driven charge transfer without loss in efficiency. The most frequently cited applications for this molecular systems are photovoltaics [4, 5] and molecular electronics [6].

There are many methods available which can be used to assemble organic molecules, such as porphyrins, into predefined molecular structures. These are e.g., thermal evaporation, SAMs [7], electrostatic layer-by-layer deposition [8] and Langmuir-Blodgett method [9]. In this chapter we discuss applications of Langmuir-Blodgett (LB) and Langmuir-Schäfer (LS) methods to assemble porphyrin base layer structures with vectorical PET function. These methods are relatively simple and versatile affording preparation of mono- and multi-layer films of lipid-like compounds with high degree of ordering of the molecules inside the layers [9]. Important advantages of the LB and LS methods are as follows: (1) the deposition is done in mild condition and does not require compounds with outstanding chemical or thermal stability, (2) rather big molecules can be deposited, e.g., proteins, (3) one can achieve an uniform orientation of molecules in the layer, which is of primary importance for deposition of donor-acceptor compounds, and (4) layers of molecules with difference functionality, e.g., donors and acceptors, can be deposited in desired sequence to form a more complex structures, as is discussed later.

Using the LB method for film deposition one can consider two options for producing solid samples with the desired vectorial PET functions:

(1) donors and acceptors are placed in two separate layers which are deposited one on top of the other, e.g., porphyrin layer deposited on top of fullerene layer; this relies on interlayer ET and requires donor and acceptor compounds suitable for LB or LS film deposition, but the orientation of the molecules in the layer can be rather random;
(2) covalently linked DA dyads can be deposited as a mono-layer with uniform orientation of the donor and acceptor moieties perpendicular to the layer plane; this requires DA compounds which can orient themselves in the layer; the result will be intra-layer charge transfer with electrons and holes appearing at the opposite sides of the layer.

The above-mentioned two approaches can be combined by depositing DA layer and then a layer of secondary donors or acceptors, which would create a film analogue of molecular triad system. Apparently, the layer-by-layer approach can be extended further more to produce organic films with complex structure and enhanced functionality. The attractiveness of the LB method is that well-defined molecular structures can be fabricated using relatively simple building blocks: such as single molecules, e.g., porphyrins or phthalocyanines, dyads, in particular porphyrin-fullerene dyads, and polymers, e.g., polythiophene.

The subject of this chapter is vectorial photoinduced electron transfer in LB layered films of porphyrin derivatives. It includes design of suitable compounds, fabrication of the films, and investigation of the film functionality. The first part of this chapter gives a short overview of LB technology in application to porphyrin films deposition. Then specific methods of electron transfer studies in LB films are examined. After that the impact of placing porphyrins in LB films on their fluorescence properties is discussed. Three following sections are devoted to inter-layer DA systems, monolayers of DA compounds, and combined multilayer structures, respectively. This order reflects the development history in the field. At first the efforts were undertaken to fabricate donor-acceptor bi-layers. It was realized later that a better control over distance and mutual orientation of donor and acceptor are required in order to achieve an efficient charge separation in solid films. The known porphyrin-fullerene dyads were redesigned to be suitable for LB films deposition as well as new dyads were engineered to be specifically used in films. When desired functionality of simple DA layers was achieved, more complex systems including secondary donors and acceptors, and light harvesting molecules were proposed, implemented, and studied.

11.2 Langmuir–Blodgett and Langmuir–Schäfer Films

11.2.1 *Layers on Water Surface*

Originally the Langmuir-Blodgett (LB) and Langmuir-Schäfer (LS) methods were developed for deposition of monomolecular layers of lipid-like molecules. The molecules should be insoluble in water, but must have hydrophilic anchor, which will determine their orientation on the water surface. The process starts with carefully dropping a small amount of the compound dissolved in a solvent non-miscible with water, e.g., chloroform or hexane, onto the water surface. While solvent evaporates, the molecules are spread on the water surface. The amount of molecules should be small enough so that there is enough room on the surface for each molecules (Fig. 11.1). In this state the molecules can freely move on the surface with almost no interaction with each other. This resembles motion and behavior of molecules in gas phase, thus this state of the film is called gas phase.

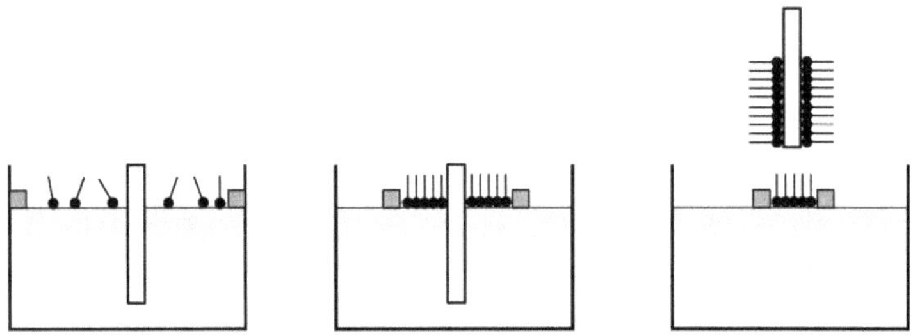

Figure 11.1. Scheme of the LB film deposition: (a) molecules are spread on to the water surface; (b) the layer is compressed to form mono-molecular layer on the water surface; (c) the layer is transferred to a solid substrate by slow lifting the substrate.

The water troughs for the LB and LS film deposition are equipped by flowing barriers, which allow to change smoothly the surface area occupied by the molecules, as schematically presented in Fig. 11.1. After spreading the molecules onto the surface and solvent evaporation, the barriers are moved slowly to decrease the surface area. This process is called compression. To monitor the compression process the surface tension is measured. While the water-free area decreases the surface tension also decreases. This process is usually presented by plotting decrease in the surface tension, which is usually referred to as surface pressure or π, as function of mean molecular area (MMA). A typical series of surface pressure–MMA isotherm is presented in Fig. 11.2 by example of porphyrin-fullerene dyad (DHD, see Fig. 11.7(a)) in octadecylamine (ODA) matrix.

At the beginning, after spreading the molecules over the surface, the density of molecules is relatively low, and they have no or little effect on surface pressure, which in indicated by zero surface pressure at large MMAs. During compression at some value of MMA the surface pressure starts to increase indicating that the lipid-like molecules on water surface are already in close contact with each other similar to molecules in liquids, therefore this state is called liquid or liquid-expended phase, and it corresponds to the state when almost no free water surface area is left (Fig. 11.1). At this stage the surface pressure is increasing sharply with decrease in MMA. For lipids with saturated hydrocarbon tails further compression

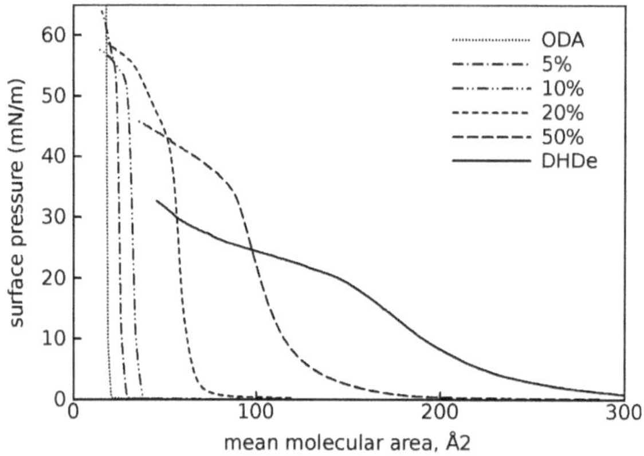

Figure 11.2. Surface pressure (π)—mean molecular area (MMA) isotherms of DHD dyad (see Fig. 11.7(a)) in octadecylamine (ODA) matrix. Molar concentration of the dyad is indicated in the plot.

results in a very sharp rise of the pressure, indicating formation of crystal-like packing of the molecules on the surface. Finally, further compression may lead to slower pressure rise or even a drop of the pressure with following irregular behavior of the isotherm. This happens when the MMA is too small to afford a mono-layer and a bi-layer and more irregular structures are formed.

The best layer deposition conditions are usually achieved at the steepest part of the isotherm. Therefore the layer is compressed till such pressure is reached, and then the surface pressure is maintained constant by moving the barrier in proper direction, i.e., compressing the layer if pressure decreases and expending the area if pressure increases over the desired value. Finally, the layer can be transferred onto a solid substrate, as presented in Fig. 11.1 by the example of LB deposition method.

The Langmuir–Blodgett (LB) and Langmuir–Schäfer (LS) techniques differ in the method of transferring the layer onto the solid substrate. In case of the LB method the substrate is a flat plate, e.g., a quartz plate, and oriented perpendicular to the surface. To deposit the layer the plate is moved slowly through the water-air interface in downward or upward direction. During the process the surface pressure is maintained constant by moving the barriers. This method is also called vertical dipping. The film is deposited on both surfaces of the plate. During down stroke deposition the orientation of the lipid-like compounds in the deposited film is with hydrophilic groups on the outer side of the film, and in case of up stroke deposition the hydrophilic groups are at the inner side of the deposited layer. Therefore the orientation of the molecules in the layer is switched by changing the deposition direction. This is an important property of the LB methods since it allows us to deposit layers of donor-acceptor compounds with predefined orientation in respect to the substrate, as is shown later.

In the case of the Langmuir–Schäfer method the substrate must have only one flat surface. The substrate surface is oriented parallel to the water surface and it is gently brought in contact with the surface and then withdrawn from the surface with layer attached to it if the layer adhesion to the substrate surface is stronger than that to the water. This method is also called horizontal dipping.

In both cases the deposition takes place only when the properties of the surface afford specific interaction between the surface and the layer. For the LS method and downward deposition direction of the LB method this implies that the surface must be hydrophobic, but in the case of the upward LB deposition the surface must be hydrophilic, though these are not the only important conditions to achieve successful deposition.

Deposition of the monolayers can be repeated to grow a multilayer LB or LS films. Also different functional layers can be combined in a specific order to gain new functionality of multilayer films. For example, an electron donor layer can be deposited on top of an electron acceptor layer to achieve a charge separation between the layers.

To monitor film deposition process, the transfer ratio is calculated as the ratio of the substrate area passed through the water interface (in case of LB method) to the decrease of the layer surface area at the water surface at constant surface pressure. Ideally the transfer ratio must be one throughout the whole deposition process. Lower ratio indicates incomplete film transfer from the water surface to the substrate.

Another routine method to verify the film deposition on to transparent substrate is measurement of the films absorption spectrum.

Assuming that the molar absorption, ε, of the compound of interest is known and the absorption spectrum of the compound in film is essentially unchanged, i.e., there is no strong aggregation or aggregation does not change the spectrum significantly, the absorbance of a mono-layer can be calculated as

$$A = \frac{\varepsilon}{sN_A} \tag{11.1}$$

where s is the mean molecular area of chromophores in the layer,[1] and N_A is the Avogadro constant. If units of s and ε are nm^2 and $\text{M}^{-1}\text{cm}^{-1}$, respectively, then

$$A \simeq 1.66 \times 10^{-7}\frac{\varepsilon}{s} \tag{11.2}$$

For example, absorbance of a porphyrin layer with mean molecular area $s = 1\ \text{nm}^2$ at the Soret band ($\varepsilon = 4 \times 10^5\ \text{M}^{-1}\text{cm}^{-1}$) is expected to be $A \simeq 0.066$.

Both methods mentioned above allow one to estimate an average coverage of the substrate surface by the monolayer, but do not provide direct information about actual layer structure. More advanced microscopy methods should be used to establish the layer topology.

11.2.2 *Complex Compounds and Mixed Films*

Originally, both the LB and LS methods were used to deposit layers of fatty acids and similar lipids. However the subject of interest of this chapter is organized layers of photoactive functional compounds, in particular porphyrin and phthalocyanine derivatives. The functional compounds can be engineered to be water insoluble and to hold specific hydrophilic side groups, e.g., carboxyl. Such compounds will stay on the water surface with predominant orientation determined by the location of the hydrophilic group, which resembles behavior of lipid, but the mono-layer formation properties of bulky

[1] In case of mixed films, such as porphyrin in stearic acid matrix, s must be the area per porphyrin but not per molecule in the layer. If the total mean molecular area is s_t and molecular fraction of porphyrins is f, then $s = \frac{s_t}{f}$.

molecules with big chromophore are not as good as for e.g., fatty acids or phospholipids. As the result, even though the desired monolayer is formed on the water surface, its transfer to a solid substrate can be a challenge task especially when a multilayer structure with particular order of the functional groups has to be fabricated. Yet another typical problem for densely packed films of porphyrin and phthalocyanine derivatives is aggregation, which causes an efficient quenching of the excited state and thus competes with the photoinduced electron transfer [10, 11].

A widely used approach to improve the layer deposition properties and to reduce the aggregation effect is to mix the functional molecules with lipids which are known to form good LB or LS films. Typical examples of such lipids are stearic acid (SA) and octadecylamine (ODA) [9]. These lipids are usually called matrix molecules, since their role is to form a monolayer in which the functional molecules will be embedded. In some cases it is even possible to use matrix molecules to deposit functional molecules which do not form a mono-layer of their own on the water surface [10, 11].

For mixed films the question of primary importance is miscibility of the functional molecules and matrix molecules in the film. Typically the lipids with saturated hydrocarbon tails form rigid monolayers with regular crystal-like molecular packing on the water surface. These layers are relatively easy to transfer onto a solid substrate and also multilayer film deposition can be achieved by both LB and LS methods. Unfortunately these lipids are mostly immiscible with molecules with shape different from their own shape. As the result domain-like structures are formed with islands of functional molecules in layer of matrix molecules [12]. However, it was shown that this type of structures can be deposited even as multilayer films.

The lipids with unsaturated hydrocarbon tails have better miscibility and can be used to reduce significantly undesired aggregation effects, but deposition of multilayer films of such lipids is not possible in most cases. These lipids do not form rigid monolayers on the water surface, which is beneficial for miscibility but strongly limits their deposition properties. Therefore application of such lipids is limited to deposition of monolayers or requires some special tricks to deposit multilayer films [13].

11.3 Measurements of Charge Transfer Dynamics in Films

11.3.1 *Time Resolved Optical Spectroscopy*

The time scale of photoinduced electron transfer in molecular systems is from fraction of picosecond to nanosecond. This phenomenon is usually studied by ultra-fast time resolved spectroscopy methods, such as femtosecond pump-probe. The same techniques can be used to study ET in LB films if the films can be deposited as repeating multilayer structures. The absorbance of a single molecular layer is usually as small as 0.01 even at Soret band for porphyrins in a supporting matrix molecules, whereas typical absorbance required for pump-probe measurements is 0.5 or higher. Therefore tens of layers have to be deposited to make use of transient absorption measurements to study kinetics of the photoinduced reactions in the film. This limits applicability of the method to study the ET in films.

Fluorescence methods, and in particular time-correlated single photon counting (TCSPC), are known to provide high sensitivity, which can be scaled down to single molecule spectroscopy [14, 15]. TCSPC and steady state spectroscopy in photon counting mode are the methods widely used to study monolayer molecular films, and can be employed to investigate the charge transfer in porphyrin containing films. Unfortunately two practical problems arise when the TCSPC is used. Firstly, the fluorescence methods can provide information on the evolution of the excited state, but cannot discriminate between different mechanisms of the excited state quenching. In other words, the fluorescence decay measurements can be used to estimate the rate of the electron transfer but cannot prove the fact of the electron transfer. The second specific problem of TCSPC method is its limited time resolution. The most of porphyrin-fullerene donor-acceptor systems of interest have time constant of the photoinduced charge separation in the range 0.5–10 ps, which is much shorter than the time resolution of the currently available TCSPC instruments. Therefore, being generally sensitive and accurate the fluorescence methods have to be complemented by other research methods to obtain information on the charge-transfer reactions in films.

11.3.2 *Transient Photo-Voltage and Photo-Current*

As was discussed in Introduction, the ultimate goal for donor-acceptor films is to achieve the vectorial charge transfer, which means, that the light flash triggers charge motion in one direction for all DA pairs in the sample. If such sample is placed between two electrodes, i.e., inside a flat capacitor, this unidirectional charge motion will generate Maxwell displacement charge on the electrodes. This phenomenon can be used to prove the fact of vectorial charge transfer and to study dynamics of the charge motion. In application to LB film studies the method was first developed to monitor photoinduced proton motion in films of bacteriorhodopsin [16]. It has been demonstrated that as small as 0.1 nm proton shift can be detected in a monolayer of uniformly oriented proteins.

Schemes illustrating possible sample structures and the method of the time-resolved Maxwell displacement charge (TRMDC) measurements is presented in Fig. 11.3 Since the aim of the measurements is to follow the charge motion in photoactive organic film, the active layer is protected from both electrodes by insulating layers. In practice this is done by depositing 10–20 layers of stearic acid or similar lipid on top of the electrode prior deposition of the photoactive layer, and 10–20 layers of the same lipid on top of the photoactive layer. One of the electrodes must be light transparent, which is easy achievable by using glass or quartz plates covered with indium tin oxide (ITO) layer, serving as bottom light transparent electrode for the film deposition. The other electrode can be a liquid metal drop, e.g., indium–gallium allow, or evaporated aluminum or gold layer. In the latter case the number of top insulating layers must be at least 20 as part of the layers will be destroyed during the metal electrode deposition.

A similar approach can be used to study any samples with predominant orientation of the donor-acceptor groups. The desired orientation can be achieved by using specially designed molecules which can be deposited as well-organized molecular layers or by fabricating anisotropic layer structures by alternating donor and acceptor layer, as is discussed in this chapter. However other methods can also be used to achieve necessary anisotropy, which would result in vectorial photo-induced charge transfer. For example, in solution the

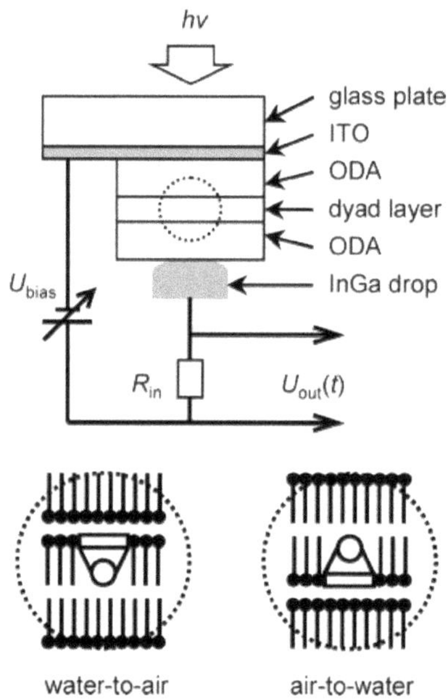

Figure 11.3. Scheme of time resolved Maxwell displacement charge measurements and possible sample structures. The latter indicates the difference in DA dyad orientation depending on the direction of the dyad layer deposition, water-to-air or air-to-water, respectively.

desired predominant orientation can be achieved by applying a high voltage to DA solution before the photo-excitation [17, 18].

The sample can be excited by a short laser flash or a continuous light. The measured signal is the voltage change between the electrodes. The sample is essentially a (flat) capacitor,[2] therefore the measurements can we done in two modes, which will be referred to as transient photo-voltage and photo-current modes. The distinction between the modes is the relation between the time scale of measurements and the time constant of the electric circuit, $\tau = R_{in}C$, where C is the capacitance of the sample (in assumption that the

[2] Ideally there is no current leakage through the insulating layers, which means that the input resistance of the measuring device, R_{in} is much smaller than the resistance of the insulating layer. Therefore the input resistance determines how fast the sample capacitor will discharge.

input capacitance of the measuring device is much smaller than the sample capacitance, C), and R_{in} is the input resistance of the measuring device.

In the time scale much shorter than τ, one can neglect by the discharge of the capacitance C via the resistor R_{in}. The potential at the capacitor is proportional to the capacitor charge, $U = q/C$, and the charge, q, is proportional to the photo-generated charge, Q, and the charge displacement in direction normal to the electrode plane, D, i.e., $q \sim Q D$. Thus the measured signal is the potential, U, which is proportional to the product of the photo-generated charge, Q and the charge displacement, D:

$$U_{PV} \sim Q D \tag{11.3}$$

This measurement mode will be referred to as transient photo-voltage mode. In case of photo-voltage measurements of a DA layer, the distance D is the donor-acceptor distance, i.e., it is constant value, and the measured signal is proportional to the charge Q, i.e., to the number of charge-separated states.

In the time scale much longer than τ the discharge of the capacitor C is fast, and the measured signal is the discharge current. Any motion of charges between the electrodes creates Maxwell displacement charge on the capacitor, which generates current through the resistor R_{in}. The measured signal, U, is proportional to the current, $U_{PC} = I R_{in}$, which, in turn, is proportional to the rate of the charges generation, $I = \frac{dq}{dt}$. As the result, the measured signal is proportional to the change in displacement charge $U_{PC} = R_{in}\frac{dq}{dt}$, thus it is proportional to the derivative of the product of the photo-generated charge and the charge displacement:

$$U_{PC} \sim \frac{d(Q D)}{dt} \tag{11.4}$$

This measurement mode will be referred to as transient photo-current mode.

An intermediate case of the measurement time scale compatible to the time constant τ also can be worked out, but requires some additional calculations [17]. In the following discussion only pure photo-voltage and photo-current will be used since by switching input resistance, R_{in}, the whole time scale can be studied without additional calculations. Typical values of the sample capacitances,

C, were 50–300 pF, and two values of input resistance, R_{in}, were used, 100 MΩ and 10 ΓΩ, which give time constants,τ, around 10 ms and 1 s, respectively. Therefore, it is possible to cover the whole time scale by carrying photo-voltage measurements with $\tau = 1$ s, i.e., in time scale $t < 1$ s, Eq. 11.3, and photo-current measurements with $\tau = 0.01$ s, i.e., in time scale $t > 0.01$ s, Eq. 11.4.

In both measurement modes the product of the charge densities and charge displacement determines the signal intensity. These two values cannot be separated one from another. The absolute value of the product can be calculates in assumption of homogeneous distribution of dielectric constant inside the capacitor. This can be used for rough estimations of the CS efficiency or distance, but validity and accuracy of such estimations need to be discussed in each individual case.

The TRMDC instrument shown in Fig. 11.3 provides also one additional method to probe the nature of the charge photogeneration in films. A static bias potential can be applied to the electrodes. For the films structure presented in Fig. 11.3 a bias of 0.5 V generates rather small disbalance energy for the ET (assuming 0.5 nm distance of the ET the disbalance energy is roughly 0.005 eV, i.e., much smaller than the thermal energy, $k_B T = 0.0026$ eV at room temperature). This disbalance energy has little to no effect on the ET in molecular DA systems, but can change gradually photo-voltage or photo-current response if the signal arises from charges "freely" migrating in the layer, since these charges have much smaller potential barrier for charge hoping [13, 19].

11.3.3 *Data Analysis, Distributed Decay*

When comparing electron transfer in one and the same donor-acceptor dyad but placed in a solid film or in solution one may notice a big difference in dynamics of the ET reactions even when macroscopic properties of the media (dielectric constant and refractive index) are the same. The difference arises from different dynamic behaviors of molecules in liquid and solid states. In solution molecules are in fast (compared with typical time scale of the ET) dynamic equilibrium with the medium, solvent molecules, and the observed behavior corresponds to that of the molecule in some averaged conformational state. In solid state isomerization and rotation do not take place at least in time scale of ET reactions. The latter is particularly important for relatively big and complex molecules such as porphyrin-fullerene dyads, since different conformers can coexist and behave differently, e.g., may have different lifetimes of the charge-separated state.

The difference in lifetimes of individual entities is usually discussed in terms of distributed decay kinetics, meaning that macroscopically observable parameters, such as emission or absorption decay, do not follow a simple exponential dependence even for simple one step spontaneous reactions, but rather presented by superposition of a large number of exponential decay with lifetime varying in a certain range [20, 21]. The practical question is how to extract useful information from experimentally measured time dependences. Two analytical approaches are used in this chapter which allow to estimate average lifetime and gives roughly the same results in case of reasonably narrow distribution width. The first is known as stretched exponential. It was found empirically and is given by the decay function [22]:

$$f(t) = e^{-\left(\frac{t}{\tau_{av}}\right)^{\beta}} \tag{11.5}$$

where τ_{av} is the average lifetime and β is the stretching parameter. When the stretching parameter $\beta = 1$ the decay is purely exponential. Distributed decay is modeled by $\beta < 1$, where smaller values of β correspond to wider distribution. A formally valid range of stretching parameter is $0 < \beta \leqslant 1$, but in practice the range is limited to $0.3 \lesssim \beta \leqslant 1$.

The main drawback of the stretch exponential model is that it has no clear physical model in its basis, and even though the stretching parameter is known to be related to the distribution width, there is no theory available to extract physically meaningful information from the parameter β. Alternative approach is based on the assumption of a narrow Gaussian distribution of the reorganization or Gibbs energy associated with the electron transfer reaction, and uses the Marcus classic electron transfer theory to calculate time dependences of the reactant or product populations [23]. The model operates with three parameters: the energy distribution width, $\triangle E$, the average lifetime, τ_{av}, which is the lifetime observed at $\triangle E = 0$, and a sensitivity parameter, a. The predicted decay function is

$$f(t) = \int_{-X}^{+X} \exp\left(-\frac{x^2}{\triangle E^2}\right) \exp\left(-\frac{t}{\tau_0} e^{\frac{ax}{kT}}\right) dx \tag{11.6}$$

where the integration interval, X, must be wide enough relative to the distribution width, i.e., $X \gg \triangle E$. The sensitivity parameter depends on the reaction Gibbs energy, $\triangle G^0$, and the reorganization energy, λ, as

$$a = -\frac{\triangle G^0 - \lambda}{2\lambda} \tag{11.7}$$

and it determines how strong is effect of the distribution on the lifetime spread.

The benefit of using decay model given by Eq. (11.6) is that in addition to the average lifetime, τ_{av}, the energy distribution width, $\triangle E$, can be estimated from the experimental data. In addition, if the Gibbs or reorganization energy is known for the reaction, the other energy can be calculated from the sensitivity parameter, a [23, 24]. A disadvantage of the model as compared to the stretch exponential is much high computational demand, since integration must be done at each iteration step.

11.4 Photophysics of Porphyrin LB Films

The lifetime of the singlet excited state is an important parameter of porphyrins intended to be used as photo-excited donor in electron transfer molecular systems. The fast quenching of the singlet excited state presents a competing process to photoinduced electron transfer (PET), and can reduce the efficiency of the desired PET. Design of compounds and film structures, which would preserve the natural lifetime of porphyrin excited state is one of important starting point in building up films with efficient ET property.

Clear distinction between the dye molecules in films and the same molecules in solutions at moderate and low concentration is enforced inter-chromophore interaction, or aggregation of the molecules. Most dyes in aggregated state are known to exhibit a strong self quenching of the singlet excited state, which may be caused by different reasons, such as accelerated non-radiative decay, faster inter-system crossing, or inter-molecular charge transfer. Porphyrins and phthalocyanines are not exception and a much lower quantum yield of the fluorescence in films as compared to that in solutions is typical for these compounds. Moreover, their plane macrocycles with conjugated π-systems have strong aggregation tendency, and a special peripheral groups are usually used to reduce the effects of aggregation.

The photophysical properties and organization of porphyrins in LB films have been subject of research for a few decades [10, 11, 25–30]. One of the ultimate goal of the studies was to find conditions and necessary porphyrin modifications to achieve stable multilayer film deposition with as less excited state quenching as possible. Considering aggregation and quenching effects tetraphenylporphyrins (TPP) have an advantage of phenyl rings turned almost perpendicular to the plane of the porphyrin macrocycle. In addition, TPP and many of its derivatives are not soluble in water and suitable for LB film deposition being mixed with supporting matrix molecules, such as stearic acid or octadecylamine [10, 11]. However, there are only few examples of mixed porphyrin LB films demonstrating the lifetime of the excited state close to that in solutions, e.g., 5-(4-acetamidophenyl)-10,15,20-tri-*p*-tolylporphyrin in dioleoylphosphatidylcholine at molar ratio 1:20 has spectral feature virtually identical to that in solutions [10]. Though it worth to notice that other porphyrins tested at the same time and in identical conditions have experienced some quenching even at lower concentrations. Also dioleoylphosphatidylcholine cannot be deposited as multilayer LB film.

Another strategy to afford LB deposition of porphyrins is to synthesize porphyrins with lipid-like side groups which can enforce monolayer formation on the water surface [25, 26, 30]. Still, noticeable changes in absorption spectra and significant reduction of the fluorescence lifetime by as much as ten times were reported for these specifically designed compounds.

Two porphyrins can be mentioned separately, for which even at high concentration the aggregation does not result in drastic quenching of the singlet excited state in LB films. The first one is 5,10,15,20-tetrakis(3,5-di-*tert*-butylphenyl)porphyrin (TBP) [11].

The *tert*-butyl groups prevent close contact between the porphyrin macrocycles in aggregated state. As the result porphyrin multilayer films at concentration up to 30 mol-% in octadecylamaine matrix have the average lifetime of 1.5 ns, and the average fluorescence lifetime is as long as 7 ns when concentration of the porphyrins is reduced to 1 mol-%. The other example is 5,10,15,20-tetrakis(pentafluorophenyl)porphyrin (PFP) [28]. Similar to TBP, PFP requires matrix molecules to stabilize monolayer on the water surface, but multilayer films can be deposited at concentrations up to 30 mol-%. The porphyrins are strongly aggregated as revealed by absorption and emission spectra, but the average fluorescence lifetime of PFP in such film is roughly 4 ns, which is only two times shorter than that of PFP monomer.

The role of matrix in porphyrin LB films can be two folded. The matrix helps to form monolayer and to deposit it as multilayer LB film. Typical lipids used for this purpose are saturated fatty acids, e.g., stearic or arachidic acids. Secondly, the matrix molecules can be used to prevent aggregation of the porphyrins. Then the important property is miscibility of the matrix molecule and porphyrin. Unsaturated fatty acids and lipids have usually better miscibility with other compounds than saturated ones [10, 13], but they cannot be deposited as multilayer films by them self. Therefore in most cases it is impossible to find matrix molecules, which would serve two goals at the same time, improve film deposition and prevent aggregation. In most practical cases the aggregation is the phenomenon, which cannot be avoided totally, and should be taken into account when designing molecular film. For most porphyrin films at concentration 10 mol-% and higher the lifetime of the singlet excited state is expected to be in the range 0.1–1 ns.

A short distance between chromophores in LB films may lead to an efficient inter-chromophore energy transfer [27]. The energy transfer in an array of identical molecules does not change the nature of the excited state, meaning that after the act of energy transfer a similar molecular entity is in excited state. However the presence of a few quenching centers in a big molecular array with efficient energy transfer may result in an efficient quenching of the excited state of the whole array. The energy transfer will not be discussed intensively here, but it is one of the phenomena, which may have drastic effect on performance of molecular electron transfer systems.

11.5 Charge Transfer Between Donor and Acceptor Layers

The most straightforward approach to achieve vectorial photoinduced electron transfer in organic film seems to be deposition of bi-layer structures with the donor and acceptor molecules placed in adjacent layers. However, very few reports were published demonstrating successful inter-layer ET. Ohta and co-workers have reported on photoinduced electron transfer between thiacyanine and viologen layers [31]. The ET was concluded from the fluorescence quenching of the thiacyanine layer, and reported electron transfer rates are $\approx 6 \times 10^8$ s^{-1}. Considering that the fluorescence lifetime of thiacyanine layer (donor) is <1 ns, the efficiency of the interlayer ET was relatively low (<0.3), though the donor and acceptor layers were separated by one layer of stearic acid.

Another example of interlayer ET was reported for polythiophene (donor)–fullerene C_{60} (acceptor) pair [32]. The fullerenes were entrapped by etoxymethylresorcinarene molecules, which has made possible deposition of alternating donor-acceptor layers, such as polythiophene layer – fullerene bi-layer and fullerene bi-layer – polythiophene layer. The inter-layer electron transfer was confirmed by transient photo-voltage measurements. The reported lifetime of the charge-separated state is rather long, a few milliseconds, which was attributed to charge migration along the polymer chain.

Nevertheless, to the best of our knowledge, there are no convincing reports on interlayer photoinduced charge transfer utilizing porphyrin derivatives as the donor, though porphyrin derivatives are very common donor in study of the photoinduced ET and in particular in design of efficient covalent linked donor-acceptor dyads [1, 33].

Obviously, the relation between the singlet excited state lifetime and the time constant of the charge separation is the important parameter determining the efficiency of the electron transfer reaction. The tendency of porphyrins to aggregate and self-quench the singlet excited states puts rather strong requirements for the rate of the electron transfer between the porphyrin layer and an acceptor layer, when the bi-layer strategy is chosen to build up films with efficient ET. Considering the above discussion on the excited state lifetime in porphyrin films, one can conclude that the electron transfer should be faster than $k_{\mathrm{ET}} \geqslant 10^{11}\ \mathrm{s}^{-1}$ to compete successfully with excitation quenching in films with high density of porphyrins, i.e., heavily aggregated films. What does this mean from the point of view of distances between the donor and acceptor? For example, typical rate constant of the photoinduced electron transfer between porphyrin and fullerene is $k_{ET} = 10^{11}\ \mathrm{s}^{-1}$ when the center-to-center distance between the chromophores is $R_{\mathrm{DA}} = 1$ nm [34]. This is not fast enough for efficient ET in the case of heavy aggregated porphyrin layer, but should be sufficiently fast for efficient ET in case of moderately aggregated porphyrin layer. However, the latter is achieved by diluting the porphyrins in a layer of matrix molecules and will result in an increased distance between the donor and acceptors in two adjacent layers if there is no specific inter-molecular coordination between the donors and acceptors.

The second question: Is 1 nm long enough distance of the charge separation to be utilized for e.g., photocurrent generation? To answer the question one need to consider the total thickness of bi-layer film and possibilities to place donors and acceptors at 1 nm distance using LB or LS methods for the film deposition. Accounting for the dimensions of the porphyrin and fullerene the latter means that the edge-to-edge distance must be 0.3–0.5 nm, which leaves almost no room for hydrophilic groups or other peripheral groups required for successful film deposition. Increasing the distance between the donor and acceptor will slow down the electron transfer. The distance dependence of the ET follows the exponential law, $k_{ET}(r) = k_0 \exp\left[-\beta(r - r_0)\right]$, where k_0 is the ET rate constant at distance r_0, and β is the decrement factor which is typically close to $\beta \simeq 4\ \mathrm{nm}^{-1}$ for through space ET [35, 36]. Thus, an increase in distance between the donor and acceptor by one nanometer will slow down the electron transfer by more than 50 times, i.e., the expected time constant of the ET is 0.5 ns, which becomes critically slow even for moderately aggregated porphyrin films.

A typical thickness of bi-layer LB film is 4–5 nm. Ideally the primary charge separation between the donor and acceptor should be followed by sequential electron and hole transfer to the layers of secondary donors and acceptors as was discussed in Introduction. If the primary distance of CS is 1 nm the distances for the secondary charge transfer steps must be longer than 2 nm. Clearly this is another weak point of the bi-layer design of the donor-acceptor films, since it is difficult to achieve an efficient CT at this distance.

The above estimation shows that to achieve an efficient photoinduced charge separation between adjacent layers of the donor and acceptor molecules is a challenging task despite of its implementation simplicity, and explains why there are very few publications demonstrating a photoinduced ET between two LB layers. Though there are numerous examples of an efficient ET between two organic layers or organic-organic interfaces formed by other deposition techniques [37, 38], which, however, goes beyond the scope of this chapter.

As was mentioned above, a pour control over distance between the donor and acceptor molecules in adjacent layers is the problem to be overcome to achieve an efficient ET in donor-acceptor films. A feasible solution is to design molecules, which would have a specific binding tendency to form desired interlayer donor-acceptor pairs, or would result in formation of an ordered donor-acceptor pairs in the layer, which is deposited then as regular LB film. The latter was achieved by using a few strategies.

A bi-layer of donor and acceptor molecules can be assembled at the water-air interface using electrostatic interaction between specifically designed donor and acceptor counterparts. In particular, tetrasulfonated phthalocyanine (an electron donor) and tetrakisdocosylphyridinium porphyrin (an electron acceptor) were used to fulfill the goal [39]. These two compounds form a bi-layer on the water surface, which can be deposited as a multi-layer LB film. Both picosecond pump-probe and transient photo-voltage measurements have confirmed the photoinduced CS in solid films. Moreover, transient photo-voltage measurements have demonstrated that direction of the ET in film can be switched by switching the deposition direction of the porphyrin-phthalocyanine layer. However the lifetime of the CS state was rather short (virtually time-unresolved in nanosecond photo-voltage measurements), which is quite expected result since the donor and acceptor were in close contact with each other providing a fast charge separation but also a fast charge recombination at the same time. Another limitation comes from the necessity to use porphyrin with long alkyl chains to allow LB film formation and deposition. These alkyl chains form an insulating layer on top of the porphyrin layer and precluded further utilization of the photo-generated CS state.

An attempt to achieve a preferable orientation of pair of donor and acceptor molecules in Langmuir layer was done by mixturing free base and zinc tertbutylphenylporphyrins placed in octadecylamine (ODA) matrix [40]. Tertbutylphenylporphyrin (TBP) is expected to form a vertically aligned dimer in ODA matrix, and since the free base and zinc porphyrins have slightly different affinity to water, one can expect preferentially uniform orientation of free base–zinc porphyrin heterodimer on the water surface. A careful study of the mixed porphyrin films absorption and emission properties, and transient photo-voltage responses have indicated that such preferential orientation does exist with zinc porphyrin situated closer to the water than the free base. However, the efficiency of the vectorial charge transfer across the layer was rather low due to the statistical nature of heterodimer formation and probably a rather low water affinity difference for the free base and zinc porphyrins.

Recently a coordination chemistry was used to form bi-layers of porphyrin and fullerene at water-air interface [41]. Imidazole functionalized fullerene derivative formed a layer on the water surface, and at the same time the imidazole moiety exposed to the water served as the binding site for zinc porphyrin (zinc tetrakis(textitN-methylpyridinium)porphyrin tetrachloride) dissolved in the water subphase. The bi-layer could be transferred onto a quartz plate or a transparent conductive substrate (ITO electrode). The photoinduced vectorial charge transfer was confirmed by measuring photo-current in a standard solar cell arrangement.

11.6 Films of Donor-Acceptor Dyads – Intralayer Charge Transfer

A more efficient and controllable charge separation can be achieved by using specially designed DA molecules for film deposition. In this case a better control over distance and mutual orientation of the donor and acceptor moieties can be achieved by engineering a proper linker between the donor and acceptor moieties. The role of the LB or LS deposition method is to provide uniform orientation of the donor and acceptor parts so that the electron transfer would take place in one direction for all the molecules and across the layer. Also the thickness of a single LB layer can be as small as 2 nm, which provides easier access to the photo-generated charges for their further utilization as will be discussed in the following section.

11.6.1 *Phytochlorin-Anthraquinone Dyad*

Technically deposition of a layer of DA molecules is a more simple task than deposition of films with the donor and acceptor placed in different layers. However, this approach requires a carefully designed DA compounds, which must be (i) efficient electron transfer pairs, (ii) be suitable for LB or LS film formation, and (iii) have donor and acceptor moieties aligned perpendicular to the film plane after deposition. One of the first examples of such DA chlorophyll-based compounds was phytochlorin-anthraquinone (PAQ) dyad shown in Fig. 11.4(b) [13, 42].

The phytyl tail of the PAQ molecule is essential part to achieve successful LB film deposition and to align donor (phytochlorin) and acceptor (anthraquinone) moieties one on top of the other. The transient photo-voltage measurements have confirmed the photoinduced CS in films of zinc PAQ derivative, and the orientation of the molecule with phytochlorin being closer to the water surface than anthraquinone. Although in addition to the vectorial CS, an undesired signal arising from aggregated phytochlorin moieties was observed. It was possible to reduce the aggregation effect by mixing PAQ molecules with so-called soft lipids, such as Soya PC and erucic acid, but the deposition properties of these lipids are rather pour and multilayer deposition was only possible after deposition of 3–5

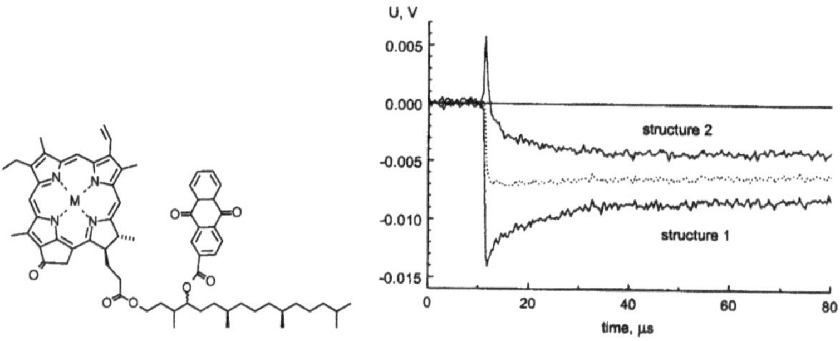

(a) (b)

Figure 11.4. (a) Structure of phytochlorin-anthraquinone (PAQ) dyad. (b)Photo-voltage responses of PAQ monolayers deposited in upward (structure 1) and downward (structure 2) directions. Note that PAQ monolayers where insulated from both electrodes by 9–10 layers of octadecylamine.

intermediate layers of stearic acid or similar lipids [13]. This limits applications of PAQ dyad films gradually.

11.6.2 *Phytochlorin-Fullerene Dyads*

Long alkyl tails, such as phytyl of PAQ, are a typical approach to syntheses of lipid-like functional molecules suitable for the deposition LB films. Though the main essential property of the alkyl tail in this application is to be strongly hydrophobic, and there are other molecular groups which can be used instead. One particularly important molecule, which was used in this type of applications, is the fullerene. It has two important properties – it is hydrophobic, and it is a good electron acceptor. Thus, designing DA dyads with fullerenes one can avoid using additional hydrophilic groups. Still the donor part should hold a hydrophilic anchor to provide a desired orientation of the dyad on the water surface and in solid film. An example of such phytochlorin-fullerene dyad is presented in Fig. 11.5(a) [43].

One another important advantage of using fullerenes in donor-acceptor applications as compared to traditional electron acceptors such as quinones, comes from its rigid ball-like shape with electron density distributed between many carbons. This has two consequences. Firstly, the reorganization energy associated with the ET reaction for dyads with fullerene acceptor is essentially smaller than

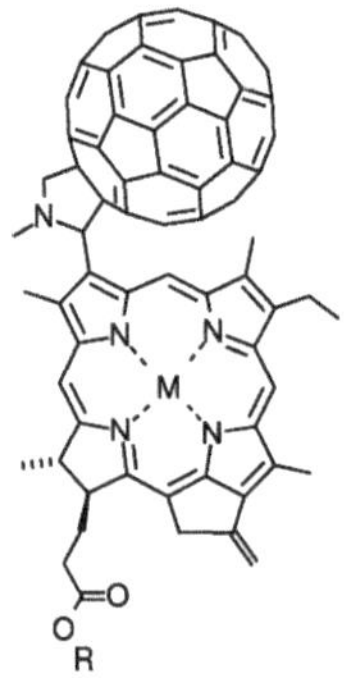

(a) PF

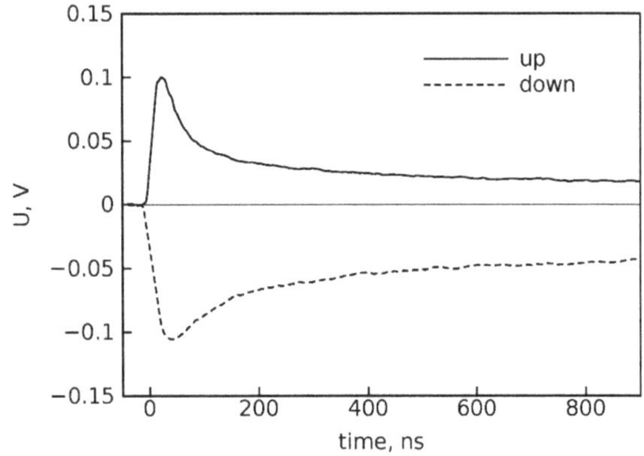

(b) Photo-voltage response

Figure 11.5. (a) Structures of phytochlorin-fullerene (PF, M=2H or Zn; R=H or CH_3), and (b) Photo-voltage responses of 20 mol-% PF (M=2H, R=H) in octadecylamine monolayers deposited in downward and upward directions. Adapted from reference [44].

that for smaller acceptor molecules with similar reduction potential, such as quinones [45, 46]. Secondly, donors of interest for this book are porphyrins and their derivatives which have extended electron π-system. At close proximity the interaction between the electronic subsystems of the porphyrin-like donors and fullerene acceptor results in formation of intra-molecular exciplex upon photo-excitation [43]. For some porphyrin-fullerene dyads the interaction can be seen already in the ground state [45, 47, 48], as will be discussed later. The important impact of the exciplex on the electron transfer properties of the dyads is that it efficiently lowers the activation energy for transition from the singlet excited state to the CS state, and thus accelerates the CS gradually [34, 47].

The PF dyad (Fig. 11.5(a)) has reasonably good film formation properties showing mean molecular area close to 1 nm^2 [44]. The area suggests vertical orientation of the dyad on the water surface. However, only a single layer was possible to deposit for pure PF film, but using octadecylamine as supporting matrix molecule the multilayer deposition was successful for PF concentrations up to 50 mol-%.

The photo-voltage response measurements have confirmed uniform dyad orientation in the film with ester moiety exposed to water and fullerene positioned at the top of the layer. Importantly, by switching the film deposition direction from downward to upward, the sign of photo-voltage response was switched to opposite, but the shape of the response remained essentially the same as illustrated in (Fig. 11.5(b)). This has proved that the response is generated by the photoinduced electron transfer in the dyad layer.

The lifetime of the charge-separated state in the film was much longer than that in solution. In non-polar solvents, such as toluene, the dyad relaxes to the ground state within a few nanoseconds, and in polar solvents the relaxation is even faster. However, in solid film the charge recombination was non-exponential with faster component around few tens of nanoseconds and longer part extended to the microsecond time domain [44]. This turned out to be a general tendency for porphyrin-fullerene dyads and is discussed later in this chapter.

Although the shape of the photo-voltage responses of PF films was virtually concentration independent, the intensity of the photovoltage signal depended strongly on the concentration showing roughly third order dependence on the surface density of the dyads. This unusual dependence suggested that the long lifetime of the CS state (as compared to that in solutions) arises from the intermolecular interaction, which has stronger effect at higher density of the dyads, i.e., at higher molar concentration of the dyad in the film [44, 49].

11.6.3 *Porphyrin–Fullerene Dyads*

Porphyrins have electron donating properties similar to that of chlorophyll derivatives, but have an advantage of providing greater flexibility in chemical engineering. This can be used to synthesize dyads with desired mutual orientation of the donor and acceptor and with hydrophilic groups at the donor or acceptor side.

The first porphyrin-fullerene dyad tested for films deposition by LB method had no specific side hydrophilic group to ensure uniform orientation of the donor and acceptor moieties in film, but it was possible to deposit even multilayer LB films using ODA as

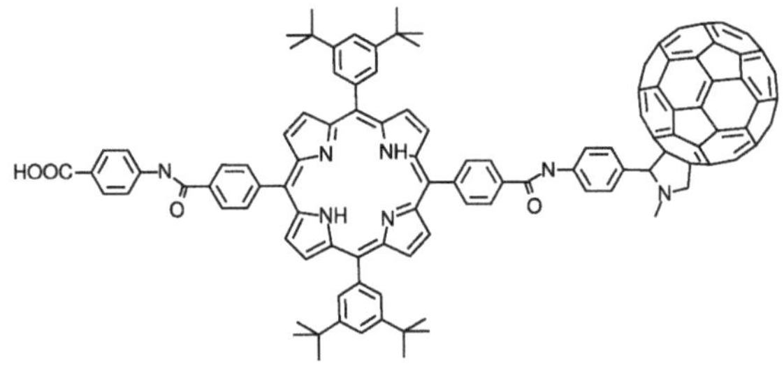

Figure 11.6. Porphyrin–fullerene dyad with hydrophilic group.

supporting matrix molecules [50]. The outstanding property of such films is the emission spectrum which has clear exciplex feature, while the same dyad in solution does not show any sign of the exciplex formation. This finding has demonstrated that conformational states of the dyads in solutions and solid films can be very different.

A dyad with carboxyl hydrophilic group presented in Fig. 11.6 was the first porphyrin-fullerene dyad designed for LB film deposition [51]. The dyad formed stable monolayers on the water surface with mean molecular area of 1.7 nm^2 at surface pressure 20 mN cm^{-1}, but stable multilayer deposition was achieved only for dyads mixed with octadecylamine supporting molecules. The donor-acceptor distance for the dyad in extended conformer is 1.8 nm, which leads to rather slow ET in solutions ($\tau_{CS} \simeq 0.5$ ns in benzonitrile). However in LB films the ET was much faster approaching $\tau_{CS} \simeq 100$ ps as estimated by the fluorescence quenching measurements. The vectorial photoinduced ET was also confirmed by photo-voltage measurements which have shown complex time profile of the charge recombination with time constants ranging from hundred nanoseconds to few microseconds. The much faster ET in films than in solutions is the clear indication that in films the distance between the porphyrin and fullerene moieties is much shorter than that in solutions, and should be less than 1 nm. It is very likely that this short distance is achieved due to a special packing of the molecules in the film with shortest distance achieved between porphyrin and fullerene moieties belonging to different molecules [51].

The above example shows that in-line arrangement of the porphyrin and fullerene moieties similar to that shown in Fig. 11.6

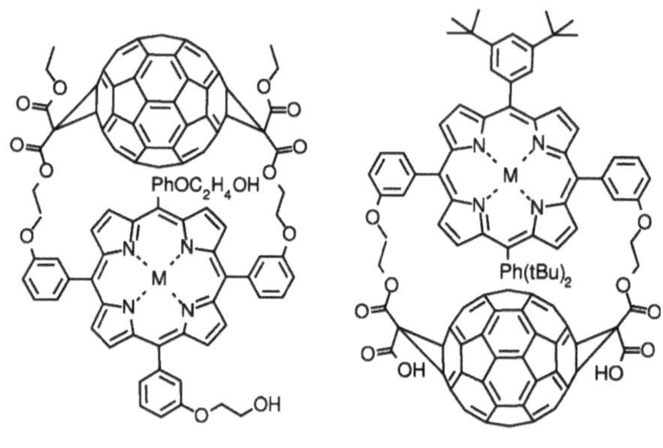

(a) DHD (b) TBD

Figure 11.7. Structures of porphyrin-fullerene dyads: (a) DHD, and (b) TBD. M = 2H or Zn.

does not provide an optimum molecular geometry for incorporation into LB films. In particular, the total length of the molecule (roughly 4 nm) exceeds the optimum thickness of Langmuir monolayer, 2–2.5 nm. Also the distance between the donor and acceptor must be shorter and better controlled for the efficient ET in films. This has lead to design of porphyrin-fullerene dyads with two linkers [47]. Example of such compounds are presented in Fig. 11.7. Both dyads have the same mutual organization of the donor and acceptor, i.e., of porphyrin and fullerene chromophores. Two linker arrangement keeps the fullerene moiety on top of the porphyrin macrocycle, which results in a fast and efficient charge separation [47]. In addition, the dyads have hydrophilic groups at different locations, which were intended to enforce opposite orientations of the dyads on water surface and LB films, respectively.

Both dyads presented in Fig. 11.7 form stable monolayers on water surface with mean molecular area around 2 nm^2 at surface pressured of 10 and 20 mN m^{-1} for DHD and TBD dyads, respectively [12]. This area agrees well with mono-molecular organization of the layer. Similarly to PF dyad only monolayer could be transferred on a solid substrate for pure dyad films, but mixed films with octadecylamine as supporting molecule can deposited as multilayers at relatively high concentrations up to 20–30 mol-%. Though

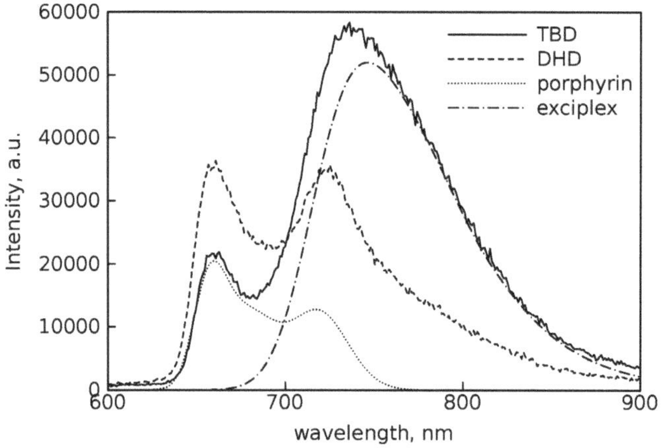

Figure 11.8. Emission spectra of 10 mol-% films of DHD and TBD, and fluorescence spectrum of porphyrin and exciplex emission spectrum. The latter was calculated assuming the exciplex energy to be 1.736 eV, the outer sphere reorganization energy 0.088 eV, the internal reorganization energy 0.053 eV, and the electronic vibrational coupling 0.76 [48].

later Langmuir-Schäfer method was adopted to achieve deposition of pure dyad layers incorporated into complex multilayer structures.

An important property of the dyads like DHD and TBD in excited state is formation of the intramolecular exciplex, which can be observed as a broad emission band in the 700–900 nm wavelength range [34, 47, 48, 52]. Interestingly, the exciplex emission was clearly seen in steady state emission spectra of TBD films, but not for DHD films, as presented in Fig. 11.8. However, the fluorescence time resolved measurements have revealed that indeed both dyads have characteristic exciplex-like emission in the near infrared part of the spectrum with lifetimes in the range 0.3–2 ns [12].

The difference in emission properties of the dyads, in particular in the appearance of the exciplex indicates that the dyads have somewhat different geometries in films despite the fact that the core parts of the dyads, the porphyrin and fullerene moieties, and the linkers between them are exactly the same for the two. Even stronger difference between the dyads was observed in photo-voltage responses, Fig. 11.9.

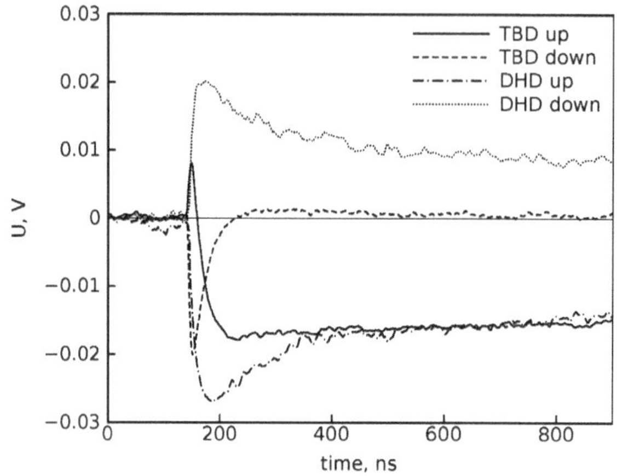

Figure 11.9. Photo-voltage responses of 10 mol-% films of DHD and TBD deposited in downward and upward directions. Adopted from reference [12].

The photo-voltage responses of DHD dyad layers deposited in opposite direction (upward and downward) are essentially the mirror images of each other. This is expected result since the only difference is orientation of the dyad in respect to the film substrate. The polarities of the responses (positive or negative) suggest that on the water surface the porphyrin moiety is located closer to the water than fullerene, which is also expected result. The formation of the charge-separated state is not time resolved in this experiment and the rise of the signal is determined by the time resolution of the instrument. The charge recombination is essentially non-exponential with the fastest decay part corresponding to roughly 100 ns lifetime and the longer decay extended into millisecond time domain.

The photo-voltage responses of TBD films have shown more complex behavior. Right after the excitation the direction of the charge motion is in agreement with the expected orientation of the dyad on the water surface, i.e., the fullerene moiety is located closer to the water than porphyrin. The orientation of the donor and acceptor in respect to the water is different for DHD and TBD dyads which results in opposite signs of the photo-voltage responses for the films deposited in one and the same direction, as seen in Fig.11.9. The photo-voltage signal of TBD film relaxes fast with lifetime close to the time resolution limit of the instrument, approximately 10 ns, which is in sharp contrast to much slower charge recombination of DHD films. Even more surprising, the photo-voltage response of TBD dyad is changing polarity at delay time 50–100 ns after the excitation. This difference in responses of DHD and TBD dyads may originate in different effect of the layer polar hydrophilic groups of the dyad and supporting matrix molecules; the phenomenon is not yet fully understood [12].

11.6.4 *Primary Photo-Reactions in Porphyrin–Fullerene Dyad Films*

The photo-voltage measurements by TRMDC method provides information on charge motion in the film in direction perpendicular to the film plane, but the time resolution of the instrument used to study films was limited by 10–20 ns. Therefore no information could be obtained on the primary reactions of the charge separation, though the CS process was expected to be a complex two-step reaction according to the dyad studies in solutions [47]. The expected time scale of the CS is from hundreds femtoseconds to few tens of picoseconds. This time scale is only accessible with femtosecond time resolved spectroscopy, and since the charge-separated state is non-emitting state the most appropriate method to study the charge transfer is femtosecond pump-probe.

The samples for pump-probe measurements should have sufficient absorption since the measured value is the change in the sample absorption. Typically absorptance of the sample must be higher than 0.4 in order to obtain a reasonable signal-to-noise ratio. In the case of LB films of DHD-like dyad this means that the sample must consist of at least 50 layers of the dyad. Deposition of such samples is much more difficult task then that for photo-voltage measurements where even a single dyad layer generates an easily detectable signal. In addition, it is highly desired to separate individual dyad layers from each other in order to reduce the interlayer interactions. The task was accomplished by depositing 60 layers of DHD dyad in octadecylamine matrix (10 mol-%), which were separated by three octadecylamine layers from each other [23, 53]. Alternative methods tested for the dyad film deposition were spin coating and drop casting. In the latter case dyads were also mixed with octadecylamine to provide similar density of the dyads for different film deposition techniques. It was found that the primary photo-reactions are essentially the same for LB and drop cast films.

The study of the excitation density dependence of the transient absorption responses of the multi-layer TBD film has shown that there is an interaction between the excited states which leads to a faster excitation relaxation [53]. This is seen as an appearance of a fast relaxation component at higher excitation densities, as illustrated in Fig. 11.10. The relaxation rate of the fast component is also excitation density dependent, and it fits the best to the second order decay model. Further analysis of the fast component dependence on the excitation density has revealed that it starts to dominate the sample transient absorption response when more than one quarter of the dyads is excited, i.e., at the excitation density >0.4 mJ cm^2 (at 420 nm).

This type of transient response dependence on the excitation energy density has been previously observed in e.g., phthalocyanine films [54], porphyrin-functionalized polypeptides [55], and

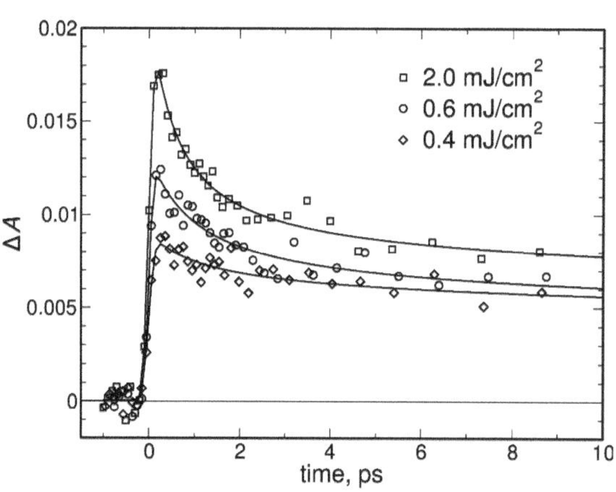

Figure 11.10. Transient absorption decays of TBD film at 700 nm recorded at different excitation densities indicated in the plot. Solid lines show the data fits. The excitation wavelength was 420 nm. Adopted from reference [53].

porphyrin dendrimers [56]. The phenomenon is well understood in terms of the singlet excited state energy transfer, or exciton migration, and exciton-exciton annihilation, which happens at event of two neighbor chromophores being in the excited state at the same time. However, in the case of studied dyad films the singlet excited state is quenched by a fast formation of the intramolecular exciplex, which takes place in less than 200 fs, whereas the observed phenomenon takes place in the time scale of few picoseconds. A typical exciplex transient absorption spectrum can also be clearly seen already at delay time of 200–300 fs, and the typical time scale of the annihilation process is 1–3 ps. Apparently the exciplex behaves in this instance as an excited state, and can be quenched by neighbor exciplex [53].

The results presented in Fig. 11.10 demonstrate that selection of the excitation energy density is important part of the ultra-fast ET dynamic measurements of the dyads in films. Since the annihilation is not the process of interest, the excitation density should be low enough to avoid the annihilation effects. For TBD dyads this means that the density must be $\leqslant$ 0.4 mJ cm^2 at excitation wavelength of 420 nm, and that the measured transient absorption signals in films will be roughly four times smaller than that in solutions for samples with the same absorption.

The excitation at 420 nm populates predominantly second singlet excited state of the porphyrin chromophore, which is known to relax quickly to the first singlet excited state. In solutions the singlet excited state relaxes by yielding the intramolecular exciplex within 200–600 fs time interval [47]. In films the exciplex formation was not time resolved, which indicates that the singlet to exciplex relaxation is faster than 200 fs (which is the time resolution of the instrument used) [23]. The exciplex can be identified due to its characteristic differential absorption spectrum (Fig. 11.11(a)), which was virtually identical to that of the dyad in solutions [47].

In the time interval of pump-probe measurements, 1.2 ns, the transient absorption decreases significantly with time, but the shape of the spectrum remained the same (Fig. 11.11(a)). This is an evidence that no new transient states are formed in the time range from 0.2 ps to 1.2 ns. At the same time the decay is clearly non-exponential (Fig. 11.11(b)), but fits well to the distributed decay

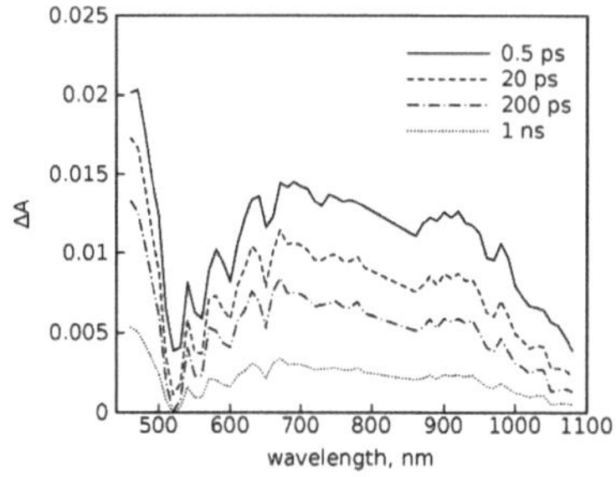

(a) time resolved spectra

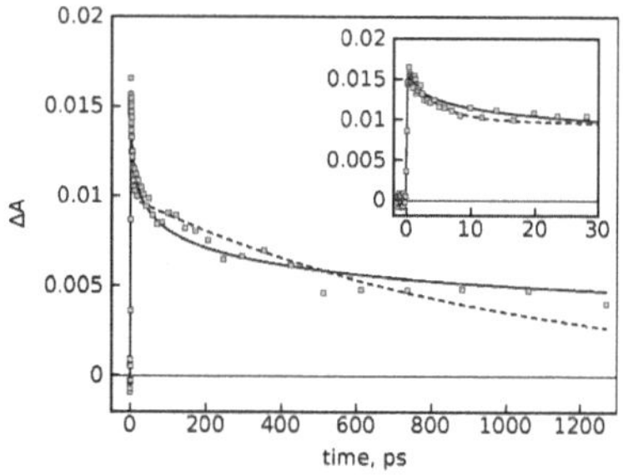

(b) decay at 700 nm

Figure 11.11. Transient absorption time resolved spectra (a) and decay at 700 nm (b) for TBD film. The delay times are indicated in the plot (a). The solid line in plot (b) is distributed decay fit, and the dashed line is bi-exponential fit. Adopted form reference [23].

model described in Section 11.3.3. The distributed decay parameters were: the average lifetime $\tau_{av} = 50$ ps, the energy distribution width $\triangle E = 0.09$ eV and the sensitivity parameter $a = 2$. Although the distribution width seems to be rather narrow, the spread of the decay lifetimes is quite large for TBD films, from few picoseconds to nanosecond.

Apparently, at a few tens of nanosecond delay a charge-separated state is formed in films of DHD and TBD dyads, as evidenced by transient photo-voltage measurements, Fig. 11.9. Most probably the charge transfer takes place in the nanosecond time domain, since the longest exciplex lifetime observed for these films is 3.7 ns and the charge transfer should take place before complete relaxation of the exciplex. Unfortunately this time scale could not be studied by pump-probe method which had maximum delay of 1.4 ns, and this time range was too fast to be resolved by the photo-voltage or flash photolysis instruments.

11.6.5 *Phthalocyanine-Fullerene Dyads*

A disadvantage of porphyrins in practical applications of photoinduced electron transfer, such as organic solar cells, is that they have relatively weak absorption in the green and red part of the spectrum. In this respect chlorophyll derivatives, e.g., phytochlorin, have an advantage of strong absorption band in the red part of the spectrum. Another type of compounds, which are structurally similar to porphyrins, but have stronger absorption in the red, are phthalocyanines. Therefore, it is not surprising that a number of phthalocyanine-fullerene dyads was synthesized and studied [57–59]. Although the photoinduced electron transfer reactions in porphyrin-fullerene and phthalocyanine–fullerene dyads are rather similar, phthalocyanines have slightly smaller oxidation potential than porphyrins, which makes the charge separation exothermic reaction even in non-polar media [58, 59]. The latter can be considered as advantage of phthalocyanines in solid film applications, since molecular solid films are typically non-polar media, especially when the dyads are placed in supporting matrix of octadecylamine or alike.

Similarly to porphyrin-fullerene dyads, the phthalocyanine–fullerene dyads designed for film deposition by LB method must be water insoluble but hold a hydrophilic group to keep orientation of the dyads uniform. A dyad with two ethyl ester groups was synthesized for this purpose (Fig. 11.12) [58]. The dyad forms a stable layer on the water surface, but multilayer film deposition by LB method was achieved only for the phthalocyanine-fullerene (PcF) dyad placed in ODA supporting matrix. Also it has to be mentioned that mean molecular area for the dyads on the water surface was only 0.8 nm^2 [24], which is smaller than expected for a monolayer of PcF and is discussed later in this section.

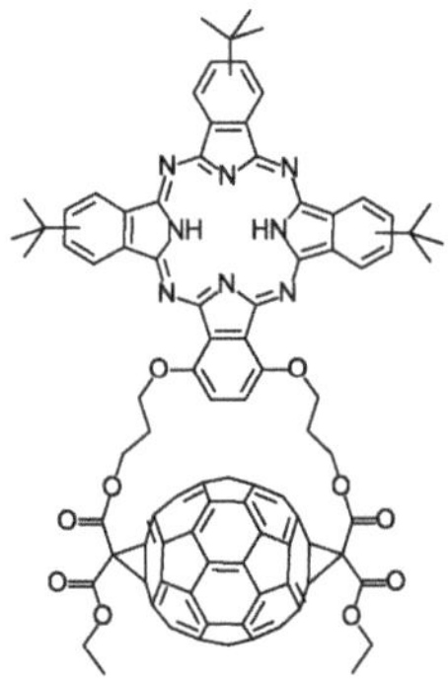

Figure 11.12. Phthalocyanine-fullerene (PcF) dyad.

By deposition of 127 PcF layers (10 mol-% in ODA) spaced by three ODA layers the absorption of 0.4 at Q-band (700 nm) was achieved. This was enough to carry transient absorption measurements by the femtosecond pump-probe method. The result of the measurements are presented in Fig. 11.13(b) [24]. To achieve a reasonable fit of the measured data a combination of two exponentials and a distributed decay (Eq. 11.6) was used. The photo-excitation populates singlet excited state of the phthalocyanine chromophore, which relaxes by yielding the charge-separated state. This behavior is similar to that observed in solutions except of the CS time constant in solution was roughly 2 ps, whereas in films it is faster being 0.8 ps. The next step in film is charge recombination, but unlike in solution it does not result in complete recombination, instead roughly 40% of dyads relaxes to the ground state whereas the rest is transformed to another charge-separated state. The transition of the other CS follows the distributed decay model with average lifetime of 35 ps and energy distribution width of 0.04 eV. This

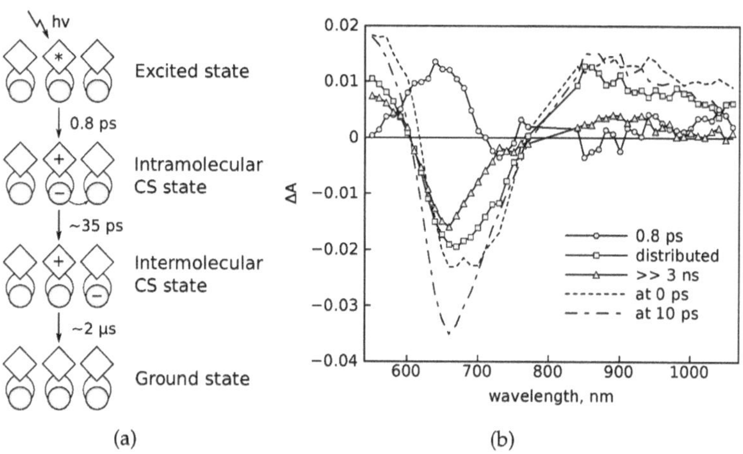

Figure 11.13. (a) Schematic presentation of photo-reactions in PcF film, and (b) transient absorption decay component spectra. The data fit model consisted of two exponential components and distributed decay component (Eq. 11.6) with average lifetime of 35 ps. The decay component spectra are shown by symbols with connecting lines. The dashed and dot-and-dashed lines present time resolved spectra right after the excitation (at 0 ps) and at 10 ps delay time, respectively. Adopted from reference [24].

part of the photo-reactions is different for solutions and films. In solutions only one CS state was observed, and it recombines directly to the ground state with much longer time constant (120 ps in benzonitrile and 2 ns in toluene).

The second CS state has lifetime much longer than the longest delay time of the pump-probe instrument. To study decay of this state nanosecond flash photolysis was used. The decay was clearly non-exponential but fitted well to the distributed decay model with average time constant of 1.6 μs and energy distribution width of 0.04 eV, which is similar to that found for the formation of the state. The formation of the long-lived CS state was also confirmed by the transient photo-voltage measurements [60]. Also reversing the deposition direction did not result in complete switching the sign of the photo-voltage response, but the result of the measurements have indicated that the photo-generated charges are migrating in the film. The latter explains the appearance of two charge separate states observed in the pump-probe measurements. The first one is the intramolecular CS state similar to that observed in solutions. The second is formed when an electron donated to fullerene migrates to another fullerene, or, respectively, a hole migrates to another phthalocyanine. This process is specific to films with tightly packed dyads.

A schematic presentation of the photoinduced reaction in PcF films is shown in Fig. 11.13(a). The first step is formation of the intramolecular CS state, which is a fast process with time constant of 0.8 ps. The charges can recombine or migrate from fullerene to fullerene or from phthalocyanine to phthalocyanine. The recombination and migration are two competing processes. In addition, the rates of both reactions depend on environment. As the result the observed decay is essentially non-exponential and follows the distributed decay model with average lifetime of 35 ps. Ones the charge has migrated an intermolecular charge-separated state is formed, which has much longer lifetime extended to the microsecond time domain, and somewhat different differential spectrum than the intramolecular CS state. The quantum yield of the intramolecular CS state was estimated to be about 60%.

Unfortunately the uniformity of the PcF film was rather poor, which can be expected from small mean molecular area (0.65 nm^2 at deposition pressure of 20 mN m^{-1}) and was confirmed by transient photo-voltage measurements. The estimated area for a PcF dyad in trans conformer standing perpendicular to the water surface is 1.2–1.4 nm^2. Two times smaller measured value means that molecules are packed in a kind of double layer rather than the desired monolayer. Probably the hydrophilicity of the methyl ester groups is too weak to enforce formation of a uniform monolayer.

The advantage of phthlocyanines, a strong absorption band in the red part of the spectrum, can be also used to increase efficiency of the light collection by adding layers of light harvesting molecules with absorption in the blue–yellow part of the spectrum and emission in the red part. The strong absorption makes phthalocyanines good energy acceptors, and in combination with efficient emitting energy donors, such as perylenediimides, the energy transfer can be as fast as 10 ps for adjacent layers of perylenediimide and phthalocyanine-fullerene dyads [61], and the interlayer distance of the energy transfer can be as long as 8 nm [62].

11.7 Interlayer Charge Transfer with Secondary Donor or Acceptor Layers

The intralayer charge separation can be efficiently achieved by using donor-acceptor molecules to form the layer. Importantly, the lifetime of the charge-separated state can be as long as hundreds of nanoseconds or even tens of microseconds. This is the time needed to achieve efficient interlayer charge transfer. To implement the interlayer charge transfer the layer of donor-acceptor (DA) molecules has to be deposited next to the layer of the secondary donors (D_2) or secondary acceptors (A_2) in a such way that an analogy of triad is formed across the layers, which is D_2-DA or DA-A_2. The choice of compounds for secondary donor and acceptor layer should be made based on their films formation and deposition properties, and on redox properties, i.e., the secondary donor should be a stronger donor than the donor in DA dyad (oxidation potential of D_2 must be lower than that of D), and secondary acceptor must be stronger acceptor than that of DA dyad.

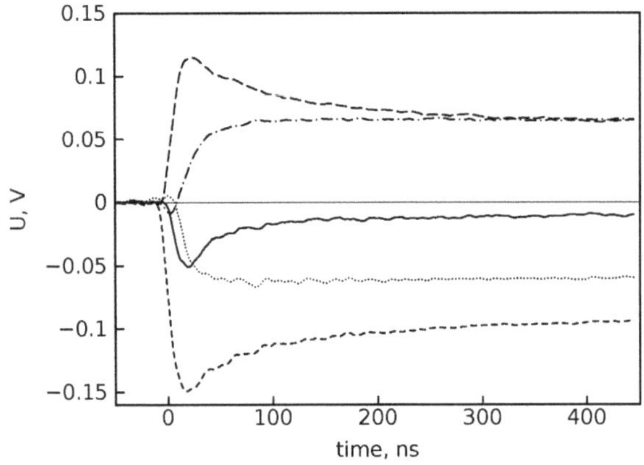

Figure 11.14. Transient photo-voltage responses of phytochlorin-fullerene (DA) – polythiophene (D_2) bi-layer films. The solid line presents response of phytochlorin-fullerene layer alone (DA film), the short-dashed line – D_2-DA bi-layer, the long-dashed line – AD-D_2 bi-layer, the dot-and-dashed line – DA-D_2 bi-layer, and the dotted line – D_2-AD bi-layer. The figure is adopted from reference [63].

The first demonstration of such type of molecular layer architectures was done with phytochlorin-fullerene dyad (PF, Fig. 11.5(a)) in DA layer and polythiophene in D_2 layer [63]. The photo-voltage response of D_2-DA bi-layer was three time stronger than that of the DA layer alone as presented in Fig. 11.14, and virtually no response of D_2 layer along was observed. Placing secondary donor, polythiophene, at the side of acceptor, fullerene, produced roughly three time smaller response than that of the D_2-DA structure with rather complex time dependence. The latter was explained by the fact that in the bi-layer DA-D_2 the electron acceptor is surrounded by two donors, D and D_2, and the observed response is superposition of two electron transfer reactions yielding two CS states with the electron shifted in opposite directions, D^+A^--D_2 and DA^--D_2^+, respectively.

The stronger photo-voltage response of the D_2-DA bi-layer indicates that the total electron transfer distance is longer in the bi-layers than in the dyad layer alone. An estimated donor-acceptor distance for PF dyad is roughly 1 nm. Three folded increase in the photo-voltage response means that in the bi-layer structure the electron transfer distance is approx. 3 nm, which agrees well with the expected distance of the interlayer electron transfer accounting for the most probable location of the acceptor, fullerene, at the side of the bi-layer structure. Another important difference between the mono-layer, DA, and bi-layer, D_2-DA, structures is the lifetime of the charge-separated state. Though the charge recombination was not exponential for both films, for mono-layer sample 80% of charges recombines in a few hundred of nanoseconds, whereas for bi-layer the charge recombination is extended to hundreds of milliseconds. This prolongation of the charge-separated state lifetime is expected since (1) the distance of the charge transfer is longer for the bi-layer structures, and (2) the secondary acceptor is conducting polymer, which makes possible charge migration in the D_2 layer, thus increasing the distance between the charges further more in lateral direction.

The same effect was observed when using the porphyrin-fullerene dyad (DHD, Fig. 11.7(a)) in DA layer [64]. The increase in the photo-voltage response was even greater, almost four folded, which is due to a smaller initial charge transfer distance in the DHD dyad compared to that in the PF dyad. For a quantitative study of the long recombination time constant the transient photo-current measurements were carried out as described in Section 11.3.2. It was shown that time constant is approaching one second, but the kinetics of the charge recombination is poorly described by a simple one step process. A better model must account for the charge migration in polymer layer. A suitable approximation is to consider a close location of the electron-hole pair, at which recombination is possible, and a long separation of the electron and hole for which the recombination is not possible, and to solve the kinetic equations with equilibrium between the states [65]. This type of analysis yields the charge recombination time constant of 0.5 s.

The advantage of the transient photo-voltage and photo-current methods is their high sensitivity, which makes possible investigation of mono- and bi-layer structures, but it does not provide direct information on the ET efficiency nor the charge location in the CS state. However, using the DHD dyad and polythiophene the multilayer film with repeating bi-layers separated by the passive ODA layers could be prepared and studied using a more conventional flash photolysis method [66, 67]. The measurements have confirmed formation of the long-lived CS in bi-layer structures. Furthermore, the transient absorption spectra of the CS state were very different for the DHD layers alone and for the DHD – polythiophene bi-layers. The former had shape typical for the charge separate state of the dyad being very similar to those observed with the femtosecond pump-probe method for the dyad in solutions and LB films. In the contrary, the spectra of the latter had shape typical for polythiophene cations with virtually no sign of porphyrin cations left already at hundred nanoseconds after excitation. This confirms the interlayer ET, though the time constant of the interlayer ET was not determined due to time resolution limit of the instrument ($\approx$ 100 ns).

The efficient and long-lived interlayer charge separation opens numerous possibilities for system extension by adding other donor and acceptor layers, as well light harvesting layers [67–69]. For example, a multi-step through layer photoinduced electron transfer was demonstrated for film consisting of two secondary donor layers: polythiophene – zinc phthalocyanine – porphyrin – fullerene dyad (DHD) [67].

The multilayer films with photoinduced charge transfer across the film are promising building blocks for organic photovoltaic devices, and in particular for organic solar cells. However, there were no reports so far demonstrating benefits of such structures compared to much more exploited dye-sensitized solar cells [70] and polymer-based bulk heterojunction solar cells [4]. A clear drawback of the bi-layer and even more complex layered structures is the rather small absorption, which, for example, for the porphyrin mono-layer is on the level of 1–3% at maximum of its absorption at Soret band (420–430 nm). For both, the dye-sensitized and bulk heterojunction solar cells this problem is overcome by non-flat surface topology at which the photoinduced electron transfer takes place. In the former case this is achieved by creating porous semiconductor electrodes. In the latter a complex topology is formed spontaneously in the bulk of mixed donor and acceptor compounds.

The LB method is not designed to be applied to micro- or nano-structured surfaces. However the fundamental results obtained with this method can be used for creating similar molecular structures with other molecular assembling methods. For example, self-assembled monolayers (SAM) by covalent, electrostatic or hydrogen bonding can be used to build up molecular layer on nano-structured surfaces [7]. Necessary surface reactive groups can be introduced to porphyrin-based compounds to allow self-assembling of porphyrin and porphyrin-fullerene dyads similar to those discussed in this chapter on glass or semi-transparent electrode surfaces [71, 72].

Another direct application of donor-acceptor dyad LB or LS monolayers in organic solar cells is the improvement of interfaces between layers. It has been demonstrated that a single phthalocyanine-fullerene dyad layer deposited between bulk layers of donors (polythiophene) and acceptors (fullerene or perylenediimide) can increase the power conversion efficiency as much as three times, though the direct photon absorption by the layer was negligible compared to that of the bulk donor and acceptor layers [73].

11.8 Conclusions

As has been shown by many examples in this chapter, inter-chromophore interactions play an important role in electron transfer in organized molecular films. A very well-known effect of such interactions is the quenching of the excited state, which is typical for almost all molecular aggregates and usually considered as undesirable phenomenon. However, many results presented above show that there is also an important benefit of aggregation – gradual increase in the lifetime of the charge-separated state. This is achieved when charges can migrated between donors or acceptors, or both. For a simple dyad molecule having the lifetime of the CS state of 1 ns in solution, i.e. isolated from other dyads, the charge migration can increase the lifetime of the CS state to 1 μs and longer in well-organized monolayer. Indeed, in aggregated films the charge separation compete with the excited state self-quenching, and it must be extremely fast process to be efficient. For the well-functioning dyads reported in this chapter, the charge separation takes place in a few picoseconds or faster, and no visible quenching was observed.

The necessity of having fast primary charge separation justifies synthesis of special donor-acceptor dyads with well-defined orientations of the donor and acceptor parts, such as DHD, and rules out the option of using separate LB layers holding donors and acceptors. On the other hand, the long lifetime of the CS states in DA layer increase the choice of compounds for secondary donor and acceptor layers, since the secondary charge transfer can be relatively slow, which means that the distance and mutual orientation of the sites involved is not as critical as for the first step. In particular, both polythiophene and phthalocyanine LB layers were demonstrated to be efficient secondary donor layers for the porphyrin-fullerene layer acting as the primary charge separation layer. This opens a wide range of possibilities for extending and utilizing photoinduced vectorial electron transfer by combining layers of photo- and elector-active organic molecules.

References

1. Wasielewski, M. (1992) Photoinduced electron transfer in supramolecular systems for artificial photosynthesis, *Chem. Rev.*, **92**, 435–461.
2. Gust, D., Moore, T. A., and Moore, A. L. (2001) Mimicking photosynthetic solar energy transduction, *Acc. Chem. Res.*, **34**, 40–48.
3. Gust, D., Moore, T. A., and Moore, A. L. (2009) Solar fuels via artificial photosynthesis, *Acc. Chem. Rev.* **42**(12), 1890–1898.
4. Hoppe, H., and Sariciftci, N. S. (2004) Organic solar cells: An overview, *J. Mater. Res.*, **19**(7), 1924–1945.
5. Kietzke, T. (2007) Recent advances in organic solar cells, *Adv. Opto Electronics*, p. 40285.
6. Heath, J. R. (2009) Molecular electronics, *Annu. Rev. Mater. Res.*, **39**, 1–23.
7. Ulman, A. (1996) Formation and structure of self-assembled monolayers, *Chem. Rev.*, **96**, 1533–1554.

8. Crespilho, F. N., Zucolotto, V., Oliveira, O. N., Jr., and Nart, F. C. (2006) Electrochemistry of layer-by-layer films: a review, *Int. J. Electrochem. Sci.*, **1**, 194–214.
9. Roberts, G. (1990) *Langmuir–Blodgett Films*, Plenum Press, New York.
10. Gust, D., Moore, T. A., Moore, A. L., Luttrull, D. K., DeGraziano, J. M., Boldt, N. J., Van der Auweraer, M., and De Schryver, F. C. (1991) Tetraarylporphyrins in mixed Langmuir-Blodgett films: steady-state and time-resolved fluorescence studies, *Langmuir*, **7**(7), 1483–1490.
11. Anikin, M., Tkachenko, N. V., and Lemmetyinen, H. (1997) Arrangement of a hydrophobically shielded porphyrin, 5,10,15,20-tetrakis(3,5-di-tert-butylphenyl)porphyrin, in octadecylamine LB multilayer, *Langmuir*, **13**(11), 3002–3008.
12. Vuorinen, T., Kaunisto, K., Tkachenko, N. V., Efimov, A., Lemmetyinen, H., Alekseev, A. S., Hosomizu, H., and Imahori, H. (2005) Photoinduced electron transfer in Langmuir-Blodgett monolayers of porphyrin-fullerene dyads, *Langmuir*, **21**, 5383–5390.
13. Tkachenko, N. V., Tauber, A. Y., Hynninen, P. H., Sharonov, A. Y., and Lemmetyinen, H. (1999) Light induced electron transfer in pyropheophytin-anthraquinone dyads: vectorial charge transfer in Langmuir-Blodgett films, *J. Phys. Chem. A.*, **103**, 3657–3665.
14. Suhling, K., French, P. M. W., and Phillips, D. (2005) Time-resolved fluorescence microscopy, *Photochem. Photobiol. Sci.*, **4**, 13–22.
15. Moerner, W. E. New directions in single-molecule imaging and analysis, Proc. Natl. Acad. Sci. **104**(31), 12596–12602.
16. Ikonen, M., Sharonov, A., Tkachenko, N., and Lemmetyinen, H. (1993) The photovoltage signals of bacteriorhodopsin in Langmuir-Blodgett films with different molecular orientation, *Adv. Mat. Opt. Electr.*, **2**, 115–122.
17. Smirnov, S., Vlassiouk, I., Kutzki, O., Wedel, M., and Montforts, F.-P. (2002) Unusual role of oxygen in electron-transfer processes, *J. Am. Chem. Soc.*, **124**, 4212–4212.
18. Smirnov, S. N., Liddell, P. A., Vlassiouk, I. V., Teslja, A., Kuciauskas, D., Braun, C. L., Moore, A. L., Moore, T. A., and Gust, D. (2003) Characterization of the giant transient dipole generated by photoinduced electron transfer in a carotene-porphyrin-fullerene molecular triad, *J. Phys. Chem. A.*, **107**, 7567–7573.
19. Tkachenko, N. V., Hynninen, P. H., and Lemmetyinen, H. (1996) Photoelectric signals of chlorophyll a in Langmuir-Blodgett films, *Chem. Phys. Lett.*, **261**, 234–240.

20. Miller, J. R., Beitz, J. V., and Huddleston, R. K. (1984) Effect of free energy on rates of electron transfer between molecules, *J. Am. Chem. Soc.*, **106**, 5057–5068.
21. Steinbach, P. J. (2002) Inferring lifetime distributions from kinetics by maximizing entropy using a bootstrapped model, *J. Chem. Inf. Comput. Sci.*, **42**(6), 1476–1478.
22. Edholm O., and Blomberg, C. (1999) Stretched exponentials and barrier distributions, *Chem. Phys.*, **252**, 221–225.
23. Lehtivuori, H., Efimov, A., Lemmetyinen, H., and Tkachenko, N. V. (2007) Distributed decay kinetics of charge separated state in solid films, *Chem. Phys. Lett.*, **437**, 238–242.
24. Lehtivuori, H., Kumpulainen, T., Efimov, A., Lemmetyinen, H., Kira, A., Imahori, H., and Tkachenko, N. V. (2008) Photoinduced electron transfer in langmuir–blodgett monolayers of double-linked phthalocyanine–fullerene dyads, *J. Phys. Chem. C.*, **112**, 9896–9902.
25. Chou, H., Chen, C.-T., Stork, K. F., Bohn, P. W., and Suslick, K. S. (1994) Langmuir-Blodgett films of amphiphilic push-pull porphyrins, *J. Phys. Chem.*, **98**, 383–385.
26. Choudhury, B., Weedon, A. C., and Bolton, J. R. (1998) Effect of molecular organization on photophysical behavior. 1. Steady-state fluorescene and fluorescene quantum yield studies of Langmuir-Blodgett monolayers of some surlactant porphyrins, *Langmuir.*, **14**, 6192–6198.
27. Vema, A. L., Zhang, Z., Tamai, N., Nakashima, K., Yoneyama, M., Iriyama, K., and Ozaki, Y. (1998) Dynamics of intra- and interlayer energy transfer in langmuir–blodgett films of 5-(4-n-octadecylpyridyl)-10,15,20-tri-p-tolylporphyrin studied by time-resolved fluorescence spectroscopy, Langmuir, **14**, 4638–4642.
28. Efimov, A. V., Anikin, M., Tkachenko, N. V., Mironov, A. F., and Lemmetyinen, H. (1998) Irregular behavior of 5,10,15,20-tetrakis (pentafluorophenyl)porphyrin in Langmuir-Blodgett films, *Chem. Phys. Lett.*, **289**, 572–578.
29. Qian, D.-J., Nakamura, C., and Miyake, J. (2001) Spectroscopic studies of the multiporphyrin arrays at the air-water interface and in Langmuir-Blodgett films, *Thin Solid Films*, **397**, 266–275.
30. Togashi, D. M., Romao, R. I. S., da Silva, A. M. G., Sobral, A. J. F. N., and Costa, S. M. B. (2005) Self-organization of a sulfonamido-porphyrin in Langmuir monolayers and Langmuir–Blodgett films, *Phys. Chem. Chem. Phys.*, **7**, 3874–3883.

31. Ohta, N., Nomura, T., and Yamazaki, I. (1997) Effect of an external electric field on photoinduced interlayer electron transfer in Langmuir-Blodgett films as revealed by the effect of the field on fluorecence: a D-A system of thiacyanine and viologen, *J. Photochem. Photobil. A.*, **106**, 37–43.
32. Vuorimaa, E., Vuorinen, T., Tkachenko, N., Cramariuc, O., Hukka, T., Nummelin, S., Shivanyuk, A., Rissanen, K., and Lemmetyinen, H. (2001) Photoinduced electron transfer between a self-assembled resorcinarene-[60]fullerene complex and poly(3-hexylthiophene) in Langmuir-Blodgett films, *Langmuir*, **17**, 7327–7331.
33. Gust, D., Moore, T. A., and Moore, A. L. (1993) Molecular mimicry of photosynthetic energy and electron transfer, *Acc Chem. Res.*, **26**(4), 198–205.
34. Tkachenko, N. V., Lemmetyinen, H., Sonoda, J., Ohkubo, K., Sato, T., Imahori, H., and Fukuzumi, S. (2003) Ultrafast photodynamics of exciplex formation and photoinduced electron transfer in porphyrin-fullerene dyads linked at close proximity, *J. Phys. Chem. A.*, **107**, 8834–8844.
35. Kang, I., Rubtsov, V., Iovine, P. M., Chen, J., and Therien, M. J. (2002) Distance dependence of electron transfer in rigid, cofacially compressed, -stacked porphyrin-bridge-quinone systems, *J. Am. Chem. Soc.*, **124**, 8275–8279.
36. Wiberg, J., Guo, L., Pettersson, K., Nilsson, D., Ljungdahl, T., Martensson, J., and Albinsson, B. (2007) Charge recombination versus charge separation in donor-bridge-acceptor systems, *J. Am. Chem. Soc.*, **129**(1), 155–163.
37. P. Peumans, A. Yakimov, and S. Forrest, (2003) Small molecular weight organic thin-film photodetectors and solar cells, *J. Appl. Phys.* **93** (7), 3693–3723.
38. Armstrong, N. R., Wang, W., Alloway, D. M., Placencia, D., Ratcliff, E., and Brumbach, M. (2009) Organic/organic' heterojunctions: organic light emitting diodes and organic photovoltaic devices, *Macromol. Rapid Commun.*, **30**, 717–731.
39. Tran-Thi, T.-H., Fournier, T., Sharonov, A. Y., Tkachenko, N., Lemmetyinen, H., Grenier, P., Truong, K.-D., and Houde, D. (1996) Photophysical, photoelectrial and nonlinear properties of porphyrin-phthalocyanine assemblies in Langmuir-Blodgett films, *Thin Solid Films.*, **273**, 8–13.
40. Kesti, T., Vehmanen, V., M¨orsky, S., Tkachenko, N., Efimov, A. A., and Lemmetyinen, H. (2000) Photoinduced vectorial charge transfer in mixed Langmuir-Blodgett films of 5,10,15,20-tetrakis-(3,5-di-tert-

butylphenyl)porphyrin and its zinc(II) derivative, *J. Mater. Chem.*, **10**, 2283–2288.

41. Marczak, R., Sgobba, V., Kutner, W., Gadde, S., D'Souza, F., and Guldi, D. M. (2007) Langmuir-Blodgett films of a cationic zinc porphyrin-imidazole-functionalized fullerene dyad: formation and photoelectrochemical studies, *Langmuir.*, **23**, 1917–1923.
42. Tkachenko, N. V., Tauber, A. Y., Grandell, D., Hynninen, P. H., and Lemmetyinen, H. (1999) Light induced electron transfer in pyropheophytin-anthraquinone and phytochlorineanthraquinone dyads: Influence of conformational exchange, *J. Phys. Chem. A.*, **103**, 3646–3656.
43. Tkachenko, N. V., Rantala, L., Tauber, A. Y., Helaja, J., Hynninen, P. H., and Lemmetyinen, H. (1999) Photoinduced electron transfer in phytochlorin-[60]fullerene dyads, *J. Am. Chem. Soc.*, **121**, 9378–9387.
44. Tkachenko, N. V., Vuorimaa, E., Kesti, T., Alekseev, A. S., Tauber, A. Y., Helaja, J., Hynninen, P. H., and Lemmetyinen, H. (2000) Vectorial photoinduced electron transfer in phytochlorin- [60]fullerene Langmuir-Blodgett films, *J. Phys. Chem. B.*, **104**, 6371–6379.
45. Imahori, H., Tkachenko, N. V., Vehmanen, V., Tamaki, K., Lemmetyinen, H., Sakata, S., and Fukuzumi, S. (2001) An extremely small reorganization energy of electron transfer in porphyrin-fullerene dyad, *J. Phys. Chem. A.*, **105**, 1750–1756.
46. Imahori, H., Tamaki, K., Guldi, D. M., Luo, C., Fujitsuka, M., Ito, O., Sakata, Y., and Fukuzumi, S. (2001) Modulating charge separation and charge recombination dynamics in porphyrin-fullerene linked dyads and triads: Marcus-normal versus inverted region, *J. Am. Chem. Soc.*, **123**, 2607–2617.
47. Chukharev, V., Tkachenko, N. V., Efimov, A., Guldi, D. M., Hirsch, A., Scheloske, M., and Lemmetyinen, H. (2004) Tuning the ground-state and excited-state interchromophore interactions in porphyrin-fullerene *p*-stacks, *J. Phys. Chem. B.*, **108**, 16377–16385.
48. Chukharev, V., Tkachenko, N. V., Efimov, A., and Lemmetyinen, H. (2005) Effect of central metal on intra-molecular exciplex of porphyrin-fullerene double linked dyad, *Chem. Phys. Lett.*, **401**, 501–505.
49. Tkachenko, N. V., Vehmanen, V., Efimov, A., Alekseev, A. S., and Lemmetyinen, H. (2003) Vectorial photoinduced electron transfer of phytochlorin-fullerene system in Langmuir-Blodgett films, *J. Porphyr. Phthalocyan.*, **7**, 255–263.
50. Tkachenko, N. V., Guenter, C., Imahori, H., Tamaki, K., Sakata, S., Fukuzumi, S., and Lemmetyinen, H. (2000) Near infra-red emission

of charge-transfer complexes of porphyrinfullerene films, *Chem. Phys. Lett.*, **326**, 344–350.

51. Tkachenko, N. V., Vehmanen, V., Nikkanen, J.-P., Yamada, H., Imahori, H., Fukuzumi, S., and Lemmetyinen, H. (2002) Porphyrin-fullerene dyad with a long linker: formation of charge transfer conformer in Langmuir-Blodgett film, *Chem. Phys. Lett.*, **366**, 245–252.
52. Kesti, T., Tkachenko, N. V., Vehmanen, V., Yamada, H., Imahori, H., Fukuzumi, S., and Lemmetyinen, H. (2002) Exciplex intermediates in photoinduced electron transfer of porphyrin-fullerene dyads, *J. Am. Chem. Soc.*, **124**, 8067–8077.
53. Lehtivuori, H., Lemmetyinen, H., and Tkachenko, N. V. (2006) Exciplex-exciplex energy transfer and annihilation in solid films of porphyrin-fullerene dyads, *J. Am. Chem. Soc.*, **128**, 16036–16037.
54. ulbinas, V., Chachisvilis, M., Valkunas, L., and Sundstr¨om, V. (1996) Excited state dynamics of phthalocyanine films, *J. Phys. Chem.*, **100**(6), 2213–2219.
55. Fujitsuka, M., Hara, M., Tojo, S., Okada, A., Troiani, V., Solladie, N., and Majima, T. (2005) Fast exciton migration in porphyrin-functionalized polypeptides, *J. Phys. Chem. B.*, **109**(1), 33–35.
56. Larsen, J., Bruggemann, B., Polivka, T., Sundstro¨m, V., Kesson, E. A., Sly, J., and Crossley, M. J. (2005) Energy transfer within Zn-porphyrin dendrimers: study of the singlet-singlet annihilation kinetics, *J. Phys. Chem. A.*, **109**(47), 10654–10662.
57. Guldi, D. M., Zilbermann, I., Gouloumis, A., Vazquez, P., and Torres, T. (2004) Metallophthalocyanines: Versatile electron-donating building blocks for fullerene dyads, *J. Phys. Chem. B.*, **108**(48), 18485–18494.
58. Isosomppi, M., Tkachenko, N. V., Efimov, A., Vahasalo, H., Jukola, J., Vainio-talo, P., and Lemmetyinen, H. (2006) Photoinduced electron transfer of double-bridged phthalocyanine–fullerene dyads, *Chem. Phys. Lett.*, **430**, 36–40.
59. Niemi, M., Tkachenko, N. V., Efimov, A., Lehtivuori, H., Ohkubo, K., Fukuzumi, S., and Lemmetyinen, H. (2008) Exciplex mediated photoinduced electron transfer reactions of phthalocyanine-fullerene dyads, *J. Phys. Chem. A.*, **112**, 6884–6892.
60. Alekseev, A. S., Efimov, A. V., and Lemmetyinen, H. (2008) Conversion of light energy into the electric potential in Langmuir–Shaefer organic molecular films, *Bull. Lebedev Phys. Inst.*, **35**(12), 363–367.
61. Lehtivuori, H., Kumpulainen, T., Hietala, M., Efimov, A., Lemmetyinen, H., Kira, A., Imahori, H., and Tkachenko, N. V. (2009) Photodynamics

of charge separation and recombination in solid alternating films of phthalocyanine or phthalocyanine–fullerene dyad and perylene dicarboximide, *J. Phys. Chem. C.*, **113**, 1984–1992.

62. Lehtivuori, H., Kumpulainen, T., Efimov, A., Lemmetyinen, H., Urthner, F. W., and Tkachenko, N. V. (2010) Interlayer energy transfer between perylene diimide and phthalocyanine monolayers, *J. Photochem. Photobiol. A.*, **211**, 26–31.
63. Alekseev, A. S., Tkachenko, N. V., Tauber, A. Y. , Hynninen, P. H., Osterbacka, R., Stubb, H., and Lemmetyinen, H. (2002) Vectorial photoinduced electron transfer in alternating Langmuir-Blodgett films of phytochlorin-[60]fullerene dyad and regioregular poly(3-hexylthiophene), *Chem. Phys.*, **275**, 243–251.
64. Vuorinen, T., Kaunisto, K., Tkachenko, N. V., Efimov, A., and Lemmetyinen, H. (2006) Photoinduced interlayer electron transfer in alternating porphyrin-fullerene dyad and regioregular poly(3-hexylthiophene) langmuirblodgett films, *J. Photochem. Photobiol. A.*, **178**(3), 185–191.
65. Vuorinen, T., Kaunisto, K., Chukharev, V., Tkachenko, N. V., Efimov, A., and Lemmetyinen, H. (2006) Kinetics of photoinduced electron transfer in polythiophene-porphyrin-fullerene molecular films, *J. Phys. Chem. B.*, **110**, 19515–19520.
66. Kaunisto, K., Chukharev, V., Tkachenko, N. V., and Lemmetyinen, H. (2008) Long-lived charge separated state in molecular films containing porphyrin-fullerene dyad, *Chem. Phys. Lett.*, **460**, 241–244.
67. Kaunisto, K., Chukharev, V., Tkachenko, N. V., Efimov, A., and Lemmetyinen, H. (2009) Energy and electron transfer in multilayer films containing porphyrin-fullerene dyad, *J. Phys. Chem. C.*, **113**, 3819–3825.
68. Kaunisto, K., Vuorinen, T., Vahasalo, H., Chukharev, V., Tkachenko, N. V., Tolkki, A., Efimov, A., Lehtivuori, H., and Lemmetyinen, H. (2008) Photoinduced electron transfer and photocurrent in multicomponent organic molecular films containing oriented porphyrin-fullerene dyad, *J. Phys. Chem. C.*, **112**, 10256–10265.
69. Kaunisto, K., Vahasalo, H., Chukharev, V., Tkachenko, N. V., Vivo, P., Niemi, M., Tolkki, A., Efimov, A., and Lemmetyinen, H. (2009) Photoinduced charge transfer through films containing poly(hexylthiophene), phthalocyanine, and porphyrin-fullerene layers, *Thin Solid Films.*, **517**, 2988–2993.
70. Grätzel, M. (2001) Photoelectrochemical cells, *Nature*, **414**(1), 338–344.

71. Chukharev, V., Vuorinen, T., Efimov, A., Tkachenko, N. V., Kimura, M., Fukuzumi, S., Imahori, H., and Lemmetyinen, H. (2005) Photoinduced electron transfer in self-assembled monolayers of porphyrin-fullerene dyads on ito, *Langmuir*, **21**, 6385–6391.
72. Isosomppi, M., Tkachenko, N. V., Efimov, A., Kaunisto, K., Hosomizu, K., Imahori, H., and Lemmetyinen, H. (2005) Photoinduced electron transfer in multilayer self-assembled structures of porphyrins and porphyrin-fullerene dyads on ito, *J. Mater. Chem.*, **15**, 4546–4554.
73. Vivo, P., Ojala, M., Chukharev, V., Efimov, A., and Lemmetyinen, H. (2009) Role of a phthalocyanine-fullerene dyad in multilayered organic solar cells, *J. Photochem. Photobiol., A.*, **203**, 125–130.

Chapter 12

Supramolecular Energy and Electron-Transfer Processes and Their Switching

Joe Otsuki

College of Science and Technology, Nihon University,
1-8-14 Kanda Surugadai, Chiyoda-ku, Tokyo, 101-8308, Japan
otsuki@chem.cst.nihon-u.ac.jp

12.1 Introduction

Molecules are the smallest entities that can have some function. Therefore, the smallest possible devices may consist of molecules, in which each individual molecule works as a functional unit [1, 2]. Molecules must be organized into well-defined architectures by means of directional intermolecular interactions, that is, non-covalent bonds, for cooperative performance [3]. Communication among functional units, i.e., molecules, may be achieved on the basis of electron- and energy-transfer processes. Therefore, properties of intermolecular bonds should be clarified as conduits for electronic communication. Further, switching functions for these processes must be installed as mechanisms for logic operation [4]. Switching may be effected by external stimuli, such as photons, electrons,

Multiporphyrin Arrays: Fundamentals and Applications
Edited by Dongho Kim

www.panstanford.com

protons, etc. Porphyrins are very attractive building blocks for such molecule-based devices due to its excellent electronic and photonic properties and their amenability to synthetic modification to afford well-defined elaborate structures [5–7].

This chapter starts with topics concerning electron-transfer processes through noncovalent bonds. Discussion on energy-transfer processes through noncovalent bonds follows. In particular, a through-bond mechanism across a noncovalent bond, specifically, an amidinium–carboxylate salt bridge, is proposed. Subsequently, supramolecular switches of energy- and electron-transfer processes will be discussed. These switches are constructed via noncovalent bonds. In addition to electronic modulation, structural modulation, particularly, reversible complexation and decomplexation, can be utilized for switching, as a unique mechanism for systems constructed via reversible intermolecular bonds. Finally, some scanning tunneling microscope images are presented for supramolecular organizations of porphyrins. These images clearly show that well-defined supramolecular organizations are formed (i) not only in solution but on surfaces, and equally importantly, (ii) not only as an averaged out architecture but on an individual molecule basis. Finally, a summary and a prospect are briefly given to conclude this chapter. This article is not intended to be comprehensive at all. Examples are mostly taken from works of my own group. Works from other groups are mentioned only when they have direct relevance to ours. This article is rather intended to convey key ideas aiming at molecular devices along the line of (i) electron- and energy-transfer processes, (ii) switching of these processes, and (iii) visualization of relevant architectures, all in supramolecular systems, in which molecules are organized by self-assembly via noncovalent interactions.

A good starting point for theories of nonradiative processes including electron- and energy-transfer processes may be Fermi's golden rule, which is derived from the time-dependent first-order perturbation theory [8, 9]:

$$k_{\mathrm{i}\to\mathrm{f}} = \frac{2\pi}{\hbar}\left|H_{\mathrm{fi}}^{1}\right|^{2}\rho\left(E_{\mathrm{i}}\right) \tag{12.1}$$

where $k_{\mathrm{i}\to\mathrm{f}}$ is the rate constant for a change from the initial state (i) to the final states (f), $\hbar$ is Planck's constant divided by 2π, and $\rho(E_{\mathrm{i}})$

is the density of *final* states at the energy E_{i}, which is the energy of the *initial* state. The Hamiltonian matrix element, H_{fi}^{1}, is defined as

$$H_{\mathrm{fi}}^{1} = \int \Psi_{\mathrm{f}}^{*} H^{1} \Psi_{\mathrm{i}} d\tau \tag{12.2}$$

where H^1 is the first-order perturbation Hamiltonian, Ψ_i and Ψ_{f} are the wavefunctions for the initial and final states, respectively, and the integration is over the whole space. The Hamiltonian H^1 is a collection of the terms such as $\Sigma e^2/4\varepsilon_o \pi \mathrm{r}$, where e is the charge of proton, ε_o is the vacuum permittivity, r is the distance between an electron in the donor and that in the acceptor, and the summation is taken over intermolecular combinations of electrons. Equation (12.1) indicates that a process occurs if there is a final state with the same energy as the initial state and the rate of the process is proportional to the interaction between the initial and final states represented by the matrix element of perturbation Hamiltonian. Specific equations for electron- and energy-transfer processes are derived on the basis of the Fermi golden rule, which will be described in the following sections.

12.2 Supramolecular Electron Transfer

12.2.1 *Driving Force Dependence of Electron-Transfer Rates Through Non-Covalent Bonds*

Electron transfer is one of the most fundamental processes in chemistry. Electron transfer among porphyrins are attracting great deal of attention because this class of compounds not only play central roles in biological electron-transfer systems including photosynthesis and respiration, but have potential to be applied in electronic and photonic devices [10, 11]. According to the Marcus theory for nonadiabatic electron transfer for weakly interacting donor–acceptor systems, the rate constant for electron transfer is given by [12]

$$k_{\mathrm{eT}} = \frac{2\pi}{\hbar} \left| H_{\mathrm{fi}}^{1} \right|^{2} \frac{1}{(4\pi \lambda k_{\mathrm{B}} T)^{1/2}} \exp\left(-\frac{(\Delta G^{\circ} + \lambda)^{2}}{4\lambda k_{\mathrm{B}} T} \right) \tag{12.3}$$

where k_{B} is the Boltzmann constant, λ is the reorganization energy, T is the temperature. Note the similarity of Eq. 12.3 to the

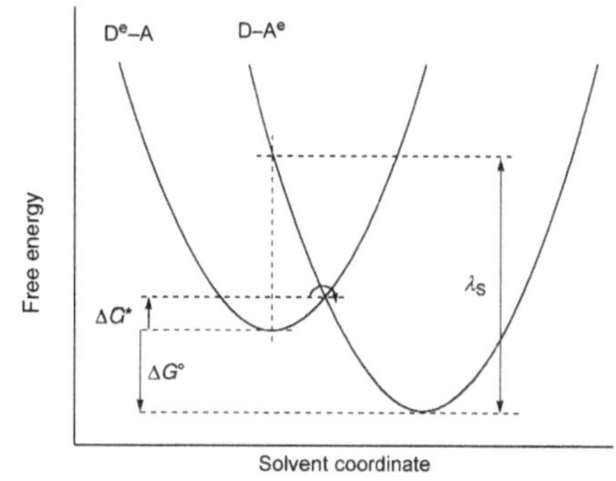

Figure 12.1. Potential energy curves before (D^e–A) and after (D–A^e) electron transfer against the solvent coordinate.

Fermi's golden rule formalism (Eq. 12.1). A diagrammatic representation corresponding this equation is given in Fig. 12.1. In this figure, only the solvent contribution to the total reorganization energy (λ_S) is included. The activation energy, $\Delta G^* = -(\Delta G^\circ + \lambda)^2/4\lambda$, is required for the electron-transfer process to occur.

If high-frequency vibration modes within the molecules are taken into account as in Fig. 12.2, the electron-transfer rate can be

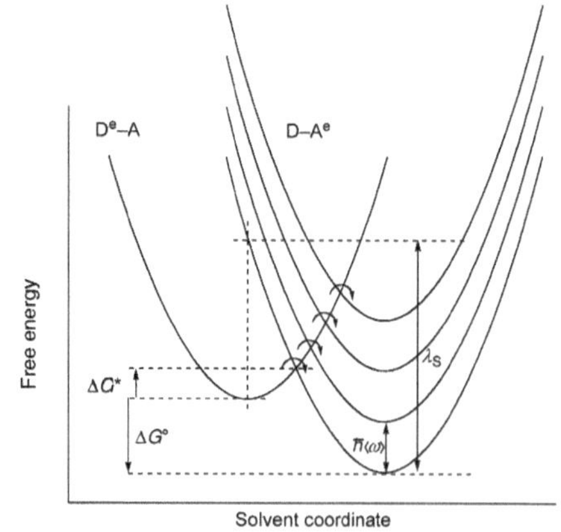

Figure 12.2. Potential energy curves including high-frequency modes.

expressed as [13]

$$k_{\mathrm{eT}} = \left(\frac{\pi}{\mathrm{h}^2 \lambda_S k_B T}\right)^{1/2} \left|H_{\mathrm{fi}}^1\right|^2 \sum_n \mathrm{e}^{-S} \frac{S^n}{n!} \times \exp\left(-\frac{(\Delta G + \lambda_S + n\mathrm{h}\langle\omega\rangle)^2}{4\lambda_S k_B T}\right) \quad (12.4)$$

where $S = \lambda_V/\hbar\langle\omega\rangle$ is the Huang-Rhys factor, λ_V is the reorganization energy associated with the averaged angular frequency $\langle\omega\rangle$ within the molecules. The sum is over integers, n.

These equations indicate that the driving force dependence of the electron-transfer rate would be bell-shaped. As the driving force increases, the rate increases as in common reactions; but this is so only as long as the driving force is less than the reorganization energy. As the driving force is further increased, the rate begins to drop, giving rise to the bell-shaped dependence. The regions where the rate increases and decreases as the driving force increases are respectively called a *normal region* and an *inverted region*.

This theoretical prediction inspired experimentalists. J. R. Miller and co-workers proved the existence of the inverted region with their experiments on a system in which donor and acceptor molecules were dispersed in a solid glass [14]. G. L. Closs and co-workers demonstrated the inverted region with their donor–acceptor conjugates linked via a rigid spacer [15]. For porphyrin derivatives, Wasielewski and co-workers investigated the rate dependence on the driving force for photoinduced charge separation processes in a series of porphyrin–triptycene molecules shown in Fig. 12.3 dispersed in a rigid glass [16]. They found that the rates were in the normal region and that the data can be fit with the parameter, $H_{\mathrm{fi}}^1 = 4.1$ meV, $\hbar\langle\omega\rangle = 0.062$ eV, $\lambda_S = 0.3$ eV, and $\lambda_V = 0.3$ eV, on the basis of Eq. 12.4. Osuka and co-workers investigated a series of porphyrin–quinone dyads, in which various quinones are attached through a common spacer as shown in Fig. 12.4 [17]. They clearly demonstrated that the rates of photoinduced charge separation are within a normal region, while the rates of charge recombination processes are in an inverted region. Parameters that best described the processes were found to be $H_{\mathrm{fi}}^1 = 3.8$ meV, $\hbar\langle\omega\rangle = 0.15$ eV, $\lambda_S = 0.18$ eV, and $\lambda_V = 0.6$ eV.

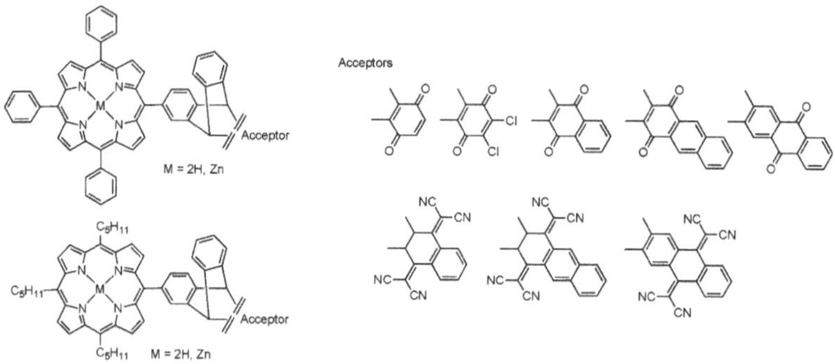

Figure 12.3. Covalent porphyrin–acceptor dyads linked via a triptycene unit [16].

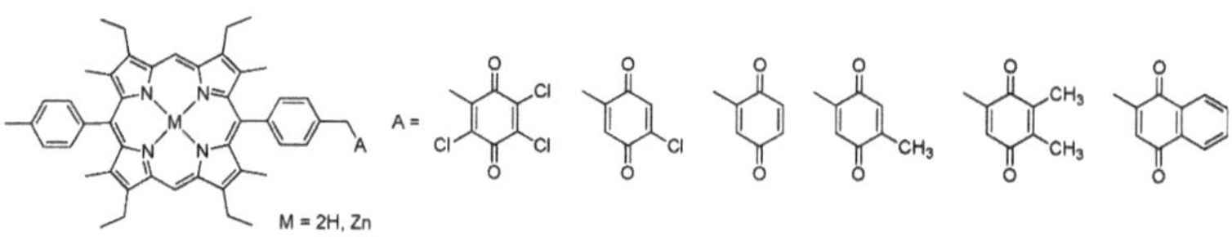

Figure 12.4. Covalent porphyrin–acceptor dyads linked via a methylene unit [17].

Hunter and co-workers reported a very simple way to assemble donor–[spacer–acceptor] supramolecules making use of axial ligation of pyridine ligands to zinc–porphyrin [18]. Zinc–porphyrin accepts only one ligand to form a well-defined supramolecules [19–21]. We employed their system to obtain donor–acceptor dyads, ZnTPP·Py-Im, in which the acceptor is systematically varied in the redox potentials, affording a set of supramolecular donor–acceptor dyads with varying driving forces for photoinduced electron transfer [22]. The structures of the supramolecules and the acceptors used are displayed in Fig. 12.5. The electron-transfer rates in CH_2Cl_2 were analyzed using Eq. 12.3. Assuming that the parameters λ and H_{fi}^1 are constant throughout the series of acceptors, which may be justified on account of their chemical and structural similarity, possible curves from some combinations of these parameters are shown in Fig. 12.6. Although it was not possible to single out parameter values to fit all the observed data, it is likely that $H_{\mathrm{fi}}^1 = 0.4$ meV and $\lambda = 0.6 - 0.8$ eV. It is interesting to note that the value of H_{fi}^1, which

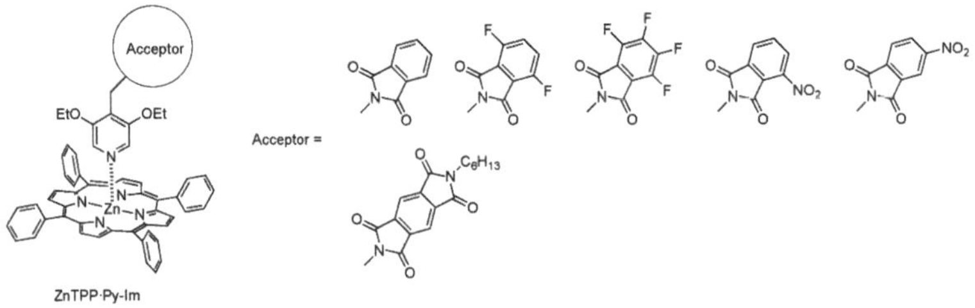

Figure 12.5. Non-covalent porphyrin–acceptor dyads [22].

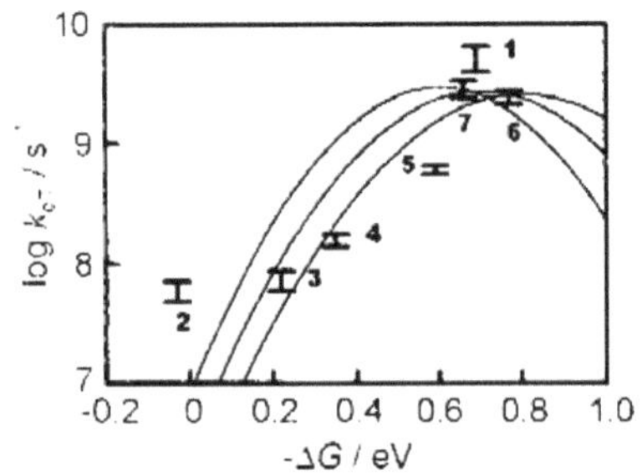

Figure 12.6. The dependence of intrasupramolecular electron-transfer rate on the energy gap. Reproduced with permission from [22].

represents electronic interactions between the donor and the acceptor, is one order of magnitude smaller in this noncovalent linkage than the cases with covalent linkages described above.

12.2.2 *Modular Construction of Supramolecular Electron-Transfer Systems*

The use of noncovalent interactions, in addition to covalent bonds to form molecules, to construct donor–acceptor electron-transfer systems is a step toward complex molecular organizations [5–7, 23, 24]. The use of an independent spacer molecule having two recognition sites, which can bind both donor and acceptor molecules to form a *ternary* complex, merits special attention, because (1) each component may be replaced and optimized independently; (2) these systems may be considered a further step towards constructing more complex molecular organizations than binary complexes such as

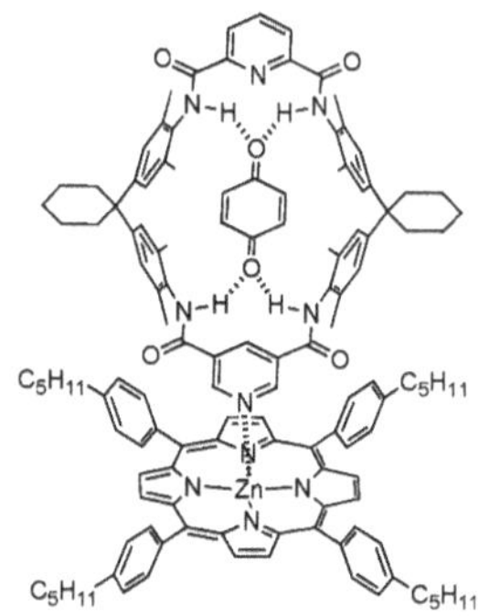

Figure 12.7. Noncovalent D–S–A ternary complex [25].

those described in the previous section; and (3) in biology, especially in photosynthetic reaction centers, proteins organize functional units such as chromophores and redox active agents, which can be regarded as spacers.

Hunter and co-workers have prepared a macrocyclic spacer molecule that binds zinc–porphyrin through axial coordination and quinones through hydrogen bonds as shown in Fig. 12.7 [25]. We prepared an amidinium derivative of pyridine that was intended as a spacer between a zinc–porphyrin and an acceptor molecule that carries a carboxylate moiety, as shown in Fig. 12.8 [26]. The

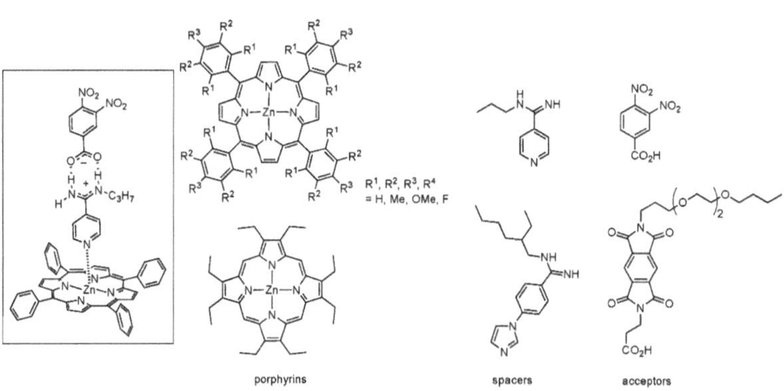

Figure 12.8. A library of porphyrins, spacers, and acceptors for ternary D–S–A complexes [26, 30]. On the left is an example of ternary complexes, ZnTPP·S1·DNBA.

amidinium–carboxylate bond is strong for a noncovalent interaction because of the electrostatic interaction in addition to hydrogen bonding. This interaction was extensively utilized in the study of proton-coupled electron-transfer processes by Nocera and coworkers [27–29].

With the ditopic spacer, S1, a ternary supramolecular complex, ZnTPP·S1·DNBA, formed with an overall association constant of 1.9×10^8 M^{-2} in CH_2Cl_2. The fluorescence of the zinc–porphyrin was nearly completely quenched. The lifetime of the excited state was estimated as shorter than 20 ps, which suggests that the spacer and the intermolecular interfaces, i.e., the axial coordination and the amidinium–carboxylate salt bridge, are effective as a conduit for electron-transfer processes.

We demonstrated the advantage of modular approach experimentally by actually constructing 20 different D–S–A ternary complexes from a library of donors, spacers, and acceptors displayed in Fig. 12.8 [30].

12.3 Supramolecular Energy Transfer

12.3.1 *Energy Transfer*

For an ensemble of molecules, both the initial and final states would have some distribution in energy. In the case of energy transfer, the distribution of the initial states may be represented by the emission spectrum of the donor and that of the final states by the absorption spectrum of the acceptor. Hence, the Fermi golden rule is recast in the form [8]:

$$k_{\mathrm{i}\to\mathrm{f}} = \frac{2\pi}{\hbar}\left|H_{\mathrm{fi}}^{1}\right|^2 \int g_{\mathrm{em}}^{D}(E)\, g_{\mathrm{abs}}^{\mathrm{A}}(E)\,\mathrm{d}E \qquad (12.5)$$

where $g_{\mathrm{em}}^{\mathrm{D}}(E)$ and $g_{\mathrm{abs}}^{\mathrm{A}}(E)$ are lineshape functions of donor emission and acceptor absorption spectra, respectively, which are normalized according to

$$\int g_{\mathrm{em}}^{\mathrm{D}}(E)\,\mathrm{d}E = 1 \text{ and } \int g_{\mathrm{abs}}^{\mathrm{A}}(E)\,\mathrm{d}E = 1 \qquad (12.6)$$

Note that these line shape functions are on an energy scale. Common spectra on a wavelength scale or on a wavenumber scale must be converted to those on an energy scale as appropriate.

The initial and final states may be represented by properly antisymmetrized wavefunctions as

$$\Psi_\mathrm{i} = 2^{-1/2}[\psi_{\mathrm{D}^*}(1)\psi_\mathrm{A}(2) - \psi_\mathrm{A}(1)\psi_{\mathrm{D}^*}(2)] \tag{12.7}$$

and

$$\Psi_\mathrm{f} = 2^{-1/2}[\psi_\mathrm{D}(1)\psi_{\mathrm{A}^*}(2) - \psi_{\mathrm{A}^*}(1)\psi_\mathrm{D}(2)] \tag{12.8}$$

respectively. Here, ψ_D and ψ_A are molecular orbitals of non-interacting isolated donor and acceptor, respectively, and the numbers 1 and 2 in parentheses represent electron 1 and electron 2, respectively. Inserting Eqs. 12.7 and 12.8 into Eq. 12.2, one obtains

$$\begin{aligned} H_\mathrm{fi}^1 &= \int \psi_\mathrm{D}^*(1)\psi_{\mathrm{A}^*}^*(2)H^1\psi_{\mathrm{D}^*}(1)\psi_\mathrm{A}(2)\mathrm{d}\tau \\ &\quad - \int \psi_\mathrm{D}^*(1)\psi_{\mathrm{A}^*}^*(2)H^1\psi_\mathrm{A}(1)\psi_{\mathrm{D}^*}(2)\mathrm{d}\tau \end{aligned} \tag{12.9}$$

The first term is the Coulomb term, from which the Coulomb mechanism in energy transfer is derived [31, 32]. The Coulomb mechanism is also called through-space mechanism, dipole–dipole mechanism, and Förster mechanism. The FRET (Förster resonance energy transfer), often used in biological probes, is based on this mechanism. The second term is the exchange term, from which the exchange mechanism is derived. The exchange mechanism is also called through-bond mechanism and Dexter mechanism.

In deriving the rate equation on the basis of the Förster mechanism, a dipole–dipole approximation is used for the Hamiltonian, H^1. The rate is given by

$$k_\text{Förster} = \frac{9 \ln 10 \kappa^2 \Phi_\mathrm{F}}{128\pi^5 N_\mathrm{A} n^4 r^6 \tau_0} \int \varepsilon_\mathrm{A}(\lambda) g_\mathrm{em}^\mathrm{D}(\lambda)\lambda^4 \mathrm{d}\lambda \tag{12.10}$$

where N_A is the Avogadro number, n is the refractive index of the medium, r is the donor–acceptor separation, τ_0 and Φ_F are the excited state lifetime and the quantum yield of the donor emission in the absence of an acceptor, respectively, $\varepsilon_\mathrm{A}(\lambda)$ is the molar absorption coefficient of the acceptor at wavelength λ. The lineshape function g_em^D is the emission spectrum of the donor in a wavelength scale, which is normalized to have a unit area. The orientation factor κ is given by

$$\begin{aligned} \kappa &= \sin\theta_\mathrm{D} \sin\theta_\mathrm{A} \cos\phi - 2\cos\theta_\mathrm{D} \cos\theta_\mathrm{A} \\ &= \cos\theta_\mathrm{T} - 3\cos\theta_\mathrm{D}\cos\theta_\mathrm{A} \end{aligned} \tag{12.11}$$

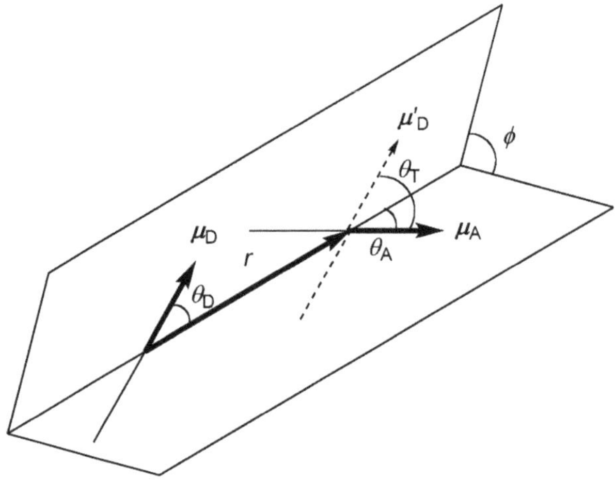

Figure 12.9. Definitions of θ_D, θ_A, θ_T, ϕ, and r in dipole–dipole interactions.

The relevant angles are defined as in Fig. 12.9, where μ_D and μ_A are the transition dipole moments of the donor and acceptor, respectively. In the case of porphyrins, the transition dipoles and the orientation factors have been analyzed thoroughly [33, 34]. Collecting numerical constants, we obtain a practical equation, with units expressed explicitly [35, 36]:

$$k_{\text{Förster}} = 8.81 \times 10^{-11} \frac{\kappa^2 \Phi_D}{n^4 (r/\text{nm})^6 (\tau_0/s)} \times \int (\varepsilon_A(\lambda)/\text{M}^{-1}\text{cm}^{-1}) g^D_{\text{em}}(\lambda)(\lambda/\text{nm})^4 \text{d}\lambda \qquad (12.12)$$

What is remarkable for this formulation is that the rate can be calculated exclusively using experimentally accessible parameters, once the distance and orientation of the donor and acceptor are established.

The formula for the rate of Dexter-type energy transfer, k_{Dexter}, is obtained in principle by inserting the second term of Eq. 12.9 into the Fermi golden rule (Eq. 12.1). In contrast to the Coulomb mechanism, however, it is difficult to predict the value of k_{Dexter} from experimentally obtainable parameters.

12.3.2 *Fast Excited Energy Transfer in Porphyrin Assemblies*

Well-defined supramolecular assemblies of zinc and free-base porphyrins were constructed through the formation of amidinium–carboxylate salt bridges, as shown in Fig. 12.10 [37]. The supramolecular 1:1 complex forms from zinc–porphyrin monoamidinate, ZnPA-1, and free-base porphyrin monocarboxylate, FbPC-1, with an association constant of 2.9×10^5 M^{-1} in toluene/DMSO (99:1). If the monocarboxylate is replaced by the tetracarboxylate, $FbPC_4$, the 4:1 complex results, as shown in Fig. 12.11. Efficient energy transfer from the zinc–porphyrin to the free-base porphyrin in these assemblies were indicated by a decreased fluorescence from the zinc–porphyrin and a sensitized fluorescence from the free-base porphyrin. For the 1:1 mixture, ZnPA-1·FbPC-1, the zinc–porphyrin fluorescence exhibited a biexponential decay with lifetime components of 0.22 and 1.92 ns. The longer lifetime was identical to that of free zinc–porphyrin (τ_0), while the shorter component was assigned as being resulted from the complex (τ). The energy-transfer rate constant was determined as 4.0×10^9 s^{-1} from the equation,

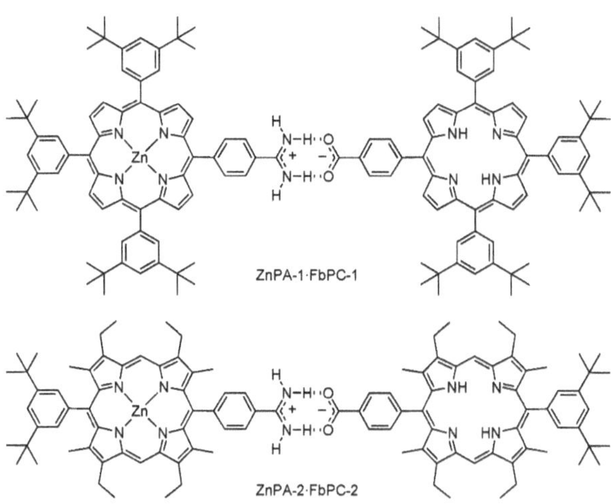

Figure 12.10. Supramolecular zinc–porphyrin–free-base porphyrin dyads linked via a common salt bridge but modified with different substituents [37, 39].

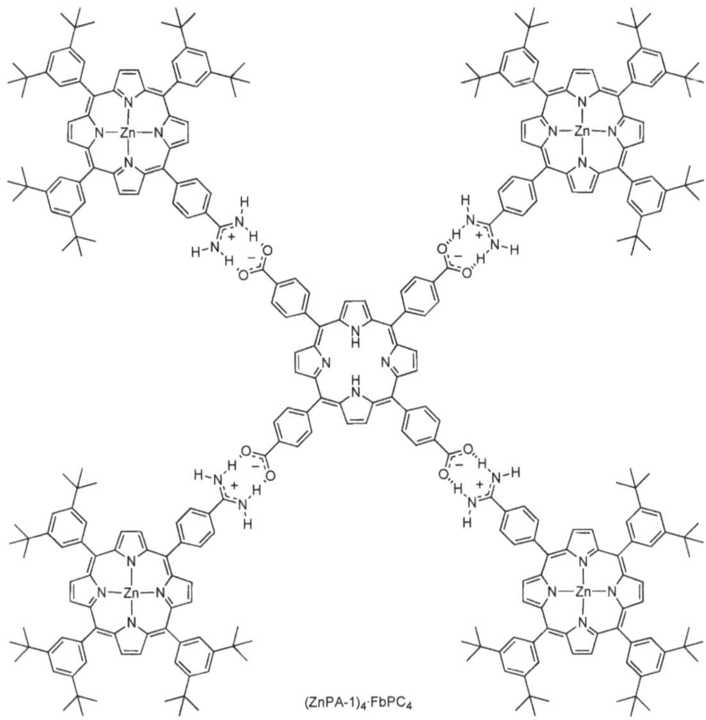

Figure 12.11. Supramolecular pentad as an energy harvesting antenna [37].

$k_{ET} = \tau^{-1} - \tau_0^{-1}$. These lifetimes gave a quantum yield for the energy transfer as 0.89. Similar analysis for the 4:1 complex gave a rate of 1.2×10^9 s^{-1} and a quantum yield of 0.96.

The rates of Förster-type energy transfer were estimated for the dyad and the pentad to be 5.1×10^8 and 6.8×10^8 s^{-1}, respectively. The estimation was based on values of $\kappa^2 = 1.01$ as an upper limit for a colinear array [33], $J_F = 3.4 \times 10^{-14}$ M^{-1} cm^3 (the 1:1 complex) or $J_F = 4.5 \times 10^{-14}$ M^{-1} cm^3 (the 4:1 complex), and $r = 2.23$ nm estimated from related crystal structures. These calculated rates are certainly too small to account for the observed rates of energy transfer. This is the case even if structural uncertainties are taken into account; if the observed rate were accounted for by the Förster mechanism alone, the distance r would have to be ~16 Å, which is improbably short. These considerations infer that a through-bond mechanism might be in operation in the excited energy transfer in these porphyrin assemblies. If this is the case, it is a rare case as most energy-transfer systems were explained only by the Förster-type mechanism [38].

The rate of Dexter-type energy transfer is sensitive to the nature of frontier orbitals, as it is the electrons in these orbitals that are exchanged during the energy-transfer process. The frontier orbitals of porphyrin are shown in Fig. 12.12 [34]. While the two e_g orbitals (LUMO) are degenerate from symmetry, the a_{2u} and a_{1u} orbitals are close in energy incidentally. Which of the a_{2u} and a_{1u} orbitals becomes the HOMO dependes on substituents on the porphyrin core. The a_{2u} orbital has a significant population on the meso carbons, whereas the a_{1u} orbital has nodes on the meso carbons. Therefore, through-bond energy transfer would benefit if the HOMO is of a_{2u} symmetry for porphyrin dyads linked via the meso

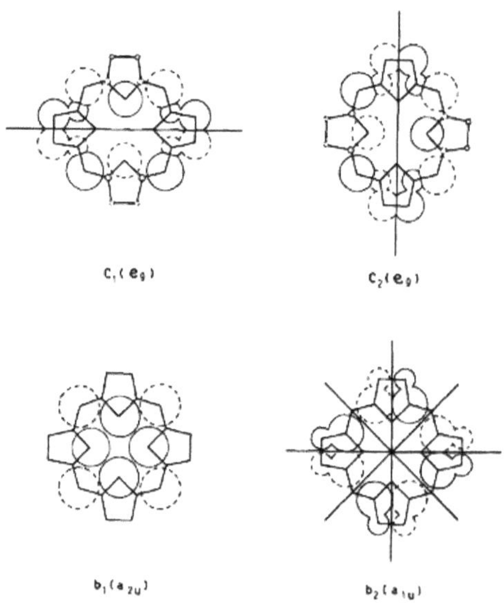

Figure 12.12. Frontier orbitals of porphyrin. The two orbitals of e_g symmetry, which are degenerate, correspond to the LUMO. The orbitals of a_{2u} and a_{1u} symmetry, corresponding to the HOMO and HOMO−1, are incidentally close in energy. Reproduced with permission from [34].

positions, because an electron or a hole in the HOMO would have a greater chance to move into the bridge. In other words, the matrix element, H_{fi}^{1}, would be larger if the HOMO is of a_{2u} symmetry. Electron donating substituents on the meso positions would raise the energy of the a_{2u} orbital, thereby making the a_{2u} orbital the HOMO. On the other hand, substitution by electron donating groups on the β positions would raise the energy of the a_{1u} orbital, thereby making the a_{1u} orbital the HOMO. These considerations suggest that the rate of Dexter-type energy transfer may be modulated with the substituent patterns on the porphyrin framework.

This notion was experimentally confirmed by comparing energy-transfer rates in the two covalent dyads, one from the Lindsey group [40] and the other from the Osuka group [41], in which a zinc–porphyrin moiety and a free-base porphyrin moiety are linked via the ethylnylene unit, as shown in Fig. 12.13. While the energy-transfer rates in the Lindsey dyad is 4.2 $\times 10^{10}$ s^{-1}, that in the Osuka dyad is 2.4 $\times 10^{9}$ s^{-1}, 17-fold smaller than the Lindsey dyad, even though these dyads have the same linker moiety. The difference has been accounted for by assuming different ordering of the

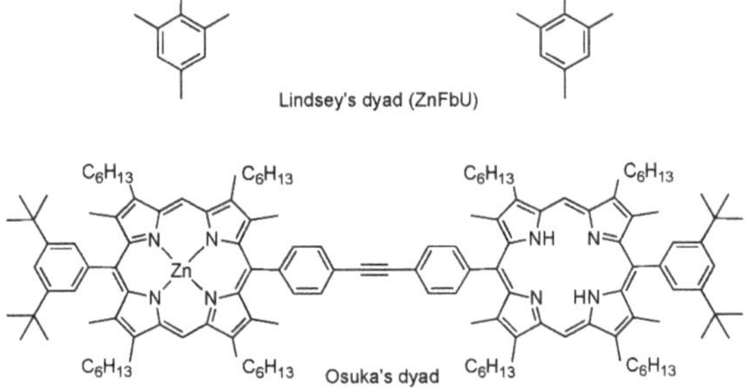

Figure 12.13. Covalent D–A dyads consisting of zinc and free-base porphyrins, linked via a common phenylene ethynylene unit but modified with different substituents [40, 41].

molecular orbitals [42, 43]. The HOMO of the meso tetraphenyl-substituted porphyrins in the Lindsey dyad is of an a_{2u} type due to the aryl substituents at the meso carbons. Therefore, the meso carbon directly linked to the bridging moiety has a significant population that may facilitate the electronic exchange processes. On the other hand, the HOMO of the β-octaalkyl substituted (and meso diphenyl-substituted) porphyrins in the Osuka dyad is of a_{1u} type and therefore has a node on the meso carbon at the linker, thereby attenuating the electron exchange processes. Faster energy-transfer processes for *meso*-tetraphenyl-type porphyrins than for 5,15-diaryl-β-octaalkyl-type porphyrins were confirmed for related covalently-linked porphyrin dyads as well [44, 45].

We examined the energy-transfer rate for our noncovalent dyads along the same line of argument. We have prepared another dyad, which were linked via the same salt bridge as in ZnPA-1·FbPC-1 but had alkyl groups on the β-positions, ZnPA-2·FbPC-2 (Fig. 12.10) [39]. The frontier orbitals of these porphyrins according to DFT calculations (B3LYP/6-31g*) are displayed in Fig. 12.14. The degeneracy of the LUMOs is lifted due to a slight deviation from a D_{4h} symmetry. However, the energies of the LUMO and LUMO+1 remain

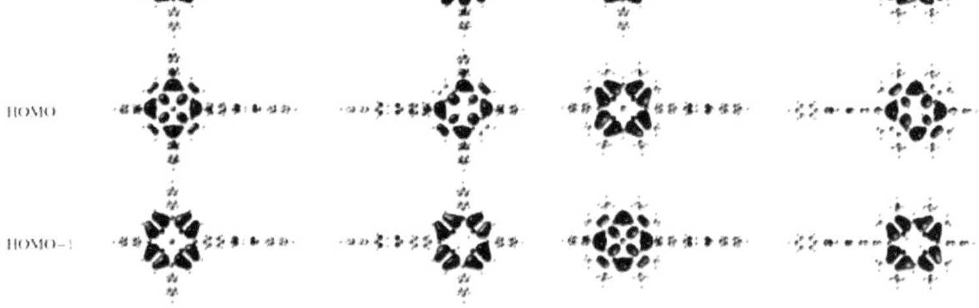

Figure 12.14. Frontier orbitals of ZnPA-1, FbPC-1, ZnPA-2, and ZnPC-2 according to DFT calculations (B3LYP/6-31g*). Reproduced with permission from [39].

almost identical. The HOMOs are of a_{2u}–a_{2u} symmetry for ZnPA-1–FbPC-1 dyad, while they are of a_{1u}-a_{2u} symmetry for ZnPA-2–FbPC-2 dyad. The involvement of an a_{1u}-type orbital would attenuate the coupling between the porphyrin and the bridge in the ZnPA-2–FbPC-2 dyad. Furthermore, the energies of a_{2u} and a_{1u} orbitals of the FbPC-2 are so close that a contribution of a_{1u} orbital may be significant. Taken together, it is expected that the exchange energy transfer would be faster for the ZnPA-1–FbPC-1 dyad than for the ZnPA-2–FbPC-2 dyad. Therefore, if the Dexter-type mechanism is involved in the energy transfer, faster energy transfer should be observed for the ZnPA-1–FbPC-1 dyad than for the ZnPA-2–FbPC-2 dyad. If the Dexter-type energy transfer is not involved, the different couplings between the porphyrin and the bridge would have little impact on the transfer rate. Time-resolved experiments showed that the energy-transfer rate for the ZnPA-2–FbPC-2 dyad was 1.3×10^9 s^{-1}, which is about a third of that in the ZnPA-1–FbPC-1 dyad. We have therefore concluded that the fast energy transfer in the noncovalently linked dyads, especially in the ZnPA-1–FbPC-1 dyad, is due to the involvement of a Dexter-type, through-bond mechanism.

The bridge must play a significant role in the through-bond mechanism for such a long-range process involving an intermolecular bond. Because direct orbital interactions between the porphyrin cores are unlikely, the matrix element, $H^1_{\rm fi}$, should be negligible if only localized excited states, D*–A and D–A* are assumed. In such cases, the involvement of the (excited states of) bridge can be invoked, giving rise to the superexchange mechanism. In a case where D and A are connected by a bridge consisting of units B_1 and B_2 in tandem, the matrix element $H^1_{\rm SE}$ to replace $H^1_{\rm fi}$ in Eq. 12.1, is represented as [46, 47]

$$H^1_{\rm SE} = \frac{H^1_{\rm DB_1} H^1_{\rm B_1B_2} H^1_{\rm B_2A}}{(\Delta E_{\rm DB})^2} \tag{12.13}$$

The pictorial illustration corresponding to this equation is given in Fig. 12.15. The matrix elements are between the localized excited states designated by the subscripts. Here, $\Delta E_{\rm DB}$ is the difference in energy between the excited states of the donor and the bridge. It is assumed in this equation that the energies of B_1* and B_2* are the same for simplicity. In the present case, benzamidinium and benzoate correspond to B_1 and B_2, respectively. The observation that

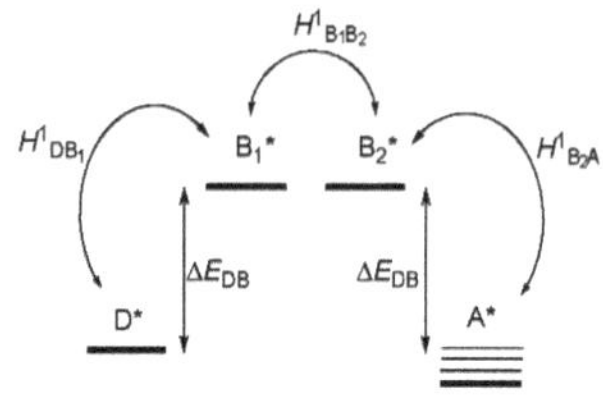

Figure 12.15. Superexchange interaction between D and A linked via a bridge, B, consisting of two identical units.

the through-bond energy transfer occurs at a significant rate suggests that the through-hydrogen-bond electronic coupling, $H^1_{B_1B_2}$, is also significant.

12.4 Switching of Electron- and Energy-Transfer Processes

12.4.1 *Supramolecular Switches*

Communication among molecular entities in supramolecular organizations is realized through electron- and energy-transfer processes. Our attempts in this respect are described in the previous sections. The next step towards molecular devices would be to realize switching of these processes with an external stimulus. Thus, molecular switches constitute an active area of research over the past 20 years [1, 2, 4, 48–55]. The simplest switch is a light switch, in which emission is switched on and off in response to a stimulus. The emitter–quencher motif is well documented [56]: when the quencher is active, no emission is exhibited, while the quencher is made inactive by some stimuli, emission appears. The situation is schematically depicted in Fig. 12.16. As a quencher, electron donor or acceptor, energy acceptor, or any other moiety may be used that facilitates a nonradiative process to effectively compete the emission process.

The emitter–quencher motif was constructed readily using zinc–porphyrin as an emitter, phenylazopyridine as a quencher, and axial coordination as an interaction for binding them [57]. Phenylazopyridine, PhNNPy, binds to ZnTPP with an association constant of

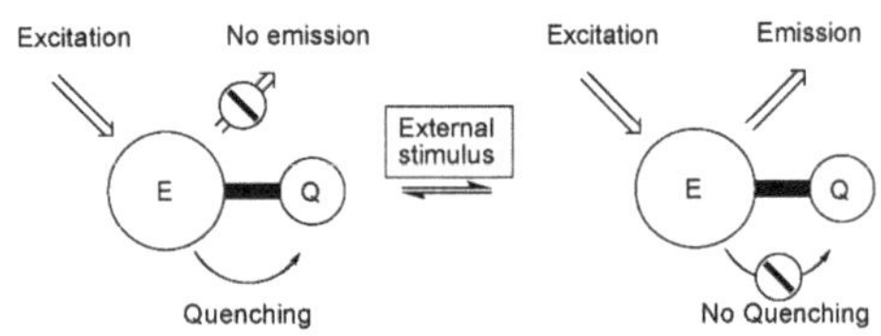

Figure 12.16. Emitter (E) and quencher (Q) motif of a molecular light switch.

3200 M^{-1} in CH_2Cl_2 and quenches the fluorescence of ZnTPP nearly completely. Assuming that the quenching is due to electron transfer from the singlet excited state of porphyrin to axially coordinated PhNNPy, reduction of PhNNPy may restore the fluorescence from the porphyrin, because the reduced form, $PhNNPy^{\bullet-}$, would not accept an additional electron when the porphyrin is excited. Electrochemical experiments confirmed this scheme (Fig. 12.17). Upon addition of PhNNPy to a solution of ZnTPP, fluorescence is largely diminished. Subsequent application of a constant potential at a −1.7 V vs. Fc^+/Fc (ferricenium/ferrocene) to an electrode immersed in the solution made the fluorescence back to the level of the initial intensity. When the solution was reoxidized with an electrode potential of −0.2 V, the fluorescence was again diminished.

That the emitter–quencher conjugate is made of noncovalent bond means that the bond can be reversibly broken and made. Upon addition of dichloroacetic acid to a mixture of PhNNPy and ZnTPP, the fluorescence of the porphyrin was restored. This is because PhNNPy was protonated and lost the ability to coordinate to ZnTPP. Addition of K_2CO_3 then diminished the fluorescence again by

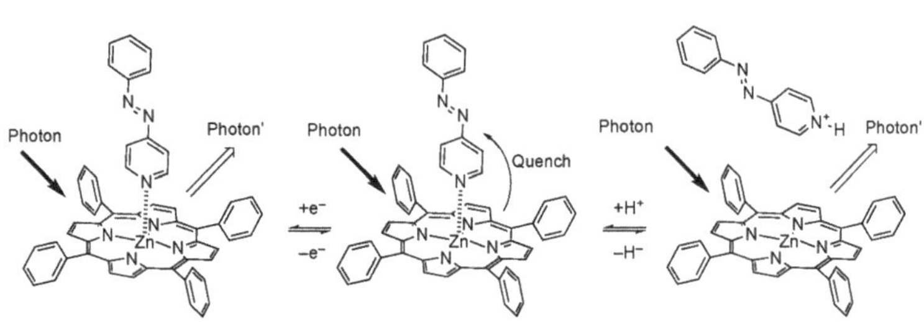

Figure 12.17. Redox- and proton-responsive supramolecular light switch [57].

deprotonating PhNNPy. Thus, this system works in a twofold manner as a redox-responsive light switch as well as a proton (acid/base)-responsive light switch.

The emitter–quencher motif was extended to an [emitter–energy acceptor]–quencher type supramolecule so that energy-transfer process may be switched. As the emitter–energy acceptor component, we used a Zn–porphyrin–free-base porphyrin dyad, ZnFbU, developed by the Lindsey group (see Fig. 12.13) [40, 58, 59]. In this dyad, energy transfer occurs from the zinc-porphyrin unit to the free-base porphyrin unit with a rate constant of $k_{\mathrm{ET}} = 4.2 \times 10^{10}\ \mathrm{s}^{-1}$ and an efficiency of 0.99. We found that the energy transfer was mostly suppressed when ZnFbU was coordinated by PhNNPy, indicating that the quenching by PhNNPy was extremely efficient with a quenching rate much larger than k_{ET} (Fig. 12.18) [60]. The energy transfer was turned on by addition of a competitive axial ligand that does not quench the fluorescence of porphyrin. Addition of *N*,*N*-dimethylaminopyridine (DMAP), having an association constant with zinc–porphyrin 20 times stronger than PhNNPy, replaces PhNNPy, thus restoring the energy-transfer process, giving rise to the fluorescence of the free-base porphyrin unit. The energy transfer was again turned off this time with addition of acid, dichloroacetic acid. Among the bases present in the solution, PhNNPy and DMAP, the latter was protonated preferentially because it is a stronger base. Resulting DMAPH^{+} lost the coordination ability, made the zinc site open for PhNNPy, which then coordinates to ZnFbU and suppresses the energy transfer again. The cycle of addition of DMAP and dichloroacetic acid can be repeated many times, demonstrating the reversible operation of the switching process, as shown in Fig. 12.18c.

The covalent linker joining the emitter and the energy acceptor in ZnFbU was further replaced by noncovalent linker to prepare all noncovalent quencher–emitter–energy acceptor triad. To make such an heterotopic A–B–C type assembly, intermolecular interactions between A and B and those between B and C must be orthogonal. We have used interactions for this purpose an axial coordination and amidinium–carboxylate salt bridge. The salt bridge is so strong that the amidinium unit and the carboxy unit exclusively bind together to form the salt bridge, leaving the zinc site open for coordination by

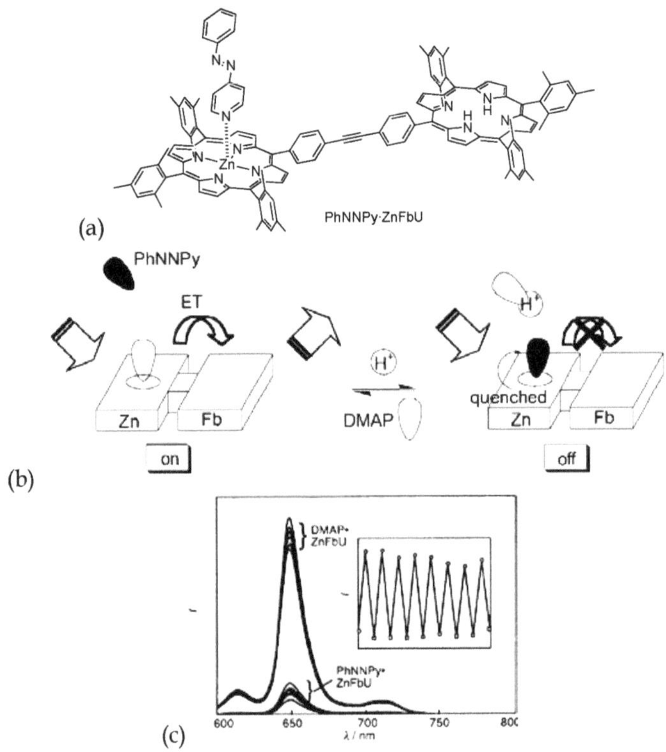

Figure 12.18. Noncovalent reversible switch for energy transfer. (a) Chemical structure of PhNNPy·ZnFbU. (b) Mode of operation of the switch. Complexation and decomplexation of PhNNPy is effected by alternate addition of acid and DMAP, respectively. (c) Fluorescence of the free-base porphyrin unit of ZnFbU as output of the switch. Reproduced with permission from [60].

a pyridine derivative. The actually constructed supramolecule was PhNNPy·ZnPA-1·FbPC-1, shown in Fig. 12.19a [61]. As described in Section 12.3.2, very fast excited energy transfer from the zinc porphyrin unit to the free-base porphyrin unit takes place at a rate of 4.0×10^9 s^{-1} in ZnPA-1·FbPC-1 likely due to the Dexter mechanism. Upon addition of PhNNPy, the fluorescence from FbPC-1 was suppressed, even though direct interaction between PhNNPy and FbPC-1 was absent. This indicates that the intra-supramolecular energy transfer is suppressed by PhNNPy. Addition of DMAP to this ternary solution resulted in recovery of the fluorescence from FbPC-1, indicating the restoration of energy transfer from ZnPA-1 to FbPC-1. The mode of operation for this system is schematically depicted in Fig. 12.19b. Also shown in the figure is a truth table corresponding to the present system, which is recast in terms of logic gates [62]. When ZnPA-1 and FbPC-1 are present in solution, they form a supramolecular dyad, in which energy transfer occurs efficiently, resulting in the fluorescence from FbPC-1; the output is 1. When only DMAP is added, the energy transfer is not affected and the fluorescence remains on; the output 1. When only PhNNPy is added, it suppresses the energy-transfer process; the output 0. When both DMAP

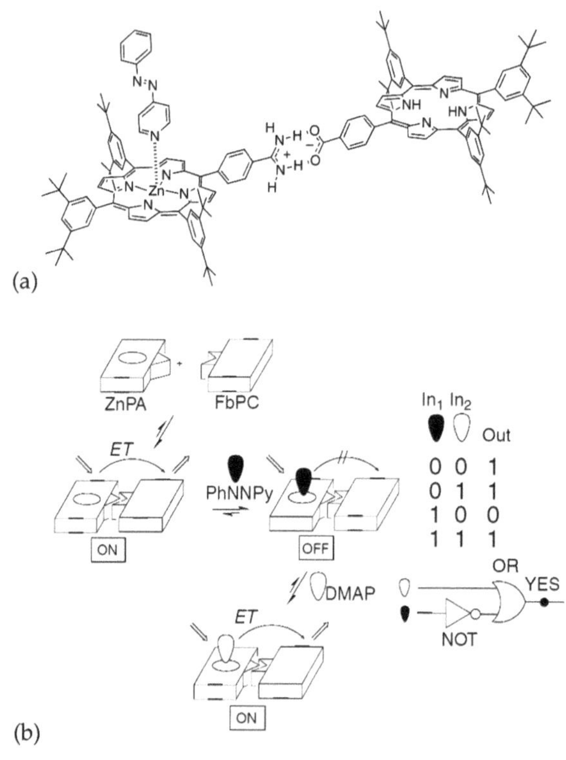

Figure 12.19. All noncovalent quencher–emitter–energy acceptor supramolecule as a switch for energy transfer. (a) Chemical structure of PhNNPy·ZnPA-1·FbPC-1. (b) A truth table and a logic circuit corresponding to the present system. Reproduced with permission from [61].

and PhNNPy is added, DMAP preferentially coordinates to ZnPA-1 so that the energy transfer is not suppressed; the output 1. The problem with this system was that further cycling of the switching processes was difficult. Addition of acid that worked for the system employing ZnFbU destroyed the assembly of the ZnPA-1 and FbPC-1.

A supramolecular switch for *electron* transfer was constructed as well using axial coordination as a key intermolecular bond [63]. In this system, ZnTPP·St-Im, a pyromellitic diimide unit, as an electron acceptor, was connected with a stilbazole moiety, as a switching unit. Stilbazole is photochromic like azobenzene is but does not work as an electron acceptor unlike azobenzene. Even with the presence of the imide group in St-Im, the fluorescence of ZnTPP was not quenched in their mixed solution (Fig. 12.20). This is in contrast to the case when Py-Im was used, as described in Section 12.2.1. This is apparently due to the longer spacer group in St-Im than in Py-Im. Light irradiation at 310 nm to this solution caused the trans-to-cis isomerization of the stilbazole group in St-Im, as well as a concomitant decrease in the fluorescence from ZnTPP. Lifetime measurements were in accord with the steady state results. The lifetime of the ZnTPP·*trans*-St-Im was 1.66 ns and that of the ZnTPP·*cis*-St-Im was 0.42 ns. Assuming that the quenching is due to electron transfer, which has been established for a variety of cases involving the combination of porphyrins and pyromellitic diimides, the electron-transfer rate is $k_{eT} = 1.83 \times 10^9\ s^{-1}$ with a quantum

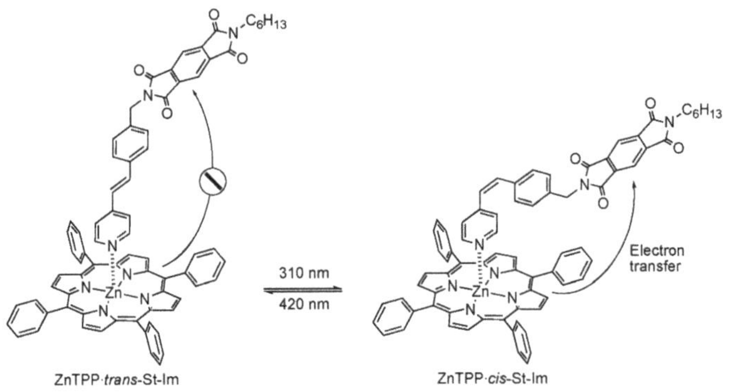

Figure 12.20. Photochromic supramolecular switch for electron transfer [63].

yield of 0.77. As the electron-transfer rate is virtually zero when the St-Im is in the trans configuration, the on–off ratio is very large. The modulation of the electron transfer is attributed to changes in the distance between ZnTPP and the imide moiety accompanying the trans-to-cis isomerization. The changes in distance were supported by ^{1}H NMR, in which the ring current effect of the porphyrin affords a good measure of geometry of the supramolecule. Molecular mechanics calculations suggested that the distance varied from 1.65 nm in ZnTPP·*trans*-St-Im to 1.0–1.3 nm in ZnTPP·*cis*-St-Im. Subsequent irradiation of 400–435 nm light to the solution including ZnTPP·*cis*-St-Im resulted in an increase of fluorescence intensity and an increase in the fraction of longer lifetime component. It is known that the cis-to-trans isomerization of stilbazole derivatives is sensitized by zinc–porphyrins through the triplet energy transfer [64–66].

12.4.2 *Supramolecular Switches by Complexation/Decomplexation*

In the previous section, I described switches based on repeated complexation and decomplexation of an axial ligand. This strategy of switching takes advantage of the reversible nature of noncovalent bond in constructing supramolecular switching systems. Acid and base were used for realizing reversible complexation/decomplexation in those systems described. Light can also be used for this purpose, if combined with photoisomerization and consequent steric changes of the ligand that lead to a modulation of the binding affinity towards zinc porphyrins. The strategy was previously employed by the Inoue group to photomodulate the catalytic reactivity of aluminum porphyrins using stilbazole derivatives [67–70].

We used 2-phenylazopyridine derivatives and 3-phenylazopyridine derivatives as detachable ligands, which double as quenchers in our case [71, 72]. In the case of 3-phenylazoderivatives (Fig. 12.21(a), the trans isomer should be more favorable for axial coordination because it has no steric hindrance upon coordination, while the cis isomer would have steric repulsion due to its phenyl group. On the other hand, in the case of 2-phenylazopyridine, the

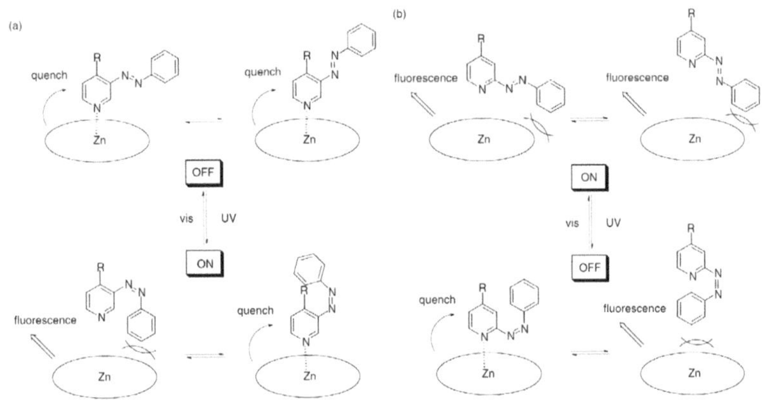

Figure 12.21. Reversibly detachable ligands that double as quenchers. (a) 3-Phenylazopyridine derivatives. (b) 2-Phenylazopyridine derivatives. Reproduced with permission from [72].

trans isomer would experience more steric repulsion than the cis isomer. The photoisomerization would provide a means to effect complexation and decomplexation that will give rise to fluorescence switches.

Chemical structures of specific ligands used for the demonstration of this strategy are shown in Fig. 12.22(a). Reversible modulation of fluorescence intensity by about 20 to 40% was realized. Note that, in the case of 3-phenylazopyridine derivatives (Fig. 12.21(a)), fluorescence intensity of zinc–porphyrin is increased by the shorter wavelength irradiation and decreased by the longer wavelength irradiation. The response is exactly opposite in the case of 2-phenylazopyridine derivatives (Fig. 12.21(b)), demonstrating the complementary nature of these two systems.

To improve the on/off ratio, we prepared two new types of photoresponsive phenylazopyridine derivatives. These are shuttlecock- (SC-Azo) and bowl-equipped (Bowl-Azo) phenylazopyridine derivatives (Fig. 12.23(a)) [73, 74]. The design principle was that, in trans configuration, the pyridyl nitrogen is protruding out so that the axial coordination to a zinc–porphyrin is possible without steric hindrance, while in the cis configuration, the pyridyl nitrogen is buried in a bulky substituent, which prevents axial coordination. A shuttlecock-shaped substituent was used in SC-Azo and a

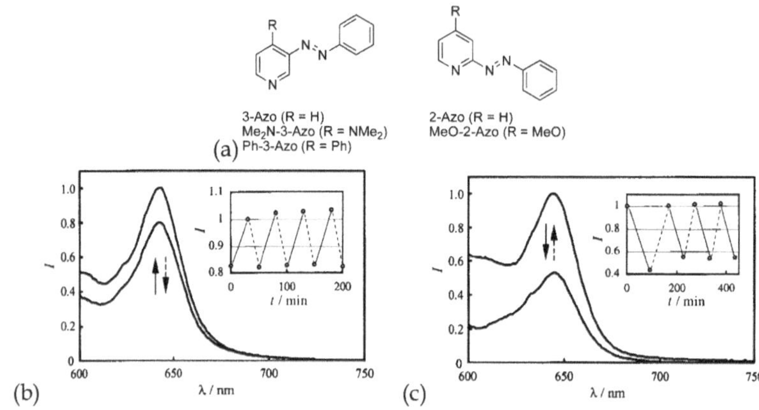

Figure 12.22. Light-induced reversible switching of the fluorescence of zinc-porphyrins with detachable ligands. (a) Chemical structures of photodetachable ligands. (b) ZnTMP·Ph-3-Azo. Solid and dotted lines represent 326- and 442-nm irradiation, respectively. (c) ZnTPP·MeO-2-Azo. Solid and dotted lines represent 315- and 436-nm irradiation, respectively. Reproduced with permission from [72].

bowl-shaped substituent was used in Bowl-Azo. The bowl was invented by Goto and co-workers to isolate reactive groups from environment [75, 76]. The crystal structure of SC-Azo (Fig. 12.23(b)) shows the pyridyl nitrogen is open for axial coordination. Although it appears that the *t*-butyl groups hinder the coordination of the pyridyl nitrogen in this particular structure obtained in the solid state, the easily rotatable phenyl–phenyl single bond would alleviate the steric repulsion. In the cis form, the structure of which was obtained from molecular mechanics calculations, the pyridyl group is in contact with one of the *t*-butylethynyl groups (Fig. 12.23(c)). The *t*-butylethynyl group may interfere with the association to zinc–porphyrin. The difference in the immediate surroundings of the pyridyl group is more clear in the *cis*- and *trans*-Bowl-Azo. In the crystal structure of *trans*-Bowl-Azo, the pyridyl group is clearly protruding out from the bowl (Fig. 12.23(e)). On the other hand, in the structure of *cis*-Bowl-Azo optimized by molecular mechanics calculations, the pyridyl group is embedded inside the bowl framework (Fig. 12.23(f)).

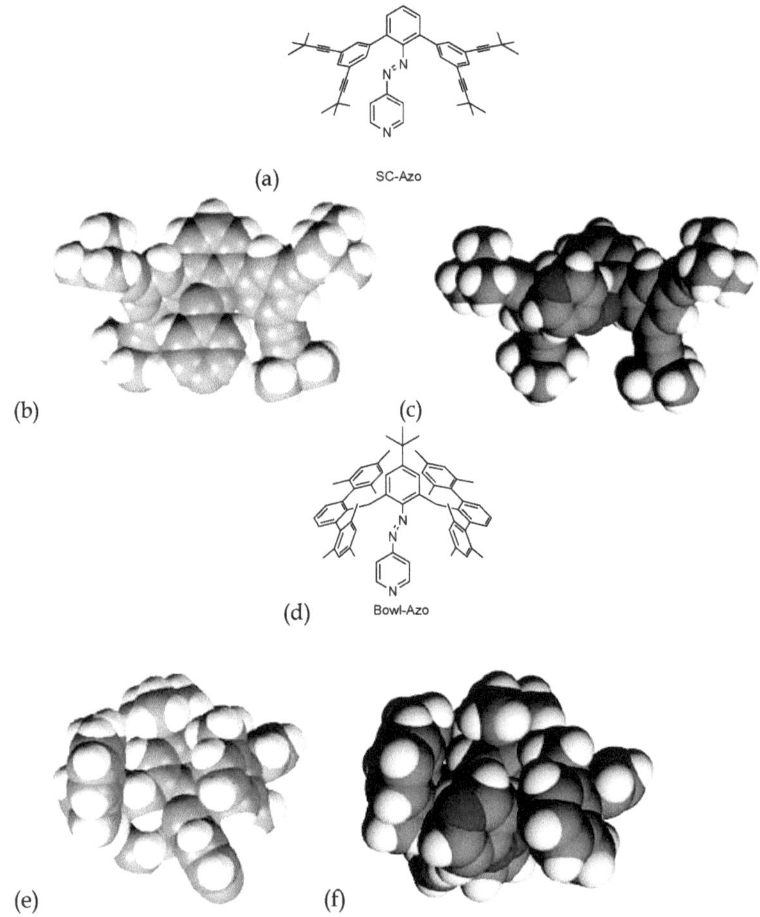

Figure 12.23. Structures of trans- and cis-configurations of ligands, which are reversibly detachable from zinc–porphyrins by light irradiation. (a) Chemical structure of SC-Azo. (b) Crystal structure of *trans*-SC-Azo. (c) Structure of *cis*-SC-Azo according to molecular mechanics. (d) Chemical structure of Bowl-Azo. (e) Crystal structure of *trans*-Bowl-Azo. (f) Structure of *cis*-Bowl-Azo according to molecular mechanics. Reproduced with permission from [74].

The photoisomerization and thermal isomerization of these sterically demanding azo compounds were investigated to examine how the bulky groups affect the cis/trans structural changes of these molecules. The quantum yields of the trans-to-cis photoisomerization of SC-Azo and Bowl-Azo were 0.013–0.022 and 0.056–0.076, respectively, depending the excitation wavelength, which are lower than those of PhNNPy, 0.10–0.21. The lower quantum yields are probably due, to a large part, to electronic factors because of the presence of aromatic systems, which are not directly related to the isomerization of the azo group. The thermal cis-to-trans isomerization of SC-Azo occurs with a rate constant of $k_{c\to t} = 0.25 \times 10^{-5}\ s^{-1}$, which is somewhat smaller than that of PhNNPy, $k_{c\to t} = 1.08 \times 10^{-5}\ s^{-1}$. Thermodynamic parameters for the process in SC-Azo, such as the activation energy, the frequency factor, the activation free energy, the activation entropy, and the activation enthalpy, all are similar to those of PhNNPy. Thus we concluded that the shuttlecock framework in SC-Azo does not affect the thermal cis-to-trans isomerization. The thermal cis-to-trans isomerization of Bowl-Azo occurs with a rate constant of $k_{c\to t} = 0.17\times 10^{-5}\ s^{-1}$, which is somewhat further smaller than that of SC-Azo. The rate reduction occurs even though the activation energy, $\Delta E_a = 70.1$ kJ mol^{-1}, is significantly lower than that of PhNNPy, $\Delta E_a = 91.7$ kJ mol^{-1}, because its frequency factor, log (A/s^{-1}) = 7.5, is much smaller than that of PhNNPy, log (A/s^{-1}) = 11.1. This is a consequence of a larger negative activation entropy of isomerization, $\Delta S^* = -110$ J K^{-1} mol^{-1}, than that of PhNNPy, $\Delta S^* = -40.7$ J K^{-1} mol^{-1}, and a smaller activation enthalpy, $\Delta H^* = 67.6$ kJ K^{-1} mol^{-1}, than that of PhNNPy, $\Delta H^* = 89.3$ kJ K^{-1} mol^{-1}. The larger activation entropy and the smaller activation enthalpy indicates that the transition state of the cis-to-trans isomerization is stabilized by the bowl framework but at the same time the conformational freedom of the molecule is severely compensated.

Figure 12.24 shows ^{1}H NMR spectra of SC-Azo and Bowl-Azo upon incremental addition of zinc-porphyrin. The azo ligands were prepared as mixtures of trans and cis configuration by prior illumination. Characteristic upfield shifts were observed on coordination to the zinc–porphyrin due to the ring current effect of the porphyrin macrocycle. The ortho and meta hydrogens on the

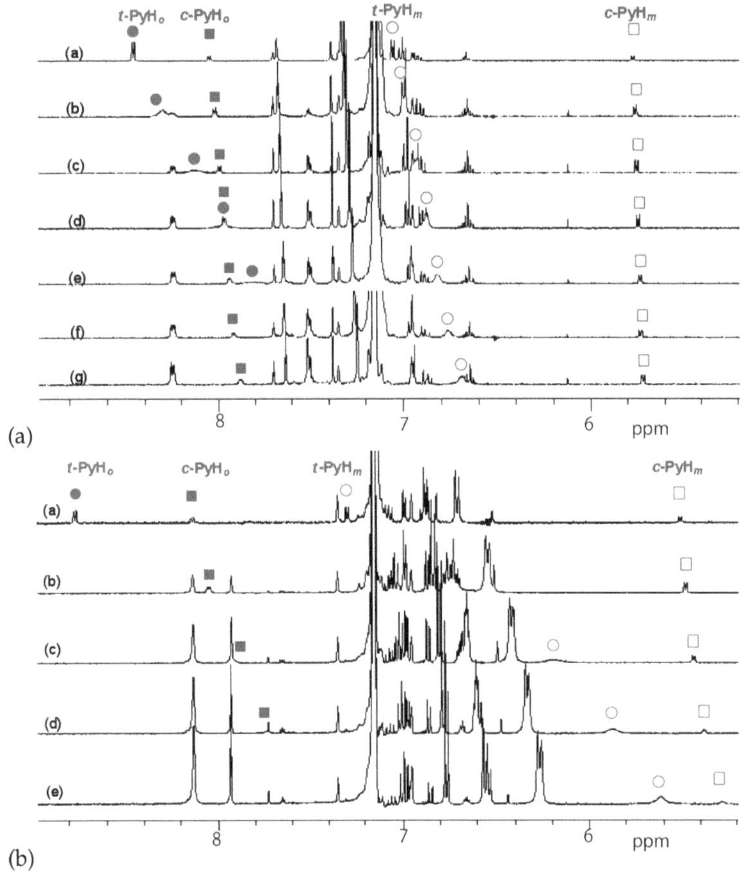

Figure 12.24. ^{1}H NMR upfield shifts in C_6D_6 for trans and cis azopyridine ligands upon addition of Zn porphyrins. (a) SC-Azo on addition of ZnTPP (0–0.6 mM). (b) Bowl-Azo on addition of zinc-5,15-bis(3′,5′-di-*tert*-butylphenyl)-2,8,12,18-tetraethyl-3,7,13,17-tetramethylporphyrin (0–2.4 mM). Reproduced with permission from [73].

pyridyl moiety with respect to the pyridyl nitrogen of the cis and trans isomers are marked on the spectra (*c*/*t*-PyH$_o$ and *c*/*t*-PyH$_m$). It is clear from the spectra that protons of the trans isomers of SC-Azo and Bowl-Azo experience large upfield shifts, while those of the cis isomers change only slightly. This demonstrates that these ligands interact much more strongly with zinc–porphyrin in the trans configuration than in the cis configuration. Quantitative analysis revealed that *trans*-SC-Azo and *cis*-SC-Azo bind with ZnTPP with an association constants of 860 and 120 M^{-1}, respectively. *Trans*-Bowl-Azo and *cis*-Bowl-Azo bind with zinc-5,15-bis(3′,5′-di-*tert*-butylphenyl)-2,8,12,18-tetraethyl-3,7,13,17-tetramethylporphyrin with association constants of 2000 and 100 M^{-1}, respectively. Thus, the bowl framework is particularly effective in the design of photocontrolled ligand systems.

12.5 Porphyrin Assemblies on an Individual Molecule Basis

In this section, we turn our attention to supramolecular structures revealed by means of scanning tunneling microscopy (STM) [77–79]. Motivation for these works are twofold. One is that the structures of molecular organization can be revealed directly in real space at the molecular level. The supramolecular porphyrin assemblies so far described are all formed in solution. The structures of the assemblies were deduced on the basis of spectroscopy including ^1H NMR spectra, UV-visible spectra, and fluorescence spectra, combined with knowledge about association motifs of related compounds with the same intermolecular interaction sites. In contrast to these indirect methods, the STM affords images directly correlated with molecular structures and molecular arrangements. The other motivation is that, to use porphyrin assemblies as electronic and/or photonic molecular devices, architectures on solid surfaces are more convenient than those in solution.

Figure 12.25 shows the modes of association of porphyrin molecules. Porphyrin molecules may be organized through peripheral association by the incorporation of intermolecular binding sites such as hydrogen bonding sites (e.g., ZnPA-1·FbPC-1) and coordination sites (No. 1). Axial coordination is a simple method to afford well-organized molecular assemblies including a metal porphyrin (e.g., ZnTPP·PhNNPy) (No. 2). As polymolecular assemblies, there are H-aggregates (No. 3), in which porphyrin disks are stacked face-to-face, and J-aggregates (No. 4), in which porphyrin disks are

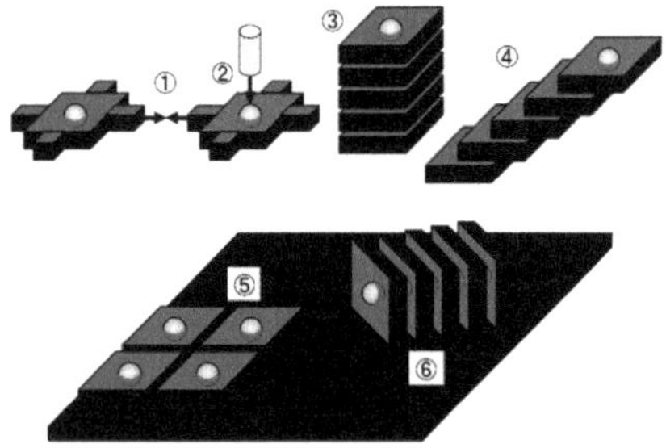

Figure 12.25. Association mode of porphyrin molecules. (1) Peripheral association through intermolecular binding sites, such as hydrogen bonding sites and coordination sites. (2) Axial coordination. (3) H-aggregates. (4) J-aggregates. (5) 2D face-on arrangements on a substrate. (6) 2D edge-on arrangements on a substrate.

associated as in 'a spread deck of cards.' As ordered assemblies on substrate surfaces, porphyrin molecules may be arranged in a face-on fashion (No. 5) or in an edge-on fashion (No. 6). STM images of porphyrin molecules organized in some of these modes are presented here to give a sense on what well-ordered assemblies are formed from porphyrin molecules. Most examples are those formed at liquid/HOPG interfaces, where HOPG stands for highly oriented pyrolytic graphite.

Incorporation of peripheral hydrogen bonding sites are exemplified by porphyrins with carboxyphenyl groups. Bai and co-workers revealed that *meso*-tetrakis(4-carboxyphenyl)porphyrin forms a square lattice with a hydrogen bonding network as shown in Fig. 12.26 [80]. 5,10-bis(4-carboxyphenyl)-15,20-bis(4-octadecyloxyphenyl)porphyrin (*cis*-R_2C_2) assembles themselves into a double row structure, in which porphyrin molecules associate head-to-head configuration due to hydrogen bond formation [81]. 5-(4-carboxyphenyl)-10,15,20-tris(4-octadecyloxyphenyl)porphyrin (R_3C_1) form a zig-zag arrangement. The kinks occur due to, again, hydrogen bond formation. These results demonstrate that well-defined hydrogen bonded assemblies form not only in solution but on surfaces as well.

There are two types of axial coordination involving matalloporphyrins. One is those having a kinetically inert bond, exemplified by rhodium and ruthenium porphyrins, axially coordinated by a

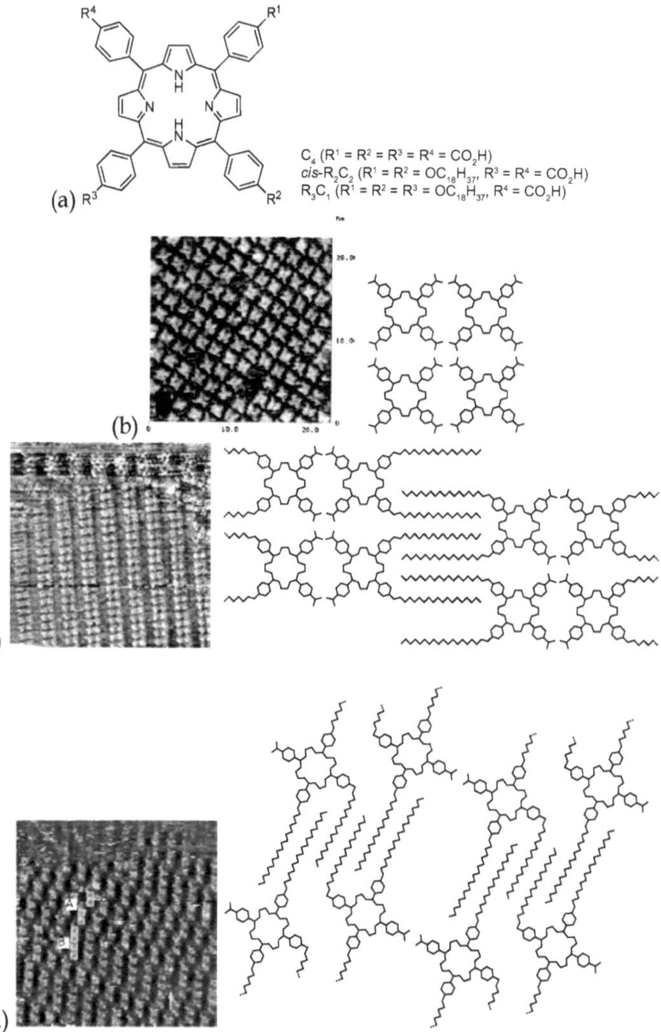

Figure 12.26. Hydrogen bonded assemblies of porphyrins bearing carboxy groups. (a) Chemical structures of porphyrins with carboxy groups. (b) C_4 at the air/HOPG interface. (c) *cis*-R_2C_2 at *o*-dichlorobenzene/HOPG interface; 50 × 50 nm^2. (d) R_3C_1 at *o*-dichlorobenzene/HOPG interface; 50 × 50 nm^2. Reproduced with permission from [80] (a) and [81]. See also Color Insert.

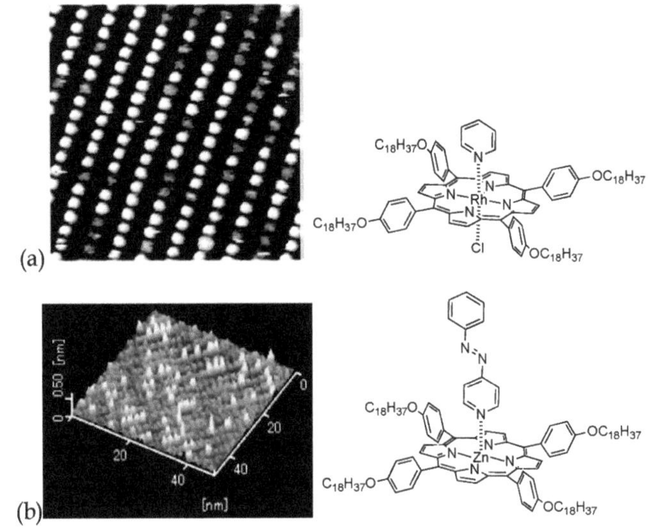

Figure 12.27. Mixed assembly of metalloporphyrin molecules with axial ligands and free-base porphyrin molecules at the 1-phenyloctane/HOPG interface. (a) Rhodium–porphyrin with pyridine; 35 × 35 nm^2. (b) Zinc–porphyrin with PhNNPy. Reproduced with permission from [82] (a) and [83] (b). See also Color Insert.

pyridyl moiety. The other is those having a kinetically labile bond, such as zinc–porphyrins, axially coordinated by a pyridyl moiety. Both types of axial coordination have been observed by STM. Figure 12.27(a) shows a STM image for a mixture of free-base porphyrin and rhodium porphyrin bearing pyridine (and chlorine) as an axial ligand, giving the lower spots and higher spots, respectively [82]. Figure 12.27(b) shows a STM image of a mixture of zinc–porphyrin with coordinated PhNNPy and free-base porphyrin [83]. In our conditions, zinc–porphyrins and free-base porphyrins in their mixture cannot be discriminated. However, upon addition of PhNNPy, tall protrusions appear on some of the spots, which we assigned zinc–porphyrin molecules coordinated by PhNNPy molecules. The axial coordination of a pyridine moiety to zinc–porphyrin is labile, reversible interaction, where attachment and detachment are occurring much faster than a STM time scale. The observed protrusions are, therefore, a kind of time-averaged image.

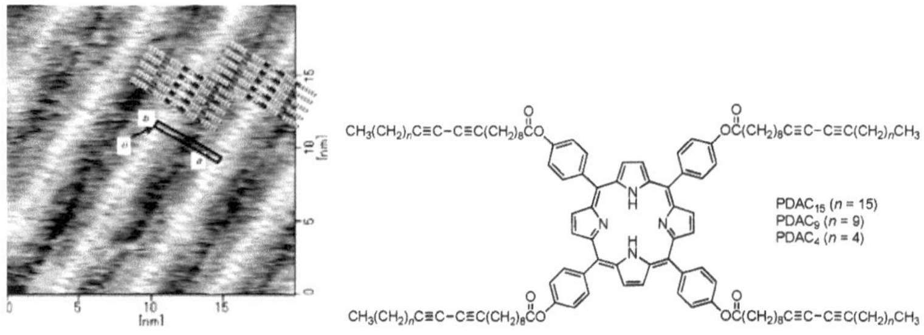

Figure 12.28. Edge-on assembly of $PDAC_9$ at the 1-phenyloctane/HOPG interface. Reproduced with permission from [84].

While most of the reported porphyrin assemblies are made up of porphyrin molecules adsorbed on a surface in a face-on manner, we have encountered assemblies, in which porphyrin molecules are stacked face-to-face forming H-type aggregates and adsorbed on the surface in an edge-on orientation (No. 6 in Fig. 12.25), as shown in Fig. 12.28 [84]. The compounds we investigated were porphyrins with long hydrocarbon chains, each of which has a diacetylene unit in the middle. Among the three molecules with different hydrocarbon lengths examined, $PDAC_n$ ($n = 4, 9, 15$), the porphyrin with the longest hydrocarbon chains, $PDAC_{15}$, adsorbed on HOPG in a face-on manner. Those with shorter hydrocarbon chains, $PDAC_9$ and $PDAC_4$, packed as highly ordered H-type aggregates and were adsorbed on the surface in an edge-on fashion. The difference in preferences for the supramolecular structures depending on the chain length may be understood in terms of the balance between molecule–molecule interactions and molecule-substrate interactions. The melting point decreases as the hydrocarbon chains becomes longer, which suggests that intermolecular association is driven by attractive force between the porphyrin cores and is alleviated by the hydrocarbon chains. On the other hand, the molecule–substrate interactions are hydrocarbon driven [85–87]. Therefore, porphyrins with longer hydrocarbon chains prefer face-on adsorption on the surface, maximizing the molecule–substrate interactions, whereas the porphyrins with shorter hydrocarbon chains prefer molecule–molecule association, giving rise to the H-aggregate formation, resulting in the edge-on adsorption. In any case, the highly ordered columnar aggregates of porphyrin molecules are of interest in view of electron- and energy-transfer properties across a number of molecules over a long distance.

12.6 Conclusions and Prospects

We have described electron- and energy-transfer processes through intermolecular bonds within supramolecular architectures. By exploiting and taking control over noncovalent interactions, more elaborate structures may become accessible than by total covalent synthetic approach. Intermolecular interactions have often been viewed as structural elements. However, to take a full advantage of noncovalent interactions, understanding of intermolecular bonds not only in structural aspects but as conduit for mediating electronic communcation is important. Electronic coupling through axial coordination was investigated for dyads consisting of a zinc–porphyrin and acceptors. Energy-transfer processes in dyads made up of a zinc–porphyrin and a free-base porphyrin linked through an amidinium–carboxylate salt bridge were investigated as well. In this particular case, fast energy transfer, which cannot be accounted for by the commonly assumed Förster mechanism was found. We propose the involvement of Dexter-type superexchange energy transfer through the hydrogen bonding interface.

From a viewpoint that these porphyrin-based supramolecular organizations will be used as components of molecular electronic and/or photonic devices, an important next step is to realize the switching of electronic processes such as electron and energy transfer, which we have demonstrated by a number of prototypes. Turning on and off of electron- or energy-transfer processes were affected by external stimuli including photons, electrons, protons (acid), and bases.

Typical types of supramolecular organizations have also been visualized at the molecular level by means of STM. The STM data showed unequivocally that highly ordered organizations of porphyrin molecules are indeed formed, not only in an averaged-out basis, as probed in solution, but on an individual molecule basis on surfaces.

Further elaboration of supramolecular architectures may be feasible in near future by judicious control over intermolecular interactions. More complex electronic operation may be realized as the complexity of the supramolecular architectures increases. Of course, as the complexity increases, difficulty increases as well, in the preparation and characterization of supramolecular organizations and also in the analysis of the electronic processes that occur within the organizations. However, living organisms have created molecule-based complex devices after trials and errors over years. Chemists may have some chance of creating artificial molecular organizations and molecular devices on a par with those found in nature.

References

1. Otsuki, J. (1998) Chemistry for molecular devices, *Recent Res. Dev. Pure Appl. Chem.*, **2**, 427–439.
2. Balzani, V., Credi, A., and Venturi, M. (2004) *Molecular Devices and Machines*, Wiley-VCH, Weinheim.
3. Lehn, J.-M. (1995) *Supramolecular Chemistry: Concept and Perspectives*, VCH, Weinheim.
4. Feringa, B. L. (ed.) (2003) *Molecular Switches*, Wiley-VCH, Weinheim.
5. Gunter, M. J. (2006) Multiporphyrin arrays assembled through hydrogen bonding, *Struct. Bonding*, **121**, 263–295.
6. Elemans, J. A. A. W., Hameren, R. V., Nolte, R. J. M., and Rowan, A. E. (2006) Molecular materials by self-assembly of porphyrins, phthalocyanines, and perylenes, *Adv. Mater.*, **18**, 1251–1266.
7. Beletskaya, I., Tyurin, V. S., Tsivadze, A. Y., Guilard, R., and Stern, C. (2009) Supramolecular chemistry of metalloporphyrins, *Chem. Rev.*, **109**, 1659–1713.
8. Yardley, J. T. (1980) *Introduction to Molecular Energy Transfer*, Academic Press, New York.
9. Atkins, P. W. (1983) *Molecular Quantum Mechanics*, 2nd edn., Oxford University Press, New York.
10. Gust, D., and Moore, T. A. (2000) Intramolecular photoinduced electron-transfer reactions of porphyrins, in *The Porphyrin Handbook*, (ed. Kadish, K. M., Smith, K. M., and Guilard, R.), Academic Press, San Diego, pp. 153–190.

11. Langford, S. J., Latter, M. J., and Woodward, C. P. (2006) Progress in charge transfer systems utilizing porphyrin donors and simple aromatic diimide acceptor units, *Photochem. Photobiol.*, **82**, 1530–1540.
12. Marcus, R. A., and Sutin, N. (1985) Electron transfers in chemistry and biology, *Biochim. Biophys. Acta*, **811**, 265–322.
13. Jortner, J. (1976) Temperature dependent activation energy for electron transfer between biological molecules, *J. Chem. Phys.*, **64**, 4860–4867.
14. Miller, J. R., Beitz, J. V., and Huddleston, R. K. (1984) Effect of free energy on rates of electron transfer between molecules, *J. Am. Chem. Soc.*, **106**, 5057–5068.
15. Closs, G. L., Calcaterra, L. T., Green, N. J., Penfield, K. W., and Miller, J. R. (1986) Distance, stereoelectronic effects, and the marcus inverted region in intramolecular electron transfer in organic radical anions, *J. Phys. Chem.*, **90**, 3673–3683.
16. George L. Gaines, I., O'Neil, M. P., Svec, W. A., Niemczyk, M. P., and Wasielewski, M. R. (1991) Photoinduced electron transfer in the solid state: Rate vs free energy dependence in fixed-distance porphyrin-acceptor molecules, *J. Am. Chem. Soc.*, **113**, 719–721.
17. Asahi, T., Ohkohchi, M., Matsusaka, R., Mataga, N., Zhang, R. P., Osuka, A., and Maruyama, K. (1993) Intramolecular photoinduced charge separation and charge recombination of the product ion pair states of a series of fixed-distance dyads of porphyrins and quinones: Energy gap and temperature dependences of the rate constants, *J. Am. Chem. Soc.*, **115**, 5665–5674.
18. Hunter, C. A., Sanders, J. K. M., Beddard, G. S., and Evans, S. (1989) A new approach to the assembly of electron donor–spacer–acceptor systems, *J. Chem. Soc. Chem. Commun.*, 1765–1767.
19. Otsuki, J. (2001) Electron and energy transfer in supramolecular complexes involving axial coordination of zinc–porphyrins, *Trends Phys. Chem.*, **8**, 61–72.
20. Sanders, J. K. M., Bampos, N., Clyde-Watson, Z., Darling, S. L., Hawley, J. C., Kim, H.-J., Mak, C. C., and Webb, S. J. (2000) Axial coordination chemistry of metalloporphyrins, in *The Porphyrin Handbook*, (ed. Kadish, K. M., Smith, K. M., and Guilard, R.), Academic Press, San Diego, pp. 1–48.
21. Bouamaied, I., Coskun, T., and Stulz, E. (2006) Axial coordination to metalloporphyrins leading to multinuclear assemblies, *Struct. Bonding*, **121**, 1–47.
22. Otsuki, J., Harada, K., Toyama, K., Hirose, Y., Araki, K., Seno, M., Takatera, K., and Watanabe, T. (1998) Energy gap dependence of electron transfer

rates in porphyrin–imide supramolecular assemblies, *Chem. Commun.*, 1515–1516.

23. Sliwa, W., and Mianowska, B. (2000) Metalloporphyrin arrays, *Transition Met. Chem.*, **25**, 491–504.
24. Satake, A., and Kobuke, Y. (2005) Dynamic supramolecular porphyrin systems, *Tetrahedron*, **61**, 13–41.
25. Hunter, C. A., and Shannon, R. J. (1996) Photoinduced electron transfer on a supramolecular scaffold, *Chem. Commun.*, 1361–1362.
26. Otsuki, J., Takatsuki, M., Kaneko, M., Miwa, H., Takido, T., Seno, M., Okamoto, K., Imahori, H., Fujitsuka, M., Araki, Y., Ito, O., and Fukuzumi, S. (2003) Formation of a supramolecular porphyrin-spacer-acceptor ternary complex and intracomplex electron transfer, *J. Phys. Chem. A*, **107**, 379–385.
27. Deng, Y., Roberts, J. A., Peng, S.-M., Chang, C. K., and Nocera, D. G. (1997) The amidinium-carboxylate salt bridge as a proton-coupled interface to electron transfer pathways., *Angew. Chem. Int. Ed.*, **36**, 2124–2127.
28. Kirby, J. P., Roberts, J. A., and Nocera, D. G. (1997) Significant effect of salt bridges on electron transfer rate, *J. Am. Chem. Soc.*, **119**, 9230–9236.
29. Roberts, J. A., Kirby, J. P., and Nocera, D. G. (1995) Photoinduced electron transfer within a donor-acceptor pair juxtaposed by a salt bridge, *J. Am. Chem. Soc.*, **117**, 8051–8052.
30. Otsuki, J., Narita, T., Tsutsumida, K., Takatsuki, M., and Kaneko, M. (2005) Modular approach toward supramolecular functional assemblies: Characterization of donor-spacer-acceptor ternary complexes, *J. Phys. Chem. A*, **109**, 6128–6134.
31. Förster, T. (1948) Zwischenmolekulare energiewanderung und fluoreszenz, *Ann. Physik*, **2**, 55–75.
32. Förster, T. (1965) Delocalized excitation and excitation transfer, in *Modern Quantum Chemistry, 3. Action of light and organic crystals. Istanbul lectures*, (ed. Sinanoglu, O.), Academic Press, New York, pp. 93–137.
33. Mårtensson, J. (1994) Calculation of the förster orientation factor for donor-acceptor systems with one chromophore of threefold or higher symmetry: Zinc porphyrin, *Chem. Phys. Lett.*, **229**, 449–456.
34. Gouterman, M. (1961) Spectra of porphyrins, *J. Mol. Spectrosc.*, **6**, 138–163.
35. Braslavsky, S. E., Fron, E., Rodríguez, H. B., Román, E. S., Scholes, G. D., Schweitzer, G., Valeur, B., and Wirz, J. (2008) Pitfalls and limitations in the practical use of förster's theory of resonance energy transfer, *Photochem. Photobiol. Sci.*, **7**, 1444–1448.

36. Klán, P., and Wirz, J. (2009) *Photochemistry of Organic Compounds*, Wiley, West Sussex.
37. Otsuki, J., Iwasaki, K., Nakano, Y., Itou, M., Araki, Y., and Ito, O. (2004) Supramolecular porphyrin assemblies through amidinium ± carboxylate salt bridges and fast intra-ensemble excited energy transfer, *Chem. Eur. J.*, **10**, 3461–3466.
38. Otsuki, J. (2009) Energy transfer in non-covalent porphyrin assemblies: Through-space or through-bond?, *J. Porphyrins Phthalocyanines*, **13**, 1069–1081.
39. Otsuki, J., Kanazawa, Y., Kaito, A., Islam, D.-M. S., Araki, Y., and Ito, O. (2008) Through-bond excited energy transfer mediated by an amidinium–carboxylate salt bridge in zn–porphyrin free-base porphyrin dyads, *Chem. Eur. J.*, **14**, 3776–3784.
40. Hsiao, J.-S., Krueger, B. P., Wagner, R. W., Johnson, T. E., Delaney, J. K., Mauzerall, D. C., Fleming, G. R., Lindsey, J. S., Bocian, D. F., and Donohoe, R. J. (1996) Soluble synthetic multiporphyrin arrays. 2. Photodynamics of energy-transfer processes, *J. Am. Chem. Soc.*, **118**, 11181–11193.
41. Osuka, A., Tanabe, N., Kawabata, S., Yamazaki, I., and Nishimural, Y. (1995) Synthesis and intramolecular electron- and energy-transfer reactions of polyyne- or polyene-bridged diporphyrins, *J. Org. Chem.*, **60**, 7177–7185.
42. Strachan, J.-P., Gentemann, S., Seth, J., Kalsbeck, W. A., Lindsey, J. S., Holten, D., and Bocian, D. F. (1997) Effects of orbital ordering on electronic communication in multiporphyrin arrays, *J. Am. Chem. Soc.*, **119**, 11191–11201.
43. Tsai, H.-H. G., and Simpson, M. C. (2002) The role of frontier molecular orbital ordering on electronic communication in porphyrin arrays, *Chem. Phys. Lett.*, **353**, 111–118.
44. Kawabata, S., Yamazaki, I., Nishimura, Y., and Osuka, A. (1997) Singlet energy transfer in bis(phenylethynyl)phenylene-bridged zinc-free base hybrid diporphyrins, *J. Chem. Soc. Perkin Trans.*, **2**, 479–484.
45. Osuka, A., Ikeda, M., Shiratori, H., Nishimura, Y., and Yamazaki, I. (1999) Accelerated singlet energy transfer in bis(phenylethynyl)phenylenebridged 5,10,15,20-tetraaryl zinc–free base hybrid diporphyrins, *J. Chem. Soc. Perkin Trans.*, **2**, 1019–1025.
46. Newton, M. D. (1991) Quantum chemical probes of electron-transfer kinetics: The nature of donor–acceptor interactions, *Chem. Rev.*, **91**, 767–792.

47. McConnell, H. M. (1961) Intramolecular charge transfer in aromatic free radicals, *J. Chem. Phys.*, **35**, 508–515.
48. Otsuki, J., Akasaka, T., and Araki, K. (2008) Molecular switches for electron and energy transfer processes based on metal complexes, *Coord. Chem. Rev.*, **252**, 32–56.
49. Otsuki, J. (2003) Molecular switches for luminescence and energy transfer, *Recent Adv. Photochem. Photobiol.*, **2**, 13–20.
50. Otsuki, J. (2009) Redox and light responsive molecular switches for energy and electron transfer processes, in *Bottom-Up Nanofabrication: Supramolecules, Self-Assemblies, and Organized Films*, (ed. Ariga, K., and Nalwa, H. S.), American Scientific Publishers, USA, pp. 93–123.
51. Callan, J. F., Silva, A. P. D., and Magri, D. C. (2005) Luminescent sensors and switches in the early 21st century, *Tetrahedron*, **61**, 8551–8588.
52. Cuisdo, J., Deniz, E., and Raymo, F. M. (2009) Fluorescent switches based on photochromic compounds, *Eur. J. Org. Chem.*, 2031–2045.
53. Raymo, F. M., and Tomasulo, M. (2005) Fluorescence modulation with photochromic switches, *J. Phys. Chem. A*, **109**, 7343–7352.
54. Raymo, F. M., and Tomasulo, M. (2005) Electron and energy transfer modulation with photochromic switches, *Chem. Soc. Rev.*, **34**, 327–336.
55. Benniston, A. C. (2004) Pushing around electrons: Towards 2-d and 3-d molecular switches, *Chem. Soc. Rev.*, **33**, 573–578.
56. Goulle, V., Harriman, A., and Lehn, J.-M. (1993) An electro-photoswitch: Redox switching of the luminescence of a bipyridine metal complex, *J. Chem. Soc. Chem. Commun.*, 1034–1036.
57. Otsuki, J., Harada, K., and Araki, K. (1999) Supramolecular electro- and proto-photoswitch, *Chem. Lett.*, 269–270.
58. Wagner, R. W., Johnson, T. E., and Lindsey, J. S. (1996) Soluble synthetic multiporphyrin arrays. 1. Modular design and synthesis, *J. Am. Chem. Soc.*, **118**, 11166–11180.
59. Seth, J., Palaniappan, V., Wagner, R. W., Johnson, T. E., Lindsey, J. S., and Bocian, D. F. (1996) Soluble synthetic multiporphyrin arrays. 3. Static spectroscopic and electrochemical probes of electronic communication, *J. Am. Chem. Soc.*, **118**, 11194–11207.
60. Otsuki, J., Yasuda, A., and Takido, T. (2003) Non-covalent switch for intramolecular energy transfer, *Chem. Commun.*, 608–609.
61. Otsuki, J., Kato, Y., Sasahara, T., Iwasaki, K., and Yasuda, A. (2005) Modular approach toward supramolecular functional assemblies: Remote control of fluorescence, *Chem. Lett.*, 1684–1685.

62. Silva, A. P. D., and McClenaghan, N. D. (2004) Molecular-scale logic gates, *Chem. eur. J.*, **10**, 574–586.
63. Otsuki, J., Suka, A., Yamazaki, K., Abe, H., Araki, Y., and Ito, O. (2004) Photocontrol of electron transfer from zn-porphyrin to an axially bound stilbazole–pyromellitic diimide conjugate, *Chem. Commun.*, 1290–1291.
64. Whitten, D. G., Wildes, P. D., and Loppla, I. G. (1969) Ligand photoisomerization in metalloporphyrin complexes. A possible case of photocatalysis, *J. Am. Chem. Soc.*, **91**, 3393–3394.
65. Wildes, P. D., and Whitten, D. G. (1970) Reversible energy transfer in metalloporphyrin complexes. A mechanism for photocatalysis, *J. Am. Chem. Soc.*, **92**, 7609–7610.
66. Brink, M., and Wennerström, O. (2001) Efficient catalysis of the photoisomerization of stilbene derivatives by zinc porphyrin complexation, *J. Photochem. Photobiol. A*, **143**, 201–208.
67. Iseki, Y., Watanabe, E., Mori, A., and Inoue, S. (1993) Photoresponsive coordination of stilbazoles to metalloporphyrins, *J. Am. Chem. Soc.*, **115**, 7313–7317.
68. Iseki, Y., and Inoue, S. (1994) Photoswitchable complexation of metalloporphyrins, *J. Chem. Soc. Chem. Commun.*, 2577–2578.
69. Sugimoto, H., and Inoue, S. (1998) A photoresponsive molecular switch. Reaction of co_2 and epoxide regulated by aluminum porphyrin-stilbazole system, *Pure Appl. Chem.*, **70**, 2365–2369.
70. Sugimoto, H., Kimura, T., and Inoue, S. (1999) Photoresponsive molecular switch to control chemical fixation of co_2, *J. Am. Chem. Soc.*, **121**, 2325–2326.
71. Otsuki, J., Narutaki, K., and Bakke, J. M. (2004) Light-triggered luminescence modulation using labile axial coordination to zinc–porphyrin, *Chem. Lett.*, **33**, 356–357.
72. Otsuki, J., and Narutaki, K. (2004) Photochromism of phynylazopyridines and its application to the fluorescence modulation of zinc-porphyrins, *Bull. Chem. Soc. Jpn.*, **77**, 1537–1544.
73. Suwa, K., Otsuki, J., and Goto, K. (2009) Syntheses of shuttlecock- and bowl-equipped phenylazopyridines and photomodulation of their coordination ability to zn-porphyrin, *Tetrahedron Lett.*, **50**, 2106–2108.
74. Suwa, K., Otsuki, J., and Goto, K. (2010) Photoisomerization and thermal isomerization of shuttlecock- and bowl-equipped (phenylazo)pyridines, *J. Phys. Chem. A*, **114**, 884–890.
75. Goto, K., and Okazaki, R. (1997) Molecular bowls and capsules with an endohedral functionality: The stabilization of highly reactive species in their inner phase, *Liebigs Ann. Recl.*, 2393–2407.

76. Goto, K., Nagahama, M., Mizushima, T., Shimada, K., Kawashima, T., and Okazaki, R. (2001) The first direct oxidative conversion of a selenol to a stable selenenic acid: Experimental demonstration of three processes included in the catalytic cycle of glutathione peroxidase, *Org. Lett.*, **3**, 3569–3572.
77. Otsuki, J. (2010) STM studies on porphyrins, *Coord. Chem. Rev.*, **254**, 2311–2341.
78. Otsuki, J. (2011) Molecular nanoarchitectures on graphite, in *Encyclopedia of Nanoscience and Nanotechnology*, (ed. Nalwa, H. S.), American Scientific Publishers, USA, in press , **16**, pp. 423–446.
79. Otsuki, J. (2011) STM studies on double- and triple-decker porphyrin and phthalocyanine complexes, *Supramol. Chem.*, in press. **23**, 169–182.
80. Lei, S. B., Wang, C., Yin, S. X., Wang, H. N., Xi, F., Liu, H. W., Xu, B., Wan, L. J., and Bai, C. L. (2001) Surface stablized porphyrin and phthalocyanine two-dimensional network connected by hydrogen bonds, *J. Phys. Chem. B*, **105**, 10838–10841.
81. Otsuki, J., Nagamine, E., Kondo, T., Iwasaki, K., Asakawa, M., and Miyake, K. (2005) Surface patterning with two-dimensional porphyrin supramolecular arrays, *J. Am. Chem. Soc.*, **127**, 10400–10405.
82. Ikeda, T., Asakawa, M., Goto, M., Miyake, K., Ishida, T., and Shimizu, T. (2004) STM observation of alkyl-chain assisted self-assembled monolayers of pyridine-coordinated porphyrin rhodium chlorides, *Langmuir*, **20**, 5454–5459.
83. Otsuki, J., Seki, E., Taguchi, T., Asakawa, M., and Miyake, K. (2007) STM observation of labile axial ligands to zinc-porphyrin at liquid/solid interface, *Chem. Lett.*, **36**, 740–741.
84. Otsuki, J., Namiki, K., Arai, Y., Amano, M., Sawai, H., Tsukamoto, A., and Hagiwara, T. (2009) Face-on and columnar porphyrin assemblies at solid/liquid interface on HOPG, *Chem. Lett.*, **38**, 570–571.
85. Cyr, D. M., Venkataraman, B., and Flynn, G. W. (1996) STM investigation of organic molecules physisorbed at the liquid–solid interface, *Chem. Mater.*, **8**, 1600–1615.
86. De Feyter, S., and De Schryver, F. C. (2003) Two-dimentional supramolecular self-assembly probed by scanning tunneling microscopy, *Chem. Soc. Rev.*, **32**, 139–150.
87. De Feyter, S., and De Schryver, F. C. (2005) Self-assembly at the liquid/solid interface: STM reveals, *J. Phys. Chem. B*, **109**, 4290–4302.

Chapter 13

Porphyrin Nanoclusters for Sensoring Chemical and Physical Stimuli

Akihiko Tsuda
Department of Chemistry, Graduate School of Science, Kobe University, 1-1 Rokkodai-cho, Nada-ku, Kobe 657-8501, Japan
tsuda@harbor.kobe-u.ac.jp

13.1 Introduction

Chromophoric π-conjugated macrocycles have potential applications such as optical, magnetic, conductive materials, and sensors for chemical and physical stimuli, originating from the unique π-electronic networks on the cyclic structure [1–12]. In a variety of macrocycles, porphyrins with 22 π-electrons, which obey Hückel's rule on aromaticity, have attracted great interest, since they play significant roles in nature such as in green leaves and red blood cells [13–16]. Light-harvesting antenna complexes with well-ordered metalloporphyrin derivatives allow efficient energy migration over many pigments within the antenna system toward a reaction center, while hemoglobin-carrying heme groups, consisting of an iron porphyrin derivative, allow oxygen transport in the body. These biological functions originate from the unique

Multiporphyrin Arrays: Fundamentals and Applications
Edited by Dongho Kim

www.panstanford.com

π-electronic structure and guest-hosting properties of the metalloporphyrin, together with well-ordered arrangements of the multi porphyrins. The characteristic π-electronic features of porphyrins are variable upon attachment of certain substituents at the periphery of the porphyrin ring, modification of the macrocyclic ring, and insertion or exchange of the metal ion of the porphyrin center [17–20]. With these interesting features of the porphyrins in mind, a variety of porphyrin-based chemical sensors, which can spectroscopically discriminate gases, metal ions, anions, pH, and molecules have been developed. These sensing systems using porphyrin and its analogues are achieved by the dynamic interaction with guest species in the molecular level. In this research background, the author has interested in sensoring not only chemical species but also physical stimuli by using supramolecular multiporphyrin architectures. The supramolecular architectures, containing the porphyrin pigments, are suitable for use as chemical sensors for direct visualization and/or indirect spectroscopic visualization of dynamic molecular motion, assembly, interaction, and orientation. This chapter presents interesting unique features of the supramolecular multiporphyrin assemblies, which can chemically sense solvent clusters in molecular recognition, and physically sense temperature and microscopic solvent fluidics induced by stirring, shaking, and sound irradiation.

13.2 Design of Multiporphyrin Supramolecular Architectures

Well-ordered architectures of self-assembled multiporphyrins [21, 22] have been attracted considerable interest in light of potential applications in material science, template-directed synthesis, reaction catalysis, and duplication of photosynthetic functions of light harvesting and charge separation. Among these, the coordination interaction between zinc porphyrin and pyridyl groups have been frequently used to construct self-assembled porphyrin architectures both in infinite and discrete forms (Fig. 13.1) [23–34]. For example, Hunter and co-workers have reported the formations of a cyclic tetramer **1** and a coordination polymer **2** from a

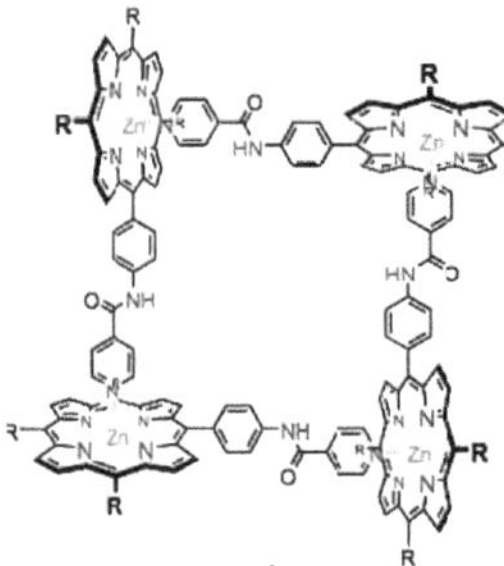

1

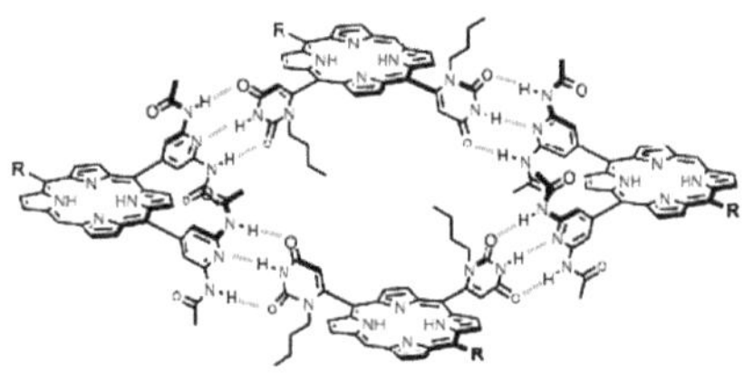

3

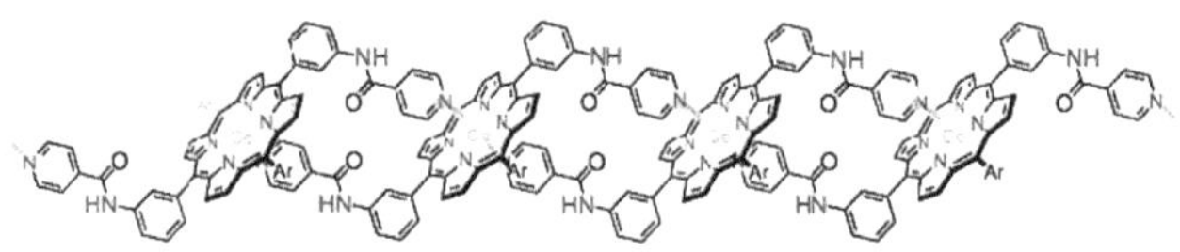

2

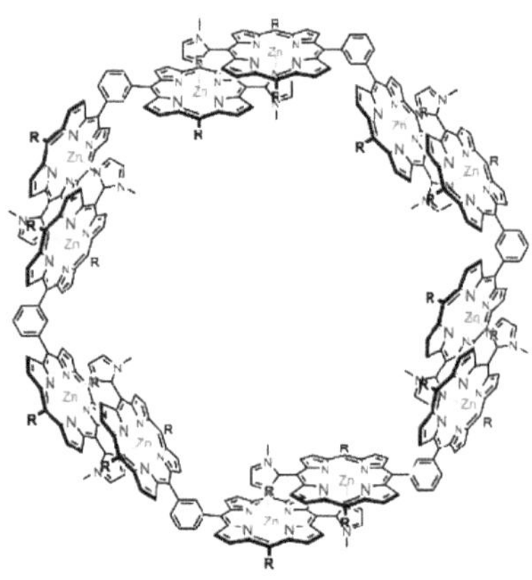

4

Figure 13.1. Representative examples of self-assembled multiporphyrin architectures.

zinc(II) porphyrin bearing a *p*-(isonicotinamide)phenyl group [35] and a cobalt(II) porphyrin bearing two *p*-(isonicotinamide)phenyl groups, respectively [36]. Drain and co-workers have reported a hydrogen-bonded porphyrin tetrameric square **3** [37]. Kobuke and co-workers have reported a giant supramolecular porphyrin ring by coordination assembly of *meso*-*N*-methyl-2-imidazolyl-substituted zinc bisporphyrin **4** [38].

With these backgrounds, the author and co-workers have synthesized self-assembled cyclic tetramers from zinc porphyrins (4-Py)P and (3-Py)P, bearing a 4-pyridyl group and 3-pyridyl group, respectively, with their X-ray crystal structures (Fig. 13.2) [39, 40]. The Zn–N coordination bonding of (4-Py)P and (3-Py)P allowed formation a cyclic tetramer with a square (1 × 1 nm) and rhombic geometry [41], respectively, both in solution and in the solid state. Based on this cyclotetrameric component, box-shaped nanoarchitectures, where zinc porphyrin components are connected at *meso* position with or without π-conjugated spacer, or by noncovalent interaction such as hydrogen bonding, were designed (Figs. 13.3, 13.4, and 13.5) [39].

Self-assembly of a directly *meso*–*meso* linked zinc bisporphyrin $[PyP]_2(I)$ gave a porphyrin nano box $\{[PyP]_2(I)\}_4$ with high stability

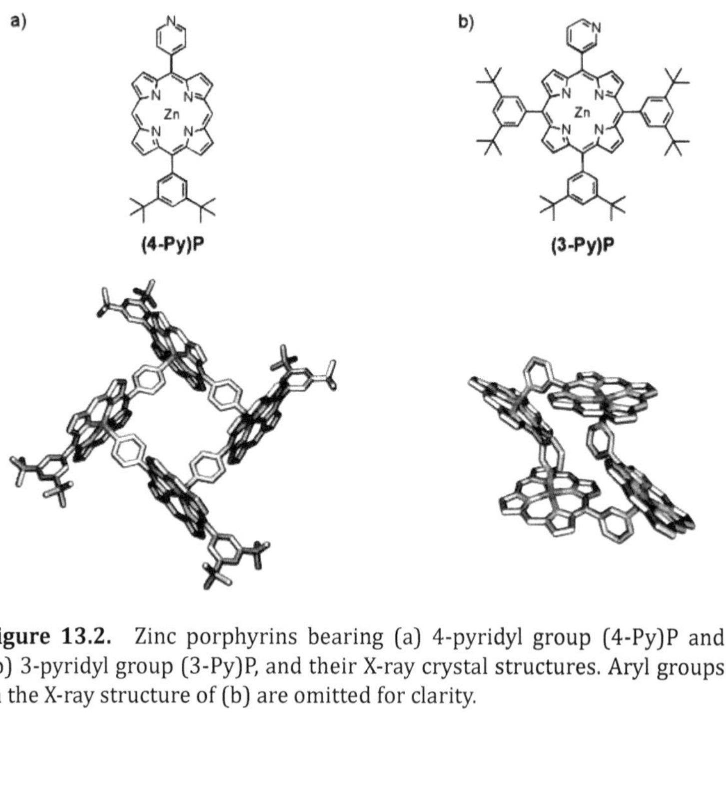

Figure 13.2. Zinc porphyrins bearing (a) 4-pyridyl group (4-Py)P and (b) 3-pyridyl group (3-Py)P, and their X-ray crystal structures. Aryl groups in the X-ray structure of (b) are omitted for clarity.

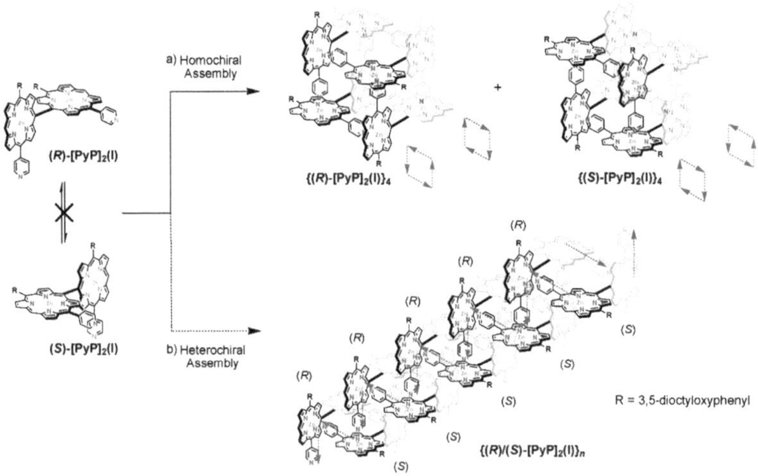

Figure 13.3. Schematic representation of a racemic mixture of *meso-meso* linked zinc bisporphyrin, bearing 4-pyridyl groups, and its self-assembling events. Red and blue arrows indicate the Py→Zn coordinating direction. See also Color Insert.

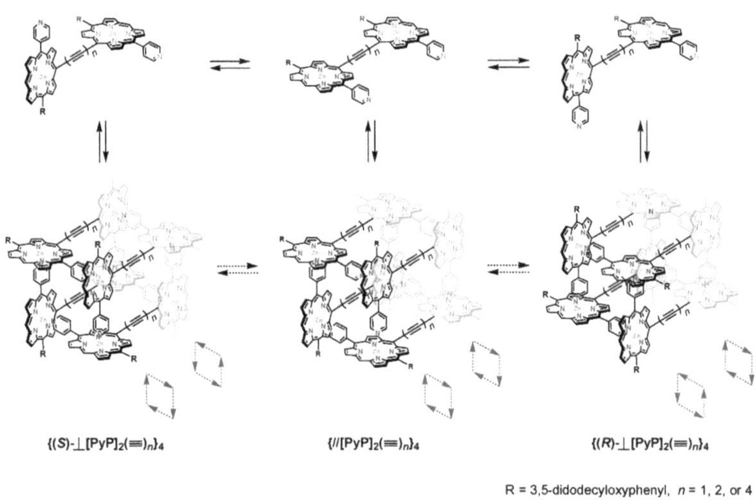

Figure 13.4. Schematic representation of alkynylene-bridged zinc bis-porphyrin rotamers, bearing 4-pyridyl groups, and their self-assembling events. Red and blue arrows indicate the Py→Zn coordinating direction. See also Color Insert.

(Fig. 13.3). The cyclic tetramerization of $[PyP]_2(I)$ could be characterized by means of size-exclusion chromatography (SEC), ^1H NMR, electronic absorption, and fluorescence spectroscopy, together with cold spray ionization mass spectrometry (CSI–MS) [42]. A THF solution of $[PyP]_2(I)$ in CSI–MS actually provided the parent ion peak of $\{[PyP]_2(I)\}_4^+$ at 6258 (calcd for $C_{376}H_{400}N_{40}O_{16}Zn_8$ 6258). However, here, it must be noted that directly *meso-meso* linked bisporphyrins have no rotational freedom of the porphyrin components around the *meso-meso* linkage due to steric hindrance of hydrogens attached at pyrrole-β positions, and thus, $[PyP]_2(I)$ is chiral, having (R)- or (S)-configurations. The racemic mixture of $[PyP]_2(I)$ has a capability to form a cyclic tetramer and/or a supramolecular ladder polymer, composed of homochiral or heterochiral bisporphyrin components, respectively. However, $[PyP]_2(I)$ forms the former discrete assembly as a sole product. Large enthalpic gain of the ring-closed system may allow dominant formation of the cyclic tetramer. Later, Osuka and coworkers succeeded in separating enantiomers of $\{(R)\text{-}[PyP]_2(I)\}_4$ and $\{(S)\text{-}[PyP]_2(I)\}_4$ by means of chiral HPLC [43].

As an extension of the above study, alkynylene-bridged zinc bispyridylporphyrin rotamers $[PyP]_2(\equiv)_n$ (n = 1, 2, and 4), having rotational freedom of the porphyrin components around the alkynylene axis have been developed (Fig. 13.4) [44–47]. $[PyP]_2(\equiv)_n$ self-assembles to form box-shaped homochiral and achiral cyclic tetramers ($\{\perp[PyP]_2(\equiv)_n\}_4$ and $\{//[PyP]_2(\equiv)_n\}_4$), where $[PyP]_2(\equiv)_n$ is conformationally frozen to adopt perpendicular and planar conformations, respectively. Further, the box-shaped assem-

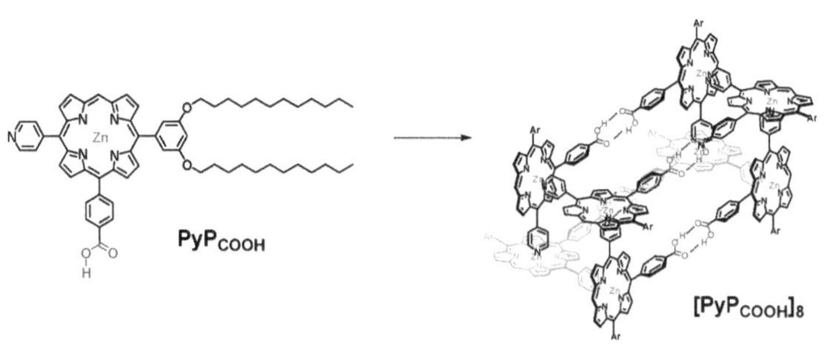

Figure 13.5. Schematic representation of zinc porphyrin PyP$_{COOH}$, bearing 4-pyridyl and carboxyphenyl groups, and its self-assembled structure formed by coordination and hydrogen bonding interactions. See also Color Insert.

bly can be also achieved in $CHCl_3$ by zinc pyridylporphyrin monomer PyP_{COOH}, bearing a carboxyphenyl group, where eight components of PyP_{COOH} self-assemble by heterogeneous interactions of coordination and hydrogen bonds to form the box-shaped architecture (Fig. 13.5) [48]. The cyclic tetramer of PyP_{COOH}, a possible intermediate of $(PyP_{COOH})_8$, can be isolated spectroscopically in the presence of acetic acid (2%), which selectively breaks the hydrogen bonding between adjacent carboxyl groups. By taking advantage of the heterogeneous interactions, a supramolecular polymer also can form from a zinc pyridylporphyrin monomer $PyP_{(COOH)2}$, bearing two carboxyphenyl groups.

13.3 Thermochromism of a π-Extended Zinc Porphyrin Nanocluster

Thermochromism is an interesting phenomenon, which can be applicable to the development of imaging and sensing materials. Classical examples include metal salts such as $CoCl_2{\cdot}6H_2O$, which displays a color change from pink to blue upon heating, as the result of a thermal-induced coordination dynamics in aqueous media [49]. π-Conjugated polymers are also candidates for thermochromic materials. For example, polythiophene displays a thermoreversible color change between red and yellow in solution, due to a change in π-electronic conjugation by a thermal-induced conformational dynamics of the backbone. Although a variety of thermochromic systems have so far been reported, there were only limited examples that are capable of developing multiple colors at ambient temperatures. With this background, an interesting example of a "supramolecular multicolor thermometer" operating at ambient temperatures have been developed [40]. The system consists of a self-assembled zinc porphyrin complex (3-Py)P$(\equiv)_1$ bearing a 3-pyridyl group and an alkynyl group, whose association/dissociation dynamics in toluene leads to a stepwise absorption spectral change with vivid color change from green to yellow to red on heating from 0 to 50 to 100°C, respectively (Fig. 13.6). The observed color change includes a contribution of the exciton coupling among self-assembled (3-Py)P$(\equiv)_1$ [50], which

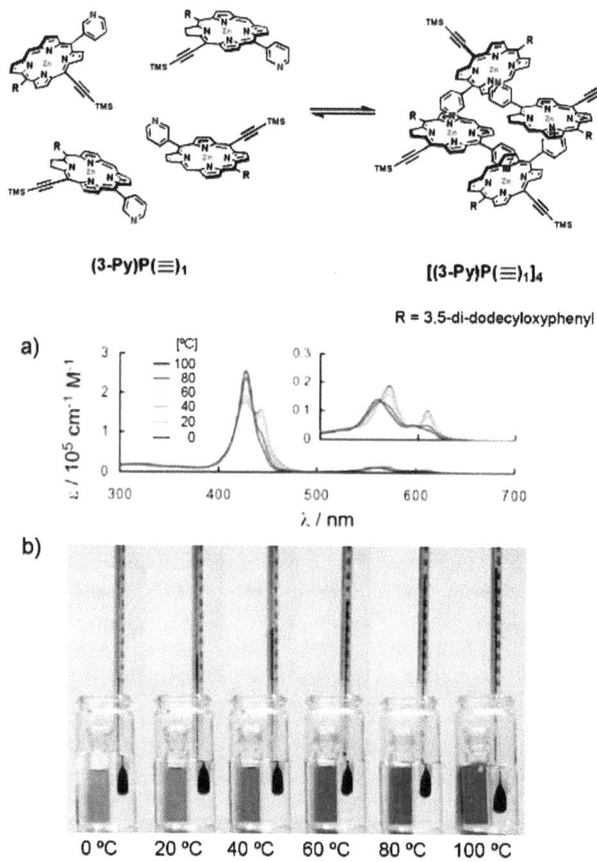

Figure 13.6. Variable-temperature a) absorption spectral profile and b) color profile of (3-Py)P(≡)$_1$ in toluene (9.8×10^{-5} M) at 0–100°C (blue: 0°C, red: 100°C). Inset shows magnified spectra in the Q-band region. See also Color Insert.

turns on and off synchronously to the above coordination dynamics on assembly and dessembly. On the other hand, the alkynyl sidearm contributes to a shift of the absorption band of the complex, so that the absorption spectral change can be detected as a vivid switching between warm and cold colors by human eyes. CSI-MS of (3-Py)P(≡)$_1$, together with ^{1}H NMR spectroscopy and X-ray crystallography of a reference complex of (3-Py)P, has indicated that (3-Py)P(≡)$_1$ preferentially adopts a cyclic tetramer with a considerable structural distortion (Fig. 13.2). The present system is a novel multicolor thermometer, which takes great advantage of the molecular design flexibility of metalloporphyrins.

13.4 Conformational Solvatochromism with a Zinc Porphyrin Supramolecular Box

Solvatochromism is the ability of chromophoric molecules to change their absorption and/or luminescence colors due to a change in solvent polarity [51]. In principle, solvatochromic shifts are caused by a change in the energy gap between the ground state and excited state having different polarities, since a change in the solvent polarity leads to differential stabilization of these electronic states. Therefore, specific interactions between chromophore and solvent molecules result in variations in wavelength, intensity, and shape of the absorption and/or luminescence spectra. Recent examples of particular interest include conjugated polymers that change their conformation and give rise to a solvatochromic response. With this general understanding on solvatochromism, one may simply assume that the color of a chromophoric molecule hardly varies with the chemical structures of solvents when their polarities are, e.g., equally very low. However, as an interesting phenomenon, a dialkynylene-bridged zinc bisporphyrin rotamer ($ZnP(\equiv)_2$) shows an explicit solvatochromic response to nonpolar solvents with low dielectric constants (2.23–2.57) (Fig. 13.7) [46]. It can discriminate even regioisomers of xylene.

As described in Fig. 13.4, alkynylene-bridged zinc bispyridylporphyrin rotamers $[PyP]_2(\equiv)_n$ (n = 1, 2, and 4), having rotational freedom of the porphyrin components around the alkynylene axis, self-assemble to form box-shaped homochiral and achiral cyclic tetramers ($\{\perp[PyP]_2(\equiv)_n\}_4$ and $\{//[PyP]_2(\equiv)_n\}_4$), where $[PyP]_2(\equiv)_n$ is conformationally frozen to adopt perpendicular and planar conformations, respectively. When $n = 1$, the planar conformer is always selected because of a strong energetic gain by the extension of a π-electronic conjugation [44]. When $n = 2$, the mole fraction of the two conformers is highly dependent on the solvent [45, 46]. Namely, $\{[PyP]_2(\equiv)_2\}_4$ with a variable extent of

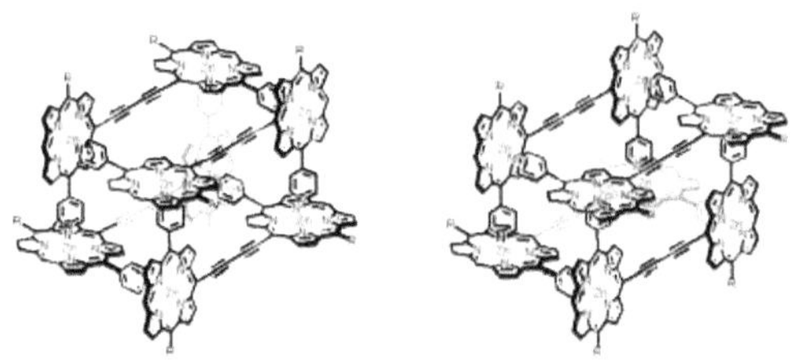

$\{\perp[PyP]_2(\equiv)_2\}_4$ $\{//[PyP]_2(\equiv)_2\}_4$

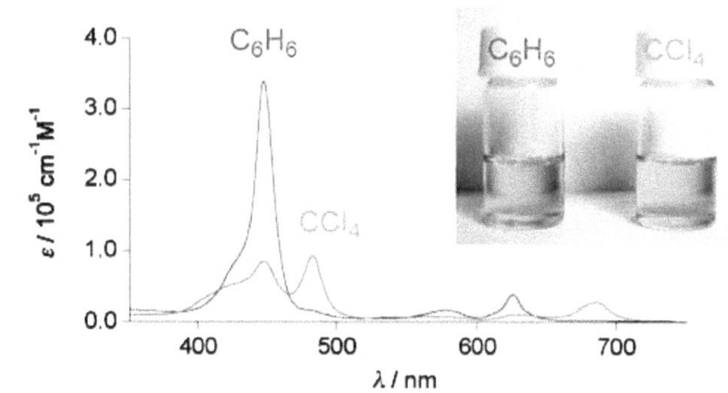

Figure 13.7. Absorption spectra of $\{[PyP]_2(\equiv)_2\}_4$ at 20°C in benzene (green) and CCl_4 (yellow). Inset shows pictures of benzene (left) and CCl_4 (right) solutions of $\{[PyP]_2(\equiv)_2\}_4$. See also Color Insert.

π-conjugation shows a unique solvatochromic response to nonpolar solvents with equally low dielectric constants (2.23–2.57) [46]. Most explicit is the case with CCl_4, where $\{//[PyP]_2(\equiv)_2\}_4$ was selected almost perfectly despite the fact that its dielectric constant is close to that of benzene with an opposite preference to $\{\perp[PyP]_2(\equiv)_2\}_4$. Further, it can spectroscopically discriminate the regioisomers of xylene, where "%-stereoisomeric excess" (%*se*) values, defined by $(\{//[PyP]_2(\equiv)_2\}_4 - \{\perp[PyP]_2(\equiv)_2\}_4)/(\{//[PyP]_2(\equiv)_2\}_4 + \{\perp[PyP]_2(\equiv)_2\}_4) \times 100$ are 39% and –78% in *o*- and *p*-xylene, respectively (Fig. 13.8), while the color of free-base $[PyP]_2(\equiv)_2$, unassembled reference, was not solvent-dependent. These observed events take advantage of a confined chromophoric cavity, given by the assembly of a π-conjugated rotamer with a variable extent of π-conjugation, and thus, can be referred to as conformational

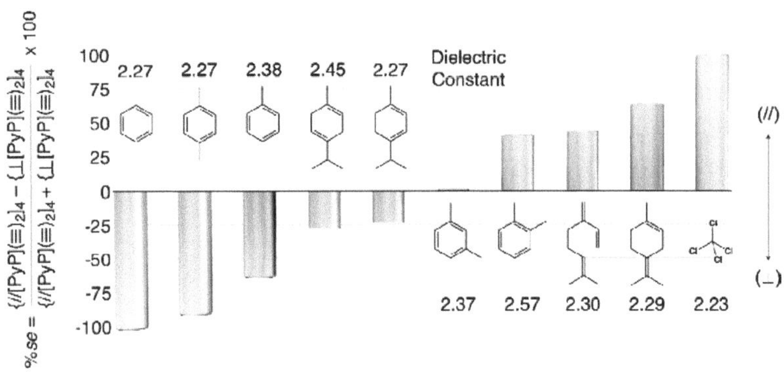

Figure 13.8. %-Stereoisomeric excess (%se) values of $\{[PyP]_2(\equiv)_2\}_4$ in nonpolar solvents at 20°C ($[PyP]_2(\equiv)_2 = 5.9\times 10^{-6}$ M).

solvatochromism. In principle, solvatochromic shifts are caused by a change in the energy gap between the ground state and excited state having different polarities, since a change in the solvent polarity leads to differential stabilization of these electronic states. However, $[PyP]_2(\equiv)_2$ shows explicit solvatochromic response to the nonpolar solvents due probably to particular electronic and/or steric effects of caged solvent molecules on their interaction with the cavity of hosting $\{[PyP]_2(\equiv)_2\}_4$.

13.5 Chiroptical Sensing of Asymmetric Hydrocarbons Using a Chiral Supramolecular Box

A variety of new concepts of chiral molecular reactions have been developed in supramolecular chemistry [52]. Certain achiral or dynamically racemic substances show a chiroptical response upon interaction with optically active species, where the chiral molecular recognition, being allowed by multipoint interactions, biases thermodynamic stability of resulting stereoisomers of the noncovalent complex. The dynamic structural conversion, induced upon chirality transfer with chemical or physical stimuli, is one of the most interesting subjects of supramolecular chiral chemistry [53–56]. In such events, porphyrins and metalloporphyrins, which possess a well understood spectral feature in both experimentally and theoretically with a strong molar ellipticity, when they are incorporated in the chiral structures or organizations, can become a spectroscopic probe for certain chiral molecular events with high sensitivity. The highly symmetric square geometry of the porphyrin allows not only the construction of well-defined supramolecular architectures, but also qualitative and quantitative evaluations of the molecular orientations spectroscopically in the architectures. These characteristic features allow finding of interesting chiroptical features of the designed porphyrin assemblies, capable of suggesting a novel concept of the chiral chemistry.

Sensing of asymmetric hydrocarbons is a challenging subject in supramolecular chemistry, since hydrocarbons do not possess functional groups appropriate for the formation of non-covalent bonds, essential for stereochemical recognition [57–60]. In the cases of $n = 4$ of box-shaped architecture $[PyP]_2(\equiv)_4$ (Fig. 13.4), the requisite for extension of the π-electronic conjugation, compared with $n = 1$ and 2, is considerably small, so that the perpendicular and planar conformers form in a 1:1 ratio in a variety of solvents [47]. The cyclic tetramers of $[PyP]_2(\equiv)_4$ are rather robust against dissociation, and the enantiomers of $\{\perp[PyP]_2(\equiv)_4\}_4$ were also successfully separated from one another by means of chiral HPLC using CH_2Cl_2/hexane (3:7) as an eluent (Fig. 13.9(a)). However, interestingly, $\{\perp[PyP]_2(\equiv)_4\}_4$ shows strong chiroptical responses due to enrichment of (*R*)- or (*S*)-enantiomer of $\{\perp[PyP]_2(\equiv)_4\}_4$ in asymmetric hydrocarbons such as limonene (Fig. 13.9(b)). By reference to the CD intensity at 458 nm of $\{\perp[PyP]_2(\equiv)_4\}_4$, limonene (= 4-isopropenyl-1-methylcyclohexane) was estimated to be at most 3%. Therefore, the successful chiroptical sensing of limonene takes great advantage of the inherently large molecular ellipticity of enantiomerically pure $\{\perp[PyP]_2(\equiv)_4\}_4$. Further, since $\{\perp[PyP]_2(\equiv)_4\}_4$ possesses also a large thermodynamic stability, a chiroptical memory, once given to $\{\perp[PyP]_2(\equiv)_4\}_4$ with (*S*)-limonene, is long-lived, even when the sample solution is diluted with (*R*)-limonene. A CPK model study predicted that $\{\perp[PyP]_2(\equiv)_4\}_4$ is likely to adopt a rectangular shape, whose inner space ($1 \times 1 \times 2$ nm) can accommodate 4–6 molecules of limonene. Nevertheless, for a better minimization of the vacancy of the solvent-including inner space, $\{\perp[PyP]_2(\equiv)_4\}_4$ is likely distorted in an orthorhombic manner, so that it displays an

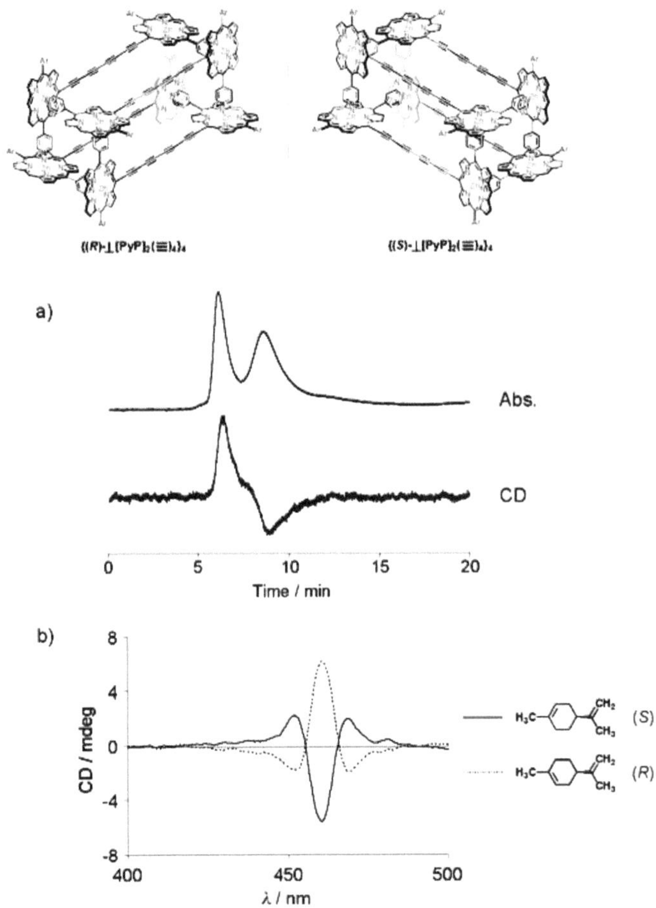

Figure 13.9. (a) Chiral HPLC profiles of $\{[PyP]_2(\equiv)_4\}_4$ using CH_2Cl_2/hexane (3:7) as an eluent at 20°C, monitored by absorption and CD responses at 455 nm. (b) Circular dichroism (CD) spectroscopy of $\{\perp[PyP]_2(\equiv)_4\}_4$ ($[PyP]_2(\equiv)_4 = 1.6 \times 10^{-6}$ M) at 20°C in (*S*)- and (*R*)-limonene (solid and broken curve, respectively).

extraordinary large molecular ellipticity. The occurrence of enantiomeric enrichment of $\{\perp[PyP]_2(\equiv)_4\}_4$ allows determination of the optical purity as well as the absolute configuration of the asymmetric hydrocarbons.

13.6 Chiroptical Visualization of Vortex Flows Using a Porphyrin Supramolecular Nanofiber

Vortex is recognized as a macroscopic chirality and one of possible origins of chiral symmetry breaking in nature [61–69]. Compared with, e.g., rheometric shear flows, the vortex is complex, involving many locally different fluidic flows. For example, upon mechanical rotary stirring of a fluid in a cuvette, a tight, torsional flow is generated at the rotary center, while the peripheral part of the vortex involves a loose, spiral flow (Fig. 13.10). The author and co-workers were prompted to investigate if vortex flows can be rendered spectroscopically. In relation to this challenging subject, certain achiral entities are known to form optically active assemblies in fluids upon rotary stirring [63–69]. Kaizu [65] and Ribó et al. [66–67] have reported one of these pioneering examples, where electrostatic *J*-aggregates of 4-sulfonylphenylporphyrin derivatives, prepared in aqueous media upon rotary stirring, are optically active, and claimed that hydrodynamic selection of either the right- or left-handed helical conformation of the aggregates is responsible. In fact, the formation of helical ribbons in such a stirred solution has recently been confirmed [67]. If such dye-containing assemblies are flexible and can readily change their conformation while flowing, they are usable as a probe for spectroscopic rendering of locally different fluidic

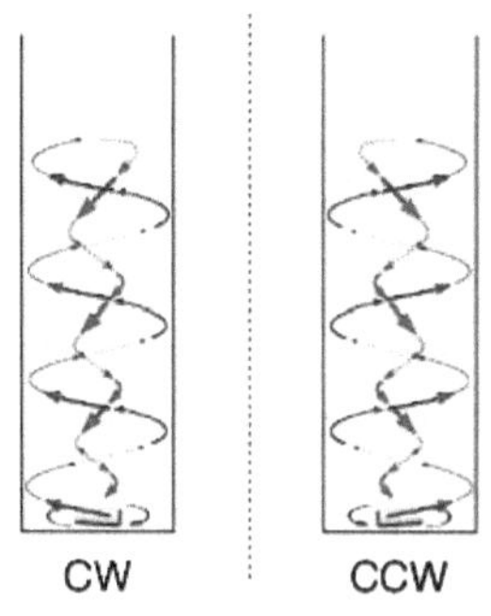

Figure 13.10. Schematic illustration of chiral vortex flows generated by rotary stirring of the solution with clockwise (CW) and counterclockwise (CCW) directions.

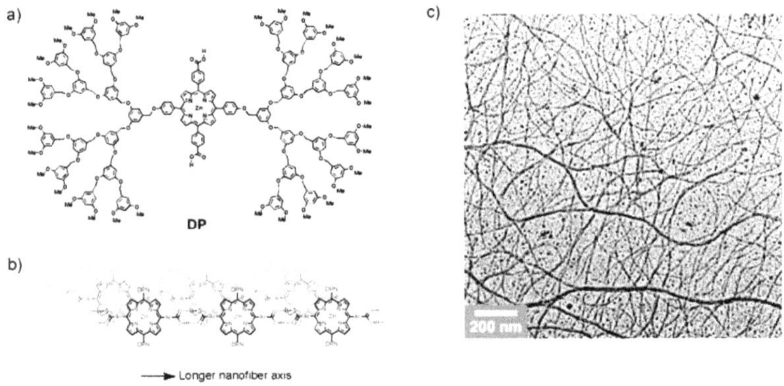

Figure 13.11. (a) Dendrimer porphyrin DP, and (b) its *J*-aggregated form of hydrogen-bonded supramolecular polymer (nanofiber). DRN represents a dendritic wedge. (c) Transmission electron micrographs (TEM) of an air-dried sample of a benzene solution of DP (1.0×10^{-5} M), deposited on a specimen grid covered with a thin carbon support film.

flows in a vortex. However, even if the stirring direction is reversed, the reported assemblies no longer invert their chiroptical sense once selected. With these background, a supramolecular polymer of *J*-aggregated zinc porphyrin dendrimer DP (Fig. 13.11) was found to visualize chiroptically the macroscopic chirality of a vortex, generated by mechanical rotary stirring of a fluid in an optical cell [70]. Although the origin of observed chiroptical activity of the sample solution could be considered as the helical twisting of the individual nanofibers of *J*-aggregated DP in the vortex flow, this is due to the consequence of a macroscopic helical alignment of the nanofibers upon rotary stirring [71].

DP forms a *J*-aggregate in benzene by a π-stacking interaction, promoted by the formation of hydrogen bonds at its carboxylic acid side groups to give a supramolecular polymer (Fig. 13.11(b)) [68]. Transmission electron microscopy (TEM) of the benzene solution of DP showed 10–20 nm wide, discrete nanofibers with a very high aspect ratio (Fig. 13.11(c)). While a benzene solution of DP (6.0×10^{-6} M; M = mol dm^{-3}) without stirring was CD silent, it became optically active upon mechanical rotary stirring (Fig. 13.12(a)). For example, when stirred at 1350 rpm in a clockwise direction (CW),

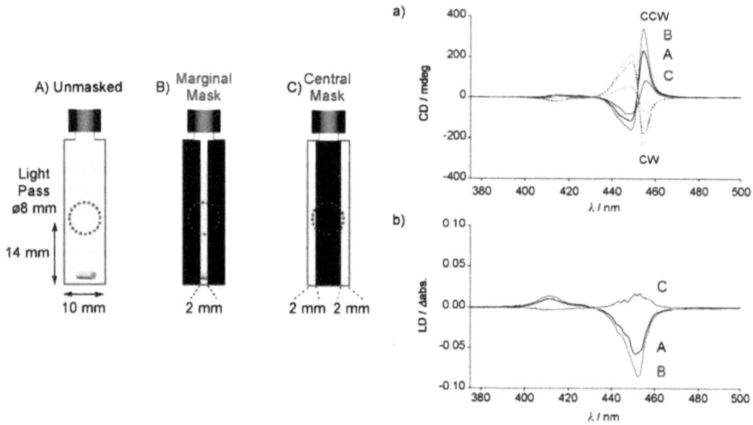

Figure 13.12. (a) Circular dichroism (CD) and (b) linear dichroism (LD) spectroscopy of a benzene solution (3.0 mL) of the nanofibers of DP (6.0×10^{-6} M), upon lower-side rotary stirring at 1350 rpm in clockwise (broken curves) or counter clockwise (solid curves) direction using a $\phi 2.0 \times 5.0$ mm Teflon-coated magnetic stirring bar at 20°C in $10 \times 10 \times 40$ mm quartz optical cells. CD and LD spectra with cells (A) unmasked (black curves), (B) masked at marginal parts with 4 mm wide black tapes to leave a 2 mm wide central slit (red curves) and (C) masked at a vertical center with a 6 mm wide black tape (blue curves). See also Color Insert.

the sample solution displayed intense CD bands at the Soret absorption region due to *J*-aggregated DP (413, 449, and 455 nm). While very weak CD bands were observed at the Q-band region (530–610 nm), no chiroptical activity emerged at the absorption band of non-assembled DP. In order to assess this intriguing macroscopic CD effect, DP was dip-coated from its benzene solution on a thick glass plate. The dip-coated thin film showed a linear dichroism (LD) activity (Fig. 13.13(a)), where the nanofibers are oriented preferentially along the dipping direction. This nanofiber-coated glass plate was then lapped over another prepared in a manner similar to the above, and observed to CD spectroscopy. Quite interestingly, when the films were lapped in such a way that their oriented directions were angled by 45° (φ), a distinct chiroptical response appeared (Fig. 13.13(b)), whose spectral shape is virtually identical to that observed for the stirred solution of *J*-aggregated DP (Fig. 13.12). When the dihedral angle for lapping (φ) was changed from 45°

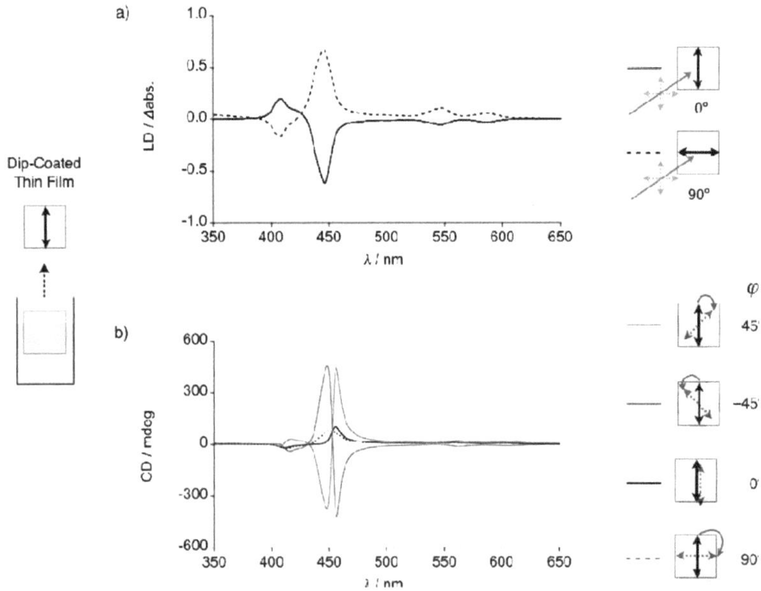

Figure 13.13. (a) Linear dichroism (LD) spectroscopy of an oriented thin film of the nanofibers of DP. The film was positioned in such a way that its oriented direction was angled by 0° (solid curve) and 90° (broken curve) relative to the vertical axis of a linearly polarized incident light for the spectroscopy. (b) Circular dichroism (CD) spectroscopy of lapped samples of two oriented thin films of the nanofibers of DP. The second film was lapped over the first one in such a way that their oriented directions were angled (φ) by 45° (red solid curve), −45° (blue solid curve), 0° (black solid curve), and 90° (black broken curve). See also Color Insert.

to −45°, a spectral inversion took place. Here, a possible major origin of the induced CD could be concluded as macroscopic chiral alignment of nanofibers upon rotary stirring. LD spectroscopy allows investigation of orientation of the nanofibers in the vortex flows. While the sample solution ([DP] = 6.0×10^{-6} M), though containing long nanofibers, was LD silent without stirring, it turned LD active upon rotary stirring. Since the spectral sign is the same as that of the dip-coated film sample, placed in such a way that its oriented direction is parallel to the vertical axis of the linearly polarized incident light, it is likely that the nanofibers preferentially align perpendicularly to the rotary direction. Nevertheless, considering that the

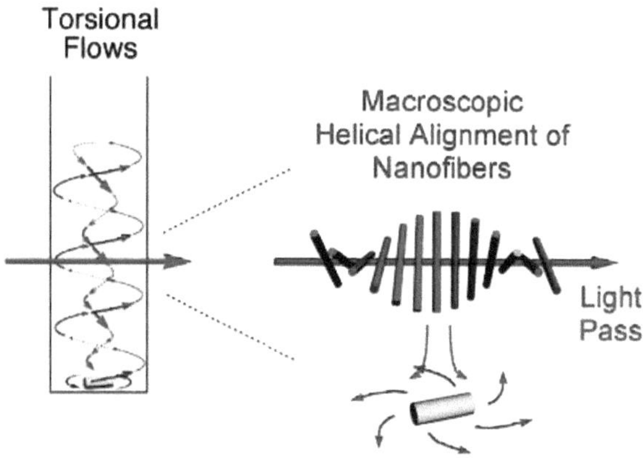

Figure 13.14. Schematic illustrations of torsional flows generated upon lower-side counterclockwise (CCW) rotary stirring and accompanying macroscopic helical alignment of nanofibers involved.

fluidic situation in a vortex, generated by rotary stirring, may change locally, point-wise LD and CD spectroscopy with optical cells selectively masked at the margins or vertical center was demonstrated (Fig. 13.12). While the LD sign observed through the central slit of the masked cell was identical to that of the entire solution, the LD spectrum observed through the margins was opposite in sign to the others. These observations suggest that the preferential alignments of the nanofibers flowing at the center and periphery of the vortex are different from each another. While the chiroptical response observed through the center slit was more intense than that with an unmasked cell, the chiroptical response at marginal parts was much less intense (Fig. 13.12(a)).

Considering the hydrodynamic model of a vortex of fluids generated in a cuvette (Fig. 13.10), they perfectly render spectroscopically local different fluidic flows in a vortex. The phenomenon observed with the nanofibers of *J*-aggregated DP is known for cholesteric liquid crystalline materials involving a macroscopic helical alignment of mesogenic molecules [72]. However, noteworthy is that in this system, such a macroscopic helical alignment occurs in non-constraint fluid media without any assistance from rheometric flows.

13.7 Spectroscopic Visualization of Sound Vibrations Using a Porphyrin Supramolecular Nanofiber

Sound is vibration, having frequencies, and can be transmitted through all forms of matter: gases, liquids, solids, and plasmas. A variety of species in nature, including humans, can sense oscillations in matter using the ears to trigger various chemical and biomolecular events. It has also been reported that the taste of alcoholic beverages such as wine and distilled spirits can change through an acoustic aging process, and an empirical technique based on this concept has been developed in industry. However, the question of whether the vibration of a medium upon sound irradiation can cause any kind of molecular or macromolecular events remains controversial. Surface acoustic waves with 1–100 MHz frequency have been used as a technique for actuation of fluids, microscale patterning of polymers and manipulation of microparticles on surfaces [73]. While, solutions containing anisotropic macroscopic objects such as polymers and colloidal particles have been known to exhibit ultrasonically induced birefringence [74–78]. The timescale of ultrasonic vibrations (>1 MHz) allows the formation of a microscopic velocity gradient or radiation pressure for macromolecules in solution, inducing them to become aligned. However, the alignment of macromolecules using audible sound, the frequency of which is much lower (20–20,000 Hz) than that of ultrasonic waves, has not been reported. This is probably due to the fact that the wavelength of audible sound is much longer than molecular length-scales. For example, the wavelengths of an audible sound with frequency of 100 Hz and an ultrasonic sound with frequency of 1 MHz in water are around 15 m and 1.5 mm, respectively. As described in the above section, a supramolecular polymer comprised of J-aggregated zinc porphyrin dendrimers can be used to chiroptically visualize the macroscopic chirality of a vortex, generated by mechanical rotary stirring of a fluid in a cuvette [70, 71]. The observed chiroptical activity is most likely due to a macroscopic helical alignment of nanofibers in torsional flows of the vortex. This finding suggests a possible visualization of sound-induced liquid vibrations using the supramolecular nanofiber. The authors found that a newly designed nanofiber, the supramoleculary polymerized zinc

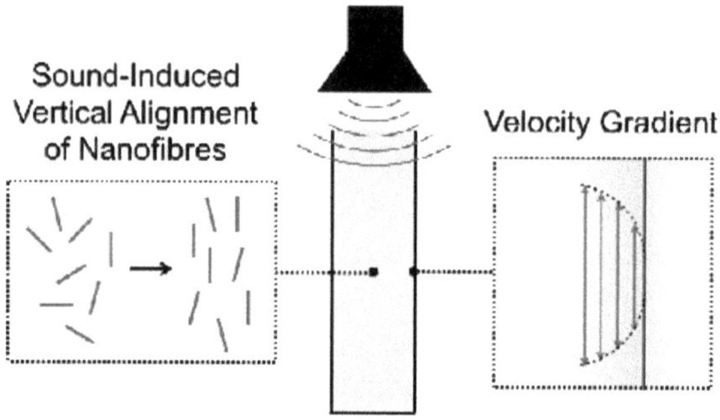

Figure 13.15. Schematic illustration of the vertical alignment of nanofibers induced by sound vibration and the resulting velocity gradient between the center and side-walls of the cuvette.

porphyrin $[PyP_{(COOH)2}]_{poly}$ (Fig. 13.16), can successfully be acoustically aligned in solution (Fig. 13.15). Zinc porphyrin $PyP_{(COOH)2}$, which bears a 4-pyridyl group and two 4-carboxyphenyl groups, occurs cyclic tetramerization due to zinc–pyridyl coordination, and following polymerization via the hydrogen bonding interactions of carboxyphenyl groups on adjacent tetramers to form the supramolecular nanofiber $[PyP_{(COOH)2}]_{poly}$ (Fig. 13.16). Dynamic light scattering (DLS) analysis showed that a solution of $PyP_{(COOH)2}$ in $CHCl_3$ (concentration 4.0×10^{-5} M at 20°C) contained self-assembled nano-objects with dimensions ranging from 425 to 1143 nm and an average radius of 667 nm. Scanning electron microscopy (SEM) performed on a solution of self-assembled $PyP_{(COOH)2}$ in $CHCl_3$ (4.0×10^{-6} M), after being air-dried, revealed the formation of linear nanofibers with an extremely high aspect ratio (Fig. 13.16(a)). Transmission electron microscopy (TEM) demonstrated that these fibers were straight; no branched structures were detected (Fig. 13.16(b)). Because the thinnest fibers in the TEM images were 3–4 nm thick, corresponding to the diameter of a molecular model of the cyclic tetramer of $[PyP_{(COOH)2}]_{poly}$, the larger diameter nanofibers, observed using both SEM and TEM, are most likely bundles of $[PyP_{(COOH)2}]_{poly}$.

Orientation of the nanofibers of $PyP_{(COOH)2}$ in solution can be characterized by using linear dichroism (LD) spectroscopy. The solution displayed a strong LD response upon irradiation with audible

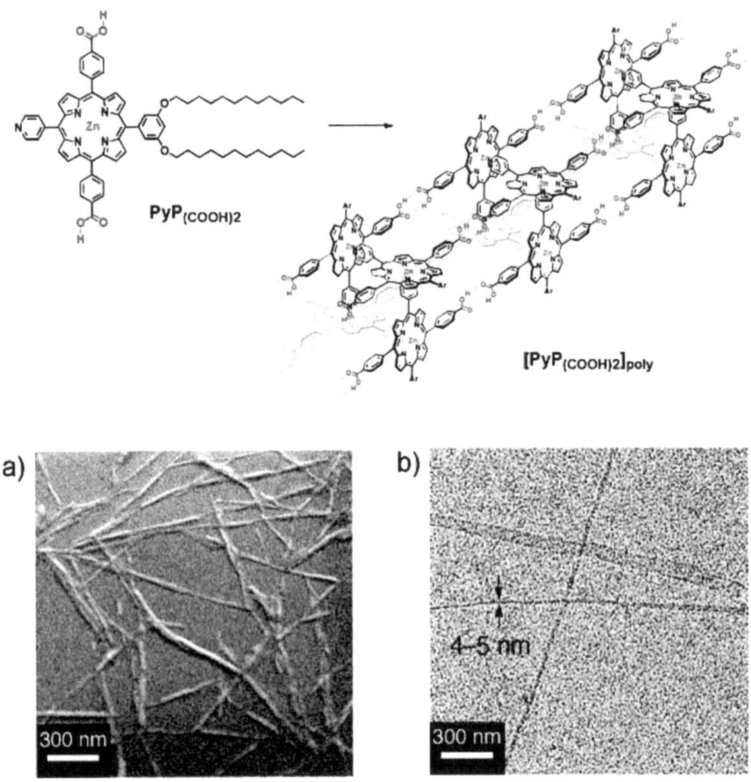

Figure 13.16. Schematic representation of zinc porphyrin $PyP_{(COOH)2}$, bearing 4-pyridyl groups, and its self-assembled supramolecular polymer formed by coordination and hydrogen bonding interactions. (a) SEM micrograph of an air-dried sample of self-assembled $PyP_{(COOH)2}$ in $CHCl_3$ solution (4.0×10^{-6} M) deposited on a silicon substrate. (b) TEM micrograph of an identical air-dried sample deposited on a specimen grid covered with a thin carbon support film.

sound in the frequency range 280 to 100 Hz, corresponding to a sound pressure of 16.6–31.6 Pa (Fig. 13.17(b)). For example, when the solution was irradiated with 120 Hz sound (31.6 Pa), intense LD bands were observed at both the Soret absorption band and the Q-band. When the sound was stopped, the solution lost LD activity immediately with a half-life of ~5 s (Fig. 13.17(c)). While, the acoustic LD response was dramatically reduced on insertion of a 12

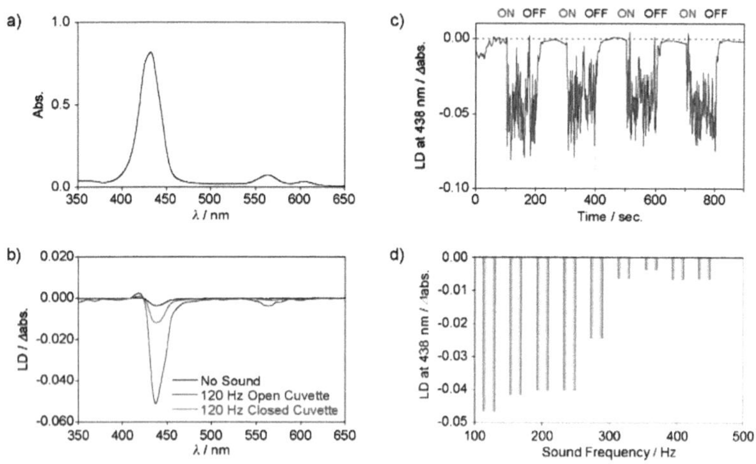

Figure 13.17. (a) Absorption spectrum of a solution of $PyP_{(COOH)2}$ in $CHCl_3$ (4.0×10^{-6} M) at 20°C. (b) LD spectra of the same solution at 20°C contained in a $10 \times 10 \times 45$ mm quartz optical cuvette, with and without 120 Hz sound irradiation (blue and black curves, respectively), and contained in a closed cuvette with a $12 \times 12 \times 2$ mm Teflon-coated plastic cap with 120 Hz sound irradiation (red curve). The sound pressure level at 120 Hz was measured as 31.6 Pa at the top of the cuvette. (c) Change in LD intensity at 438 nm in response to a repeated ON-OFF sequence of 120 Hz sound irradiation at 20°C. (d) Changes in LD intensity at 438 nm and 20°C in response to different sound frequencies for an applied voltage of 10 V on the function generator. The LD intensities were averaged for 30 s.

$\times$ 12 $\times$ 2 mm Teflon-coated plastic cap at the top of the cuvette (Fig. 13.17(b)). Because there is little propagation of sound waves from gas to liquid phases due to their extremely different acoustic impedances, the nanofibers may sense extremely weak induced vibrations of the liquid media. The dip-coated thin film, in which the nanofibers were preferentially oriented along the dipping direction, showed a virtually identical LD spectral pattern to that obtained for the solution. Hence, the nanofibers in $CHCl_3$ solution predominantly adopt a vertical orientation, parallel to the propagation direction of the irradiated sound.

The LD response was highly dependent on the frequency and amplitude of the sound (Fig. 13.17(d)). Sound with frequencies of

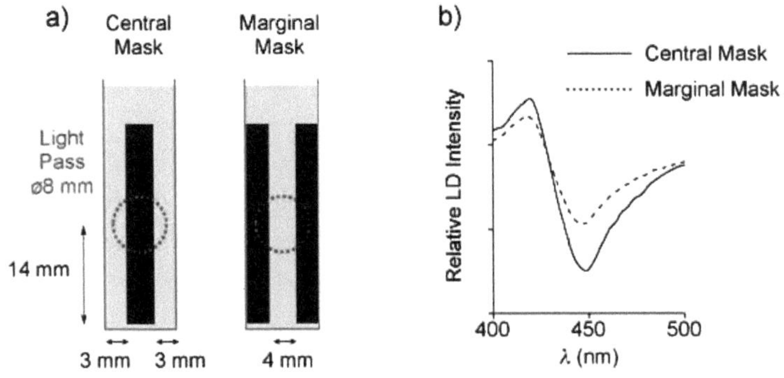

Figure 13.18. (a) Quartz optical cuvettes (12 × 12 × 45 mm) masked at a vertical center with a 4 mm wide black tape (left), and masked at marginal parts with 3 mm wide black tapes to leave a 4 mm wide central slit (right). (b) Relative LD intensity of a $CHCl_3$ solution of $PyP_{(COOH)2}$ (4.0×10^{-6} M) at 20°C with the cuvettes masked at the center and marginal parts (solid and dashed curves, respectively) upon 120 Hz sound irradiation.

less than 280 Hz (pressures in the range 21.6–31.6 Pa) allowed LD induction; the maximum value of Δabs. (−0.047) was achieved at 100 Hz (28.4 Pa). In contrast, no LD response was observed upon the irradiation of sound with frequencies in the 320–1000 Hz range (18.9–21.4 Pa). Because low frequency sound corresponds to a slow directional change of the liquid vibrations, the resulting large fluctuations of the solvent may allow alignment of the supramolecular nanofibers.

Pointwise LD spectroscopy for the local liquid vibrations in the cuvettes selectively masked at the vertical center or marginal parts revealed that the LD intensity observed through the central slit of the masked cuvette was smaller than that through the marginal parts (Fig. 13.18). The flowing nanofibers preferentially align around side wall of the cuvette. Considering the mechanism of the sound-induced alignment of nanofibers in solution, it should be noted that the wavelength of the audible sound is far longer than the length of the cuvette (for example, the wavelength is 8.3 m in the $CHCl_3$ phase for a frequency of 120 Hz at 25°C). The sound, which can be considered as periodic cycles of the condensation and rarefaction of air, will act on the sample solution in the form of a periodic increase and decrease of atmospheric pressure, respectively. This will induce a periodic change in the volume of the sample solution and give rise to vibrations, causing the solvent molecules to move parallel to the glass walls of the cuvette. Because such fluidic behavior gives rise to a velocity gradient in the solution between the side-walls and center of the cuvette (Fig. 13.15) [79], physical friction can cause the nanofibers to align hydrodynamically, becoming oriented parallel to the direction of liquid vibration, as observed for the dip-coated film. Since large hydrodynamic gradient must be occur in boundary layer of the flowing liquid around the wall surface, the larger LD intensity observed at the wall side of cuvette in the pointwise LD spectroscopy (Fig. 13.18), is quite reasonable in this mechanism.

These results suggest possible visualizations of local vibrations of acoustic liquids with qualitative as well as quantitative insights, and then, a spectroscopic visualization of the sound-induced local liquid vibration in an L-shaped cuvette was demonstrated. Because lower frequency sound is poorly directional, propagation of sound vibration to the terminal of the L-shaped cuvette via its square corner can be expected. As was also observed for the straight cuvette, a negative LD response at 438 nm was obtained at a position 20 mm below the solution surface (position (i) in Fig. 13.19). In sharp contrast, a weak LD response with a positive sign emerged in the horizontal branch of the cuvette, at position (ii) in Fig. 13.19. The nanofibres are thus able to sense weak horizontal vibrations associated with the propagation of low-frequency sound over long distances in the solution, including a directional change of 90 degrees.

The previously developed supramolecular polymer of dendritic zinc porphyrin also exhibit analogous phenomenon, though it was achieved at higher concentration. While, as controls, no acoustic LD responses were observed for solutions of conjugated polymers such as a directly linked polyporphyrin, having a long rod-shaped structure [80], and polystyrene, which allows the ultrasonically induced birefringence. Hence, the observed phenomenon is most likely a general unique function of linear supramolecular nanofibers. The high sensitivity of $[PyP_{(COOH)2}]_{poly}$ towards liquid vibrations is very attractive for sensing applications involving a variety of vibrations.

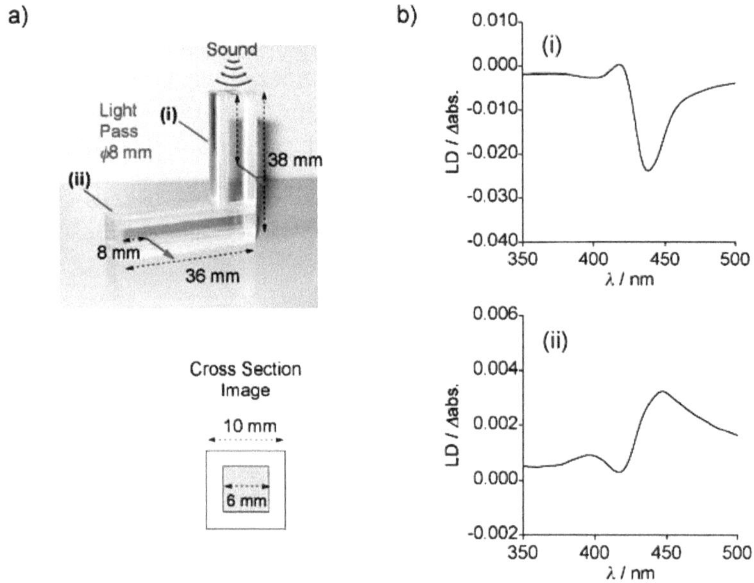

Figure 13.19. (a) A L-shaped cuvette with vertical and horizontal sections of 40 mm long, which having 10 × 10 mm widths with a glass thickness of 2 mm. (b) LD spectra for a $CHCl_3$ solution of $PyP_{(COOH)2}$ (4.0×10^{-6} M) contained in the L-shaped cuvette at 20°C are shown for position (i), located 20 mm from the top of the cuvette, and for position (ii), located in the horizontal section of the cuvette 25 mm from the center of the vertical section.

13.8 Conclusion

Design of a variety of metalloporphyrin nanoclusters, connected via covalent or noncovalent linkages with an objective to find/develop their potential properties and functions, is a very important subject in supramolecular and material sciences. π-Electrons as well as metal centers of the metalloporphyrin derivatives determine the basic electronic properties, related to optical, magnetic, and conductive properties, and assembling events. Their unique self-assembling behaviors, which can dynamically change its structure and conformation depending on the environment, allow sensing of chemical and physical stimuli such as molecular recognition, solvent polarity, light, temperature, magnetic force fields, and hydrodynamic interactions. By understanding the basic electronic structure and the self-assembling events of artificial metalloporphyrin nanoarchitectures, one may achieve their functionalization, but in most cases serendipitously found unexpected specific functions in the course of study. Authors achievements, described herein, suggest the importance of the "molecular design" from both microscopic and macroscopic viewpoints.

References

1. Huijser, A., Suijkerbuijk, B. M. J. M., Gebbink, R. J. M. K., Savenije, T. J., and Siebbeles, L. D. A. (2008) Efficient exciton transport in layers of self-assembled porphyrin derivatives, *J. Am. Chem. Soc.*, **130**, 2485–2492.
2. Gnichwitz, J.-F., Wielopolski, M., Hartnagel, K., Hartnagel, U., Guldi, D. M., and Hirsch, A. (2008) Cooperativity and tunable excited state deactivation: modular self-assembly of depsipeptide dendrons on a hamilton receptor modified porphyrin platform, *J. Am. Chem. Soc.*, **130**, 8491–8501.
3. Drain, C. M. (2002) Self-organization of self-assembled photonic materials into functional devices: photo-switched conductors, *Proc. Natl. Acad. Sci. USA*, **99**, 5178–5182.
4. Röger, C., Müller, M. G., Lysetska, M., Miloslavina, Y., Holzwarth, A. R., and Würthner, F. (2006) Efficient energy transfer from peripheral chromophores to the self-assembled zinc chlorin rod antenna: a bioinspired light-harvesting system to bridge the "green gap", *J. Am. Chem. Soc.*, **128**, 6542–6543.
5. Würthner, F. (2004) Perylene bisimide dyes as versatile building blocks for functional supramolecular architectures, *Chem. Comm.*, **14**, 1564–1579.
6. Hill, J. P., Jin, W., Kosaka, A., Fukushima, T., Ichihara, H., Shimomura, T., Ito, K., Hashizume, T., Ishii, N., and Aida, T. (2004) Self-assembled hexa-peri-hexabenzocoronene graphitic nanotube, *Science*, **304**, 1481–1483.
7. Yamamoto, Y., Fukushima, T., Suna, Y., Ishii, N., Saeki, A., Seki, S., Tagawa, S., Taniguchi, M., Kawai, T., and Aida, T. (2006) Photoconductive Coaxial Nanotubes of Molecularly Connected electron donor and acceptor layers, *Science*, **314**, 1761–1764.
8. Takase, M., Enkelmann, V., Sebastiani, D., Baumgarten, M., and Müllen, K. (2007) Annularly fused hexapyrrolohexaazacoronenes: an extended

π system with multiple interior nitrogen atoms displays stable oxidation states, *Angew. Chem. Int. Ed.* , **46**, 5524–5527.

9. Williams-Harry, M., Bhaskar, A., Ramakrishna, G., Goodson, T., Imamura, M., Mawatari, A., Nakao, K., Enozawa, H., Nishinaga, T., and Iyoda, M. (2008) Giant thienylene-acetylene-ethylene macrocycles with large two- photon absorption cross section and semishape-persistence, *J. Am. Chem. Soc.*, **130**, 3252–3253.
10. Yoon, Z. S., Kwon, J. H., Yoon, M.-C., Koh, M. K., Noh, S. B., Sessler, J. L., Lee, J. T., Seidel, D., Aguilar, A., Shimizu, S., Suzuki, M., Osuka, A., and Kim, D. (2006) Nonlinear optical properties and excited- state dynamics of highly symmetric expanded porphyrins, *J. Am. Chem. Soc.*, **128**, 14128–14134.
11. Stepien, M., Donnio, B., and Sessler, J. L. (2007) Supramolecular liquid crystals based on cyclo[8]pyrrole, *Angew. Chem. Int. Ed.*, **46**, 1431–1435.
12. Boul, P. J., Cho, D.-G., Rahman, G. M. A., Marquez, M., Ou, Z., Kadish, K. M., Guldi, D. M., and Sessler, J. L. (2007) Sapphyrin-nanotube assemblies, *J. Am. Chem. Soc.*, **129**, 5683–5687.
13. Roszak, A. W., Howard, T. D., Southall, J., Gardiner, A. T., Law, C. J., Isaacs, N. W., and Cogdell, R. J. (2003) Crystal structure of the RC- LH1 core complex from rhodopseudomonas palustris, *Science*, **302**, 1969–1972.
14. Koepke, J., Hu, X., Muenke, C., Schulten, K., and Michel, H. (1996) The crystal structure of the light- harvesting complex II (B800-850) from Rhodospirillum molischianum, *Structure*, **4**, 581–597.
15. McDermott, G., Prince, S. M., Freer, A. A., Hawthornthwaite-Lowless, A. M., Papiz , M. Z., Cogdell, R. J., and Isaacs, N. W. (1995) Crystal structure of an integral membrane light-harvesting complex from photosynthetic bacteria, *Nature*, **374**, 517–521.
16. Perutz, M. F. (1970) Stereochemistry of cooperative effects in hemoglobin. Hem-hem interaction and the problem of allostery, *Nature*, **228**, 726–737.
17. Collman, J. P., Gagne, R. R., Reed, C. A., Halbert, T. R., Lang, G., and Robinson, W. T. (1975) Picket fence porphyrins. Synthetic models for oxygen binding hemoproteins, *J. Am. Chem. Soc.*, **97**, 1427–1439.
18. Wojaczynski, J., and Latos-Grazynski, L. (2000) Poly- and oligometalloporphyrins associated through coordination, *Coord. Chem. Rev.*, **204**, 113–171.
19. Nakamura, Y., Aratani, N., and Osuka, A. (2007) Cyclic porphyrin arrays as artificial photosynthetic antenna: synthesis and excitation energy transfer, *Chem. Soc. Rev.*, **36**, 831–845.

20. Kobuke, Y., and Ogawa, K. (2003) Porphyrin supramolecules for artificial photosynthesis and molecular photonic/electronic materials, *Bull. Chem. Soc. Jpn.*, **76**, 689–708.
21. Imamura, T., and Fukushima, K. (2000) Self-assembly of metallopyridylporphyrin oligomers, *Cood. Chem. Rev.*, **198**, pp.133–156.
22. Wojaczynski, J., and Grazynski, L. L. (2000) Poly- and oligometalloporphyrins associated through coordination, *Cood. Chem. Rev.*, **204**, pp. 113–171.
23. Abrahams, B. F., Hoskins, B. F., Michail, D. M., and Robson, R. (1994) Assembly of porphyrin building blocks into network structures with large channels, *Nature*, **369**, 727–729.
24. Bhyrappa, P., Wilson, S. R., and Suslick, K. S. (1997) Hydrogen-bonded porphyrinic solids: supramolecular networks of octahydroxy porphyrins, *J. Am. Chem. Soc.*, **119**, 8492–8502.
25. Kumar, R. K., and Goldberg, I. (1998) Supramolecular assembly of heterogeneous multiporphyrin arrays - structures of $[\{Zn^{II}(tpp)\}_2(tpyp)]$ and the coordination polymer $[\{[Mn^{III}(tpp)]_2(tpyp)(ClO_4)_2\}_\infty]$, *Angew. Chem. Int. Ed.*, **37**, 3027–3030.
26. Kobayashi, K., Koyanagi, M., Endo, K., Masuda, H., and Aoyama, Y. (1998) Self- assembly of porphyrin arrays by hydrogen bonding in the solid state: an orthogonal porphyrin - bisresorcinol system, *Chem. Eur. J.*, **4**, 417–424.
27. Ikeda, C., Nagahara, N., Motegi, E., Yoshioka, N., and Inoue, H. (1999) Self-assembly of monopyrazolylporphyrins by hydrogen bonding in solution, *Chem. Commun.*, **17**, 1759–1760.
28. Ogawa, K., and Kobuke, Y. (2000) Formation of a giant supramolecular porphyrin array by self-coordination, *Angew. Chem. Int. Ed.*, **39**, 4070–4073.
29. Fujita, N., Biradha K., Fujita M., Sakamoto, S., and Yamaguchi, K. (2001) A porphyrin prism: structural switching triggered by guest inclusion, *Angew. Chem. Int. Ed.*, **40**, 1718–1721.
30. Mines, G. A., Tzeng, B.-C., Stevenson, K. J., Li, J., and Hupp, J. T. (2002) Microporous supramolecular coordination compounds as chemosensory photonic lattices, *Angew. Chem. Int. Ed.*, **41**, 154–157.
31. Li, J., Ambroise, A., Yang, S. I., Diers, J. R., Seth, J., Wack, C. R., Bocian, D. F., Holten, D., and Lindesy, J. S. (1999) Template-directed synthesis, excited-state photodynamics, and electronic communication in a hexameric wheel of porphyrins, *J. Am. Chem. Soc.*, **121**, 8927–8940.

32. Anderson, S., Anderson, H. L., Bashall, A., McPartlin, M., and Sanders, J. K. M. (1995) Assembly and crystal structure of a photoactive array of five porphyrins, *Angew. Chem. Int. Ed.*, **34**, 1096–1099.

33. Fleischer, E. B., and Shachter, A. M. (1991) Coordination oligomers and a coordination polymer of zinc tetraarylporphyrins, *Inorg. Chem.*, **30**, 3763–3766.

34. Hoffmann, M., Kärnbratt, J., Chang, M.-H., Herz, L. M., Albinsson, B., and Anderson, H. L. (2008) Enhanced π conjugation around a porphyrin[6] nanoring, *Angew. Chem. Int. Ed.*, **47**, 4993–4996.

35. Chi, X., Guerin, A. J., Haycock, R. A., Hunter, C. A., and Sarson, L. D. (1995) Self-assembly of macrocyclic porphyrin oligomers, *J. Chem. Soc., Chem. Commun.*, **24**, 2567–2569 .

36. Michelsen, U., and Hunter, C. A. (2000) Self-assembled porphyrin polymers, *Angew. Chem. Int. Ed.* **39**, 764–767.

37. Shi, X., Barkigia, K. M., Fajer, J., and Drain, C. M. (2001) Design and synthesis of porphyrins bearing rigid hydrogen bonding motifs: highly versatile building blocks for self-assembly of polymers and discrete arrays, *J. Org. Chem.* **66**, 6513–6522.

38. Shoji, O., Tanaka, H., Kawai, T., and Kobuke, Y. (2005) Single molecule visualization of coordination-assembled porphyrin macrocycles reinforced with covalent linkings, *J. Am. Chem. Soc.*, **127**, 8598–8599.

39. Tsuda, A., Nakamura, T., Sakamoto, S., Yamaguchi K. and Osuka, A. (2002) A self-assembled porphyrin box from *meso-meso*-linked bis[5-*p*-pyridyl-15-(3, 5-di-octyloxyphenyl) porphyrinato zinc(II)], *Angew. Chem. Int. Ed.*, **41**, 2817–2821.

40. Tsuda, A., Sakamoto, S., Yamaguchi, K., and Aida, T. (2003) A Novel Supramolecular multicolor thermometer by self- assembly of a πextended zinc porphyrin complex, *J. Am. Chem. Soc.*, **125**, 15722–15723.

41. Balaban, T. S., Goddard, R., Linke-Schaetzel, M., and Lehn, J.-M. (2003) 2-Aminopyrimidine directed self-assembly of zinc porphyrins containing bulky 3,5-di-*tert*-butylphenyl groups, *J. Am. Chem. Soc.*, **125**, 4233–4239.

42. Yamaguchi, K. (2003) Cold-spray ionization mass spectrometry: Principle and applications, *J. Mass. Spectrom.*, **38**, 473–490.

43. Hwang, I.-W., Kamada, T., Ahn, T. K., Ko, D. M., Nakamura, T., Tsuda, A., Osuka, A., and Kim, D. (2004) Porphyrin boxes constructed by homochiral self-sorting assembly: optical separation, exciton coupling, and efficient excitation energy migration, *J. Am. Chem. Soc.*, **126**, 16187–16198.

44. Tsuda, A., Hu, H., Watanabe, R., and Aida, T. (2003) π-Conjugated multiporphyrin box via self-assembly of an ethynylene-bridged zinc porphyrin dimer, *J. Porphyrin Phthalocyanine*, **7**, 388–393.

45. Tsuda, A., Hu, H., Tanaka, R., and Aida, T. (2005) Planar or perpendicular? Conformational preferences of π-conjugated metalloporphyrin dimers and trimers in supramolecular tubular arrays, *Angew. Chem. Int. Ed.*, **44**, 4884–4888.

46. Aimi, J., Nagamine, Y., Tsuda, A., Muranaka, A., Uchiyama, M., and Aida, T. (2008) "Conformational" solvatochromism: spatial discrimination of nonpolar solvents by using a supramolecular box of a π-conjugated zinc bisporphyrin rotamer, *Angew. Chem. Int. Ed.*, **47**, 5153–5156.

47. Aimi, J., Oya, K., Tsuda, A., and Aida, T. (2007) Chiroptical sensing of asymmetric hydrocarbons using a homochiral supramolecular box from a bismetalloporphyrin rotamer, *Angew. Chem. Int. Ed.*, **46**, 2031–2035.

48. Tsuda, A., Nagamine, Y., Watanabe, R., Nagatani, Y., Ishii, N. and Aida, T. (2010) Spectroscopic visualization of sound-induced liquid vibrations using a supramolecular nanofibre, *Nature Chem*, **2**, XXXX–XXXX.

49. Cotton, F. A., and Wilkinson, G. (1972) *Advanced Inorganic Chemistry, A Comprehensive Text*, 4th ed. (John Wiley & Sons, New York).

50. Osuka, A., and Maruyama, K. (1988) Synthesis of naphthalene-bridged porphyrin dimers and their orientation-dependent exciton coupling, *J. Am. Chem. Soc.*, **110**, 4454–4456.

51. Reichardt, C. (1994) Solvatochromic dyes as solvent polarity indicators, *Chem. Rev.*, **94**, 2319–2358.

52. Furusho, Y., Kimura, T., Mizuno, Y., and Aida, T. (1997) Chirality-memory molecule: a D_2-symmetric fully substituted porphyrin as a conceptually new chirality sensor, *J. Am. Chem. Soc.*, **119**, 5267–5268.

53. Takeuchi, M., Imada, T., and Shinkai, S. (1998) A strong positive allosteric effect in the molecular recognition of dicarboxylic acids by a cerium(IV) bis[tetrakis(4-pyridyl) porphyrinate] double decker, *Angew. Chem. Int. Ed.*, **37**, 2096–2099.

54. Huang, X., Rickman, B. H., Borhan, B., Berova, N., and Nakanishi, K. (1998) Zinc porphyrin tweezer in host-guest complexation: determination of absolute configurations of diamines, amino acids, and amino alcohols by circular dichroism, *J. Am. Chem. Soc.*, **120**, 6185–6186.

55. Borovkov, V. V., Lintuluoto, J. M., Fujiki, M., and Inoue, Y. (2000) Temperature effect on supramolecular chirality induction in bis(zinc porphyrin), *J. Am. Chem. Soc.*, **122**, 4403–4407.

56. Yashima, E., Maeda, K., and Nishimura, T. (2004) Detection and amplification of chirality by helical polymers, *Chem. Eur. J.*, **10**, 42–51.
57. Kawasaki, T., Tanaka, H., Tsutsumi, T., Kasahara, T., Sato, I., and Soai, K. (2006) Chiral discrimination of cryptochiral saturated quaternary and tertiary hydrocarbons by asymmetric autocatalysis, *J. Am. Chem. Soc.*, **128**, 6032–6033.
58. Paolesse, R., Monti, D., La Monica, L., Venanzi, M., Froiio, A., Nardis, S., Di Natale, C., Martinelli, E., and D'Amico, A. (2002) Preparation and self-assembly of chiral porphyrin diads on the gold electrodes of quartz crystal microbalances: a novel potential approach to the development of enantioselective chemical sensors, *Chem. Eur. J.*, **8**, 2476–2483.
59. Prince, R. B., Barnes, S. A., and Moore, J. S. (2000) Foldamer-based molecular recognition, *J. Am. Chem. Soc.*, **122**, 2758–2762.
60. Kobayashi, K., Asakawa, Y., Kikuchi, Y., and Aoyama, Y. (1993) CH-π interaction as an important driving force of host-guest complexation in apolar organic media. Binding of monools and acetylated compounds to resorcinol cyclic tetramer as studied by proton NMR and circular dichroism spectroscopy, *J. Am. Chem. Soc.*, **115**, 2648–2654.
61. Mason, S. F. (1984) Origins of biomolecular handedness, *Nature*, **311**, 19–23.
62. Link, D. R., Natale, G., Shao, R., Maclennan, J. E., Clark, N. A., Körblova, C. E., and Walba, D. M. (1997) Spontaneous formation of macroscopic chiral domains in a fluid smectic phase of achiral molecules, *Science*, **278**, 1924–1927.
63. Kondepudi, D. K., Kaufman, R. J., and Singh, N. (1990) Chiral symmetry breaking in sodium chlorate crystallization, *Science*, **250**, 975–976.
64. Kondepudi, D. K., Laudadio, J., and Asakura, K. (1999) Chiral symmetry breaking in stirred crystallization of 1,1'-Binaphthyl Melt, *J. Am. Chem. Soc.*, **121**, 1448–1451.
65. Ohno, O., Kaizu, Y., and Kobayashi, H. (1993) *J* -aggregate formation of a water-soluble porphyrin in acidic aqueous media, *J. Chem. Phys.*, **99**, 4128–4139.
66. Ribó, J. M., Crusats, J., Sague, F., Claret, J., and Rubires, R. (2001) Chiral sign induction by vortices during the formation of mesophases in stirred solutions, *Science*, **292**, 2063–2066.
67. Escudero, C., Crusats, J., Díez-Pérez, I., El-Hachemi, Z., and Ribó, J. M. (2006) Folding and hydrodynamic forces in *J* -aggregates of 5-phenyl-10,15,20-tris- (4-sulfo-phenyl) porphyrin, *Angew. Chem. Int. Ed.*, **45**, 8032–8035.

68. Yamaguchi, T., Kimura, T., Matsuda, H., and Aida, T. (2004) Macroscopic spinning chirality memorized in spin- coated films of spatially designed dendritic zinc porphyrin *J* -aggregates, *Angew. Chem. Int. Ed.*, **43**, 6350–6355.

69. Dzwolak, W., Loksztejn, A., Galinska-Rakoczy, A., Adachi R., Goto, Y., and Rupnicki, L. (2007) Conformational indeterminism in protein misfolding: chiral amplification on amyloidogenic pathway of insulin, *J. Am. Chem. Soc.*, **129**, 7517–7522.

70. Tsuda, A., Alam, M. A., Harada, T., Yamaguchi, T., Ishii, N., and Aida, T. (2007) Spectroscopic visualization of vortex flows using dye-containing nanofibers, *Angew. Chem. Int. Ed.*, **46**, 8198–8202.

71. Wolffs, M., George, S. J., Tomovic, Z., Meskers, S. C. J., Schenning, A. P. H. J., and Meijer, E. W. (2007) Macroscopic origin of circular dichroism effects by alignment of self-assembled fibers in solution, *Angew. Chem. Int. Ed.*, **46**, 8203–8205.

72. Saeva, F. D., Sharpe, P. E., and Olin, G. R. (1973) Cholesteric liquid crystal induced circular dichroism (LCICD). V. Mechanistic aspects of LCICD, *J. Am. Chem. Soc.*, **95**, 7656–7659.

73. Shilton, R., Tan, M. K., Yeo, L. Y., and Friend, J. R. (2008) Particle concentration and mixing in microdrops driven by focused surface acoustic waves, *J. Appl. Phys.*, **104**, 149101–149109.

74. Kawamura, H. (1938) *Kagaku* (Iwanami Shoten, Publisher), **7**, 6–7 (in Japanese) .

75. Lipeles, R., and Kivelson, D. (1980) Experimental studies of acoustically induced birefringence, *J. Chem. Phys.*, **72**, 6199–6208.

76. Yasuda, K., Matsuoka, T., Koda, S., and Nomura, H. (1994) Dynamics of V_2O_5 sol by measurement of ultrasonically induced birefringence, *Jpn. J. Appl. Phys.*, **33**, 2901–2904.

77. Yasuda, K., Matsuoka, T., Koda, S., and Nomura, H. (1997) Dynamics of entanglement networks of rodlike micelles studied by measurements of ultrasonically induced birefringence, *J. Phys. Chem. B*, **101**, 1138–1141.

78. Nomura, H., Matsuoka, T., and Koda, S. (2004) Ultrasonically induced birefringence in polymer solutions, *Pure Appl. Chem.*, **76**, 97–104.

79. Yoshimura, T., Sakashita, H., and Wakabayashi, N. (1983) Real-time measurements of spatial velocity distribution with a laser Doppler imaging system, *Appl. Opt.*, **22**, 2448–2452.

80. Yoshida, N., Aratani, N., and Osuka, A. (2000) Poly(zinc(II)-5,15-porphyrinylene) from silver(I)-promoted oxidation of zinc(II)-5,15-diarylporphyrins, *Chem. Commun.*, 197–198.

Chapter 14

Energy- and Electron-Transfer Processes in Porphyrin Dendrimers

Myung-Seok Choi[a] and Woo-Dong Jang[b]

[a]*Department of Materials Chemistry and Engineering, Konkuk University, 1 Hwayang-dong, Gwangjin-gu, Seoul, 143-701, South Korea*

[b]*Department of Chemistry, Yonsei University, 262 Seongsanno, Seodaemun-gu, Seoul 120-749, South Korea*

mchoi@konkuk.ac.kr; wdjang@yonsei.ac.kr

A recent development of dendritic macromolecules has provided a new methodology for the design of well-controllable three-dimensional nano-object. Dendrimers generally consist of three different topological parts, i.e., focal core, monomeric building blocks, numbers of terminal functional groups. For the design of novel photofunctional nano-devices, many efforts have been paid for the incorporation of porphyrin derivatives into each topological parts of dendrimer structure. In this chapter, we are going to discuss about energy and electron transfer processes in dendritic architecture with porphyrin derivatives.

14.1 Introduction

Nature developed the photosynthetic processes several billion years ago to utilize solar energy and store it in the fuels that are essential

Multiporphyrin Arrays: Fundamentals and Applications
Edited by Dongho Kim

www.panstanford.com

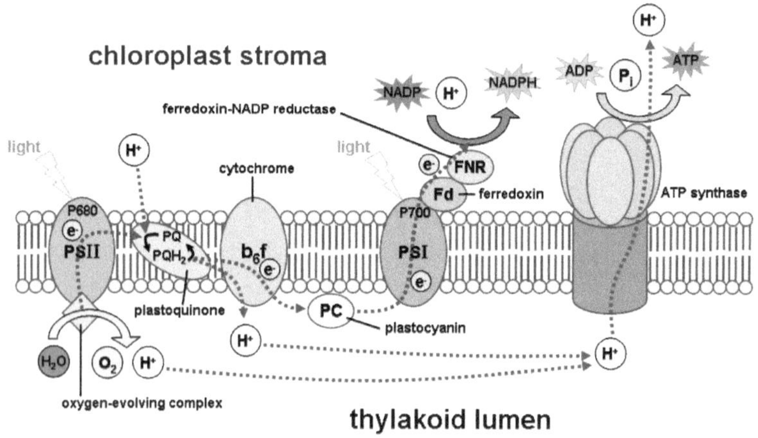

Figure 14.1. Photosynthesis of purple bacteria (from Wikipedia).

for life on earth. Interestingly, photosynthesis convert only 0.02–0.05% of the incident solar energy of about 10^{22} kJ/year into biological material. This is about 100 times more than the food needed for mankind [1]. Thus, a batter understanding of photosynthesis is very important due to its potential uses of solar energy.

In nature, some bacteria and green plants have such a photosynthetic work. In particular, photosynthesis of purple bacteria has much in common with its analogue in green plants but is simpler and better understood (Fig. 14.1).

In fact, bacterial photosynthesis efficiently use sunlight by employing supramolecular machines, where the absorbed photon energy is converted into the chemical potential through creation of a charge-separated state via photoinduced electron transfer with a high efficiency. The resulting potential energy is used to pump protons across the membrane, generating an osmotic and charge imbalance which in turn powers the synthesis of biological high-energy molecules in the form of adenosine triphosphate (ATP) by ATP synthase. The ATP fills the majority of the energy needs of the bacterium.

Artificial model mimicking photosynthesis is a hot issue recently in the field of material science and nano-chemistry [2, 3, 4]. What are the design requirements for an artificial photosynthesis? As shown in Fig. 14.1, at least two molecular complexes, light-harvesting antennae to collect sunlight and reaction center to convert photon energy into chemical energy are essential. To utilize sunlight in a similar fashion, it is reasonable to borrow some design concepts from structures and functions of natural systems. For this reason, synthetic porphyrins are particularly promising building blocks for artificial systems, since they share many of the general structural, chemical, redox, and photophysical properties with the natural components such as chlorophyll and bateriochlorophyll. A common objective in the fabrication of many of these supramolecular systems like photosynthetic artificial systems has been the facile organization of the individual porphyrin molecules by using covalent or noncovalent approaches.

In particular, many efforts to mimic and understand the functional and mechanistic features of these natural photosynthetic processes has inspired by X-ray crystal analysis of the light-harvesting antenna complexes [5] and reaction center RC [6] of photosynthetic purple bacteria (Fig. 14.2).

Photosynthetic bacteria evolved a pronounced energetic hierarchy in the light-harvesting system, which have two types of antenna complexes LH-1 and LH-2, consisting of wheel-like assemblies of bateriochlorophyll a (porphyrin derivatives) supported by α, β-apoproteins with distinct biochemical and spectroscopic

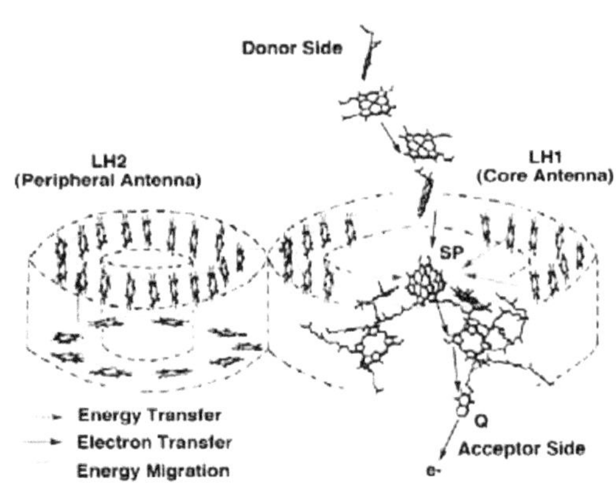

Figure 14.2. Photophysical processes in photosynthetic purple bacteria.

properties [7]. The hierarchy furnishes a cascade-like system of excited states that funnels electronic excitation from the peripheral antenna LH-2 through core antenna LH-1 to the reaction center (RC) [8]. Such well-studied structural information has inspired considerable research efforts into the design, synthesis of novel light-harvesting antennae that could be created different forms with precise molecular structures, based on transition-metal complexes [9], dye-labeled polymer [10], multi-chromophoriccyclodextrins [11], anthryloligothienyl porphyrins [12], boron-dippyrrins [13], dendrimer [14], or porphyrin assemblies [15]. Most effective synthetic approach for the natural light-harvesting antennae is dependent upon achieving following several distinctive features: (1) fabricating systems that have large absorption throughout the visible light, (2) vectorial energy transfer from the antenna molecules to the energy trap. Along this line, the authors wish to describe a recent progress relating to the light-harvesting multi-porphyrin arrays with dendritic architecture, consisting of many peripheral energy-donating units with or without a focal energy-accepting unit. Most of synthetic antenna systems described herein are also comparable, in their sizes and shapes, to the LH2 of purple bacteria.

The popularity of dendrimers in science is not merely due to aesthetics. Years of intense research activity have shown that dendrimers can be considered as promising nano-sized architectures for use in materials science and biological applications [16].

Dendrimers are highly branched three-dimensional macromolecules with a branch point at each monomer unit (Fig. 14.3). As a result of this high degree of branching, large dendrimer adopt a globular shape in which all bonds converge to a focal point or central core unit. What makes dendrimers special is the combination of several interesting properties, particularly in term of light-harvesting activity [17]. Two basic properties make these macromolecules promising candidates for the photosynthetic molecular mimic: the large number of absorbing units that grow exponentially with their generation number (high probability for the capture of light) and the relatively short distance of the periphery from the center, where a fluorescent trap, reaction center, or a chemical sensor can be located. Along the line, this chapter focuses on the molecular design of

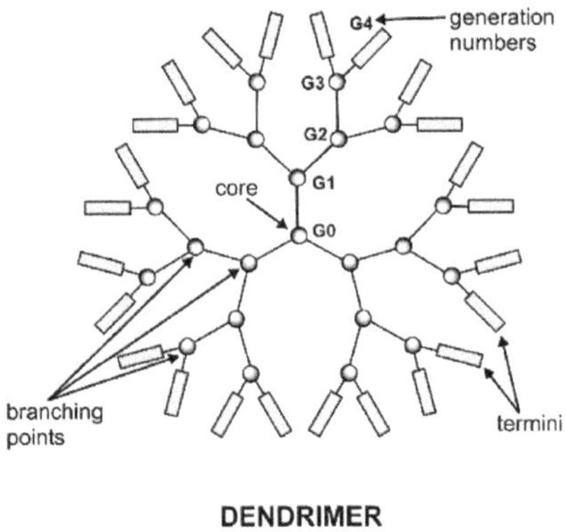

Figure 14.3. Structure of dendrimer (From Wikipedia).

donor–acceptor arrays using porphyrin dendrimers, and their photoinduced energy/electron transfer events.

14.2 Energy Transfer in Porphyrin Dendrimers

14.2.1 *Design of Dendritic Light Harvesting Antennae for Effective Excitation Energy Transfer*

Owing to their unique structural and topological features, dendritic architectures are often utilized for various bio-inspired applications from bio-mimicry to biomedicine. The incorporation of porphyrin units to the dendritic architecture also widely explored to aim various applications. Well-organized three-dimensional architecture of dendrimer provides a variety of unique properties, which cannot be achievable using traditional linear polymers. Typically, dendrimers can provide multivalent modification of peripheral functional groups and site isolation effect of core functionality. Large dendrimers eventually have density gradient because the number of branching unit gradually increase from core to surface functional groups. A variety of dendrimers have been designed as mimicry of natural light harvesting system to achieve efficient energy transfer. Photosynthesis is one of the most important chemical events in nature which provides vital energy for all living things on earth.

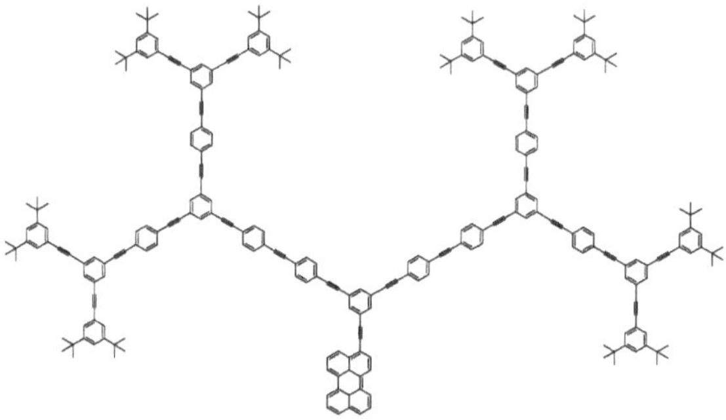

1

Figure 14.4. Structure of phenyleneethynylene dendrimer.

The photosynthetic process includes the absorption of light energy and successive conversion of light energy to chemical energy. A large number of chromophores form supramolecular array system, which absorbs photons from the sunlight and then transfer to the reaction center. The density gradient of dendritic structure is one of powerful candidates for the achievement of directional energy transfer from large number of chromophores to focal energy acceptor. As a pioneering work of directional energy transfer, phenyleneethynylene dendrimers (**1**) have been synthesized (Fig. 14.4) [18]. The phenyleneethynylene branching units provide large absorption cross-section of UV light at the range of 250–350 nm, and the absorbed energy transported to core chromophore with high efficiency. The energy transfer efficiency was increased with generation increment. At similar period, the theoretical consideration of energy migration on the dendrimers was also carried out by Kopelman et al. for the direction of designing controlled artificial energy antennae [19]. They emphasized that the geometric bias and energetic funnel in dendrimers should be considered for the successful design of the designing artificial antennae. On the basis of above indication, porphyrin bearing poly(benzyl ether) dendrons would be one of the simple but successful examples. The focal porphyrin works as energetic funnel to accept excited energy of benzyl ether groups in the dendritic wedges.

14.2.2 *Porphyrin-Centered Dendrimers*

Porphyrin-centered poly(benzyl ether) dendrimers (**2a**–**2e**) were firstly synthesized by Aida et al., and their energy or electron transfer phenomenon has been extensively studied (Fig. 14.5) [17b]. As mentioned above, the focal porphyrin unit works as the acceptor for excited energy of dendritic wedges. Under 280 nm UV irradiation, large numbers of benzene rings in dendrimers frameworks absorb light energy and transfer their excitation energy to the focal porphyrin unit, which emit strong fluorescence at 656 and 718 nm. The energy-transfer quantum efficiency is over 80% in fourth generation dendrimer (**2a**). In sharp contrast, excitation of partially defected dendrimers (**2b**–**2e**) results in only weak fluorescence emission from focal porphyrin. This phenomenon can be explained by excitation energy migration among aromatic building units of dendritic wedges. In the globular shape dendrimer, the aromatic building units become close and have homogenous density distribution around the focal core. Therefore, the excitation energy migration process can be preceded continuously to enhance the probability of energy transfer to focal porphyrin. Alternatively, the density of aromatic units in the partially defected dendrimers becomes looser and discontinuous, so that the excitation energy migration becomes less effective. The continuous energy migration process within dendritic architecture might be closely related to the

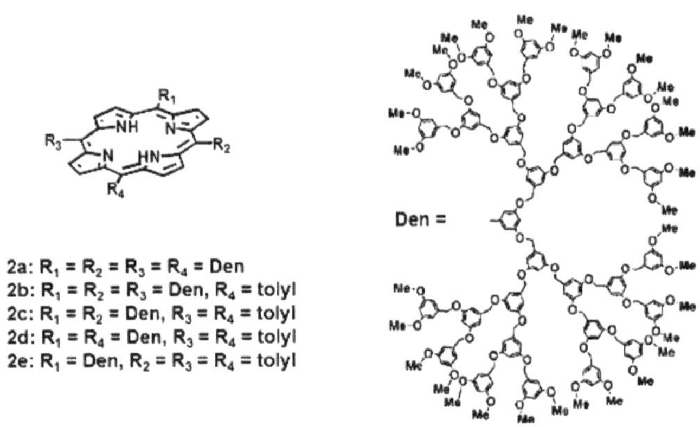

Figure 14.5. Structure of porphyrin-centered poly(benzyl ether) dendrimers.

energy transfer events in bacterial light harvesting antenna in photosynthetic bacteria, where the excitation energy migration proceeded very rapidly and efficiently along the wheel-like arrangement of pigments to initiate the photosynthetic process.

An interesting and unique application of dendrimer porphyrin is sensitization of oxygen. The excitation energy of focal porphyrin can be transferred to oxygen molecules to produce highly toxic singlet oxygen. This mechanism can be utilized to photodynamic cancer therapy. Kataoka et al. have extensively researched the biomedical application of porphyrin-centered poly(benzyl ether) dendrimers [20]. Photodynamic therapy is based on the accumulation of a photosensitizer in malignant tissue after its administration usually through intravenous route. Most conventional photosensitizers for the application in photodynamic therapy are porphyrin derivatives due to their large absorption cross-section and high quantum yield for the singlet oxygen generation. However, porphyrin derivatives are easy to form aggregates in aqueous medium due to the strong π–π interactions and their hydrophobic nature. By the introduction of large dendritic wedges to porphyrin, the formation of aggregates can be perfectly prevented. Furthermore, polyion complex (PIC) micelles can be obtained by using ionic dendrimer porphyrin. The PIC micelles showed a higher stability with increasing ionic strength and highly enhanced photocytotoxicity, because the dendritic architecture may prevent collisional quenching of photo-excited core functionality. The large hydrodynamic volume of micellar formulation is expected to provide high localization into the tumor tissue due to the Enhanced Permeation and Retention (EPR) effect. Indeed, dendrimer porphyrin-incorporated PIC micelle exhibited selective accumulation onto the neovascular lesion of eye disease models. In addition to the porphyrin-centered dendrimer, ionic Zinc phthalocyanine dendrimer also developed to achieve improved tissue-penetration of light. As another approach to achieve deeper tissue penetration of light energy, Fréchet et al. developed a porphyrin-centered dendrimer (**3**) bearing two photon absorption dyes (Fig. 14.6) [21]. This system is based on the energy transfer from the two photon absorption dyes to a focal porphyrin, and has been shown to efficiently generate singlet oxygen after irradiation at 780 nm.

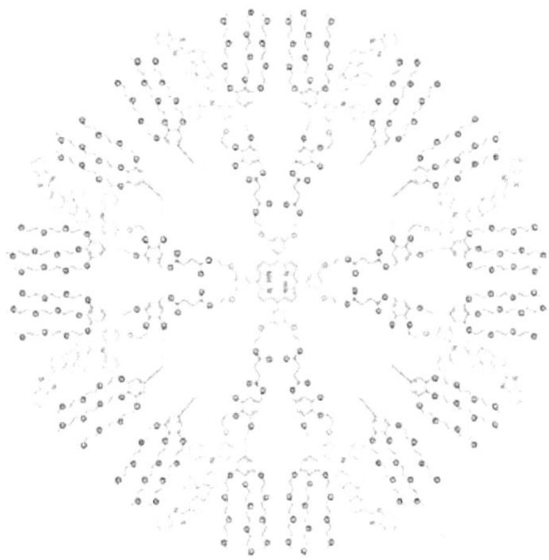

3

Figure 14.6. Porphyrin-centered dendrimer bearing two photon absorption dyes.

14.2.3 *Porphyrins in Periphery of Dendrimers*

Dendrimers can provide large numbers of peripheral functional groups. Therefore, the dendritic architecture is one of the most powerful motifs for the integration of functional moieties into limited spaces. Globular shape multi-porphyrin arrays can be prepared by the introduction of porphyrins onto the surface of dendrimers. As an example, poly(L-lysine) dendrimer (**4**) having 32 porphyrin units as surface functional groups has been synthesized (Fig. 14.7) [22]. Dendrimer **4** bears 16 zinc porphyrin units (P_{Zn}) on one hemisphere and 16 free base porphyrins (P_{FB}) on the other hemisphere. A steady-state fluorescence study has shown the occurrence of energy transfer from P_{Zn} to P_{FB} with an efficiency of 43%, where the energy transfer from P_{Zn} to P_{FB} occurs mainly at the hemispherical boundaries. Aida et al. have synthesized a series of poly(benzyl ester) dendrimers with P_{Zn} units (**5a–5d**) as end functional groups to evaluate the excitation energy migration within the porphyrin arrays (Fig. 14.8) [23]. A fluorescence anisotropy decay and transient absorption of the two-branched type porphyrin dendrimers revealed that tightly packed three-dimensional structures

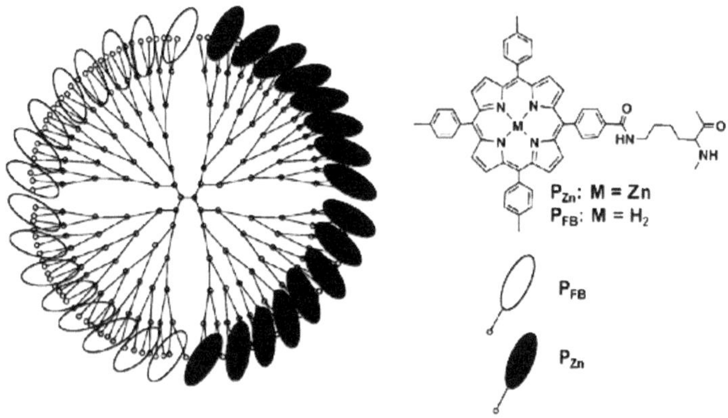

Figure 14.7. Poly(L-lysine) dendrimer having 32 porphyrin units.

of porphyrins in dendrimers enhance the efficiency of excitation energy transfer. For example, the experimental excitation energy transfer times between the porphyrin units in **5a** were faster than those of **5b**. On the other hand, three-branched type dendrimers **5c** and **5d** exhibits biexponential anisotropy decays with similar time constants to those of structural constituent unit **3P**$_{\mathbf{Zn}}$, suggesting that **3P**$_{\mathbf{Zn}}$ constituent units acts as a cluster in the excitation energy-transfer process. Because the porphyrins in structural constituent unit **3P**$_{\mathbf{Zn}}$ are already in closely contacted, fast local excitation energy migration processes might be occurred and then slow excitation energy migration processes between the clusters. More interestingly, addition of bipyridyl guest molecule to **5a** and **5b**, which is possible to intercalate between two porphyrin units, results in suppression of the P_{Zn}–P_{Zn} interaction and an enhancement of the fluorescence quantum yield. From the above observation, it can be concluded that geometrical homogeneity and packing density of chromophores is significantly important for the effective excitation energy transfer.

Multi-porphyrin dendrimers with focal porphyrin core were also synthesized. As a model study of light-harvesting antennae, several types of star-shaped pentameric porphyrin arrays with focal P_{FB} have been extensively studied. The first example of dendritic porphyrin array was a vinylene-linked nonamericninkel porphyrin

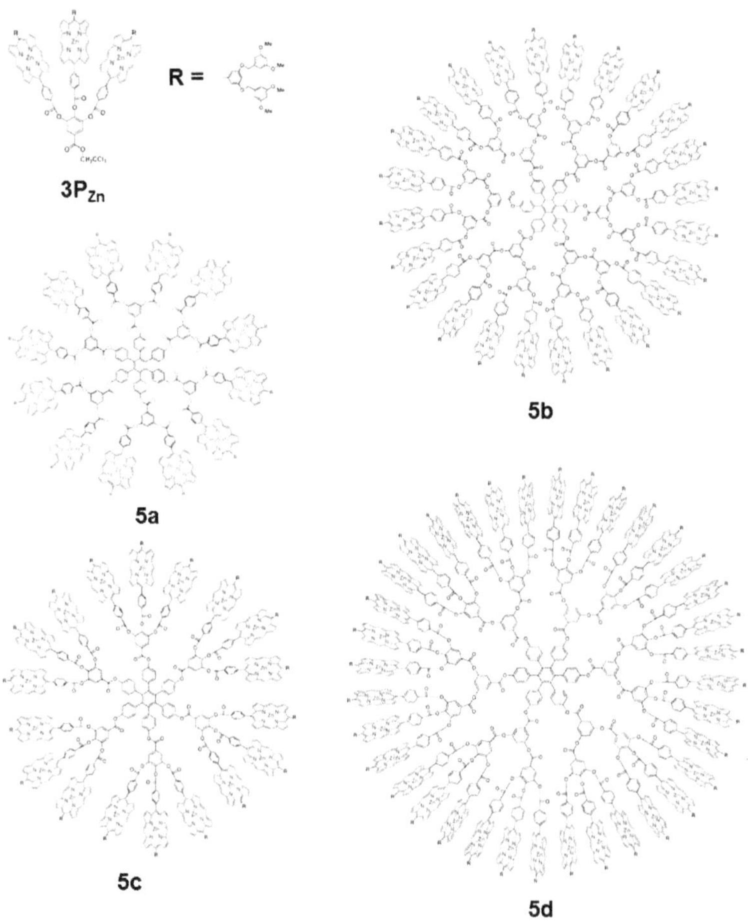

Figure 14.8. Poly(benzyl ester) dendrimers with P_{Zn} units as end-functional groups.

dendrimer with P_{FB} core [24]. However, their energy-transfer properties were not investigated because the nickel porphyrin derivatives are inherently inactive for photochemical energy transfer phenomena. Nonameric dendritic porphyrin array with eight peripheral P_{Zn} with P_{FB} (**6**) was synthesized by Sanders et al, and evaluated their photochemical properties (Fig. 14.9) [25]. Excitation of the P_{Zn} units of **6** results in fluorescence emission from the

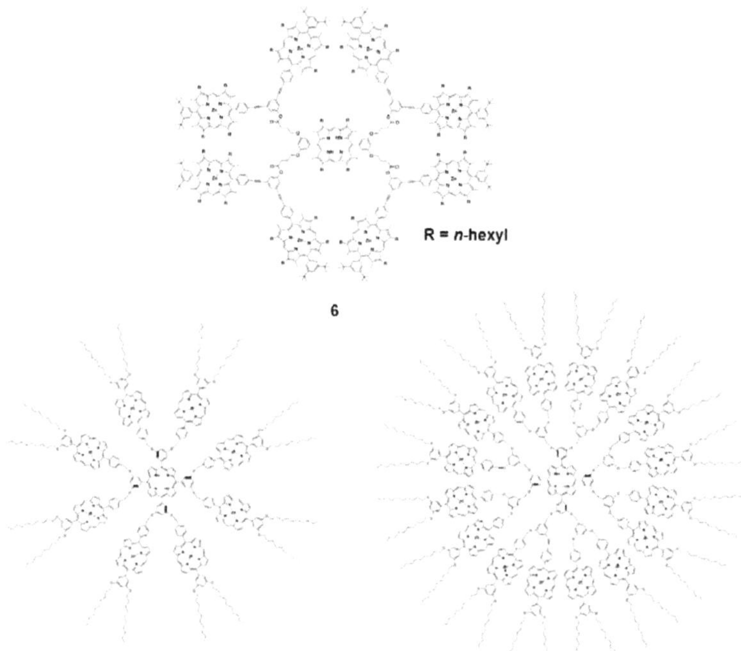

7a 7b

Figure 14.9. Multiporphyrin dendrimers with focal porphyrin core.

P_{FB} focal core as a result of a P_{Zn}-to-P_{FB} energy transfer. Evaluation of the fluorescence decay profile of the P_{Zn} units shows that the energy transfer takes place with a quantum efficiency (FENT) of 60% (kENT $= 9 \times 10^{-8}$ s^{-1}). Poly(benzyl ether) dendrimers with peripheral P_{Zn} units and P_{FB} core have been reported (Fig. 14.9) [26]. The quantum efficiencies of energy transfer from P_{Zn} to P_{FB} were obtained by steady state fluorescence measurement. Because the distance from P_{Zn} to P_{FB} of **7a** is little longer than that of **7b**, the quantum efficiency of **7a** was slightly lower than that of **7b**. However, both **7a** and **7b** showed relatively high quantum efficiency compare to cone-shaped control models, indicating that efficient energy migration processes between P_{Zn}s can be preceded to enhance the probability of energy transfer to focal P_{FB}. Considering the high quantum efficiency and large absorption cross-section, these

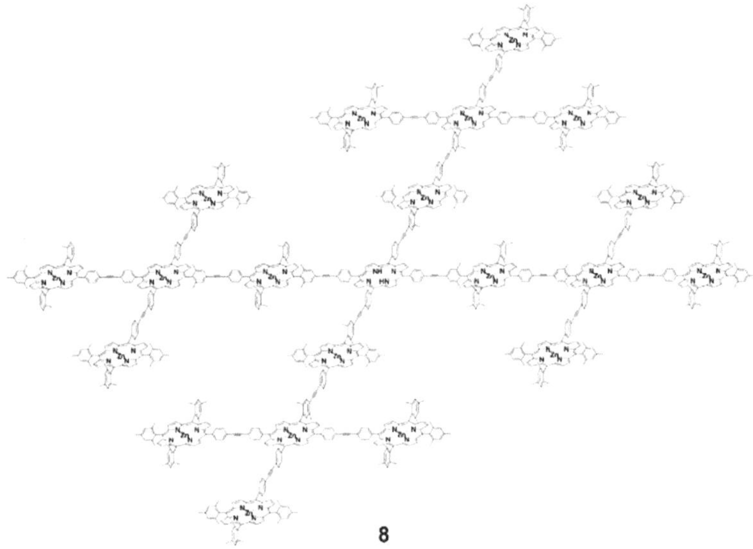

8

Figure 14.10. Dendritic metalloporphyrin array consisting of 20 P_{Zn} units conjugated with P_{FB} core through diarylethyne linker.

dendritic multiporphyrin arrays would be one of the best candidates for artificial light harvesting antennae.

14.2.4 *Porphyrins as Building Block of Dendrimers*

Owing to the density gradient of dendritic architecture, vectorial energy migration within dendrimer was also investigated. To achieve the vectorial energy migration, building blocks of dendrimers should be composed of porphyrin units. The synthesis of a dendritic metalloporphyrin array consisting of 20 nickel porphyrin units with P_{FB} core has been reported, but photochemical properties were not investigated [27]. For the investigation of photofunctional properties, another type of dendritic multi-porphyrin array, in which 20 P_{Zn} units attached to a focal P_{FB} core through diarylethyne linkers, has been synthesized (Fig. 14.10). Time-resolved absorption spectroscopy of compound **8** has shown a decay of the excited singlet state of the P_{Zn} units, followed by an increase in the absorption of the P_{FB} core. The P_{Zn}-to-P_{FB} energy transfer has a FENT value of 92%.

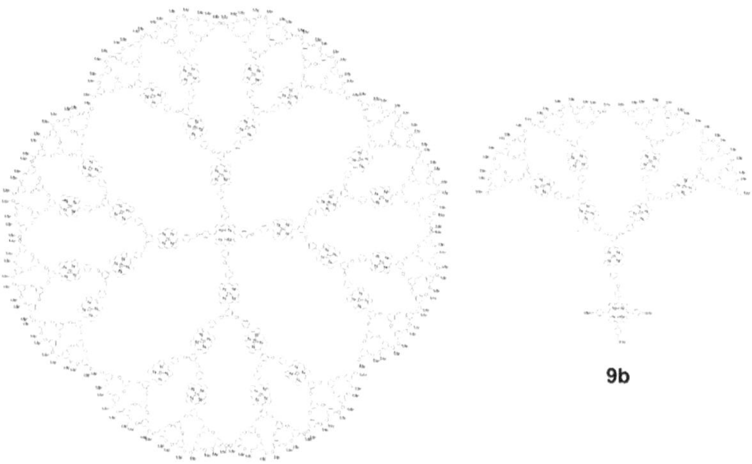

9b

9a

Figure 14.11. Dendritic multiporphyrin array consisting of four $28P_{Zn}$ units and one focal P_{FB} core (**9a**) and conical model (**9b**).

Because dendrimer **8** is composed of fully conjugated diarylethyne linkage, the energy transfer processes mainly take place through bond mechanism.

Aida et al. have synthesized a large dendritic multi-porphyrin array to investigate energy transfer phenomenon, which consists of four dendritic wedges having heptameric P_{Zn} units and one focal P_{FB} core (Fig. 14.11) [28]. Excitation of the P_{Zn} units in **9a** results in a highly efficient P_{Zn}-to-P_{FB} energy transfer; thereby fluorescence is predominantly emitted from the focal P_{FB}. However, compound **9b**, a conical control model of **9a**, displays an emission mostly from the P_{Zn} units, with only a weak emission from the PFB focal core. Time-dependent fluorescence change also indicates efficient P_{Zn}-to-P_{FB} energy transfer. The time resolved fluorescence spectroscopy of **9a** initially shows an increase in the fluorescence from the P_{Zn} units, which then decays progressively to allow the generation of a new fluorescence from the PFB core. The energy transfer rate constant (kENT) for **9a** and **9b** have been estimated to be 1.04×10^9 and 0.10×10^9 s^{-1}, respectively. The rate constant kENT for **9a** is one order of magnitude greater than that for **9b**.

The energy-transfer efficiency (FENT) for **9a** (71%), calculated from these kENT values, is also significantly greater than that for **9b** (19%).This observation again indicates the importance of the density and morphology of the chromophores for the effective excitation energy transfer. A fluorescence depolarization study has shown that the energy transfer in **9a** is facilitated by a cooperative energy migration among the dendritic P_{Zn} units.

14.2.5 *Porphyrins in Supramolecular Dendrimer Assembly*

To achieve dendritic architecture, several supramolecular assembly systems were developed. To a P_{FB} bearing pyrazine arms, P_{Zn} dimers formed by hydrogen-bonding interaction have coordinated to form nonameric and heptadecameric multi-porphyrin arrays (Fig. 14.12 **10a**) [29]. Upon excitation of the P_{Zn} units, energy transfer takes place with 80% of FENT, which results in a fluorescence emission from the P_{FB} focal core. Alternatively, serial-type compound **10b** becomes less effective for energy transfer than **10a**, where the FENT of **10b** is only 55%. The difference of FENT between **10a** and **10b** indicates the importance of geometry between donor and acceptor.

Platinum porphyrin bearing three poly(benzyl ether) dendrons and one carboxylate has been utilized as a ligand for Er^{3+} ion for

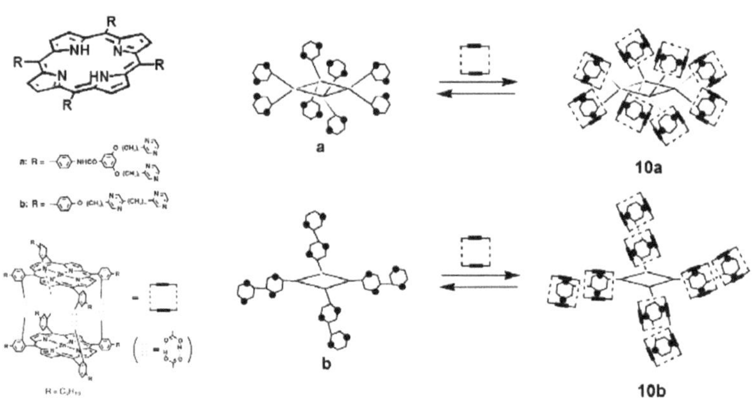

Figure 14.12. P_{Zn} dimers formed by hydrogen-bonding interaction have coordinated to form supramolecular heptadecameric multiporphyrin arrays.

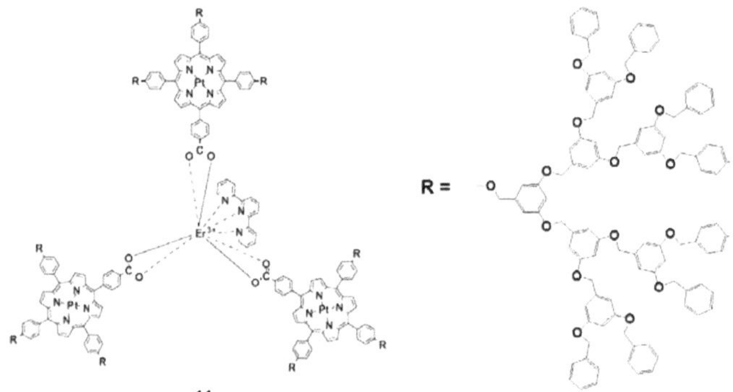

11

Figure 14.13. Supramolecular Er^{III} complex dendrimer.

the design of near-IR(NIR) emitting devices (Fig. 14.13) [30]. For the development of planar waveguide optical-emplifier-integrated wavelength division multiplexing devices (WMD), strong NIR emission is especially important. Typically, Er^{III} complexes are currently recognized to be essential for the successful photonic devices using WDM technology. However, Er^{3+} ion itself has very low absorption coefficient, and shows poor solubility in conventional inorganic media. The supramolecular Er^{III}-cored dendrimer complexes (**11**) show strong NIR emission at 1530 nm via efficient energy transfer from the excited triplet state of the PPt ligand to Er^{3+} ions. Notably, the NIR emission intensity of thin films of Er^{III} complexes is drastically enhanced by generation increment of poly(benzyl ether) dendron due to the site-isolation as well as light harvesting effects. Because the Er^{III}-cored dendrimers exhibit excellent thermal and photochemical stability with good solubility against various organic solvents, such kinds of molecules have strong potential for the design of highly efficient photonic nano-devices.

14.3 Electron Transfer in Porphyrin Dendrimers

14.3.1 *Star-shaped Porphyrin-C_{60} Hexad*

A star-shaped porphyrin antenna array, consisting of three peripheral zinc porphyrins (P_{ZP}), a central zinc porphyrins (P_{ZC}), and an

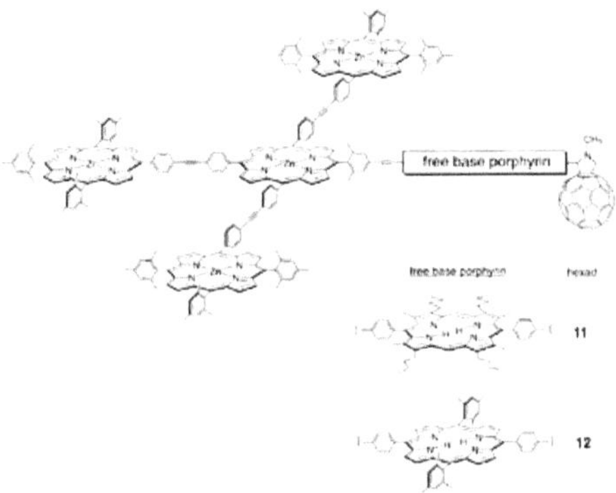

Figure 14.14. Molecular mimics of photosynthetic antenna-reaction center.

energy-accepting P_{FB}, has been joined to a fullerene, to mimic a photosynthetic antenna-reaction center complex **11** and **12** represented by $(P_{ZP})_3$-P_{ZC}-$P_{FB}^{\bullet+}$-$C_{60}^{\bullet-}$ (Fig. 14.14) [31]. Using time-resolved absorption and emission spectroscopic studies in 2-methyltetrahydrofuran solution, it is revealed that excitation energy of any energy-donating P_{ZP} moiety in **12** is transferred to the P_{ZC} to give $(P_{ZP})_3$-$^1P_{ZC}$-P_{FB}-C_{60} with a time constant of 50 ps. In addition to exchanging excitation with the P_{ZP} moieties, this state donates excitation energy to the P with a time constant of 32 ps to yield $(P_{ZP})_3$-P_{ZC}-$^1P_{FB}$-C_{60}, which decays by electron transfer to the fullerene with a time constant of 25 ps. Thermodynamically favorable migration of positive charge over P_{Zn} moieties transforms the initial state of charge separation into a long-lived $((P_{ZP})_3$-$P_{ZC})^{\bullet+}$-P-$C_{60}^{\bullet-}$ with a time constant of 380 ps, which finally decay to the ground state with a lifetime of 240 ns. The first step in this transformation is presumably electron transfer from the P_{ZC} to give $(P_{ZP})_3$-$P_{ZC}^{\bullet+}$-P_{FB}-$C_{60}^{\bullet-}$. The peripheral and central zinc porphyrin moieties differ slightly in oxidation potential. The quantum yields of $(P_{ZP})_3$-P_{ZC}-$P_{FB}^{\bullet+}$-$C_{60}^{\bullet-}$ and $((P_{ZP})_3$-$P_{ZC})^{\bullet+}$-P_{FB}-$C_{60}^{\bullet-}$ are 0.95–0.98 and 0.86, respectively.

It is interesting to compare the photochemical properties of antenna-reaction center complex **12** with **11**, as shown in Table 14.1. The rate constants for exchange of singlet excitation energy among the zinc porphyrins of the antenna are identical in **11** and **12**, as one would expect given the structural similarity. On the other hand, the

Table 14.1. Comparison of photochemical properties of **11** and **12**

Property	Hexad 11	Hexad 12
time constant for P_{ZC} to P energy transfer ($1/k_3$).ps	244	32
time constant for photoinduced electron transfer (($1/k_3$).ps)	3	25
lifetime of final charge-separated state, ns	1.3	240
quantum yield[a] of $(P_{ZP})_3$-P_{ZC}-$P^{\bullet+}$-$C_{60}^{\bullet-}$	0.70	0.98
quantum yield[a] of $((P_{ZP})_3$-$P_{ZC})^{\bullet+}$-P-$C_{60}^{\bullet-}$		0.86
^{1}P energy, eV	1.96	1.91
P_z oxidation potential,[b] V	0.77	0.77
P oxidation potential,[b] V	0.84	1.05
energy of $(P_{ZP})_3$-P_{ZC}-$P^{\bullet+}$-$C_{60}^{\bullet-}$, eV	1.46	1.67

[a]Based on ^{1}Pz. [b]In V vs SCE.

rate constant k_3 for energy transfer from the excited state of P_{ZC} to the P_{FB} is only 4.1×10^9 s^{-1} in **11** but is 3.2×10^{10} s^{-1} in **12**. This result leads to an increase in the quantum yield of charge separation from 0.69 in 1 to 0.98 in **12**; the slower rate in **11** allows competition with energy transfer from the usual photophysical pathways that depopulate the excited singlet state of zinc porphyrin moiety. Synthetic antenna-reaction center complex **12** has been optimized to collect excitation energy efficiently and use the resulting energy to generate a charge separated state in high quantum yield. The only difference is that the P_{FB} is a β-octaalkyl porphyrin for **11** and *meso*-tetraaryl porphyrin for **12**. Basically, these complexes potentially mimic the structure and function of natural photosynthetic antenna systems and reaction center complexes.

14.3.2 *Fullerene-Terminated Dendritic Multiporphyrin Arrays*

As shown in Fig. 14.15, light-harvesting dendritic P_{Zn} arrays have been integrated into C_{60} moiety [32, 33]. The arrays **13**, **14**, and **15** were synthesized from hydroxy-terminated P_{Zn} dendrons (nP_{Zn}, $n =$ 1, 3, and 7) and a carboxylic acid appended C_{60} derivative by using azodicarboxylic acid diethyl ester as a condensation agent [34]. Excitation of a solution of **5** in PhCN at 544 nm resulted in a P_{Zn} fluorescence with two major bands at 590 and 645 nm, while no fluorescence was detected from the C_{60} terminus. The fluorescence

13 14 15

Figure 14.15. Schematic structures of fullerene-terminated dendritic multiporphyrin arrays.

of **15** was clearly quenched by 84 ± 1% relative to the reference system of C_{60}-free $1P_{Zn}$. As expected, the one-generation-higher **14** showed a lower quenching efficiency (60 ± 2%) than **15**. On the other hand, the quenching efficiency of **13**, the largest homologue of the series, stayed at the same level as that of **14** (62 ± 2%). A similar trend was observed for the analysis by time-resolved spectroscopy. Excitation of solutions of arrays in PhCN under argon at 22°C at 410 nm resulted in the fluorescence monitored at 610 nm having dual-exponential decay characteristics, where the relatively long-lived fluorescing components are assignable to residual components after the photochemical communication between the P_{Zn} and C_{60} units (Fig. 14.16(a)). The average rate constant for the fluorescence decay (k_{CS}) became smaller by a factor of four as the zinc porphyrin array increased from **5** (1.55×10^9 s^{-1}) to **14** (0.40×10^9 s^{-1}). On the other hand, the k_{CS} value (0.43×10^9 s^{-1}) of the largest homologue of the family **3** was comparable to that of **14**, in spite of the fact that **13** has a greater number of the P_{Zn} units located away from the C_{60} terminus.

Nanosecond time-resolved absorption spectroscopy has been employed to investigate the charge recombination (CR) process at 600–1200 nm. Excitation of a solution of **13** in PhCN at 532 nm with a 6-ns laser pulse produced an absorption band at 1020 nm, which correspond to the anion radical of C_{60}($C_{60}^{\bullet-}$), and bands centered at 850 and 740 nm, which correspond to the triplet-excited $^3P_{Zn}^*$ and $^3C_{60}^*$ species, respectively. The time profile of the

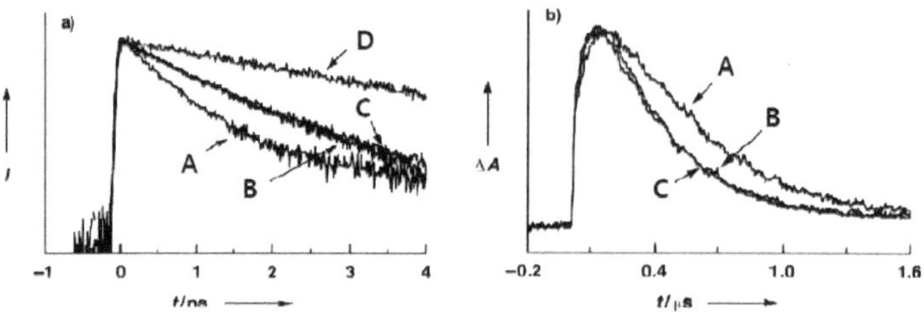

Figure 14.16. (a) Fluorescence decay profiles of **13**(A), **14**(B), **15**(C), and reference(D) at 610 nm upon excitation at 410 nm, and (b) absorption decay profiles of anion radical of C_{60}($C_{60}^{\bullet-}$) at 1020 nm, generated upon excitation at 532 nm, in PhCN at 22°C under argon.

absorption corresponding to $C_{60}^{\bullet-}$ at 1020 nm was analyzed for determination of the lifetime of the charge-separated state(τ_{CS}), and the decay curve nicely fitted with dual-exponential components (Fig. 14.16(b)). The slow-decaying component corresponded to the triplet-excited species, while a radical ion pair ($P_{Zn}^{\bullet+}$-$C_{60}^{\bullet-}$) was identified as the fast-decaying component, whose τ_{CS} value was evaluated to be as long as 0.66 ms in PhCN at 22°C. Very interestingly, **14** and **15** also showed similar transient absorption spectral characteristics in PhCN, but their τ_{CS} values (0.41 and 0.35 μs, respectively) were clearly shorter than that of **13**. Accordingly, **13** showed a smaller CR rate constant ($k_{CS} = (1.5 \pm 0.2) \times 10^6\ s^{-1}$) than **15** ($(2.9 \pm 0.3) \times 10^6\ s^{-1}$) and **14** ($(2.4 \pm 0.3) \times 10^6\ s^{-1}$).

The electron transfer reaction most likely takes place from the P_{Zn} unit directly connected to the C_{60} terminus, while the energy for the electron transfer is acquired by the dendritic antenna. From the slowest back-electron-transfer process in **13**, the role of the large dendritic antenna is proposed that the cation radical of P_{Zn}($P_{Zn}^{\bullet+}$) species, generated at the focal point upon electron transfer, move away from the $C_{60}^{\bullet-}$ towards the periphery by an intramolecular hole hopping mechanism. In relation to this proposal, the first oxidation potential of **13** (0.29 V) was lower than those of **15** (0.33 V) and **14** (0.31 V). Temperature dependence of k_{CR} was investigated in Fig 14.17 which gave activation energy (E_a) for the charge recombination process. Quite interestingly, the E_a value (slope of curve) for **13** (12.4 kJmol^{-1}) was much greater than those for the lower generation homologues (7.5 and 7.8 kJmol^{-1} for **15** and **14**, respectively).

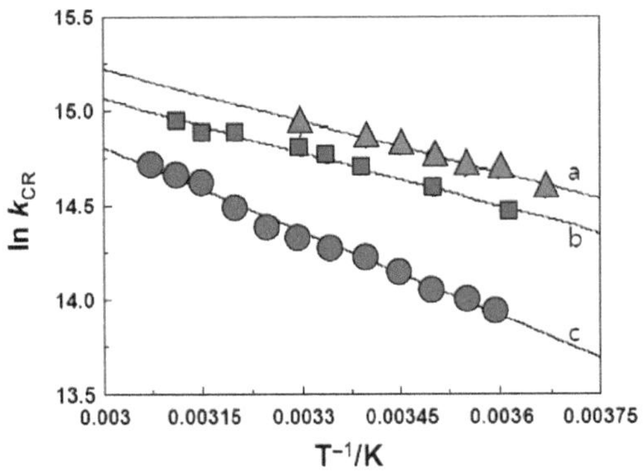

Figure 14.17. Temperature dependence of charge recombination processes (k_{CR}) of **13**(c), **14**(b), and **15**(a).

In studies on the photoinduced electron transfer properties, the large dendritic antenna not only harvests visible light for the electron transfer to the C_{60}- terminus, but also retards the back-electron-transfer process. These results indicate a new potential of dendronized dye molecules for the efficient conversion of solar energy into chemical potentials.

The intermolecular electron transfer property of the triplet excited states of dendritic multi-porphyrins arrays (nP$_{Zn,}$ n (number

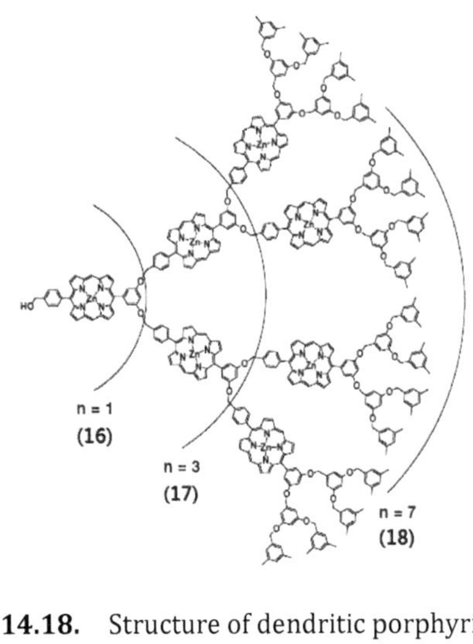

Figure 14.18. Structure of dendritic porphyrin arrays.

Table 14.2. Quenching rate constants (k_q) of $^3nP_{Zn}$* by C_{60}, quantum yields (Φ_{et}) and electron-transfer rate constants (k_{et}^{obs}) from $^3nP_{Zn}$* to C_{60}, and back electron-transfer rate constants (k_{bet}^{2nd}) from $C_{60}^{\bullet-}$ to $P_{Zn}^{\bullet+}$ in PhCN

	E_{ox}^a/V	ΔG_{et}^{0b}/ eV	$k_q \times 10^{-9}$/ mol^{-1} dm^3 s^{-1}	Φ_{et}^c	$k_{et}^{obs} \times 10^{-9}$/ mol^{-1} dm^3 s^{-1}	$k_{bet}^{2nd\,c} \times 10^{-9}$/ mol^{-1} dm^3 s^{-1}
16	0.33	−0.26	1.6 ± 0.1	0.63 ± 0.03	1.0 ± 0.1	5.5 ± 0.3
17	0.31	−0.28	1.4 ± 0.1	0.36 ± 0.02	0.5 ± 0.1	6.0 ± 0.3
18	0.29	−0.31	2.1 ± 0.1	0.56 ± 0.03	1.2 ± 0.1	5.2 ± 0.3

(a) vs Fc/Fc$^+$. (b) The values of Δ Get0 were evaluated from the Rehm-Weller equation, using Ered(-0.93 V) and ET(1.54 eV) for C_{60}. (c) ε RA = 12100 mol^{-1}dm^{3}cm^{-1} for C60$^{\bullet-}$.

of P_{Zn}) = 1, 3, and 7), shown in Fig. 14.18, have been also investigated with free C_{60} by measuring the nanosecond transient absorption spectra in the visible and near-IR regions with changing the generation number [35]. Intermolecular triplet–triplet annihilation rates decrease with the dendrimer generation, which was interpreted on a proposed kinetic model assuming that the excited triplet energy almost localizes in one P_{Zn} unit in the arrays. In the presence of C_{60}, intermolecular electron-transfer takes place via the excited triplet states of P_{Zn} ($^3P_{Zn}$*), yielding the cation radical of P_{Zn} ($P_{Zn}^{\bullet+}$) and the anion radical of C_{60} ($C_{60}^{\bullet-}$) in PhCN. Electron transfer properties $^3P_{Zn}$*-to-C_{60} were summarized in Table 14.2. Deceleration of the electron-transfer rate-constants from **16** to **17** and the acceleration from **17** to **18** were observed, in which the latter tendency was interpreted by considering an increase in effective encounter radius for excited triplet state of **18**. The observed small change of the rate constants for back electron transfer between the oppositely charged species with the dendrimer generation was also reasonably interpreted by taking a smaller effective radius due to electrostatic attraction into consideration. Dendrimer generation effect was also observed for the intermolecular hole-transfer process.

14.3.3 *Segregated Arrays of Multiple Porphyrins and C_{60} Units*

As shown in Fig. 14.19, porphyrin dendrimers (**19**, **20**, and **21**) are able to form coordination complexes with multiple C_{60} (F) acceptors having bipyridine ligands (Py_2F_n, *n*(number of C_{60}) = 1, 2, and

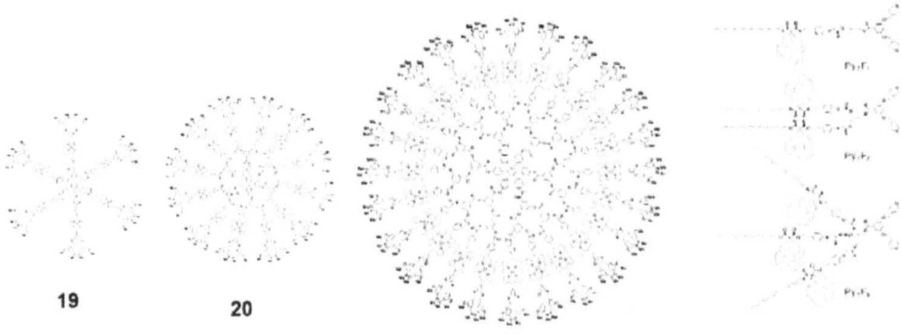

19 20 21

Figure 14.19. Structures of multiporphyrin dendrimers and fullerene ligands Py_2F_1, Py_2F_2, and Py_2F_3.

3) on the surface of dendrimers. The complexes are stable enough in chloroformat 25°C to be isolated by gel permeation chromatography. UHV-STM microscopy enables clear visualization of a petal-like structure of the arrays. Photoinduced electron transfer in the complexes was confirmed by means of steady-state emission spectroscopy and nanosecond flash photolysis measurements. Excitation of a $CHCl_3$ solution of **21** (1.5×10^{-7} M) at 550 nm resulted in a fluorescence emission from P_{Zn} units at 591 and 635 nm (Fig. 14.20(a)). Upon titration with Py_2F_3, the fluorescence stepwise decreased in intensity and was quenched almost completely in the final stage. By means of transient absorption spectroscopy, we confirmed the occurrence of an electron transfer from the photoexcited P_{Zn} units to the fullerene units in the complexes. Excitation of a CH_2Cl_2 solution of **9** (1.7×10^{-5} M) at 532 nm in the presence of 3 equiv of Py_2 F_3 resulted in a transient absorption spectrum (Fig. 14.20(b)) with bands at around 680 and 1000 nm assignable to the cation and anion radicals of the P_{Zn} and C_{60} units, respectively. Since the fluorescence spectral profiles of complexes did not show any sign of P_{Zn}-to-C_{60}energy transfer [36] the quenching of the P_{Zn} fluorescence observed for the complexes is mostly due to the intracomplex photoinduced electron transfer between the P_{Zn} and the C_{60} units.

Upon increment of the numbers of these donor and acceptor units, the electron transfer reaction was remarkably facilitated, while the recombination of the resulting charge-separated state remained virtually intact. Consequently, the complex between **21**

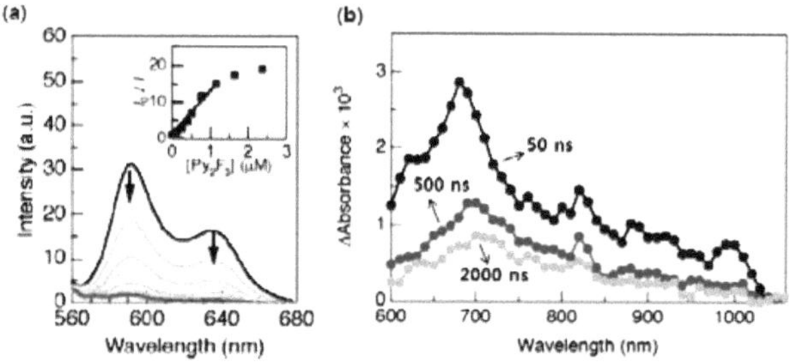

Figure 14.20. (a) Fluorescence spectral change and Stern-Volmer plot (inset) of **21** (1.5×10^{-7} M) upon excitation at 550 nm in the presence of Py_2F_3($[Py_2F_3]/[$**21**$]$) 0, 0.85, 1.7, 2.6, 3.4, 5.1, 7.7, 11, and 16 in $CHCl_3$ at 25°C under Ar. (b) Nanosecond transient absorption spectra at 20°C of a CH_2Cl_2 solution of **19** (1.7×10^{-5} M) containing 3 equiv of Py_2F_3 upon photoexcitation at 532 nm.

and Py_2F_3, accommodating roughly 30 fullerene units on the **21** surface, achieved the largest ratio of the charge-separation to charge-recombination rate constants (3400), which is even 1 order of magnitude greater than those of precedent examples. Application of this molecular design strategy to the development of optoelectronic materials is one of the subjects worthy of further investigations.

14.3.4 *SWNT-Porphyrin Conjugates*

The functionalization of single-walled carbon nanotubes (SWNTs) with P_{Zn} has been afforded a series of SWNTs-P_{Zn} (**22** and **23**) electron donor-acceptor conjugates (Fig. 14.21) [37]. SWNT-P_{Zn} conjugates were fully characterized by Raman, X-ray photoelectron (XPS), absorption and emission spectroscopy, atomic force microscopy (AFM), and electrochemistry.

Insight into electronic interactions between SWNT and P_{Zn} moieties came from fluorescence spectroscopy. The high fluorescence quantum yield of P_{Zn} allows monitoring energy or electron transfer property even at relatively low concentrations. Upon excitation at 420 nm in DMF solution, the P_{Zn} fluorescence over 570–700 nm is quantitatively quenched via electron transfer to the C_{60} moiety, namely 99 and 96% for **22** and **23**, respectively, as shown

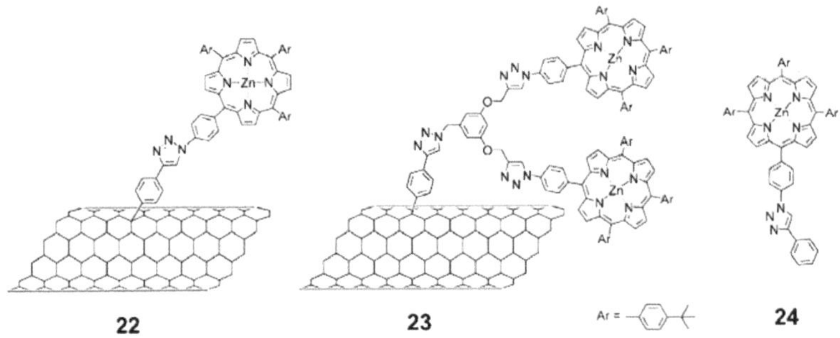

Figure 14.21. Schematic structures of SWNT-porphyrin conjugates.

in Fig. 14.22(a). Excitation of P_{Zn} is accompanied by the rapid formation (i.e., 2 ps) of a transient that exhibits the following features: minima evolve at 425 and 560–600 nm for the Soret- and Q-band absorptions of P_{Zn}, respectively, while maxima are seen at 455, 580, and 630 nm which are due to the singlet-singlet absorptions of P_{Zn}. Within the time window of 3.0 ns these singlet excited-state features transform (i.e., 2.4 ns) into those of the triplet excited state at 475 and 840 nm. The triplet excited state is long-lived with lifetimes of up to hundreds of micro-seconds in the absence of molecular oxygen.

Following femtosecond excitation at 387 nm we note two sets of features in Fig. 14.22(b). The first set in the visible range corresponds to the singlet excited state of ZnP and attests to the successful excitation of P_{Zn} (minima at 455, 580, and 630 nm). The second set in the near-infrared region is attributed to the SWNT centered excited states. The P_{Zn} singlet excited state decay rather rapidly in the presence of covalently linked SWNT with 26 ps for **22**, to afford a newly developed band arising from 600 and 850 nm, which are due to the one-electron oxidized form of P_{Zn} [38]. This result clearly demonstrates that electron transfer occur from P_{Zn} singlet excited state to the nanotubes. It is important to notice that the 980 and 1605 nm minima of SWNT transform within the same time window into new features (i.e., minima at 940 and 1580 nm as well as a maximum at 1085 nm). An overall blue-shift of the SWNT-centered transitions is observed, a feature that has been seen previously in spectro electrochemical and pulse

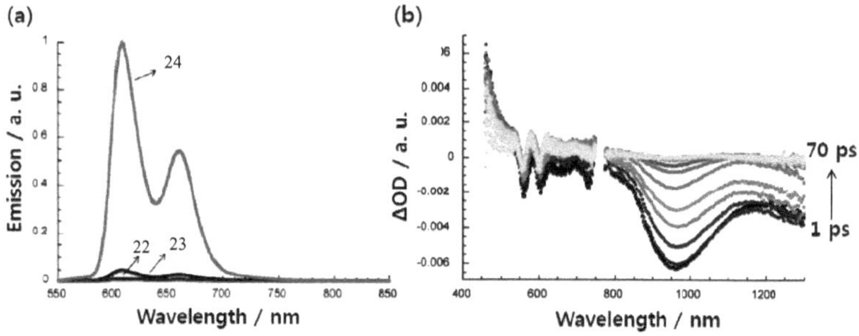

Figure 14.22. (a) Fluorescence spectra of **22** and **23** with that of reference **24**. (b) Differential absorption spectra obtained upon femtosecond flash photolysis (387 nm, 150 nJ) of **22** in DMF with time delays between 1 ps and 70 ps at room temperature. See also Color Insert.

radiolytical reduction experiments [39] and which is linked to the injection of electrons into the conduction band of the semiconducting SWNT. Of interest, the life time of charge separation state for **23** (200 ps) was shorter than that of **22** (820 ps), because the inherent flexibility of the dendritic moiety in **23** favors a configuration, in which SWNT and P_{Zn} are in close proximity to each other.

14.3.5 *Polyoxometallate-Appended Multiporphyrin Array*

A multi-porphyrin cluster has been covalently attached to a polyoxometallate (POM) catalyst to form **25**, a molecular model for the photosynthetic reaction complex (Fig. 14. 23) [40]. The system comprises a 4-membered porphyrin-based cluster (Zn_3Fb) fitted with three P_{Zn} covalently linked to a P_{FB}. The design rationale is that the ancillary P_{Zn} units will function as photon collectors and transfer excitation energy to the nearby P_{FB} [41]. The resultant excited-singlet state of FbP unit is expected to donate an electron to the polyoxometallate (POM), since there is a reasonable driving force for this reaction [42]. The various parameters relating to photoinduced electron transfer are summarized in Table 14.3.

Furthermore, solely in terms of thermodynamics, two electrons could be added successively to the POM from the excited-singlet state of the P_{Zn}-based donor. In DMF solution, the fluorescence quantum yield ($\Phi_F = 0.053$) of P_{FB} in **25** was notably decreased with

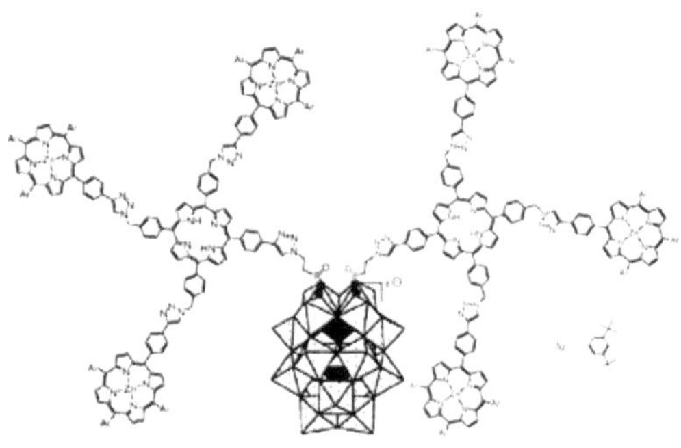

25

Figure 14.23. Molecular formula for the polyoxometallate-appended multiporphyrin array.

respect to Zn_3Fb (0.071). There is a concomitant decrease in the fluorescence lifetime, which drops from 6.3 ns to 5.4 ns, allowing estimation of the quenching rate constant of 2.6×10^7 s^{-1} for **25**.

Since the thermodynamic driving force for PET from the P_{Zn} unit to the appended POM is somewhat larger ($\Delta G^\circ = -1.07$ eV) than that from the P_{FB} unit, thus, electron and energy transfer events are competing. PET might involve diffusion contact between the porphyrin and POM units and/or through-bond super-exchange interactions between remote reactants. Time-resolved fluorescence decay curves recorded for the P_{FB} unit in **25** could be well described in terms of exponential kinetics while the agreement between quenching probabilities derived from steady state and time-resolved

Table 14.3. Derived parameters for photoinduced electron transfer (PET) for **25**

Process	ΔG^0 / ev	Rate constant/s^{-1}
PET[a]	-0.68^b	$2.6 \times 10^{7\,b}$
PER[a]	-0.20^c	$5.3 \times 10^{5\,c}$
HT[d]	-0.24	5×10^4
CR[e]	-1.00	$36.7 \times 10^{9\,f}$

[a]PET from FbP to POM.
[b]Singlet excited state.
[c]Triplet excited state.
[d]Hole transfer.
[e]Charge recombination.
[f]M^{-1} s^{-1}.

modes argues against the presence of multiple conformations in the ground state. The very low rate constant of electron transfer (k_{ET}) is fully consistent with through-bond electron transfer along a poor conductor [43]. Transient absorption spectra using laser flash photolysis techniques were made for the triplet state of the P_{FB} unit in a deoxygenated DMF solution of **25** at room temperature. For Zn_3Fb, this metastable species decays with a lifetime (τ_T) of 370 μs but this is lowered to 1.9 μs for **25**. Nonetheless there is a modest thermodynamic driving force for electron transfer from the triplet state of the P_{FB} to the POM($\Delta G^\circ = -0.20$ eV), the k_{ET} value (5.3×10^5 s^{-1}) is smaller than from the corresponding singlet excited state, due to both the thermodynamics and unfavorable geometry. After decay of the triplet state, transient absorption spectral records show the corresponding cation radical of P_{FB} ($P_{FB}^{\bullet+}$) in the solution state. As shown in Fig. 14.24, the most diagnostic feature of this triplet state PET process is the relatively fast decay of the triplet state at 770 nm and the evolution of the spectral changes between 510 and 650 nm. Initially, the signal is negative, this being characteristic of the triplet state, but becomes positive as the $P_{FB}^{\bullet+}$ is formed. Over some tens of μs, the spectrum evolves further until it matches that of the cation radical of P_{Zn} (c with contamination from the reduced POM at longer wavelength. The most diagnostic features of this spectral evolution relate to the transformation of the bleaching of four Q-bands, this being characteristic of the FbP triplet state, to the two Q-bands appropriate of the ZnP. The $ZnP^{\bullet+}$ is responsible for the strong absorption band seen at around 650 nm. The reduced POM

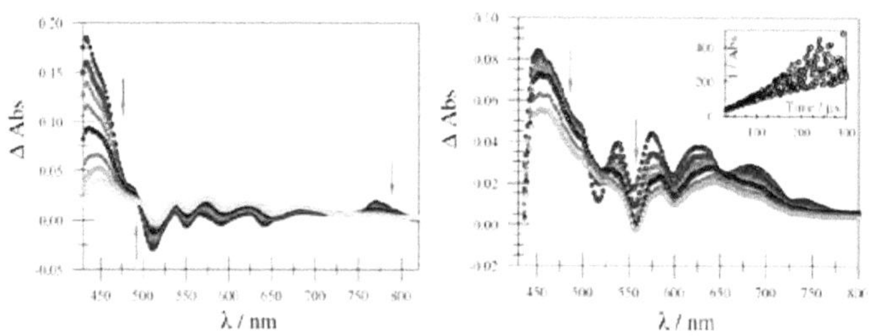

Figure 14.24. Transient differential absorption spectra recorded after 532 nm excitation of **25**. Spectra were recorded at delay time of 0.1–5 μs (left) and 10–100 μs (right).

contributes to the signal in the near-IR region. This charge-shift reaction occurs via first-order kinetics, having a lifetime of 19 μs in deoxygenated DMF, and leads to further spatial isolation of the redox products. There is a thermodynamic driving force of 0.24 eV for this charge-shift step. Charge recombination (CR) now occurs primarily by way of second-order kinetics with a bimolecular rate constant of 3.7×10^9 M^{-1} s^{-1}, implying that the $P_{Zn}^{\bullet+}$ is sterically blocked from encountering the reduced POM by way of an intramolecular process.

14.3.6 *Dendritic Porphyrin–Fullerene Hybrid*

A novel dendritic porphyrin-fullerene hybrid **27** (C_{60}-P_{FB}-$(P_{Zn})_3$) have been introduced as the mimicry of the primary events in photosynthesis (Fig. 14.25). Owing to the flexibility of the linkers that connect the C_{60}, P_{FB}, P_{Zn} units, the outcome depends strongly on the rigidity/viscosity of the environment. In the conjugates **26** and **27**, the P_{FB} fluorescence is quantitatively quenched with the fluorescence equantum yields in the range $2.0 \times 10^{-3} \sim 0.8 \times 10^{-3}$, in which the P_{Zn} fluorescence in **27** is negligible. Mostly, the P_{FB} and P_{Zn} fluorescence dominates the spectral region around 720 nm where no C_{60} fluorescence is detectable. In contrast to **26**, the photophysics of **27** follows a different deactivation pathway. Upon photoexcitation in THF, the P_{Zn} singlet excited-state features are discernable, which convert rapidly ($k = 2.7 \times 10^{10}$ s^{-1}) into a new transient. Interestingly, the new transient is characteristic of the C_{60} radical anion ($C_{60}^{\bullet-}$) that evolves at 1035 nm, indicating the formation of the charge-separation (CS) state of $C_{60}^{\bullet-}$-P_{FB}-$(P_{Zn})_3^{\bullet+}$ (Fig. 14.26(a)). Concurrently, the corresponding P_{Zn} radical cation ($P_{Zn}^{\bullet+}$) is registered in the visible range between 580 and 800 nm. The CS state of **27** recombines in THF unexpectedly quickly ($k = 2.7 \times 10^9$ s^{-1}) and reinstates the singlet ground state. The rate constant of CS and CR processes reveal a weak solvent dependence with values of $k_{CS} = 1.7 \times 10^{10}$ (59 ps) and $k_{CR} = 2.0 \times 10^9$ s^{-1}(489 ps) in toluene, and $k_{CS} = 8.3 \times 10^{10}$(12 ps) and $k_{CR} = 1.0 \times 10^{10}$ s^{-1} (100 ps) in benzonitrile. However, no spectroscopic/kinetic evidence was found in favor of forming the CS state of $C_{60}^{\bullet-}$-$P_{FB}^{\bullet+}$-$(P_{Zn})_3$. Two reasons are plausible; (1) a larger driving force (ΔG°) for the formation of $C_{60}^{\bullet-}$-P_{FB}-$(P_{Zn})_3^{\bullet+}$ than for the formation of $C_{60}^{\bullet-}$-$P_{FB}^{\bullet+}$-$(P_{Zn})_3$, (2) a flexible linker facilitate a

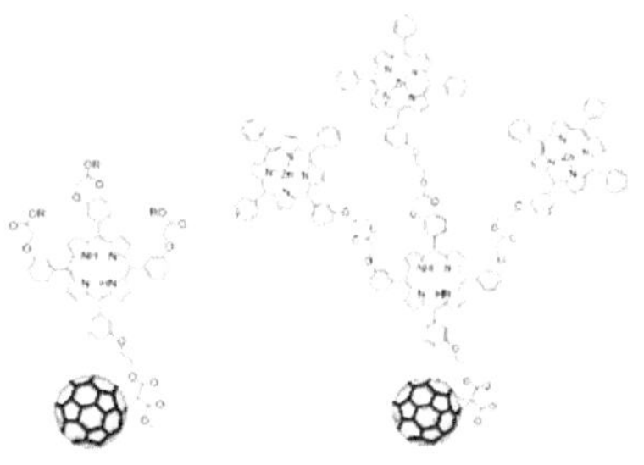

Figure 14.25. Structure of porphyrin-fullerene hybrid **26** and **27**.

very close approach between the triad termini so that a direct charge separation between the C_{60} and P_{Zn} units is feasible. In an effort to disrupt such potentially folded conformations of the triad, the addition of pyridine (i.e., 1 % (v/v) in toluene) to coordinate the P_{Zn} was investigated. The result showed a red-shift of 10 nm in absorption and a longer lived $C_{60}^{\bullet-}$-P_{FB}–$(P_{Zn})_3^{\bullet+}$ species (790 ps) in toluene, due to a strong perturbation of the distance and interactions between the C_{60} and P_{Zn} units upon coordination of pyridine.

Transient absorption measurements carried out after embedding **27** in an agar matrix led to drastic changes (Fig. 14.26(b)). Agar (0.75 vol%) was added to an aqueous buffer solution (pH 7) and heated till the agar dissolved completely, the compound was dissolved in Triton X-100 and added to the solution at 60°C, and a transparent jelly was formed after cooling to room temperature.

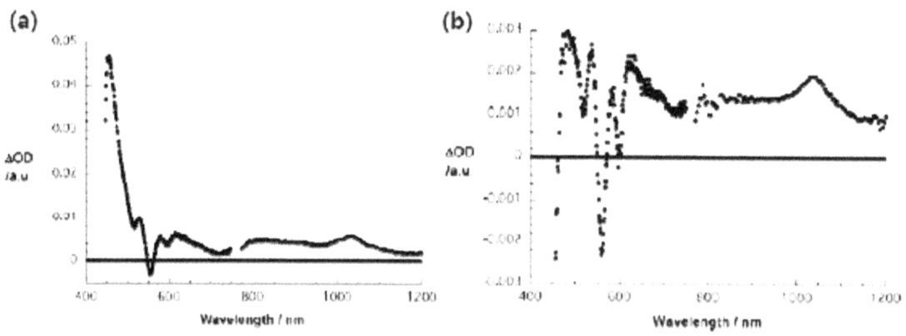

Figure 14.26. Differential absorption spectra obtained upon femtosecond flash photolysis at 387 nm of **27** (10^{-4} M) in argon-saturated THF with a time relay of 50 ps (a) and embedded in agar with a time relay of 100 ps (b) at room temperature.

Under such experimental conditions, the P_{Zn} singlet excited-state features decay through energy transfer to afford the corresponding H_2P singlet excited state. The latter is then susceptible to charge transfer: the stable $C_{60}^{\bullet-}$-$P_{FB}^{\bullet+}$-$(P_{Zn})_3$ is seen within 657 ps (1.5×10^9 s^{-1}) on the femtosecond scale. Complementary nanosecond experiments disclose a radical ion pair state with a lifetime of 100 ns (1.0×10^7 s^{-1}) corresponding to $C_{60}^{\bullet-}$-P_{FB}-$(P_{Zn})_3^{\bullet+}$. This result indicates a cascade of charge-transfer reactions that convert the initially formed $C_{60}^{\bullet-}$-$P_{FB}^{\bullet+}$-$(P_{Zn})_3$ into $C_{60}^{\bullet-}$-P_{FB}-$(P_{Zn})_3^{\bullet+}$. The underlying charge shift reaction must have taken place on a timescale between 3 and 10 ns (i.e., the time resolution of the femtosecond and nanosecond setups, respectively).

14.3.7 *Ru(II)/Sn(IV) Multiporphyrin Arrays*

The photophysical behavior of a series of heterometallic arrays made of a central Sn(IV) porphyrin connected, respectively, to two (**28**), four (**29**), or six (**30**) Ru(II) porphyrin units (Fig. 14.27) has been studied in dichloromethane [44]. Two different motifs connect the ruthenium porphyrin units to central tin porphyrin core, axial coordination via ditopic bridging ligands and/or coordination to peripheral pyridyl groups of the central porphyrin ring.

All the arrays showed very similar absorption pattern of metal porphyrin with two Q-bands and the Soret bands in the visible region, in which the Q-band region is dominated by the Ru(II) porphyrin absorption features with an increase of the 534 nm band,

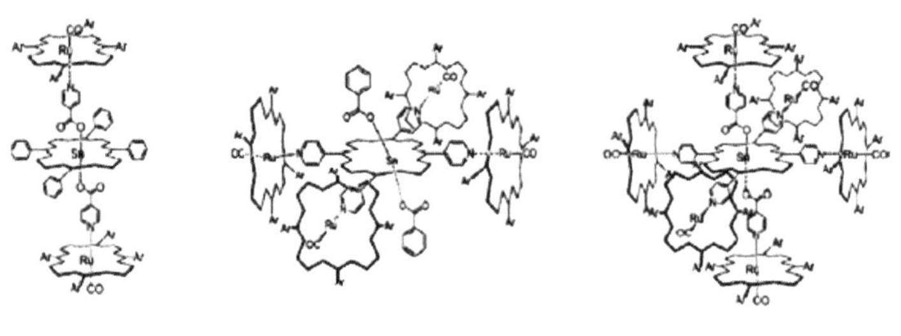

28 **29** **30**

Figure 14.27. Structural formulas of Ru(II)/Sn(IV) multiporphyrin arrays.

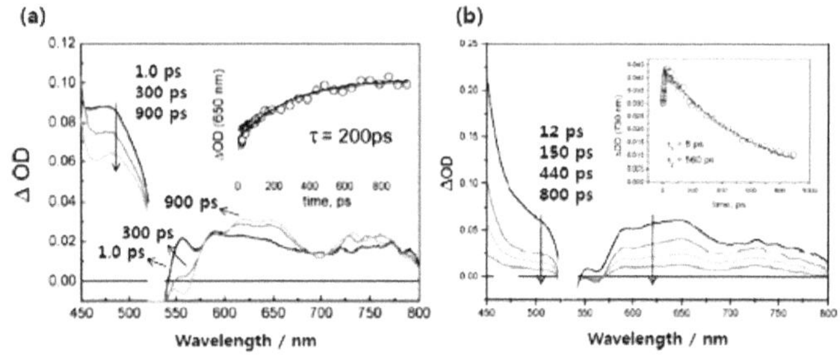

Figure 14.28. Ultrafast spectroscopy of **28**(a) and **29**(b) in CH_2Cl_2 upon excitation at 530 nm. See also Color Insert.

proportional to the number of ruthenium units. Basically, the fluorescence emission of Sn(IV) porphyrin usually appears at 600 (0,0) and 660 nm(0.1), but the fluorescence of Ru(II) porphyrin is not detectable at room temperature. Emission spectra of **28**, **29**, and **30** were also displayed with similar profiles. They exhibit an exceedingly weak emission, excited at 595 nm, both in aerated and deaerated solution. This weak emission is fluorescence from the tin chromophore, strongly quenched with respect that of the single Sn(IV) porphyrin.

For **28**, transient behavior has been also investigated by selective excitation of the Ru(II) porphyrin unit at 530 nm (Fig. 14.28(a)). The initial spectrum (1 ps) is very similar to that exhibited by the single Ru(II) porphyrin as reference, showing a positive absorption with a maximum at 460 nm. A clear evolution from this spectrum to a final spectrum with features practically identical to those found upon 555 nm excitation (broad absorption at $\lambda > 560$ nm with a maximum at 730 nm) is observed in the 1–500 ps time scale. The time for formation of this transient is by far slower than that observed for excitation at 550 nm. Kinetic analysis of spectral changes at 650 nm yields a time constant of 200 ps (inset of Fig. 14.28(a)). As in the case of tin unit excitation, these spectral changes remain constant over the time window of the experiment.

For **29**, the transient spectral changes obtained by ultrafast spectroscopy in CH_2Cl_2 are shown in Fig. 14.28(b). The behavior is clearly biphasic. In the shorter time scale (1–16 ps), the spectral changes (formation of a positive broad absorption at $\lambda > 550$

peaking at 730 nm) are very similar to those observed for **28** under the same experimental conditions. Contrary to what happened for **28**, however, this spectrum decays in the time window of the ultra-fast experiment. Upon excitation of **29** at 530 nm, where the excitation light is 100% absorbed by the Ru(II) porphyrin unit, the spectral changes in the initial time scale are qualitatively the same as observed for **28** in the same experimental conditions. In contrast the kinetic behavior was found to be quite different. The initial spectrum, taken immediately after the excitation pulse, evolves in the early timescale (<20 ps) to a new spectrum featuring broad absorption at $\lambda > 560$ nm with a maximum at 730 nm. The formation of this transient is much faster than in the case of **28**. A time constant of 8 ps is obtained from the kinetic analysis at 730nm. On a longer time scale this transient decays to ground state uniformly at all wavelengths with a time constant of 560 ps (inset of Fig. 14.28(b))., while in the case of **28** the transient spectral changes remain constant over the time window of the experiment. For **20**, the transient changes observed upon 530 nm excitation were very similar (spectral shape and time scale) to those detected for **29** in the same experimental conditions. A transient spectrum featuring broad absorption at $\lambda >$ 560 nm with maxima at 650 and 730 nm is formed in 7 ps and decays to ground state in 650 ps. The time constant values and the corresponding driving forces are summarized in Table 14.4.

Table 14.4. Free energy-change and electron-transfer-rate constants for **28** and **29**

array	process	ΔG_{eb} eV[b]	$1/k$, ps
28	^{1*}Sn–Ru→^{1}Sn$^{(-)}$–Ru$^{(+)}$	−0.52	25
	Sn–^{3*}Ru→^{3}Sn$^{(-)}$–Ru$^{(+)}$	−0.18	200
	^{1}Sn$^{(-)}$–Ru$^{(+)}$→Sn–Ru	−1.53	$2-8\times10^{3}$
	^{3}Sn$^{(-)}$–Ru$^{(+)}$→Sn–Ru	−1.53	$2-8\times10^{3}$
29	^{1*}Sn–Ru→^{1}Sn$^{(-)}$–Ru$^{(+)}$	−0.80	4.3
	Sn–^{3*}Ru→^{3}Sn$^{(-)}$–Ru$^{(+)}$	−0.46	8
	^{1}Sn$^{(-)}$–Ru$^{(+)}$→Sn–Ru	−1.25	510
	^{3}Sn$^{(-)}$–Ru$^{(+)}$→Sn–Ru	−1.25	560

[a]All measurements were made in dichloromethane solution at 298 K.
[b]Free energy changes calculated on the basis of the potophysical data and electrochemical potentials

The main interesting features of this study include the following: (i) using a combination of emission spectroscopy and femtosecond and nanosecond time-resolved absorption techniques, a very detailed picture of the photophysics of these systems has been obtained; (ii) a common (except for spin multiplicity) charge-separated state, where the ruthenium porphyrin is oxidized and the tin porphyrin is reduced, is reached by intercomponent electron transfer following selective excitation of both chromophoric components; (iii) changing the solvent polarity from dichloromethane to toluene has the effect of lifting the energy of the charge-separated state, and thus of switching electron transfer off and triplet energy transfer on following excitation of the Ru-based units; and (iv) the kinetics of the various intercomponent processes has been rationalized on the basis of standard electron transfer theory.

References

1. Barber J. and Anderson, B. (1994) Revealing the blueprint of photosynthesis, *Nature*, **370**, 31–34.
2. Wasielewski, M. R. (1992) Photoinduced electron transfer in supramolecular systems for artificial photosynthesis, *Chem. Rev*, **92**, 435–461.
3. Gust, D. and Moore, T. A. (1999) *The Porphyrin Handbook*, (ed. Kadish, K., Smith, K. M., and Guilard, R.), volume 8 Academic Press, New York pp. 153.
4. Gust, D., Moore, T. A., and Moore, A. L. (2001) Mimicking photosynthetic solar energy transduction, *Acc. Chem. Res*, **34**, 40–48.
5. (a) Mcdermott, G. Prince, S. M., Freer, A. A., Hawthornthwaite-Lawless, A. M., Papiz, M. Z., Cogdell, R. J., and Isaacs N. W. (1995) Crystal structure of an integral membrane light-harvesting complex from photosynthetic bacteria, *Nature*, **374**, 517–521. (b) Kühlbrandt W. (1995) Many wheels make light work, *Nature*, **374**, 497–498.
6. Deisenhofer, J. and Michel H. (1989) The photosynthetic reaction center from the purple bacterium rhodopseudomonasviridis, *Science*, **245**, 1463–1473.
7. Pullerits, T. and Sundstrom V. (1996) Photosynthetic light-harvesting pigment-protein complexes: Toward understanding how and why, *Acc. Chem. Res.* **29**, 381–389.

8. Hu, X. Damjanovic, A. Ritz, T., and Schulten K. (1998) Architecture and mechanism of the light-harvesting apparatus of purple bacteria, *Proc. Natl. Acad, Sci, USA* **95**, 5935–5941.
9. (a) Cleary, R. L. Byrom, K. J., Bardwell, D. A., Jeffery, J. C., Ward, M. D., Calogero, G. Armaroli, N. Falmigni, L., and Barigelletti F. (1997) Intercomponent electronic energy transfer in heteropolynuclear complexes containing ruthenium- and rhenium-based chromophores bridged by an asymmetric quaterpyridine ligand, *Inorg. Chem.*, **36**, 2601–2609. (b) Balzani, V. Campagna, S. Denti, G. Juris, A. Serroni, S., and Venturi M. (1998) Designing dendrimers based on transition-metal complexes: Light-harvesting properties and predetermined redox patterns, *Acc. Chem. Res.*, **31**, 26–34. (c) Zhou, X. Tyson, D. S., and Castellano F. N. (2000) First generation light-harvesting dendrimers with a $[Ru(bpy)_3]^{2+}$ core and aryl ether ligands functionalized with coumarin 450, *Angew. Chem. Int. Ed.*, **39**, 4301–4305. (d) Fleming, C. N. Maxwell, K. A. DeSimone, J. M. Meyer, T. J., and Papanikolas J. M. (2001) Ultrafast excited-state energy migration dynamics in an efficient light-harvesting antenna polymer based on Ru(II) and Os(II) polypyridyl complexes, *J. Am. Chem. Soc.***123**, 10336–10347. (e) MaClenaghan N. D., Loiseau, F. Puntoriero, F., Serroni, S., and Campagna S. (2001) Light-harvesting metal dendrimers appended with additional organic chromophores: a tetranuclear heterometallic first-generation dendrimer exhibiting unusual absorption features, *Chem. Commun.*, **37**, 2634–2635.
10. (a) Webber S. E. (1990) Photon-harvesting polymers, *Chem. Rev.*, **90**, 1469–1482. (b) Fox M. A. (1992) Polymeric and supramolecular arrays for directional energy and electron transport over macroscopic distances, *Acc. Chem. Res.*, **25**, 569–574. (c) Watkins, D. M., and Fox M. A. (1994) Rigid, well-defined block copolymers for efficient light harvesting, *J. Am. Chem. Soc.*, **116**, 6441–6442. (d) Schultze, X. Serin, J. Andronov, A., and Fréchet J. M. J. (2001) Light harvesting and energy transfer in a ruthenium–coumarin-2 copolymer, *Chem. Commoun.*, 1160–1161.
11. (a) Nowakowska, M. Loukine, N. Gravett, D. M. Burke, N. A. D., and Guillet J. E. (1997) A novel antenna cyclodextrin: Synthesis and photosensitized reaction of an included guest, *J. Am. Chem. Soc.*, **119**, 4364–4368 (b) Tamura, M. Gao, D., and Ueno A. (2001) A polyrotaxaneseries containing α-cyclodextrin and naphthalene-modifiedcyclodextrin as a light-harvesting antenna system, *Chem. Eur. J.* **7**, 1390–1397.
12. Vollmer, M. S. Effenberger, F. Stümpfig, T. Hartschuh, A., Port, H., and Wolf, H. C. (1998) Steroid-bridged anthryloligothienyl porphyrins: Syn-

thesis and study on the intramolecular energy transfer, *J. Org. Chem.* **63**, 5080–5087.

13. Li, F. Yang, S. I. Ciringh, Y. Seth, J. Martin III C. H. Singh, D. L., Kim, D. H. Birge, R. R., Bocian, D. F., Holten, D., and Lindsey J. S. (1998) Design, synthesis, and photodynamics of light-harvesting arrays comprised of a porphyrin and one, two, or eight boron-dipyrrin accessory pigments, *J. Am. Chem. Soc.*, **120**, 10001–10017.
14. (a) Stewart G. M. and Fox M. A. (1996) Chromophore-labeled dendrons as light harvesting antennae, *J. Am. Chem. Soc.*, **118**, 4354–4360. (b) Devadoss, C. Barathi P., and Moore J. S. (1996) Energy transfer in dendritic macromolecules: Molecular size effects and the role of an energy gradient, *J. Am. Chem. Soc.*, **118**, 9635–9644.
15. (a) Vincente, M. G. H. Jaquinod, L., and Smith K. M. (1999) Oligomeric porphyrin arrays, *Chem. Commun.*, 1771–1782. (b) Anderson H. L. (1999) Building molecular wires from the colours of life: conjugated porphyrin oligomers, *Chem. Commun.*, 2323–2330. (c) Burrell, A. K., Officer, D. L. Plieger, P. G., and Reid D. C. W. (2001) Synthetic routes to multiporphyrin arrays, *Chem. Rev.*, **101**, 2751–2796.
16. (a) Tonalia, D. A. Baker, H. Dewald, J. Hall M., Kallos, G. Martin, S. Roeck, J., Ryder, J., and Smith P. (1985) A new class of polymers: Starburst-dendritic macromolecules, *Polym. J.*, **17** 117–132. (b) Fréchet J. M. J. (1994) Functional polymers and dendrimers: reactivity, molecular architecture, and interfacial energy, *Science*, **263**, 1710–1715. (c) Frey H. (1998) From random coil to extended nanocylinder: Dendrimer fragments shape polymer chains, *Angew. Chem. Int. Ed.*, **37**, 2193–2197.
17. (a) Jiang, D.L. and Aida, T. (1997) Photoisomerization in dendrimers by harvesting of low-energy photons, *Nature* **388**, 454–456. (b) Jiang, D.L., and Aida T. (1998) Morphology-dependent photochemical events in aryl ether dendrimer porphyrins: Cooperation of dendron subunits for singlete transduction, *J. Am. Chem. Soc.*, **120**, 10895–10901. (c) Adronov, A., and Fréchet J. M. J. (2000) Light-harvesting dendrimers, *Chem. Commun.*, 1701–1710.
18. Devadoss C., Bharathi P, and Moore J S. (1996) Energy transfer in dendritic macromolecules: molecular size effects and the role of an energy gradient *J Am Chem Soc.*, **118**, 9635–9644.
19. Bar-Haim, A., Klafter, J., and Kopelman R. (1997) Dendrimers as controlled artificial energy antennae *J. Am. Chem. Soc.*, **119**, 6197–6198.
20. (a) Nishiyama, N., Morimoto, Y., Jang, W-D., and Kataoka, K. (2009) Design and development of dendrimer photosensitizer-incorporated

polymeric micelles for enhanced photodynamic therapy *Adv Drug Delivery Rev.*, **61**, 327–338. (b) Li, Y., Jang, W-D., Nishiyama, N., Kishimura, A., Kawauchi, S., Morimoto, Y., Miake, S., Yamashita, T., Kikuchi, M., Aida, T., and Kataoka, K. (2007) Dendrimer generation effects on photodynamic efficacy of dendrimer porphyrins and dendrimer-loaded supramolecular nanocarriers *Chem Mater.*, **19**, 5557–5562. (c) Jang, W.-D., Nishiyama, N., Kataoka, K. (2007) Suparmolecular assembly of photofunctional dendrimers for biomedical nano-devices. *Supramol. Chem.*, **19** 309–314 (d) Arnida, Nishiyama, N., Kanayama, N., Jang, W.-D., Yamasaki, Y., and Kataoka, K. (2006) PEGylatedgene nanocarriers based on block catiomers bearing ethylenediamine repeating units directed to remarkable enhancement of photochemical transfection, *J. Control Rel.,***115** 208–215. (e) Jang, W.-D., Nakagishi, Y., Nishiyama, N., Kawauchi, S., Morimoto, Y., Kikuchi, M., and Kataoka K. (2006) Polyion complex micelles for photodynamic therapy: incorporation of dendritic photosensitizer excitable at long wavelength relevant to improved tissue-penetrating property, *J Control Rel*, **113**, 73–79. (f) Nishiyama, N., Iriyama, A., Jang, W.-D., Miyata, K., Itaka, K., Inoue, Y., Takahashi, H., Yanagi, Y., Tamaki, Y., Koyama, H., and Kataoka, K. (2005) Light-induced gene transfer from packaged DNA enveloped in a dendrimeric photosensitizer *Nat Mater.*, **4** 934–941 (g) Ideta, R., Tasaka, F., Jang, W.-D., Nishiyama, N., Zhang, G.-D., Harada, A., Yanagi, Y., Tamaki, Y., Aida, T., and Kataoka, K. (2005) Nanotechnology-based photodynamic therapy for neovascular disease using a supramolecular nanocarrier loaded with a dendritic photosensitizer. *Nano Lett.*, **5**, 2426–2431. (h) Jang, W.-D., Nishiyama, N., Zhang, G.-D., Harada, A., Jiang, D.-L., Kawauchi, S., Morimoto, Y., Kikuchi, M., Koyama, H., Aida, T., and Kataoka, K. (2005) Supramolecular nanocarrier of anionic dendrimer porphyrins with pegylated cationic block copolymer to enhance intracellular photodynamic efficacy, *Angew Chem Int Ed.*, **44**, 419–423.

21. (a) Oar, M A. Serin, J M., Dichtel, W. R., Frechet, J. M. J., Ohulchanskyy T. Y., and Prasad, P. N. (2005) Photosensitization of singlet oxygen via two-photon-excited fluorescence resonance energy transfer in a water-soluble dendrimer, *Chem. Mater.*, **17** 2267–2275. (b) Oar, M A. Dichtel, W. R., Serin, J M., Frechet, J. M. J., Ohulchanskyy T. Y., Prasad, P. N. (2006) Light-harvesting chromophores with metalated porphyrin cores for tuned photosensitization of singlet oxygen via two-photon excited FRET, *Chem. Mater*, **18**, 3682–3692.
22. Maruo, N., Uchiyama, M., Kato, T., Arai, T., Akisada, H., Nishino, N. (1999) Hemispherical synthesis of dendritic poly(l-lysine) combining

sixteen free-base porphyrins and sixteen zinc porphyrins, *Chem. Commun.*, 2057–2058.

23. (a) Yang, J., Cho, S., Yoo, H., Park, J., Li, W.-S., Aida, T., and Kim, D. (2008) Control of molecular structures and photophysical properties of Zinc(II) porphyrin dendrimers using bidentate guests: utilization of flexible dendrimer structures as a controllable mold, *J. Phys. Chem. A.* **112**, 6869–6876. (b) Cho, S., Li, W.-S., Yoon, M.-C., Ahn, T. K., Jiang, D.-L., Kim, J., Aida, T., and Kim, D. (2006) Relationship between incoherent excitation energy migration processes and molecular structures in Zinc(II) porphyrin dendrimers, *Chem. Eur. J.* **12** 7576–7584.
24. Officer, D. L., Burrell, A. K., and Reid, D. C. W. (1996) Building large porphyrin arrays: pentamers and nonamers *Chem. Commun.*, 1657–1658.
25. Mak, C. C., Pomeranc, D., Montalti, M., Prodi, L., and Sanders, J. K. M. (1999) A versatile synthetic strategy for construction of large oligomers: binding and photophysical properties of a nine-porphyrin array *Chem. Commun.*, 1083–1084.
26. Jang, W.-D., Lee, C.-H., Osada, M., and Choi, M.-S. (2009) Synthesis of multi-porphyrin dendrimer as artificial light harvesting antennae *J Porphyr Phthalocya*, **13**, 787–793.
27. Sugiura, K., Tanaka, H., Matsumoto, T., Kawai, T., and Sakata, Y. (1999) A mandala-patterned bandanna-shaped porphyrin oligomer, C_{1244} H_{1350} N_{84} Ni_{20} O_{88}, having a unique size and geometry, *Chem. Lett.* 1193–1194.
28. Choi, M.-S. Aida, T., Yamazaki, I., and Yamazaki T. (2001) A large dendritic multiporphyrin array as a mimic of the bacterial light-harvesting antenna complex: molecular design of an efficient energy funnel for visible photons, *Angew. Chem* 113, 3294–3298; *Angew. Chem. Int. Ed.*, 40, 3194–3198.
29. (a) Sugou, K., Sasaki, K., Kitajima, K., Iwaki, T., and Kuroda, Y. (2002) Light-harvesting heptadecameric porphyrin assemblies *J. Am. Chem. Soc.*, **124** pp 1182–1183. (b) Kuroda, Y., Sugou, S., and Sasaki, K. (2000) Nonameric porphyrin assembly: antenna effect on energy transfer *J. Am. Chem. Soc.*, **122**, 7833–7834.
30. Oh, J. B., Nah, M.-K., Kim, Y. H., Kang, M. S., Ka, J.-W., and Kim, H. K. (2006) Er^{III} cored complexes based on dendritic Pt^{II} porphyrin ligands: synthesis, near-IR emission enhancement, and photophysical studies, *Adv. Funct. Mater.*, **17** 413–424.
31. Kodis, G. Liddell, P. A., de la Garza, L., Clausen, P. C., Lindsey, J. S., Moore, A. L., Moore, T. A., and Gust D. (2002) Efficient energy transfer and electron

transfer in an artificial photosynthetic antenna-reaction center complex, *J. Phys. Chem. A* 106, 2036–2048.

32. Choi, M.-S. Aida, T. Yamazaki, I., and Yamazaki T. (2002) Dendritic multiporphyrin arrays as light-harvesting antennae: effects of generation number and morphology on intramolecular energy transfer, *Chem. Eur. J.*, 8, 2667–2678.
33. Choi, M.-S. Yamazaki, T. Yamazaki, I., and Aida, T. (2003) Bioinspired molecular design of light-harvesting multiporphyrin arrays, *Angew. Chem. Int. Ed.*, **43**, 150–158.
34. Choi, M.-S., Aida, T. Luo, H. Araki, Y., and Ito, O. (2003) Fullerene-terminated dendritic multiporphyrin arrays: dendrimer effect on photoinduced charge separation, *Angew. Chem. Int. Ed.*, **42**, 4060–4063.
35. Luo, H., Choi, M-S, Araki, Y, Ito, O., and Aida, T. (2005) Dendrimer effects on intermolecular energy-transfer of photoexcited triplet states of dendritic multiporphyrin arrays and electron transfer vs fullerene[60], *Bull. Chem. Soc. Jpn.*, **78**, 405–412.
36. Kuciauskas, D. Lin, S. R. Seely, G. Moore, A. L. Moore, T. A., Drovetskaya, D. G. Christopher A., and Boyd R. D. W. (1996) Energy and photoinduced electron transfer in porphyrin-fullerene dyads, *J. Phys. Chem.* **100**, 15926–15932.
37. Thomas, P., Khang, H. L., Jousselme, B., Jegou, P., Filoramo, A., Ehli, C., Guldi, D. M., and Campidelli, S. (2009) Efficient functionalization of carbon nanotubes with porphyrin dendrons via click chemistry, *J. Am. Chem. Soc.*,**131**, 15394–15402.
38. Fukuzumi, S. Imahori, H. Yamada, H., El-Khouly, M. E. Fujitsuka, M. Ito, O., and Guldi D. M. (2001) Catalytic effects of dioxygen on intramolecular electron transfer in radical ion pairs of zinc porphyrin-linked fullerenes, *J. Am. Chem. Soc.*, **123**, 2571–2575.
39. Herranz, M. A. Ehli, C. Campidelli, S. Gutie'rrez, M. Hug, G. L. Ohkubo, K. Fukuzumi, S. Prato, M. Mart'in, N., and Guldi, D. M. (2008) Spectroscopic characterization of photolytically generated radicalion pairs in single-wall carbon nanotubes bearing surface-immobilized tetrathiafulvalenes, *J. Am. Chem. Soc.*, **130**, 66–73.
40. Elliott, K. J., Harriman, A., Pleux, L. L., Pellegrin, Y., Blart, E., Mayer, C. R., and Odobel, F. (2009) A porphyrin-polyoxometallate bio-inspired mimic for artificial photosynthesis, *Phys. Chem. Chem. Phys.*, 11, 8767–8773.
41. Brookfield, R. L. Ellul, H., Harriman A., and Porter G. (1986) Luminescence of porphyrins and metalloporphyrins. Part 11.—Energy transfer

in zinc–metal-free porphyrin dimmers, *J. Chem. Soc., Faraday Trans.* 2, 82, 219–233.

42. Harriman, A. Elliott, K. J. Alamiry, M. A. H., Le Pleux, L. Se'verac, M., Pellegrin, Y., Blart, E., Fosse, C. Cannizzo, C. Mayer C. R., and Odobel F. (2009) Intramolecular electron transfer reactions observed for Dawson-type polyoxometalates covalently linked to porphyrin residues, *J. Phys. Chem. C* **113**, 5834–5842.
43. Albinsson, B., and Martensson J. (2008) Long-range electron and excitation energy transfer in donor–bridge–acceptor systems, *J. Photochem. Photobiol. C*, **9**, 138–155.
44. Indelli, M. T., Chiorboli, C., Ghirotti, M., Orlandi, M., Scandola, F., Kim, H. J., and Kim, H-J, Photoinduced electron transfer in ruthenium(II)/tin(IV) multiporphyrin arrays, *J. Phys. Chem. B*, in press.

Chapter 15

Push-Pull Porphyrins for Efficient Dye-Sensitized Solar Cells

Chen-Yu Yeh,[a] Ching-Yao Lin,[b] and Eric Wei-Guang Diau[c]

[a] *Department of Chemistry, National Chung Hsing University, Taichung, Taiwan.*
[b] *Department of Applied Chemistry, National Chi Nan University, Puli, Nantou, Taiwan.*
[c] *Department of Applied Chemistry and Institute of Molecular Science, National Chiao Tung University, Hsinchu, Taiwan.*
cyyeh@dragon.nchu.edu.tw; cyl@ncnu.edu.tw; diau@mail.nctu.edu.tw

15.1 Introduction

Dye-sensitized solar cells (DSSC) have attracted much attention because they present a highly promising alternative to conventional photovoltaic devices based on silicon [1–3]. In nanocrystalline TiO_2 solar cells sensitized with a dye, efficiencies of conversion of light to electric power up to 11% have been obtained with polypyridyl ruthenium complexes [4, 5]. The advantages of using such ruthenium complexes are that they exhibit broad absorption in the near UV and visible regions and appropriate excited-state oxidation potentials for electron injection into the conduction band of TiO_2 [6]. The cost, rarity, and environmental issues of ruthenium complexes limit their wide application and encourage exploration of cheaper and safer sensitizers.

Multiporphyrin Arrays: Fundamentals and Applications
Edited by Dongho Kim

www.panstanford.com

Considerable effort has been devoted to the development of new and efficient sensitizers suitable for practical use; among these, organic sensitizers have drawn great interest because of their modest cost, ease of synthesis and modification, large molar absorption coefficients, and satisfactory stability. Organic dyes such as coumarin [7–9], indoline [10–12], oligoene [13, 14], thiophene [15–17], triarylamine [18–20], perylene [21–23], cyanine [24–26], and hemicyanine [27, 28] derivatives have been investigated as sensitizers in DSSC. Some organic dyes with conversion efficiencies in a range 5–9% have been prepared [29–31].

In the photosynthetic cores of bacteria and plants, solar energy is collected by chromophores based on porphyrin [32], and the captured radiant energy is converted efficiently to chemical energy. Various artificial photosynthetic model systems have been designed and synthesized to elucidate the factors that control the photo-induced electron-transfer reaction [33–40]. Valuable knowledge has been acquired from these artificial systems. Inspired by the efficient energy transfer in naturally occurring photosynthetic reaction centers, numerous porphyrin [41–49] and phthalocyanine [50–54] dyes have been synthesized and used for photovoltaic solar cells. The intrinsic advantages of porphyrin-based dyes are their rigid molecular structures with large absorption coefficients in the visible region and their many reaction sites, i.e., four *meso* and eight β positions, available for functionalization. Fine tuning of their optical, physical, and electrochemical properties thus becomes feasible.

Some recent reviews of DSSC based on porphyrin-related dyes have been published elsewhere [55–57] In this chapter, we focus mainly on advances in our monomeric porphyrin dyes and describe the structure-property relationship that may be useful for future molecular design to enhance the conversion efficiency of a DSSC. The dimeric porphyrin dyes developed by our and others' groups are also discussed.

15.2 Porphyrins with a Variety of Bridges

It is well known that the efficiency of power conversion of a DSSC device depends on how the sensitizer is attached to the surface of a semiconductor [58–60]. Therefore, the first strategy on design of porphyrin sensitizers for DSSC applications is to examine the effect of the length of the spacer with respect to the photovoltaic performance of the device. For this purpose, we have studied a series of porphyrins with length-controlled anchoring groups [61]. In our design, 10,20-biphenylporphinato zinc(II) (**ZnBPP**) served as a light-harvesting center, and a carboxylic acid as an anchoring group. To control the distance between the porphyrin and the anchoring carboxylic group, we utilized a phenylethylnyl (PE) unit as a bridging moiety to connect a *meso* position of **ZnBPP** and the carboxylic acid end. We chose **ZnBPP** because of its ease of synthesis, the stability of zinc porphyrins against irradiation and the efficiency of zinc porphyrins superior to that of copper and free-base porphyrins [62]. 4-carboxylic acid groups were used to ensure efficient adsorption on TiO_2 surfaces and to promote electronic coupling between the donor levels of the excited porphyrins and the acceptor levels of TiO_2 [63]. PE units were employed to control the distance because of their rigidity, their linear structure, and the possible π-interactions between the porphyrin core and the anchoring end [64, 65].

Introduction of PE units as spacers has been reported for other systems [66–68]. For example, Piotrowiak, Galoppini, and Meyer reported "tripod" Ru^{II} complexes using PE groups as rigid links to connect the sensitizers and TiO_2. These sensitizers consisted of a tripod-shaped base with three COOR groups to attach to TiO_2 nanocrystalline surfaces, a rigid (4-ethynyl)phenylethyne spacer, and a Ru^{II} complex as a chromophoric center. The base design provides a stable and well defined orientation of the sensitizers with respect to the nanoparticulate surfaces, whereas the rigidity of the (4-ethynyl)phenylethyne spacer ensures the control of the distance between the light-harvesting centers and the TiO_2 surfaces. By a similar design, we have prepared novel porphyrins to investigate the interfacial electron transfer and for DSSC applications: 5-(4-carboxy-phenylethynyl)-10,20-biphenylporphinato zinc(II), 5-[4-(4-carboxy-phenylethynyl)-phenylethynyl]-10,20-biphenylporphinato zinc(II), 5-{4-[4-(4-carboxy-phenylethynyl)-phenylethynyl]-phenylethynyl}-10,20-biphenylporphinato zinc(II), and 5-(4-{4-[4-(4-carboxy-phenylethynyl)-phenylethynyl]-phenylethynyl}-phenyleth-

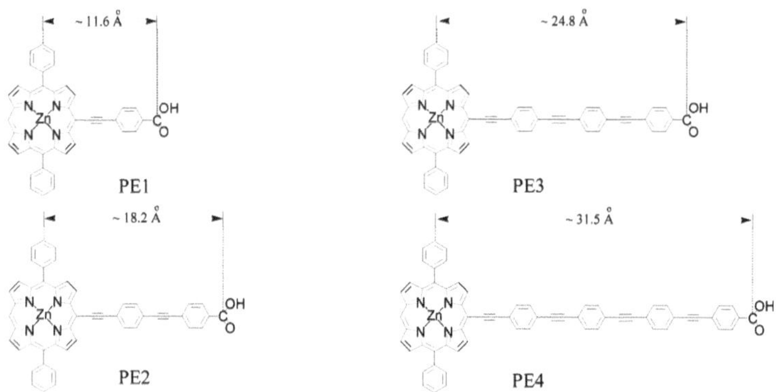

Figure 15.1. Molecular structures of PE1-PE4.

ynyl)-10,20-biphenylporphinato zinc(II), abbreviated as **PE1**, **PE2**, **PE3** and **PE4**, respectively. As shown in Fig. 15.1 these porphyrins share the same light-harvesting unit (**ZnBPP**) with rigid, linear, edgewise and fully conjugated substituents of varied lengths. When these porphyrins were fabricated into DSSC devices, the efficiencies of power conversion of these devices decreased systematically from 2.5 $\pm$ 0.2% for **PE1** to 0.25 $\pm$ 0.02% for **PE4** [61]. Measurements of femtosecond fluorescence of the porphyrin-sensitized TiO_2 films showed that electron injection from the porphyrin core to the surface of TiO_2 was equally rapid for **PE1-PE4**, consistent with electrochemical results showing their similar reduction potentials. Because a long link did not slow the electron injection at the dye/TiO_2 interface, other factors, such as dye aggregation and charge recombination, should be considered to account for the great degradation of performance in the devices with a long PE link [69, 70]. For this reason, the following discussions on the design and characterizations of porphyrin sensitizers for DSSC applications are mainly focusing on the molecular structures similar to that of **PE1**.

The second strategy is to consider the positions of the anchoring group attached to the surface of TiO_2. In general, there are two possible sites, the *meso*- and β-positions, available for the attachment of the anchoring group. To investigate how the structure of porphyrins affects the cell performance of devices, we designed and synthesized a series of porphyrin sensitizers, in which the porphyrin ring and the

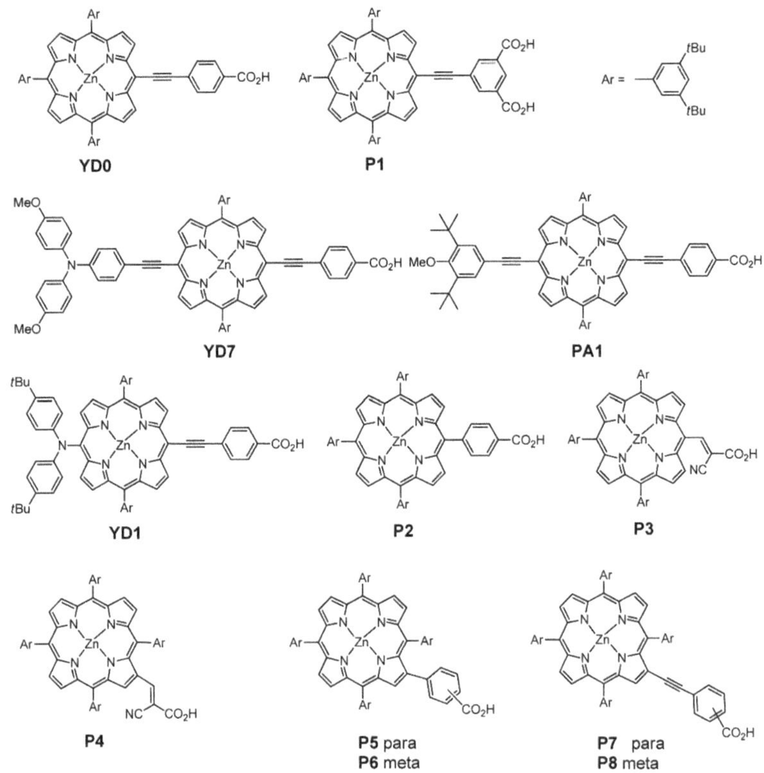

Figure 15.2. Structures of carboxylated porphyrin sensitizers with a variety of bridging units.

carboxyl anchoring group are connected with a vinyl, phenyl or PE bridging unit at the *meso*- or β-position (Fig. 15.2) [71].

The major factor in decreasing the efficiency of sensitized photocurrent generation is the formation of dye aggregates on the semiconductor surface [72, 73]. To decrease that aggregation, 3,5-di-*tert*-butylphenyl groups were introduced at the *meso*-positions of the porphyrin ring. One of approaches to the enhancement of absorption by porphyrin dyes in the visible region is to expand the π-conjugation system, which causes a red shift and broadening of both Soret and Q bands. In research on porphyrin arrays and porphyrin-based push-pull chromophores, bridges of ethyne type have been shown to allow efficient conjugation and strong electronic interaction between chromophores, and to provide a well defined and rigid structural arrangement [74–81]. We thus designed porphyrins having an ethyne bridging unit, which might facilitate electron transfer from the excited dye to the TiO_2 surface, and which are expected to show an improved efficiency of energy conversion. An alternative approach to increase the domain of absorption in the visible region, thus to increase the conversion efficiency, is to use a unit of ethenyl type as the link [82]. The efficient porphyrin-based sensitizers, exhibiting efficiencies of energy conversion $\geq$5%, are those with a β-substituted cyanoacetic or malonic acid, reported by Grätzel and co-workers [83–85]. Thus, porphyrin **P3** and **P4** having a cyanoacetic acid substituted at *meso*- and β-positions, respectively, were designed and synthesized.

The position and nature of the link between the porphyrin and the carboxyl group are expected to influence significantly the efficiency of energy conversion. Compounds **P5-P8** having a phenyl or PE bridge at the β-position were designed and synthesized. Most highly efficient dyes used in DSSC have a push-pull structure [86, 87]. To increase the electron-donating ability of the porphyrin ring, compound **YD7** with a triarylamine moiety and porphyrin **YD1** having a diarylamine group directly attached to the *meso*-position were synthesized.

The UV-visible spectra of these compounds in solution exhibit the features typical of a porphyrin ring, with an intense Soret band in the range 400–500 nm and less intense Q bands in a range from 550 to 750 nm. As expected, the electronic absorption bands of porphyrins are sensitive to substituents on the periphery of the porphyrin ring. Examples of absorption spectra for porphyrins **YD0**, **YD7**, **YD1**, **P2**, and **P5** appear in Fig. 15.3; UV-visible data for each compound are summarized in Table 15.1. Comparison of the spectra of **P5** and **P6** with **P2** reveals that substitution of carboxyphenyl at the β-position causes no significant red shift of either Soret or Q band because of an orthogonal orientation between the carboxyphenyl and porphyrin rings. For **P7** and **P8**, in which ethyne serves as the bridge between the porphyrin ring and anchoring group, the Soret and Q bands are more red-shifted than those of **P5** and **P6**. Perturbation of the energy of the HOMO and LUMO orbitals of the porphyrin ring is pronounced when *meso*- or β-ethenyl is employed as the linker, as indicated by

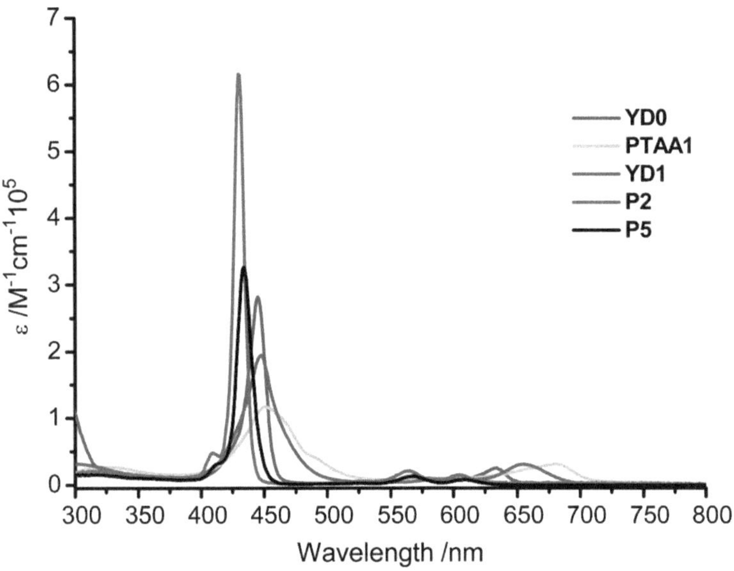

Figure 15.3. UV-visible absorption spectra of **YD0**, **YD7**, **YD1**, **P2**, and **P5** in CH_2Cl_2/Pyridine (100/1). See also Color Insert.

significant broadening and a red shift of the absorption bands of porphyrins **P3** and **P4** relative to **P5**-**P8**. Compounds **YD0** and **P1** show a similar spectral feature and substitution of carboxyphenyl at their *meso*-position via an ethyne linker that results in red shifts of the absorption bands larger than for those of β-substituted analogues **P7** and **P8**. Compound **YD1** shows broadening of the Soret band and a red shift of the Q band relative to the spectrum of **YD0** due to electronic interaction between the diarylamino group and the porphyrin ring of **YD1**. Comparison of the absorption spectra of porphyrins **YD7** and **PA1** with **YD0** shows that red shifts and broadening increase systematically with increasing π-conjugation [88, 89]. A similar trend was observed in the fluorescence spectra.

With compound **YD0** as our standard, we compared the cell performance for each porphyrin sensitizer. Compound **P4** was designed to mimic the molecular structure of **Zn-3** [82], of which the carboxyl is attached to the β-position via a vinylene group, but incorporating the *tert*-butyl substituents to avoid dye aggregation. As shown in Figure 8c, compound **P4** with a short linker attached at the

Table 15.1. Electronic absorption and emission data for porphyrins **YD0, P1, PTAA1, PA1, YD1**, and **P2-P8**

Porphyrin	Absorption, λ_{max}/nm ($\varepsilon/10^3 M^{-1}cm^{-1}$)	Emission, λ_{max}/nm
YD0	445 (282), 579 (9.5), 636 (24.8)	653
P1	445 (231), 582 (8.2), 632 (19.4)	651
PTAA1	451 (117), 680 (30.6)	707
PA1	454 (283), 668 (51.0)	687
YD1	448 (194), 601 (8.3), 654 (29.7)	687
P2	430 (616), 565 (20.7), 605 (14.7)	617, 660
P3	455 (106), 571 (7.1), 636 (8.4)	659
P4	451 (129), 564 (11.7), 613 (11.1)	666
P5	434 (326), 567 (12.9), 607 (7.7)	631
P6	433 (409), 566 (16.3), 609 (10.7)	630
P7	442 (348), 574 (21.0), 618 (13.0)	635
P8	444 (375), 575 (25.1), 618 (16.3)	635

β-position of the porphyrin ring exhibits poor adsorption on the TiO_2 film that causes poor cell performance; compounds **P5** and **P6** similarly showed small efficiencies of power conversion for the same reason—that the steric hindrance of the bulky *tert*-butyl groups impedes adsorption of the dye on TiO_2.

For compound **P7** with a longer linker at the β-position that promoted sufficient dye becoming adsorbed on the TiO_2 film, the efficiency of power conversion of **P7** became comparable to that of compound **YD0** (η = 2.7% vs. 2.4%), indicating that the cell performance of a porphyrin-sensitized solar cell was insensitive to the position of the PE linker attached at the β- or *meso*-position. This result, although contrary to reported data [57], provides new evidence that *meso*-substituted porphyrins might serve as potential photosensitizers in DSSC applications. In what follows, we discuss the photovoltaic performance of devices fabricated with various *meso*-substituted porphyrins.

Comparison first of **YD0** and **P1** shows that the anchoring group (−COOH) in the *para* position performs better than that in the *meta* position (η = 2.4% vs. 1.2%), consistent with literature results [57]. Relative to compound **YD0**, the structure of **P2** lacks a triple bond in the linker so that its π-conjugation is decreased. The Q band of **P2** thus became blue-shifted and the photocurrent density of **P2** decreased slightly relative to **YD0**. As a result, the power-conversion efficiency of **P2** became slightly less than that of **YD0** (2.2%

vs. 2.4%). In contrast, the Q band of **P3** is red-shifted relative to **YD0** because the structure of **P3** has an additional cyano group. The poor performance of **P3** in both J_{SC} and V_{OC} leads to the power conversion efficiency being much less than that of **YD0**. We expect that the superior electron-pulling capability of the cyano group in **P3** might first pull the electrons and then transfer them back into the porphyrin core through space; this process is impracticable for **P4** or **Zn-3** in which the linker is attached to the β-position.

To improve the charge separation in the sensitizer, we modified the structure of compound **YD0** on adding an efficient electron-pushing group at the *meso*-position of the porphyrin ring, and designed porphyrins **YD7**, **PA1**, and **YD1** accordingly. The additional triple bond in **YD7** and **PA1** extends the absorption to the near IR region, which might facilitate harvesting sunlight toward a more efficient region, but **YD7** and **PA1** are poorly soluble in ethanol, leading to molecular aggregation in solution. The Q band of compound **YD7** adsorbed on TiO_2 became broad, an indication of serious dye aggregation on the surface of the TiO_2 films. As a result, the cell performances of both compounds **YD7** and **PA1** were poorer than that of **YD0**. In contrast, porphyrin **YD1** with a diarylamino group attached directly at the *meso*-position exhibited an excellent cell performance with $J_{SC} = 13.60$ mA cm^{-2}, $V_{OC} = 0.701$ V, $FF = 0.629$, and overall $\eta = 6.0\%$ obtained for a TiO_2 film of thickness ~9 μm. The photocurrent density of **YD1** is more than twice that of **YD0**, indicating the superior capabilities of light harvesting and charge separation of that dye. The open-circuit voltage of **YD1** is notably the largest among all porphyrins under investigation.

The cell performance of **YD1** was further optimized in comparison with that of N719 on a film-to-film basis of different thickness (L) of the TiO_2 films [89]. The corresponding photovoltaic parameters are summarized in Table 15.2. The results indicate that the short-circuit photocurrent densities of these **YD1** devices are significantly greater than those of the N719 devices as compared on the basis of the same film thickness. Even though the open-circuit photovoltages and fill factors for the former are smaller than for the latter, the net effects of these variations make the overall efficiencies of power conversion of the **YD1** devices almost equal to those of the N719 devices; the maximum efficiency occurs at L ~12 μm (three active layers) in both cases.

Table 15.2. Photovoltaic parameters of DSSC with photosensitizers **YD1** and **N719** as a function of TiO_2 film thickness (L) under simulated AM-1.5 illumination (power 100 mW cm^{-1}) and active area 0.16 cm^2

Dye	$L/\mu m$	J_{SC}/mA cm^{-1}	V_{OC}/mV	FF	η (%)
YD1	4	9.2	718	0.69	4.6
	8	12.9	714	0.69	6.4
	12	14.3	712	0.69	7.0
	16	14.3	708	0.68	6.9
	12 + 4	15.1	717	0.70	7.6
N719	4	8.6	766	0.72	4.7
	8	11.9	751	0.72	6.4
	12	13.9	737	0.69	7.1
	16	13.5	735	0.70	6.9
	12 + 4	14.9	771	0.69	7.9

The superlative cell performance of **YD1** reflects its extraordinarily great J_{SC} value that arises from the large IPCE values broadly extending beyond 700 nm. The diarylamino group attached at the *meso*-position of the porphyrin core not only *spectrally* extends the absorption to a region of greater wavelength but also *spatially* pushes the excited electrons toward TiO_2 for an improved charge separation. In particular, we found in absorption spectra of the **YD1**/TiO_2 films an additional band about ~800 nm, which is not visible in the IPCE spectrum shown in the inset of Fig. 15.3. This 800 nm absorption band thus makes no contribution to photocurrent generation. In our work on the spectroelectrochemistry of **YD1**, a characteristic band at ~800 nm was observed for the oxidized species of **YD1** [60]. The presence of such a band at ~800 nm for the **YD1**/TiO_2 film thus indicates the formation of the **YD1**$^+$ species. This result implies the excellent character of charge separation in the **YD1**/TiO_2 film with protracted charge recombination.

The observed retardation of charge recombination might be due to an optimal spatial separation between the amino moiety and the TiO_2 surface reported for a Ru-dye system [90]. For the porphyrin system, Durrant and co-workers found that the period for charge recombination of a triarylamino porphyrin/TiO_2 film is 20 times that of a free-base porphyrin counterpart (80 vs. 4 ms) [41]. Mozer et al. further noted that the lower V_{OC} of porphyrin-sensitized solar cells is due to the decreased electron lifetime related to either a more rapid recombination of electrons with dye cations or I_3^- ions [91]. In our case, the long-lived charge separation in the **YD1** film implies that the interception of conduction band electrons is controlled solely by recombination of electrons with I_3^-. The observed V_{OC} in **YD1**-based solar cells significantly larger than other porphyrin-based devices [41–49] is thus inferred to be due to the diminished recombination between I_3^- and conduction band electrons because I_3^- might be attached to the positively charged diarylamino moiety far from the TiO_2 surface. The femtosecond fluorescence decays (Fig. 15.4) show that the quantum yield for electron injection of **YD1** on TiO_2 surface becomes $\Phi_{inj} = 0.86$. For a Ru-based DSSC, Φ_{inj}

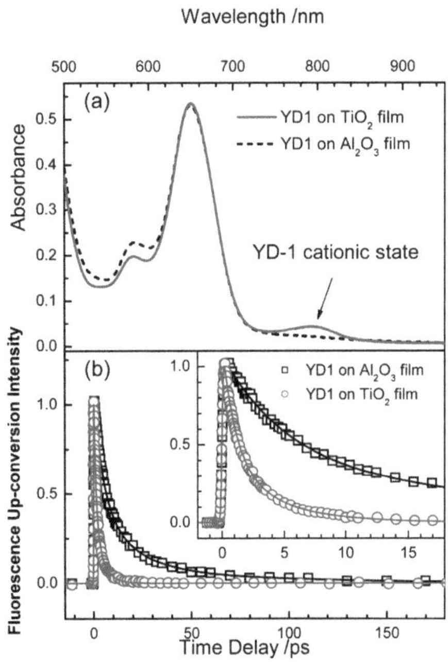

Figure 15.4. (a) Absorption spectrum of YD-1 adsorbed on a TiO_2 film (solid curve) compared with that adsorbed on an Al_2O_3 film (dashed curve); (b) the corresponding femtosecond fluorescence decay curves of the two thin-film samples obtained at the emission maxima, i.e., $\lambda_{em} = 680$ nm for the TiO_2 film and $\lambda_{em} = 692$ nm for the Al_2O_3 film, the inset shows the same temporal profiles at the smaller delays ($\lambda_{ex} = 430$ nm).

was expected to be ~1.0 [5, 92–94], but a recent study reported a smaller value (~0.9) [95]. For porphyrin sensitizers such as **YD1**, even though several bulky *tert*-butyl groups were incorporated in the aryl substituents around the porphyrin macrocycle and the dye was co-sensitized with chenodeoxycholic acid (CDCA) on the TiO_2 surface [60], the effect of aggregation cannot be completely eliminated.

To get insight into the electron contribution in the frontier orbitals, we performed quantum-chemical calculations on **YD1** using density-functional theory (DFT) at the B3LYP/6-31G(d) level [60]. As shown in Fig. 15.5, the electron density of **YD1** is significantly distributed to the π-system of the porphyrin ring, the diphenylamino moiety and the phenylene linker at the HOMO and HOMO−1, but the π-conjugation is extended to only the porphyrin ring and the PE link at the LUMO, facilitating the electron transfer from the dye to TiO_2 through the link; LUMO+1 shows the localization of electron density solely on the porphyrin ring. The electron density distribution of the frontier orbitals is appropriate for an efficient dye used in DSSC.

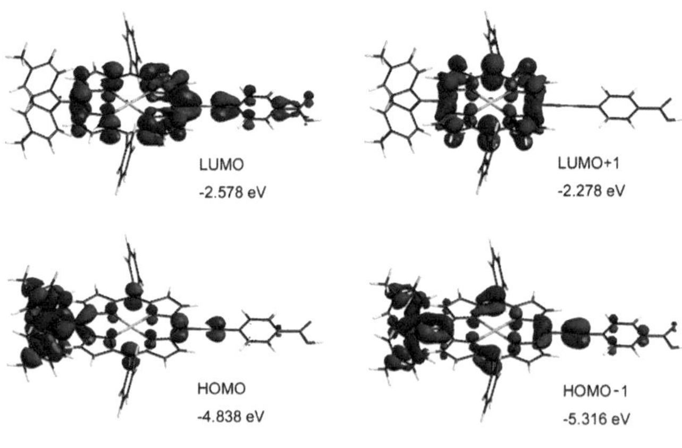

Figure 15.5. Surfaces of frontier molecular orbitals of **YD1** predicted with DFT calculations. To simplify computations, *tert*-butyl groups at *para*- and *meta*-positions were replaced with methyl groups and hydrogen atoms, respectively. See also Color Insert.

15.3 Porphyrins with a Push-Pull Framework

Numerous organic dyes have been synthesized for the use in DSSC. Organic dyes with large conversion efficiencies are typically composed of a donor-π-conjugated unit-acceptor (D-π-A) structure with a well defined architecture. As mentioned, a novel zinc porphyrin dye **YD1** with the molecular design based on a concept of a push-pull structure, by simply adding an electron-donating diarylamino group attached at the *meso*-position of porphyrin opposite the phenylethynyl–carboxyl anchoring group (Fig. 15.6). We found that a device made from this **YD1** dye has a cell performance similar to that of a Ru-based DSSC.

Previous work showed that introduction of hydrophobic alkyl chains onto organic dyes suppresses the electron transfer from TiO_2 to the electrolyte [31, 96]. Compound **YD2** having two hexyl chains on the diphenylamine is expected to act as a more efficient sensitizer in a DSSC than **YD1**. Alkoxyl groups are considered to be stronger electron-donating groups than alkyl groups; accordingly, porphyrins **YD3** and **YD4** with methoxyl and pentoxyl groups, respectively, have been synthesized. Another promising strategy to increase the electron-donating ability of the porphyrin dye is to introduce an N-substituent onto the diarylamine moiety; compound **YD5** incorporated with a triamine group to the *meso*-position has also been synthesized. To extend the charge separation between porphyrin and TiO_2, porphyrins with a triarylamino group, **YD6** and **YD8**, have been designed and prepared [97].

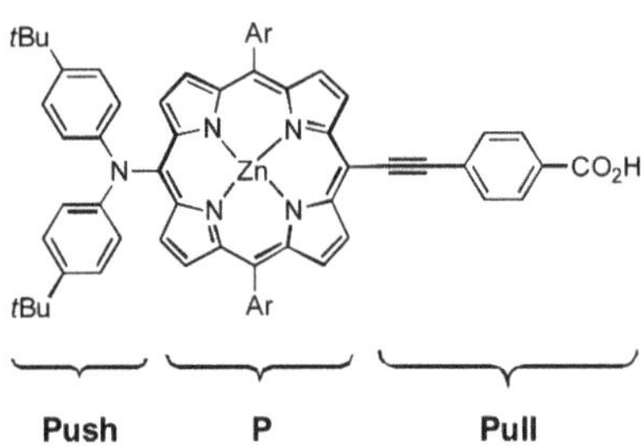

Figure 15.6. The porphyrin structure with a push-pull framework.

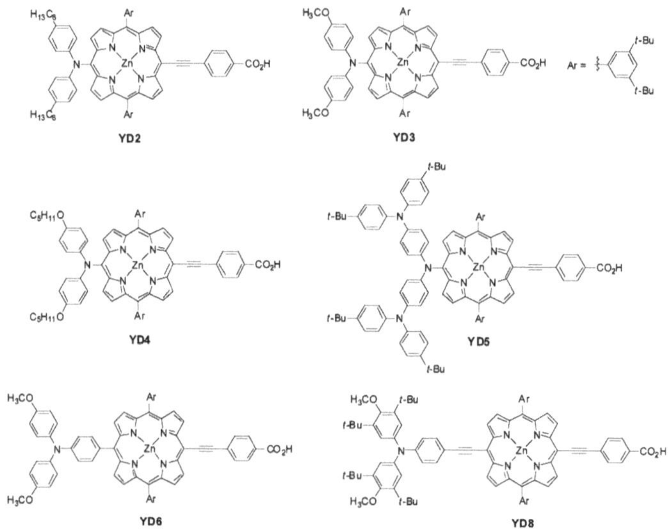

Figure 15.7. Molecular structures of porphyrin dyes with various electron-donating groups opposite the anchoring group.

The absorption spectra of these compounds are summarized in Table 15.3. All porphyrin dyes in this work exhibit strong Soret band and moderate Q bands. The molar absorption coefficients/10^5 M^{-1} cm^{-1} for the Soret band of these porphyrin dyes range from 1.36 to 4.98, so fulfilling one requirement for a dye usable in a DSSC,

Table 15.3. Electronic absorption and emission data for porphyrins **YD0-YD8**

Dye	FWHM B-band/cm^{-1}	f_B	FWHM, Q-band/cm^{-1}	f_Q	Absorption, λ_{max}/nm (ε/103 $M^{-1}cm^{-1}$)	Emission, λ_{max}/nm
YD0	672	1.46	540	0.15	442(498), 579(16.8), 627(39.8)	634
YD1	1708	1.36	914	0.15	442(207), 587(11.1), 644(31.2)	672
YD2	1700	1.42	934	0.15	444(217), 589(10.8), 648(33.7)	676
YD3	1366	0.88	1170	0.10	440(141), 592(5.9), 664(16.9)	701
YD4	1313	0.86	1169	0.10	438(143), 591(5.4), 665(17.6)	701
YD5	1088	1.55	1592	0.20	441(307), 659(27.2)	687
YD6	1217	0.76	634	0.07	443(216), 579(8.0), 628(20.8)	641
YD7	2042	1.05	898	0.20	449(141), 672(52.7)	689
YD8	1940	0.95	876	0.17	457(136), 671(49.9)	689

whereas those/10^3 M^{-1} cm^{-1} of the Q(0,0) band are in the range 16.9–52.7. The Soret and Q bands for **YD1–YD8** are broadened and red-shifted relative to those of **YD0**. Both Soret and Q(0,0) bands of **YD1–YD8** have full widths at half maximum (FWHM) height two or three times as large as those for the reference compound **YD0**; this effect is ascribed to an electronic interaction between the porphyrin core and the amino group. Both the Soret and Q bands for **YD1** and **YD2** show oscillator strengths comparable with that of **YD0**, although their absorption coefficients are much smaller than those of **YD0**. **YD5** exhibits a larger oscillator strength for both Soret and Q bands than for **YD0**. Among these porphyrins, **YD8** exhibits the most pronounced broadening and bathochromic shift of the Soret band, indicating that the acetylenic link effectively mediates the electronic coupling between the porphyrin and triarylamine units. Similar to the absorption, the emission is red-shifted on incorporation of an amino group onto the porphyrin macrocycle (Table 15.3).

The energy levels of these porphyrins are depicted in Fig. 15.8. The excited-state oxidation potentials (E_{0-0}*) are obtained from the relation $E_{0-0}* = E_{ox1} - E_{0-0}$, in which E_{ox1} is the first oxidation potential of a porphyrin dye and E_{0-0} is the zero-zero excitation energy obtained from the absorption edge [98, 99]. The calculated E_{0-0}* values are all more negative than the conduction edge (−0.50 V vs. NHE) of TiO_2, which is compatible with electron injection from the excited state of the dye to the conduction band (CB) of TiO_2. As the HOMO levels are more positive than the oxidation

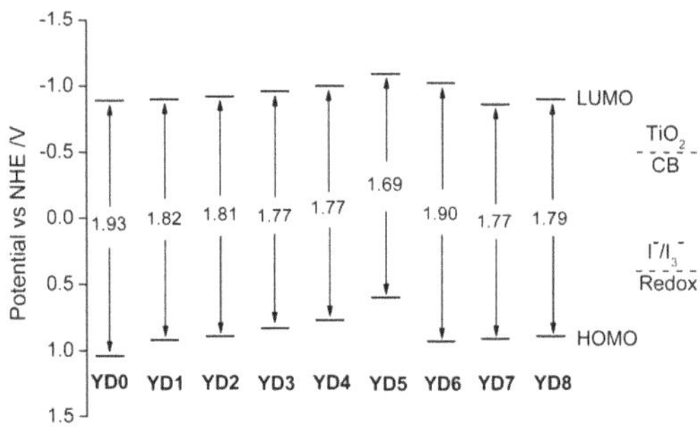

Figure 15.8. A schematic energy-level diagram of porphyrins **YD0–YD8**. HOMO = E_{ox1} and LUMO = E_{0-0}*.

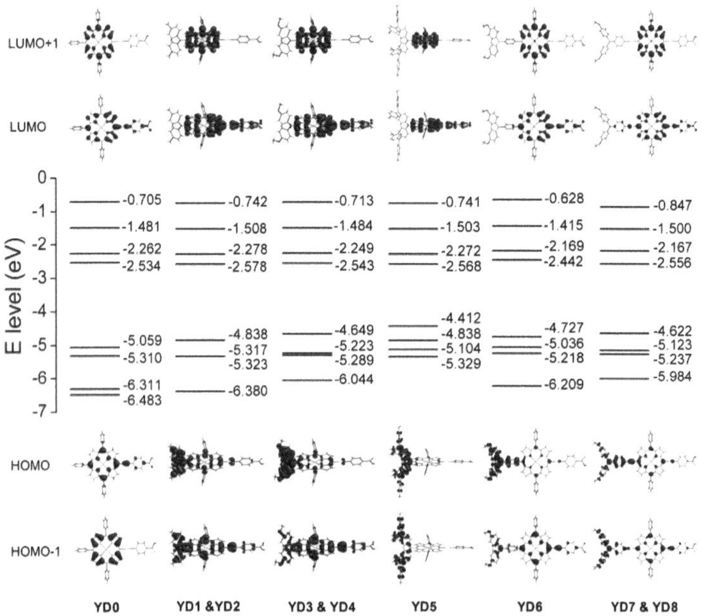

Figure 15.9. Energy-level diagram and the corresponding molecular orbitals of porphyrins **YD0–YD8** calculated at the B3LYP/6-31G(d) level of theory.

potential for I^-/I_3^- (+0.40 V vs. NHE), the energy levels for **YD1–YD8** all fulfill the requirement for effective electron injection and dye regeneration in a DSSC system. The quantum-chemical calculations on some porphyrins show a reduced HOMO-LUMO gap upon incorporation of an electron-donating group to the porphyrin ring (Fig. 15.9), which is in accordance with the tendency on the change of the HOMO-LUMO gap calculated from electrochemical and absorption data as shown in Fig. 15.8.

In the electronic absorption for a porphyrin, both Soret and Q bands arise from π–π^* transitions, which can be explained by considering the Gouterman's four-orbital model: two π orbitals (a_{1u} and a_{2u}) and two degenerate π^* orbitals (e_{gx} and e_{gy}) [100]. In compounds **YD1–YD8**, there is considerable electronic coupling between the electron-donating group and the porphyrin core, thus decreasing the HOMO-LUMO energy gaps relative to **YD0**. This effect is consistent with the red shift in absorption and emission upon introduction of an electron-donating group onto the porphyrin ring. The separation between the a_{1u} and a_{2u} orbitals increased upon introduction of a donor group whereas the energy splitting of the e_{gx}/e_{gy} pair was only slightly perturbed.

As mentioned, the electronic density of **YD1** is significantly distributed into the π-system of the porphyrin ring and the diphenylamino moiety at the HOMO and HOMO-1. Introduction of a more strongly electron-donating group would increase the electronic density on the diarylamino moiety. The HOMO of **YD3** shows an increased electron distribution located on the electron-donating diarylamine relative to **YD2**. Similar to the LUMO of **YD2**, the π-conjugation is extended to only the porphyrin ring and the phenylene linker at the LUMO of **YD3**. Comparison of **YD1**, **YD3,** and **YD5** with **YD0** shows that increasing the electron-donating ability of the amino substituents results in a decreased HOMO-LUMO gap, which is ascribed to the significantly elevated HOMO as the LUMO is moderately altered. The HOMO electronic densities of **YD5** and **YD6** are significantly distributed on the triamine and triarylamine moieties, respectively; the LUMO shows an electronic distribution on both the porphyrin and phenylene units. In the case of **YD7**, for which the triarylamine is attached at the *meso*-position through an acetylene link, the distribution of electronic density of the molecular orbitals is similar to that of **YD6**.

Porphyrins **YD0–YD8** were sensitized onto TiO_2 nanoparticulate films to serve as working electrodes of DSSC devices for photovoltaic characterization. Figure 15.10(a) shows the transmittance absorption spectra of the sensitized thin-film samples in three parts; Figures 15.10(b) and 15.10(c) show the corresponding action spectra and current–voltage curves of the devices, respectively. The thin-film samples show much broader absorption (Fig. 15.10(a)) than their solution counterparts because of strong intermolecular interactions of the molecules aggregated on TiO_2 surface. The large absorbances of the thin-film spectra indicate that the amounts of dye loading on TiO_2 films were sufficient for all porphyrin sensitizers. The IPCE action spectra (Fig. 15.10(b)) reflect the photoelectric conversion efficiency at each wavelength. Although there is a large gap between the Soret and the Q bands of the absorption spectra of porphyrins,

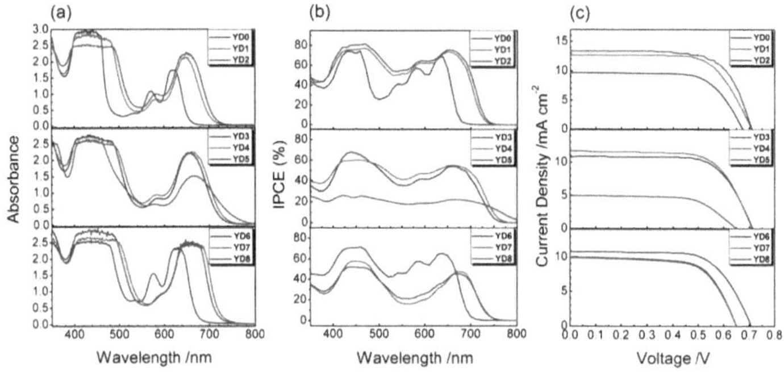

Figure 15.10. (a) Absorption spectra of films with no added scattering layer (film thickness ~10 μm), (b) incident photon-to-current conversion efficiency (IPCE) spectra of devices, and (c) current–voltage characteristics of devices fabricated from **YD0-YD8** sensitized on TiO_2 films (film thickness ~ (10 + 4) μm). See also Color Insert.

this feature is not evident in the IPCE spectra because the scattering by TiO_2 nanoparticles increases the photocurrents for the weak absorption in that region.

Table 15.4 summarizes the photovoltaic parameters for devices **YD0–YD8**. The overall efficiencies of power conversion of the devices exhibit a systematic trend for porphyrins with varied *meso*-substituents according to three classes. First, the diarylamino groups substituted directly on the *meso*-position, i.e., **YD1–YD4,** show a cell performance much better than for our reference cell,

Table 15.4. Photovoltaic parameters of porphyrin-based dye-sensitized solar cells under AM-1.5 illumination (power 100 mW cm^{-2}) and active area 0.16 cm^2

Dye	J_{SC}/mA cm^{-1}	V_{OC}/mV	FF	η (%)
YD	9.45	0.675	0.68	4.34
YD1	12.73	0.710	0.68	6.15
YD2	13.40	0.710	0.69	6.56
YD3	10.85	0.713	0.69	5.34
YD4	11.68	0.711	0.68	5.65
YD5	5.05	0.651	0.64	2.10
YD6	10.81	0.708	0.67	5.13
YD7	10.05	0.650	0.67	4.38
YD8	9.94	0.651	0.66	4.27

YD0. Second, the triarylamino-substituted porphyrins, **YD6–YD8**, exhibit cell performance comparable to that of **YD0**. Third, the triamine-substituted porphyrin, **YD5**, displays cell performance much poorer than that of **YD0**.

Because the π-conjugation of the diarylamino groups extends the spectral absorption towards long wavelength, the photocurrents of **YD1–YD4** are significantly greater than that of **YD0**; the long-chain hydrocarbons further extend the spectral region of light harvesting and make the performance of **YD2** slightly larger than **YD1** and that of **YD4** slightly greater than **YD3**. Although both **YD3** and **YD4** have alkoxyl groups on the diarylamino substituent, the additional electron-donating groups increased no further the device performance. The action spectra indicate that the efficiency values of **YD3** and **YD4** are less than those of **YD1** and **YD2**, which accounts for the J_{SC} values of the former being smaller than those of the latter. This decrease might reflect that the additional electron-donating groups raise the energy level of the HOMO (Figs. 15.8 and 15.9) so that dye regeneration of the porphyrin cations from the electrolyte becomes slower. Such an effect became more pronounced when a more strongly electron-donating group was present. For example, **YD5** has a triamine substituent, which is a stronger electron-donating group than a diarylamine, so that the HOMO level is the highest among all porphyrins under investigation. As a result, **YD5** shows a small J_{SC} value that is confirmed by its significantly smaller efficiency through the entire action spectrum shown in the middle panel of Fig. 15.10(b).

YD7 and **YD8** also have excellent π-conjugation between the porphyrin core and the phenyl group through the CC triple-bond bridge so that their thin-film absorption spectra are similar to those of **YD1–YD4**, but their device performances are poorer than those of **YD1–YD4**. **YD7** suffered from dye aggregation in ethanol solution [60]; additional *tert*-butyl groups in **YD8** failed to decrease the tendency of aggregation. The smaller I_{SC} and V_{OC} of **YD7** and **YD8** are thus expected to be due to the effect of dye aggregation so that fewer electrons were injected into the conduction band of TiO_2 after photoexcitation. In contrast, the structure of **YD6** has a triarylamino group directly attached at the *meso*-position of porphyrin core without a triple bond as a bridge; rotation about the bridged C–C single bond is thus feasible so that effective π-conjugation between the porphyrin core and the triarylamine substituent was unattainable. As a result, the thin-film absorption spectrum of **YD6** is similar to that of **YD0**, but the electron-donating ability of the triarylamino group increases the efficiency of the former relative to the latter to improve the cell performance of the former.

The electron-donating feature of amino substituents in **YD1–YD4** and **YD6** seems to be responsible for the V_{OC} value (~0.71 V) being larger than that (0.675 V) of the reference cell (**YD0**). Among evidence from other work, Durrant and coworkers found that the period for charge recombination of a triarylamino porphyrin/TiO_2 film is 20 times that of a free-base porphyrin counterpart (80 vs. 4 ms) [41]; Mozer et al. noted that the smaller V_{OC} of porphyrin-sensitized solar cells is due to the decreased electron lifetime related to either a more rapid recombination of electrons with dye cations or I_3^- ions [91]. We therefore expect that the observed larger V_{OC} in **YD1–YD4** and **YD6** than in **YD0** is due to a diminished recombination between I_3^- and conduction-band electrons, because I_3^- might be attached to the positively charged diarylamino moiety far from the TiO_2 surface for the former case. For **YD7** and **YD8**, although the amino group is farther from TiO_2 than in the cases of **YD1–YD4** and the electron distribution of the frontier orbitals is appropriate for their use in DSSC, the V_{OC} values of **YD7** and **YD8** (~0.65 V) are smaller than that of **YD1–YD4**, and even smaller than that of **YD0** because of the effect of aggregation.

We found that the tuning of the HOMO-LUMO energy gap, thus the optical and electrochemical properties, is achievable on varying the structure of the donor group. The electronic interaction between the donor and the porphyrin units in **YD1–YD8** significantly affects the spectral features of absorption and electrochemical properties. Our results reveal that direct attachment of an alkyl-substituted diarylamino group to the porphyrin ring result in significant improvement in solar-to-electrical conversion efficiency.

To elucidate the influences of the electron-donating group (EDG) on their optical, electrochemical and photovoltaic properties, porphyrins **YD14-YD17** (Fig. 15.11) with more than one EDG on various positions of the porphyrin ring were designed and synthesized [101].

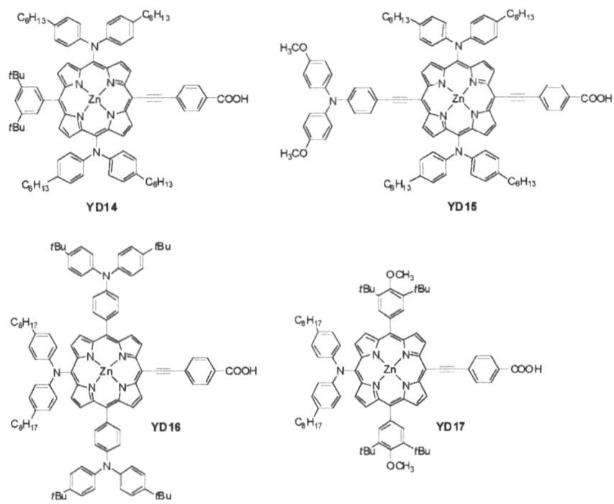

Figure 15.11. Molecular structures of porphyrin dyes with various electron-donating groups at *meso*-positions.

In sharp contrast to typical tetraphenylporphyrins, porphyrin **YD14** shows split Soret bands, and broadened and red-shifted Q bands, indicating strong electronic interaction between the porphyrin and diphenylamino units. Such an absorption property has been observed also in 5,15-bis(di-*p*-anisylamino)-10,20-diphenylporphyrin by Sakamoto et al [102]. The absorption band of **YD14** at ~490 nm fills the blue part of the absorption gap between the Soret and Q bands for a typical tetraphenylporphyrin. The light-harvesting efficiency is thus expected to improve in this region for the DSSC made with **YD14**. To enhance further the absorption in the visible region and even in the near IR region we expanded the π-conjugation system by introducing a triphenylamine at the *meso*-position via an ethynylene bridge (**YD15**). This condition causes an even more pronounced red shift and broadening of both Soret and Q bands relative to **YD14**. In compound **YD16**, two strongly electron-donating triphenylamino units are directly attached at the 5,15-*meso*-positions together with one diarylamino unit attached at the 10-*meso*-position. The absorption spectrum of **YD16** shows that both Soret and Q bands are slightly broadened and red-shifted relative to those of **YD2**, indicating that only mild electronic communication is involved between the triphenylamino moieties and the

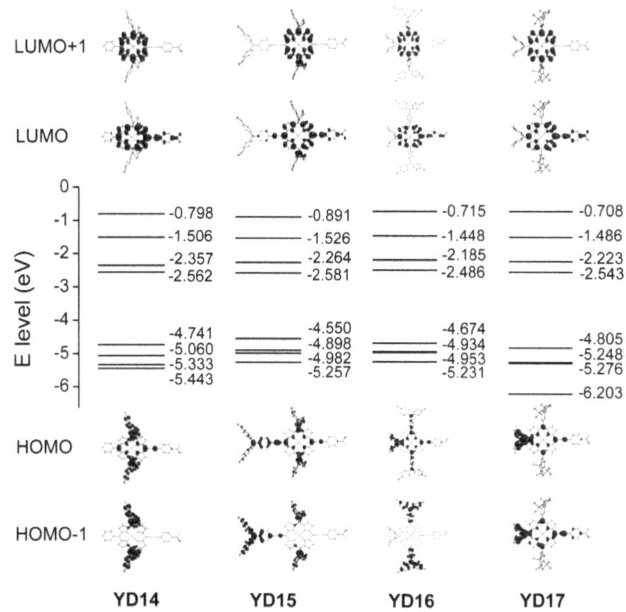

Figure 15.12. Energy-level diagram and the corresponding molecular orbitals of porphyrins **YD14–YD17** calculated at the B3LYP/6-31G(d) level of theory. bands.

porphyrin ring because of the steric interactions that cause the triphenylamino units to rotate from the porphyrin plane [103] Porphyrin **YD17** has additional methoxyl groups on the phenyl ring that would increase the electron-donating ability relative to **YD2**. The spectral features of **YD17** resemble those of **YD2** except that the Soret and Q bands exhibit a slightly bathochromic shift.

Figure 15.12 shows the energy levels and the corresponding molecular orbitals for these porphyrin dyes. Comparison of **YD14–YD17** with **YD2** shows that the HOMO-LUMO gap is decreased upon incorporation of additional electron-donating groups on the porphyrin ring because there is considerable electronic coupling between the electron-donating groups and the porphyrin core. This phenomenon is more pronounced for **YD15** because of the extended π-system. These results are consistent with the red-shift and broadening in the absorption bands.

As shown in Fig. 15.9, the electronic density of **YD2** is mainly located on the π-system of the porphyrin and the diarylamine at the HOMO and HOMO-1. The electronic distribution of the frontier orbitals for compound **YD17** resembles that for **YD2** because of the structural similarity. The electron densities of the HOMO of **YD14** and **YD16** are also greatly delocalized over the porphyrin and diphenylamine moieties despite the structural diversity of these two porphyrins. For **YD15** the electron density of the HOMO is delocalized across not only the porphyrin and diphenylamine units but also the triphenylamine via the conjugated ethynylene link. Similar to the LUMO of **YD2**, the π-conjugation is extended to only the porphyrin, ethynylphenylene link and carboxyl anchoring group at the LUMO of **YD14–YD17**.

Porphyrins **YD14–YD17** along with **YD2** and **YD0** were sensitized onto TiO_2 films (with active layer of thickness 10 μm and scattering layer 4 μm) to serve as working electrodes of a DSSC. With the TiO_2 films soaked in dye solutions (concentration $\sim 2 \times 10^{-4}$ M) for 6 h for each dye, the amounts of dye loading on TiO_2 films (Table 15.5), in order **YD2**>**YD17**>**YD0**>**YD14**>**YD15**>**YD16**, are sufficient for photovoltaic measurements. Figures 15.13a and 15.13b show the current–voltage characteristic curves and the corresponding IPCE action spectra of the devices, respectively; the derived photovoltaic parameters are summarized in Table 15.5. The device performance shows an order **YD2**>**YD17**>**YD14**>**YD16**>**YD0**>**YD15**. The top three dyes perform similarly with $\eta \sim$ 7%, whereas the performances of **YD15** and **YD16** are comparable to only that of **YD0** ($\eta \sim$ 5%). For **YD16**, the poor cell performance

Table 15.5. Photovoltaic parameters and dye loading of DSSC with TiO_2 films sensitized with **YD0**, **YD2** and **YD14-YD17** under simulated AM-1.5 illumination (power 100 mW cm^{-2}) and active area 0.16 cm^2

Dye	Dye-loading/nmol cm^{-2}	J_{SC}/mA cm^{-1}	V_{OC}/mV	FF	η (%)
YD	140	10.42	697	0.69	5.0
YD2	148	14.80	714	0.67	7.1
YD14	131	14.27	712	0.67	6.8
YD15	128	9.42	623	0.71	4.2
YD16	111	12.21	704	0.64	5.5
YD17	142	13.99	722	0.69	7.0

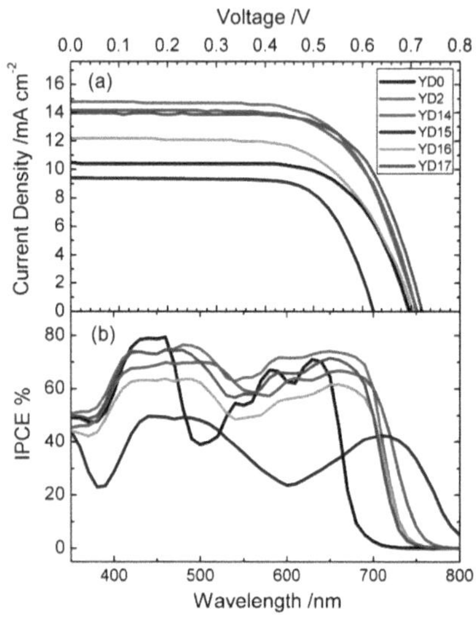

Figure 15.13. (a) Current–voltage characteristics of devices fabricated from **YD0**, **YD2**, and **YD14-YD17**; (b) incident photon-to-current conversion efficiency (IPCE) spectra of devices. See also Color Insert.

is consistent with the smaller amount of dye loading, which gives a J_{SC} value substantially less than for the top three dyes. However, the absorption spectra of all porphyrins adsorbed on TiO_2 films indicate that the light-harvesting efficiencies of all porphyrins are close to one for both major absorption bands. Therefore, electron injection and/or charge collection efficiencies might play a role for the observed cell performance. Apparently the two bulky triphenylamino units on the 5,15-*meso*-position hinder the dye adsorption on the TiO_2 surface. The poor performance of **YD15** reflects its smaller J_{SC} and V_{OC} values, which are understood to be due to the effect of aggregation. We encountered difficulty in dissolving **YD15** in ethanol; the cell performance reported for **YD15** was obtained from immersion of the TiO_2 films in a CH_2Cl_2 solution.

The IPCE action spectra (Fig. 15.13(b)) reflect the efficiency of photoelectric conversion at each wavelength. Although there is a large gap between the Soret and the Q bands of the absorption spectra of porphyrins (Fig. 15.12), this feature is not evident in the IPCE spectra because the effect of light scattering by TiO_2 nanoparticles increases the photocurrents for the weak absorption in that region. There still exists a small gap near 500 nm for **YD0**, but such a feature becomes obscure for other push-pull porphyrins. The reported push-pull porphyrins all exhibit a relatively flat IPCE spectral feature covering the entire visible region. For **YD16** and **YD17**, the IPCE spectral features are similar to that of **YD2** but the corresponding efficiencies are **YD2**>**YD17**>**YD16**, consistent with the variation of J_{SC} showing the same order (J_{SC}/mA cm^{-2} = 14.80, 13.99, and 12.21 for **YD2**, **YD17** and **YD16**, respectively) because integrating the IPCE over the AM 1.5G solar spectrum would give a calculated J_{SC} similar to the collected value. In contrast **YD14** and **YD15** exhibit a broader IPCE spectral feature than that of **YD2**. For **YD14**, even though the IPCE values are smaller than those of **YD2** by 5–10%, the extended IPCE spectrum makes its J_{SC} value less than that of **YD2** by only 3.6%. For **YD15**, the IPCE spectrum extends beyond 800 nm. The problem of aggregation for this porphyrin makes the average IPCE value significantly less than those of the other porphyrins under investigation, which results in much smaller J_{SC} (9.42 mA cm^{-2}).

We conclude that, increasing the number of diarylamino and/or triarylamino groups at the *meso*-positions increases the light-harvesting efficiency of the porphyrin through the extension of π-conjugation. The electrochemical measurements show that the involvement of an EDG increases the stability of the oxidized porphyrin, which is an essential requirement for their application in a DSSC. The DSSC devices based on the **YD14** and **YD17** sensitizers have achieved the efficiencies 6.8% and 7.0% of power conversion, respectively, under AM-1.5 illumination (100 mW cm^{-2}). In view of the prospective light-harvesting ability of **YD15**, from the approach of molecular design it is desirable to eliminate the effect of dye aggregation on the TiO_2 surface so as to enhance its cell performance significantly.

With a similar approach, we also designed a series of push-pull porphyrin sensitizers with added electron-withdrawing or -donating groups [104]. As shown in Fig. 15.14, the new porphyrins (denoted as X-PE1) are based on the molecular structure of PE1 with additional 4-dimethylamino (**Me$_2$N-PE1**), 4-methoxy (**MeO-PE1**), 4-cyano (**CN-PE1**), or 4-nitro (**NO$_2$-PE1**) phenylethynyl

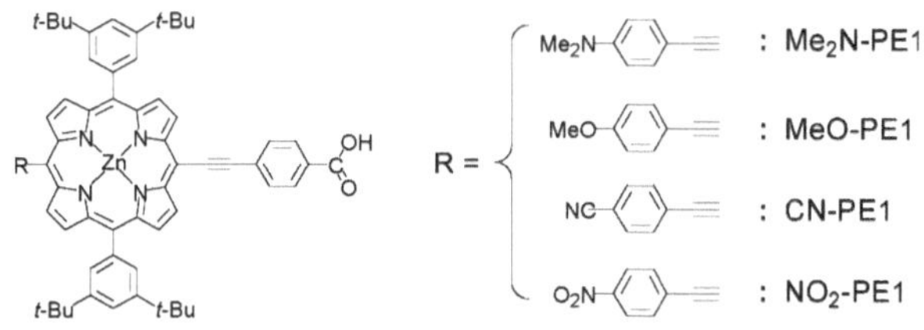

Figure 15.14. Molecular structures of PE1 and X-PE1 porphyrins.

groups in the *meso*-position opposite the anchoring group. For these porphyrins, zinc porphine acts as the primary chromophore. 3,5-di-*tert*-butyl-phenyl groups serve to increase the solubility, to lessen aggregation, and to protect the porphine core structure. 4-dimethylamino-phenyl, 4-methoxy-phenyl, 4-cyano-phenyl, and 4-nitro-phenyl groups are employed to represent substituents with varied electron-donating or withdrawing properties. Using electron-donating or withdrawing groups to tune the photophysical properties of a porphyrin is well documented by Therien and co-workers [105–107]. Recently, a theoretical screening by Grätzel and co-workers showed that preparing porphyrin sensitizers in a suitable donor-acceptor fashion might create a highly efficient porphyrin sensitizer [108].

As shown in Fig. 15.15, the MO patterns of X-PE1 porphyrin are consistent with those of Gouterman's four-orbital model [100], i.e., HOMO-1 and HOMO resemble those of the a_{1u} and a_{2u} orbitals whereas the LUMO and LUMO+1 are similar to those of the e_g orbitals. The patterns are also consistent with the suggestion that the first reduction and oxidation of X-PE1 are reactions centered on the porphyrin ring. However, the HOMO and LUMO patterns exhibit a feature with slight delocalization to the anchoring group and the electron-donating or withdrawing substituents. This minor deviation from Gouterman's four-orbital model is rationalized through a decreased symmetry of the complexes and an extended π-conjugation of the substituent. This condition is consistent with the potential shifts of X-PE1 relative to PE1 porphyrin. The HOMO pattern of **Me_2N-PE1** shows a considerable electron distribution at the Me_2N-substituent whereas its LUMO pattern shows an electron distribution at the porphine core and the anchoring group. Because

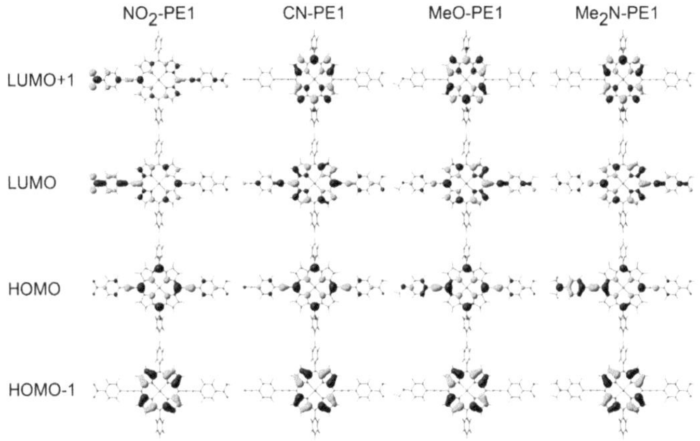

Figure 15.15. Molecular-orbital patterns of X-PE1 porphyrins with geometries of each molecule optimized at the B3LYP/LanL2DZ level of theory. These MO patterns serve to assist our understanding of the porphyrins qualitatively. See also Color Insert.

the anchoring group is attached to TiO_2, the above MO feature indicates that there exists a driving force for an efficient electron injection from the dye to the surface of TiO_2 upon excitation. In contrast, the LUMO pattern of **NO_2-PE1** exhibits an electron distribution localized mainly on the NO_2-substituent group, so that the photoexcitation provides a driving force in the opposite direction. The NO_2-substituent hence pulls the electron in competition with TiO_2 for electron injection from the excited state of the dye.

Figure 15.16 shows an energy-level diagram of the X-PE1 system, comparing the HOMO/LUMO of each porphyrin with the valence band (VB) and conducting band (CB) of TiO_2. This diagram is estimated based on our spectral and electrochemical data: the first reduction potentials were used to determine the LUMO levels and the UV-visible and the fluorescence spectra were used to determine the gaps between the HOMO and LUMO levels. This diagram shows that the LUMO levels of X-PE1 porphyrins become increasingly stabilized as the substituents are varied from the electron-donating Me_2N-phenylethynyl group to the electron-withdrawing NO_2-phenylethynyl group. Although the LUMO level of **NO_2-PE1** is the least in the series, it is still above the CB level of TiO_2. This

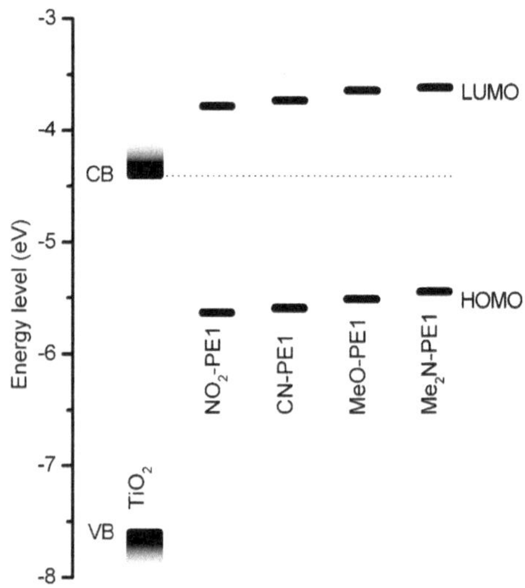

Figure 15.16. Energy-level diagram of X-PE1 porphyrins and TiO_2.

condition indicates that (i) LUMO levels of X-PE1 porphyrins should all be capable of injecting electrons to the CB of TiO_2, and (ii) electron injection from the dye to TiO_2 might be more favorable for **Me_2N-PE1** than for **NO_2-PE1**.

Figure 15.17 shows the IPCE spectra of the DSSC devices fabricated using these porphyrins. To demonstrate the reproducibility of the data, three independent measurements were performed for each dye using TiO_2 films fabricated with an identical procedure; the corresponding averaged photovoltaic parameters are listed in Table 15.6. The overall efficiencies of the devices display a systematic trend of increase from **NO_2-PE1** (1.09%), **CN-PE1** (4.05%), **MeO-PE1** (4.76%) to **Me_2N-PE1** (6.12%). In particular, **Me_2N-PE1** is the most efficient dye in the series and the corresponding device has reached 90% of the cell performance of a **N719**-sensitized solar cell (7.03%) fabricated under the same experimental conditions.

The variation of the overall efficiencies of power conversion of the system is consistent with their variation of short-circuit current density (Table 15.6). The IPCE spectra mimic the corresponding UV-visible spectra but with broader features. The two major responses of the IPCE spectra represent the energy conversion of the porphyrin

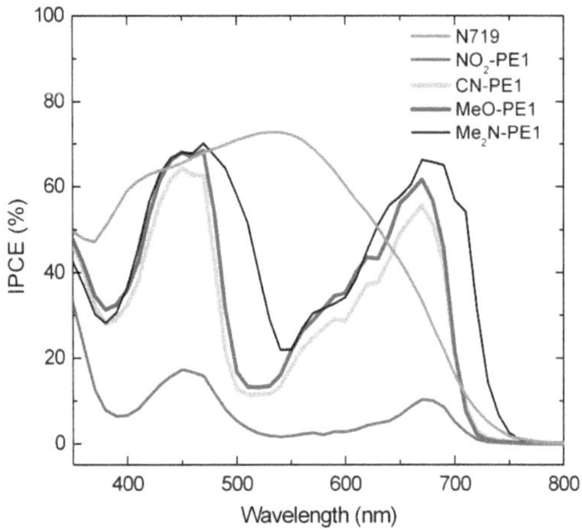

Figure 15.17. Photocurrent action spectra of X-PE1-sensitized solar cells. Corresponding overall efficiencies are **N719** 6.89%, **Me_2N-PE1** 5.67%, **MeO-PE1** 4.20%, **CN-PE1** 3.68%, and **NO_2-PE1** 0.84% in these figures (not a mean value). See also Color Insert.

B and Q bands with maximum efficiencies above 60% for the **Me_2N-PE1**, **MeO-PE1** and **CN-PE1** devices, but the efficiencies of the **NO_2-PE1** device are significantly smaller. The IPCE responses increase as the electron-donating abilities increase from NO_2-, CN-, MeO- to Me_2N-substituents. Both B and Q bands of the IPCE spectrum of the **Me_2N-PE1** device are significantly broader than those of the other porphyrin devices, especially in the regions 500–550 and 700–750 nm. As a result, the greater light-harvesting ability of **Me_2N-PE1** contributes the larger J_{SC} value of the SC (13.08 mA cm^{-2}). The greater efficiencies of **Me_2N-PE1** are also attributed to the greater

Table 15.6. Photovoltaic parameters of X-PE1-sensitized solar cells under AM-1.5 illumination (power 100 mW cm^{-2}) with active area 0.16 cm^2.

Dye[b]	J_{SC}/mA cm^{-2}	V_{OC}/V	FF	η (%)
Me_2N-PE1	13.08	0.68	0.68	6.12
MeO-PE1	10.13	0.56	0.69	4.76
CN-PE1	8.59	0.65	0.72	4.05
NO_2-PE1	2.53	0.57	0.74	1.09
N719[c]	13.48	0.75	0.70	7.03

efficiency of electron injection as predicted by the driving force to give an efficient electron injection inferred from the electron distribution of the frontier orbitals aforementioned. An electron-donating group such as Me_2N-substituent hence facilitates an electron injection from the dye to TiO_2 whereas an electron-withdrawing group such as NO_2-substituent obstructs that function to deteriorate the cell performance.

The molecular structure of **Me_2N-PE1** is similar to a porphyrin (**YD7**) discussed earlier [97], but the device performance of **Me_2N-PE1** (6.12%) is much better than that of **YD7** (4.38%). We note that separate immersing solvents were used for the two systems: the dye-soaking of **YD7** was made in ethanol whereas that of **Me_2N-PE1** was made in THF. Protic solvents might influence the adsorption and geometry of the porphyrins on a TiO_2 surface [109], and in some cases [97, 109, 110] ethanol or methanol is a superior immersing solvent whereas in other cases [61, 111, 112] THF is superior. As **YD7** does not dissolve well in ethanol, it suffered from the problem of dye aggregation when adsorbed on the TiO_2 surface [97]. Because dye aggregation is greatly suppressed in THF, the IPCE values of **Me_2N-PE1** are significantly greater than those of **YD7**. As for the dye adsorption time, we used two hours for dye uptake in order to generate the most satisfactory photovoltaic performance. We noticed incomplete dye-adsorption at a shorter immersing time and observed poorer overall efficiencies of the solar cells. On the other hand, prolonged immersing time did not yield higher overall efficiencies of the devices. This is consistent with the recent report by Imahori and co-workers [109].

In summary, we prepared four porphyrins bearing NO_2-, CN-, MeO-, or Me_2N-phenylethynyl substituents in order to examine the electron-donating/-withdrawing effects on the photovoltaic properties of the corresponding porphyrin-sensitized solar cells. **Me_2N-PE1** yields the most red-shifted and broadened UV-visible absorption and fluorescence bands in the series. Electrochemical tests show that **Me_2N-PE1** is the most easily oxidized whereas **NO_2-PE1** is the most easily reduced. The photovoltaic performances of X-PE1-sensitized solar cells exhibit a systematic trend: **Me_2N-PE1** > **MeO-PE1** > **CN-PE1** > **NO_2-PE1**. The superior cell performance of **Me_2N-PE1** is rationalized to be due to the broader and more red-shifted spectral feature and the more efficient electron-injection ability. Furthermore, the IPCE spectra of X-PE1 sensitizers exhibit a complementary spectral feature compared with those of the Ru-complex sensitizers. Co-sensitization of the two dyes can thus be employed in a tandem solar system to enhance further the overall performance.

15.4 Push-Pull Porphyrins with an π-Extended Linker

The overall solar energy conversion efficiency is proportional to the short-circuit current density, J_{SC}. The most viable way to enhance J_{SC} is to harvest a broader region of the solar spectrum. In general, porphyrins show a Soret band at 400–450 nm and Q bands at 500–650 nm. To extend the absorption of porphyrin dyes to the near IR region, the optical gap between HOMO and LUMO needs to be lowered. Our approach is to introduce a highly conjugated π-extended link between the porphyrin ring and the anchoring group. As shown in Fig. 15.18, the bridge between acetylene and carboxyl groups is varied from phenylene for **YD11** to nathylene for **YD12**, and to anthracenylene for **YD13** [113].

Figure 15.19 shows the absorption spectra of **YD11-YD13** in ethanol solution; the Soret and Q bands shift toward longer wavelengths as the π system in **B** is expanded, because of the electronic coupling between the aromatic substituent and the porphyrin ring. This coupling effect was exceptionally pronounced in **YD13**, for which both Soret and Q bands become significantly broad and red-shifted.

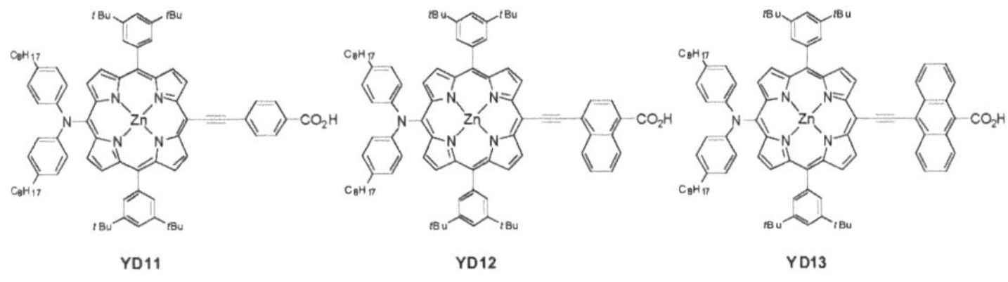

Figure 15.18. Molecular structures of porphyrin dyes **YD11-YD13**.

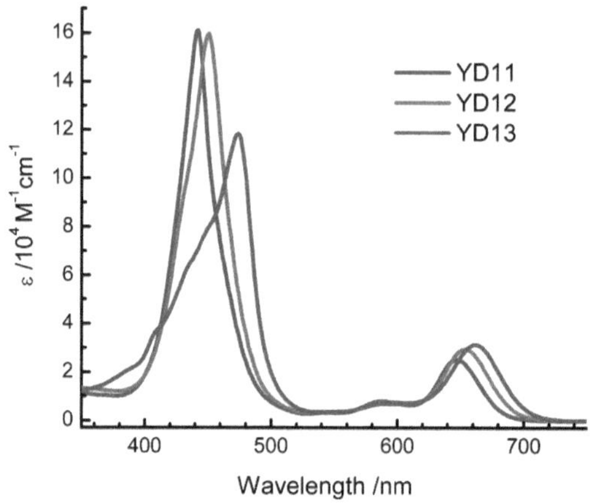

Figure 15.19. UV-visible absorption spectra of **YD11-YD13** in ethanol. See also Color Insert.

The photovoltaic measurements were carried out with at least three identical working electrodes for each porphyrin adsorbed on each TiO_2 film (A-C) at the same experimental conditions. The corresponding averaged photovoltaic parameters are summarized in Table 15.7. Figures 15.20(a)–(c) show one set of typical $J-V$ curves of the porphyrin-based DSSC devices for TiO_2 films A-C, respectively; for comparison, the cell performances of the devices made of **N719** dye with the same TiO_2 films as for **YD11-YD13** are shown as gray curves in each plot. Our results indicate that both **YD11** and **YD12** exhibit exceptionally superior performance relative to **N719** dye; the poor performance of **YD13** is remarkable and is discussed below. The short-circuit photocurrent densities (J_{SC}) of the two promising porphyrin-based devices are significantly greater than those of the **N719** devices, in particular for those of film B (Fig. 15.20(b)). Even though the open-circuit photovoltages (V_{OC}) and fill factors (FF) for the former are smaller than for the latter, the net effects of these variations make the overall efficiencies of power conversion of the **YD11** and **YD12** devices outperform those of the **N719** devices at $L \sim 10$ μm without a scattering layer (Fig. 15.20(b)). **YD11** ($\eta = 6.7\%$) and **YD12** ($\eta = 6.8\%$) are thus two remarkable green sensitizers for their outstanding cell performances relative to that of **N719** ($\eta = 6.1\%$) without an added scattering layer (film B) for

Table 15.7. Photovoltaic parameters of DSSC with photosensitizers **YD11–YD13** and **N719** as a function of TiO_2 film thickness (L) under simulated AM-1.5 illumination (power 100 mW cm^{-2}) and active area 0.16 cm

L/μm	Dye	Dye-loading /nmol cm^{-2}	J_{SC}/mA cm^{-1}	V_{OC}/mV	FF	η (%)
~ 5	**YD11**	75	10.54 ± 0.33	0.723 ± 0.013	0.73 ± 0.01	5.54 ± 0.11
(Film A)	**YD12**	82	10.75 ± 0.34	0.724 ± 0.009	0.72 ± 0.02	5.60 ± 0.09
	YD13	62	3.30 ± 0.16	0.633 ± 0.006	0.71 ± 0.01	1.49 ± 0.08
	N719	90	9.27 ± 0.13	0.794 ± 0.006	0.74 ± 0.01	5.47 ± 0.03
~ 10	**YD11**	154	12.99 ± 0.83	0.715 ± 0.006	0.71 ± 0.03	6.56 ± 0.09
(Film B)	**YD12**	160	13.77 ± 0.40	0.714 ± 0.005	0.68 ± 0.02	6.69 ± 0.09
	YD13	129	3.97 ± 0.10	0.618 ± 0.002	0.72 ± 0.01	1.76 ± 0.04
	N719	178	10.97 ± 0.42	0.769 ± 0.002	0.73 ± 0.02	6.16 ± 0.14
−(10 + 4)	**YD11**	a	14.01 ± 0.14	0.716 ± 0.003	0.68 ± 0.01	6.79 ± 0.12
(Film C)	**YD12**	a	14.23 ± 0.82	0.717 ± 0.008	0.68 ± 0.03	6.91 ± 0.15
	YD13	a	4.12 ± 0.08	0.630 ± 0.002	0.72 ± 0.01	1.86 ± 0.04
	N719	a	13.08 ± 0.31	0.786 ± 0.007	0.71 ± 0.01	7.27 ± 0.14

[a]The dye-loading amounts are similar to those of the corresponding films of the same thickness without a scattering layer (film B).

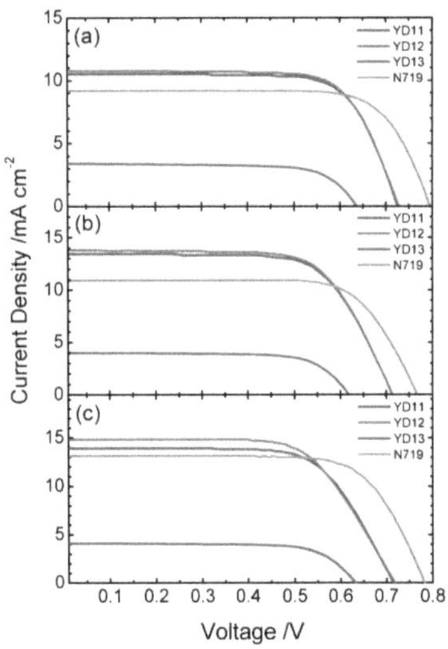

Figure 15.20. Current–voltage characteristics of devices fabricated from **YD11-YD13**, under illumination of simulated AM1.5 full sunlight (100 mW cm^{-2}) with an active area 0.16 cm^2 of three film thicknesses: (a) ~5 μm (film A); (b) ~10 μm (film B); (c) ~(10 + 4) μm (film C). See also Color Insert.

light-penetrable DSSC applications. When the ~10 μm TiO_2 films were covered with a scattering layer (film C), we found that the cell performance of **N719** was significantly improved to $\eta = 7.3\%$ whereas the performances of the porphyrin dyes increased only slightly ($\eta = 6.8\%$ and 7.0% for **YD11** and **YD12**, respectively, Fig. 15.20c). Our results indicate that a substantial increase in J_{SC} for the **N719** device is a key factor for the improvement of the cell performance with the addition of a scattering layer.

To understand why the scattering layer was insensitive to the cell performances of the porphyrin-based devices, measurements of the incident photon-to-current conversion efficiency (IPCE) for each device were performed [114]. Figures 15.21(a)–(c) show the efficiency spectra of the same DSSC devices of which the corresponding J–V characteristics are shown in Figs. 15.20(a)–(c). Integrating the IPCE over the AM 1.5G solar spectrum gives a calculated J_{SC} similar to the collected value for all devices under investigation. There are three important points deduced from our IPCE results. First, the efficiency spectra of both **YD11** and **YD12** sensitizers are similar for all three TiO_2 films, but the spectra of **YD12** have a Q-band

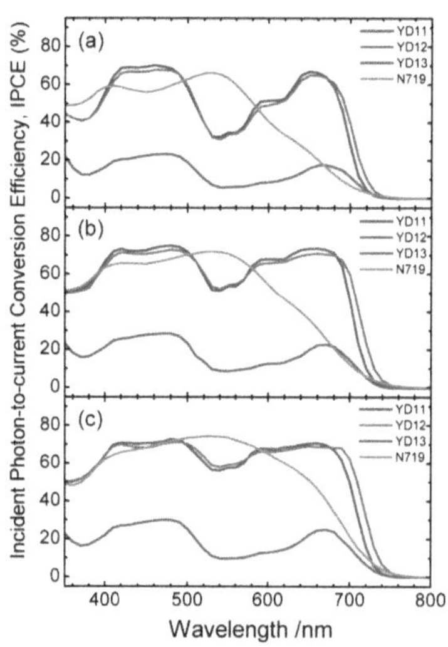

Figure 15.21. Corresponding IPCE action spectra of the same DSSC devices as those shown in Fig. 15.20. See also Color Insert.

shoulder slightly extended to longer wavelengths, which increases slightly J_{SC} for **YD12** relative to **YD11**. This effect is consistent with the absorption spectral feature shown in Fig. 15.19. Second, the efficiency spectra of **YD13** show smaller values than those of **YD11** and **YD12**, which explains its poor performance. This effect is inconsistent with the absorption feature shown in Fig. 15.19 and is discussed later. Third, the efficiency spectra of **YD11** and **YD12** involve a large gap between the Soret and the Q bands in film A (Fig. 15.21(a)), but this gap became smaller when thicker TiO_2 films (B) were applied (Fig. 15.21(b)). On addition of another scattering layer (film C), the gaps in **YD11** and **YD12** became even smaller so that the efficiency spectra display a nearly flat nature in the entire visible region, 400–700 nm (Fig. 15.21(c)). The shoulders of the efficiency spectra on the red side extended no further beyond the edge of the Q band in the presence of a scattering layer; for this reason only a slight improvement in J_{SC} was found for the porphyrin-based DSSC with a scattering layer.

In contrast, a significant improvement in cell performance was found for a **N719**-based DSSC with a scattering layer, because of the effective scattering effect in the red shoulder of the efficiency spectrum. Based on the above observations, we conclude that the involvement of the partially allowed triplet MLCT states of ruthenium complexes [115] is responsible for the enhanced efficiency in the red shoulder of the IPCE spectrum of **N719**, whereas the effect of spin-orbit coupling in zinc porphyrins was not great enough for the $S_0 \rightarrow T_1$ transitions to occur so that the additional scattering layer provides no improvement in the efficiency spectra of **YD11-YD13** beyond the Q-band absorptions.

Femtosecond excitation of the thin-film samples was performed at 430 nm using a fluorescence up-conversion system [116]. The emissions at the maximum intensity were optically gated with the fundamental pulse (860 nm) to yield the emission decays of the **YD11-YD13**/TiO_2 films shown in Figs. 15.22(a)–(c); those of the **YD11-YD13**/Al_2O_3 films are also shown for comparison. The temporal profiles of all samples show multi-exponential decay, and the corresponding time coefficients were obtained on analyzing the data with a parallel kinetic model [113, 116]. With the time coefficients weighted by their relative amplitudes (shown in parentheses), the

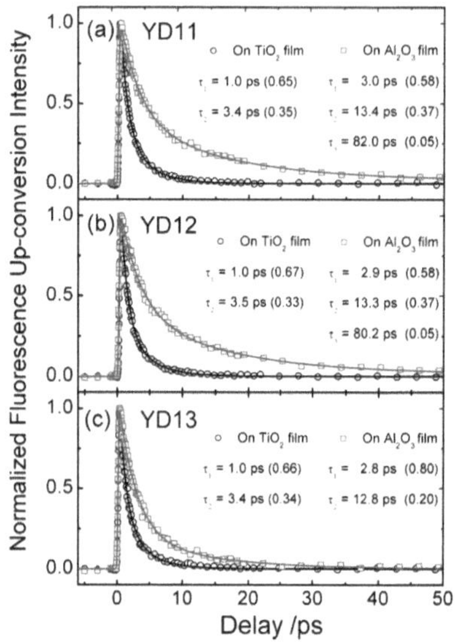

Figure 15.22. Femtosecond emission decay curves of the thin-film samples excited at the Soret band ($\lambda_{ex} = 430$ nm) and probed at the wavelength of maximum emission intensity. (a)–(c) represent the transients of **YD11-YD13** on TiO_2 (circles) and Al_2O_3 (squares) films. Solid curves represent theoretical fits with the corresponding time coefficients and relative amplitudes (in parenthesis) as indicated. See also Color Insert.

average time coefficients of the TiO_2 films are determined to be all similar (τ_{TiO_2} ~1.8 ps) for **YD11-YD13**; those of the Al_2O_3 films are determined to be $\tau_{Al_2O_3} = 10.8$, 10.6, and 4.8 ps, respectively. The emission decays of the Al_2O_3 films reflect only the intermolecular energy transfer due to aggregation of the dye on the Al_2O_3 surface, but the emission decays of the TiO_2 films contain not only the aggregate-induced energy transfer but also rapid electron injection from the excited state of a porphyrin into the conduction band of TiO_2. If we assume that the extent of dye aggregation on both TiO_2 and Al_2O_3 films is similar, based on the same amount of dye molecules adsorbed on the films, the quantum yields of **YD11**, **YD12** and **YD13** for electron injection on a TiO_2 surface become evaluated to be $\Phi_{inj} = 0.83$, 0.83, and 0.62, respectively [113].

In Fig. 15.22, the fluorescence decays are similar for all three porphyrins sensitized on TiO_2 films, but the fluorescence decay of

YD13 was much more rapid than that of **YD11** or **YD12** sensitized on Al_2O_3 films. Hence the presence of the anthracene group in the bridge from **YD13** to TiO_2 did not hamper the rate of interfacial electron transfer for the observed small injection yield. We consider two other possibilities responsible for the small injection yield of **YD13**: one is the anthracene-induced rapid intramolecular relaxation due to effective vibronic coupling, and the other is the anthracene-induced rapid relaxation of intermolecular energy due to aggregation. To examine the first possibility, we measured time-correlated single-photon counting (TCSPC) to determine the lifetimes of excited state of **YD11-YD13** in dilute solutions. All three porphyrins in ethanol (2×10^{-5} M) have similar lifetimes in a nanosecond range (1.3–1.4 ns). Ultrarapid non-radiative relaxation through intramolecular channel is thus excluded. For the other possibility, we performed photovoltaic measurements for **YD11-YD13** co-adsorbed with chenodeoxycholic acid (CDCA) on TiO_2 (film B) in a ratio [porphyrin]:[CDCA] = 1:2. The presence of CDCA slightly decreases J_{SC} for **YD11** and **YD12** through a slight reduction of the amount of dye loading, but CDCA plays a role to improve dye aggregation to some extent so as to significantly enhance the J_{SC} for **YD13**. As a result, the efficiency of **YD13** greatly increased from 1.8% to 2.9% in the presence of CDCA. Anthracene thus induces much more rapid intermolecular energy transfer due to dye aggregation, leading to a cell performance for **YD13** poorer than for **YD11** and **YD12**.

The electron injection yield is an important factor to be considered to improve further the cell performance of the devices for organic dyes with a tendency to aggregate. For porphyrin sensitizers such as **YD11** and **YD12**, even though several bulky *tert*-butyl groups and hydrophobic long alkyl chains were incorporated around the porphyrin macrocycle, the effect of aggregation cannot be completely eliminated.

The same idea of designing **YD11-YD13** was further extended based on the molecular structure of **PE2** [111]. As shown in Fig. 15.23, the new porphyrins (denoted as **LAC-1** to **LAC-5**) are essentially **PE1** added one unit of acenyl-ethynyl group. For these porphyrins, zinc porphine acts as the primary chromophore. 3,5-di-*tert*-butyl-phenyl groups are used to increase the solubility, to lessen aggregation, and to protect porphine core structure.

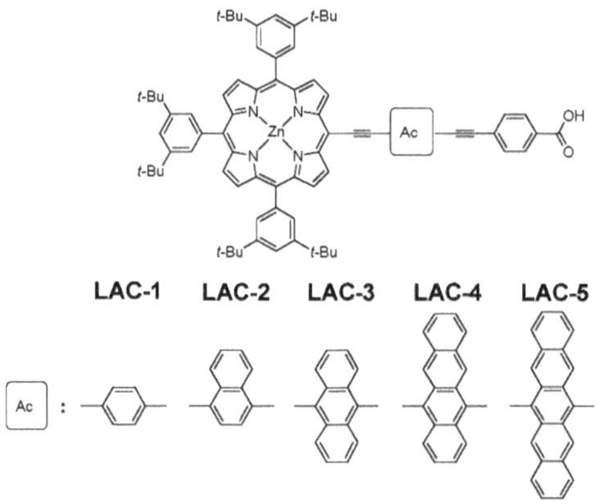

Figure 15.23. Structural diagram of LAC porphyrins.

Acenes (benzene to pentacene) are put between porphine and carboxylic anchoring group as the secondary chromophores in order to broaden porphyrin absorption bands. Scattered examples in the literature have shown that porphyrin absorption bands can be significantly affected by attaching phenyl-ethyne [117, 118], naphthylenyl-ethyne [117], anthracenyl-ethyne [118], and pentacenyl-ethyne [119] to the porphyrin *meso*-positions. More importantly, the use of benzene to pentacene allows us to adjust absorption wavelengths of the secondary chromophore to observe the impacts of the acenes to porphyrin absorption spectra, i.e., a titration of porphyrin absorption bands by acenes.

Figure 15.24 compares the UV-visible spectra of **PE1**, acenyl-ethynyl precursors (denoted as ⟨1⟩ to ⟨5⟩) and LAC porphyrins. For the acenyl-ethynyl precursors, the lowest-energy absorptions gradually red-shift from 293 to 639 nm as the conjugation system expands. This is consistent with the report by Anthony and co-workers [120]. Note that the wavelengths of ⟨3⟩ and ⟨5⟩ match those of **PE1**'s B and Q bands, respectively, whereas the wavelengths of ⟨4⟩ locate between **PE1**'s B and Q bands. For LAC porphyrins, absorption bands are affected in each different fashion upon incorporating acenyl-ethynes into the π-conjugation system. These spectrum changes are consistent with the literature reports [117–119]. For **LAC-1**, the absorption bands are slightly red-shifted

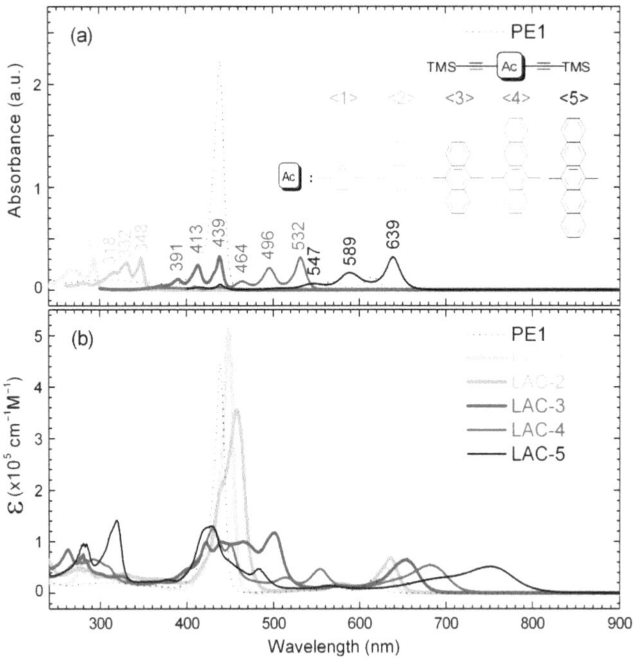

Figure 15.24. UV-visible spectra of **PE1**, acene-ethynyl precursors and LAC porphyrins in THF. See also Color Insert.

and intensified from those of **PE1**, indicating the effect of the additional phenyl-ethynyl conjugation. For **LAC-2**, further red-shifts of the absorption bands and a splitting pattern of the B bands are observed. These spectrum changes are consistent with the expanded π-conjugation and the lowered molecular symmetry of the complex. For **LAC-3**, additional red-shifts of the absorption bands are observed for both B and Q bands. Remarkably, the split B bands spread out widely in the 400 to 500 nm region. For **LAC-4**, the B bands are not as red-shifted but the Q bands are red-shifted to 682 nm. In addition, the tetracene group gives rise to absorption bands at 513 and 554 nm, complementing the porphyrin B and Q bands. For **LAC-5**, the B bands are at 429 and 482 nm and the extremely broadened Q bands locate in the 500–800 nm region. The lowest-energy absorption band of **LAC-5** might be an intramolecular charge transfer band because of the significant red-shift from the

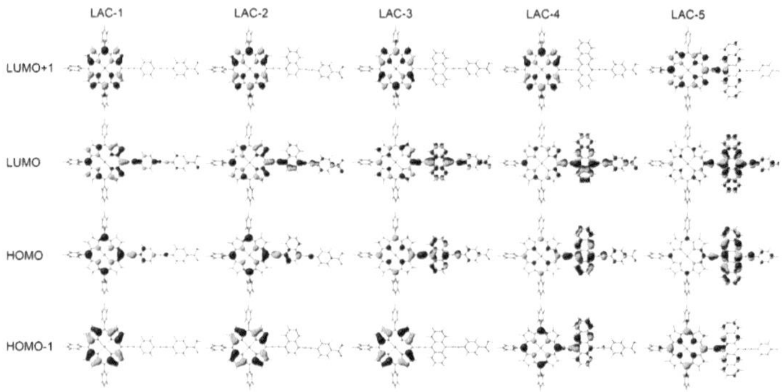

Figure 15.25. MO patterns of LAC porphyrins with geometries optimized at the B3LYP/LanL2DZ level of theory. Note that these MO patterns are used to qualitatively assist our understanding to the porphyrins. See also Color Insert.

component bands. It is noteworthy that **LAC-3**'s B bands and **LAC-5**'s Q bands are enormously broadened while the absorption wavelengths of ⟨3⟩ and ⟨5⟩ match those of **PE1**'s B and Q bands, respectively. This implies that, for a fully conjugated system, matching the absorption wavelengths of the primary and secondary chromophores may result in significantly broadened absorption bands of the total system.

Figure 15.25 depicts molecular orbital (MO) patterns of **LAC-1** to **LAC-5** at HOMO-1 to LUMO+1 levels. As shown in the figure, the MO patterns of **LAC-1** are consistent with those of Gouterman's four-orbital model discussed earlier [100]. Minor deviations are observed: for **LAC-1**, the HOMO and LUMO patterns are calculated to be mainly at the porphine core with slight delocalization to the anchoring group, suggesting the decreased symmetry of the complex and the extended π-conjugation of the substituent. For **LAC-2**, the MO patterns are similar to those of **LAC-1** with very small MO patterns residing at the second ring of naphthalene. This is consistent with the small positive-shift of **LAC-2** reduction potential. For **LAC-3**, the delocalization is calculated to be much more extensive at the HOMO and LUMO levels. As shown in the figure, the HOMO and LUMO patterns populate at both the porphine core and the anthracenyl-ethynyl anchoring group. The more delocalized LUMO pattern of **LAC-3** is consistent with the further positive-shift of the

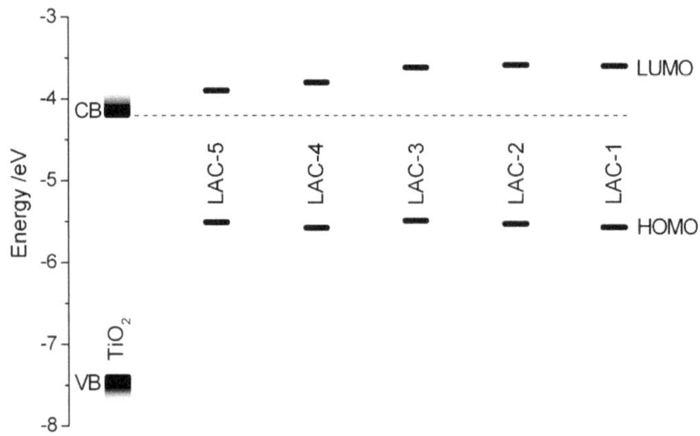

Figure 15.26. Energy-level diagram of LAC porphyrins and TiO_2.

reduction potentials. For **LAC-4** to **LAC-5**, the MO patterns at the HOMO and LUMO levels become increasingly localized at the tetracenyl and pentacenyl groups, respectively.

Figure 15.26 shows the energy-level diagram of LAC porphyrins and TiO_2, comparing the HOMO/LUMO of each porphyrin with the valence band (VB) and conducting band (CB) of TiO_2. This diagram shows that the LUMO levels of LAC porphyrins become more and more stabilized as the sizes of the acenes increase. Although the LUMO level of LAC-5 is the lowest in the series, it is still above the conducting bands of TiO_2. This suggests that the LUMO levels of LAC porphyrins should all be capable of injecting electrons to the conducting bands of TiO_2.

For the photovoltaic measurements, averaged photovoltaic parameters of LAC-sensitized solar cells are listed in Table 15.8. Figure 15.27 compares photo-current action spectra of LAC-sensitized solar cells. In short, the overall efficiencies increase from **LAC-1** (2.95%), **LAC-2** (3.31%) to **LAC-3** (5.44%); however, overall efficiencies rapidly decrease to 2.82% of **LAC-4** and 0.10% of **LAC-5**. Overall efficiency of **LAC-3**-sensitized solar cell is nearly twice of that of **LAC-1**-sensitized solar cell (2.95%) and is about 81% overall efficiency of **N719**-sensitized solar cells under the same experimental conditions. As shown in Fig. 15.27, the photo-current action spectra are consistent with the UV-visible spectra. From **LAC-1** to

Table 15.8. Photovoltaic parameters of LAC-sensitized solar cells under AM-1.5 illumination (power 100 mW cm^{-2}) with an active area of 0.16 cm^2

Dye	J_{SC}/mA cm^{-2}	V_{OC}/V	FF	η (%)
LAC-1[a]	6.13 ± 0.04	0.67 ± 0.02	0.72 ± 0.01	2.95 ± 0.02
LAC-2	7.27 ± 0.41	0.65 ± 0.01	0.70 ± 0.02	3.31 ± 0.06
LAC-3	12.67 ± 0.46	0.67 ± 0.01	0.64 ± 0.03	5.44 ± 0.06
LAC-4	6.68 ± 0.05	0.61 ± 0.01	0.68 ± 0.01	2.82 ± 0.14
LAC-5	0.33 ± 0.06	0.49 ± 0.02	0.62 ± 0.05	0.10 ± 0.02
N719[b]	12.67 ± 0.91	0.74 ± 0.04	0.72 ± 0.02	6.73 ± 0.10

Note: Three independent measurements were performed for all dyes using the TiO_2 films fabricated with an identical procedure in order to demonstrate the reproducibility of the data.
[a]Typical dye-loads of LAC-sensitized solar cells are estimated to be (in nmol/cm^2): 120 (**LAC-1**), 135 (**LAC-2**), 132 (**LAC-3**), 140 (**LAC-4**), and 140 (**LAC-5**).
[b]As a reference, overall efficiency of **N719** sensitized solar cell was also determined. The value is smaller than the literature value because of thinner TiO_2 films and without adding the scattering layer in our case.

LAC-3, the responses in the B- and Q-band regions gradually broaden, intensify and red-shift, contributing to the higher photo-currents and overall efficiencies. Note that the gap between B- and Q-band regions of **LAC-3**-sensitized solar cell is minimized possibly owing to the broadened absorption bands. This is consistent with the broadened absorption bands of **LAC-3**/TiO_2 films. For **LAC-4**, the presence of tetracene group complements porphyrin B and Q bands, and the combined absorption bands generate photo-currents beyond 800 nm. However, covering more wavelengths does not compensate the general drop of the IPCE and J_{SC} values. For **LAC-5**, the device also generates photo-current beyond 800 nm. However, the IPCE values and J_{SC} are negligible in the visible region.

In summary, we systematically demonstrate that porphyrin absorption spectra can be significantly affected by including acenes into the π-conjugation system. Among the acene-modified porphyrins, anthracene and pentacene groups substantially broaden porphyrin B (**LAC-3**) and Q (**LAC-5**) bands, respectively. On the other hand, tetracene group complements porphyrin absorption bands (**LAC-4**). For the performance of the DSSC, the overall efficiencies first increase then decrease for LAC porphyrins: **LAC-1** $<$ **LAC-2** $<$ **LAC-3** $>$ **LAC-4** $\gg$ **LAC-5**. **LAC-3** outperforms other porphyrins in the series and overall efficiency of the DSSC is about 81%

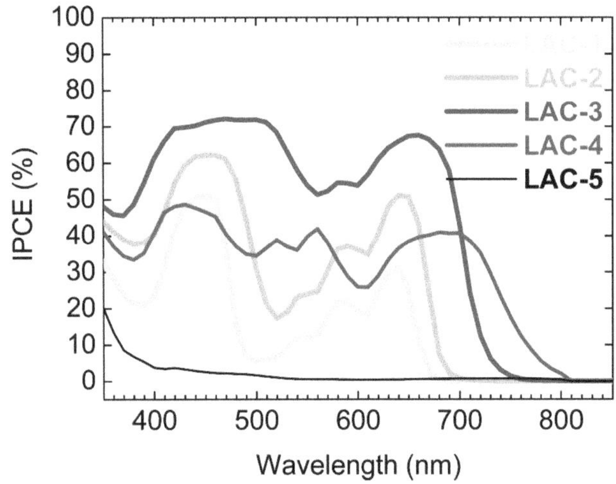

Figure 15.27. Photo-current action spectra of LAC-sensitized solar cells. Corresponding overall efficiencies are: 2.43% (**LAC-1**), 3.38% (**LAC-2**), 5.49% (**LAC-3**), 2.67% (**LAC-4**), and 0.12% (**LAC-5**) in these figures (not an averaged value). See also Color Insert.

overall efficiency of N719-sensitized solar cells under the same experimental conditions. For **LAC-3**, IPCE and TiO_2 film spectrum show that broadening the absorption bands by anthracene effectively minimizes the gap between porphyrin B- and Q-band regions, contributing to the higher photo-current and improved photovoltaic performance. Although **LAC-5** also exhibits very broad absorption bands, the DSSC performs poorly. The very poor performance of DSSC using **LAC-5** is suggested to be caused by (1) aggregation of the dye and/or (2) non-radiative relaxation of the molecule in the singlet excited state. As a result, electron injection from **LAC-5** to TiO_2 may not efficiently take place, rendering **LAC-5** a poor photo-sensitizer.

15.5 Dimeric Porphyrin Dyes

To generate a large photocurrent response, organic dyes in an efficient DSSC must have broad and intense absorption in the visible and near IR regions [121–123]. Porphyrin sensitizers are dominant candidates for this purpose because of their intense absorption in

Soret and Q bands to harvest solar energy efficiently in a broad spectral region [56], but the existence of a gap between the Soret and Q bands in monomeric porphyrins limits their cell performances.

Officer and coworkers reported the photovoltaic properties of porphyrin dimers, comprising two monoporphyrin units connected in either a linear anti or a 90° syn fashion [124]. The dimeric porphyrin dyes exhibited significantly improved light-harvesting efficiency compared to the corresponding monomeric porphyrin dyes when the devices were fabricated on a thin TiO_2 film. The improved performance for the dimer is due to the higher absorption coefficients. They found that no major difference in the device performance has been observed between the linear and angled dimers, suggesting both of the arrangements can be used to build larger 3D light harvesting porphyrin arrays for efficient DSSC.

Porphyrin arrays linked with conjugated acetylene bridges exhibit strong electronic coupling between porphyrin rings, resulting in splitting of the Soret band and broadening of the Q bands [125, 126]. Electronic absorption spectra of *meso-meso*-linked porphyrin arrays and their corresponding doubly and triply fused porphyrin arrays also show wide absorption covering the visible and near IR region [127]. By dint of such spectral features, these porphyrin arrays are prospectively efficient sensitizers for application in DSSC. Osuka and coworkers developed a series of doubly β-functionalized directly linked porphyrin dimers for applications in DSSC [128]. The devices based on the porphyrin dimers showed broad IPCE action spectra in the Soret and Q-band regions due to the broadened absorption induced by exciton coupling. The maximum power conversion efficiency has reached 4.2% in a dimer-based DSSC, which is much higher than the corresponding monomers.

Recently, we reported the synthesis and the spectral, electrochemical and photovoltaic properties of four porphyrin dimers; their molecular structures show diverse connectivity between the two porphyrin macrocycles, as displayed in Fig. 15.28 [129].

Absorption spectra of these porphyrin dyes are shown in Fig. 15.29. All porphyrin dimers exhibit much broader absorption than that of reference compound **YD0**. Compound **YDD0** shows split

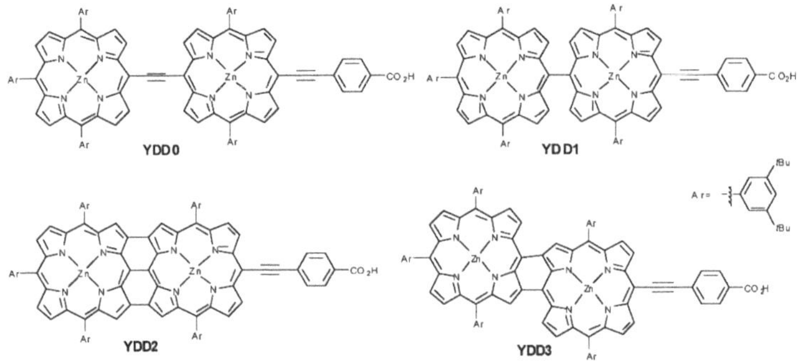

Figure 15.28. Molecular structures of porphyrin dyes **YDD0-YDD3**.

Soret bands in the range 400–500 nm, and red shifts and broadening of the Q bands due to interporphyrin electronic coupling. Dimer **YDD1** also exhibits a split Soret band ascribed to excitonic coupling. The absorption spectra of **YDD2** and **YDD3** exhibit a typical feature for fused porphyrin dimers with three major bands. Bands I and II of **YDD2** and **YDD3** feature a range across almost the entire visible region. Bands III appears at 756 and 845 nm for **YDD3** whereas

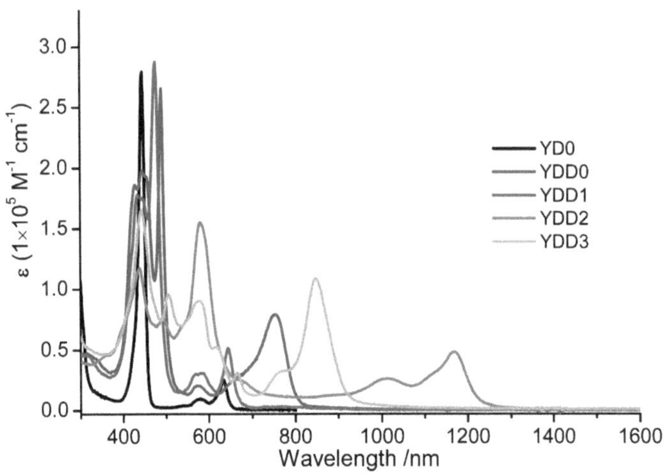

Figure 15.29. UV-visible absorption spectra of **YD0**, and **YDD0-YDD3** in ethanol. See also Color Insert.

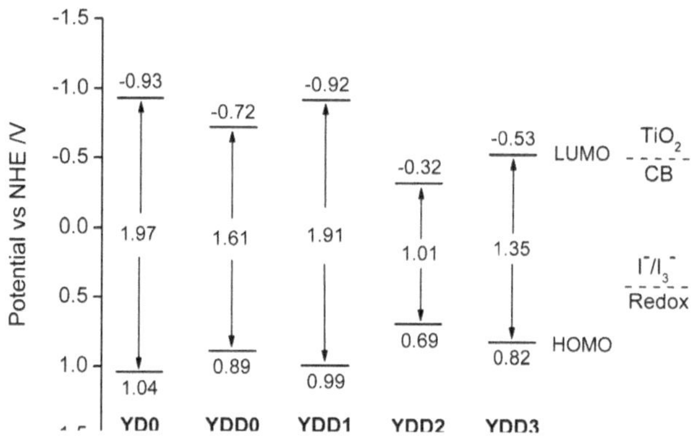

Figure 15.30. Schematic energy levels of porphyrins **YD0** and **YDD0–YDD3** based on absorption and electrochemical data.

those for **YDD2** are much wider and range from 900 to 1300 nm. The molar absorption coefficients at the maximum absorption wavelengths for these dimers are all large and fall in the range 0.3–2.9 10^5 dm^3 mol^{-1} cm^{-1}. Fluorescence spectra of **YDD2** and **YDD3** were unobtainable because of the sensitivity limit of our detector for this region, but they are expected to fall in the near IR region [130].

Figure 15.30 shows energy levels of **YDD0-YDD3**, with **YD0** for comparison. The HOMO levels of these porphyrin dimers are all more positive than the oxidation potential for I^-/I_3^-, indicating that dye regeneration might be feasible for all sensitizers; the LUMO energy levels of only **YDD0** and **YDD1** are more negative than the conduction-band (CB) edge of TiO_2, whereas those of **YDD2** and **YDD3** are not negative enough for effective injection of the excited-state electrons into the CB of TiO_2

Figure 15.31 shows the current–voltage characteristics and efficiencies of conversion of incident photons to current (IPCE) of porphyrin-based DSSC; the corresponding photovoltaic parameters are summarized in Table 15.9. Both porphyrin dimers **YDD0** and **YDD1** perform similarly to the reference compound (**YD0**), but the fused porphyrins (**YDD2** and **YDD3**) exhibit poor cell performance. In particular, the photocurrents of **YDD2** and **YDD3** are small, consistent with the potential feature shown in Fig. 15.29. For

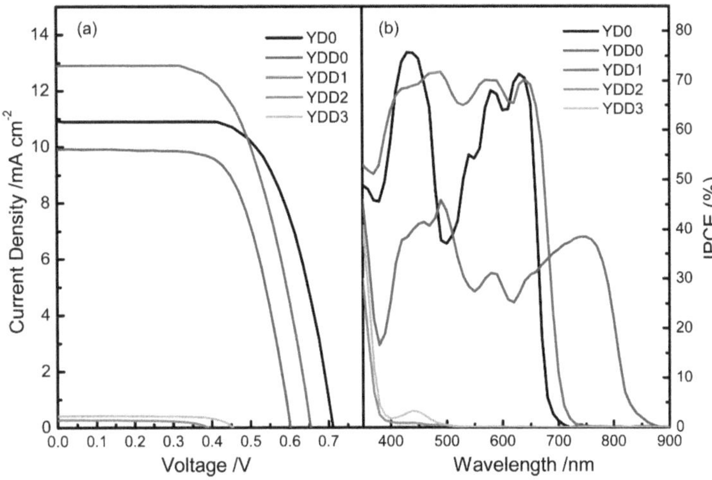

Figure 15.31. The current–voltage characteristics and efficiencies of conversion of incident photons to current (IPCE) of porphyrin-based DSSC. See also Color Insert.

YDD2, essentially no injected electrons were observed (Fig. 15.31b), because the energy level of LUMO is substantially lower than the CB edge of TiO_2. For **YDD3**, a small response in the IPCE action spectrum corresponds to the contribution of broad Bands I and II of the fused porphyrin, but no injected electrons were observed for the broad Band III in region 700–900 nm. We infer that electron injection from the excited states of **YDD3** to TiO_2 competed with energy relaxation from higher excited states (Bands I/II) to the lowest excited state (Band III), and there was insufficient kinetic energy for the electrons to inject from Band III of **YDD3** to the CB of TiO_2.

Table 15.9. Photovoltaic parameters of porphyrin-based dye-sensitized solar cells under AM1.5 illumination (power 100 mW cm^{-2}) and active area 0.16 cm^2

Dye	J_{SC}/mA cm^{-2}	V_{OC}/mV	FF	η (%)
YD0	10.955	0.711	0.66	5.14
YDD0	9.936	0.602	0.68	4.07
YDD1	12.929	0.653	0.62	5.23
YDD2	0.265	0.389	0.64	0.07
YDD3	0.417	0.454	0.71	0.13

Integrating the IPCE over the AM 1.5G solar spectrum yields a calculated J_{SC} similar to the collected value for devices of **YD0, YDD0** and **YDD1**, confirming the accuracy of the current–voltage results shown in Fig. 15.31(a). As shown there, the short-circuit photocurrent density (J_{SC}) of **YDD1** is greater than that of **YD0**, whereas the open-circuit voltage (V_{OC}) of the former is smaller than the latter; the net effect produces a slightly greater power-conversion efficiency of the former than that of the latter ($\eta = 5.23$ vs. 5.14%). The fact that the J_{SC} value of **YDD1** was significantly greater than that of **YD0** is understood to be due to the effective excitonic coupling of the two porphyrin macrocycles in the dimer whereas such a character was absent in the monomer. As a result, the IPCE spectrum of **YDD1** exhibits a flat response over the entire visible region whereas that of **YD0** shows a large gap between the Soret and Q bands (Fig. 15.31(b)). In contrast, the overall efficiency of **YDD0** was significantly smaller than that of **YDD1** (4.07 vs. 5.23%) because of the smaller J_{SC} and V_{OC} values of the former. Even though the IPCE spectrum of **YDD0** shown in Fig. 15.31(b) displays a broad feature covering spectral range 400–800 nm, the IPCE values of **YDD0** were much smaller than those of **YDD1** in the region of spectral response.

We notice that the absorption coefficients of **YDD0** are in general greater than those of **YD0** and **YDD1** (Fig. 15.29), but the values of IPCE of the former are substantially lower than those of the latter (Fig. 15.31(b)). As the results of dye-loading experiments show similar amounts of dye molecules ($\sim$75 nmol cm^{-2}) being sensitized on TiO_2 films for **YD0-YDD1**, the smaller external quantum efficiencies of **YDD0** relative to those of **YD0** or **YDD1** are thus inferred to arise from the efficiency of electron injection into TiO_2, which was smaller for the former than for the latter. Of two reasons to explain these effects, one is that the LUMO energy level of **YDD0** is much lower than that of **YD0** or **YDD1**, which might impede electron injection for the former. The other is that the nearly planar structure of **YDD0** facilitates π-conjugation between two porphyrin macrocycles and provides a decreased driving force to push electrons toward TiO_2. The planar geometry of **YDD0** might also facilitate the formation of dye aggregates that significantly decrease the efficiency of electron injection.

15.6 Conclusion

Inspired by the efficient energy transfer in naturally occurring photosynthetic reaction centers, numerous potential porphyrin sensitizers have been introduced in this chapter for highly efficient dye-sensitized solar cells. The most promising porphyrin, **YD2**, has a molecular structure with the integration of a prophyrine chromphore as π-bridge in a D-π-A system exhibiting an unprecedented efficiency of 11% when used as a photo-sensitizer on a double-layer TiO_2 film under standard illumination test conditions [131]. To our best knowledge, this is for the first time such high efficiency of power conversion has been achieved with a ruthenium-free sensitizer. Based on the molecular design of **YD2**, other porphyrin sensitizers such as **YD12**, **YD15**, **Me_2N-PE1**, and their derivatives, are expected to be promising for DSSC applications. We believe that the current study has opened new avenues to the improvement of photovoltaic performance through an appropriate design of the donor–acceptor substitution on porphyrin dyes towards a new milestone for the development of DSSC.

Acknowledgements

We are indebted to Prof. Michael Grätzel of EPFL for helpful discussions. National Science Council of Taiwan and Ministry of Education of Taiwan, under the ATU program, provided support for this project.

References

1. Nazeeruddin M. K., Péchy P. and Grätzel M. (1997) Efficient panchromatic sensitization of nanocrystalline TiO_2 films by a black dye based on a trithiocyanato–ruthenium complex, *Chem. Commun.*, 1705–1706.
2. Nazeeruddin M. K., Zakeeruddin S. M., Humphry-Baker R., Jirousek M., Liska P., Vlachopoulos N., Shklover V., Fischer C.-H., and Grätzel M. (1999) Acidbase equilibria of (2,2'-bipyridyl-4,4'-dicarboxylic acid)ruthenium(II) complexes and the effect of protonation on charge-transfer sensitization of nanocrystalline titania, *Inorg. Chem.*, **38**, 6298–6305.

3. Nazeeruddin M. K., Péchy P., Renouard T., Zakeeruddin S. M., Humphry-Baker R., Comte P., Liska P., Cevey L., Costa E., Shklover V., Spiccia L., Deacon G. B., Bignozzi C. A., and Grätzel M. (2001) Engineering of efficient panchromatic sensitizers for nanocrystalline TiO2-based solar cells, *J. Am. Chem. Soc.*, **123**, 1613–1624.
4. O'Regan B. and Grätzel M. (1991) A low-cost, high-efficiency solar cell based on dye-sensitized colloidal TiO_2 films, *Nature*, **353**, 737–740.
5. Nazeeruddin M. K., Kay A., Rodicio I., Humphry-Baker R., Müller E., Liska P., Vlachopoulos N., Grätzel M. (1993) Conversion of light to electricity by cis-X_2bis(2,2'-bipyridyl-4,4'-dicarboxylate)ruthenium(II) charge-transfer sensitizers (X = Cl^-, Br^-, I^-, CN^-, and SCN^-) on nanocrystalline TiO_2 electrodes, *J. Am. Chem. Soc.*, **115**, 6382–6390.
6. Kalayansundaram K. and Grätzel M. (1998) Applications of functionalized transition metal complexes in photonic and optoelectronic devices *Coord. Chem. Rev.*, **177**, 347–414.
7. Hara K., Sayama K., Ohga Y., Shinpo A., Suga S., and Arakawa H. (2001) A coumarin-derivative dye sensitized nanocrystalline TiO_2 solar cell having a high solar-energy conversion efficiency up to 5.6%, *Chem. Commun.*, 569–570.
8. Hara K., Kurashige M., Danoh Y., Kasada C., Shinpo A., Suga S., Sayama K., and Arakawa H. (2003) Design of new coumarin dyes having thiophene moieties for highly efficient organic-dye-sensitized solar cells, *New J. Chem.*, **27**, 783–785.
9. Wang Z.-S., Cui Y., Dan-oh Y., Kasada C., Shinpo A., and Hara K. (2007) Thiophene-functionalized coumarin dye for efficient dye-sensitized solar cells: electron lifetime improved by coadsorption of deoxycholic acid, *J. Phys. Chem. C*, **111**, 7224–7230.
10. Horiuchi T., Miura H., and Uchida S. (2003) Highly-efficient metal-free organic dyes for dye-sensitized solar cells, *Chem. Commun.*, 3036–3037.
11. Horiuchi T., Miura H. Sumioka, K., and Uchida S. (2004) High efficiency of dye-sensitized solar cells based on metal-free indoline dyes *J. Am. Chem. Soc.*, **126**, 12218–11219.
12. Ito S., Zakeeruddin S. M., Humphry-Baker R., Liska P., Charvet R., Comte P., Nazeeruddin M. K., Péchy P., Takata M., Miura H., Uchida S., and Grätzel M. (2006) High-efficiency organic-dye-sensitized solar cells controlled by nanocrystalline-TiO2 electrode thickness, *Adv. Mater.*, **18**, 1202–1205.

13. Kitamura T., Ikeda M., Shigaki K., Inoue T., Anderson N. A., Ai X., Lian T., and Yanagida S. (2004) Phenyl-conjugated oligoene sensitizers for TiO2 solar cells, *Chem. Mater.*, **16**, 1806–1812.
14. Hara K., Sato T., Katoh R., Furabe A., Yoshihara T., Murai M., Kurashige M., Ito S., Shinpo A., Suga S., and Arakawa H. (2005) Novel conjugated organic dyes for efficient dye-sensitized solar cells, *Adv. Funct. Mater.*, **15**, 246–252.
15. Kim S., Choi H., Kim D., Song K., Kang S. O., and Ko J. (2007) Novel conjugated organic dyes containing bis-dimethylfluorenyl amino phenyl thiophene for efficient solar cell *Tetrahedron*, **63**, 9206–9212.
16. Kim S., Choi H., Baik C., Song K., Kang S. O., and Ko J. (2007) Synthesis of conjugated organic dyes containing alkyl substituted thiophene for solar cell, *Tetrahedron*, **63**, 11436–11443.
17. Jung I., Lee J. K., Song K. H., Song K., Kang S. O., and Ko J. (2007) Synthesis and photovoltaic properties of efficient organic dyes containing the benzo[*b*]furan moiety for solar cells, *J. Org. Chem.*, **72**, 3652–3658.
18. Velusamy M., Thomas K. R. J., Lin J. T., Hsu Y., and Ho K. (2005) Organic dyes incorporating low-band-gap chromophores for dye-sensitized solar cells, *Org. Lett.*, **7**, 1899–1902.
19. Hagberg D. P., Edvinsson T., Marinado T., Boschloo G., Hagfeldt A., and Sun L. (2006) A novel organic chromophore for dye-sensitized nanostructured solar cells, *Chem. Commun.*, 2245–2247.
20. Liang M., Xu W., Cai F., Chen P., Peng B., Chen J., and Li Z. (2007) New triphenylamine-based organic dyes for efficient dye-sensitized solar cells, *J. Phys. Chem. C* **111**, 4465–4472.
21. Ferrere S., Zaban A., and Greg B. A. (1997) Dye sensitization of nanocrystalline tin oxide by perylene derivatives, *J. Phys. Chem. B*, **101**, 4490–4493.
22. Ferrere S. and Greg B. A. (2002) New perylenes for dye sensitization of TiO_2, *New J. Chem.*, **26**, 1155–1160.
23. Shibano Y., Umeyama T., Matano Y., and Imahori H. (2007) Electron-donating perylene tetracarboxylic acids for dye-sensitized solar cells, *Org. Lett.*, **9**, 1971–1974.
24. Ehret A., Stuhl L., and Spitler M. T. (2001) Spectral sensitization of TiO2 nanocrystalline electrodes with aggregated cyanine dyes, *J. Phys. Chem. B*, **105**, 9960–9965.
25. Ushiroda S., Ruzycki N., Lu Y., Spitler M. T., and Parkinson B. A. (2005) Dye sensitization of the anatase (101) crystal surface by a series of dicarboxylated thiacyanine dyes, *J. Am. Chem. Soc.*, **127**, 5158–5168.

26. Tatay S., Haque S. A., O'Regan B., Durrant J. R., Verhees W. J. H., Kroon J. M., Vidal-Ferran A., Gaviña P., and Palomares E. (2007) Kinetic competition in liquid electrolyte and solid-state cyanine dye sensitized solar cells *J. Mater. Chem.*, **17**, 3037–3044.
27. Yao Q.-H., Shan L., Li F.-Y., Yin D.-D., and Huang C.-H. (2003) An expanded conjugation photosensitizer with two different adsorbing groups for solar cells *New J. Chem.*, **27**, 1277–1283.
28. Chen Y.-S., Li C., Zeng Z.-H., Wang W.-B., Wang X.-S., and Zhang B.-W. (2005) Efficient electron injection due to a special adsorbing group's combination of carboxyl and hydroxyl: dye-sensitized solar cells based on new hemicyanine dyes, *J. Mater. Chem.*, **15**, 1654–1661.
29. Hara K., Sato T., Katoh R., Furube A., Ohga Y., Shinpo A., Suga S., Sayama K., Sugihaa H., and Arakawa H. (2003) Molecular design of coumarin dyes for efficient dye-sensitized solar cells, *J. Phys. Chem. B*, **107**, 597–606.
30. Morandeira A., Boschloo G., Hagfeldt A., and Hammarstrolm L. (2005) Photoinduced ultrafast dynamics of coumarin 343 sensitized p-type-nanostructured NiO films, *J. Phys. Chem. B*, **109**, 19403–19410.
31. Koumura N., Wang Z.-S., Mori S., Miyashita M., Suzuki E., and Hara K. (2006) Alkyl-functionalized organic dyes for efficient molecular photovoltaics, *J. Am. Chem. Soc.*, **128**, 14256–14257.
32. Deisenhofer J., and Norris J. R. (1993) *Photosynthetic Reaction Center* Academic Press, New York.
33. Collin J.-P., Harriman A., Heitz V., Odobel F., and Sauvage J.-P. (1994) Photoinduced electron- and energy-transfer processes occurring within porphyrin-metal-bisterpyridyl conjugates, *J. Am. Chem. Soc.*, **116**, 5679–5690.
34. Seth J., Palaniappan V., Wagner R. W., Johnson T. E., Lindsey J. S., and Bocian D. F. (1996) Soluble synthetic multiporphyrin arrays. 3. static spectroscopic and electrochemical probes of electronic communication, *J. Am. Chem. Soc.*, **118**, 11194–11207.
35. Jiang D.-L., and Aida T. (1998) Morphology-dependent photochemical events in aryl ether dendrimer porphyrins: cooperation of Dendron subunits for singlet energy transduction, *J. Am. Chem. Soc.*, **120**, 10895–10901.
36. Gust D., Moore T. A., and Moore A. L. (2001) Mimicking photosynthetic solar energy transduction, *Acc. Chem. Soc.*, **34**, 40–48.
37. Cho H. S., Rhee H., Song J. K., Min C. K., Takase M., Aratani N., Cho S., Osuka A., Joo T., and Kim D. (2003) Excitation energy transport

processes of porphyrin monomer, dimer, cyclic trimer, and hexamer probed by ultrafast fluorescence anisotropy decay, *J. Am. Chem. Soc.*, **125**, 5849–5860.

38. Duncan T. V., Wu S. P., and Therien M. J. (2006) Ethyne-bridged (porphinato)zinc(II)(porphinato)iron(III) complexes: phenomenological dependence of excited-state dynamics upon (porphinato)iron electronic structure, *J. Am. Chem. Soc.* **128**, 10423–10435.
39. Imahori H. (2007) Creation of fullerene-based artificial photosynthetic systems, *Bull. Chem. Soc. Jpn.*, **80**, 621–636.
40. Winters M. U., Kärnbratt J., Blades H. E., Wilson C. J., Frampton M. J., Anderson H. L., and Albinsson B. (2007) Control of electron transfer in a conjugated porphyrin dimer by selective excitation of planar and perpendicular conformers, *Chem. Eur. J.*, **13**, 7385–7394.
41. Clifford J. N., Yahioglu G., Milgrom L. R., and Durrant J. R. (2002) Molecular control of recombination dynamics in dye sensitized nanocrystalline TiO_2 films, *Chem. Commun.*, 1260–1261.
42. Hasobe T., Imahori H., Kamat P. V., Ahn T. K., Kim S. K., Kim D., Fujimoto A., Hirakawa T., and Fukuzumi S. (2005) Photovoltaic cells using composite nanoclusters of porphyrins and fullerenes with gold nanoparticles, *J. Am. Chem. Soc.*, **127**, 1216–1228.
43. Hasobe T., Kamat P. V., Troiani V., Solladié, N., Ahn T. K., Kim S. K., Kim D., Kongkanand A., Kuwabata S., and Fukuzumi S. (2005) Enhancement of light-energy conversion efficiency by multi-porphyrin arrays of porphyrin-peptide oligomers with fullerene clusters, *J. Phys. Chem. B*, **109**, 19–23.
44. Borgström M., Blart E., Boschloo G., Mukhtar E., Hagfeldt A., Hammarström L., and Odobel F. (2005) Sensitized hole injection of phosphorus porphyrin into NiO: Toward new photovoltaic devices *J. Phys. Chem. B*, **109**, 22928–22934.
45. Luo L., Lo C.-F., Lin C.-Y., Chang I-J., and Diau E. W.-G. (2006) Femtosecond fluorescence dynamics of porphyrin in solution and solid films: The effects of aggregation and interfacial electron transfer between porphyrin and TiO_2, *J. Phys. Chem. B*, **110**, 410–419.
46. Huijser A., Savenije T. J., Kotlewski A., Picken S. J., and Siebbeles L. D. A. (2006) Efficient light-harvesting layers of homeotropically aligned porphyrin derivatives, *Adv. Mater.*, **18**, 2234–2239.
47. Hagemann O., Jørgensen M., and Krebs F. C. (2006) Synthesis of an all-in-one molecule (for organic solar cells), *J. Org. Chem.*, **71**, 5546–5559.

48. Hasselman G. M., Watson D. F., Stromberg J. R., Bocian D. F., Holten D., Lindsey J. S., and Meyer G. J. (2006) Theoretical solar-to-electrical energy-conversion efficiencies of peryleneporphyrin light-harvesting arrays, *J. Phys. Chem. B*, **110**, 25430–25440.
49. Eu S., Hayashi S., Umeyama T., Oguro A., Kawasaki M., Kadota N., Matano Y., and Imahori H. (2007) Effects of 5-membered heteroaromatic spacers on structures of porphyrin films and photovoltaic properties of porphyrin-sensitized TiO_2 cells, *J. Phys. Chem. C*, **111**, 3528–3537.
50. O'Regan B C., López-Duarte I., Martínez-Díaz M. V., Forneli A., Albero J., Morandeira A., Palomares E., Torres T., and Durrant J R. (2008) Catalysis of recombination and its limitation on open circuit voltage for dye sensitized photovoltaic cells using phthalocyanine dyes, *J. Am. Chem. Soc.*, **130**, 2906–2907.
51. Cid J-J, Yum J-H, Jang S-R, Nazeeruddin M K., Martínez-Ferrero E, Palomares E, Ko J, Grätzel M., and Torres T (2007) Molecular cosensitization for efficient panchromatic dye-sensitized solar cells, *Angew. Chem., Int. Ed. Engl.*, **46**, 8358–8362.
52. Reddy P Y, Giribabu L, Lyness C, Snaith H J., Vijaykumar C, Chandrasekharam M, Lakshmikantam M, Yum J-H, Kalyanasundaram K, Grätzel M., and Nazeer M K. (2007) Efficient sensitization of nanocrystalline TiO_2 films by a near-IR absorbing unsymmetrical zinc phthalocyanine *Angew. Chem., Int. Ed. Engl.*, **46**, 373–376.
53. He J, Benkö G, Korodi F, Polívka T, Lomoth R, Åkermark B, Sun L, Hagfeldt A., and Sundström V (2002) Modified phthalocyanines for efficient near-IR sensitization of nanostructured TiO_2 electrode *J. Am. Chem. Soc*, **124**, 4922–4932.
54. Palomares E, Martínez-Díaz M. V, Haque S A., Torres T., and Durrant J R. (2004) State selective electron injection in non-aggregated titanium phthalocyanine sensitised nanocrystalline TiO_2 films *Chem. Commun.*, 2112–2113
55. Rio, Y., Vazquez, P., and Palomares, E. (2009) Extended π-aromatic systems for energy conversion: phthalocyanines and porphyrins in molecular solar cells *J. Porphyrins Phthalocyanines*, **13**, 645–651.
56. Imahori H., Umeyama T., and Ito S (2009) Large π-aromatic molecules as potential sensitizers for Highly Efficient dye-sensitized solar cells, *Acc. Chem. Res.*, **42** 1809–1818.
57. Campbell W. M., Burrell A. K., Officer D. L., and Jolley K. W. (2004) Porphyrins as light harvesters in the dye-sensitised TiO_2 solar cell, *Coord. Chem. Rev.* **248**, 1363–1379.

58. Hasobe T., Fukuzumi S., Hattori S., and Kamat P. V. (2007) Shape- and functionality-controlled organization of TiO_2–porphyrin–C_{60} assemblies for improved performance of photochemical solar cells *Chem. Asian J.* **2**, 265–272.
59. Rochford J., Chu D., Hagfeldt A., and Galoppini E. (2007) Tetrachelate porphyrin chromophores for metal oxide semiconductor sensitization: Effect of the spacer length and anchoring group position *J. Am. Chem. Soc.*, **129**, 4655–4665.
60. Lee C.-W., Lu H.-P., Lan C.-M., Huang Y.-L., Liang Y.-R., Yen W.-N., Liu Y.-C., Lin Y.-S., Diau E. W.-G., and Yeh C.-Y. (2009) Novel zinc porphyrin sensitizers for dye-sensitized solar cells: synthesis, spectral, electrochemical and photovoltaic properties, *Chem. Eur. J.* **15**, 1403–1412
61. Lin C-Y, Lo C-F, Luo L, Lu H-P, Hung C-S., and Diau E W-G (2009) Design and characterization of novel porphyrins with oligo(phenylethylnyl) links of varied length for dye-sensitized solar cells: synthesis and optical, electrochemical, and photovoltaic investigation, *J. Phys. Chem. C*, **113**, 755–764.
62. Campbell W. M. Burrell A. K. Officer D. L., and Jolley K. W. (2004) Porphyrins as light harvesters in the dye-sensitised TiO_2 solar cell *Coord. Chem. Rev.*, **248**, 1363.
63. Bignozzi C. A. Argazzi R., and Kleverlaan C. J. (2000) Molecular and supramolecular sensitization of nanocrystalline wide band-gap semiconductors with mononuclear and polynuclear metal complexes, *Chem. Soc. Rev.*, **29**, 87.
64. Siemsen P. Livingston R. C., and Diederich F. (2000) Acetylenic coupling: a powerful tool in molecular construction." *Angew. Chem. Int. Ed.*, **39**, p. 2632.
65. Bunz U. H. F. (2000) Poly(aryleneethynylene)s: syntheses, properties, structures, and applications *Chem. Rev.*, **100**, 1605.
66. Piotrowiak P. Galoppini E. Wei Q. Meyer G. J., and Wiewior P. (2003) Subpicosecond photoinduced charge injection from "molecular tripods" into mesoporous TiO_2 over the distance of 24 angstroms *J. Am. Chem. Soc.*, **125**, 5278.
67. Myahkostupov M. Piotrowiak P. Wang D., and Galoppini E. (2007) Ru(II)-bpy complexes bound to nanocrystalline TiO_2 films through phenyleneethynylene (OPE) linkers: effect of the linkers length on electron injection rates *J. Phys. Chem. C*, **111**, p. 2827.
68. Rochford J. Chu D. Hagfeldt A., and Galoppini E. (2007) Tetrachelate porphyrin chromophores for metal oxide semiconductor sensitization:

effect of the spacer length and anchoring group position *J. Am. Chem. Soc.*, **129**, 4655.

69. Luo L., Lin C.-J., Tsai C.-Y., Wu H.-P., Li L.-L, Lo C.-F., Lin C.-Y., and Diau, E. W.-G. (2010) Effects of aggregation and electron injection on photovoltaic performance of porphyrin-based solar cells with oligo(phenylethynyl) links inside TiO_2 and Al_2O_3 nanotube arrays, *Phys. Chem. Chem. Phys.*, **12**, 1064–1071
70. Luo L., Lin C.-J., Tsai C.-Y., Hung C.-S., Lo C.-F., Lin C.-Y., and Diau, E. W.-G. (2010) *Phys. Chem. Chem. Phys.*, **12**, in press (DOI: 10.1039/C0CP00458H)
71. Liu D., Fessenden R. W., Hug G. L., and Kamat P. V. (1997) Light-induced charge separation across Ru(II)-modified nanocrystalline TiO_2 interfaces with phenothiazine donors, *J. Phys. Chem. B*, **101**, 2583–2597.
72. Kim S., Lee J. K., Kang S. O., Ko J., Yum J.-H., Fantacci S., Angelis F. D., Censo D. D., Nazeeruddin M. K., and Grätzel M. (2006) Molecular engineering of organic sensitizers for solar cell applications *J. Am. Chem. Soc.* **128**, 16701–16707.
73. Lin V. S.-Y., DiMagno S. G., and Therien M. J. (1994) Highly conjugated, acetylenyl bridged porphyrins: new models for light-harvesting antenna systems *Science*, **264** 1105–1111.
74. Lin V. S.-Y., and Therien M. J. (1995) The Role of Porphyrin-to-Porphyrin Linkage Topology in the Extensive Modulation of the Absorptive and Emissive Properties of a Series of Ethynyl- and Butadiynyl-Bridged Bis- and Tris(porphinato)zinc Chromophores *Chem. Eur. J.* **1**, 645–651.
75. Taylor P. N., Huuskonen J., Rumbles G., Aplin R. T., Williams E., and Anderson H. L. (1998) Conjugated porphyrin oligomers from monomer to hexamer, *Chem. Commun*, 909–910.
76. Nakamura K., Fujimoto T., Takara S., Sugiura K.-I., Miyasaka H., Ishii T., Yamashita M., and Sakata Y. (2003) Systematic synthesis of porphyrin dimers linked by conjugated oligoacetylene bridges *Chem. Lett.*, **32**, 694–695.
77. Anderson H. L., Martin S. J., and Bradley D. D. C. (1994) Synthesis and third-order nonlinear optical properties of a conjugated porphyrin polymer, *Angew. Chem., Int. Ed. Engl.*, **33**, 655–657.
78. Osuka A., Tanabe N., Kawabata S., Yamazaki I., and Nishimura Y. (1995) Synthesis and intramolecular electron- and energy-transfer reactions of polyyne- or polyene-bridged diporphyrins, *J. Org. Chem.*, **60**, 7177–7185.

79. Maruyama K. and Kawabata S. (1990) Synthesis and characterization of polyyne porphyrins *Bull. Chem. Soc. Jpn.*, **63** 170–175.
80. Taylor P. N. and Anderson H. L. (1999) Cooperative self-assembly of double-strand conjugated porphyrin ladders, *J. Am. Chem. Soc.*, **121** 11538–11545
81. Li S.-L., Jiang K.-J., Shao K.-F., and Yang L.-M. (2006) Novel organic dyes for efficient dye-sensitized solar cells *Chem. Commun.*, 2792–2794.
82. Wang Q., Campbell W. M., Bonfantani E. E., Jolley K. W., Officer D. L., Walsh P. J., Gordon K., Humphry-Baker R., Nazeeruddin M. K., and Grätzel M. (2005) Efficient light harvesting by using green Zn-porphyrin-sensitized nanocrystalline TiO_2 films, *J. Phys. Chem. B*, **109**, 15397–15409.
83. Schmidt-Mende L., Campbell W. M., Wang Q., Jolley K. W., Officer D. L., Nazeeruddin M. K., and Grätzel M. (2005) Zn-porphyrin-sensitized nanocrystalline TiO_2 heterojunction photovoltaic cells *Chem Phys Chem*, **6**, 1253–1258.
84. Campbell W. M., Jolley K. W., Wagner P., Wagner K., Walsh P. J., Gordon K. C., Schmidt-Mende L., Nazeeruddin M. K., Wang Q., Grätzel M., and Officer D. L. (2007) Highly efficient porphyrin sensitizers for dye-sensitized solar cells, *J. Phys. Chem. C*, **111**, 11760–11762.
85. Hara K., Sayama K., Ohga Y., Shinpo A., Suga S., and Arakawa H. (2001) A coumarin-derivative dye sensitized nanocrystalline TiO_2 solar cell having a high solar-energy conversion efficiency up to 5.6%, *Chem. Commun.* 569–570.
86. Hara K., Kurashige M., Ito S., Shinpo, A., Suga S., Sayama K., and Arakawa H. (2003) Novel polyene dyes for highly efficient dye-sensitized solar cells, *Chem. Commun.*, 252–253.
87. Chang J.-C., Ma C.-J., Lee G.-H., Peng S.-M., and Yeh C.-Y. (2005) Porphyrin–triarylamine conjugates: strong electronic communication between triarylamine redox centers via the porphyrin dication *Dalton Trans.*, 1504–1508.
88. Yen W.-N., Lo S.-S., Kuo M.-C., Mai C.-L., Lee G.-H., Peng S.-M., and Yeh C.-Y. (2006) Synthesis, structure, and optical and electrochemical properties of star-shaped porphyrintriarylamine conjugates *Org. Lett.*, **8**, 4239–4242.
89. Lu H.-P., Tsai C.-Y., Yen W.-N., Hsieh C.-P., Lee C.-W. Yeh C.-Y., and Diau E. W.-G. (2009) Control of dye aggregation and electron injection for highly efficient porphyrin sensitizers adsorbed on semiconductor

films with varying ratios of coadsorbate, *J. Phys. Chem. C*, **113**, 20990–20997.

90. Haque S. A., Handa S., Peter K. Palomares E. Thelakkat M., and Durrant J. R. (2005) Supermolecular control of charge transfer in dye-sensitized nanocrystalline TiO_2 films: Towards a quantitative structure–function relationship *Angew. Chem. Int. Ed.*, **44**, 5740–5744.
91. Mozer A. J., Wagner P., Officer D. L., Wallace G. G. Campbell W. M. Miyashita M. Sunahara K., and Mori S. (2008) The origin of open circuit voltage of porphyrin-sensitised TiO_2 solar cells *Chem. Commun.*, 4741–4743.
92. Hamann T. W. Jensen R. A. Martinson A. B. F. Ryswyk H. V., and Hupp J. T. (2008) Advancing beyond current generation dye-sensitized solar cells *Energy Environ. Sci.*, **1**, 66–78.
93. Tachibana Y. Moser J. E. Grätzel M., Klug D. R., and Durrant J. R. (1996) Subpicosecond interfacial charge separation in dye-sensitized Nanocrystalline titanium dioxide films, *J. Phys. Chem.*, **1996**, **100**, 20056–20062.
94. Wenger B. Grätzel M., and Moser J. E. (2005) Rationale for kinetic heterogeneity of ultrafastlight-induced electron transfer from Ru(II) complex sensitizers to nanocrystalline TiO_2, *J. Am. Chem. Soc.*, **127**, 12150–12151.
95. Barnes P. R. F. Anderson A. Y. Koops S. E. Durrant J. R., and O'Regan B. C. (2009) Electron injection efficiency and diffusion length in dye-sensitized solar cells derived from incident photon conversion efficiency measurements *J. Phys. Chem. C*, **113**, 1126–1136.
96. Wang Z.-S., Koumura N., Cui Y., Takahashi M., Sekiguchi H., Mori A., Kubo T., Furube A., and Hara K. (2008) Hexylthiophene-functionalized carbazole dyes for efficient molecular photovoltaics: Tuning of solar-cell performance by structural modification *Chem. Mater.*, **20**, 3993–4003.
97. Hsieh C.-P., Lu H.-P. Chiu C.-L., Lee C.-W. Chuang S.-H Mai C.-L., Yen W.-N. Hsu S.-J., Diau E. W.-G., and Yeh C.-Y. (2010) Synthesis and characterization of porphyrin sensitizers with various electron-donating substituents for highly efficient dye-sensitized solar cells, *J. Mater. Chem.*, **20**, 1127–1134.
98. Hagberg D P., Yum J-H, Lee H, Angelis F D, Marinado T, Karlsson K M, Humphry-Baker R, Sun L, Hagfeldt A, Grätzel M., and Nazeeruddin M. K. (2008) Molecular engineering of organic sensitizers for dye-sensitized solar cell applications *J. Am. Chem. Soc*, **130**, 6259–6266.

99. Park J. K., Lee H. R., Chen J., Shinokubo H., Osuka A., and Kim D. (2008) Photoelectrochemical properties of doubly β-functionalized porphyrin sensitizers for dye-sensitized nanocrystalline-TiO_2 solar cells *J. Phys. Chem. C*, **112**, 16691–16699.
100. Anderson H L. (1999) Building molecular wires from the colours of life: conjugated porphyrin oligomers, *Chem. Commun.* 2323–2330.
101. Wu S.-L., Lu H.-P., Yu H.-T., Chuang S.-H. Chiu C.-L., Lee C.-W., Diau E. W.-G., and Yeh C.-Y. (2010) Design and characterization of porphyrin sensitizers with a push-pull framework for highly efficient dye-sensitized solar cells *Energy Environ. Sci.*, **3**, 949–955.
102. Sakamoto R., Sasaki T., Honda N., and Yamamura T. (2009) 5,15-Bis(di-p-anisylamino)-10,20-diphenylporphyrin: distant and intense electronic communication between two amine sites *Chem. Commun.* 5156–5158.
103. Huang C-W, Chiu K Y., and Cheng S.-H. (2005) Novel spectral and electrochemical characteristics of triphenylamine-bound zinc porphyrins and their intramolecular energy and electron transfer *Dalton Trans.*, 2417–2422
104. Lo C-F, Hsu S-J, Wang C-L, Cheng Y-H, Lu H-P, Diau E W-G., and Lin C-Y (2010) Tuning spectral and electrochemical properties of porphyrin-sensitized solar cells" *J. Phys. Chem. C*, **114**, 12018–12023.
105. LeCours S M. DiMagno S G., and Therien M J. (1996) Exceptional electronic modulation of porphyrins through *meso*arylethynyl groups electronic spectroscopy, electronic structure, and electrochemistry of [5,15-bis[(aryl)ethynyl]-10,20-diphenylporphinato]zinc(II) complexes. X-ray crystal structures of [5,15-bis[(4′-fluorophenyl)ethynyl]-10,20-diphenylporphinato]zinc(II) and 5,15-bis[(4′-methoxyphenyl) ethynyl]-10,20-diphenylporphyrin *J. Am. Chem. Soc.* **118**, 11854–11864.
106. LeCours S. M. Guan H. W. DiMagno, S. G. Wang C. H., and Therien, M. J (1996) Push-pull arylethynyl porphyrins: new chromophores that exhibit large molecular first-order hyperpolarizabilities *J. Am. Chem. Soc.*, **118**, 1497–1503.
107. LeCours S. M., Philips C. M., Paula J. C. D., and Therien, M. J. (1997) Synthesis, transient absorption, and transient resonance raman spectroscopy of novel electron donor-acceptor complexes: [5,15-bis[(4'-nitrophenyl)ethynyl]-10,20-diphenylporphinato]copper (II) and [5-[[4'-(Dimethylamino)phenyl]-thynyl]-15-[(4"-nitrophenyl) ethynyl]-10,20-diphenylporphinato] copper(II), *J. Am. Chem. Soc.*, **119**, 12578–12589.

108. Ma R., Guo P., Cui H., Zhang X., Nazeeruddin M. K., and Grätzel M. (2009) Substituent effect on the meso-substituted porphyrins: theoretical screening of sensitizer candidates for dye-sensitized solar cells, *J. Phys. Chem. A*, **113**, 10119–10124.
109. Imahori H., Hayashi S., Hayashi H., Oguro A., Eu S., Umeyama T., and Matano Y. (2009) Effects of porphyrin substituents and adsorption conditions on photovoltaic properties of porphyrin-sensitized TiO_2 Cells *J. Phys. Chem. C*, **113**, 18406–18413.
110. Imahori H., Hayashi S., Hayashi H., Oguro A., Eu S., Umeyama T., and Matano Y. (2009) Effects of porphyrin substituents and adsorption conditions on photovoltaic properties of porphyrin-sensitized TiO_2 Cells *J. Phys. Chem. C*, **113**, 18406–18413.
111. Lin C-Y, Wang Y-C, Hsu S-J, Lo C-F., and Diau E W-G (2010) Preparation and spectral, electrochemical, and photovoltaic properties of acene-modified zinc porphyrins, *J. Phys. Chem. C*, **114**, 687–693.
112. Nazeeruddin Md. K., Humphry-Baker R., Officer D. L., Campbell W. M., Burrell A. K., and Grätzel M. (2004) Application of metalloporphyrins in nanocrystalline dye-sensitized solar cells for conversion of sunlight into electricity *Langmuir*, **20**, 6514–6517.
113. Lu H.-P., Mai C.-L., Tsia C.-Y., Hsu S.-J., Hsieh C.-P., Chiu C.-L., Yeh C.-Y., and Diau E. W.-G. (2009) Design and characterization of highly efficient porphyrin sensitizers for green see-through dye-sensitized solar cells, *Phys. Chem. Chem. Phys.*, **11**, 10270–10274.
114. Chen C.-C., Chung H.-W., Chen C.-H., Lu H.-P., Lan C.-M., Chen S.-F., Luo L., Hung C.-S., and Diau E. W.-G. (2008) Fabrication and characterization of anodic titanium oxide nanotube arrays of controlled length for highly efficient dye-sensitized solar cells *J. Phys. Chem. C*, **112**, 19151–19157.
115. Nazeeruddin M. K., Angelis F. D., Fantacci S., Selloni A., Viscardi G., Liska P., Ito S., Takeru B., and Grätzel M. (2005) Combined experimental and DFT-TDDFT computational study of photoelectrochemical cell ruthenium sensitizers *J. Am. Chem. Soc.*, **127**, 16835–16847.
116. Luo L., Lo C.-F., Lin C.-Y., Chang I.-J., and Diau E. W.-G. (2006) Femtosecond fluorescence dynamics of porphyrin in solution and solid films: The effects of aggregation and interfacial electron transfer between porphyrin and TiO_2, *J. Phys. Chem. B*, **110**, 410–419.
117. Kuo M.-C., Li L.-A. Yen W.-N. Lo S.-S. Lee C.-W., and Yeh C.-Y. (2007) New synthesis of zinc tetrakis(arylethynyl)porphyrins and substituent effects on their redox chemistry *Dalton Trans.* 1433–1439.

118. Drobizhev M., Stepanenko Y. Dzenis Y. Karotki A. Rebane A. Taylor P. N., and Anderson H. L. (2005) Extremely strong near-IR two-photon absorption in conjugated porphyrin dimers: quantitative description with three-essential-states model, *J. Phys. Chem. B* **109**, 7223–7236.

119. Susumu K., Duncan T. V., and Therien M. J. (2005) Potentiometric, electronic structural, and ground- and excited-State optical properties of conjugated bis[(porphinato)zinc(II)] compounds featuring proquinoidal spacer units, *J. Am. Chem. Soc.,* **127**, 5186–5195.

120. Payne M. M., Parkin S. R., and Anthony J. E. (2005) Functionalized higher acenes: hexacene and heptacene, *J. Am. Chem. Soc.,* **127**, 8028–8029.

121. He J., Benkö G, Korodi F, Polívka T., Lomoth R, Åkermark B., Sun L., Hagfeldt A., and Sundström V. (2002) Modified phthalocyanines for efficient near-IR sensitization of nanostructured TiO_2 electrode *J. Am. Chem. Soc.,* **124**, 4922–4932.

122. Hao Y., Yang X., Cong J., Tian H., Hagfeldt A., and Sun L. (2009) Efficient near infrared D–π–A sensitizers with lateral anchoring group for dye-sensitized solar cells *Chem. Commun.,* 4031–4033.

123. Tian H., Yang X., Chen R., Hagfeldt A., and Sun L. (2009) A metal-free "black dye" for panchromatic dye-sensitized solar cells *Energy Environ. Sci.,* **2**, 674–677.

124. Mozer A. J., Griffith M. J., Tsekouras G., Wagner P., Wallace G. G. Mori S., Sunahara K., Miyashita M., Earles J. C., Gordon K. C., Du L., Katoh R., Furube A., and Officer D. L., (2009) ZnZn Porphyrin Dimer-Sensitized Solar Cells: Toward 3-D Light Harvesting *J. Am. Chem. Soc,* **131**, 15621–15623.

125. Angiolillo P. J., Lin V. S.-Y., Vanderkooi J. M., and Therien M. J. (1995) EPR spectroscopy and photophysics of the lowest photoactivated triplet state of a series of highly conjugated (porphinato)Zn arrays, *J. Am. Chem. Soc.,* **117**, 12514–12527.

126. Martinson A. B. F., Hamann T. W., Pellin M. J., and Hupp J. T. (2008) New architectures for dye-sensitized solar cells *Chem. Eur. J.,* **14**, 4458–4467.

127. Tsuda A., and Osuka A. (2001) Fully conjugated porphyrin tapes with electronic absorption bands that reach into infrared Science, **293**, 79–82.

128. Park J. K., Chen J., Lee H. R., Park S. W., Shinokubo H. Osuka A., and Kim D. (2009) Doubly β-functionalized mesomeso directly linked porphyrin dimer sensitizers for photovoltaics, *J. Phys. Chem. C,* **113**, 21956–21963.

129. Mai C.-L. Huang W.-K. Lu H.-P Lee C.-W., Chiu C.-L. Liang Y.-R., Diau E. W.-G., and Yeh C.-Y. (2010) Synthesis and characterization of diporphyrin sensitizers for dye-sensitized solar cells, *Chem. Commun.*, **46**, 809–811.
130. Cho H. S., Jeong D. H., Cho S., Kim D., Matsuzaki Y., Tanaka K., Tsuda A., and Osuka A. (2002) Photophysical properties of porphyrin tapes *J. Am. Chem. Soc.*, **124**, 14642–14654.
131. Bessho T, Zakeeruddin S M., Yeh C-Y, Diau E. W.-G., and Grätzel M (2010) Donor-acceptor Substituted Porphyrins Achieve Eleven Percent Power Conversion Efficiency in Mesoscopic Dye-sensitized Solar Cells, *Angew. Chem. Int. Ed.*, **49**, 6646–6649.

Index

Color Insert

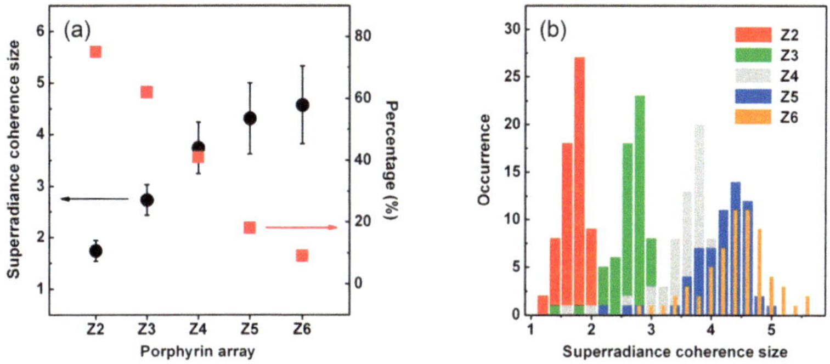

Figure 1.5. (a) A plot of the averaged superradiance coherence sizes (L_S) with standard deviations (black circles) and the percentage of FITs showing the same number of steps as porphyrin units (red squares). (b) Histograms for L_S of $\mathbf{Z}_N$, which were constructed by collecting 64 single-molecule datasets for each array. Reprinted with permission from [38].

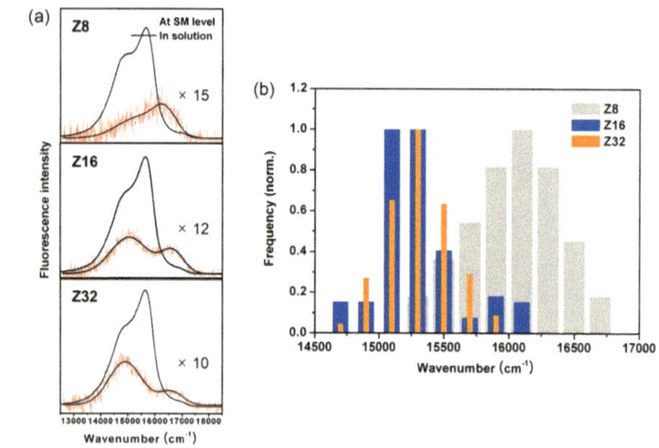

Figure 1.7. (a) Representative single-molecule fluorescence spectra of **Z8**, **Z16**, and **Z32**, with their ensemble-level fluorescence spectra taken in THF. (b) Histograms for fluorescence maximum positions of single-molecule fluorescence spectra of **Z8**, **Z16**, and **Z32**. Reprinted with permission from [47].

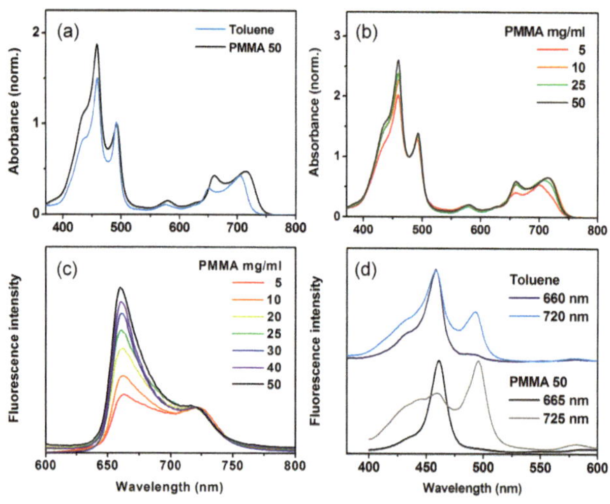

Figure 1.9. (a) Steady-state absorption spectra of **Z2B** in toluene and concentrated film prepared with 50 mg/ml PMMA. Steady-state absorption (b) and emission (c) spectra of concentrated films of **Z2B** prepared with a couple of different density of PMMA. (d) Fluorescence excitation spectra of **Z2B** in toluene and concentrated film prepared with 50 mg/ml PMMA. In the excitation spectra, the detection wavelengths of 660 (or 665) and 720 (or 725) nm correspond to the emission from the twisted and planar conformers, respectively.

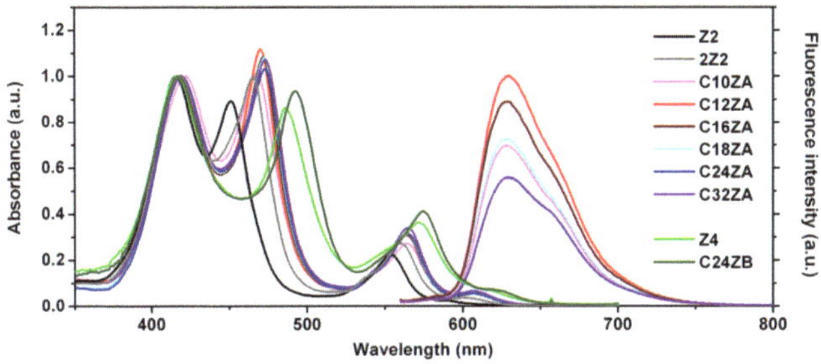

Figure 1.16. Steady-state absorption and emission spectra of **Z2**, **2Z2**, C_N**ZA**, **Z4**, and **C24ZB** in toluene. The absorption spectra were normalized to the high-energy Soret bands around 420 nm and the emission spectra were recorded with optical density of 0.1 at the excitation wavelength of 550 nm. Reprinted with permission from [94, 101].

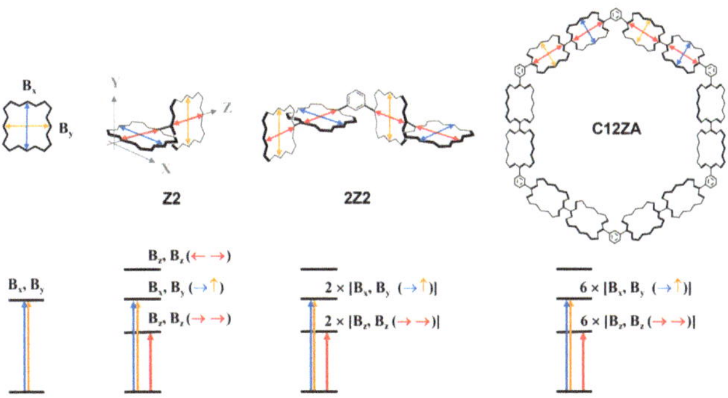

Scheme 1.3. Exciton coupling in **C12ZA**.

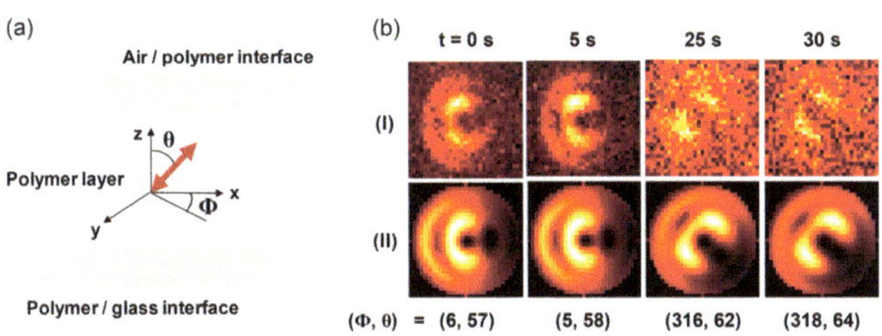

Figure 1.18. (a) Definition of angles: Φ indicates the in-plane (in the *xy* plane) angle, and θ indicates the out-of-plane angle (inclination angle between the optical dipole (bold arrow in red) and excitation optical axis (*z* axis)). (b) Snapshots of experimentally observed emission patterns of **C12ZA** (sequence I) and corresponding calculated patterns (sequence II) as a function of time. Reprinted with permission from [94].

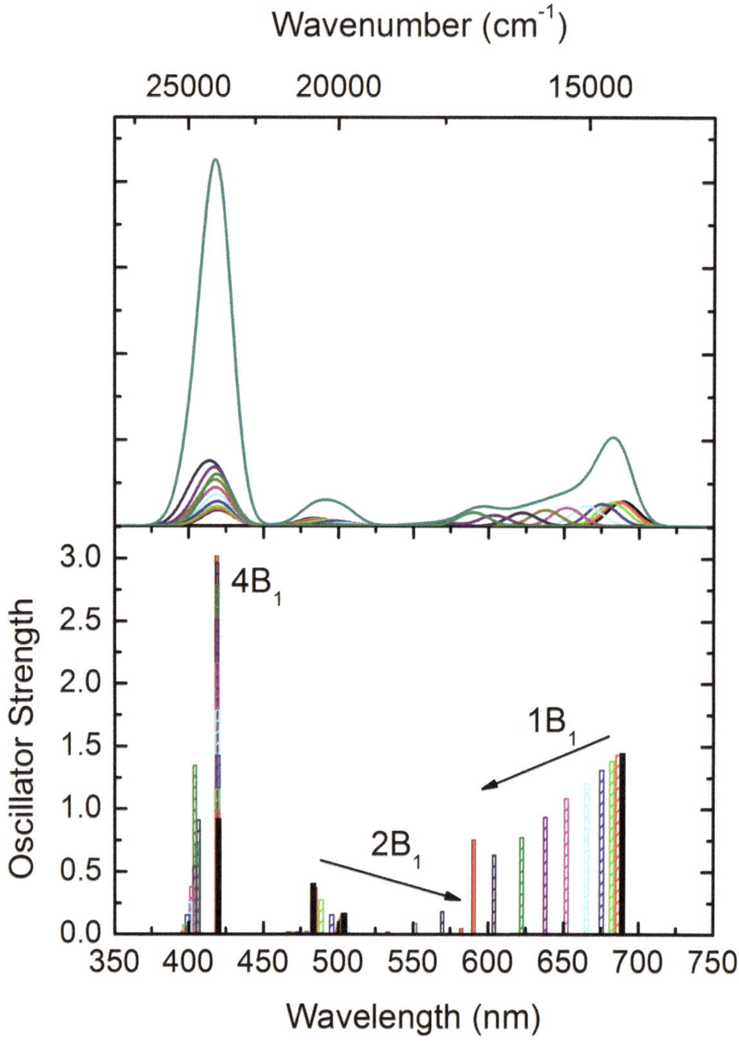

Figure 2.4. (Bottom) Calculated ground-state stick spectra at different dihedral angles; 0° (black, solid) and 90° (red, solid) represent the border cases and the intermediate dihedral angles are represented by hatched bars (10° (red), 20° (green), 30° (blue), 40° (cyan), 50° (magenta), 60° (yellow), 70° (dark green), 80° (navy blue)). (Top) Calculated spectra fitted to Gaussian components with half-widths at half-max arbitrarily set to 1000 cm^{-1}; 0°–90° in various colors and a simulated room temperature spectrum (dark cyan) based on the sum of the 10 equally weighted component spectra.

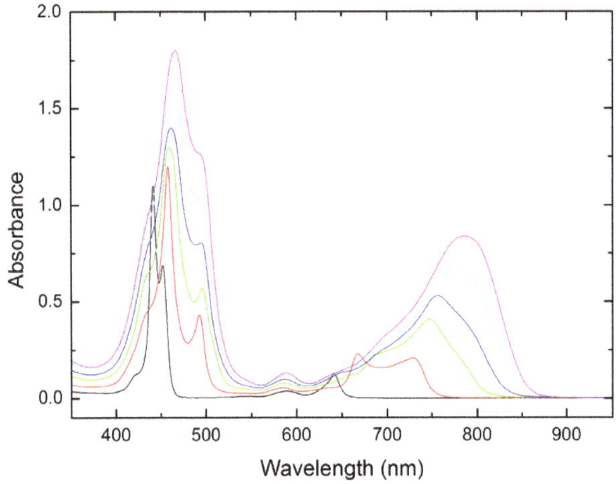

Figure 2.8. Ground-state absorption spectra of $\mathbf{P}_1$ (black), and $\mathbf{P}_2$ (red), $\mathbf{P}_3$ (green), $\mathbf{P}_4$ (blue), $\mathbf{P}_6$(cyan), and $\mathbf{P}_8$ (magenta) in THF with 1% pyridine.

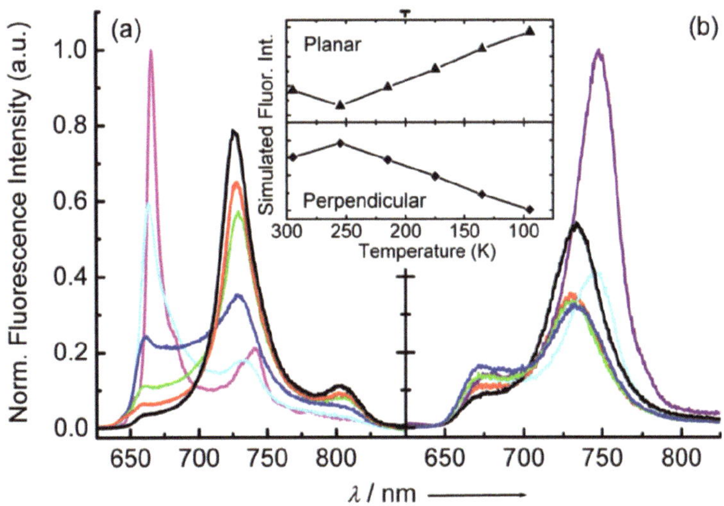

Figure 2.10. Emission spectra recorded at several temperatures obtained by exciting (a) $\mathbf{P}_2$ (2-MTHF) and (b) $\mathbf{P}_2$-$\mathbf{C}_{60}$ (2-MTHF) at 460 nm. The spectra were measured at 295 K (black), 255 K (red), 215 K (green), 175 K (blue), 135 K (cyan), 115 K (purple, $\mathbf{P}_2$-$\mathbf{C}_{60}$ only), and 95 K (magenta, $\mathbf{P}_2$ only). The inset shows the temperature variation of the integrated concentration of the planar and perpendicular species of $\mathbf{P}_2$-$\mathbf{C}_{60}$ (top and bottom, respectively), estimated from the extracted rate constants.

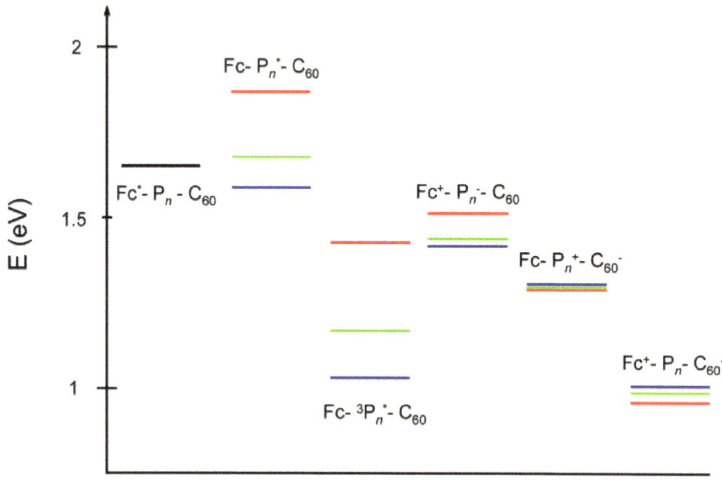

Figure 2.14. State energies estimated from electrochemical and spectroscopic data. Red lines are for $n = 1$, green lines for $n = 2$, and blue lines for $n = 4$.

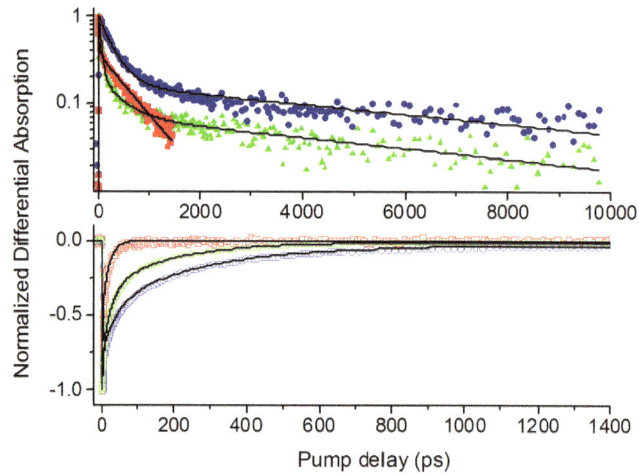

Figure 2.15. Transient absorption decays measured at 1000 nm (top), shown in logarithmic scale to emphasize the fact that they are single exponential decays at long delay times. The bottom panel shows the ground-state recovery of the oligomers (660 nm, 732 nm, and 756 nm). **Fc-P$_1$-C$_{60}$** is shown by red squares, **Fc-P$_2$-C$_{60}$** by green triangles, and **Fc-P$_4$-C$_{60}$** by blue circles. Note the different scales of the abscissa in the two panels. All measurements are in THF.

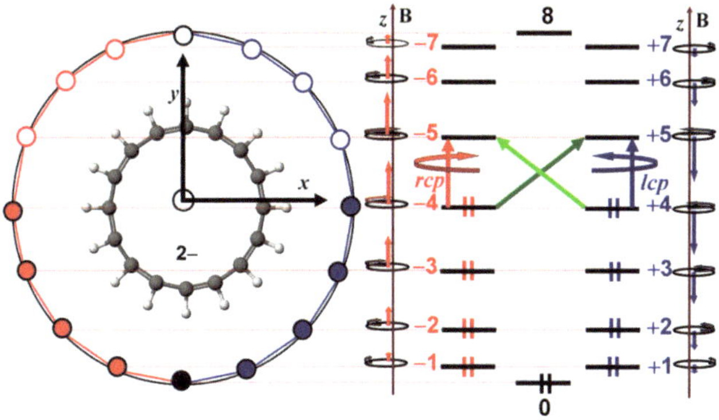

Figure 3.6. Michl's perimeter model (left) [15]. The circle is a diagrammatic representation of the clockwise (red) and counterclockwise (blue) motion of π-system electrons on the $C_{16}H_{16}^{2-}$ parent perimeter of porphyrinoid compounds based on the 16 $2p_z$ atomic orbitals of the inner ligand perimeter, generating the $M_L = 0, \pm1, \pm2, \pm3, \pm4, \pm5, \pm6, \pm7, 8$ value for each complex wavefunction MO. The perspective is towards the light source. The magnitudes of the magnetic moments induced in the applied field within each MO (right) are plotted diagrammatically, based on the results of a set of LCAO calculations carried out by Michl [15]. The origin of the forbidden L and allowed B transitions linking the HOMO ($M_L = \pm4$) and the LUMO ($M_L = \pm5$) are indicated using arrows with the same color scheme adopted in Fig. 3.5 for the Q and B bands in Gouterman's 4-orbital model.

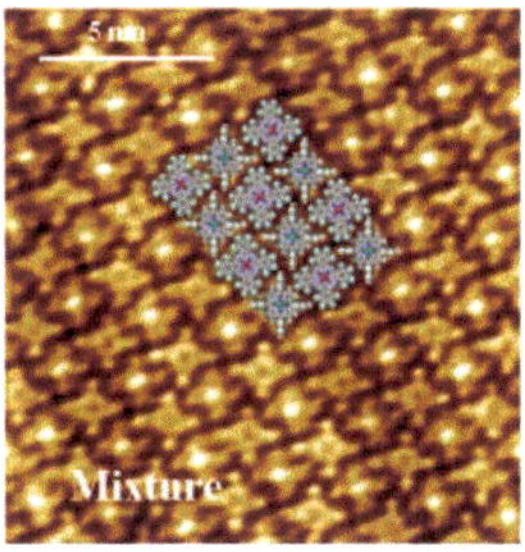

Figure 4.12. High resolution STM image of hexadecafluoro Co(II)Pc and tetraphenyl Ni(II)Por adsorbed on Au(111) imaged under ultra high vacuum (UHV) conditions. The space-filling models of both the Pc and the Por compounds have been superimposed to the STM image. Reproduced from Ref. [64] with permission of the American Chemical Society.

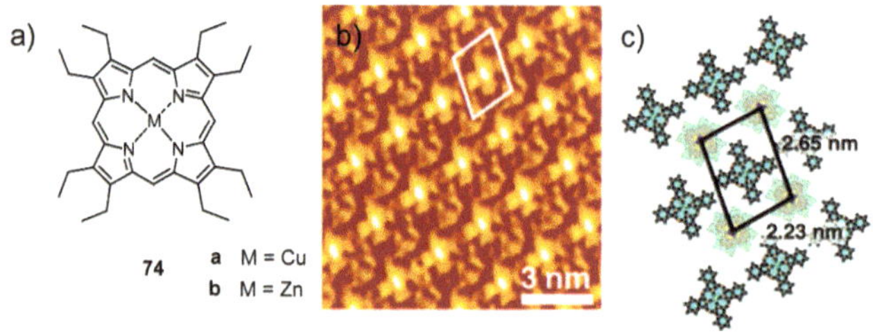

Figure 4.13. (a) Molecular structures of octaethyl Pors **74a,b**. (b) High-resolution STM image of alternate binary arrays of Co(II)Pc and Cu(II)Por **74a** on Au(111) in 0.1 M $HClO_4$ and (c) corresponding proposed model. Figure 4.13b reproduced from Ref. [68] with permission of the American Chemical Society.

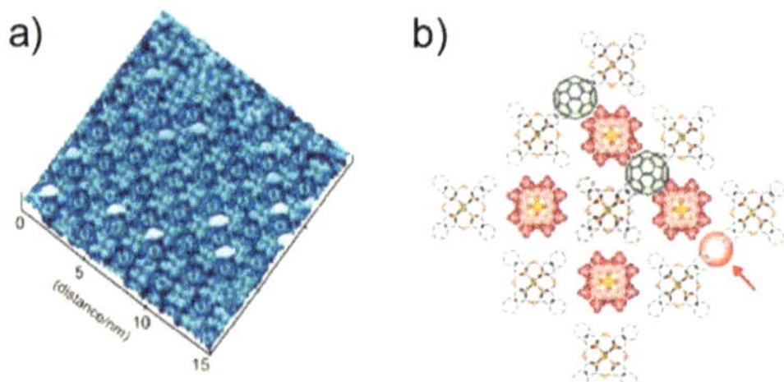

Figure 4.14. (a) Height shaded view of a STM image and (b) proposed model of C_{60} array in the bimolecular Pc·Por "chessboard" on Au(111). Reproduced from Ref. [69] with permission of the American Chemical Society.

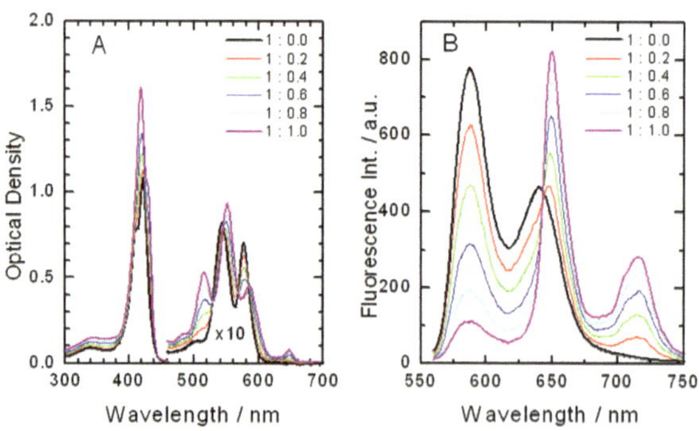

Figure 5.9. Absorption (A) and fluorescence (B, λ_{ex} = 546 nm, isosbestic point) spectra of the dimer $(ZnOEP)_2Ph$ with increasing amounts of extra-ligand $H_2P(m{^\wedge}Pyr)_2$-(iso-PrPh)$_2$ (toluene, 295 K). The molar ratio x = [ligand]/[dimer] varies from $x = 0.0$ to 1.0 (0.0, 0.2, 0.4, 0.6, 0.8, 1.0).

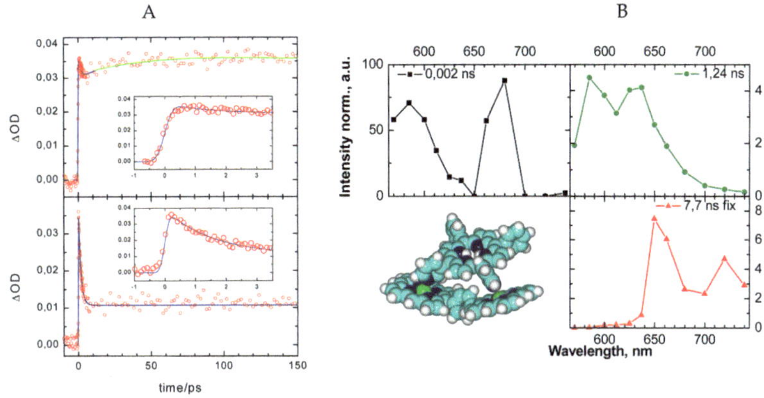

Figure 5.11. Time-resolved femtosecond pump-probe (A) and picosecond fluorescence (B) data for the triad $(ZnOEP)_2Ph \otimes H_2P(m^\wedge Pyr)_2$-(iso-PrPh)$_2$ in toluene at 293 K. A: Time evolution of transient absorbance at λ_{pump} = 555 nm and λ_{probe} = 670 nm (top), 510 nm (bottom). B: Decay-associated spectra (λ_{ex} = 546 nm) derived from a global analysis of 12 time correlated single photon counting (TCSPC) fluorescence measurements.

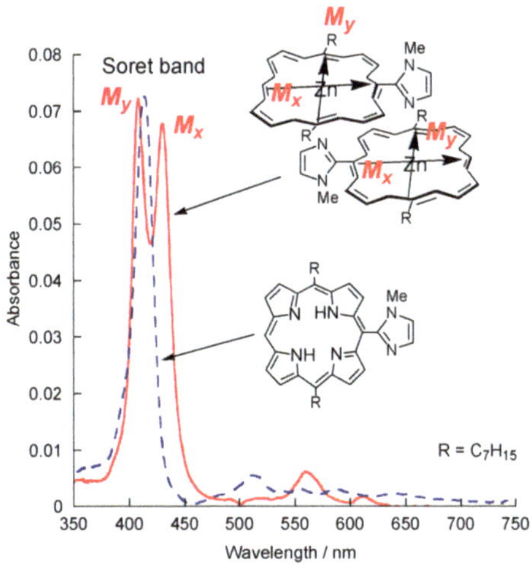

Figure 6.6. UV-vis absorption spectra of a slipped cofacial porphyrin dimer and interaction modes of the transition dipole moment.

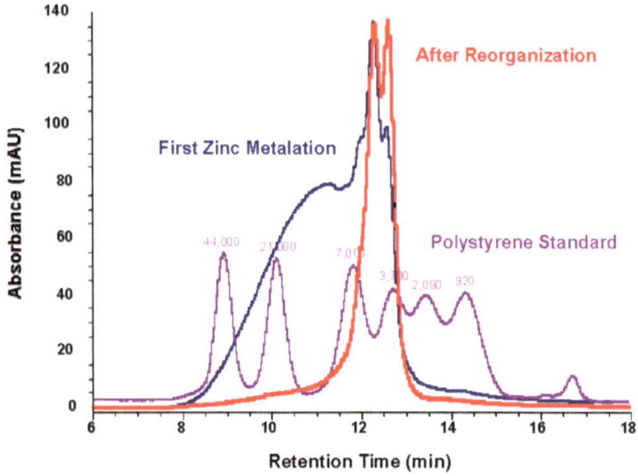

Figure 6.8. GPC elution curves immediately after Zn(II) insertion (blue), after reorganization (red), and of polystyrene standards (purple). Reprinted with permission from Takahashi, R., and Kobuke, Y. (2003) *J. Am. Chem. Soc.*, **125**, 2372–2373. Copyright 2003 American Chemical Society.

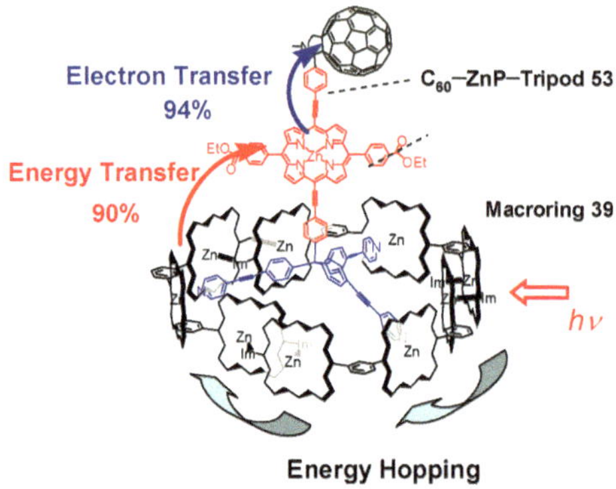

Figure 6.32. Energy transfer from the antenna ring to the acceptor porphyrin followed by electron transfer to the fullerene unit.

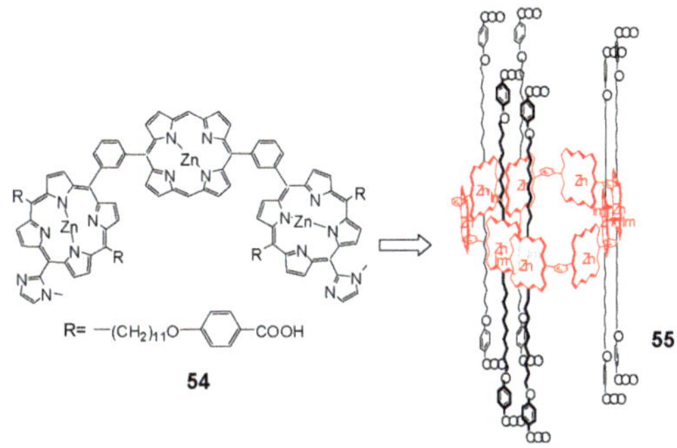

Figure 6.33. Membrane incorporation model of light-harvesting antenna.

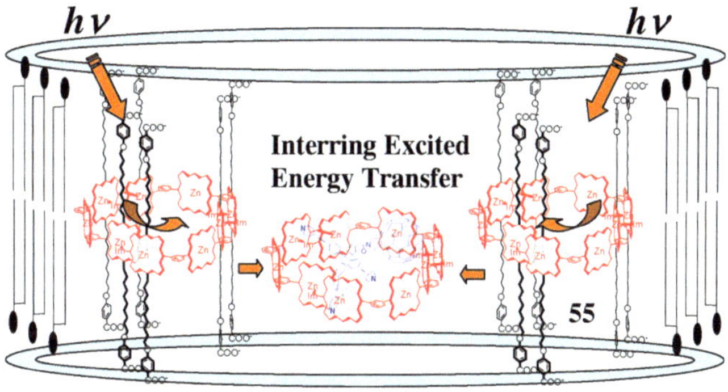

Figure 6.35. Schematic diagram of interring excitation-energy transfer among antenna rings in a bilayer membrane.

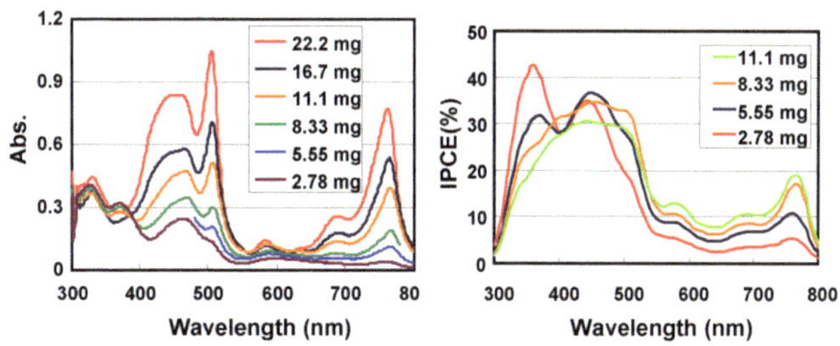

Figure 6.37. (a) Absorption spectra and (b) IPCE from the butadiynylene-linked bisporphyrin array **31** at various concentrations.

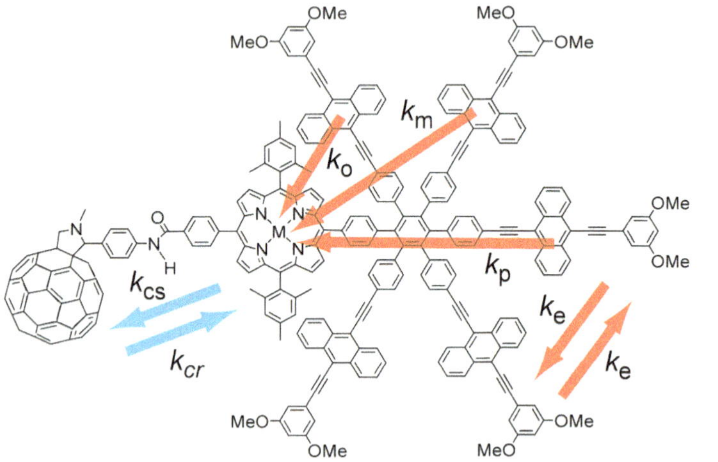

Figure 7.9. Singlet energy (red) and electron (blue) transfer pathways in heptads **3** and **4**.

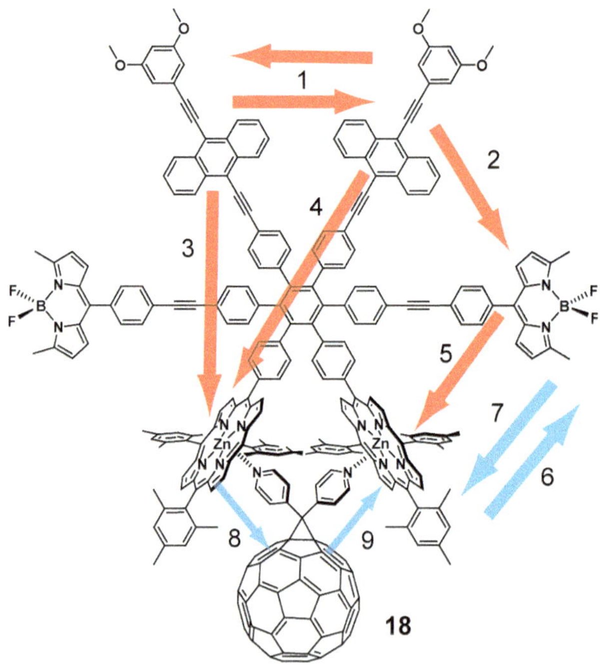

Figure 7.15. Singlet excitation energy (red arrows) and electron (blue arrows) transfer pathways observed in heptad **18** and model compounds. The relevant time constants are shown in Table 7.2.

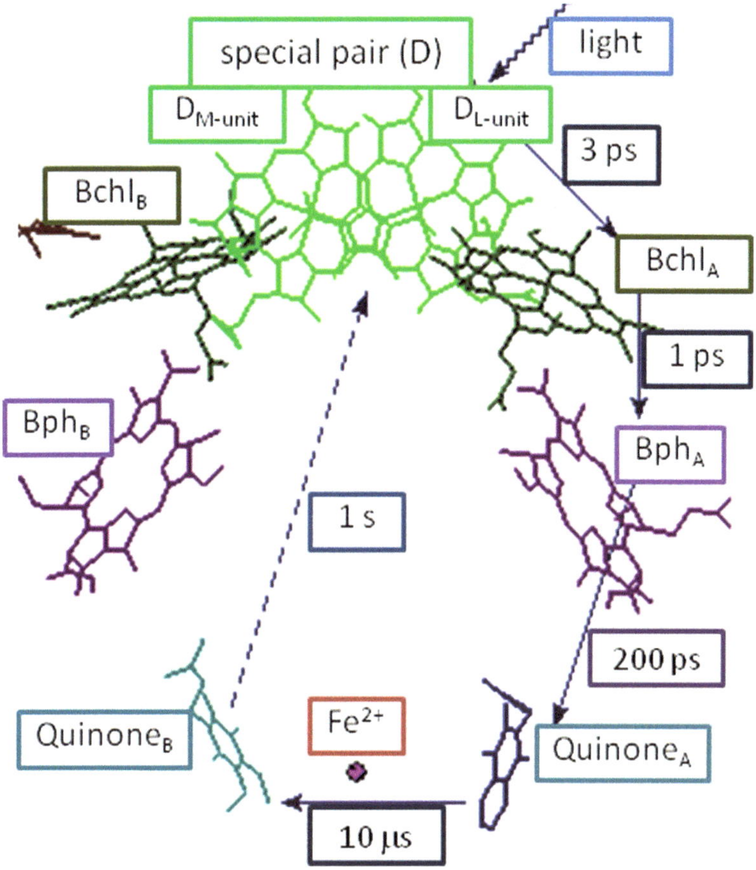

Figure 8.1. Photosynthetic reaction centers of purple bacteria. Adapted from Ref. [4].

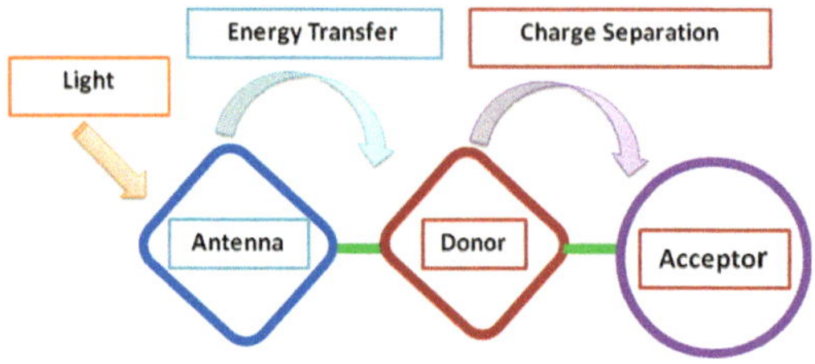

Figure 8.2. Schematic of a covalently linked triad for mimicking photosynthetic antenna-reaction center functionality.

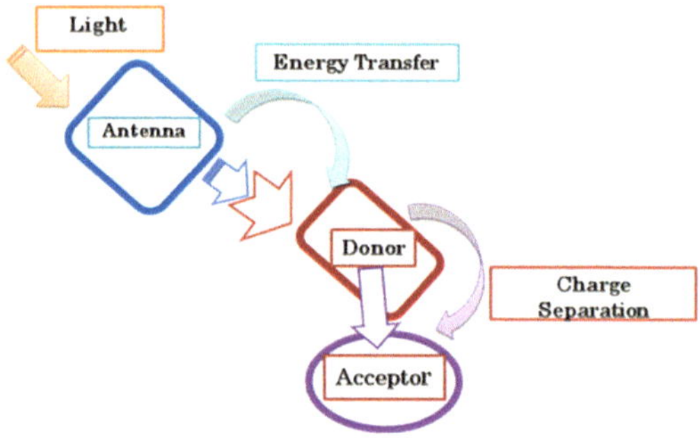

Figure 8.3. Schematic example of a non-covalently linked supramolecular triad system.

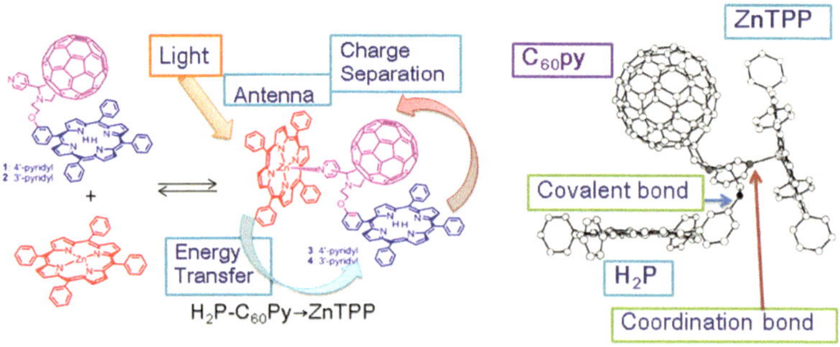

Figure 8.4. (Left) Self-assembly of H_2P-C_{60}Py→ZnTPP to result in a triad, and the expected roles of the components. (Right) Optimized structure of H_2P-C_{60}Py→ZnTPP. Modified from Ref. [65].

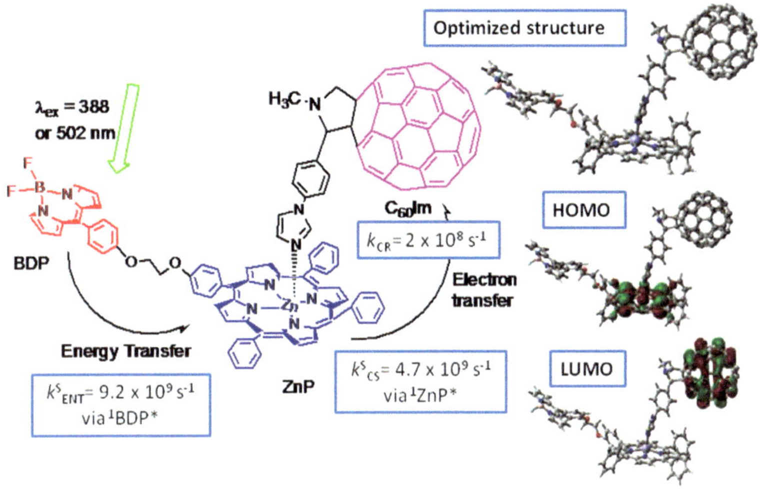

Figure 8.8. Structure of C_{60}Im→ZnP-BDP triad, its optimized structure, the HOMO and the LUMO. Modified from Ref. [66].

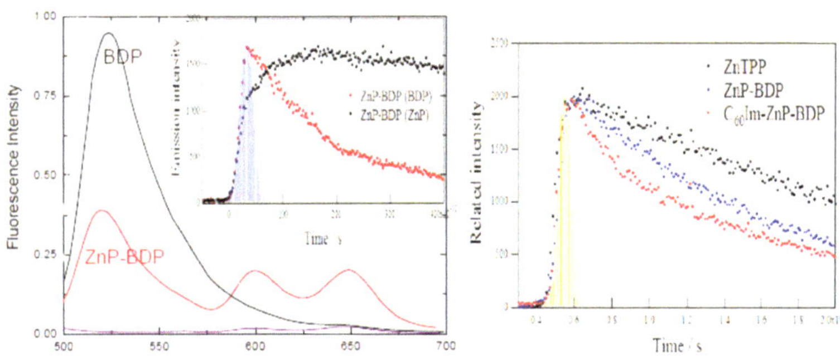

Figure 8.9. (Left) Steady-state fluorescence spectra in *o*-DCB; (black) BDP, (red) ZnP-BDP, and (purple) C_{60}Im→ZnP-BDP. Inset: Fluorescence time profiles of ZnP-BDP monitored at ZnP and BDP fluorescence wavelengths. (Right) Fluorescence time-profiles at 600 nm of ZnP; (black) ZnTPP, (blue) ZnP-BDP, and (red) C_{60}Im→ZnP-BDP in *o*-DCB. Hatched area shows the laser pulse profile of the excited light at 550 nm. Modified from Ref. [66].

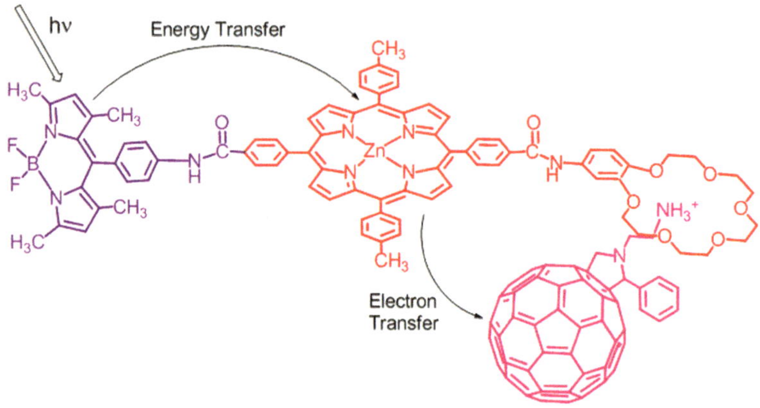

Figure 8.11. Structure of the C_{60}-NH_3^+; (Crown)ZnP-BDP) supramolecular triad. Modified from Ref. [67].

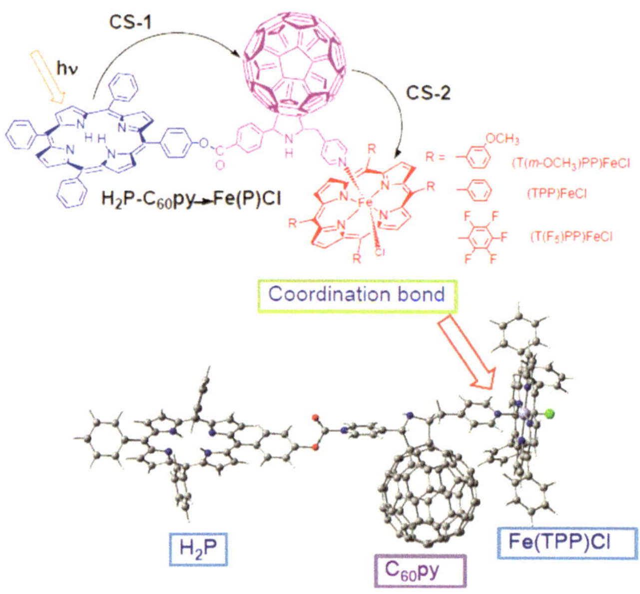

Figure 8.12. Structures of H_2P-$C_{60}Py \rightarrow Fe(P)Cl$ triads and the optimized structure of H_2P-$C_{60}py \rightarrow Fe(TPP)Cl$ triad. Modified from Ref. [68] .

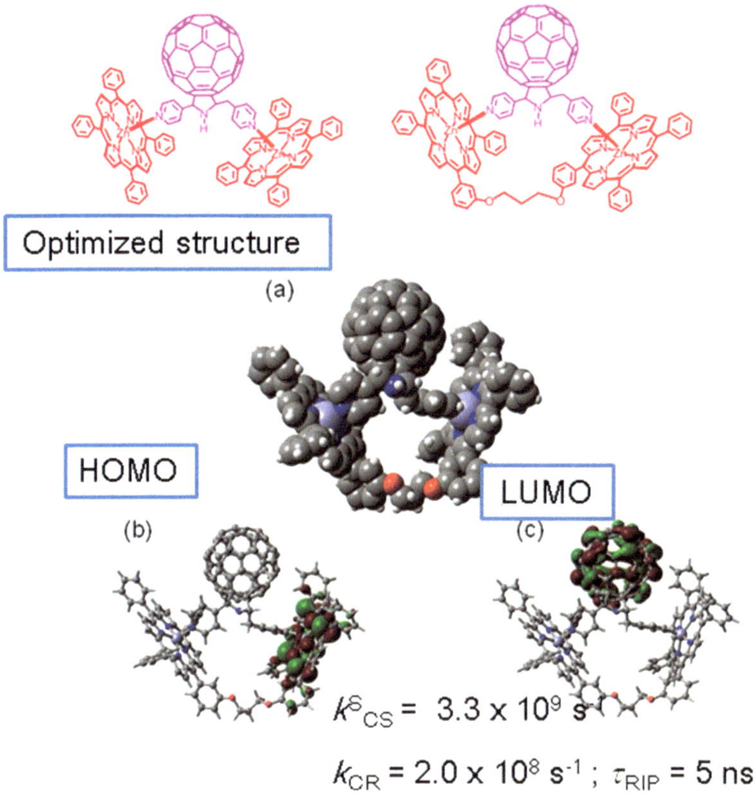

Figure 8.15. Two bisporphyrins-fullerene supramolecular complexes; (a) optimized structure, (b) the HOMO, and (c) LUMO of the ZnP dimer complex with PyC_{60} Py. Modified from Refs. [69, 70].

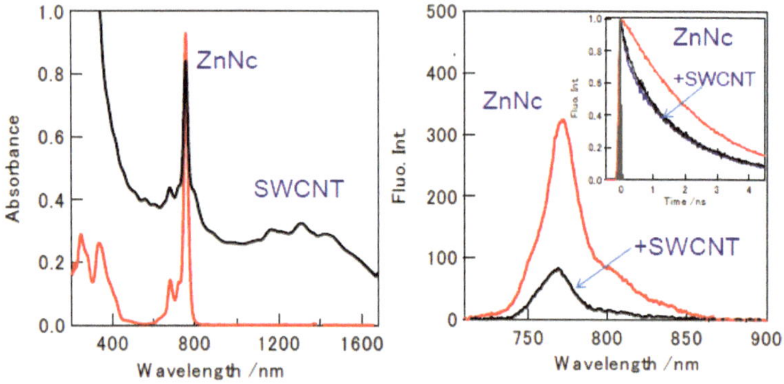

Figure 8.30. (Left) Steady-state absorption spectra of (ZnP←ImPyr)$_n$/SWCNT and ZnNc-fluorescence quenching of (ZnP←ImPyr)$_n$/SWCNT in DMF. Modified from Ref. [100].

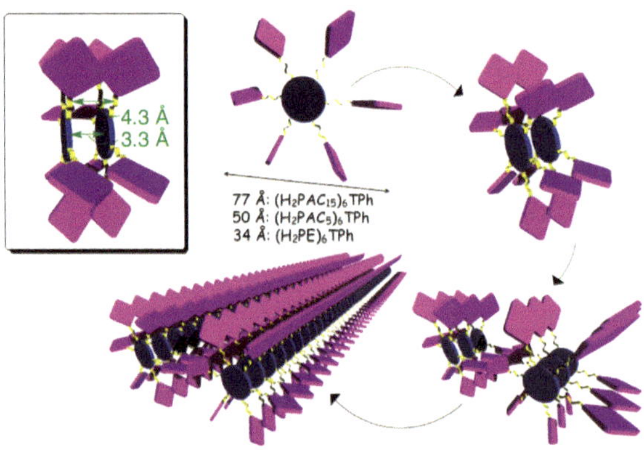

Figure 10.16. A schematic illustration of organization process. The purple square, yellow wavy line and blue round shapes stand for porphyrin, alkyl chain, and triphenylene units, respectively. The molecular sizes were estimated by DFT calculation. The insertion: proposed stacking structure of $(H_2PAC_{15})_6TPh$ units. Reprinted with permission from [74].

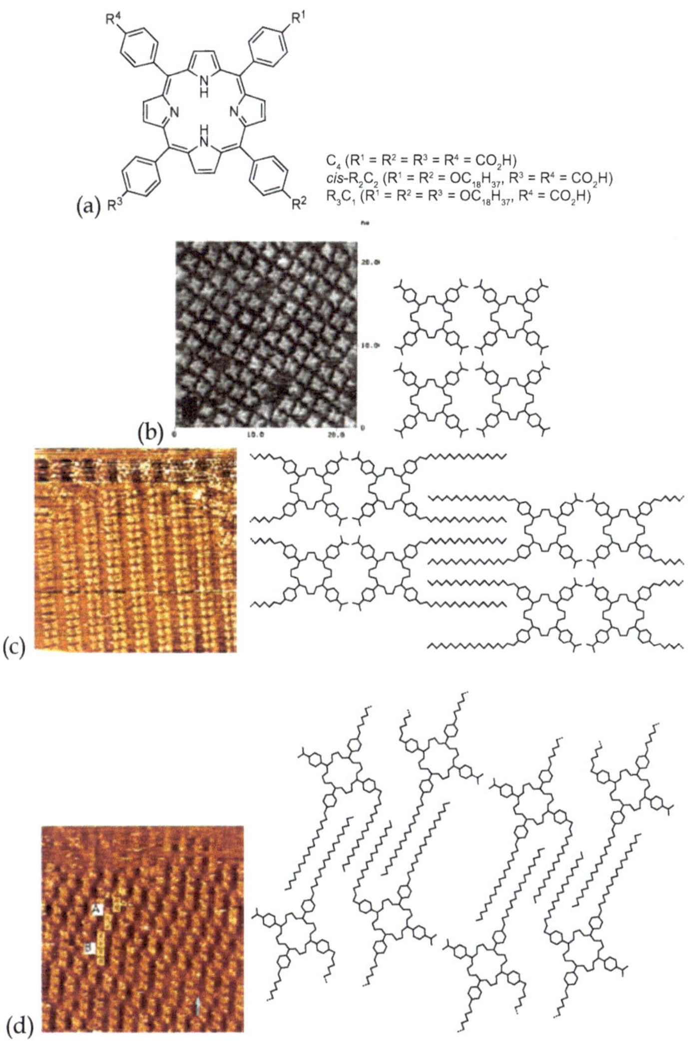

Figure 12.26. Hydrogen bonded assemblies of porphyrins bearing carboxy groups. (a) Chemical structures of porphyrins with carboxy groups. (b) C_4 at the air/HOPG interface. (c) *cis*-R_2C_2 at *o*-dichlorobenzene/HOPG interface; $50 \times 50\ nm^2$. (d) R_3C_1 at *o*-dichlorobenzene/HOPG interface; $50 \times 50\ nm^2$. Reproduced with permission from [80] (a) and [81].

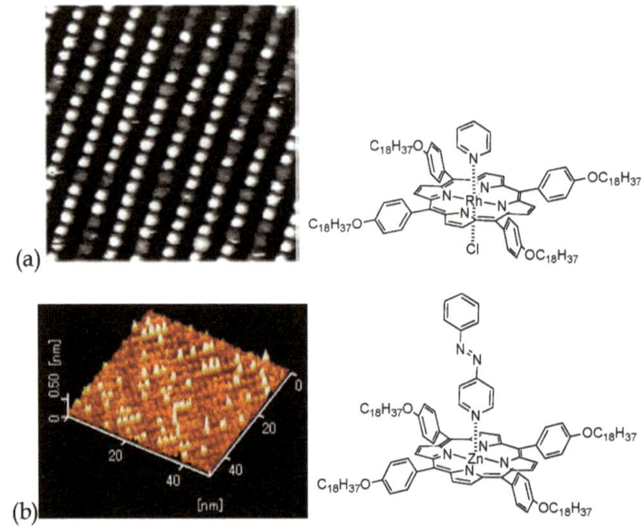

Figure 12.27. Mixed assembly of metalloporphyrin molecules with axial ligands and free-base porphyrin molecules at the 1-phenyloctane/HOPG interface. (a) Rhodium–porphyrin with pyridine; 35 × 35 nm^2. (b) Zinc–porphyrin with PhNNPy. Reproduced with permission from [82] (a) and [83] (b).

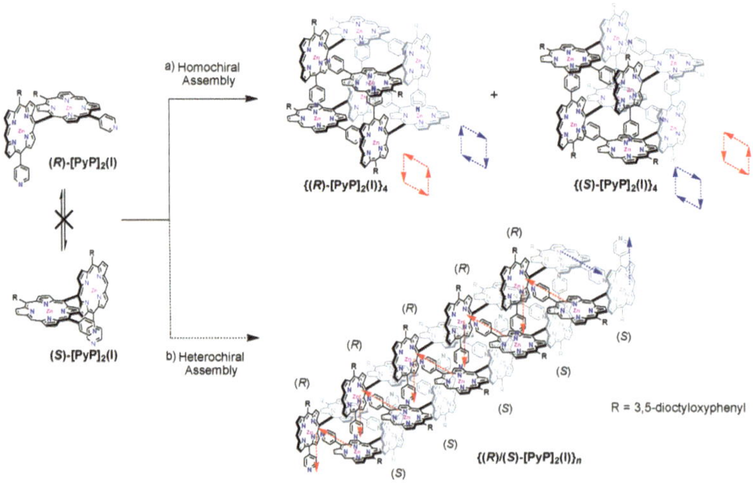

Figure 13.3. Schematic representation of a racemic mixture of *meso-meso* linked zinc bisporphyrin, bearing 4-pyridyl groups, and its self-assembling events. Red and blue arrows indicate the Py→Zn coordinating direction.

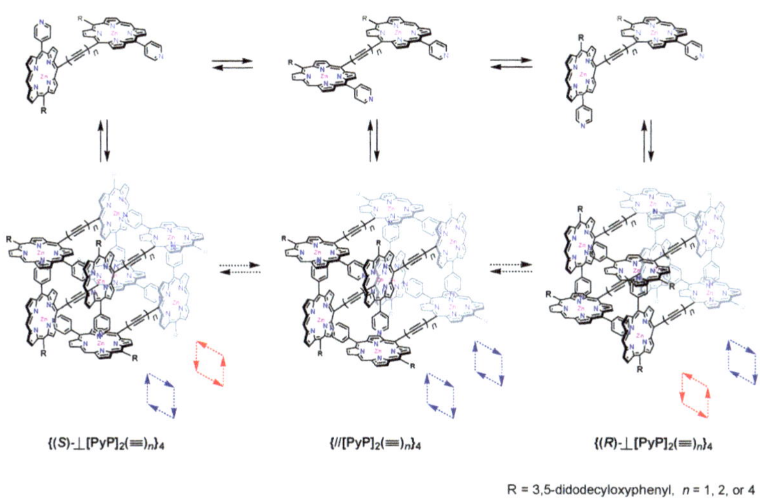

Figure 13.4. Schematic representation of alkynylene-bridged zinc bisporphyrin rotamers, bearing 4-pyridyl groups, and their self-assembling events. Red and blue arrows indicate the Py→Zn coordinating direction.

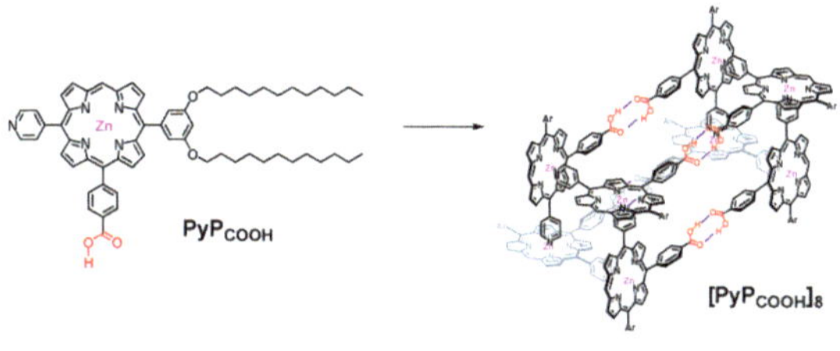

Figure 13.5. Schematic representation of zinc porphyrin PyP_{COOH}, bearing 4-pyridyl and carboxyphenyl groups, and its self-assembled structure formed by coordination and hydrogen bonding interactions.

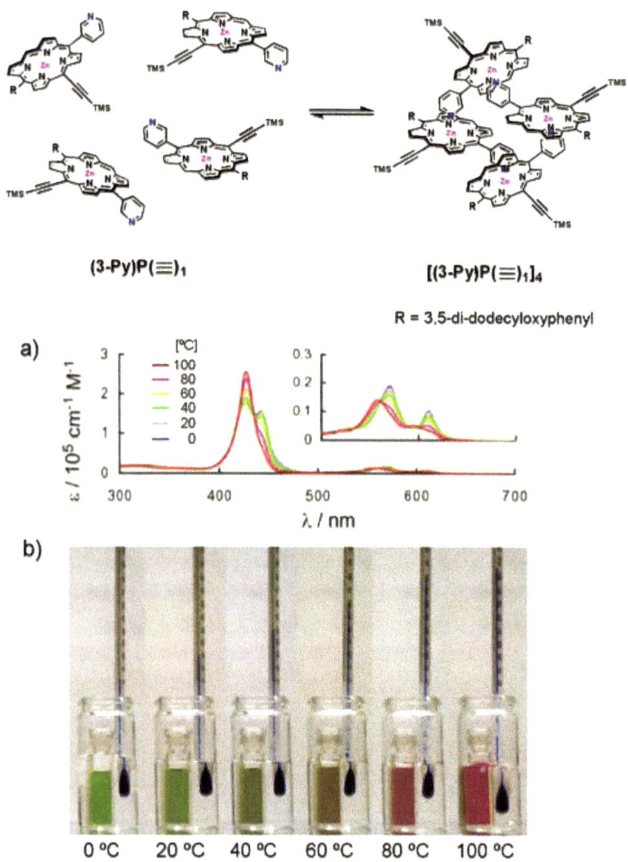

Figure 13.6. Variable-temperature a) absorption spectral profile and b) color profile of (3-Py)P(≡)$_1$ in toluene (9.8×10^{-5} M) at 0–100°C (blue: 0°C, red: 100°C). Inset shows magnified spectra in the Q-band region.

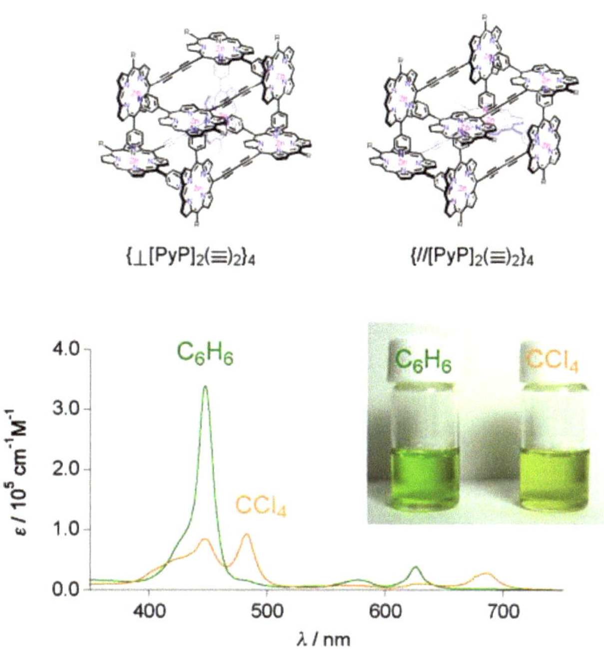

Figure 13.7. Absorption spectra of $\{[PyP]_2(\equiv)_2\}_4$ at 20°C in benzene (green) and CCl_4 (yellow). Inset shows pictures of benzene (left) and CCl_4 (right) solutions of $\{[PyP]_2(\equiv)_2\}_4$.

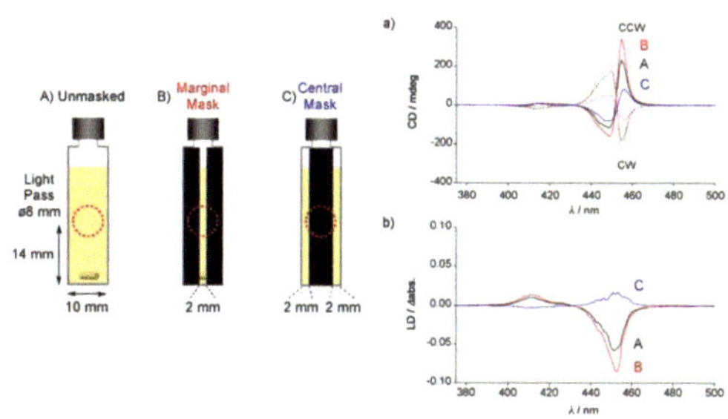

Figure 13.12. (a) Circular dichroism (CD) and (b) linear dichroism (LD) spectroscopy of a benzene solution (3.0 mL) of the nanofibers of DP (6.0×10^{-6} M), upon lower-side rotary stirring at 1350 rpm in clockwise (broken curves) or counter clockwise (solid curves) direction using a $\phi 2.0 \times 5.0$ mm Teflon-coated magnetic stirring bar at 20°C in $10 \times 10 \times 40$ mm quartz optical cells. CD and LD spectra with cells (A) unmasked (black curves), (B) masked at marginal parts with 4 mm wide black tapes to leave a 2 mm wide central slit (red curves) and (C) masked at a vertical center with a 6 mm wide black tape (blue curves).

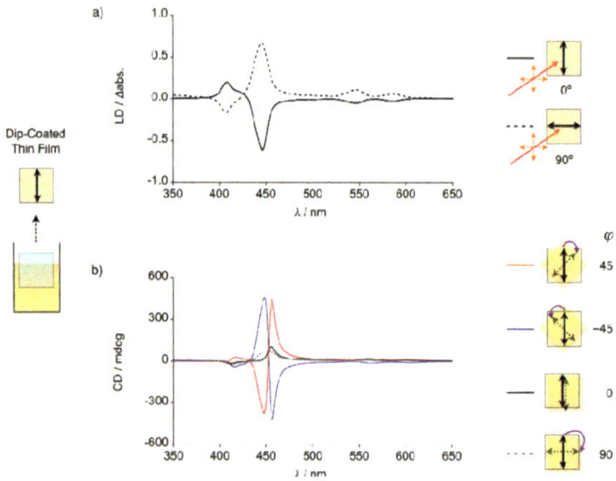

Figure 13.13. (a) Linear dichroism (LD) spectroscopy of an oriented thin film of the nanofibers of DP. The film was positioned in such a way that its oriented direction was angled by 0° (solid curve) and 90° (broken curve) relative to the vertical axis of a linearly polarized incident light for the spectroscopy. (b) Circular dichroism (CD) spectroscopy of lapped samples of two oriented thin films of the nanofibers of DP. The second film was lapped over the first one in such a way that their oriented directions were angled (φ) by 45° (red solid curve), –45° (blue solid curve), 0° (black solid curve), and 90° (black broken curve).

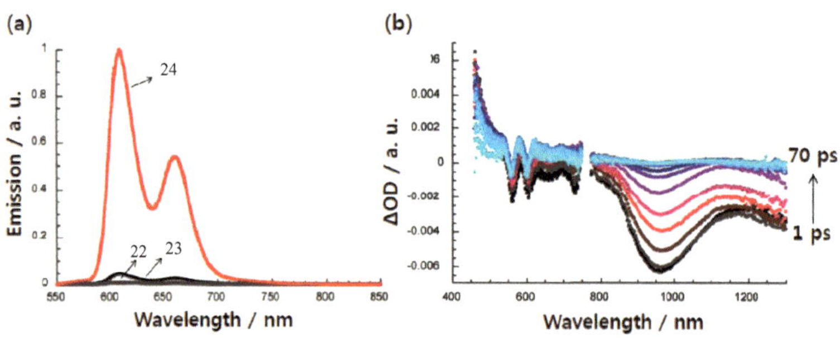

Figure 14.22. (a) Fluorescence spectra of **22** and **23** with that of reference **24**. (b) Differential absorption spectra obtained upon femtosecond flash photolysis (387 nm, 150 nJ) of **22** in DMF with time delays between 1 ps and 70 ps at room temperature.

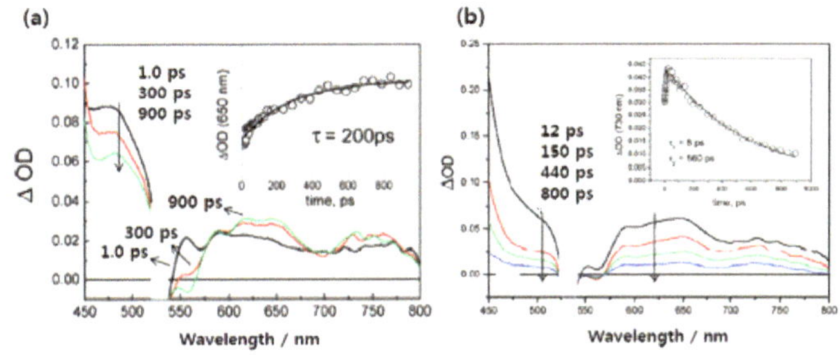

Figure 14.28. Ultrafast spectroscopy of **28**(a) and **29**(b) in CH_2Cl_2 upon excitation at 530 nm.

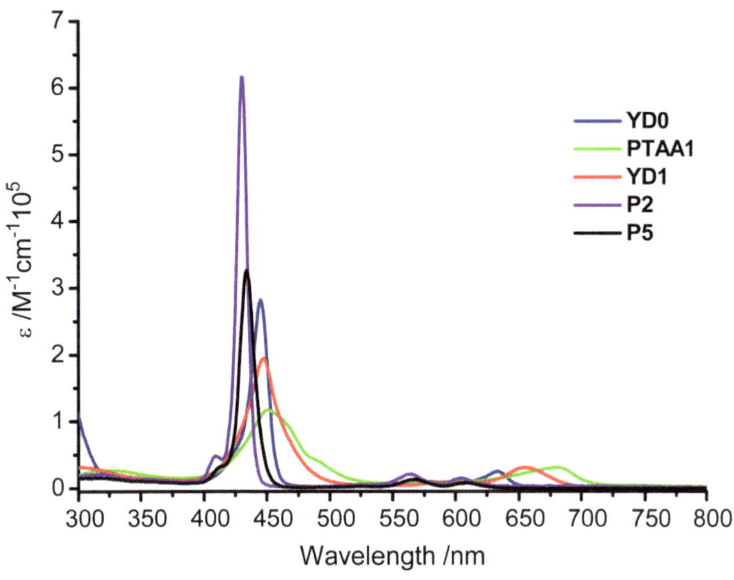

Figure 15.3. UV-visible absorption spectra of **YD0**, **YD7**, **YD1**, **P2,** and **P5** in CH_2Cl_2/Pyridine (100/1).

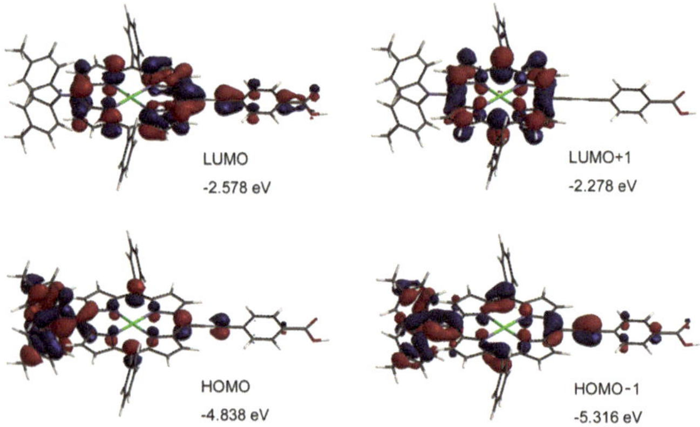

Figure 15.5. Surfaces of frontier molecular orbitals of **YD1** predicted with DFT calculations. To simplify computations, *tert*-butyl groups at *para*- and *meta*-positions were replaced with methyl groups and hydrogen atoms, respectively.

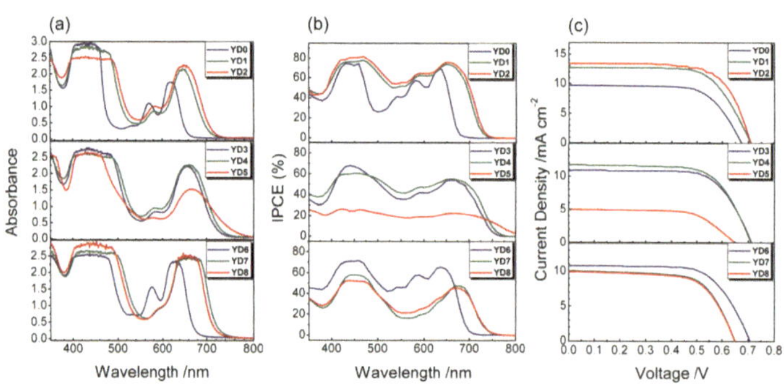

Figure 15.10. (a) Absorption spectra of films with no added scattering layer (film thickness ~10 μm), (b) incident photon-to-current conversion efficiency (IPCE) spectra of devices, and (c) current–voltage characteristics of devices fabricated from **YD0-YD8** sensitized on TiO_2 films (film thickness ~ (10 + 4) μm).

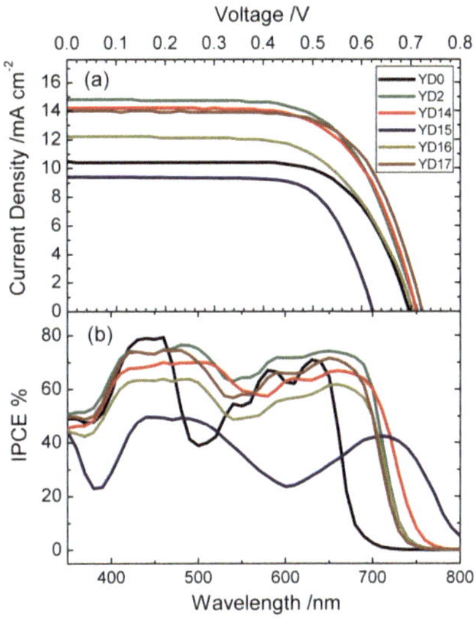

Figure 15.13. (a) Current–voltage characteristics of devices fabricated from **YD0**, **YD2**, and **YD14-YD17**; (b) incident photon-to-current conversion efficiency (IPCE) spectra of devices.

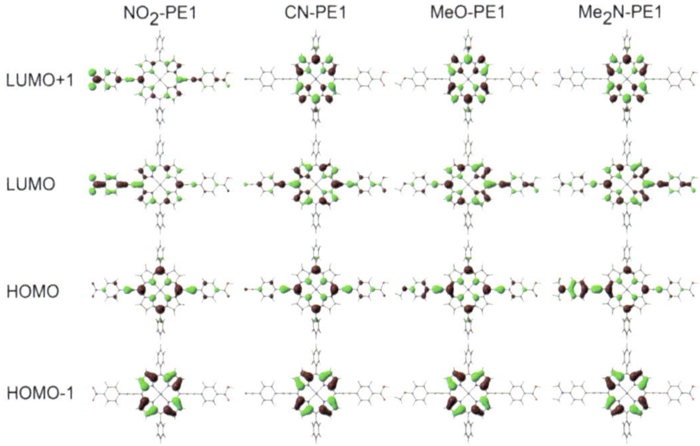

Figure 15.15. Molecular-orbital patterns of X-PE1 porphyrins with geometries of each molecule optimized at the B3LYP/LanL2DZ level of theory. These MO patterns serve to assist our understanding of the porphyrins qualitatively.

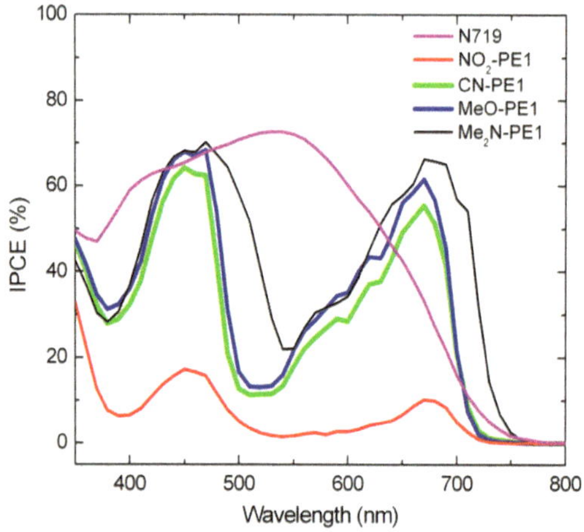

Figure 15.17. Photocurrent action spectra of X-PE1-sensitized solar cells. Corresponding overall efficiencies are **N719** 6.89%, **Me_2N-PE1** 5.67%, **MeO-PE1** 4.20%, **CN-PE1** 3.68%, and **NO_2-PE1** 0.84% in these figures (not a mean value).

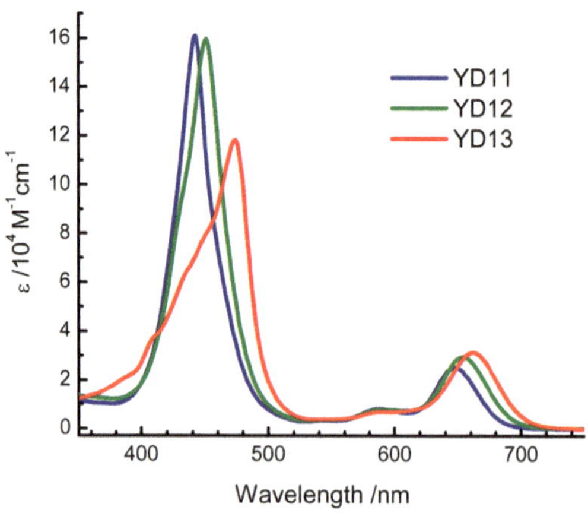

Figure 15.19. UV-visible absorption spectra of **YD11-YD13** in ethanol.

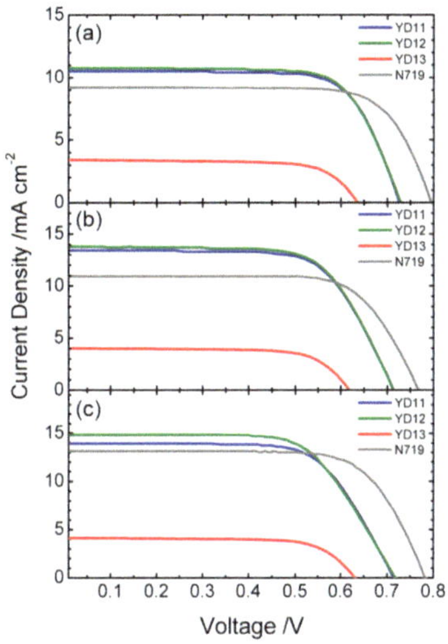

Figure 15.20. Current–voltage characteristics of devices fabricated from **YD11-YD13**, under illumination of simulated AM1.5 full sunlight (100 mW cm^{-2}) with an active area 0.16 cm^2 of three film thicknesses: (a) ~5 μm (film A); (b) ~10 μm (film B); (c) ~(10 + 4) μm (film C).

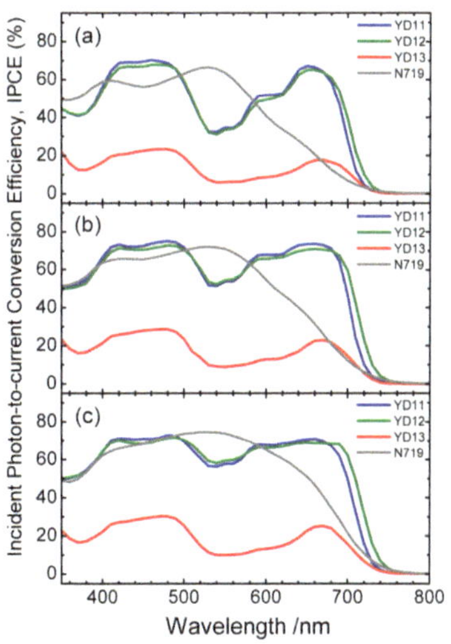

Figure 15.21. Corresponding IPCE action spectra of the same DSSC devices as those shown in Fig. 15.20.

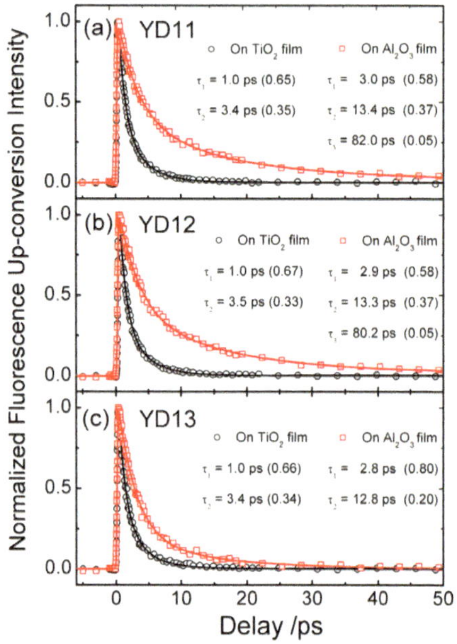

Figure 15.22. Femtosecond emission decay curves of the thin-film samples excited at the Soret band ($\lambda_{ex} = 430$ nm) and probed at the wavelength of maximum emission intensity. (a)–(c) represent the transients of **YD11-YD13** on TiO_2 (circles) and Al_2O_3 (squares) films. Solid curves represent theoretical fits with the corresponding time coefficients and relative amplitudes (in parenthesis) as indicated.

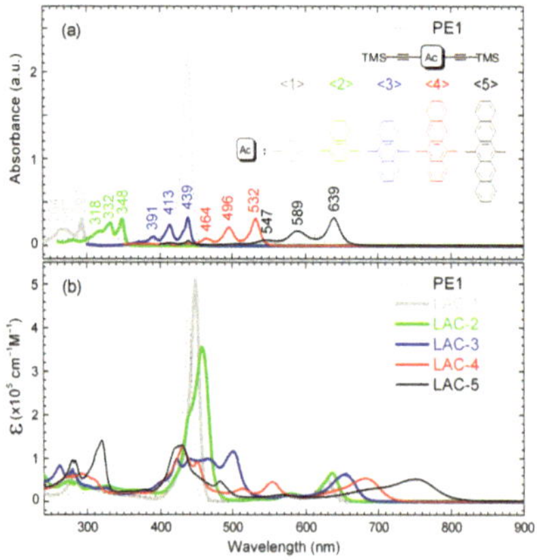

Figure 15.24. UV-visible spectra of **PE1**, acene-ethynyl precursors and LAC porphyrins in THF.

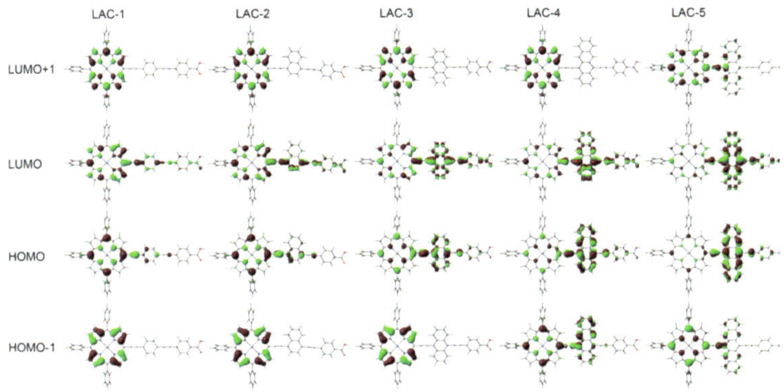

Figure 15.25. MO patterns of LAC porphyrins with geometries optimized at the B3LYP/LanL2DZ level of theory. Note that these MO patterns are used to qualitatively assist our understanding to the porphyrins.

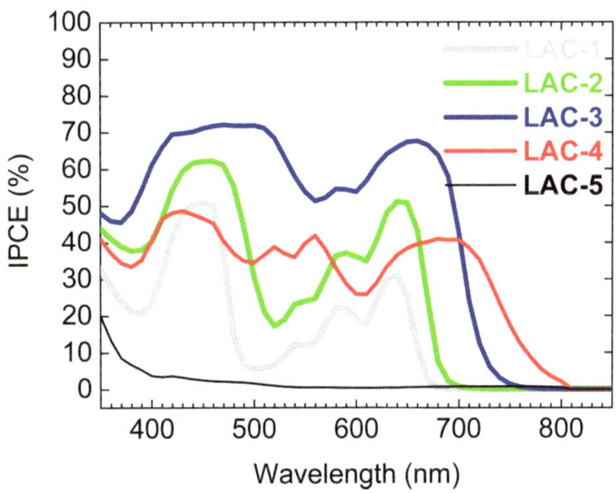

Figure 15.27. Photo-current action spectra of LAC-sensitized solar cells. Corresponding overall efficiencies are: 2.43% (**LAC-1**), 3.38% (**LAC-2**), 5.49% (**LAC-3**), 2.67% (**LAC-4**), and 0.12% (**LAC-5**) in these figures (not an averaged value).

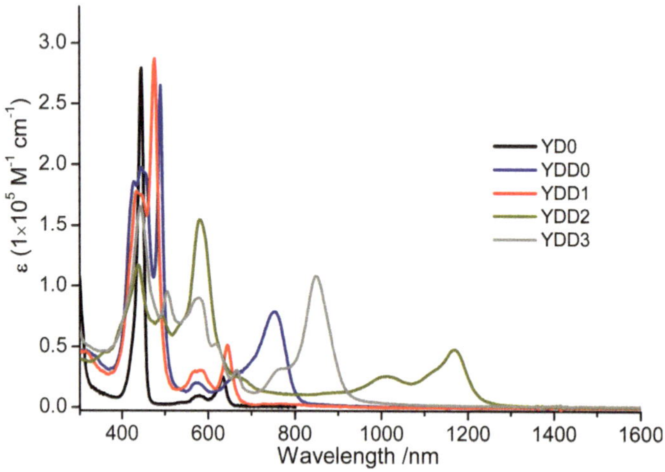

Figure 15.29. UV-visible absorption spectra of **YD0**, and **YDD0-YDD3** in ethanol.

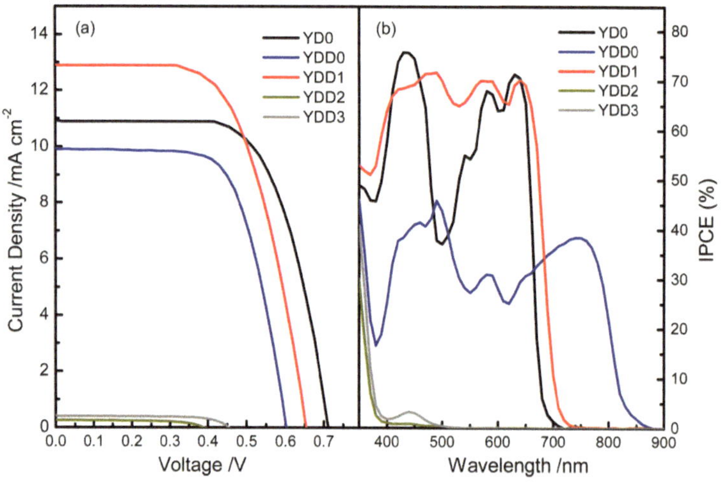

Figure 15.31. The current–voltage characteristics and efficiencies of conversion of incident photons to current (IPCE) of porphyrin-based DSSC.